MW01065730

Famous First Facts™

About the Environment

Other H.W. Wilson titles by New England Publishing Associates

Facts About the World's Languages
Facts About the American Wars
Wilson Calendar of World History
Facts About China (forthcoming)

Other titles in the Wilson Facts Series

Famous First Facts About Sports
Famous First Facts About American Politics
Famous First Facts, Fifth Edition
Famous First Facts, International Edition

Facts About the Presidents, Seventh Edition
Facts About American Immigration
Facts About Retiring in the United States
Facts About the 20th Century
Facts About Canada, Its Provinces and Territories
Facts About the British Prime Ministers
Facts About the Cities, Second Edition
Facts About the Congress
Facts About the Supreme Court of the United States
Facts About the World's Nations

Famous First Facts™

About the Environment

Edited by Ronald J. Formica

Contributors:

Victoria S. Chase
Lee Davis
Michael Golay
Edward W. Knappman
Lisa Paddock
Martin Manning
Tom Smith
John Wright

A New England Publishing Associates Book

The H.W. Wilson Company
New York • Dublin
2002

Copyright © 2002 by the H.W. Wilson Company

All rights reserved. No part of this work may be reproduced or copied in any form or by any means, including but not restricted to graphic, electronic, and mechanical—for example, photocopying, recording, taping, or information and retrieval systems—without the express written permission of the publisher, except that a reviewer may quote and a magazine, newspaper, or electronic information service may print brief passages as part of a review written specifically for inclusion in that magazine, newspaper, or electronic information service.

Library of Congress Cataloging-in-Publication Data

Formica, Ronald J.
Famous first facts about the environment / Ronald J. Formica.
p. cm.
Includes bibliographical references and index.
ISBN 0-8242-0974-5
1. Environmental sciences—Miscellanea. I. Title

GE 105 .K53 2001
363.7—dc21 2001017704

Printed in the United States of America

The H. W. Wilson Company
950 University Avenue
Bronx, NY 10452

Visit H.W. Wilson's Web site: www.hwwilson.com

Contents

Preface

Famous First Facts About the Environment follows in the footsteps of Joseph Nathan Kane's *Famous First Facts*, first published by H.W. Wilson in 1933, with the fifth edition appearing in 1997. In the nearly 70 years since its initial publication, *Famous First Facts* has come to be regarded as the quintessential reference book on notable American "firsts." Its diverse content includes entries on numerous topics, including agriculture, crime, entertainment, medicine, and weapons, and nearly two hundred firsts on environmental issues. *Famous First Facts About the Environment* uses Kane's well-established formula to expand the record of important and interesting environmental firsts and presents them from a global perspective.

The entries in *Famous First Facts About the Environment* demonstrate that the environment has always been a topic of concern and a point of contention for people around the world. Some claim that too much money and resources are spent protecting the environment, while others argue that not nearly enough is being done. The entries in *Famous First Facts About the Environment* do not espouse either view. Instead, as in Joseph Nathan Kane's *Famous First Facts*, the items in this book are presented impartially.

The goal of *Famous First Facts About the Environment* is to cover not only the obvious environmental firsts, but to also include entries on topics not found in other environmental reference works. The Expanded Contents indicates the diversity of the subjects covered here. Indeed, perhaps the most difficult task in assembling this book was deciding under which category each entry should be placed, since many entries could easily fall under two or more categories. In the end, the contributors and I decided to place entries where we believed readers would be most likely to look them up in the Subject Index.

Famous First Facts About the Environment uses the same system as *Famous First Facts* to organize the approximately 4,000 entries. They are first listed under a main subject category; then, if needed, the main subject category is divided into subcategories. Within each category or subcategory, the entries are organized chronologically. There are five indexes in the back of the book to help readers find specific entries with ease. This organizational strategy is explained in greater detail in the How to Use This Book section which follows.

The contributors to this book went to great lengths to find the most interesting environmental firsts. They spent thousands of hours researching in a wide variety of sources. While not every environmental first is covered within these pages (an impossible task), we are confident that no other reference work available today can match the number of environmental firsts presented here.

I was fortunate to have a great deal of assistance in editing this book. I would like to thank Edward W. Knappman of New England Publishing Associates for the opportunity to work on this book and for his guidance. The rest of the staff of New England Publishing—Elizabeth Frost-Knappman, Kristine Schiavi, and Victoria Harlow—were very supportive throughout this project. I would like to thank the many editors at H.W. Wilson who worked on this book, especially Hilary Claggett for her guidance during the early stages of organizing the entries, as well as Beth Levy, Gray Young, Lynn Messina, and Norris Smith.

The support of my family, not just during the course of editing this book, but throughout my life, has been very important. I want to extend my gratitude and love to my father and mother, Joseph and Lucy Formica, as well as to my brothers Sebastian and Steven. A special thanks must go to my niece and nephew, Sabrina and Nicholas, for just being who they are and for the way they can always bring a smile to my face.

Finally, this book would not have been possible without the work done by all of the contributors. They worked diligently over the course of several months researching and writing these entries. They were always willing to go that extra mile to make sure that all of the items were factually accurate.

Ronald J. Formica
Haddam, Connecticut
March 2002

How to Use This Book

Entries in *Famous First Facts About the Environment* are grouped together under main subject categories. The subject categories are arranged alphabetically, starting with **Activist Movements—Animal Rights** and ending with **Zoos, Aquariums and Museums—Planetariums**. When the situation is appropriate, main subject categories are divided into subcategories. These subcategories are also arranged alphabetically. For example, the main category of **Air Pollution** is divided into five subcategories—**Health**, **Indoor**, **Legislation and Regulation**, **Public Opinion**, and **Research**. A complete list of subject categories and subcategories appears in the Expanded Contents.

Within each main subject category or subcategory, the entries are arranged chronologically. Each entry begins with a four-digit indexing number, staring with 1001, and is followed by an introductory phrase in boldface type. This introductory phrase is the entry head. Please note that the four-digit indexing number does not indicate the year of an event. The date on which an entry took place is found in the main text of the entry.

There are five indexes in the back of this book—Subject Index, Index by Year, Index by Month and Day, Personal Name Index, and Geographical Index. This system of indexing eliminates the need for cross-referencing within the main text. In addition, it will allow the reader to quickly find specific entries according to subject, date, name, or place. In all five indexes, entries are identified by the four-digit indexing number—not by page number.

The Subject Index lists alphabetically all entry heads, as well as words representing key subjects. The Index by Year lists entries according to the year when they took place. Under each year, entries are listed alphabetically. Entries in the Index by Month and Day are listed first by month, then by the specific day of the month on which they took place. For example, all of the entries that took place in the month of January but do not have a specific day are listed first. Then the entries that took place on January 1 are listed, and so forth. In this index, the entries are listed by year under each day and not alphabetically. The Personal Name Index is an alphabetical listing of the key personal names found in the main body of the text. The names are listed in bold, with the entry heads related to each name listed below. Finally, the Geographical Index lists the key locations of the entries. The entries are listed according to country, state or province, and then city. For example, an event that took place in Los Angeles would be found by looking up **UNITED STATES**, then **California**, and finally ***Los Angeles***. The entry heads in this index are listed alphabetically.

Expanded Contents

Activist Movements—Animal Rights . . . 1
Activist Movements—Anti-Environmentalism . . . 2
Activist Movements—Conservationism . . . 3
Activist Movements—Ecofeminism . . . 4
Activist Movements—Environmentalism . . . 4
Activist Movements—Grassroots . . . 8
Activist Movements—Green Politics . . . 9
Agriculture and Horticulture . . . 10
Agriculture and Horticulture—Animal Husbandry and Livestock . . . 11
Agriculture and Horticulture—Cash Crops . . . 13
Agriculture and Horticulture—Diseases and Pests . . . 13
Agriculture and Horticulture—Food Crops . . . 14
Agriculture and Horticulture—Irrigation . . . 14
Agriculture and Horticulture—Legislation . . . 15
Agriculture and Horticulture—Methods and Equipment . . . 16
Agriculture and Horticulture—Organic Farming . . . 17
Agriculture and Horticulture—Organizations . . . 18
Agriculture and Horticulture—Pest and Weed Control . . . 18
Agriculture and Horticulture—Preservation . . . 21
Air Pollution . . . 21
Air Pollution—Health . . . 22
Air Pollution—Indoor . . . 23
Air Pollution—Legislation and Regulation . . . 24
Air Pollution—Public Opinion . . . 28
Air Pollution—Research . . . 28
Alien Species and Species Migration . . . 31
Automotive Industry . . . 34
Automotive Industry—Highways . . . 36
Automotive Industry—Legislation and Regulation . . . 37
Automotive Industry—Technology . . . 38
Biodiversity . . . 41
Birds . . . 43
Birds—Legislation and Regulation . . . 46
Birds—Observations . . . 47
Birds—Organizations . . . 47
Birds—Refuges and Reservations . . . 48
Birds—Research . . . 48
Birds—Treaties . . . 49
Botany . . . 50
Climate and Weather . . . 53
Climate and Weather—Global Warming . . . 56
Climate and Weather—Meteorology . . . 57

Climate and Weather—Ozone Layer . . . 65
Climate and Weather—Paleoclimatology . . . 68
Coal and Natural Gas . . . 69
Dams . . . 72
Dams—Construction . . . 73
Dams—Disasters . . . 75
Droughts and Water Shortages . . . 76
Droughts and Water Shortages—Pluviculture . . . 80
Earthquakes . . . 80
Ecotourism . . . 85
Education and Research . . . 86
Education and Research—Expositions . . . 89
Extinct and Endangered Species . . . 90
Extinct and Endangered Species—Extinctions . . . 93
Extinct and Endangered Species—Lists . . . 96
Extinct and Endangered Species—Reintroductions . . . 99
Fish and Fishing . . . 101
Fish and Fishing—Commercial . . . 103
Fish and Fishing—Hatcheries and Fish Farming . . . 104
Fish and Fishing—Legislation and Regulation . . . 106
Fish and Fishing—Sport Fishing . . . 106
Floods and Flood Control . . . 107
Floods and Flood Control—Levees . . . 110
Floods and Flood Control—Mudslides . . . 110
Forests and Trees . . . 111
Forests and Trees—Deforestation . . . 111
Forests and Trees—Diseases and Pests . . . 111
Forests and Trees—Fires and Fire Prevention . . . 113
Forests and Trees—Legislation and Regulation . . . 114
Forests and Trees—Logging . . . 115
Forests and Trees—Management . . . 116
Forests and Trees—Organizations . . . 117
Forests and Trees—Preservation and Reserves . . . 118
Forests and Trees—Rain Forests . . . 119
Forests and Trees—Reforestation and Planting . . . 119
Forests and Trees—Research . . . 120
Genetic Engineering and Biotechnology . . . 121
Geology and Geophysics . . . 123
Hazardous Waste . . . 127
Hunting and Trapping . . . 131
Hunting and Trapping—Game Management . . . 133
Hunting and Trapping—Legislation and Regulation . . . 133
Hunting and Trapping—Organizations . . . 135
Industrial Pollution . . . 135
International Law . . . 139
International Law—Conventions, Agreements, and Treaties . . . 140

International Law—Organizations 144
Land Use and Development 145
Land Use and Development—Legislation and Regulation 145
Land Use and Development—Planning 146
Land Use and Development—Preservation 147
Land Use and Development—Property Rights 148
Land Use and Development—Settlement 148
Land Use and Development—Wetlands 149
Land Use and Development—Zoning 150
Literature and Arts—Arts and Artists 150
Literature and Arts—Books and Writers 153
Literature and Arts—Photography and Television 156
Littering 156
Mines and Mining 159
Natural Resources 162
Noise Pollution 171
Noise Pollution—Legislation and Regulation 171
Nuclear Power 173
Oceans and Oceanography 180
Oil and Petroleum Industry 185
Oil and Petroleum Industry—Fires, Spills, and Blowouts 186
Oil and Petroleum Industry—Offshore Drilling and Production 187
Oil and Petroleum Industry—War and Diplomacy 188
Oil and Petroleum Industry—Wells and Refineries 188
Parks 191
Parks—Legislation and Regulation 192
Parks—National Monuments 192
Parks—National Parks 193
Parks—State and Urban Parks 197
Polar Regions 197
Population Growth 201
Population Growth—Legislation and Regulation 202
Population Growth—Organizations and Conferences 203
Public Health and Safety 204
Scenery 209
Soil Resources 214
Soil Resources—Conservation 215
Soil Resources—Desertification 215
Soil Resources—Erosion 216
Soil Resources—Fertility and Fertilizers 216
Soil Resources—Research 217
Solar Energy 217
Solid Waste 221
Solid Waste—Disposal 222
Solid Waste—Recycling 223
Solid Waste—Street Cleaning 225

Storms 225
Storms—Blizzards and Snowstorms 226
Storms—Hailstorms 228
Storms—Hurricanes, Cyclones, and Typhoons 228
Storms—Ice Storms 231
Storms—Thunderstorms 231
Storms—Tornadoes 231
Sustainable Development 233
Synthetic Fuels 235
Volcanoes 238
Water Pollution 241
Water Pollution—Destruction of Species 242
Water Pollution—Health 242
Water Pollution—Industrial Discharge 243
Water Pollution—Legislation and Regulation 244
Water Pollution—Marine Pollution 246
Water Pollution—Organic Pollution 247
Water Pollution—Research and Technology 247
Water Pollution—Sewers and Human Waste 248
Water Power 249
Water Supply and Purification 251
Water Supply and Purification—Reservoirs 256
Whaling 256
Wilderness 257
Wildlife 261
Wildlife—Breeding 266
Wildlife—Conservation 267
Wildlife—Control 270
Wildlife—Preserves and Restoration 271
Wind Power 274
Zoos, Aquariums, and Museums 276
Zoos, Aquariums, and Museums—Planetariums 282

Famous First Facts About the Environment

A

ACTIVIST MOVEMENTS—ANIMAL RIGHTS

1001. Evidence of bullfighting as a sport dates from 2000 B.C. in Crete. A wall painting excavated at Knossos shows male and female acrobats confronting a bull. Bullfighting was introduced to Spain in the 11th century by the Moors, and later spread to other Spanish-speaking countries. Bullfighting, which generally leads to the death of a bull, has been strongly criticized by animal rights activists.

1002. Recorded evidence of the use of animals in research dates from the 3rd century B.C. The Greek physician and anatomist Erasistratus of Alexandria in Egypt (fl. c. 250 B.C.) used animals in his studies of circulation and nerves.

1003. Published presentation of the philosophy that humans should not be cruel to animals because animals can feel pain was presented by English philosopher Jeremy Bentham in 1780. Bentham wrote about his philosophy in *Introduction to the Principles of Morals and Legislation.*

1004. Use of the phrase "The question is not, Can they reason?, not, Can they talk?, but, Can they suffer?" was by English philosopher Jeremy Bentham in 1780. It was in his book *Introduction to the Principles of Morals and Legislation.* The phrase sums up the philosophy that we should be kind to animals because they can feel pain.

1005. Animal protection legislation was passed by the British Parliament in 1822. The push for the measure grew out of city-dwellers' gradual recognition that animals share many traits with humans, such as the ability to feel pain and the capacity for love, loyalty, and grief.

1006. Animal welfare organization started on a national level was the Society for the Protection of Animals, founded in England in 1824. The Society for the Protection of Animals was a response in part to Parliament's passage of animal protection legislation two years before. Similar private societies were established in the United States later in the 19th century. Though the movement spread more slowly in Europe, private animal protection organizations had appeared in several European countries by 1900.

1007. Society for animal welfare in the U.S. was the Society for the Prevention of Cruelty to Animals, established in 1866. Its founder was Henry Bergh of New York City, a pioneer in humane treatment of animals. The society was chartered in New York State in April 1866.

1008. Antivivisection legislation was introduced in the British House of Lords on May 4, 1875. Vivisection is the practice of cutting open living, and often unanesthetized, animals. Parliament in 1876 approved a measure regulating painful animal experiments.

1009. National act regulating painful experimentation on animals was approved in Britain in 1876. The Cruelty to Animals Act allowed painful experiments only if they furthered physiological knowledge and prohibited the use of animals for the practice of surgical skills.

1010. National animal rights organization in the United States was the American Anti-Vivisection Society, organized on February 23, 1883, at Philadelphia, PA. According to its charter, the goal of the Society was to restrict "the practice of vivisection within proper limits" and "the prevention of the injudicious and needless infliction of suffering upon animals under the pretense of medical and scientific research." The founder of the society was Caroline Earle White and its first president was Thomas George Morton. The first annual meeting was held on January 30, 1884, in Philadelphia.

ACTIVIST MOVEMENTS—ANIMAL RIGHTS—*continued*

1011. Fur farm was established on Prince Edward Island in Canada in 1887. Fur farmers used controlled breeding to achieve characteristics of size, color, and texture. Fur farms have since become a controversial issue. Animal rights advocates strongly oppose raising animals in captivity simply for their fur.

1012. Use of the term *animal rights* came in a work by India-born British naturalist and classical scholar Henry Salt in 1892. His *Animals' Rights Considered in Relation to Social Progress* argued the notion that animals should live free of human interference and humans should not use animals for their own purposes: for eating, for clothing (as in fur coats), or for scientific experimentation.

1013. Law in the United States to make shipment of wild animals a federal offense if the animals were taken in violation of state laws was the Lacey Game and Wild Birds Preservation and Disposition Act, approved by Congress in 1900.

1014. Mass-membership animal rights organization in the United States was the Humane Society of the United States, established in 1954. The society, with 3.5 million members, preserves wildlife and wilderness and intervenes on behalf of endangered species; it has campaigned for protection of whales, dolphins, elephants, bears, and wolves, and to curtail use of animals in medical research.

1015. Bill to require humane slaughter of animals in the United States was introduced by Rep. Martha Griffiths (D, MI) in 1956. Although that original bill was not passed, Congress in 1958 enacted the Humane Slaughter Law for animals in slaughterhouses and trapped animals on federal land.

1016. Broad-based animal welfare organization to oppose the slaughter of fur seals was the Humane Society of the United States, established in 1960. Other groups such as People for the Ethical Treatment of Animals (PETA) and the Fund for Animals soon joined the campaign against seal killings.

1017. Federal law in the United States closely regulating the use of animals in medical and commercial research was the Laboratory Animal Welfare Act of 1966. A grassroots lobby campaign led by Christine Stevens, founder of the Animal Welfare Institute, was credited with leading Congress to approve the measure.

1018. Law in the United States to require laboratories to furnish adequate food and shelter for test animals was enacted in 1966. The Animal Welfare Act of that year did not, however, limit the types of experiments that could be done on lab animals.

1019. Activist to popularize the term *speciesism* was the author Peter Singer in his book *Animal Liberation*, published in 1975. The term describes an attitude that places human interests above those of all other species. Opposition to speciesism is a foundation of the modern animal rights movement.

1020. Legislation in the United States making animal fighting a federal offense was the Animal Welfare Act Amendments, enacted in 1976. The Humane Society of the United States lobbied hard for the measure.

1021. Underground animal rights organization to operate in the United States was the Animal Liberation Front, an offshoot of an English group, established in 1979. Activists broke into government and university laboratories, medical research facilities, slaughterhouses, hatcheries, and furriers and claimed responsibility for setting fires, destroying equipment, and stealing research materials.

1022. Boycott of a major U.S. automaker in protest of the use of animals in crash tests began on October 1, 1991, against some General Motors dealerships. People for the Ethical Treatment of Animals (PETA), the radical animal rights group, launched the "Heartbreak of America" boycott against General Motors dealerships in 20 states.

1023. State in the United States to call for a moratorium on planned breeding of dogs and cats was Washington in February 1994. The one-year moratorium came in response to a Humane Society of the United States campaign to reduce the population of unwanted pets.

ACTIVIST MOVEMENTS—ANTI-ENVIRONMENTALISM

1024. Publication that outlined the goals of the "wise-use" movement was *The Wise Use Agenda: The Citizen's Policy Guide to Environmental Resource Issues: A Task Force Report*, by Ron Arnold, leader of the "wise-use" movement, and Alan M. Gottlieb. It appeared in 1989, advocating what the authors called "a middle way between extreme environmentalism and extreme industrialism." The "wise-use" movement was a coalition of anti-environmentalist interests that had its beginnings at the Multiple-Use Strategy Conference in Reno, NV, in August 1988.

1025. Organized resistance to environmental regulations was signaled by publication of the essay "An Anti-Environmentalist Manifesto," by Llewellyn H. Rockwell, a conservative critic and propagandist who took a critical look at the environmental movement. His essay was published in a special quarterly report of conservative journalist Patrick J. Buchanan's newsletter, *Patrick J. Buchanan . . . From the Right*, in 1990.

ACTIVIST MOVEMENTS—CONSERVATIONISM

1026. Conservationist of note in the United States was the Scottish-born naturalist John Muir. Born in 1838, he arrived in the United States at age 11. He spent many years traveling around the United States and Canada as a forester and botanist before settling in California, where he campaigned for the establishment of national forests and parks, including Yosemite National Park, formed in 1890 by an act of Congress. He was also the founder of the Sierra Club. California's Muir Woods National Monument and Alaska's Muir Glacier were named after him.

1027. Wildlife conservation group in the U.S. was the New York Association for the Protection of Game formed in 1844.

1028. Society emphasizing conservation in the United States was the American Forestry Association. It was established in 1875 to help protect forests and to promote the planting of new trees.

1029. Private and nonprofit land conservation organization in the United States was the Appalachian Mountain Club, established in 1876. The Appalachian Mountain Club, organized to promote the conservation of the Appalachian Mountains, is still in existence today.

1030. Boone and Crockett Club meeting occurred in January 1888 in New York City. Theodore Roosevelt and *Field and Stream* editor George Bird Grinnell founded the sportsmen's conservation group. The group, now located in Alexandria, VA, lobbies for conservation laws and wildlife refuges.

1031. Irrigation advocacy group in the United States was the National Irrigation Association, begun in a hotel room in Wichita, KS, on June 2, 1889. George Hebard Maxwell formed the association to educate the nation about using irrigation to reclaim western lands. The group's efforts helped pass two major acts, the Carey Irrigation Act in 1894 and the Newlands Reclamation Act in 1902.

1032. National Conservation Commission in the United States was formed in 1908. President Theodore Roosevelt appointed 50 members from government, industry, and science to the new commission, which was charged with working with the states to inventory U.S. natural resources.

1033. Governors' conference on U.S. conservation issues convened in Washington, DC, on May 13, 1908. President Theodore Roosevelt called the Conference on the Conservation of Natural Resources to discuss land and forest management, irrigation, livestock grazing on public lands, and other issues.

1034. Tri-national conference on conservation in North America was the North American Conservation Conference, convened on February 18, 1909. An outgrowth of President Theodore Roosevelt's U.S. governors' conference of 1908, the North American Conservation Conference included representatives of Canada, Mexico, and the United States.

1035. Conservation group to focus on the eastern United States in addition to western states was the National Parks Association (NPA), founded by Stephen T. Mather and Robert Sterling Yard in May 1919. The first meeting was held at the Cosmos Club in Washington, DC. The NPA promoted parks to tourists and lobbied for more parks to be established. Since the 1970s it has been called the National Parks Conservation Association.

1036. Noted "back to the land" conservationist in the United States was Scott Nearing, whose life and works helped inspire the modern environmental movement. Nearing and his wife Helen bought a run-down farm in Vermont in 1932 and set out "to live sanely and simply in a troubled world." Several books, including *Living the Good Life* (1954), recount their experiences and offer a radical critique of the wastefulness of modern consumer society.

1037. International organization of significance dedicated to protecting and conserving global resources was established at Fontainebleau, France, in 1948. Governmental agencies and private groups of 33 countries were represented initially in the World Conservation Union. By the mid-1990s, the union claimed about 800 members from more than 100 countries.

ACTIVIST MOVEMENTS—CONSERVATIONISM—*continued*

1038. Private environmental organization of significance in the United States to purchase and preserve substantial natural habitats was the Nature Conservancy, established in 1951. Now the fourth largest environmental organization in the United States, with 600,000 members, the Conservancy owns 1,000 nature preserves totaling 5 million acres.

1039. Woman to head a major U.S. conservation organization and receive a salary was Kathryn Fuller, who became the president and CEO of the World Wildlife Fund (WWF) in 1989. In 1997 she launched the WWF's Living Planet Campaign.

ACTIVIST MOVEMENTS—ECOFEMINISM

1040. Ecological protests by women occurred in the 18th century when Amrita Devi, her daughters, and more than 350 Indian women gave their lives to protect a forest in the district of Rajasthan, India. They were members of the Bishbios sect and believed trees to be sacred.

1041. Nonviolent protest of significance by the modern Chipko movement in India occurred on April 4, 1973, when women of Gopeshwar in Uttar Pradesh confronted loggers. The women, who were educated by Mira Behn and Sarrala Behn, put the principles of nonviolence to work by hugging trees to stop the loggers from cutting them down. They eventually saved more than 4,633 square miles of forest in India.

1042. Use of the term *ecofeminism* occurred in 1974. French novelist, essayist, and journalist Françoise d'Eaubonne coined the term to show the relationship between ecology and feminist theory; both embraced the notion of the interconnectedness of all things. She used the term in her 1974 book *Le feminisme où la mort* (*Feminism or Death*).

1043. Environmental group in Kenya to tie deforestation to poverty was the Green Belt Movement, founded by Wangari Maathai in 1977. The group taught Kenyan women agricultural survival tips such as intercropping and agroforestry and planted several hundred trees in Kenya.

1044. WorldWIDE Network Inc. meeting was held in Washington, DC, in 1981. WorldWIDE Network Inc. is an international environmental forum started by Joan Martin-Brown. It provides a forum for women to share information and identifies women to speak to government bodies about local environmental issues.

1045. Environmental activist to become a member of a women's hall of fame was Mary Sinclair, inducted into the Michigan Woman's Hall of Fame in the fall of 1990. Mary Sinclair led the battle against Dow Chemical and Consumer Power in Midland, MI, in the 1960s, 1970s, and 1980s. As a result of her activism the proposed nuclear power plant in Midland is now a natural-gas power plant. Women's halls of fame honor women of outstanding achievement in all fields of study.

1046. Women's international conference on the environment was held in Miami in 1991. Joan Martin-Brown and the Women's Environment and Development Organization cosponsored this conference where conferees shared information on environmental progress in their countries.

ACTIVIST MOVEMENTS—ENVIRONMENTALISM

1047. Environmental movement of significance in the United States dates from about 1836 in New England, the locus of literary and philosophical transcendentalism. Ralph Waldo Emerson, Henry David Thoreau, and other transcendentalists expressed reverence for the natural world and believed that human beings and nature shared a divine spirit. Most of the first meetings in transcendentalism took place in Concord, MA.

1048. Research showing that human activity could cause irretrievable damage to the earth was contained in George Perkins Marsh's *Man and Nature*. Written while Marsh was in Italy, it was published in the United States in 1864. He discussed how deforestation, loss of wetlands, species extinction, and changes in weather patterns are related to human activity.

1049. Use of the term *ecology* was attributed to the German naturalist Ernst Haeckel. A supporter of Charles Darwin and his evolutionary theory, Haeckel in 1869 coined the term to describe "the body of knowledge concerning the economy of nature." He was also the first to draw a "family tree" illustrating the relationships between various animal groups.

1050. Expression of the concept that the earth is a living organism came in 1892 in *The Land Problem* by ethnologist Otis T. Mason, an early environmentalist. Ecologist James Lovelock later used the name of the Greek earth goddess Gaia to symbolize complex and changing biological behavior on the planet.

1051. National environmental organization in the United States was the Sierra Club, a product of the conservation movement, founded in 1892. The naturalist and conservationist John Muir and his allies established the organization in response to logger, miner, and rancher efforts to shrink the boundaries of the newly created Yosemite National Park. With more than 650,000 members, the Sierra Club is today probably the most powerful environmental group in the world.

1052. Use of the term *oekology* (ecology) in the United States occurred during a speech given by Ellen Swallow on December 1, 1892, at Boston's Vendome Hotel. She said "Oekology is the worthiest of all the applied sciences which teaches the principles on which to found healthy and happy homes."

1053. Introduction of the discipline of animal ecology came in 1927, when British zoologist Charles S. Elton published *Animal Ecology*. Elton was a key figure in the founding of the Nature Conservancy in England in 1948.

1054. "Land Ethic" concept was introduced by U.S. naturalist Aldo Leopold in 1933. His plea to care for land and its biological complex, instead of considering it a commodity, was first stated in a *Journal of Forestry* article. He later expanded and clarified his ideas in his book *A Sand County Almanac* (1949). The core of the "Land Ethic" concept argues that "a thing is right when it tends to preserve the integrity, stability and beauty of the biotic community. It is wrong when it tends otherwise."

1055. National organization dedicated specifically to preserving wilderness (as opposed to wildlife) in the United States was the Wilderness Society. It grew out of a discussion between founder Robert Marshall and others at an American Forestry Conference in Knoxville, TN, on October 19, 1934. The society organized its first conference in January 1935.

1056. Use of the term *ecosystem* was in 1935. The English plant ecologist Arthur G. Tansley coined the term to define a community of plants, animals, and bacteria and its interrelated physical and chemical environment.

1057. Sierra Club chapter formed outside of California was started by Polly and John Dyer in Washington State during the 1950s. Polly Dyer later became the first Sierra Club national board member who was not from California.

1058. Exploits of "The Fox," the anonymous Chicago-area environmentalist, were in the mid-1960s. The Fox jammed chimneys, plugged effluent pipes, and hung signs such as one that read: "We're involved—in killing Lake Michigan. U.S. Steel." His main targets were large industries with poor environmental records. The Fox was ultimately identified as James Phillips, a middle-school science teacher from Aurora, IL, after his death on October 3, 2001.

1059. World Charter for Nature was adopted by the United Nations in 1982. The charter proclaimed that "nature shall be respected and its essential processes not impaired."

1060. International conference to discuss global environmental issues was the Biosphere Conference in Paris in 1968. However, the UN-sponsored conference made no recommendations and took no other action.

1061. Assembly of significance of environmental lawyers, law school faculty and environmental leaders in the United States occurred in 1969 at Airlee House in Virginia. The gathering, sponsored by the Conservation Foundation, discussed the emerging field of environmental law.

1062. Comprehensive national policies for protecting the environment in the United States were introduced in 1969 with the National Environmental Policy Act. The act created a Council on Environmental Quality and required federal agencies to take environmental consequences into account in their plans and activities. The council's influence was sharply restricted during the Reagan administration in the 1980s.

1063. Environmental organization to pursue direct action as a primary strategy was Greenpeace, founded in Vancouver, British Columbia, in November 1969. Greenpeace operations have included plugging industrial effluent pipes, interfering with commercial whale fisheries, and sailing into nuclear test zones to protest scheduled test explosions. Greenpeace had offices in around 30 countries and a world membership of 4 million in the late 1990s.

1064. Requirement for an environmental impact statement for development projects in the United States was introduced in 1969. In most cases, the requirement applies to individual projects such as housing developments and shopping centers.

ACTIVIST MOVEMENTS—ENVIRONMENTALISM—*continued*

1065. Environmental movement of significance in the Soviet Union arose in the late 1960s in opposition to development and industrialization that threatened South Siberia's natural treasure, Lake Baikal. Such organizations as the Limnological Institute in Irkutsk and such prominent authors as Valentin Rasputin were at the forefront of the campaign to protect the lake. Their efforts influenced the Soviet Council of Ministers to pass a resolution in January 1969 mandating strict pollution controls on new industry constructed in the Baikal watershed and directing various national and regional Soviet authorities to make recommendations for the basin's sustainable use.

1066. Appearance of the environmental movement known as bioregionalism was in California in the 1970s. The movement, which spread through the United States in the 1980s, regarded the earth as life territory defined by topography and biota rather than by human beings and their governmental apparatus.

1067. Cabinet-level regulatory environmental agency in the United States came into being in December 1970. The Nixon administration created the Environmental Protection Agency to consolidate federal regulatory functions in a single bureau. William D. Ruckelshaus was the agency's first administrator.

1068. Organizational expression of the bioregional movement was Planet Drum, established in San Francisco, CA, in the 1970s. It produced a biannual publication, *Raise the Stakes*, and introduced the "Green City Program" to promote bioregionalism in cities.

1069. Earth Day to be celebrated nationwide in the United States was held on April 22, 1970. Twenty million Americans, including students at some 2,000 colleges and 10,000 high schools, participated in marches, educational programs, and rallies intended to increase public awareness of the world's environmental problems. In subsequent years some groups and officials have celebrated Earth Day on April 22, while others have chosen the Vernal Equinox to mark the occasion. An earlier Earth Day celebration had been organized on March 21, 1970, in San Francisco, CA.

1070. Global environmental conference to spur activism was the United Nations Conference on the Human Environment, held in Stockholm, Sweden, in 1972. Its most important result was the revolutionary precedent it set for international cooperation in addressing environmental degradation. All nations, the conference concluded, have a shared responsibility for the quality of the environment, particularly the oceans and the atmosphere. The immediate result was the formation of the United Nations Environment Programme, which gave rise to a growing group of national environmental agencies and private groups dedicated to the salvation of the environment through public awareness and protective legislation.

1071. Successful conservation organization to use paying volunteers to sponsor and assist scientists on research trips throughout the world was Earthwatch, established in Watertown, MA, in 1972. By the mid-1990s, Earthwatch had 70,000 members and more than 150 research projects underway.

1072. United Nations organization to establish its headquarters in an underdeveloped country was the United Nations Environment Programme, launched in Nairobi, Kenya, in 1972. UNEP's mission was to encourage conservation policies in member nations and to monitor environmental issues worldwide.

1073. Use of the phrase "think globally, act locally" dates to 1972, when ecologist René Dubos coined it at the United Nations Conference on the Human Environment in Stockholm, Sweden. Dubos meant to convey the idea that environmental consciousness begins at home.

1074. *EPA Journal* a publication of the Environmental Protection Agency, was issued in January 1975 by Joan Martin-Brown, the EPA director of public affairs. Published between 1975 and 1995, the *EPA Journal* was used by government leaders and activists alike.

1075. Environmental organization to substantially practice "ecotage" was Earth First!, a controversial radical group founded in Missoula, MT, in 1979. Earth First! founders took inspiration from naturalist/author Edward Abbey's novel *The Monkey Wrench Gang* (1975), in which the main characters plotted "monkey wrenching" in environmental causes and fantasized about blowing up the Glen Canyon Dam. Ecotage tactics include putting sand in bulldozer fuel tanks, spiking trees, and sabotaging drilling equipment. Earth First! has been heavily involved in challenging logging operations in the U.S. Pacific Northwest.

1076. Patron saint of ecology was Francis of Assisi, proclaimed by Pope John Paul II in 1979. The pope urged the faithful to practice the Italian saint's respect for all creation.

1077. Use of the acronym SLAPP dates from the mid-1980s. SLAPP is the tactic employed by polluters and developers who file suit in U.S. courts against their environmentalist opponents. It stands for "Strategic Lawsuit Against Public Participation."

1078. Group of Ten meeting took place in a Washington, DC, restaurant on January 21, 1981. Leaders of the nine (the tenth executive joined later) largest national environmental organizations hosted potential contributors who wanted to donate money to their cause and to discuss common goals for political action. Deliberately held the day after Ronald Reagan was inaugurated, the Group of Ten ushered in the modern era of environmentalism. The organization disbanded at the end of the 1980s but the name Group of Ten continued to be used as a symbol of mainstream environmentalism.

1079. National protest of significance sponsored by Earth First! was on March 21, 1981, at the Glen Canyon Dam on the Colorado River along the Utah-Arizona border. Earth First!, which considered itself a radical environmental organization, was founded in April 1980 and operated under the slogan "No Compromise in the Defense of Mother Earth."

1080. Substantial cuts in the budget of the U.S. Environmental Protection Agency were made in 1981 and 1982, the first two years of the Reagan administration. The agency, established in 1970, lost 29 percent of its budget and a quarter of its staff during those two years.

1081. Joint document from major U.S. environmental groups to spell out a national environmental policy was "An Agenda for the Future," published in 1984. The Sierra Club, the Audubon Society, and other major national organizations collaborated to produce the document.

1082. Radical "deep ecology" platform was developed in 1984 and published the following year. Norwegian philosopher Arne Naess and Bill Devall, an American sociologist, laid out a set of eight principles underpinning the need to protect the nonhuman world from human beings. Naess called the compromises of mainstream environmentalists "shallow ecology."

1083. Handbook on ecological sabotage was published in 1985. Dave Foreman's *Ecodefense: A Field Guide to Monkeywrenching*, established guidelines for sabotage in defense of the environment. Foreman, a cofounder of Earth First!, gave detailed instructions for destroying or disabling things that damage the wilderness.

1084. Radical environmental activist arrested and convicted for "monkeywrenching" in the United States was Howie Wolke, an Earth First! cofounder, in 1985. A fellow cofounder, Dave Foreman, the author of *Ecodefense: A Field Guide to Monkeywrenching* (i.e., environmental sabotage), was later arrested on federal conspiracy charges. Though Foreman pleaded guilty, he refused to disavow monkeywrenching as a tactic.

1085. Student Environmental Action Coalition in the United States was organized at the University of North Carolina in 1988. Within four years, the coalition claimed 33,000 members at 1,500 college campuses and 750 high schools throughout the United States.

1086. Joint United Nations–World Council of Churches "Environmental Sabbath" was declared for June 18, 1989. The council's convention, meeting in San Antonio, TX, in June 1989 urged clergy of all denominations to preach on the integrity of all creation that day.

1087. Introduction of the concept of "ecotheology" came in a message from Pope John Paul II on December 5, 1989. In his message for the World Day of Peace, the pope told Christians it was their moral duty to protect the environment. The papal message had a major impact on Roman Catholic teaching on ecology.

1088. Mobilization and training of youth to participate in nonviolent environmental protest was organized by Judi Bari. The Redwood Summer of 1990 was modeled on the Mississippi civil rights campaign of 1964. More than 3,000 people came to Northern California to chain themselves to logging equipment and hug trees to protect the Headwater Forest from logging.

1089. Earth Day to be celebrated around the world took place on April 22, 1990, the 20th anniversary of the first environmental awareness Earth Day in the United States. In Kenya, celebrants planted 1.5 million trees; activists in Munich, Germany, released 10,000 balloons carrying environmental messages.

1090. Environmental coalition of small island nations was the Alliance of Small Island States (AOSIS), organized in November 1990. AOSIS has a membership of 43 small island nations, coastal nations, and observers from all regions of the world. It was organized by developing countries concerned about climate change and the impact it might have on low-lying maritime areas. It operates mainly as a lobby and consultant within the United Nations system.

ACTIVIST MOVEMENTS—ENVIRONMENTALISM—*continued*

1091. Widely acknowledged environmentalist to be chosen as a major-party U.S. vice presidential candidate was Tennessee senator Albert Gore. Democratic presidential candidate Bill Clinton tapped Gore as his running mate on July 9, 1992. Both were elected in November.

1092. Guidelines issued in the United States by the Federal Trade Commission for environmental marketing claims were adopted on July 28, 1992. The guidelines gave examples of legitimate "earth friendly" environmental product claims as well as ones that could be false or misleading.

1093. Scholarly article to identify a legal trend curbing the powers of environmental groups and private citizens to sue polluters in U.S. courts appeared in *The Environmental Law Forum* in 1999. In the article, two Georgetown University scholars traced the gradual weakening of citizen powers to a 1983 law review article by Antonin Scalia, who won appointment to the U.S. Supreme Court in 1986.

ACTIVIST MOVEMENTS—GRASSROOTS

1094. Mass death of "tree huggers" occurred during the 17th century in the district of Rajasthan in India. Amrita Devi, her daughters, and more than 350 Indian women died to protect the trees in that district. Amrita is the inspiration for the modern Chipko Movement in India.

1095. Land purchased for what would become Everglades National Park in Florida was bought by the Florida Federation of Women's Clubs in 1927. Twenty years later the Everglades National Park became a reality.

1096. Citizen lawsuit provisions allowing people to file legal actions in U.S. courts to protect the environment were contained in the Clean Air Act, the Clean Water Act, and other environmental legislation of the 1970s. Citizens could seek injunctions to stop pollution as well as penalties against polluters to be paid into the U.S. Treasury.

1097. Valhalla Wilderness Society meeting was in 1975. Colleen McCrory founded the group to publicize the value of the Valhalla Range in British Columbia, Canada, and to lobby for its preservation. In 1983, the British Columbian government created the Valhalla Provincial Park.

1098. Trees planted by the Green Belt Movement were planted in Nyeri, Kenya, on World Environment Day in June, 1977. A few women gathered in Wangari Maatahi's backyard and planted seven trees. These trees were the first of over 17 million trees the group has planted worldwide as of 1995.

1099. Love Canal Homeowners Association meeting was organized by Lois Gibbs in the spring of 1978. Love Canal was a section of Niagara Falls, NY, that was once a chemical disposal site. Between 1942 and 1953, Hooker Chemicals and Plastics dumped tons of toxic wastes in poorly sealed metal drums into a nearby empty canal. The canal was eventually filled in and sold by the company to the city of Niagara Falls. Housing and an elementary school were later built on the site. When a high number of birth defects, miscarriages, cancers, and other illnesses began to plague the area in the late 1970s, it was determined that toxins were seeping from the buried drums and making their way to the surface. The Love Canal Homeowners Association was organized in response to the growing concern over the toxins. In late 1978, Love Canal inhabitants had to move from their homes due to industrial pollutants. President Jimmy Carter declared Love Canal a federal disaster area in1978.

1100. Widespread use of the acronym NIMBY occurred in the United States in the 1980s. NIMBY stands for "not in my backyard," and it became a standard response to proposals for environmentally unsound or otherwise unwanted development, from nuclear waste disposal sites to low-income public housing.

1101. National Toxics Campaign in the United States was launched in Boston, MA, by the Sierra Club in 1984. The campaign coordinated grassroots movements pressing for cleanups of military waste sites.

1102. National organization formed to assist U.S. local, regional, and state grassroots environmental groups with their projects was the Environmental Support Center, established in Washington, DC, in 1990. The center helped local groups find free legal advice, get technical aid, and raise funds.

1103. Person in England to chain herself to a bulldozer to protest a highway was Emma Must, in June 1993. Later this tactic became a hallmark of the British Anti-Roadway Movement. Emma Must founded the Twyford Down Alert! in 1993 to protest the building of a major road through Twyford Down.

ACTIVIST MOVEMENTS—GREEN POLITICS

1104. State governor in the United States to implement a doctrine of conservation in balance with economic concerns was Oregon's Thomas McCall, who served from 1967 to 1975. Under the leadership of McCall, a Republican, Oregon introduced statewide land use planning, coastal management, and strict industrial pollution standards.

1105. Local environmental political parties were established in Sweden and Switzerland in 1971. These were the precursors of large and sometimes influential national parties in New Zealand, Germany, Great Britain, the Netherlands, and elsewhere. The first regional environmental political party emerged in the Australian state of Tasmania in 1972.

1106. Use of the term *green* in the context of environmental activism was recorded in 1971 with the founding of the radical environmental group Greenpeace. *Green* is now widely used to denote movements and political parties that campaign on environmental issues. Greenpeace got its start in the early 1970s with widely publicized direct action against nuclear weapons testing.

1107. National environmental political party is commonly regarded as the New Zealand Values party, established in 1972. The first national environmental party in Great Britain emerged in 1973; French and Belgian green parties were founded later in the 1970s.

1108. Political party to take an environmental issue to the polls was the United Tasmanian Group, which entered the Australian state elections in April 1972. The group mobilized to oppose a hydroelectric project that would flood Lake Pedder.

1109. National political party in Europe committed primarily to environmental and ecological issues was founded in Britain in February 1973. Known first simply as "People" and later as the Ecology Party, it became the Green Party in 1985. The party's electoral success has been modest, though it has influenced Britain's major political alignments on some environmental issues.

1110. Ecological candidate elected to a national parliament took office in Switzerland in 1979. Perhaps the best-known European environmental political party, Germany's Greens, emerged on a local and regional level in West Germany in 1977 but did not send their first representatives to the national legislature until 1983. The Swiss and German parties today gather between 5 and 10 percent of the national vote.

1111. Politician from a Green party to be elected to a national parliament was Daniel Brelaz in Switzerland in 1979. Two years later, four Greens were elected to the Belgian parliament.

1112. Parliament to have a substantial Green bloc represented was the West German parliament in 1982. Twenty-seven members of Germany's Green Party were elected. Petra Kelly was chosen as one of the German Greens' speakers.

1113. Establishment of what would become the U.S. Greens was in Minneapolis, MN, in 1984. The organization, modeled on the German Green Party, grew to 200 local chapters in all 50 states and changed its name to the U.S. Greens.

1114. Person to be elected the head of a government after serving as a minister of the environment was Gro Harlem Brundtland, elected prime minister of Norway in 1986. As prime minister, Bruntland became a world ecological leader who advocated a new emphasis on prevention instead of cleanup.

1115. State in which the U.S. Green political party gained ballot status was Alaska, in 1990. California followed in 1992.

1116. Woman to head the U.S. Environmental Protection Agency was Carol Browner, who was appointed in 1993. Browner believed that "economic progress and environmental protection can and must go hand in hand."

1117. Green politician to join a national government as a minister was Pekka Haavisto in Finland in March 1995. Haavisto won appointment as minister for the environment and planning.

1118. U.S. Green Party presidential campaign was launched in 1995. Candidate Ralph Nader's name appeared on the ballot in 22 states and he polled about 1 percent of the vote in the 1996 election.

ACTIVIST MOVEMENTS—GREEN POLITICS—*continued*

1119. National Green party to become part of a governing coalition in Europe was Germany's Greens, in December 1998. The Greens were the junior partner in Social Democrat Gerhard Schroeder's government.

AGRICULTURE AND HORTICULTURE

1120. Domestication of the grain barley occurred in Persia circa 7000 B.C.

1121. Domestication of the root vegetable potato occurred in South America circa 2500 B.C.

1122. Introduction of drought-resistant rice occurred in China and Southeast Asia circa A.D. 1000. The strain of rice was known as Champa, after the region (now Vietnam) where it first appeared.

1123. Appearance in Europe of American corn and potatoes occurred in the 16th century. Europeans fed corn to livestock but ate most varieties of potatoes themselves. Compared to traditional crops, potatoes provided a dramatically higher yield for the labor involved, and a higher standard of nutrition in many areas.

1124. Introduction of tobacco plants into Europe occurred beginning in 1559, when Spanish sailors brought seeds home to Spain from Santo Domingo in the Caribbean. Tobacco plants reached Italy in 1561.

1125. Commercial cultivation of tobacco began in the Jamestown Colony in present-day Virginia, in 1612. It proved highly profitable and a substantial export trade had developed with England by 1620.

1126. Agricultural experimental farm in an American colony was established on a 10-acre plot set aside by Savannah, GA, in 1735. A botanist was appointed "to collect the seeds of drugs and dying-stuffs in other countries in the same climate, in order to cultivate such of them as shall be found to thrive well in Georgia."

1127. Synthetic fertilizer was a superphosphate of lime made from charred bone (the waste products of sugar refineries), to which were added sulfate of ammonia and Peruvian guano, developed by James Jay Mapes, who experimented with fertilizers in 1847 on his 20-acre farm in Newark, NJ. He applied for a patent in 1849, which was granted on November 22, 1859.

1128. Widespread use of South American guano in European agriculture occurred after 1850. Use of this natural fertilizer, used for centuries in South America, led to a significant increase in agricultural productivity in Europe.

1129. Head of the Department of Agriculture to be a member of the President's Cabinet was Norman Jay Colman in 1889. On February 8, 1889, the title Commissioner of Agriculture was changed to Secretary of Agriculture.

1130. Introduction of blended tobacco occurred in the United States during the first decade of the 20th century. Together with the development of popular name brands such as Camel (1913), milder blended tobaccos touched off a great expansion of cigarette smoking.

1131. Model conservationist farm in the United States was Malabar Farm started by Louis Bromfield in 1938 outside Mansfield, OH. Bromfield documented his experiments at Malabar Farm in four books, *Pleasant Valley* (1945), *Malabar Farm* (1948), *Out of the Earth* (1950), and *From My Experience* (1955). He was awarded the Audubon Medal in 1952 for his research at Malabar Farm, which included soil saving techniques, conservation, and cheaper farming techniques.

1132. High-yield crops were developed beginning in September 1944 when U.S. plant pathologist Norman Borlaug took up his duties with a team of agricultural researchers in Mexico City, Mexico. Borlaug's high-yield wheat and corn strains made him the father of the Green Revolution. By 1953, as a result of his work, Mexico had doubled its output of wheat.

1133. Variety of high-yield rice was introduced in 1964. Building on plant pathologist Norman Borlaug's work on wheat in Mexico in the 1940s, scientists at the International Rice Research Institute at Los Banos in the Philippines bred a "miracle rice" from Indonesian and Taiwanese strains. The new crops proved susceptible, however, to diseases and pests.

1134. Proposal for a national network of living historical farms in the United States was advanced in 1965. Agricultural economist Marion Clawson's idea caught on, and within a decade the living farm network was attracting 600,000 visitors a year.

1135. National memorial dedicated to farmers was the National Farmers Memorial in the Agricultural Hall of Fame in Bonner Springs, KS. It was built by the National Ideals Foundation in 1986.

1136. Farmland of significance in New York State to be permanently protected was the David and Margaret Rockefeller farm in the Hudson Valley. The 2,000 acres were set aside for preservation on June 23, 1992.

AGRICULTURE AND HORTICULTURE—ANIMAL HUSBANDRY AND LIVESTOCK

1137. Human herding of semi-domesticated gazelles dates from circa 18,000 B.C. in the Levant.

1138. Animal to be domesticated was probably the dog, in Persia, circa 12,000 B.C. The first dogs domesticated in the United States were in the northwest, a thousand years later. Early societies domesticated canines for use in hunting other animals.

1139. Domestication of sheep and goats occurred in Mesopotamia circa 9000 B.C.

1140. Domestication of pigs occured in Mesopotamia circa 6750 B.C.

1141. Domestication of cattle occurred in Greece circa 6000 B.C.

1142. Domestication of the chicken occurred circa 6000 B.C. in China. Domestic birds are thought to have derived from the red jungle fowl native to Southeast Asia.

1143. Domestication of the horse occurred in the Ukraine circa 3500 B.C. Researchers base this belief upon evidence of skeletal remains found in that region. Some authorities, however, speculate domestication may have happened somewhat earlier in Central Asia.

1144. Known use of honey from honeybees as a food dates from 3500 B.C. in Egypt. Honey was humankind's first source of sweets.

1145. Domestication of the common goose occurred in southeastern Europe circa 3000 B.C.

1146. Evidence that dromedary (one-humped) camels were domesticated in Arabia dates from 3000 B.C. Bactrian (two-humped) camels were domesticated in Persia and Turkestan around 2500 B.C.

1147. Use of cattle for dairy products dates from circa 2500 B.C., probably in Greece or neighboring areas of the eastern Mediterranean.

1148. Large animals domesticated in the Western Hemisphere were the llama and the alpaca, in what is now Peru. Domestication is thought to have occurred about 4,000 years ago. Llamas and alpacas provided early Andeans with meat, hides, milk, and wool; male llamas were also valued as beasts of burden.

1149. Evidence of the use of shoes for horses dates from circa 900 A.D. in western Europe. Along with the development of more effective harness equipment, shoeing allowed humans to extract more work from draft horses.

1150. Introduction of rabbits into England occurred in the 12th century. Domesticated rabbits were imported from southern Europe; they only later escaped into the wild.

1151. Widespread use of the horse as a plough animal in western Europe dates from circa 1100. In England and some other places, oxen, cheaper to maintain, remained dominant for several more centuries.

1152. Introduction of cattle and reintroduction of horses into the Americas occurred in 1493. Columbus carried the domesticated animals on his second voyage to the New World. Horses had once lived in the Americas, during prehistoric times, but had become extinct there for unknown reasons.

1153. Wild horses appeared in America in the 16th century. The horses were descendants of tame animals introduced by Spanish explorers and settlers. By 1900, as many as 2 million wild horses, known as mustangs, roamed the west. The mustang population is around 20,000 today.

1154. Sheep imported into an American colony were shipped to the Jamestown Colony in present-day Virginia, in 1609, by the London Company.

1155. Honey bees carried into the Western Hemisphere came from England to Virginia in 1622. Bees were common in the eastern American colonies by the late 18th century and honey production became an important element of the colonial economy.

1156. Cattle imported into a North American colony arrived in March 1624 from Dover, England, brought over by Edward Winslow, governor of the Plymouth Colony in Massachusetts. At the time, cows were valued principally for their hides, secondly for meat, and very incidentally for their milk.

1157. Appearance of modern sedentary ranching in the Western Hemisphere came in the mid-18th century, when Spanish explorers and missionaries introduced herds of horses, cattle, sheep, and goats into the open rangelands of what would be Mexico and the United States.

1158. Naturalization of the honeybee in North America occurred circa 1800. Europeans introduced the bees to the continent and soon they were used to produce domesticated honey throughout the United States and Canada.

1159. Widespread use of oil cake as an animal feed in England occurred during the 1820s. Its general use led to an increase in livestock productivity.

AGRICULTURE AND HORTICULTURE—ANIMAL HUSBANDRY AND LIVESTOCK—*continued*

1160. Camels imported into the United States arrived in the late 1840s. They were used for a short time along mail and express routes across arid western territories recently acquired from Mexico. The trials were not a success and the animals were destroyed.

1161. Fur-bearing animals raised commercially were minks reared in Oneida County, NY, by H. Ressegue in 1866. Live animals for breeding stock sold for $30 a pair.

1162. Pekin duck brought to the United States was in 1873, when a clipper ship sailing from China delivered nine to Long Island. Millions of members of the Long Island duckling variety descended from this founding stock.

1163. Barbed wire product on the market was introduced in 1874 by Joseph Farwell Glidden. Glidden's product made such a hit that by 1890 most privately owned Great Plains rangeland had been fenced with barbed wire.

1164. Introduction of larger, meatier British cattle on U.S. ranches occurred in the 1880s. The new breeds displaced traditional longhorn cattle from Texas and California north to Wyoming, Montana, and Canada.

1165. Recommendation to sell U.S. public lands in large parcels for grazing was made in the 1880s by the U.S. Public Lands Commission. The proposal called for selling off lands to ranchers and stockmen in 2,560-acre parcels. Congress ignored the recommendation.

1166. Ban on fencing the public domain in the United States came via Congressional action in 1885. Ranchers and stockmen in the West had been fencing in public rangelands and limiting access to water to protect their own "range rights."

1167. Grazing fees assessment for use of U.S. federal lands came in 1906, when the government began charging for grazing use of national forest reserves.

1168. Major exposé of unsanitary conditions in the U.S. meatpacking industry was Upton Sinclair's novel *The Jungle*, published in 1906. A convert to socialism, Sinclair invested his earnings from this best-selling investigation of the Chicago stockyards in a cooperative community.

1169. Federal management policy for public grazing lands in the United States was contained in the Taylor Grazing Act of 1934. The act sought to protect grazing lands and to provide for their "orderly use, development and improvement" and to "stabilize the livestock industry dependent on the public range."

1170. Grazing district on public lands in the United States was established in Wyoming in 1935 under the terms of the Taylor Grazing Act of 1934.

1171. Screwworm infestations in Florida and Georgia occurred in 1937. The larvae of this fly feeds off living flesh; the plague devastated livestock herds, with as many as 1 million cases a year reported.

1172. Use of antibiotics as an animal feed additive was in 1949. In experiments, Lederle Laboratories researcher Thomas Jukes found that antibiotics added weight to animals and improved the quality of meat. By the 1950s, feed additives were widespread, even though there were suspicions from the first that they might contribute to the development of resistant bacteria.

1173. Comprehensive report to conclude that subtherapeutic use of antibiotics in animal feed posed a significant health risk to humans was issued in Great Britain in 1969. The Swann Committee's report followed an epidemic of antibiotic-resistant salmonella that began in cattle and spread to humans.

1174. Large-scale seizure and auction of cattle grazing on U.S. land without a permit occurred in Arizona on June 26, 1992. The U.S. Bureau of Land Management seized 84 head of cattle belonging to an Arizona rancher.

1175. Ranchers' organization to promote "predator-friendly" wool was the Growers' Wool Cooperative, formed in 1997 in Montana. Its members do not trap, gas, poison, or shoot the predators (principally coyotes) that threaten their sheep; instead, they rely on guard animals, such as llamas, donkeys, and dogs, to protect the flocks.

1176. Substantial revisions in U.S. guidelines for approving new antibiotics for livestock and for monitoring the effects of existing drugs were proposed in 1999. The U.S. Food and Drug Administration's aim was to curb misuse of antibiotics, which were suspected of being a factor in the emergence of drug-resistant bacterial infections in humans and animals.

AGRICULTURE AND HORTICULTURE—CASH CROPS

1177. Introduction of bananas into the Americas occurred in 1516. Spanish colonizers brought the plant from the Canary Islands off the coast of Africa.

1178. Coffee plantations in Brazil were developed near Rio de Janeiro in 1774. By the end of the 19th century, Brazil produced three-quarters of the world's coffee.

1179. Introduction of cocoa into West Africa occurred in the late 1870s. The region's British colonial rulers promoted cocoa, figuring it would produce export revenue; by 1911 Ghana had become the world's largest cocoa producer.

1180. Experimental rubber plantations in Malaya were started in 1877 with rubber tree seeds taken from Brazil. By 1919, the British colony was growing half the world's rubber.

1181. Banana plantations in Central America started by a U.S. company were established in 1889 by the America United Fruit Company. With plantations in Panama, Nicaragua, and elsewhere, the American firm soon came to dominate the world banana trade.

AGRICULTURE AND HORTICULTURE—DISEASES AND PESTS

1182. Colorado potato beetles to feed on the potato began to do so in the American West in the early 19th century. The beetles originally fed on native toxic weeds like the buffalo burr, but developed a taste for the leaves of plants in the nightshade family, like potatoes, tomatoes, and eggplant, as well as other vegetables. With abundant supplies of food, the beetles quickly spread across the United States, reached Europe in the 1920s, and have become one of the world's most destructive and insecticide-resistant insect pests.

1183. Famine as a result of potato blight occurred in Ireland from 1846 to 1851. The failure of Irish monoculture caused the Great Hunger, in which more than a million people died and hundreds of thousands were forced to emigrate to England, the United States, Canada, and Australia.

1184. Introduction of the citrus-damaging cottony-cushion scale insect in California occurred in 1868. The pest probably came in on oranges imported from Australia. By the 1880s, it had caused major damage to Southern California's citrus industry.

1185. Appearance of the potato-destroying Colorado beetle on the east coast of the U.S. was in 1874. The pest had moved rapidly eastward from Colorado, ravaging potato crops in many states along the way.

1186. Massively destructive eastward migration of the Rocky Mountain grasshopper occurred from 1874 to 1877. Grasshopper swarms devastated Great Plains grain fields before the plague burned itself out in 1878.

1187. Widespread damage from coffee leaf disease in Ceylon (Sri Lanka) occurred in 1874. Over the following 30 years, the spread of the disease severely reduced Ceylon's coffee production.

1188. Appearance of the alfalfa weevil in the United States was in 1900. The pest arrived from Europe, probably in imported nursery stock. By the 1940s, it had become a major threat to the alfalfa crop in the eastern United States

1189. Japanese beetles in the United States appeared in Riverton, NJ, in 1916. Their grubs were believed to have arrived in the roots of imported nursery stock.

1190. Appearance of the cotton-destroying boll weevil in Virginia occurred in 1922. Thirty years after its first appearance in Texas in 1892, the pest had migrated through the entire cotton-growing South.

1191. Infestation in the United States of the Mediterranean fruit fly occurred in Florida in 1929. State authorities moved rapidly to eradicate the fly, which attacks citrus and other fruit crops.

1192. Probable appearance of the cereal leaf beetle in the United States was in 1959 in Michigan. The pest, a native of Europe, seriously damaged Midwestern grain fields in the 1960s and spread as far eastward as Massachusetts.

1193. Appearance of the wooly whitefly in California was in 1966, in the San Diego area. This citrus-damaging pest migrated northward from Mexico.

1194. Appearance of the alfalfa blotch leaf miner in the United States came in 1969. The pest, a native of Europe, spread rapidly in the northeastern United States and Canada, attacking fields of alfalfa, the most important North American livestock forage crop.

AGRICULTURE AND HORTICULTURE—FOOD CROPS

1195. Domesticated plants genuinely cultivated were beans and chili peppers in Peru, squash in Mexico, and wheat and barley in Jericho in the Jordan Valley, all from circa 8000 B.C., according to archeological evidence. Cultivation of these crops marked the beginning of sedentary agriculture.

1196. Domestication of millet, a type of cereal grass occurred on the North China plain circa 6000 B.C. The first settled communities appeared there around the same time.

1197. Cultivation of vines, olives, and figs dates from circa 4000 B.C. in the Mediterranean region.

1198. Appearance of wet-rice cultivation in paddy fields occurred circa 500 B.C. Developed in Southeast Asia, the technique spread over the next thousand years to China, Korea, Japan, India, and Java.

1199. Introduction of orange, lemon, and lime trees into Spain and the western Mediterranean occurred by the 10th century. Islamic traders brought the first trees to the region from Southeast Asia.

1200. Introduction of sugar cane from Mesopotamia into the Levant, Egypt, and Cyprus occurred during the 10th century. Large numbers of slaves were used to cultivate the crop.

1201. Introduction of tomatoes to Europe occurred in the 1550s. Spanish priests brought the New World plant from Mexico to Spain and Italy.

1202. Widespread cultivation of buckwheat in northern Europe occurred in the Netherlands from 1550. Farmers found buckwheat attractive because it was hardy and had a short growing season.

1203. Introduction of the potato into Europe occured when it was introduced into Spain in 1570. A native of South America, the potato then spread rapidly throughout Europe.

1204. Introduction of manioc into Africa from Brazil occurred in the early 17th century. The high yield and the drought- and pest-resistance of manioc, a root similar to a sweet potato, made it a staple food in equatorial Africa.

1205. Introduction of the potato into North American colonies is believed to have occurred in 1613, when the potato was brought to Bermuda from Europe. A native of South America, the potato had been taken to Europe by the Spanish in 1570 and widely adopted, in Spain and elsewhere, as a nutritious and economical crop. It returned to the New World via Bermuda (1613) and Canada (1621); the first records of its cultivation in what is now the United States date from 1685 in Pennsylvania.

1206. Cultivation of tea in India for commercial use developed after 1833, when the East India Company lost its monopoly on the tea trade. Forests were cleared and large plantations established on the hills of Assam; by 1900, 375,000 acres were in production.

1207. Development of the Burbank or Idaho potato came in the early 1870s through the work of U.S. plant breeder Luther Burbank. His better potato was estimated to have added $1 billion to the world's wealth over a 50-year period.

1208. Introduction of a chemical "on-off switch" that would make seeds sterile in a generation occurred in the 1990s. The new seed touched off a biotechnology debate in 1999. Seed companies were interested in the development as a means of preventing farmers from growing valuable genetically engineered crops without buying the seeds each year. Farmers opposed it, fearing higher costs.

1209. Food irradiation plant in the United States received a license to operate in Florida on December 11, 1991. The irradiation process kills bacteria and insects on food.

AGRICULTURE AND HORTICULTURE—IRRIGATION

1210. Irrigation, as distinguished from dry farming, first appeared circa 5500 B.C. in southwest Khuzistan, on the eastern border of Mesopotamia. It consisted of the digging from larger waterways and creating small earth dams to supplement the watering of crops by rainfall alone.

1211. Environmental damage of significance resulting from irrigation occurred in Sumeria in Mesopotamia as early as 2500 B.C. Rising salt levels in the soil caused crop yields to decline and eventually made the cultivation of wheat impossible.

1212. Agricultural collapse caused by irrigation began in Mesopotamia around 900 B.C. with the first recorded large-scale building of irrigation canals to boost food production. Dug between the Tigris and Euphrates Rivers, they led to a rapidly rising water table and salinization that destroyed farm lands.

1213. Automatic water-driven irrigation wheel was developed in Egypt circa 100 B.C.

1214. Irrigation project of significance in what is now the United States dates from 5th century in the Hohokam Indian communities in Arizona. By the 14th century, the Hohokams had a network of irrigation canals exceeding 150 miles.

1215. Large-scale irrigation of land in the modern era began in British India in the 1840s. Civil engineers constructed irrigation systems that converted vast sections of Northwest India (now Pakistan) into farmland.

1216. Irrigation law passed by U.S. Congress was the act of July 26, 1866, which ruled states could control their waterways subject to "local customs, laws and decisions of the court."

1217. Legislation in the United States granting a homesteader land at bargain rates in return for a promise to irrigate was the Desert Land Act of 1877. A settler could claim 640 acres at 25 cents an acre and a commitment to irrigate the tract.

1218. Water resources inventory of the arid western U.S. was authorized by Congress in 1888. The U.S. Geological Survey measured water supplies, surveyed potential dam and canal sites, and calculated the area of potentially irrigable land.

1219. Federal land-grant program to promote irrigation in western U.S. states was introduced in 1894. The Carey Irrigation Act made available one million acres of public land to each western state to promote irrigation. Funding limitations meant few irrigation projects were actually started under the land grant program.

1220. Law to provide for federal development of irrigated agriculture in the United States was the Reclamation Act (also known as the Newlands Act) of 1902. Approved June 17, 1902, the measure authorized federal funding of irrigation projects, with financing through the sale of arid lands in 16 western states.

1221. Discovery of large numbers of dead and deformed birds in the Kesterson Refuge and other ponds containing irrigation runoff from farms in the Central Valley of California came in 1983. Government biologists blamed the problems on high levels of the heavy metal selenium in the drainage waters; evaporation of stored water had led to dangerous concentrations of selenium, other metals, and salts. Some 16,000 birds died during the breeding season of 1984. In February 1985, the state ordered the federal Bureau of Reclamation to clean up the Kesterson Refuge or drain it.

AGRICULTURE AND HORTICULTURE—LEGISLATION

1222. Agricultural bureau established by the U.S. government was the Department of Agriculture, established in 1862. The department was raised to cabinet rank in 1889.

1223. Federal land grants for colleges that would further the "agricultural and mechanical arts" were made under the terms of the Morrill Act in 1862. The measure gave U.S. public universities responsibility for scientific research on agricultural problems. There are now some 70 land grant institutions in the United States.

1224. Agricultural experiment stations established by the U.S. government in individual states were started through the Hatch Act of 1887. The first Secretary of Agriculture, Norman Jay Coleman, wrote the Act, which established stations at land-grant colleges and instructed them to conduct research and experiments to help improve and maintain the agriculture industry in the United States. The Act provided federal subsidies as long as the states matched the contribution.

1225. Legislation in the United States to withdraw substantial amounts of land from cultivation was the Agricultural Adjustment Act of 1933. This New Deal measure was designed to raise farmers' income and reduce overproduction. More than 40 million acres were taken out of production during the first year of the act and farmers received several hundred million dollars in benefit payments.

1226. Legislation to regulate Indian grazing ranges in the United States was the Indian Reorganization Act of 1934. The measure led to new grazing regulations that emphasized conservation.

1227. Extensive U.S. government standards for quality and labeling of foods were set by the Federal Food, Drug, and Cosmetic Act of 1938. The U.S. Food and Drug Administration (established in 1927) enforced the act.

AGRICULTURE AND HORTICULTURE—LEGISLATION—*continued*

1228. Government-backed crop insurance in the United States was offered in 1938. The Crop Insurance Act provided farmers with protection against the loss of growing crops.

1229. Legislation in the United States to remove ecologically fragile farmland from production was the Food Security Act of 1985. It set aside a 45-million acre conservation reserve, the largest such reserve ever established in the United States up to then.

AGRICULTURE AND HORTICULTURE—METHODS AND EQUIPMENT

1230. Human practice of sedentary agriculture developed between 9000 and 5000 B.C. The first farming did not occur in any one locale but developed gradually in Egypt, Mesopotamia, India, and other places.

1231. Use of the shaduf or bucket-and-pole system of watering fields in the Nile Valley of Egypt dates from circa 1340 B.C. The more efficient system of watering led to an increase in cultivable land of about 10 percent.

1232. Introduction of the animal-drawn water wheel in the Nile Valley of Egypt dates from circa 300 B.C. An improved watering system allowed for an increase in the amount of cultivable land.

1233. Development of the heavy ox-drawn plow occurred in northern Europe during the 6th century. Its use spread slowly throughout Europe over the next four centuries.

1234. Adoption of the three-field rotation system occurred in France circa 800. One field was sown with winter wheat or rye in the autumn, a second field was planted with oats or barley in the spring, and a third field was left fallow. The system spread gradually throughout Europe.

1235. Sugar cane cultivation in the Madeiras Islands of the Atlantic began in the 1450s. The islands' Portuguese colonizers introduced the plantation system of the eastern Mediterranean and brought in thousands of African slaves to work the plantations.

1236. Widespread use of the cotton gin occurred in the United States in the late 18th century. By 1820, Eli Whitney's machine for processing cotton, invented in 1793, had made America the world's largest cotton producer.

1237. Cast-iron plow was patented on June 26, 1797, by Charles Newbold, a Burlington County, NJ, farmer. Except for the handles and beam, the plow was made of solid cast iron and consisted of a bar, a sheath, and a moldplate. It was the first cast-iron plow made. Few farmers adopted the plow, as many believed that iron poisoned the land.

1238. Successful harvesting machine was demonstrated by U.S. inventor Cyrus McCormick in July 1831 in a wheatfield near Walnut Grove, VA. McCormick's harvesting machine, also known as a reaper, would revolutionize farming in the mid-19th century.

1239. Patent on a practical reaping machine was given to Obed Hussey of Cincinnati, OH, in 1833. Cyrus McCormick invented the first successful reaper in 1831 but did not obtain a patent for it until 1834. Still, he eventually won out over Hussey and other competitors.

1240. Use of artificial irrigation in the Nile Valley of Egypt dates from the 1840s. The systems were built not to grow food but to grow cotton to supply European textile mills.

1241. Plow for pulverizing the soil was patented by George Page of Washington, DC, on August 7, 1847. The design used a revolving disk on the side of the plow.

1242. Appearance of the two-wheeled sulky plow occurred in the United States during the Civil War era, 1861–65. The sulky plow had a seat for the driver.

1243. Large-scale practice of dry farming methods in the United States occurred in the arid Great Plains in the 1880s. Dry farming involves planting crops that make the best use of available moisture in areas of light rainfall such as Montana, Wyoming, and the Dakotas of the north central U.S.

1244. Large-scale dredging effort to reclaim land in Florida's Everglades for farming began in 1906. Within a year, two large dredging vessels were carving out canals to drain the land. The canals drastically altered the natural drainage of the Everglades.

1245. Crop dusting by an airplane occurred circa 1918 in the United States.

1246. Small all-purpose tractor in the United States was introduced 1924. It led to the widespread mechanization of American farms.

1247. Description of hydroponic agriculture was in William Frederick Gericke's article "Aquaculture, a Means of Crop Production," published in December 1929 in the *American Journal of Botany*. The first use of the term *hydroponics* was in Gericke's article "Hydroponics—Crop Production in Liquid Culture Media," published on February 12, 1937, in *Science*. Early attempts at hydroponic growing of crops had used sand beds, mounted over tanks containing nutrient solutions, to hold the plantings.

1248. Private hydroponic garden was created in 1931 by William Frederick Gericke at his home in Berkeley, CA. He grew vegetables and flowers in nutrient-enriched water.

1249. Commercial production of plants in water instead of soil was begun in February 1934 by the firm of Vetterle and Reinelt in a Capitola, CA, 100-tank greenhouse (100 by 33 feet). The first planting consisted of about 2,000 begonias, which grew more rapidly than if planted in soil, thanks to the precise regulation of humidity and food supply. On October 12, 1935, the growers set out tomato plantings that grew to 15 feet in height within six to eight months.

1250. Commercial hydroponicum on a large scale was established on December 5, 1935, in Montebello, CA, by Ernest Alfred Brandon and Frank Farmington Lyon, who installed a water circulating system. They obtained a patent on December 1, 1936, on a "system of water culture" and incorporated the Chemi-Culture Company on October 19, 1937.

1251. Use of the term *hydroponics* came in 1936. University of California scientist William Frederick Gericke, who developed commercially viable hydroponically grown tomatoes, coined the term. Hydroponics involves cultivating plants by placing the roots in nutrient-rich liquids instead of soil.

1252. Government programs in the United States to assist migrant workers and their families were introduced in the mid-1960s. Within 20 years, government agencies were spending $600 million a year on aid for migrants and their dependents.

1253. "Miracle rices" were developed in several countries during the early 1960s. These new rices, richer in protein and providing higher yields, proved to be a boom to underdeveloped countries.

1254. Introduction of the concept of "regenerative farming" was by Pennsylvania publisher Robert Rodale in the 1970s. Rodale argued that farmers could produce safe and profitable crops and protect the environment at the same time by maximizing the use of resources such as the sun, soil, water, and natural enemies, and by minimizing the use of chemical fertilizers, insecticides, and pesticides.

1255. Federal guidelines governing research with genetically altered material in the United States were issued in 1977. The National Institutes of Health guidelines specifically prohibited the release of genetically altered organisms. The standards were revised in 1978 to permit release under some circumstances.

1256. Deliberate release of genetically altered organisms in the United States occurred in California on April 24, 1987. Scientists from the Advanced Genetics Corporation of Oakland sprayed strawberry plants with genetically altered bacteria to improve the plants' freeze resistance.

1257. Use of the term *genetic pollution* occurred on May 20, 1999. Cornell University scientists reported in the journal *Nature* that pollen from genetically engineered corn containing a toxin gene had killed monarch butterflies feeding on leaves dusted with the pollen.

AGRICULTURE AND HORTICULTURE—ORGANIC FARMING

1258. Important scientific discussion of the negative impact of chemical fertilizers on food crops came in the work of Austrian scientist Rudolf Steiner in the 1890s. In Paris Steiner experimented successfully with close plant spacings and animal manures, an approach that became known as the "French method" of organic cultivation.

1259. Instances of modern organic farming as a conscious rejection of traditional chemical methods dates from the mid-1930s. British agricultural scientist Sir Albert Howard introduced natural animal and plant husbandry in which wastes were returned to the soil as nutrients.

1260. Comparative study of organic and conventional farming was begun by Lady Evelyn Balfour in the 1940s. She began her 34-year experiment on her farm in Haughley, England. The results of the Haughley experiment are still referred to today.

AGRICULTURE AND HORTICULTURE—ORGANIC FARMING—*continued*

1261. Serial publication in the United States devoted to organic agriculture was *Organic Gardening and Farming*, published by J. I. Rodale in 1945. Rodale created an organic farm in Emmaus, PA, and organized organic gardening clubs throughout the United States. He published results of his research on his farm in his magazine.

1262. Organic farming association in the United Kingdom was the Soil Association, formed in 1946 as a result of Lady Evelyn Balfour's book *The Living Soil.* One of its goals was to research the loss of soil through erosion and depletion. The association is based at Haughley Farm in England.

1263. Organic farming organization in Australia was Biodynamic Farming Association, founded by Alex Podolinsky in the 1950s. Podolinsky's farm near Powelltown was the model for his applications of Rudolf Steiner's theory of biodynamics, a system of farming that uses only organic materials for soil conditioning and fertilizing.

1264. Law in the United States to establish federal and state standards for organic certification was enacted in 1990. The federal law set standards for organic farming, marketing, and distribution. California and Minnesota were among the first states to adopt certification programs.

1265. Major apparel maker to make its entire line of clothing from organically grown cotton and nontoxic dyes was the VF Corporation, in 1992. The U.S.-based company marketed Wrangler and Lee jeans.

AGRICULTURE AND HORTICULTURE—ORGANIZATIONS

1266. Law setting up the Cooperative Extension Service in the United States was the Smith-Lever Act of 1914. The service, which operates in all 50 states and in U.S. dependencies, offers agriculture and forestry programs.

1267. Conference to discuss the scope of the newly established international Food and Agriculture Organization convened in Hot Springs, VA, in April 1943. Representatives of 43 nations discussed ways to improve agricultural conditions and food supplies around the world.

AGRICULTURE AND HORTICULTURE—PEST AND WEED CONTROL

1268. Chemical insecticides were used in biblical times in Palestine. Inorganic materials such as sulfur, arsenic, and mercury were used to kill insect pests on crops and vines.

1269. Use of Chrysanthemum powder as an insecticide was in China circa A.D. 100. Many consider this to be the first attempt at using an insecticide to help crop growth.

1270. Known case of the importation of a natural enemy to control an alien pest occurred in 1762 in India. Farmers brought in mynah birds to feed on crop-damaging red locusts.

1271. American vine aphids or *Phylloxera vastatrix* in Europe arrived in the mid-19th century by ship. The parasite devastated European vineyards before it was checked by the grafting of vulnerable European vines onto resistant American rootstocks.

1272. Entomologist hired by the U.S. government was Townend Glover, who was commissioned on June 14, 1854, to collect "statistics and other information on seeds, fruits, and insects of the United States." His first report, entitled *Insects Injurious and Beneficial to Vegetation,* was published by the U.S. Patent Office in 1854.

1273. Use of biological control against an unwanted plant occurred in Ceylon (Sri Lanka) in 1856. Cochineal insects were imported from India to attack the prickly-pear cactus.

1274. Insecticide that was effective in the United States was Paris green, a very poisonous copper-based powder. It was first successfully used in 1867 to combat the Colorado potato beetle. Within the next decade, Paris green was widely used in the United States against several types of chewing and sucking insects.

1275. Recognition of the citrus-attacking California red scale insect as a serious pest in the United States came in the 1870s. Various biological methods, including the use of parasitic wasps, gradually brought the scale under control.

1276. Fruit spraying in the United States was done in 1878 in Niagara County, NY, when a grower sprayed his apples with Paris green to control canker worms.

1277. Quarantine law for plants by a state was enacted in California on March 4, 1881. The law was particularly designed to protect cultivated grapevines in the Sonoma Valley against an insect known as *Phylloxera vastatrix*, which had attacked the vineyards in

1873, and against the San Jose scale insect and codling moth, which in 1875 had caused serious damage to tree fruits. Quarantine rules, covering both intrastate and interstate shipments of fruit and fruit trees, were issued on November 12, 1881.

1278. Fungicide considered effective was bordeaux mixture, developed in 1882 in France. Made of slaked lime and copper sulfate, it was used for decades on plants and fruit trees.

1279. Introduction of the scale insect–destroying Vedalia beetle in California occurred in the autumn of 1888. This method of biological control conquered a scale epidemic that had seriously damaged Southern California's citrus industry.

1280. Use of copper sulphate to destroy the charlock weed occurred in the 1890s. A weed in the mustard family, charlock growing uncontrolled can choke out cereal crops.

1281. Law in the United States that protected the public from hazardous substances was the Biologics Control Act of 1902. Congress passed a related measure, the Pure Food and Drug Act, in 1906.

1282. National insect quarantine program in the United States was enacted in 1905. California had its own quarantine in place as early as 1881.

1283. Quarantine law for plants enacted by the U.S. Congress was passed on August 20, 1912. It was directed against dangerous plant diseases and insect pests "new to or not theretofore widely prevalent or distributed within and throughout the United States." The immediate targets were plants that could transmit white-pine blister rust or potato wart, and plants that might harbor the Mediterranean fruit fly. The first quarantine imposed under the act was issued on September 16, 1912.

1284. Successful control of the cotton-destroying boll weevil came in the mid-1920s with the introduction of the pesticide calcium arsenate in the southern United States. Initial successes spurred the widespread use of chemical pesticides in the 1950s.

1285. Law in the United States to require warning labels on hazardous substances was the Caustic Poison Act of 1927. It called for warning labels on a dozen poisons.

1286. Use of organic chemicals for weed control occurred in 1935. Nitrophenols were used as selective herbicides in the United States

1287. DDT (Dichlorodiphenyl-Trichloroethane) synthesized for use as an insecticide was in 1939. Although DDT was first synthesized in 1874, its insecticidal properties were not known until discovered by the Swiss chemist Paul Müller in 1939. DDT is an odorless, white crystalline poison, with no apparent taste, that disrupts the central nervous systems. The use of DDT as an insecticide became widespread during the late 1940s and early 1950s. Although DDT was hailed as a "miracle insecticide" when it was first introduced, during the 1960s it became apparent that many insects had developed an immunity to DDT. Other species, such as fish, birds, and some plant life, were found to contain dangerously high levels of DDT. The U.S. Environmental Protection Agency banned DDT in 1972, but many countries around the world still use it as an insecticide.

1288. Commercial use of the pesticide DDT in the U.S. began in 1945. DDT (dichlorodiphenyl-trichloroethane) had proven effective in controlling carriers of malaria and other diseases during World War II. Use declined in the 1960s amid concerns about harm to ecosystems, plant and animal life, and human health. Rachel Carson's classic book *Silent Spring*, published in 1962, exposed DDT's dangers.

1289. Modern herbicide in the United States was 2,4-D, introduced in 1945. The herbicide's full name is 2,4-dichlorophenoxyacetic acid. It was developed to kill broad-leafed plants but not grass and grain crops. Unlike DDT, 2,4-D does not linger in the environment for long. However, its effects are not fully known. When mixed with its relative 2,4,5-T, it becomes the potent and potentially carcinogenic defoliant Agent Orange.

1290. Law of importance in the United States regulating pesticide use was the Federal Insecticide, Fungicide, and Rodenticide Act, passed in 1947 and much amended and strengthened over the years. Since 1972, the U.S. Environmental Protection Agency has administered the act, which monitors health and environmental consequences of pesticides.

1291. Widely used chemical herbicide regarded as benign in the environment was diquat, developed in England in 1955. Diquat proved highly effective in controlling weeds and was said to be harmless to animals and humans.

AGRICULTURE AND HORTICULTURE—PEST AND WEED CONTROL—*continued*

1292. Large-scale epidemic of fungicide poisoning from treated grain seeds occurred in Iraq in 1956. New outbreaks, caused when Iraqi peasants consumed rather than planted seeds treated with alkyl-mercurial fungicides, were reported in 1960 and 1971. The death toll reached into the hundreds, and many thousands became seriously ill.

1293. Eradication of the flesh-eating screwworm in the southeastern U.S. occurred in 1958–59. Government scientists used the release of millions of sterile screwworms to wipe out the pest over a period of 18 months.

1294. Public outcry of significance in the United States about cancer-causing chemicals in food occurred in 1959, when the federal government declared a herbicide found in small amounts in cranberry sauce to be carcinogenic to animals.

1295. Tests by the military of the defoliant Agent Orange were conducted at Fort Drum in New York State and in Texas and Puerto Rico in 1959. U.S. military aircraft sprayed Agent Orange, a suspected carcinogen, in Southeast Asia beginning in 1961. The defoliant destroyed more than one-third of the mangrove forest areas in South Vietnam.

1296. Ban on mercury fungicides in agriculture came in Sweden, in 1966. Swedish conservationists in the 1950s were the first to raise the alarm about the health effects of agricultural materials treated with mercury compounds.

1297. Mass fish kill attributed to an insecticide spill occurred in the Rhine river downstream of Bingen, Germany, beginning on June 19, 1969. Within a week, estimates of the number of dead fish and eels reached 50 million. By the end of June, there were virtually no fish in the Rhine.

1298. Move to halt all fungicide uses of mercury in the United States came on March 22, 1972. The U.S. Environmental Protection Agency determined the compounds posed a serious health hazard.

1299. Mediterranean fruit fly ("medfly") infestation in California occurred in 1975. State and federal authorities managed to eradicate the flies temporarily, but they returned in greater numbers in subsequent years.

1300. Major release of the pesticide dioxin into the atmosphere occurred as a result of the explosion of a chemical plant in Seveso, Italy, in July 1976. The highly toxic chemical spread in the form of a dense white cloud. Dioxin is known to cause liver disease and other ailments.

1301. Appearance of the bayberry whitefly in California citrus groves was in 1978. The introduction of a natural enemy, a parasite, brought the citrus-damaging whitefly under control by the mid-1980s.

1302. Introduction of early biologically based alternatives to chemical pesticides occurred in the 1980s. Biotechnology researchers created biologically based pesticides that were regarded as safer than chemical alternatives

1303. Massive spraying of the insecticide malathion to control the Mediterranean fruit fly in California commenced on July 10, 1981. The campaign cost the state and federal governments $100 million; farmers claimed $100 million in crop losses. The spraying caused an outcry among environmentalists, who opposed the use of the insecticide, as well as among farmers, who complained the government waited too long to attack the medfly.

1304. Large settlement of Agent Orange lawsuits came in 1984, when manufacturers of the herbicide agreed to a $180 million settlement with 9,000 U.S. veterans of the Vietnam War. The manufacturers did not, however, admit guilt or concede any connection between exposure to Agent Orange and health problems.

1305. National Resources Defense Council study to show that the chemical Alar is a carcinogen was in 1989. Apple growers had used the chemical to enhance the shelf life of their product.

1306. All-natural pesticide made from the seeds of the neem tree was introduced in the United States in 1991. The Ringer Corporation, a lawn products company, marketed the pesticide as Neem.

1307. Massively lethal pesticide spill in the Sacramento River occurred on July 14, 1991, when a railroad car carrying 19,500 gallons of the pesticide metam sodium went off the rails and spilled most of its contents into the river near Dunsmuir, CA. The pesticide killed all aquatic life in the river, from ducks and fish to insects and algae, for a distance of 45 miles downstream. A state study estimated a full recovery would take 50 years.

1308. Ban on some U.S. uses of the EBDC fungicide class was imposed on February 13, 1992. The U.S. Environmental Protection Agency barred the fungicides' use on 11 food crops, but said they could still be used on 45 others. EBDCs had been used on nearly all U.S.-grown vegetable and fruit crops.

1309. Reported large-scale bird kills traced to the lawn care pesticide Diazinon occurred in Virginia in July 1992.

1310. Confirmation that peach oil kills fungi and other soil pests came in 1999. Scientists at the U.S. Agricultural Research Service reported that the peach compound called benzaldehyde could replace methyl bromide a pesticide that is toxic to people and damages the earth's ozone layer.

1311. Use of an ancient weed-control technique by a modern American city occurred in 1999, when Denver authorities employed a herd of 100 cashmere goats to nibble weeds on 25 to 35 municipal acres. Goats prefer broadleafed plants to grass and have been used since ancient times to keep down weeds in fields and pasturelands.

AGRICULTURE AND HORTICULTURE—PRESERVATION

1312. Living historical farm in the United States was established in Iowa in 1967. Living History Farms near Des Moines grew out of agricultural economist Marion Clawson's proposal for a national network of historic farms.

1313. Program for public acquisition of farmland development rights was established in Suffolk County, NY, in 1974. The $50 million program aimed to preserve farmland on eastern Long Island.

AIR POLLUTION

1314. Use of sulfur in an industrial process dates to Egypt in 2000 B.C., when textile workers bleached their linens with fumes of burning sulfur. The burning of fossil fuels forms sulfur dioxide, a principal contributor to air pollution.

1315. Known use of the substance later called asbestos dates from the 2nd century B.C. The Romans made cremation cloths and wicks from the substance. It has since been shown that airborne asbestos fibers can cause lung cancer and other diseases.

1316. Published mentions of serious air pollution in classical Rome were in the 1st century A.D. Authors wrote of the blackening of building surfaces in Rome from air pollution.

1317. Recorded use of the word *asbestos* was by the Roman scientist Pliny the Elder in the 1st century A.D.

1318. Documented worldwide climate change from volcanic eruptions occurred in 1815 and caused the "year without summer." Mt. Tambora in Sumbawa, Indonesia, erupted, spewing about 150 cubic kilometers of ash. Average temperatures around the globe were 0.6 to 3.2 degrees lower the next summer.

1319. Campaign of significance against air pollution in London was launched on June 8, 1819. The campaign of Michael Angelo Taylor, a member of the British Parliament, led to the establishment of a select parliamentary committee to investigate the problem.

1320. Widespread deaths of trees in Moscow attributed to air pollution were reported in the 1930s. During this time, the Soviet Union, under Joseph Stalin, greatly increased its industrial output. The Soviet Union's rapid industrialization was accompanied by substantially increased pollution, which in turn proved to be a great threat to plants and trees across the country.

1321. Aerosol spray containers for insecticide that were portable were developed in 1941 by U.S. Department of Agriculture scientists Lyle D. Goodhue and William N. Sullivan Jr. The aerosol "bug bombs" were important in controlling disease on the fighting fronts during World War II. Aerosols came into widespread commercial use in the 1950s. They were gradually banned as their harmful effects on the environment, particularly the atmosphere, became known.

1322. Atmospheric tests of a new, more powerful atomic weapon, the hydrogen or thermonuclear bomb, were conducted by the United States in the Marshall Islands of the Pacific in 1952. Hydrogen bombs generated significantly greater radioactive fallout, and the contaminants were dispersed globally.

1323. Major public outcry on atmospheric nuclear-weapons testing occurred in 1954, when the United States detonated a thermonuclear device over the Marshall Islands in the Pacific. Inhabitants of nearby islands were exposed to radioactive pollution and large areas of Pacific fisheries were contaminated. In 1983, the U.S. gave almost $184 million to the Marshall Islands to pay for damages caused by the many nuclear and atomic tests conducted on the islands in the years after World War II.

AIR POLLUTION—*continued*

1324. North American country to recognize acid rain as a serious environmental and transboundary problem was Canada, in 1977. In a report, the Canadian Environmental Ministry described acid rain as an environmental time bomb and called for a bilateral U.S.-Canadian solution.

1325. Recorded rainfall with a pH of 2 occurred in Wheeling, WV. The pH scale measures the acidity or alkalinity of a solution, with the number 7 representing neutrality. Lower numbers on the scale indicate increasing acidity, and higher numbers, increasing alkalinity. When rainwater registers a pH of 5, it is officially considered acid rain; a pH of 2 indicates an acidity between that of lemon juice and that of battery acid.

1326. Radioactive fallout of significance from a space program occurred when a Soviet nuclear-powered satellite, *Cosmos 954*, disintegrated over western Canada in January 1978. The Soviet Union paid Canada $3 million to assist with recovery costs.

1327. Call from a U.S. governors' group for remedial action on acid rain came in 1985. The bipartisan Alliance for Acid Rain Control campaigned for a public/private partnership to seek solutions to the problem.

1328. Immediate ban on major uses of asbestos in the United States was proposed by the U.S. Environmental Protection Agency in 1986. A U.S. appeals court later partially overturned the ban.

AIR POLLUTION—HEALTH

1329. Smelting of copper occurred in Anatolia (present-day Turkey) circa 6000 B.C. Smelter emissions today are recognized as a source of air pollution.

1330. Large-scale air pollution problems in London date from the 16th century with a great increase in the burning of coal. A huge smoke pall could be seen from great distances hanging over the city.

1331. Use of London's weekly Bills of Mortality to establish a link between death rates and the burning of coal was in 1662. Statistician John Graunt claimed that his research proved such a link between the air pollution caused by burning coal and the death rate.

1332. Portland cement patent was taken out in England in 1796. A large cement industry developed along the north coast of the county of Kent; the kilns caused a severe air pollution problem, with smoke, dust and odor.

1333. Use of asbestos as an insulator on a large scale commenced in the 1870s, when mining for chrysolite, one of four types of asbestos, began in Quebec, Canada. Asbestos has been recognized for some time as a hazardous air pollutant that can cause lung scarring and lung cancer.

1334. Smog-related deaths reported in London occurred in 1873. Authorities reported unusually high death rates from bronchitis, pneumonia, tuberculosis and other respiratory diseases.

1335. Naming of the disease asbestosis was by a London physician in 1900. Asbestosis is a chronic, debilitating disease caused by inhaling asbestos fibers, which damage the lungs. It is often eventually fatal.

1336. Major private organization to campaign against air pollution in the United States was the Smoke Prevention Association (later, the Air Pollution Control Association), founded in 1907. The group, now known as the Air and Waste Management Association, has 11,000 members in 50 countries.

1337. Warning of air pollution dangers from coal-fired electrical generation came from Charles P. Steinmetz in 1908. Steinmetz, a German-born electrical engineer, issued the warning in a series of lectures in New York City. He said burning coal to generate electrical power threatened to "poison nature and ourselves with smoke and coal gas." Steinmetz called for development of hydroelectric power as a cleaner alternative.

1338. Deaths caused by air pollution in Scotland occurred in Glasgow in 1911. More than 1,000 people died as a result of smog from coal smoke pollution.

1339. Chemist to add tetraethyl lead to gasoline was Thomas Midgley of General Motors, in 1921. The lead kept the enginefrom knocking, but it also made auto exhaust a more dangerous pollutant. Midgley went on to develop chlorofluorocarbons (CFCs) in 1928; these nontoxic, non-flammable compounds were used as refrigerants, solvents, and aerosol propellants. In the mid-1970s, scientists identified CFCs released into the atmosphere as the major cause of the depletion of the ozone layer.

1340. Effective refrigerant for mass consumer use was Freon. The Du Pont Chemical Company and General Motors introduced Freon, which contained chlorofluorocarbons (CFCs), in Wilmington, DE, in 1930. Freon was later shown to harm the earth's ozone layer and contribute to global warming—the "greenhouse effect."

1341. Smog-related deaths recorded in the United States occurred in October 1947 in Donora, PA, which is set in a natural basin. Pollutants were trapped over the heavily industrialized town by a thermal inversion, which lasted four days, leaving 20 people dead and 6,000 in need of medical treatment.

1342. Nuclear-weapons tests in the atmosphere in the Pacific were conducted by the United States in 1946. Between then and 1980, the United States and four other nuclear powers detonated more than 500 nuclear devices above ground, raising widespread fears of radioactive air pollution.

1343. Casualties of the great London killer fog of December 1952 were cattle at the Smithfield livestock show. An Aberdeen Angus died and 12 others had to be slaughtered; another 160 were treated for respiratory problems.

1344. Smog episode to cause widespread loss of life occurred in London from December 4 to 8, 1952. Cold weather, fog, and the burning of coal and gasoline created a killer smog that claimed at least 4,000 lives. The tragedy led to Britain's first strict clean air standards.

1345. Recorded blanket of killer smog to cover New York City occurred in 1953. Death toll for the episode ranged from 200 to 300.

1346. Substantial radiation "hot spot" to be observed far from the site of an atmospheric nuclear test occurred near Troy, NY, in April 1953. A rainstorm delivered a high concentration of radioactive contaminants from a Pacific Ocean test.

1347. Smog emergency alert plan in Los Angeles County was adopted in 1955. The county also has a smog forecasting system that warns people of smog dangers on particular days; it has been in place since 1969.

1348. Conclusive link between asbestos and lung cancer came from studies done by Dr. Irving Selikoff and others in the 1960s. Their research showed the incident of death was six times higher in wartime shipbuilders. Some 632 industry workers were randomly chosen for the study.

1349. Air pollution index broadcast on television was aired in 1964. As a result of a letter writing campaign by Hazel Hunderson, WABC-TV in New York City began including the air pollution index in daily weather reports. Other stations and publications followed in New York City and nationwide soon after.

1350. Air Pollution Control Alert Warning System for New York City was established on October 31, 1969. The four stages of the alert were triggered by weather forecasts as well as readings of concentrations of carbon monoxide, sulfur dioxide, and particulates.

1351. Disabling smogs in Tokyo began in 1970 and lasted into 1972. Polluted air made some 50,000 people seriously ill throughout the city.

1352. Recognition of radon as a dangerous indoor household pollutant in the United States came in the early 1980s. Radon is a naturally occurring gas that is odorless, colorless, and almost inert. High levels, however, present a significant health hazard and can cause lung cancer. The Radon Gas and Indoor Air Quality Research Act of 1986 spurred studies to find means to curb radon exposure.

1353. Law of importance in the United States requiring government and industry to inform the public about toxic chemicals was the Emergency Planning and Community Right-to-Know Act of 1986. Congress enacted the law after the accidental release of chemicals at a Union Carbide plant in India killed thousands of people.

AIR POLLUTION—INDOOR

1354. Identification of some disabling industrial diseases was made by the Italian medical scholar Bernardino Ramazzini between 1682 and 1714. He discovered that potters suffered from paralysis and loss of teeth from lead exposure, and that glassblowers experienced lung damage from breathing in borax and antimony.

1355. Building code to be adopted in the United States was the code in New York City in 1867. The code specified, among other things, that all tenement buildings be equipped with ventilators and water closets.

1356. Formaldehyde synthesization took place in 1867 in Germany. Used improperly as a foam insulator in the United States in the 1970s, it caused severe indoor pollution problems.

1357. Mass production of cigarettes began in the 1880s in Durham, NC. During that decade, as many as 500 million ready-made cigarettes were sold annually. Cigarettes have been shown to be a significant health hazard; second-hand smoke is a form of indoor air pollution.

AIR POLLUTION—INDOOR—*continued*

1358. Ban on lead-based paint in the United States was enacted in 1971. It barred the use of paint with lead pigment in federally financed construction. The paint was considered hazardous because scraping or burning it released toxic lead into the air; also because children might eat paint chips.

1359. Free investigations of indoor air pollution in U.S. workplaces were offered in 1971. The National Institute for Occupational Safety and Health (NIOSH), a federal agency, conducted the surveys.

1360. Indoor air quality conference in the United States convened in Maine in 1972. The conference considered such sources of indoor air pollution as wood-burning and gas stoves.

1361. Recognition of "sick building syndrome" came in the United States as early as 1974 when employees in climate-controlled office buildings complained of fatigue, headaches, dizziness, and other ailments. New building materials, overinsulation in response to rising energy costs, and windows sealed against fresh air trapped allergens and other irritants inside for continual recirculation via air conditioning, heating, and ventilation systems.

1362. State in the United States to require nonsmoking sections in restaurants was Connecticut, in 1974. Other states followed suit, and soon smoking restrictions in hotel and motel rooms were common as well.

1363. Indoor ventilation guidelines for commercial buildings, from the American Society of Heating, Refrigeration and Air-Conditioning Engineers, were published in 1977. The standards were intended to assure acceptable air quality for 80 percent of a building's occupants.

1364. Urea-formaldehyde foam insulation ban in the United States came in 1982. When improperly installed, it caused nosebleeds, dizziness, and vomiting. The ban was later overturned on a legal challenge.

1365. Law in the United States mandating that all school systems check their buildings for asbestos-containing materials was the Asbestos Hazard Emergency Response Act of 1986. The measure required school officials to develop a plan for control or removal where asbestos was found.

1366. Recognition by the U.S. government of the health hazards of passive smoking came in a U.S. surgeon general's report in 1986. The report spurred passage of local and state ordinances that restricted smoking in public and private places.

1367. Radon control legislation in the United States was the Indoor Radon Abatement Act of 1988. The measure provided $45 million to attack the problem.

1368. Report from the U.S. Environmental Protection Agency indicating a strong link between second-hand smoke and lung cancer was published in 1991. The report led to tougher workplace and public-place restrictions on smoking in the United States.

1369. Airline flight attendants' class action suit against tobacco companies was filed on November 1, 1991. Seven flight attendants claimed they had contracted cancer, heart disease, or respiratory ailments from breathing in smoke from passengers smoking cigarettes on airplanes.

1370. U.S. government declaration of cigarettes as a "drug delivery device" came in 1996. The U.S. Food and Drug Administration declared nicotine in cigarettes addictive. Tobacco causes indoor air pollution and is a leading cause of illness and death.

AIR POLLUTION—LEGISLATION AND REGULATION

1371. Laws to locate sources of objectionable odors and smoke downwind of city walls were adopted in the eastern Mediterranean region around the time of Christ, early in the 1st century A.D.

1372. Regulations designed to curb the ill effects of coal smoke were enacted in England in the 13th century. The laws regulated fuel type, chimney height, and time of use.

1373. Statute in England against air pollution was promulgated in 1273 by King Edward I. The law sought to stem the high rate of pollution, especially in London.

1374. Commission of inquiry to investigate complaints about smoke levels in London was established in 1287.

1375. Royal proclamation in England to ban the burning of coal was issued in 1307. The edict sought to curtail the amount of pollution caused by the burning of coal, especially in the growing city of London. This attempt to curb air pollution was widely ignored.

1376. Parliamentary committee to investigate air pollution in England was empanelled in 1819. The committee took no effective action, however.

1377. British government inquiry into the practicability of laws to control smoke was launched in 1845. The resulting De la Beche–Playfair report of 1846 concluded that smoke abatement would be possible with vigorous enforcement of the laws.

1378. Air pollution prosecutions under the smoke clause of the City of London Sewers Bill were in 1851. Some 150 compulsory notices were sent to polluters during the first year the bill was law.

1379. Air pollution law of significance in England aimed at actually ridding London of air pollution rather than making it easier to prosecute polluters was the Smoke Nuisance Abatement (Metropolis) Act. Parliament approved the measure in 1853.

1380. English police force given power to enforce smoke abatement laws was the London police. The London police were granted the authority by the British Parliament in 1853.

1381. Formal government antipollution agency in England was the Alkali Inspectorate, established in the early 1860s. The bureau attempted to regulate chemical industry emissions but lacked enforcement powers.

1382. Air pollution law in England granting British inspectors the right to enter factories during their investigations was the Alkali and Works Act. Parliament approved the measure in 1863.

1383. Law to establish limits on emissions of hydrochloric acid was England's Alkali and Works Act. Parliament enacted legislation regulating the highly poisonous gas, which is used in metallurgy, in 1863.

1384. Legislation explicitly barring industries in England from emitting "black smoke" was the Public Health Act of 1875. The law was not, however, strictly enforced.

1385. Local ordinance in the United States to regulate chimney height as a means of curbing air pollution was adopted in St. Louis, MO, in 1876. It required factory chimneys to be at least 20 feet higher than adjacent buildings.

1386. Local ordinances curbing air pollution in the United States were passed in Chicago, IL, and Cincinnati, OH, in 1881. The local laws attempted to control factory smoke and soot.

1387. Alkali Act (1863) revisions of significance were enacted in 1906 by the British Parliament. The measure for the control of "noxious vapour" remained essentially unchanged until 1975.

1388. Building code in the United States to require new factories to reduce smoke production was passed in Chicago in 1907. The regulation stated that builders must use the most current engineering techniques to reduce emissions from smokestacks.

1389. Arbitration ruling to assign national responsibility for air pollution damage was handed down on April 15, 1935, when jurors ordered Canada to compensate the United States for damage to crops from emissions from a smelter in Trail, British Columbia. The ruling and $350,000 award were landmarks in international law, affirming a country's responsibility for the pollution it generates.

1390. Effective control measures for solid-fuel stationary sources of air pollution in the United States were initiated in St. Louis, MO, in 1940. The local regulations were designed to curb smog that sometimes became so thick in St. Louis during winter that streetlights and auto headlights were needed during the middle of the day.

1391. Air pollution control director position for a U.S. city was created for Los Angeles, CA, in February 1945. The director enforced pollution laws and mounted public education campaigns about the causes and consequences of air pollution.

1392. Municipal smokeless zones in England were created in Manchester in 1946. Householders in the zones were required to convert from coal-burning stoves to smokeless stoves.

1393. Air pollution control districts in a U.S. state were established in California by state legislative act in 1948. Los Angeles formed its own pollution control district in that year.

1394. Significant air pollution control program to be adopted in the United States went into effect in Oregon in 1952. The program sought to cut down on the amount of pollutants emitted by businesses and homes.

1395. National Air Pollution Control Act in the United States was approved in 1955. Congress expanded this initiative with the Clean Air Act of 1963, providing federal aid for research and for the development of state air pollution control agencies.

AIR POLLUTION—LEGISLATION AND REGULATION—*continued*

1396. Air pollution ordinance enacted in Chicago was in 1958. However, the ordinance exempted steelmakers, the city's largest single industry and a major source of pollution.

1397. Multilateral treaty to address an air pollution problem was the Nuclear Test Ban Treaty of 1963, also known as the Moscow Agreement. It prohibited nuclear tests in the atmosphere, the oceans, and outer space. More than 100 countries became parties to the treaty.

1398. Air pollution law of importance enacted by the U.S. Congress was the Clean Air Act of 1963, signed by President Lyndon B. Johnson on December 17, 1963. The law authorized the expenditure of $93 million in matching grants for state-funded air pollution prevention and control programs.

1399. Federally sponsored interstate air pollution abatement conference took place November 9–10, 1965. The Interstate Air Pollution Abatement Conference of Bishop, MD, and Selbyville, DE, sought to stop noxious and stinking emissions from the Bishop Processing Co., a rendering plant for fish and chicken parts.

1400. Large U.S. industrial state to ban the burning of high-sulfur fuels was New York, in March 1967. Several east coast states, including Maryland, Virginia, and Florida, followed suit.

1401. Law in the United States calling for ways of controlling jet aircraft emissions was contained in the Air Quality Act of 1967. It was estimated that jet aircraft produced, in Los Angeles alone, the daily particulate equivalent of one million automobiles.

1402. Large U.S. states to file civil suits charging airlines with violating air pollution control were New Jersey and Illinois, in 1969. New Jersey threatened a $1,600 fine for each landing and takeoff.

1403. State in the United States to enact legislation to curtail aircraft emissions was California, in 1969.

1404. Local private grassroots organization to successfully pressure governmental agencies to take steps to curb air pollution was the Group Against Smog and Pollution (GASP), established in Pittsburgh, PA, on October 20, 1969. The group mobilized strong public support for pollution abatement measures in the Pennsylvania industrial center, long known as the "smoky city."

1405. Regular and systematic monitoring of air pollution levels in the United States began with congressional passage of the Clean Air Act Amendments of 1970. These established limits on the amounts and types of pollutants that could be released into the atmosphere.

1406. National standards for air polluting motor vehicle emissions in the United States were set in the Clean Air Act Amendments of 1970, signed into law on December 31, 1970.

1407. Presidential order prohibiting U.S. government assistance to concerns that violate air emissions standards was issued in 1971 by President Richard M. Nixon. Nixon's order barred federal agencies from doing business with companies convicted of intentionally violating emissions standards.

1408. Clean Indoor Air Act adopted by a U.S. state was in Minnesota, in 1975. Among other things, the act banned smoking in many public and some private places.

1409. State in the United States to ban the sale of aerosol products containing chlorofluorocarbons (CFCs) was Oregon, in June 1975. Other states soon followed suit.

1410. Ban on aerosol chlorofluorocarbons (CFCs) was enacted in the United States in 1978. CFCs are responsible for depletion of the earth's stratospheric ozone layer. However, CFCs remained in widespread use in refrigeration, air conditioning, and cleaning solvents and as a foaming agent in plastics.

1411. International agreement on cross-border air pollution was the Convention on Long-Range Transboundary Air Pollution, adopted in Geneva, Switzerland, in 1979.

1412. Introduction of the "bubble concept" in U.S. air pollution policy was in December 1979. The U.S. Environmental Protection Agency announced it would allow a business to increase pollution in some areas if it reduced pollution in other areas, so long as the total amount remained within the clean air requirements.

1413. Law of significance to control acid rain in the United States was passed in 1979. New York senator Daniel Patrick Moynihan introduced the measure; acid rain had seriously damaged lakes in the Adirondack Mountains of Moynihan's home state.

1414. Effective international action to curb acid rain was taken in 1984 with the founding of the "30 Percent Club." The club consisted of industrialized countries (including Austria, France, and Switzerland) that agreed to reduce their sulfur dioxide emissions 30 percent by 1993.

1415. International accord requiring a reduction on nitrogen oxide levels was the United Nations Sofia Agreement, signed in 1989 in Sofia, Bulgaria. Nitrogen oxide emissions, formed when motor vehicles, power plants, and factories burn fossil fuels, cause acid rain.

1416. Amendments of the U.S. Clean Air act to regulate emissions of pollutants, including sulfur dioxide and nitrogen oxides were added in 1990. Previously, the Clean Air Act punished polluters but did not regulate emissions of pollutants.

1417. Comprehensive effort to introduce preventive pollution control in the United States was launched on November 5, 1990, with the passage of the Pollution Prevention Act of 1990. The act, a major policy shift, established pollution prevention as a national objective and mandated efforts to reduce or prevent pollution at its source rather than merely treating or disbursing pollutants.

1418. Comprehensive upgrades of the 1970 and 1977 Clean Air Act and Amendments came with the Clean Air Act Amendments of 1990. Approved on November 15, 1990, the amendments toughened emission standards, set up controls for clorofluorocarbons, air toxics, and acid rain, and established deadlines for meeting the new standards.

1419. Antipollution device installed as a result of the 1977 law imposing special restrictions on pollution near national parks occurred at the Navajo Generating Station in Page, AZ. The power consortium's agreement with the U.S. Environmental Protection Agency resulted in the consortium installing $30 million worth of new antipollution equipment.

1420. U.S. government program to reduce air pollution in the Grand Canyon was launched in 1991. It was one of several limited pre-election year environmental initiatives by the Bush administration. Many environmentalists believe that air pollution is slowly destroying many of the natural elements of the Grand Canyon.

1421. National rules for reducing acid rain in the United States were proposed by the U.S. Environmental Protection Agency on October 30, 1991. The standards were adapted from the 1990 Clean Air Act Amendments, which called for a halving of sulfur dioxide emissions by 2000.

1422. Purchase of emissions allowances in the United States as permitted under the Clean Air Act Amendments occurred in May 1992, when the Tennessee Valley Authority (TVA) bought the right to emit 10,000 tons of sulfur dioxide from Wisconsin Power and Light. The two utilities were the first to disclose a trade under the new free market system that gave companies flexibility in seeking cost-effective ways of reducing pollution.

1423. State in the United States in which all electric utilities joined the U.S. Environmental Protection Agency's "Green Lights" energy efficiency program was New Jersey, in January 1992. The program encouraged the voluntary installation of energy efficient lighting as a means of curbing air pollution.

1424. Cooperative effort by northeastern states to curb nitrogen oxide emissions from coal-burning power plants was announced on March 11, 1992. The states agreed to work toward implementation of a two-stage plan for reducing emissions.

1425. Cooperative effort by 10 northeastern U.S. states to develop plans to reduce emissions of greenhouse gases was announced on March 30, 1992. The states also agreed to plan for a potential global warming trend and a rise in sea level.

1426. Federal provision requiring companies that service air conditioning and refrigeration equipment to "capture and recycle" chlorofluorocarbons (CFCs) went into effect on July 1, 1992. The Clean Air Act Amendments of 1990 mandated CFC recycling; previously, the ozone-damaging gases were released into the air.

1427. Recommendations in California to shift responsibility, from businesses to drivers, for paying for Southern California air pollution abatement was made on July 8, 1992. The South Coast Air Quality Management District in Los Angeles said private automobile operators should bear the main economic burden of fighting

AIR POLLUTION—LEGISLATION AND REGULATION—*continued*

1428. Edict issued by the U.S. Environmental Protection Agency requiring the states to clean up air pollution in national parks was in April 1999. The federal agency set a 2064 deadline to return air quality in the parks and wilderness areas to what amounts to a pre-industrial purity. The order targeted the main source of air pollution affecting the parks—power plants that burn coal or other fossil fuels.

AIR POLLUTION—PUBLIC OPINION

1429. Major anti–air pollution tract was *Fumitugium: or the Inconvenience of the Aer and Smoake of London*, by John Evelyn. He submitted it to King Charles II of England in 1661.

1430. Anti–air pollution activities by the English public interest group known as the Fog and Smoke Committee were in 1880. Among other things, the group enlisted public support for its work by organizing an exhibition of equipment and design for smoke abatement.

1431. Smoke reduction exhibition was held in London in 1881. The Smoke Abatement Exhibition was sponsored by the Fog and Smoke Committee and displayed more than 230 exhibits on how to reduce smoke in industry and home.

1432. Rankings by the U.S. government of the nation's most polluted cities were initiated in the early 1960s. The list was regarded as an effective means of building public support for air pollution control.

AIR POLLUTION—RESEARCH

1433. Patent on a device to enable a steam engine to consume its own smoke was registered by James Watt in 1785. Similar devices were patented by others around the same time.

1434. Effective treatment for stack emissions of hydrochloric acid in the alkali industry was developed in 1836 in England. William Gossage's method involved washing the gases with water as they ascended.

1435. Reference to the problem of acid rain appeared in a Swedish publication in 1848.

1436. Scientist to systematically study dispersal of industrial age air pollutants was the English alkali inspector Robert Angus Smith. In 1852, he issued a report on the chemistry of rainwater near the industrial city of Manchester, England.

1437. Treatise on the phenomenon of acid rain was published in England in 1872. Robert Angus Smith, one of Britain's first pollution inspectors, coined the term in his book *Acid and Rain.*

1438. Mathematical equation to calculate atmospheric warming from carbon dioxide concentrations was developed by Swedish chemist Svante Arrhenius in 1896. He figured that a doubling of carbon dioxide in the air might raise the earth's temperature by 9 degrees Fahrenheit.

1439. Use of the word *smog* occurred in London at a meeting of the Public Health Congress in 1905. The word, a combination of *smoke* and *fog*, describes noxious urban air containing gases and aerosols. H. A. Des Voeux of the Coal Smoke Abatement Society is said to have coined the word.

1440. Smokeless coalite patent was registered in London, in 1906. Supplies were not sufficient, however, to make coalite a commercial and environmental success at the time.

1441. Research showing that the acidity of rain decreased the farther one traveled from a city center appeared in 1911. Scientists in Leeds, England, argued that this proved a direct link between industrial and power plant emissions and acid rain.

1442. Federal research on ways to limit the release of smoke into the atmosphere in the United States was undertaken in 1912. Scientists at the U.S. Bureau of Mines were assigned the project.

1443. Air pollution measurements taken in London were begun in 1914. In 1917, the private Coal Smoke Abatement Society turned over the monitoring project to the British government meteorological office.

1444. Introduction of the pH scale to measure acidity in rain came in 1923. The pH scale measures the acidity or alkalinity of a solution based on its hydrogen ion concentration.

1445. Federal studies of the effects on the atmosphere of carbon monoxide from automobile exhaust in the United States were launched in 1925. U.S. Public Health Service scientists undertook the research.

1446. Electrostatic precipitation application was at the Willesden, England, power station in 1929. The technique evolved from the experiments of the physicist Oliver Lodge.

1447. Successful application of electrostatic precipitation to scrub pollutants from cement kiln stacks in Britain was in 1933. The technique was successfully used at the British Portland Cement Manufacturers plant in North Kent, England.

1448. Strong claim that industrial burning of fossil fuels increased carbon dioxide levels in the atmosphere came from English coal engineer George Callendar in a report to the Royal Meteorological Society on February 16, 1938. Callendar observed that high carbon dioxide concentrations in the air caused measurable effects—in particular, warming—in the weather.

1449. Indication of a new type of air pollution occurred in Los Angeles, CA, in 1943. On September 8, emissions of butadiene from a synthetic rubber plant reacted with sunlight to create a heavy smoke that caused eye irritation and other problems. The day became known as "Black Wednesday." By the early 1950s, scientists generally agreed that a photochemical reaction had created the Black Wednesday smog.

1450. Acid rain measurement monitoring systems were established in Sweden in 1948. Monitoring programs in the United States were set up around the same time. Sources of acid rain include the burning of fossil fuel, industrial smelting operations, automobile exhaust, and nitrogen fertilizer in agriculture.

1451. United States National Air Pollution Symposium was held in Pasadena, CA, on November 10–11, 1949. The Stanford Research Institute (now SRI International), the California Institute of Technology, and the University of California sponsored the conference so authorities in the field could exchange ideas and technology on pollution research. It was the first of three symposiums sponsored by SRI. The next two followed in 1952 and 1955.

1452. Instrument to measure carbon dioxide in the atmosphere in parts per million was designed by California chemist Charles Keeley in the 1950s.

1453. Careful scientific monitoring of Lumsden Lake in Ontario, Canada, for acid rain began in the early 1950s. When monitoring commenced, there were eight species of fish in the lake. By 1978 all eight species were extinct. The cause of their extinction is believed to be the high amounts of acid rain that fell over the lake.

1454. Scientific paper demonstrating that hydrocarbons and nitric oxide react in sunlight to form ozone pollutants was published by Prof. A. J. Haagen-Smit of the California Institute of Technology in 1951. Smit argued seasonal photochemical reactions accounted for thick summer smogs in the Los Angeles area, including the notorious Black Wednesday "gas attack" of September 8, 1943.

1455. Air pollution study sponsored by the U.S. government began in 1955, when the Public Health Service started researching air pollution.

1456. Research showing acid-forming pollution could travel great distances was published by Canadian scientist Eville Gorman in 1955. Gorman's research showed that, when the wind blew from the industrial area in the south, the rain over the Lake District in northern England contained high concentrations of sulfuric acid.

1457. Smoke density standard measurement in England was issued in 1956. Earlier, it had been proposed to call a unit of smoke a "murk."

1458. Carbon dioxide monitoring station was set up in Mauna Loa, HI, in 1958. C. D. Keeling of the Scripps Institution of Oceanography set up the station to provide data about the carbon dioxide content of the lower atmosphere.

1459. Measurement of carbon dioxide levels in the atmosphere occurred at Mauna Loa Research Station in Hawaii in 1958. The result showed a level of 314 parts per million. It reached 358 parts per million in 1994.

1460. Carefully documented reports of a link between acid rain and damage to fish populations came from Scandinavia in 1959. Fisheries inspectors blamed acid rain for the disappearance of fish in lakes and streams of southern Norway and southwestern Sweden.

1461. International conference on air pollution was held in London in 1959. In response to growing national concerns about air pollution and recognizing the transbroundary nature of air pollution, countries gathered for the First International Air Pollution Conference. At the conference, experts, scientists, and government leaders shared research ideas and technology.

1462. Smog chamber for air pollution research built by an industrial organization went into operation in July 1962 at the General Motors Research Laboratories in Warren, MI. The cylindrical chamber contained 300 cubic feet of space bathed in simulated noontime created sunshine by 247 fluorescent lights.

AIR POLLUTION—RESEARCH—*continued*

1463. Serious evidence of acid rain in the United States was observed in 1963 in the Hubbard Brook Experimental Forest in New Hampshire.

1464. Report on the effects of acid rain on fish populations was presented by Dr. Harold Harvey of Ontario, Canada, in 1966.

1465. International monitoring system for air pollution was the Background Air Pollution Monitoring Network, established in 1968 with 20 reporting stations. The system was set up by the World Meteorological Organization in Geneva, Switzerland.

1466. Direct measurements of chlorofluorocarbon (CFC) levels at the earth's surface were taken in 1970. British scientist James Lovelock, using a detector of his own make, suggested that CFC concentrations would survive in the atmosphere for decades without change.

1467. Evidence of significant acid rain damage in the Great Britain emerged in the late 1970s in the lakes of Scotland and northern England. By the early 1980s, British forests were showing signs of disease.

1468. Notice of the wintertime phenomenon known as "Arctic haze" was in the 1970s. Scientists made the chance discovery of large quantities of wind-borne pollutants in the Arctic during the winter and early spring.

1469. Reintroduction of the term *acid rain* occurred in 1972. The term was originally coined by British scientist Robert Angus Smith in 1872. However, the work of Cornell University ecologist Gene Likens in the early 1970s reintroduced the term into common usage.

1470. Evidence of pollution damage to the earth's stratospheric ozone layer emerged in 1974. The layer, 9–30 miles above ground, makes the planet habitable by blocking 95–99 percent of the sun's potentially deadly ultraviolet radiation. Chemical compounds known as chlorofluorocarbons (CFCs) were blamed for the damage.

1471. Precipitation sampling network in Canada was established in 1976. Canadian researchers conducted a national assessment of acid deposition from 1985 to 1990.

1472. Activities of the Arctic Chemical Network occurred in 1977. The group of scientists from Arctic rim countries investigated transboundary pollution affecting the northern reaches of the planet.

1473. National network in the United States to analyze the effects of air pollution on precipitation chemistry was set up in 1978. The National Atmospheric Deposition Program was privately operated and funded.

1474. Long-term study of the causes and effects of acid rain in the United States was proposed in 1979. President Jimmy Carter called for a 10-year, $100-million survey of the problem. By the 1980s, some 25,000 North American lakes had suffered acid rain damage; acid rain pollution was a suspect, too, in the "dieback" of North American forests.

1475. Widespread damage to and deaths of trees in Germany from air pollution were seen in the early 1980s. By 1984, 50 percent of conifers in West Germany were believed to have been damaged or killed by severe air pollution.

1476. Epidemiological evidence of the association between passive smoking and lung cancer appeared in a study, by Japanese researcher T. Hirayama, released in 1981. A Greek study confirmed the findings two years later. Additional studies were carried out subsequently in the United States, Germany, and Hong Kong.

1477. Government report issued jointly by the United States and Canada to rate acid rain a serious environmental and transboundary problem for both countries was issued in January 1986. The release also marked the first time U.S. President Ronald Reagan acknowledged acid rain as a serious problem. The report led to the 1991 Air Quality Accord between the United States and Canada.

1478. Proof that thunderstorms can push pollutants as high as the lower stratosphere occurred in 1987 when U.S. scientists aboard a Saberliner jet observed levels of carbon monoxide and nitric acid at high elevations around thunderstorm clouds.

1479. Reformatted gasoline tested in the United States was EC-1 Regular, produced in 1989 by ARCO Petroleum in Los Angeles, CA. Scientists at Arco reduced the smog precursors by 37% after 90 days of research. They published their results in technical journals and waived the patent rights.

1480. Evidence that the "ozone hole" had expanded outward to include New Zealand, Australia, Argentina, and Chile emerged in the early 1990s. This led to the growing concern that the hole in the ozone layer would soon spread even farther.

1481. U.S. government research suggesting nitric acid (rather than sulfuric acid) is the chief cause of acid rain was released on May 4, 1992. U.S. Geological Survey scientists recommended that the government rethink its forest management practices designed to attack the problem.

1482. Recorded appearance of a large haze of air pollution over the Indian Ocean was in 1999. Scientists said the brownish blanket of dirty air, which covers an area from the Bay of Bengal and the Arabian Sea south to the equator and is about the size of the United States, could have implications for the global climate and for the regional environment on the Indian subcontinent.

ALIEN SPECIES AND SPECIES MIGRATION

1483. Rabbit infestation due to introduction occurred on Porto Santo in the Madeira Islands. After the animals were introduced by Portuguese settlers in the 1420s, the absence of natural predators allowed the rabbits to flourish, destroying the island's plant life and causing subsequent erosion problems. Within a few years, the island was unsuitable for human habitation and did not recover for decades.

1484. Cattle in the Western Hemisphere were introduced to Hispaniola by Christopher Columbus on his second voyage in 1493. In less than a century, herds of cattle could be found roaming throughout Central, North, and South America.

1485. Horses reintroduced in the Americas were those brought to Hispaniola by Christopher Columbus on his second voyage in 1493. Horses had become extinct in the Americas during prehistoric times, for unknown reasons.

1486. Pigs in the Western Hemisphere were introduced by Christopher Columbus, who brought eight pigs to Hispaniola on his second voyage in 1493. The scarcity of natural predators allowed pig populations to spread throughout Central America after their arrival on the mainland.

1487. Sheep in the Western Hemisphere were introduced by Christopher Columbus on his second voyage to Hispaniola in 1493. During the 16th century, herds of domestic sheep escaped to become feral and spread throughout Central and South America.

1488. Horses reintroduced to the American mainland were those used by Hernán Cortés during his conquest of Mexico, beginning in 1519. Escaped horses soon became wild and migrated throughout the hemisphere, profoundly changing Native American society.

1489. Wheat plants grown in the Americas were brought by the Spaniards in 1520. The plants were grown in parts of South America.

1490. Silkworms in the Western Hemisphere were brought by Spaniard Hernán Cortés to Mexico in 1522. He also brought mulberry trees to Mexico, to provide food for the silkworms.

1491. Magnolias to arrive in England were sent by John Banister to Bishop Henry Compton in 1688. Magnolias have flourished there ever since. Bishop Compton was said to have grown more than 100 exotic plants under glass in his home in London.

1492. Cattle and sheep in Australia and New Zealand were introduced by Europeans in the late 18th century. The livestock reproduced rapidly and had a profound effect on native vegetation.

1493. Mangos in the Western Hemisphere were planted by the Portuguese in Bahia, Brazil, in the 18th century.

1494. Hydrangea in England was planted by Peter Collinson in 1736 in his garden in Peckham. Collinson was an avid collector of exotic plants and seeds that could flourish in the English climate. He used his business connections to collect specimens from all over the world.

1495. Australian pine in the United States was introduced to coastal areas of Florida in the late 19th century for use in windbreaks and as a shade tree. By cutting off sunlight for smaller plant species and blanketing the ground with its needles, the tree displaced native vegetation, creating wildlife displacement, erosion, and other problems.

1496. Brazilian pepper in the United States is thought to have been imported to Florida during the mid- to late 19th century. Its ripened red berries, which earned it the nickname "Florida holly," made it suitable for use as an ornamental tree. Yet it has become one of southern Florida's most troublesome alien species because it overwhelms native plant life, particularly mangroves and pine, thus reducing habitat and species diversity.

ALIEN SPECIES AND SPECIES MIGRATION—*continued*

1497. Leafy spurge in the United States is thought to have been introduced accidentally in seeds imported from Europe in the early 19th century. It reached the plains states by the early 20th century. Leafy spurge is a major pest which displaces other vegetation, including millions of acres of native grassland important for livestock grazing and to wildlife habitats.

1498. Purple loosestrife in North America was imported from Europe in the early 19th century for use as an ornamental plant. It is now found nearly everywhere in North America and is a listed pest in some states. Purple loosestrife seriously damages habitat diversity in wetlands by outcompeting native vegetation, thus destroying plants upon which other species rely for survival.

1499. Introduction of goats on St. Helena Island in the South Atlantic occurred in 1810. The herds eventually destroyed 22 of 33 native island plant species.

1500. Person to establish a vineyard in San Francisco, CA was Frenchman Jean Louis Vignes during the 1830s. Vignes imported vines from France and brought cuttings when he immigrated to San Francisco.

1501. Sea lamprey positively identified in Lake Ontario was seen in 1835. The salt-water fish reached the lake via the Hudson/Mohawk/Erie Canal system and flourished as agricultural and industrial development led to sharp increases in the water temperature of streams where it spawned. A predatory fish, the lamprey had wiped out Lake Ontario's Atlantic salmon population and seriously damaged whitefish and lake trout stocks by 1900; in 1921 it reached Lake Erie through the Welland Canal and from there infested the other Great Lakes, doing great damage to commercial fisheries.

1502. Prickly pear in Australia was introduced in 1839 for use as hedging. The spreading dominance of wild prickly pear in the subsequent century made the growth of any other vegetation impossible on millions of acres of land.

1503. Saltcedar or tamarisk in the United States was introduced in arid western states during the 1840s to prevent stream banks from eroding. During the mid-20th century, saltcedar began to spread throughout the West, becoming a listed pest by consuming more water than native vegetation, and altering water supplies and flow patterns, endangering wildlife.

1504. Forsythia in England was sent there from China by Robert Fortune in 1844. He sent his specimen, found in the mandarin's garden in Zhoushan, to the Horticultural Society of London.

1505. Appearance of the American vine aphid in Europe was in the 1850s. The tiny insect spread quickly through European vineyards, causing massive damage in France, Germany, and Italy.

1506. English sparrows imported into the United States arrived in 1850. Eight pairs of English, or house, sparrows (*Passer domesticus*) were imported under the auspices of Nicholas Pike and other directors of the Brooklyn Institute in Brooklyn, NY, for the purpose of protecting shade trees from the foliage-eating caterpillars. The birds were kept in cages until freed in the spring of 1851. The first group did not thrive, but a larger number were imported in 1852 and thereafter multiplied rapidly.

1507. Cinchona trees in India were introduced by Clements R. Markham, an Englishman, in 1859. Markham and Richard Spruce succeeded in growing the tree from which quinine, an antimalarial drug, is produced. An antimalarial was not available in India at the time.

1508. Rabbits in Australia were imported from England for sport hunting by Thomas Austin. On Christmas Day of 1859, Austin released 24 rabbits at Barwon Park near Geelong, Victoria. Because Australia had no native predators to contain them, within twenty years wild rabbits became a national plague, decimating agriculture and plant life. They remain the worst vertebrate pest on the continent despite more than a century of eradication schemes.

1509. California quail in New Zealand were released by M. W. Hay at Papakura, New Zealand in 1862. The quail became so abundant by 1880 that they were exported by the thousands to England and other parts of Europe.

1510. *Begonia sutherlandii* to flower in England bloomed in James Backhouse's nursery in Acomb in 1867. A Quaker missionary and botanist, Backhouse also collected seeds and bulbs in Australia in 1813 and in South Africa between 1838 and 1840, for William Hooker.

1511. Introduction of the gypsy moth in the United States occurred by accident at Medford, MA, near Boston in 1869. E. Leopold Truvelot, a French-born artist with an interest in entomology, was responsible for importing the foliage-eating pest.

1512. Ostrich in southeastern Australia was introduced in 1869 for the feather industry. Later in 1882 the South Australian government supported ostrich farming. Many farm birds were released after World War I when the industry failed. There are still ostriches in Red Cliff and near Port Augusta.

1513. Rabbits in Antarctica were released during 1874 on the Kerguelen Islands. By 1955, the rabbit population was so large that a rabbit disease, myxomatosis, was released to control the population. Due to cold and distance, the disease spread so slowly that resistance was built up to it and it did not significantly reduce the rabbit population.

1514. Successful introduction of the ringneck pheasant in the United States was in 1882 by Judge O. N. Denny to the Willamette Valley in Oregon. Since the 1730s unsuccessful attempts had been made to introduce the ringneck to the United States. In 1882, Judge Denny released several dozen birds, and ten years later Oregon hunters harvested over 250,000.

1515. Boll weevils in the United States migrated into Texas during the 1890s from Mexico and quickly spread across the South. The boll weevil measures just a quarter of an inch and uses its long snout to bore into a cotton boll. The pest has caused billions of dollars in losses to the U.S. cotton industry despite a century of sporadically successful eradication efforts.

1516. European starlings introduced to the United States were imported in 1891 to serve as ornamental birds in Central Park in New York City. A few of these birds escaped, and by the mid-1950s the European starling was established across the continental United States, taking over ecological niches formerly occupied by bluebirds and flickers.

1517. Australian melaleuca in the United States was imported during the early 1900s. Its appetite for water made its introduction an effective method of drying Florida wetlands. Unfortunately, its spread has reduced the diversity of animal species by severely displacing native vegetation, including cypress.

1518. Magellan Goose on South Georgia Island was introduced to the South Antarctic Ocean in 1910 as fresh food for whale and seal hunters. They increased in numbers by the 1920s but suffered a setback and were extirpated by 1950.

1519. Appearance of the Colorado potato beetle in Europe was in 1920, in France. The potato-destroying pest spread throughout Europe, reaching the Soviet Union in 1955.

1520. Common quail introduced to the Hawaiian Islands were released on Maui and Lanai in 1921. They established themselves quickly and are still present on the islands.

1521. Emus on Kangaroo Island off the coast of South Australia were introduced between 1926 and 1929. During that period, eight emus were brought from mainland Australia to the island. Their few remaining descendants have been known to damage crops, leading Kangaroo Islanders to build emu-proof fences.

1522. Mediterranean fruit fly infestation in North America was in Florida in 1929. The "medfly" is a major agricultural pest in tropical and subtropical areas throughout the world. It became established in Hawaii during the early 20th century. Sporadic infestations in California are thought to have resulted from fruit shipped to the mainland from Hawaii.

1523. Dutch elm disease in the United States was detected in 1930 in Cleveland and Cincinnati, OH. The blight, caused by the Asiatic fungus *Ceratocystis ulmi*, was brought to the United States on elm logs shipped from Europe. It spread rapidly, killing vast numbers of American elm trees, *Ulmus americanus*, and changing the appearance of many American cities.

1524. Nile perch in Lake Victoria were introduced during the 1950s to compensate for depletion of native species caused by overfishing in the lake, located on the Ugandan-Tanzanian-Kenyan border in central east Africa. During the next two decades, the predatory perch decimated the remaining native fish in the lake, causing numerous species of haplochromines to become extinct. This disruption to the lake's ecosystem is thought to be one of several major problems—including pollution and invasive alien plant species—that are contributing to a rapid decline in the lake's oxygen level.

1525. Sri Lankan hydrilla in the United States was introduced to Florida in the 1950s. Initially, it was intended to be used in aquariums, but the weed has since spread throughout the South, congesting waterways and displacing native vegetation. It can ruin human recreation areas and imperils plant and fish species by preventing sunlight from reaching below the surface to sustain other life.

ALIEN SPECIES AND SPECIES MIGRATION—*continued*

1526. TFM chemical releases into the lamprey-infested waters of Lake Superior were in the spring of 1958. The chemical killed spawning lampreys without apparent harm to other aquatic life.

1527. Asian tiger mosquitoes in the United States arrived in Texas during the 1980s. Their eggs are suspected of having been transported and hatched in Asian tires sent to the United States for retreading.

1528. Zebra mussels introduced in the United States came from Europe in the mid-1980s, probably in the bilge water of ships. First identified in 1988 in Lake St. Clair, near Detroit, MI, the prolific shellfish (*Dreissena polymorpha*) quickly spread through Lake Erie into Lake Ontario and other waterways, replacing native species and damaging underwater structures and marine equipment.

1529. "Killer bees" in the United States arrived from Mexico in October 1990, threatening agriculture, the honey industry, and public safety in southern Texas. Properly called Africanized honeybees, they were members of an aggressive new variety created in 1956, when African bees escaped into the wild from a Brazilian laboratory and hybridized with previously established colonies of European honeybees. The new bees bred and spread more quickly than the European types, produced less honey, and defended their hives much more fiercely. By the late 1990s, "killer bees" had caused more than 1,000 human deaths, mostly in Mexico and Argentina.

1530. Aquatic Nuisance Species Task Force was created by the U.S. Congress on November 29, 1990, with the passage of the Nonindigenous Aquatic Nuisance Prevention and Control Act. The U.S. Fish and Wildlife Service, the U.S. Coast Guard, the Environmental Protection Agency, the Army Corps of Engineers, and the National Oceanic and Atmospheric Administration were all assigned roles in preventing new nuisance species from being introduced into the Great Lakes through the ballast of foreign vessels.

AUTOMOTIVE INDUSTRY

1531. Person to imagine the possibilities of power-driven vehicles was Leonardo da Vinci, the Italian inventor, in the 15th century. A wide-ranging genius, he also envisioned manned flight.

1532. Automobile race in the United States was run in Chicago in 1895 under sponsorship of the Chicago Times-Herald newspaper. A car with a four-cycle engine built by Charles and Frank Duryea won the race.

1533. Mass-produced automobile in the United States was manufactured by Charles and Frank Duryea in 1896. Their small firm, the Duryea Motor Wagon Company, produced 13 cars that year, but failed in 1898 for financial reasons.

1534. Successful U.S. passenger car manufacturer was the Olds Motor Vehicle Company, established in Lansing, MI, in 1897. Olds joined the General Motors Company in 1908.

1535. Agreement to finance development of a new high-performance car on the condition that the model be named for the financier's daughter dates from 1900. Emil Jellinek agreed to back the venture of the Daimler Engine Company so long as Daimler agreed to name the car after his daughter, Mercedes.

1536. Long-lasting partnership to build luxury cars in Great Britain was formed in 1906 between Charles Stewart Rolls and Frederick Henry Royce. They produced their Silver Ghost car in time for the Paris Motor Show of 1906. The first Rolls-Royce commercial model became available in 1907 at a cost of $1,744.

1537. Carmaking company to produce a range of models was the General Motors Company, established in Flint, MI, in 1908. The founder, William C. Durant, believed a large company making a number of models would be better protected from market forces than a smaller manufacturer producing only a single model line.

1538. Car produced by the firm that would become the Nissan Motor Company of Japan was the DAT made in Tokyo in 1914. DAT was an acronym formed from the first letters of the first names of the Kwaishinsha Motor Car Works' three principal investors. It later became known as Datsun.

1539. National car industry to produce one million units a year was that of the United States, in 1916. The second country to reach that level of production was Great Britain, in 1956.

1540. Ethyl gasoline with tetraethyl lead was sold in 1923 at a gasoline station of the Refiners Oil Company in Dayton, OH.

1541. Year in which General Motors vehicles outsold those of the Ford Motor Company was 1927. By then, GM founder William C. Durant had been ousted and Pierre S. Du Pont had taken charge of the company. The Ford Motor Company had dominated the automobile industry around the world for nearly two decades. However, by the early 1920s, Ford's reliance on one model and reluctance to introduce new models led other companies to seize on the opportunity to present to consumers a wider variety of car models to choose from.

1542. Automobile safety advocate of note in the United States was Ralph Nader. Born in 1934 in Winsted, CT, he studied law at Harvard and published the book *Unsafe at Any Speed* in 1965. This indictment of safety standards within the automobile industry, along with Nader's testimony before the U.S. Congress, resulted in the passage of the National Traffic and Motor Vehicle Safety Act of 1966. Nader and his young activists, "Nader's raiders," established the Center for Auto Safety and other consumer-protection organizations, such as the Center for the Study of Responsive Law, Congress Watch, and the Public Interest Research Group.

1543. Mass-produced FIAT Topolino came off the assembly line at Mirafiori, Italy, in 1936. Using production techniques adapted from the Ford Motor Company in the United States, FIAT made many thousands of the two-seater Topolino.

1544. Sit-down strikes to cripple General Motors car production occurred at GM's assembly plants in Flint, MI, in December 1936 and January 1937. The strikes spread, and in February the company relented and agreed to recognize the United Auto Workers union as the collective bargaining agent for its laborers.

1545. Automobile rotary engine was invented in 1954 by Felix Wankel, an East German mechanical engineer, who developed a prototype in 1956. He then worked for several automobile manufacturers at the Technische Entwicklungstelle in Lindow, near Berlin. His rotary engine had great potential for controlling automobile pollution because its small size allowed more space under the hood for emission-control devices. The Mazda Motor Corporation of Japan used it for several models and, in 1970, the General Motors Corporation paid $50 million for the patent. The Wankel engine has been used in both automobiles and aircraft, but problems with the rotor seals have limited its success.

1546. Exports of Datsun cars from Japan to the U.S. were in 1958. The company formed a U.S. subsidiary in 1960. Datsun would change its name to the Nissan Motor Company during the 1980s.

1547. Compact car built by General Motors was the Chevrolet Corvair, introduced in 1959. Sales plunged after Ralph Nader exposed the model's flaws in his 1965 book *Unsafe at Any Speed*. Soon after, Chevrolet stopped producing the Corvair.

1548. Federal government bailout of a major U.S. carmaker was arranged for the Chrysler Corporation in 1980. Chrysler chairman Lee Iacocca engineered with the U.S. government a deal worth $1.5 billion in loans and tax concessions. Under Iacocca, Chrysler rebounded from its financial woes, and Chrysler was able to pay back the $1.5 billion loans in 1983, seven years ahead of schedule.

1549. Unleaded gasoline campaign in Great Britain was by the Campaign for Lead-free Air (CLEAR), launched on January 25, 1982, in London. Among its objectives that were realized were the introduction of unleaded gasoline and a tax on leaded gasoline to create a price advantage for drivers using unleaded gas. The chairman of CLEAR was Des Wilson, a journalist and former member of the Liberal Party Council.

1550. Electric automobile battery research group in the United States was the United States Advanced Battery Consortium, established in 1991. The group was formed as a partnership between the three major domestic automobile companies—the Chrysler Corporation, Ford Motor Company, and General Motors Corporation—together with the U.S. Department of Energy, the Electric Power Research Institute, and battery manufacturers. The goal was to produce batteries that would advance the market potential of electric vehicles. Research was conducted on such versions as the nickel-metal and lithium-ion batteries.

1551. Leaded gasoline total ban by the United States Congress came into effect on January 1, 1996. Although all cars manufactured in the United States were required to run on unleaded gasoline, until this ban, leaded gasoline was still available for older cars. As of January 1, 1996, it became unlawful to sell leaded gasoline for use in any motor vehicles in the United States.

AUTOMOTIVE INDUSTRY—HIGHWAYS

1552. Highway built in the United States entirely with federal funds was the paved National Road, or Cumberland Road (now U.S. 40), approved by the U.S. Congress in 1806. Construction began on November 20, 1811, in Cumberland, MD, and progressed—through Pennsylvania, Virginia, Ohio, and Indiana—until the money ran out in 1840 at Vandalia, IL. Congress passed a bill on May 4, 1822, to make it a toll road, but President James Monroe vetoed it. The road was the main one used by settlers moving to the Midwest and beyond. The early part of the road closely followed the military road laid out in 1754–1755 by Lieutenant Colonel George Washington and British General Edward Braddock.

1553. Use of asphalt for street paving in the United States was in 1870. Asphalt derived from petroleum today is used to surface 90 percent of U.S. roads.

1554. State highway department in the United States was established in New Jersey in 1891. The agency had responsibility for local road construction and maintenance.

1555. Census of public roads in the United States was taken in 1904. It showed two million miles of unpaved roads and 154,000 miles of pavement in the country. The rapid development of motor traffic beginning about this time exposed the inadequacy of the U.S. road network.

1556. Road inventory in the United States was taken in 1904 by the federal Office of Public Roads. Its report, published in 1907, listed 1,598 miles of stone-surfaced toll roads and 2,151,570 miles of rural public roads. Most of these rural roads were not paved.

1557. Highway program of significance in the United States granting federal funds for state roads was created by the Federal Aid Road Act, signed on July 11, 1916, by President Woodrow Wilson. It provided $75 million for rural roads, rather than long-distance highways, in order to bring about "social betterment." The act also earmarked $10 million ($1 million for each of 10 years) for roads and trails in, or partly in, national forests.

1558. Federal highway system in the United States was created by the Federal Highway Act, passed by the U.S. Congress on November 9, 1921. It required that the selection of official federal highways be made from up to seven percent of each state's highways that connected with neighboring states. It also funded the state highway departments to help them maintain their part of the system.

1559. Modern tollways were constructed in 1924 in Italy by private companies that owned them. Called autostrade, the highways had limited access.

1560. Parkway in the United States was the Bronx River Parkway opened in 1925 in New York City. It was the first example of the parkway system that circles a city but has intersections. It had trees and shrubs planted along the sides.

1561. Highway-marking system in the United States was approved by the states on November 11, 1926. The system of highway numbers on black-and-white shield markers is still used today. It was devised in 1924 by the American Association of State Highway Officials (AASHO). Prior to this, no uniform numbering system existed, and highways were often given auto club names.

1562. Network of high-speed freeways in Germany was the autobahn, developed in the 1930s. Limited-access autobahns now cover more than 6,800 miles of the reunified country.

1563. Limited-access highway in the United States built like an interstate was the Pennsylvania Turnpike, opened to the public on October 1, 1940. This first section ran 160 miles from Middlesex to Irwin, PA. After its plans were completed in October 1938, 155 construction companies began the work, using 1,100 engineers and 15,000 workers from 18 states. The all-concrete highway was used by the U.S. Army before being opened to the public. Now part of Interstate 76, the turnpike has been called "America's First Superhighway" and "The Granddaddy of the Pikes."

1564. Road link between Alaska and the lower 48 states was the Alcan Military Highway, built in 1942 during World War II. After the war, it opened to the general public and was renamed the Alaska (Alcan) Highway. While a boon to motorists and a boost to the automobile industry and the Alaskan economy, it caused an unmitigated environmental debacle. Permafrost was disturbed, producing thermokarts—collapse of the land into subsurface voids left by a loss of water—and extensive aquatic habitats were degraded by erosion and or sedimentation.

1565. Express nationwide highway system in the United States was the National System of Interstate and Defense Highways, first approved in 1944 by the U.S. Congress. World War II delayed the program, but in 1956 Congress passed the Federal Aid Highway Act, approved construction of more than 40,000 miles of interstate highways and also the Highway Revenue Act, which authorized the first funds of $25 billion. Costs were shared with states, with the federal government financing 90 percent. The system, originally built to last 20 years, also included intercity highways and farm-to-market roads. It was the largest public works project in U.S. history.

1566. Interstate section in the United States was opened on November 14, 1956, a few miles west of Topeka, KS, with the paving of U.S. Highway 40, now Interstate 70. It was posted with a sign calling it the "first project in the United States completed under the provision of the new Federal-Aid Highway Act of 1956."

1567. British four-lane limited-access highway was the London-to-Leeds "motorway," the M1, whose first section opened on November 1, 1959. The interstate-type road had been proposed in the 1920s.

1568. Expressway in France built with private funds opened in 1973. It stretched 42 miles, linking Paris and Chartres. Before that, expressways were funded by either the French government or local city governments.

1569. Annual highway information for the U.S. government was collected in 1978 by the Federal Highway Administration (FHWA). Its Highway Performance Monitoring System, a cooperative federal-state program, continually collects data on the entire highway and public road system in order to estimate current and future needs. This includes the conditions, performance, and future needs of the highway system. It is used to help apportion federal aid. The U.S. Congress in 1965 had first required biennial reports from the FHWA.

AUTOMOTIVE INDUSTRY—LEGISLATION AND REGULATION

1570. Highway law in the United States for federal-state funding was the Federal Aid Road Act, passed by the U.S. Congress in 1916. It required states to establish their own highway departments and provided matching funds for five years in advance. Each state received funds in relation to its area, population, and post-road mileage.

1571. Gasoline tax levied by the U.S. Congress was enacted on June 6, 1932, as part of the Revenue Act of 1932. A tax of one cent per gallon was imposed on gasoline and other motor fuel.

1572. Automobile emissions standards set by a U.S. state were imposed on cars in California in 1959. They required that crankcase emissions be recycled through a "blow-by" valve.

1573. Automobile crankcase emissions law in the United States was passed in 1961 in California. It required that the devices, which control hydrocarbon emissions, be installed on all cars sold in the state.

1574. Automobile emissions bill approved by the U.S. Congress was passed on October 1, 1965. The Secretary of Health, Education and Welfare was given the power to set emission standards for new automobiles with gasoline or diesel engines. The antipollution law also gave the federal government the authority to prohibit the sale of any automobile that failed to meet the new standards.

1575. Automobile safety legislation of significance passed by the U.S. Congress was the National Traffic and Motor Vehicle Safety Act of 1966. It was mainly brought about by Ralph Nader's book *Unsafe at Any Speed* (1965) and his testimony before a congressional committee. The act established federal authority to create and enforce safety standards for automobiles.

1576. Laws requiring catalytic converters for automobiles in the United States were the Clean Air Act Amendments of 1966. This was the first federal legislation to mention automobile pollution. It established stricter automobile emission standards to become effective in 1968 cars. The act had the U.S. Environmental Protection Agency regulate gasoline additives, if these were needed to develop the catalytic technology.

1577. Automobile emissions control law of significance in the United States was passed in California in 1970. The California Resources Board required 2 percent of the new cars sold in the state by 1998 to release no harmful emissions and 10 percent by 2003, when at least 70 percent must have lower hydrocarbon emissions than do 1993 models.

1578. Automobile emissions requirements in Japan were set in 1972 by that nation's Environment Agency. The policy was modeled on U.S. laws and required that dangerous emissions be reduced by one-tenth by 1975. All automobile manufacturers met the new standards by 1978.

1579. Automobile and truck ban in Greece to control emissions was in June 1982 in Athens. Because of intense smog, those vehicles were not allowed to enter the center of the city from 8 A.M. to 4 P.M., Monday through Friday. In Athens's outer areas, vehicles (divided by odd and even numbers on license plates) could operate on alternate days.

AUTOMOTIVE INDUSTRY—LEGISLATION AND REGULATION—*continued*

1580. Limit on lead in gasoline initiated by the Italian government was included in the Presidential Decree of March 10, 1982. The maximum lead content in gas was set at .40 milligrams per liter. The decree also promised support to introduce unleaded gas.

1581. Transportation act in the United States with significant mass transit funding was the Intermodal Surface Transportation Efficiency Act, passed by the U.S. Congress on November 27, 1991. It provided $31 billion over six years to create and encourage programs of public transportation, although it gave states more power to decide how the funds would be spent. The act also allocated a record $124 billion for a highway and bridge program.

1582. Automobile emissions standards in Canada were detailed in an agreement between the Canadian government and officials of the automobile industry. The requirements paralleled those set out in the United States Clean Air Act but were even more stringent.

1583. Mandatory automobile-emissions tests in Canada were announced on March 24, 1992. The first annual testing of vehicle exhausts began at the end of summer in the Vancouver area.

1584. Tax on "gas-guzzling" automobiles by a U.S. state was passed on April 12, 1992, by Maryland. Tax credit was also given to cars that were fuel efficient. However, on June 17, the National Highway Traffic Safety Administration ordered the state to rescind the new tax.

1585. Transportation act for the 21st century passed by the United States was the Transportation Equity Act for the 21st Century (TEA-21), signed by President Bill Clinton on June 9, 1998. It has many ecological measures, including the reduction of the environmental impact of transportation, the conversion of public-fleet vehicles to cleaner fuel, support for less-polluting trucks, and projects to reduce vehicle-caused wildlife deaths. Overall, the act strengthens the government's strategies—such as its continual support of the Congestion Mitigation and Air Quality Improvement Program—to safeguard public health and the environment.

AUTOMOTIVE INDUSTRY—TECHNOLOGY

1586. Automobile was a steam-powered three-wheeled artillery vehicle built in 1769 by a French military engineer, Nicholas Joseph Cugnot, in the province of Lorraine in northeast France. It carried four people for 20 minutes, running at a little more than two miles per hour. He built two vehicles, but lack of financial support ended production.

1587. Automobile using steam power in England was a "road carriage," completed on December 12, 1801, at Camborne, Cornwall, by Richard Trevithick, an engineer. He drove passengers in it on Christmas Eve, but this first model burned up three days later after he forgot to add water to the boiler. Trevithick continued building steam cars until 1815 but was not commercially successful. He also constructed the first steam railroad locomotive.

1588. Automobile with a practical internal-combustion engine was built in 1860 by Jean-Joseph Étienne Lenoir, a French engineer born in Belgium. He also built a boat propelled by the engine in 1886.

1589. Automobile that proved practical with an internal-combustion engine was the Motorwagen, built in 1885 by the German engineer Karl Benz at his factory in Mannheim, Germany. It was a three-wheel vehicle powered by a two-cycle, one-cylinder engine. Within three years, he had 50 employees building the cars, and with an improved design produced the first commercially successful car in the world, 4,000 being manufactured between 1897 and 1900.

1590. Automaker to use an assembly line was Ransom Eli Olds at his Olds, Motor Vehicle Company in 1899. The company later became part of General Motors, and Olds moved on to form the Reo Motor Car Company.

1591. U.S. manufacturer to make cars in great quantities to reduce costs was Henry Ford, during the early 20th century. Ford's Model F sold for $1,200 in 1904. By 1924, with the techniques of mass production, the cost of a sturdy, reliable new Model T was $290.

1592. Automatic starter for automobiles was invented in Dayton, OH, in 1911. The inventor, Charles F. Kettering, was founder of the Dayton Engineering Company, Delco. With Kettering's device, cars no longer had to be started with a hand crank.

1593. Gasoline additive tried was iodine in 1916. It was added to kerosene automobile fuel in an experiment at a laboratory in Dayton, OH. The iodine successfully stopped engine knocking. However, idodine proved impractical because of its expense.

1594. Use of tetraethyl lead for automobiles was introduced in 1923 by the General Motors Corporation as an antiknock gasoline additive. It was produced by GM's former subsidiary, the Ethyl Corporation. Tetraethyl lead allowed cars to run more efficiently on lower-octane gas. Two years later the Public Health Service began to investigate it as an air pollutant. In 1996, all lead-based gasolines were banned in the United States.

1595. Car model with a high-compression six-cylinder engine was the Chrysler Six, which became available in 1924. The Maxwell Motor Corporation's Chrysler Six, designed by Walter Chrysler, was an immediate success, with 32,000 sold the first year. Maxwell became Chrysler Corporation in 1925.

1596. Subcompact car designed to be fuel-efficient was the Volkswagen Beetle, designed in Germany by Ferdinand Porsche in 1935 and imported into the United States beginning in 1949. The Beetle was noted for its rear-mounted, air-cooled engine. A redesigned Beetle was introduced in 1998.

1597. Mass-produced car with a fiberglass body was the Chevrolet Corvette. Chevrolet, a division of the General Motors, introduced the Corvette in 1953.

1598. Pollution-control data swapped by U.S. car companies was after a 1955 "cross-licensing" agreement signed by the manufacturers. This was to help inform smaller companies of technical progress made by the major corporations. Informal swapping had begun a year earlier. Critics of the agreement said it actually slowed technical breakthroughs by dulling competition.

1599. Advanced pollution-control system introduced by a U.S. car manufacturer was the Chrysler Clean Air Package in 1962. It involved a new manifold heat valve, new cylinder gaskets, and control valves for the carburetor and distributor. By 1970, about 80 percent of U.S. automobiles used the Chrysler system.

1600. Automobile seat-belt buzzers were required by the U.S. government to be installed on cars built after January 1, 1972. Some companies had already added them in 1971. The buzzers warn people in the front seats to fasten their seat belts.

1601. Personal rapid transit (PRT) system in the United States began in 1975 at Morgantown, WV. Costing $119 million, it connected the downtown with two university campuses on a route of 1.4 miles. The PRT uses small 12-seat minibuses that are controlled by a computer in a guideway at speeds of up to 30 miles per hour.

1602. Automobile pollution device of significance in China was the PCV (positive crankcase ventilation) valve introduced in 1991 in Beijing. The device draws crankcase fumes into the cylinders to be burned.

1603. Automobile technology organization formed by large U.S. car companies was the United States Council for Automotive Research (USCAR), established in 1992 by the Chrysler Corporation, the Ford Motor Company, and General Motors Corporation. Their goal was to strengthen the technological base of the domestic automobile industry through cooperative, precompetitive research into such areas as cleaner emissions and fuel efficiency.

1604. Electric-car "solar carpark" opened in 1992 in Diamond Bar, CA. It was built by Southern California Edison for the headquarters of the South Coast Air Quality Management District. The 3,000 square feet of solar cells were placed above the parking lot to provide electricity to car batteries through simple plugs. It was hoped that solar-recharging stations would eventually be found throughout California cities in parking lots next to offices and shopping centers.

1605. Low-emission automobiles were introduced in May 1992 by the Ford Motor Company. The two 1993 models appeared in California showrooms and were four years ahead of the state's stringent requirements for low-emission automobiles. Ford's new technology involved increasing the size of the catalytic converter and shifting it near the engine. The device also heats more quickly to eliminate dangerous exhaust fumes when the vehicle is started.

1606. Automobile technology agreement between the U.S. government and industry was the establishment of the Partnership for a New Generation of Vehicles (PNGV), announced on September 29, 1993, by President Bill Clinton. The public-private program was between seven U.S. government agencies and 20 federal laboratories with the Chrysler Corporation, Ford Motor Company, and General Motors Corporation to develop future technologies for a "Supercar." A year earlier the three companies

AUTOMOTIVE INDUSTRY—TECHNOLOGY—*continued*

had formed the United States Council for Automotive Research (USCAR). The long-term goal was to develop an environmentally friendly car with triple fuel efficiency, obtaining 80 miles per hour with low emissions. Concept vehicles were to be built by 2000 by each company and production prototypes by 2004. Federal funding for the research in 1998 was $227 million.

1607. Vehicle recycling research center in the United States was the Vehicle Recycling Development Center (VRDC) established in January 1994 in Highland Park, MI. It was created by a recycling consortium formed in 1991 by the United States Council for Automotive Research (USCAR), which is the cooperative organization of the Chrysler Corporation, the Ford Motor Company, and General Motors Corporation. About 75 to 85 percent of a vehicle's weight is recycled. The center dismantles up to 500 vehicles a year to study ways to recycle the materials. It works with such organizations as the Automotive Recyclers Association, the American Plastics Council, Aluminum Association, the Institute for Scrap Recycling Industries, and the Argonne National Laboratory.

1608. Automobile powered by fuel cells was introduced in 1996 by the Daimler-Benz company of Germany. It could run at speeds greater than 100 miles per hour and go about 155 miles before refueling.

1609. Electric car to be mass-produced using modern technology was marketed by General Motors, in Detroit, MI, beginning on December 4, 1996, under the name Electric Vehicle One, or EV-1. Intended to comply with California's tough antipollution regulations, the first EV-1 models were leased to customers in Los Angeles, CA, based on a retail price of $34,000. The two-seater coupe was powered by a 137-horsepower, three-phase induction motor that ran on stored energy in a lead-acid battery pack and could travel 70 to 90 miles between chargings, which took 3 to 12 hours.

1610. Cars designed using computer-aided tools were designed at the Chrysler Corporation in the United States in 1997. These cars became known as "paperless cars" because they were not designed in the conventional way. Design teams used digital model assembly for 1998 and 1999 full-size sedans.

1611. Test drive of the "veggie van," an experimental motor home powered by a diesel engine that ran on fuel made from vegetable oil was in 1997. The vehicle reached speeds of 65 mph on its 10,000-mile trip around the United States. The exhaust was said to smell like French fries.

1612. Electric car from a major U.S. company was the EV-1 introduced in January 1997 by General Motors in 24 dealerships in Arizona and Southern California. Made by its Saturn division, the EV-1 has two seats, accelerates from 0 to 60 miles per hour in less than 9 seconds, and on a full battery can go 70 miles in the city and 90 on a highway. The original price was about $35,000.

1613. Automobile gasoline converted to hydrogen for future cars was announced in February 1997 by the Chrysler Corporation. Company research had developed a process to extract hydrogen from low-octane gasoline and convert the hydrogen into electricity. This was done with an on-board fuel processor and platinum catalyst. Chrysler estimated thay by 2005 the company could introduce fuel-cell cars that would give off virtually no dangerous emissions and be 50 percent more fuel-efficient than internal-combustion engines. The technology had been originally developed for spacecraft.

1614. Electric automobile to convert gas to hydrogen fuel was developed in the United States by the Chrysler Corporation. The company announced the breakthrough in February 1997 and said the car, which will seat six, could be available by 2015. It will emit virtually no pollution and require no maintenance.

1615. Automobile emission cleaner than city air was announced on March 16, 1998, by Honda. Honda's experimental Zero-Level Emission Vehicle (ZLEV) was built by the Honda Research and Development Center in Tochigi, Japan. The company said its 2.3-liter, 4-cylinder Accord engine produces only one-tenth of the maximum amount of dangerous emissions allowed by California, which has the world's most stringent requirements: ZLEV carbon monoxide emissions are 0.17 gram per mile, compared to California's 1.7 gram maximum allowance. This is done using exhaust catalysts, to trap and clean hydro-carbon emissions when the engine starts up, and also a computer to control the combustion timing.

1616. Electric-vehicle project by the Canadian government and industry was the Montreal 2000 Electric Vehicles Project, announced January 25, 1999. At a cost of more than $3 million for a two-year trial, it provided one or more of 40 available electric vehicles for up to 20 commercial or institutional organizations in the greater Montreal area. The vehicles, delivered in the spring of 1999, were Ranger compact pickup trucks and Force four-door sedans, provided by Ford of Canada, and Citivans by Solectria. The joint project involved the Canadian government, providing funding through its Climate Change Action Fund (CCAF); the Quebec government; Hydro-Quebec; Norvik Traction Inc.; and the Centre d'experimentation des véhicules electriques du Quebec (Center for Experimentation on Electric Vehicles).

B

BIODIVERSITY

1617. Use of taxonomy to classify living organisms was conceived by Swedish scientist Carolus Linnaeus (Carl von Linné). The influential system Linnaeus devised for naming, ranking, and classifying organisms was first published in 1735 in his *Systema Naturae.*

1618. Scientist to use the term *biosphere* was Austrian geologist Hans Eduard Suess in the 1890s. The term, however, was not elucidated until 1929, when Russian geochemist Vladimir Vernadsky used it to define the part of the earth which supports living matter.

1619. "Life zones" concept was developed by American naturalist Clinton Hart Merriam during his tenure as chief of the U.S. Biological Survey. While doing field work in Arizona in the 1890s, Merriam observed that temperature and altitude corresponded to particular distributions of plants and animals. Merriam created an identification system that divided habitats in the western United States into six "life zones," each with distinctive natural characteristics—Alpine, Hudsonian, Canadian Forest, Transition, Upper Sonoran, and Lower Sonoran. In 1898, Merriam described his concept in *Life Zones and Crop Zones of the United States.* The "life zones" concept was refined by later naturalists.

1620. "Ecological niche" concept was developed by U.S. ornithologist Joseph Grinnell. He began using the term in 1917 to describe how each animal species has its own feeding habits and habitat within environments shared by other species. Other zoologists refined the "niche" concept to define how species interact within larger ecological communities, especially in matters of competition, environmental conditions, and predation.

1621. Scientist to use the term *ecosystem* was English botanist Arthur George Tansley. In July 1935, Tansley used the term to describe ecological units composed of both organisms and their surrounding environment.

1622. United Nations conference linking biodiversity and sustainable development was the UNESCO (United Nations Educational, Scientific and Cultural Organization) Conference on the Conservation and Rational Use of the Biosphere, held in Paris in 1968. The conference resulted in UNESCO's Man and the Biosphere (MAB) program. More than 85 countries attempted to balance the competing needs of biodiversity, economic development, and the cultural values of indigenous people living within designated protected MAB zones.

1623. Biosphere reserve in Asia is central Thailand's Sakaerat Biosphere Reserve, which was designated in 1976. Reforestation and resource management of, and biodiversity in, the reserve's variety of Thailandian monsoon forests are administered as the Plualuang National Forest Reserve.

1624. Biosphere reserve in South America is the Bañados del Este of Uruguay's pampas. The reserve was designated in 1976. Its conservation and development programs attempt to regulate the use of water and grasslands as pastures in sustainable ways.

1625. Biosphere reserves in Europe were designated in 1976. Their programs included study of mountain systems in Yugoslavia's Tara River Basin, environmental monitoring of North East Svalbard tundra in Norway, and forest-related studies in four Polish reserves, Babia Gorn, Bialowieza, Lukajno Lake, and Slowinski. The various programs in the Polish reserves range from climate monitoring to the ecology of the region's rare bison.

1626. Biosphere reserves in the Middle East were established in Iran in 1976. Nine reserves were created for preservation and study sites in the country's northern mountains and southern deserts. All the reserves are administered by Iran's Department of the Environment.

BIODIVERSITY—*continued*

1627. Biosphere reserves in the United States were designated in 1976. The 25 protected areas in a dozen continental U.S. states and Puerto Rico included a variety of biomes—mountains, tundra, deserts, grasslands, rain forests, and different kinds of forests. The Man and the Biosphere Secretariat for U.S. reserves is located in Washington, DC. It serves as the national committee of UNESCO's worldwide Man and the Biosphere Progamme.

1628. Biosphere reserves in Africa were designated in 1977 in five countries. Biomes protected included rain forests in the Congo, Côte d'Ivoire (Ivory Coast), and Nigeria, and on island dependencies of Mauritius. Reserves designated for study and protection in Tunisia included evergreen forests, grasslands, and the steppes of the Atlas Mountains.

1629. Biosphere reserves in Australia were designated in 1977. The continent's first eight reserves included a variety of biomes, including arctic desert and tundra, evergreen forests (Croajingolong National Park), tropical forests, rain forests, deserts (Uluru National Park), and grasslands.

1630. Biosphere reserves in Mexico were designated in 1977. The Michilia Reserve researches ornithology, botany, and crop development in the mountains of central Mexico. The Mapimi Reserve focuses on climatology, rare and endangered species, and resource management in the Chihuahuan desert.

1631. Biosphere reserve in Canada is the Mont-St.-Hilaire Biosphere Reserve in Quebec. The broadleaf forest was designated a biosphere reserve in 1978 and is the site of studies in zoology, meteorology, geology, and applied geophysics.

1632. Modern study linking rain forest deforestation with accelerating species extinction due to habitat loss was contained in English ecologist Norman Myers's book *The Sinking Ark* (1979). Myers's influential work helped focus the attention of international scientific community on the disappearance of plant and animal species that accompanies the rapid destruction of the planet's rain forests.

1633. Biosphere reserve in Central America is the Rio Platano Biosphere Reserve, which was designated in northeastern Honduras in 1980. The biosphere program for the forested reserve protects endangered species, develops tourism, and is attempting to stem pollution and deforestation caused by farming and the timber industry.

1634. Biosphere reserve in Hawaii was the Hawaiian Islands Biosphere Reserve, designated in 1980 and managed from Hawaii's Volcanoes National Park. Monitoring and research programs are focused on rare and endangered species, ecosystem restoration, vegetation, and inventorying Hawaiian invertebrate species.

1635. Biosphere reserve in Ecuador is located in the Archipiélago de Colón (the Galápagos Isands). The reserve was designated as such under UNESCO's Man and the Biosphere Programme in 1984. Its mission includes ecosystems research, resources management, and the study and preservation of the islands' rare species.

1636. Use of the term *biodiversity* was established at the 1986 National Forum on BioDiversity in Washington, DC. The term was coined by National Academy of Sciences administrator Walter Rosen and his colleagues, who used it to describe the varietyor biological diversity of all organisms and ecosystems on earth. The origin of the term is frequently misattributed to influential biologist Edward O. Wilson.

1637. Joint biodiversity program between Nicaragua and Costa Rica is the Si a Paz, or "Yes to Peace," Park situated on their common border. The two countries initiated the park program in 1988 after years of guerrilla warfare in the region. The Si a Paz strategy promotes ecologically sound agricultural practices, in hopes of preventing deforestation from causing wildlife habitat loss and irreversible damage to the area ecosystem.

1638. Biosphere reserve in Madagascar is the Manara Nord Biosphere Reserve, which was designated in 1990. The protected zone on Madagascar's northeastern coast includes both rain forest and a marine park. Programs range from land management, to reverse the effects of deforestation, to educating regional residents in fisheries management education.

1639. "Bioprospecting" agreement was reached in September 1991 between the U.S.-based pharmaceutical firm Merck and Costa Rica's National Biodiversity Institute (INBio). INBio agreed to provide Merck with drug-screening from wild plants, insects, and micro-organisms, in return for a two-year research and sampling budget of $1.14 million and royalties on any resulting commercial products. Merck also agreed to provide technical assistance and training to help establish drug research in Costa Rica. INBio agreed to contribute a portion of the proceeds to Costa Rica's

National Park system. This biodiversity prospecting or "bioprospecting" contract marked a significant departure from an earlier pattern whereby commercial biotechnology companies would exploit the resources of developing countries without compensation.

1640. Canadian biodiversity government agency was the Biodiversity Convention Office, established in September 1991. The office became responsible for guiding national policy relating to the 1992 United Nations Conference on the Environment and Development, which committed Canada to the conservation of biodiversity and sustainable use of biological resources.

1641. Biodiversity governmental agency in Mexico was the National Commission for the Knowledge and Use of Biodiversity (CONABIO). The commission was created by a presidential decree signed by Ernesto Zedillo Ponce de León on March 16, 1992. CONABIO centralized the activities of numerous governmental and non-governmental agencies to promote biodiversity research, sustainable use policies, and public education about biodiversity issues.

1642. Trilateral biosphere reserve was created in 1998 when Ukraine added a nature park to a transfrontier biosphere reserve administered by Poland and Slovakia. Wildlife conservation, pollution monitoring, and land management programs within the new entity, the East Carpathians Biosphere Reserve, are concentrated on mountainous forestland shared by the three countries.

BIRDS

1643. Recorded use of pigeons as messengers was in 1200 B.C. in Egypt, when four birds were released to carry the message that Pharaoh Ramses III had assumed the throne.

1644. Known exploitation of migratory swallows' homing instincts was in approximately A.D. 1, when a Roman noble took swallows with him from his home in Volterra to Rome, 134 miles away. After watching the chariot races, he marked the birds with colors indicating the winners and then released them to carry the news back to his friends and family.

1645. Commerce in cave swiftlet nests was during the T'ang Dynasty (618–907), when Chinese traders traveled to Borneo to trade for the valuable nests. The nests are an essential ingredient in bird's nest soup, an Asian delicacy.

1646. Discovery of the spectacled cormorant was in 1741, when Georg W. Steller observed the bird while he was marooned on the Komandorskiye Islands with explorer Vitus Bering.

1647. Oilbird was discovered in 1799 in Venezuela by explorer Alexander von Humboldt.

1648. Commercial ostrich farm was started in 1838 in South Africa. Such farming came about because of the plume trade that flourished during the 19th century. Many experts today believe that ostrich farming probably saved the large bird from extinction.

1649. Emperor penguin egg seen and collected by humans was collected on an ice floe in Antarctica between 1839 and 1843 by a member of a French expedition commanded by Dumont d'Urville. Little was known about the bird's breeding habits at the time, and the significance of the egg went unrecognized.

1650. Bird species native to North America to be extinguished was the great auk, a flightless bird once common on North Atlantic coastlines. The great auk became extinct circa 1844 as a result of human impacts, such as expanding settlements and hunting pressure.

1651. House (English) sparrows to thrive in the United States were imported in 1852 into Brooklyn, NY. This was the second batch to be brought to the New World, an introduction two years earlier having failed. During the remainder of the 19th century, the species grew to become the most abundant one in the country. However, the number of house sparrows in the United States declined with the advent of the automobile in the 20th century.

1652. Description of the Virginia warbler was in 1858 by Dr. W. W. Anderson, an assistant army surgeon who discovered the bird in New Mexico, naming it after his wife, Virginia.

1653. Eurasian tree sparrow in the United States was introduced in 1870 in St. Louis, MO, when 20 birds were brought over from Germany. For the next 100 years, the species was confined to that area, but it began spreading in the 1960s, occupying adjacent areas of Missouri and Illinois.

1654. Lincoln's sparrow was discovered in 1883, during John James Audubon's travels in Labrador, Newfoundland.

1655. Successful introduction of a tinamou was in 1885, when the Chilean tinamou was introduced onto Easter Island. Many other attempts had been made elsewhere to introduce various species of tinamous, but this was the only successful one.

BIRDS—*continued*

1656. Extinction of a native American bird in modern times was that of the American passenger pigeon in 1914. Martha, the last known specimen, died on September 1, 1914, in the Cincinnati Zoo in Cincinnati, OH.

1657. Report of the reappearance of the cahow, or Bermuda petrel was made in 1916, when several were sighted on the Castle Rocks off Bermuda. The species was thought to have become extinct in the early 17th century. In 1951, 18 pairs were sighted at the same location. Although the species continued to be completely protected, probably no more than 50 breeding pairs remained at the end of the twentieth century.

1658. Bird of a new type discovered in the 20th century in the United States was the Cape Sable sparrow, an endangered sparrow discovered in 1918 at Cape Sable, FL. Although once thought to be a distinct species, the bird was later determined to be a race of the seaside sparrow.

1659. Imperial pheasants collected from the wild were a pair captured in 1924 in Annam, in what was then known as French Indochina, by the ornithologist Jean Delacour. The pair were then taken to France. The limited number of imperial pheasants still existing in captivity are all descended from that pair. Modern surveys carried out in Vietnam indicate that small numbers of the endangered birds still exist there in the wild.

1660. Chestnut-winged chachalaca introduced in the United States was in 1928, when some members of the species were brought into Hawaii from Panama. However, the species did not establish itself in this new habitat.

1661. Crested guan introduced in the United States was in 1928, when some members of the species were brought to Hawaii from Panama. However, the species did not establish itself in this new habitat.

1662. Great currasows introduced into the United States were brought from Panama to Hawaii in 1928. However, the species did not establish itself in this new habitat.

1663. Albatross colony located on a mainland was a Northern Royal Albatross nest site located at Taiaroa Head on the Otago Peninsula on the South Island of New Zealand. The first authenticated breeding took place here in 1920, but the resulting egg disappeared from the nest. The first chick hatched at this sight in 1938, but it was killed. Only in 1938 did a chick fully fledge at this location. The colony remains small, consisting of only 17-20 breeding pairs.

1664. Utilization of the Federal Aid in Wildlife Restoration Act (Pittman-Robertson Act) in Hawaii was in 1945. The state used federal funds to finance a contract for the study of game birds.

1665. Kiwi bred in captivity was a North Island brown kiwi hatched on September 19, 1945, near Napier, New Zealand. The small, flightless kiwi is native to New Zealand.

1666. Open hunting season on the chukar in the United States was in Nevada in 1947. A native of Asia, the chukar is a kind of partridge. It was widely introduced in the United States, but flourished only in arid western regions.

1667. Rediscovery of the takahe took place in 1948 in the Murchison Mountains of Fiordland in New Zealand, near Lake Te Anu. Prior to that time, the bird was known only from four specimens and, having not been seen since 1898, was thought to be extinct.

1668. Scarlet ibis introduced into the United States was in 1960 in Florida, when scarlet ibis eggs were placed under incubating white ibis parents. The pure-bred foster scarlet ibis disappeared after a few years, but hybridization occurred.

1669. Endangered species list to include birds was the first endangered species list released by the U.S. Department of the Interior, in 1966. It listed a total of 78 species of endangered plants, reptiles, birds, and mammals.

1670. Reported new sighting of the taiko was in 1967, when several were seen near the Chatham Islands in New Zealand. Also known as the magenta petrel, the bird had previously been known only from a single specimen collected in 1867.

1671. Wild turkey restoration program in Georgia began in 1972 when the Georgia Game and Fish Department initiated the project with funds received through the Federal Aid in Wildlife Restoration Act (Pittman-Robertson Act)

1672. Hooded grebe to be discovered was found in 1974 in Laguna Las Escarchadas in Patagonia in Argentina by ornithologist Mauricio Rumboll.

1673. Report of the reappearance of the Campbell Island flightless brown teal was in 1975, when the supposedly extinct species was sighted on Dent Island, a small, predator-free islet adjacent to Campbell Island, New Zealand. The wild population of this rare duck probably consists of no more than 30 to 50 individuals. The teal disappeared from Campbell Island—their primary home—after it was overrun by Norway rats and feral cats, descendants of animals that came ashore from whaling ships and sealers in the 19th century. In 2001 the New Zealand Conservation Department began a drive to eliminate non-native species from the island and restore its pre–19th century ecology.

1674. Pheasant restoration program in South Dakota began in 1976 as a cooperative effort, between the state's Game, Fish and Parks Department and the South Dakota Pheasant Congress, to restore wildlife habitat on private land. Participating landowners are paid to plant and maintain, for a minimum of five years, up to 40 acres of retired cropland as prime dense nesting cover.

1675. Goose to lay an egg weighing 1.5 pounds was a white domestic goose, named Speckle, that on May 3, 1977, laid this substantial egg in Goshem, OH. Normally, goose eggs weigh between 6 and 7 ounces.

1676. Known bird with a wingspan of 25 feet was a new soaring-type species whose remains were found in Argentina in 1979. The bird could have weighed as much as 250 pounds and measured as long as 11 feet. Because of its size, this extinct species would not have been capable of prolonged flapping while flying.

1677. Year Oklahoma required the purchase of a state duck stamp for hunting waterfowl was in 1980. Revenues from these sales were used to match funds authorized by the Federal Aid in Wildlife Restoration Act (Pittman-Robertson Act) to create an intensive statewide waterfowl management program.

1678. Emperor penguins bred outside of Antarctica were born at Sea World in San Diego, CA, on September 16, 1980. The birds have bred sporadically ever since.

1679. Bird known to have lived to a documented age of 80 was a sulphur-crested cockatoo named Corky, acquired as an adult by the London Zoo at the beginning of the 20th century. Because Corky had been raised by an individual owner before being acquired by the zoo, when the bird died on October 28, 1982 his keepers were able to establish that he had lived for at least 80 years.

1680. Chatham Islands black robin to continue breeding to the age of 13 was Old Blue, who was the only female of the species breeding during the period 1979–1983. Without her longevity and vitality, the species probably would have become extinct: all 50 such robins alive in 1987 were descendants of Old Blue.

1681. Fossils of *Protoavis* a hollow-boned bird-like creature, were discovered in the badlands of western Texas in 1983 by the paleontologist Sankar Chatterjee. These fossils are believed to be 225 million years old, about 75 million years older than fossils of *Archaeopteryx*, long identified as the earliest bird. Since the *Protoavis* fossils do not show any distinct feathers, some paleontologists believe that *Protoavis* was not the "first bird" but a tree-dwelling reptile, possibly a bird ancestor.

1682. Rediscovery of MacGillivray's petrel was in 1983, when a specimen flew ashore on Gua Island, Fiji, striking an ornithologist in the head. The species had previously been known only from one specimen, a fledgling collected on Gua in 1855 and named for the Scottish ornithologist William MacGillivray, a friend of John James Audubon. In 1985, another fledgling was found on Gua.

1683. Reported sighting of Jerdon's courser since 1900 was on January 12, 1986, in India. The shorebird had long been thought extinct and was discovered accidentally when a local hunter retrieved one specimen that had been blinded by the light of his torch. The bird remains very rare, and its actual status is unknown.

1684. California condors to be bred in captivity were four chicks successfully reared at the San Diego Wild Animal Park in 1989. Within four years a second generation was being bred, and in January 1992 a captive-raised pair were released to the wild, with six more released nine months later. Within a year, half of these birds had died, and the remaining four were removed to remote Lion Canyon in the Los Padres National Forest, where they were released along with five more captive-raised condors.

BIRDS—*continued*

1685. Hybridization of the Norfolk Island boobook owl was in 1990, when, after the population of this bird had been reduced to a single female, the island was stocked with several owls from a separate subspecies that had been imported from New Zealand. Several hybrid chicks were thus produced.

1686. Hybridization of a whooping crane and a sandhill crane was in 1992, when a male whooping crane mated with a female sandhill crane. They wintered with their chick at Bosque del Apache Refuge in New Mexico. The hybrid chick is referred to as a "whoopbill."

1687. New bird species established solely by DNA tests and photographs was the Bulo Burti boubou, in Somalia in 1992. Feathers and a blood sample taken from a mist-netted bird, together with photographs, enabled scientists to establish the existence of a previously unknown species of bush shrike. The captured bird was released; the practice of describing a new species without collecting a type specimen has come in for much criticism.

BIRDS—LEGISLATION AND REGULATION

1688. Bird officially designated as the national bird of the United States was the bald eagle, so designated on June 20, 1782.

1689. Legislation to protect migratory game and insectivorous birds was authorized by the Weeks-McLean Act of 1913. Also known as the Migratory Bird Act, it gave the U.S. Department of Agriculture authority to regulate the hunting season of a wide range of bird species. The act, signed on March 4, 1913, by President William Howard Taft, placed under U.S. protection all migratory game and insectivorous birds with northern or southern migration paths that passed through any U.S. state or territory.

1690. Successful challenge to state-ownership doctrine was the Migratory Bird Treaty Act, signed by President Woodrow Wilson on July 3, 1918. The act established federal ownership of economically important migratory birds. This shifted away from the states the power to regulate wildlife. The legislation implemented the provisions of a 1916 treaty, which the United States signed with Great Britain and Canada, that established the international framework for managing and protecting migratory waterfowl in North America

1691. Government commission to rent or purchase lands to be set aside for migratory birds in the United States was the Migratory Bird Conservation Commission, established on February 18, 1929, with the passage of the Migratory Bird Conservation Act. The commission considers areas of land and/or water recommended by the U.S. Secretary of the Interior and fixes prices to be paid for purchase or lease of those areas. Since 1989, the commission has also been responsible for establishing new waterfowl refuges under the North American Wetlands Conservation Act.

1692. Duck stamps for waterfowl hunters in the United States were issued in 1934 by the federal government. Purchase of the stamps, which are offered by the U.S. Fish and Wildlife Service and made available through most U.S. post offices, is required of every waterfowl hunter age 16 or older. Proceeds of these sales go toward the purchase of wetlands. The duck stamp program has proven to be one of the government's most successful conservation methods, adding more than four million acres of wetlands habitat to the National Wildlife Refuge System since 1934.

1693. Legislation that provided regular federal funding for waterfowl management in the United States was the Migratory Bird Hunting Stamp Act of 1934, also known as the Duck Stamp Act. The act, signed by President Franklin D. Roosevelt, provided critical funds for wetlands and waterfowl conservation programs. Until this act, no stable funding source was available for this conservation work.

1694. Federal protection for bald eagles in the lower 48 states of the United States was granted in 1940, when Congress passed the Bald Eagle Protection Act. The law has been amended twice, most recently in 1972, when Wyoming ranchers were discovered to be poisoning eagles and shooting them from helicopters.

1695. Programs in Georgia authorized by the Federal Aid in Wildlife Restoration Act (Pittman-Robertson Act) began in 1944. Five wildlife restoration projects were funded by federal and state money through the Federal Aid in Wildlife Restoration Act.

1696. Law passed by U.S. Congress specifically aimed at preserving endangered birds was the Endangered Species Act of 1966, signed on October 15, 1966. The law aimed to protect species of birds, fish, plants, and other wildlife in danger of extinction.

BIRDS—OBSERVATIONS

1697. Identification of the Congo peacock occurred in 1928, when James Chapin of the American Museum of Natural History was visiting the Congo Museum in Brussels, where he found two birds, labeled as imported juvenile blue peafowls, bearing feathers that matched those of a bird he had received 25 years earlier while visiting the Belgian Congo.

1698. Blue goose nest discovered was on Baffin Island in Canada in 1929.

1699. Discovery of a bristle-thighed curlew's nest was in 1948 in western Alaska, the only known breeding area for this curlew. It is one of the few North American birds whose nests and eggs remained undiscovered until well into the 20th century. Little is known about its exact habits, as the bird has seldom been studied.

1700. Colony of puna flamingos to be discovered was found at Laguna Colorado in Bolivia in January 1957. The puna flamingo is also known as James' flamingo, named for the naturalist and businessman H. B. James.

1701. Bird known to fly at an altitude of 21,000 feet was a mallard, which on July 9, 1963, collided with a Western Airlines L-188 aircraft over Elko, NV.

1702. Discovery of the Hutton's shearwater breeding grounds was in 1965 on the slopes of the Kaikoura Mountains on New Zealand's South Island.

1703. Wandering albatross recorded ashore in the United States was in July 1967, when a specimen was discovered grounded 60 miles north of San Francisco, CA.

1704. Report of a shorebird at latitude 71 degrees south was in October 1968, when the carcass of a Wilson's phalarope was found on Alexander Island in Antarctica. It was the most southerly sighting of a shorebird ever recorded.

1705. Manx shearwater nest discovered in the United States was in 1973 in Massachusetts. The Manx shearwater is a pelagic bird that spends most of its time at sea, coming to land only to breed. Certain small islands off the coast of Newfoundland customarily serve as nesting sights for this species.

1706. Bird flight recorded at 37,000 feet was on November 29, 1973, when a Rueppell's griffon was sucked into the jet engine of a plane flying over Abidjan, Côte d'Ivoire, at 37,000 feet. This was the highest recorded altitude ever achieved by a bird. The plane's engine was damaged so badly that it was shut down.

1707. Discovery of a marbled murrelet's nest occurred on August 7, 1974, when a tree surgeon at Big Basin Park in California climbed up a large Douglas fir to cut off a limb that threatened to fall on tourists. Unfortunately, he saw the murrelet's nest and the chick in it only after the tree limb had been severed. He managed to catch the chick, but it escaped and fell to the ground, dying two days later. The murrelet, a coastal bird that prefers to nest in forests, was one of the few North American–breeding species whose nests had never been identified.

1708. Female golden-fronted bowerbird seen was in 1981, when the species, once thought to be extinct, was rediscovered in the Foya Mountains of New Guinea. Previously, the species had been known only from three or four male trade skins.

1709. Bird recorded at the North Pole was a snow bunting which was observed in May 1987 by a Royal Navy submarine that had surfaced at the Pole.

1710. Swans sighted in the Antarctic were five black-necked swans observed in various locations along the Antarctic Peninsula, approximately 500 miles south of the tip of South America, during the austral summer of 1988–1989. Swans were again sighted in Antarctica during the austral summer of 1993–1994.

BIRDS—ORGANIZATIONS

1711. International Ornithological Congress was held in Vienna, Austria, in 1884.

1712. Bird preservation organization in the United States was founded in 1886 by George Bird Grinnell in New York City. Named for the famed American naturalist and wildlife painter John James Audubon, the Audubon Society attempted to put a stop to the wholesale slaughter—and even extinction—of various bird species hunted for their plumes, eggs, and meat. The response to Grinnell's efforts was so overwhelming—38,000 people joined in only three months—that he was obliged to disband the organization in 1888. A group of women then founded the Massachusetts Audubon Society in 1896, and groups in other states followed suit, forming a loose alliance in 1901 and incorporating in 1905 as the National Association of Audubon Societies for the Protection of Wild Birds and Animals.

1713. Avicultural organization founded was the Avicultural Society, founded in Great Britain in 1894. *Avicultural Magazine*, its journal, has been published continuously since then.

BIRDS—ORGANIZATIONS—*continued*

1714. Warden hired by a bird conservation society in the United States was in 1900, when the first Audubon warden was hired, even before the National Audubon Society was officially founded. By 1904, 34 such wardens were employed in 10 states.

1715. National Audubon Society meeting was held in 1905. The Society took its name from the American naturalist and wildlife painter John James Audubon. Audubon societies were formed in several states during the 1890s. They made a loose alliance called the National Committee of the Audubon Societies in 1901, before becoming incorporated in 1905 as the National Association of Audobon Societies for the Protection of Wild Birds and Animals, today's National Audobon Society.

1716. National U.S. bird banding association was founded as the American Bird Banding Association, by 30 charter members on December 8, 1909, in New York City. When it dissolved in 1980, its records were turned over to the Bureau of Biological Survey of the Department of Agriculture in Washington, DC.

1717. Parrot protection and welfare organization was established in the United Kingdom in 1989 as the World Parrot Trust. Growing out of aviculture, the trust also focuses on conservation of parrots in the wild. The trust has developed branch offices in Australia, Belgium, Canada, Denmark, France, and the United States, where an IRS-recognized, tax-exempt, nonprofit organization, the World Parrot Trust USA Inc., has been set up. The trust publishes a newsletter, *PsittaScene*, and backs a "Parrot Portfolio" project that records work carried out in 18 countries on behalf of 20 parrot species.

BIRDS—REFUGES AND RESERVATIONS

1718. Bird refuge established by a U.S. state was created at Lake Merritt in Oakland, CA, by authority of a law enacted on February 14, 1872.

1719. Bird reservation established by the U.S. government was created under an executive order of President Theodore Roosevelt on March 14, 1903, at Pelican Island, located in the Indian River near Sebastian, FL, to protect a nesting colony of pelicans and herons. It was enlarged in 1909 to include adjacent mangrove islands and swamps.

1720. Bird sanctuary to protect migrating hawks and eagles was Hawk Mountain Sanctuary, established in 1934 on the Kittatinny Ridge near Harrisburg, PA.

1721. Raptor sanctuary was founded in 1934 on the Kittatinny Ridge near Harrisburg, PA. The Hawk Mountain Sanctuary was the first in the world to protect migrating hawks and eagles.

1722. Project in Utah approved under the Federal Aid in Wildlife Restoration Act (Pittman-Robertson Act) was the Ogden Bay Waterfowl Management Area, established in 1937.

1723. California condor sanctuary was the Sisquac Condor Sanctuary, founded in 1937. Sisquac's 1,200 acres were later dwarfed by the 35,000-acre size of the Sespe Condor Sanctuary, which was established in the Los Padres National Forrest in Ventura County, CA, in 1947.

1724. Refuge for whooping cranes was established with the Aransas National Wildlife Refuge, at Aransas Bay on the Gulf Coast of Texas, in 1938.

1725. Waterfowl management area developed by the Missouri Conservation Commission was the Fountain Grove Wildlife Management Area, purchased in 1947 with Federal Aid in Wildlife Restoration Act (Pittman-Robertson Act) funds.

1726. Whooping crane artificial colony was established in Grays Lake, ID, in 1975, when whooping crane eggs were placed under incubating sandhill cranes. The resulting young whoopers wintered with their sandhill "foster parents" in central New Mexico, but have yet to reproduce in Idaho. A second attempt to create an artificial colony of the endangered birds occurred in February 1993, when 14 captive-bred whoopers were released in south central Florida, but within a month five had been killed by bobcats. These five were replaced in November 1993.

BIRDS—RESEARCH

1727. Bird band recovery was in 1710 in Germany, when a gray heron was found to be wearing metal rings placed on it several years earlier in Turkey.

1728. Bird banding in the United States was done in 1803 by John James Audubon at Mill Grove Farm, in Montgomery County, PA, 24 miles northwest of Philadelphia. He used silver wire to band a brood of phoebes and later recovered two of the birds.

1729. Breeding bird census was conducted by Scottish immigrant Alexander Wilson in 1806 in the garden of his friend, William Bartram, near Philadelphia, PA. As a result of this research and other avian observations, Wilson was able to correct earlier errors in bird taxonomy.

1730. Brandt's Cormorant description came in 1837 by the German zoologist Johann Friedrich Brandt. Brandt's is the only surviving cormorant to employ an individual's name as part of the bird's common name.

1731. Towhee named after an individual was Abert's towhee, named for Major James W. Abert, who collected the first specimen in 1852 in New Mexico.

1732. Discovery of *Archaeopteryx*, a fossil proto-bird was in 1861, in a limestone quarry located in Solenhofen, Bavaria, in southern Germany.

1733. National Christmas Bird Count in the United States was sponsored in 1900 by Frank Chapman, an ornithologist with the American Museum of Natural History and publisher of *Bird Lore* magazine. It was an attempt to substitute bird counting for bird killing. By the end of the century, more than 42,000 people were participating in the event, providing valuable information to ornithologists.

1734. Bird banding by a U.S. government agency was done during the summer of 1914 by the United States Biological Survey at the Bear River marshes in Utah. Alexander Wetmore, who was investigating an outbreak of duck sickness, attached bands to different species of ducks and other waterfowl.

1735. Emperor penguin egg collected for science was taken in the winter of 1911 at Cape Crozier in Antarctica by Dr. Edward Wilson, Birdie Bowers, and Apsley Cherry-Gerrard.

1736. Detailed life-history investigation of a major wildlife species was done on the bobwhite quail by Herbert L. Stoddard beginning in 1924 in Georgia in a project funded by a group of sportsmen in cooperation with the Bureau of Biological Survey in the U.S. Department of Agriculture. The project lasted until 1928.

1737. Species to be given a scientific name incorporating the first and last names of a naturalist was the Zapata wren (*Ferminia cerverai*), named for Fermin Z. Cervera, who was in Cuba in 1926 when the bird was first discovered. The Zapata wren is found only in a small marsh, the Zapata Swamp, located in the southern region of the island.

1738. Ross's goose breeding grounds were discovered on June 31, 1940, in the Perry River Delta, Northwest Territories, Canada.

1739. Comprehensive study on wild turkeys was published in 1943 as *The Wild Turkey in Virginia, Its Status, Life History and Management*. It was based on investigations by the Virginia Cooperative Wildlife Research Unit and the Federal Aid in Wildlife Restoration Project 2-R of the Virginia Commission of Game and Inland Fisheries. The study reported habitat reqirements for Turkeys and described specific management recommendations for restoring wild turkeys. wild

1740. Hibernating bird discovered was the common poorwill, found in a rock crevice in the Chuckwalla Mountains of the Colorado Desert in California in December 1946. Its temperature registered 64.4–68 degrees Fahrenheit, markedly below its normal temperature of 106. The bird was banded and subsequently found in the same spot in later years.

1741. Institution to breed all 15 species of cranes was the International Crane Foundation in Baraboo, WI. In 1993 it hatched its last species, the Wattled Crane.

1742. Documentation of a *Mononychus* (one claw), a birdlike dinosaur was in April 1993, when the turkey-sized dinosaur was described in the Gobi Desert of Mongolia. The fossilized remains resembled those of a flightless bird and included feathers. The animal's skeletal structure included both birdlike and dinosaur-like elements, adding weight to the theory that birds evolved from dinosaurs.

BIRDS—TREATIES

1743. International treaty to protect birds was the Migratory Bird Treaty of 1916. It sought to protect migratory birds in the United States and Canada. The Treaty was signed by representatives of the United States and Great Britain in Washington, DC, on August 16, 1916. It was signed by President Woodrow Wilson on September 1, 1916 and later officially ratified by Great Britain on October 20, 1916. The treaty went into effect December 8, 1916.

1744. Environmental treaty dealing with a particular ecosystem was adopted in Ramsar, Iran, in 1971, and implemented in 1975. The mission of the Convention on Wetlands of International Importance Especially as Waterfowl Habitat, also known as the Ramsar Convention, is to conserve wetlands through international cooperation and thereby work toward sustainable development throughout the world. As of January 1, 1998, 106 states were signatories.

BOTANY

1745. Written description of the medical uses of plants was found in the works of Hippocrates in the 5th century B.C. In his writings, Hippocrates mentions more than 240 plants and their uses as medicine.

1746. Botanical study of significance was the work of Theoprastus, Aristotle's student and successor as head of the Lyceum. His *De historia plantarum*, written circa 320 B.C., described and classified more than 500 plants.

1747. Catalog of Roman plants was Cato the Censor's *De re rustica*, from the 2nd century B.C. He lists 125 plants, including eight types of vines, found exclusively in Rome.

1748. Catalog of significance of plants in the ancient world was contained in Pliny's *Historia naturalis*, composed in Rome during the first century. Sixteen of the 37 sections of this encyclopedic work dealt with more than 1,000 different kinds of plants. Not until 1583, when Caesalpinus published *De Plantis*, covering more than 1,500 species, was there a more extensive work on the subject.

1749. Use of the word *pollen* is attributed to German botanist Valerius Cordus in the 16th century.

1750. Botanical garden in Europe was cultivated in Italy at the University of Pisa in 1543. While the garden was devoted to studying plants for their medical properties, by the end of the 16th century the emerging recognition of botany as a science distinct from medicine transformed the work done in Pisa.

1751. Herbarium was created by Italian botanists Caesalpinus and Aldrovani in 1563. The herbarium contained 768 plants dried, mounted, and labeled with Latin and Italian names.

1752. Western botanist to note the sexual identity of plants was Venetian physician Prospero Alpini in 1580. While working for the Venetian government in Egypt, he discovered that plants exist in male and female forms. Alpini later taught botany and was the first European to describe the coffee plant.

1753. Plant taxonomist was Italian botanist Caesalpinus. In his work *De Plantis*, published in 1583 in Florence, he classifies more than 1,500 plants by various attributes such as seeds, fruit, and leaf formation.

1754. Scientific botanical garden in Western Europe was founded in Holland at the University of Leiden in 1590. Its first director, Carolus Clusius (Charles de L'Écluse), and first curator, Cornelis Clutius, developed the "hortus botanicus" into a center for teaching and research, as well as a showplace for many varieties of world plant life. In 1989 the hortus was merged with Leiden University's Rijksherbarium, which had been founded in 1829 to study plant life in Dutch colonies in the East and West Indies.

1755. Tulips in Holland were introduced by botanist Carolus Clusius (Charles de L'Écluse) in 1593. Head botanist of the University of Leiden's "hortus" or botanical garden, Clusius planted tulips given to him by Austria's ambassador to Turkey. Tulips had been popular in Persia for centuries, but before Clusius planted his garden in 1593 the Dutch were unfamiliar with the flower that was to become a popular symbol of their country.

1756. Botanic garden in England was founded at Oxford in 1621 by Lord Henry Danvers, the earl of Danby. The garden's plant life was collected for medical, botanical, and horticultural study. Originally known as the Oxford Physic Garden, it was the foundation for the later University of Oxford Botanic Garden.

1757. Flora of the British Isles was compiled by William How. In 1650, the English physician published his systematic descriptions of plant life in his *Phytologia Britannica.*

1758. Use of the term *cell* to describe microscopic units of plants occurred in 1665 in *Micrographia*, authored by English scientist Robert Hooke.

1759. Recognition of the correlation between cellular rings in trees and age was made by Italian anatomist Marcello Malpighi, who had discovered capillaries and numerous human tissues for the first time with the aid of a microscope. Malpighi's discoveries in human anatomy led him to apply similar research to plants. His research, entitled *Anatome plantarum idea*, was published by the Royal Society of London in 1675, while Malpighi was a professor at the University of Bologna.

1760. Botanical account, in detail, of the plants of North America was found in *Historia Plantarum*, by English naturalist John Ray, whose volumes were published from 1686 to 1704. Ray's work contained the botanical observations of the Reverend John Banister, an English cleric who studied and described the plants he found while living in the Virginia colony.

1761. Female botanist of significance in the United States was Jane Colden, who studied the local flora in the Hudson River Valley in the mid-18th century. In one of America's earliest manuscripts of local flora she described more than 400 species of plants, and made leaf prints and drawings. It was not until June 1963 that the Garden Clubs of Orange and Dutchess counties of New York State published some of her botanic manuscripts.

1762. Botanical garden in the United States was planned and made in 1728 by John Bartram in Philadelphia, PA. The garden consisted of five or six acres located at 43rd and Eastwick streets on the banks of the Schuylkill River. Bartram at one time acted as botanist to King George III of England.

1763. Japanese flora description printed in English appeared in *The History of Japan and Siam*, published in London in 1728 as the posthumous translation of a work by the Dutch physician and botanist Engelbert Kämpfer. The book contained the first Western description of a ginkgo. Kämpfer made his observations while working in Japan as a physician from 1690 to 1692. His 1712 *Amoenitatum Exoticarum* (not translated into English) included the first Western description of any Japanese flora.

1764. Systematic use of binary nomenclature to describe plants appeared in 1753 in *Species Plantarum* (Species of plants) by Swedish scientist Carolus Linnaeus (Carl von Linné). Linnaeus's binomial system of describing a plant—by its genus and its species—became the standard form in which all plant life is described.

1765. Artificial plant hybrid was produced by German botanist Josef Gottlieb Kölreuter in 1761.

1766. Scientific discovery of the role of bees in pollination was made in 1761 by Joseph Gottlieb Kölreuter, who realized that plant fertilization can occur with the help of pollen-carrying insects.

1767. Botanical classification based on the natural relationships between plants was developed by French naturalist Michel Adanson, who rejected artificial taxonomic categories. The system used in Adanson's *Familles des Plantes* (1763) closely prefigured modern classification of plant life. Adanson was also one of the first European experts on the flora of West Africa.

1768. Botanical expedition in the United States was made by Manasseh Cutler, who set out on Ipswich, MA, on July 19, 1784, to study and classify botanical species in New England. At Mount Washington, NH, he examined 350 species and classified them according to the Linnaean method.

1769. Botanical society formed that is still in existence today was the Linnean Society of London, founded in March 1788 by Sir James Edward Smith and others.

1770. Female botany experimentalist of significance in England was Agnes Ibbetson in the late 18th and early 19th centuries. In her home in Exeter, England, she conducted experiments, dissecting plants and viewing them under a microscope. From 1790 to 1823 she published more than 50 articles in the scientific journals of the day. *Ibbetsonia genistoides*, a yellow-flowered shrub discovered in South Africa, was named in her honor.

1771. Introduction to the botanical sciences written by a woman was published in London in 1796. Priscilla Wakefield's *An Introduction to Botany in a Series of Familiar Letters*, or, as it was commonly called, *Wakefield's Botany*, consisted of letters written between two teenage sisters. The letters discussed lectures given by their governess on the Linnaean system of classification.

1772. Demonstration of the alternation of generations was done by Wilhelm Hofmeister in the mid–19th century. Alternation of generations forms the basis for understanding the plant life cycle.

1773. Use of the term *protoplasm* to describe the contents of plant cells was credited to German botanist Hugo Von Mohl in the 19th century.

1774. Use of the term *taxonomy* to describe the science of classification appeared in 1813. Swiss botanist and educator Augustin Pyrame de Candolle coined the term in his *Théorie élémentaire de la Botanique*

1775. Use of *chlorophyll* as a term to describe the green substance that colors plants was in 1817 by Pierre-Joseph Pelletier and Joseph Bienaimé Caventou, two French pharmacists, who were the first to isolate chlorophyll from green leaves.

BOTANY—*continued*

1776. Botanist to explore the American West was English-born naturalist Thomas Nuttall. After describing plant life in the northeastern and midwestern United States in his *Genera of American Plants* (1818) and teaching botany at Harvard University, Nuttall undertook a series of botanical expeditions into the far West and Hawaii. He wrote extensively about his botanical work in the West before shifting his attention to ornithology.

1777. Discovery and naming of the process of osmosis was by French naturalist Henri Dutrochet between 1824 and 1830. Dutrochet discovered the movement of water through cell membranes and named it *osmosis*.

1778. Professor of botany at the University of London was John Lindley. He began teaching botany there in 1829. Lindley was one of the first persons to advocate botany as a science.

1779. Terrarium was invented by English surgeon and botanist Nathaniel B. Ward, who noticed that moisture inside a closed glass bottle containing soil sustained the growth of plant life. After Ward publicized his experiments, terrariums or "Wardian cases" were widely used to transport plants on voyages, thus increasing the international availability of plant life for botanical study and agricultural transplanting.

1780. Presentation of the cell theory was given at the Beiträge zur Phytogenesis (Conference on Photogenesis) in Germany in 1838. German botanists Matthias Jakob Schleinden and zoologist Theodore Schwann proposed that all living matter was made up of cells.

1781. Botanical chart to be published was created by Elizabeth Warren in Falmouth, England. Her children's governess wanted a chart to teach botany, so Warren created an introductory chart. After it was published around 1839, botanist William Hooker endorsed it, but only a few sold. Warren donated the remaining charts to schools in 1844.

1782. Director of Kew Gardens in England was William Hooker, who was appointed to the post by Parliament in 1841. Hooker and his son and successor, Joseph, began the transformation of Kew Gardens from a royal horticultural showplace to a scientific botanical facility, run by the government and open to the public.

1783. Complete flora of Leicestershire, England, was published by Mary Kirby in 1850. In *A Flora of Leicestershire* she organized more than 900 local flowering plants and ferns according to their natural classes.

1784. Systematic study of *Sequoia gigantea*, the California sequoia, was undertaken by Albert Kellogg in 1852. In 1882 he published *Forest Trees of California*, the first publication on California forests.

1785. Research showing that lichen are two organisms was completed in 1867 by Swiss botanist Simon Schwendener. It is now known that lichen are a distinct type of organism composed of both fungal and algal cells working in symbiotic association.

1786. Public arboretum affiliated with a university was the Arnold Arboretum of Harvard University, which is also part of Boston's park system in the city's Jamaica Plain section. It was established in 1872.

1787. National Herbarium in the United States was established in Washington, DC, in 1894. It consolidated the botanical collections of the National Museum and the U.S. Department of Agriculture and included flora collected in expeditions to the western territories and states. It is now administered by the Department of Botany of the Smithsonian Institution, where it is located in the Museum of Natural History.

1788. Meeting to create the New England Botanical Club occurred on December 10, 1895, in Cambridge, MA. The botanist William G. Farlow and colleagues from Harvard University gathered at his house to discuss forming a club for those interested in the study of local flora. The NEBC was officially organized on February 5, 1896. Its main publication was the monthly journal *Rhodora*, whose first editor was Dr. Benjamin Lincoln Robinson.

1789. Female pioneer in the field of mycology, the study of mushrooms, was children's author Beatrix Potter. She experimented in spore germination and recorded her experiments in detailed drawings. Her paper *On the Germination of the Spore of Agaricineae* was read by proxy (women were not admitted at the time) at the April 1, 1897, meeting of the Linnean Society of London.

1790. Person to suggest starting the New York Botanical Garden was Elizabeth Knight Britton in 1899. She was the unofficial curator of mosses at the Columbia College Herbarium starting in 1899 and was also an environmental activist. She helped to found the Wild Flower Preservation Society of America in 1902.

1791. Woman on the science staff of Manchester University was Marie Carmichael Stopes, appointed as junior lecturer and demonstrator in botany in 1904. She specialized in paleobotany, the study of fossil plants, and published several books on the subject. Later she became famous (or notorious) for advocating birth control and sex education.

1792. Botanical garden dedicated to the native flora of a single country was Kirstenbosch Gardens in Cape Town, South Africa. Kirstenbosch was founded in 1913. It is one of eight national botanical gardens, each of which specializes in flora indigenous to southern Africa. Since its foundation, Kirstenbosch has been administered by the Botanical Society of South Africa.

1793. Use of the term *genecology* to describe the study of genetic variations within individual plant species as they relate to the environment came in 1923. The term was coined by Swedish botanist Göte Turesson.

1794. Person to lobby for a botanical garden in Montreal was Canadian botanist Brother Marie-Victorin in the early 1930s. Due in large part to his efforts, the garden was established in 1932. Brother Marie-Victorin also founded the Young Naturalists Club and the Awakening School to instruct children in natural history.

1795. Plant Materials Center (PMC) in the United States was established in 1939 in Beltsville, MD. It was the first of more than 25 PMCs, where plants and plant technology are developed to address specific regional needs, such as preventing erosion, restoring wetlands, controlling invasive plant species, and providing ground cover on environmentally damaged land. PMCs are administered by the U.S. Department of Agriculture.

1796. Female botanist to become a member of the Royal Society in England was Agnes Roberston Arber. Arber was elected to the society in 1948.

1797. Vegetation inventory in the U.S. national parks system was inaugurated on June 30, 1994. The five-year program of data gathering was undertaken by the National Park Service, the National Biological Survey, and numerous private nature-research institutions. The resulting standardized inventory and mapping of plant communities was intended to improve management and preservation programs throughout the park system.

C

CLIMATE AND WEATHER

1798. European to systematically record a North American winter was the French explorer Jacques Cartier, in 1535–1536. On his second voyage to the New World, Cartier wintered near present-day Quebec. He described ice more than six feet thick and snow four feet deep.

1799. Appearance of the "Little Ice Age," the coldest regime since the end of the last major ice age 100,000 years before, dates from about 1550. The cold era in Europe and North America lasted until about 1850. Average temperatures around the world fell significantly during this period.

1800. Severe winter said to have introduced a new artistic approach occurred in Europe between 1564–1565. Peter Brueghel the Elder painted his famous picture *Hunters in the Snow* in February 1565 in the midst of the coldest and most severe winter in more than a century. Brueghel made the landscape the center of interest of the picture, rather than merely the background.

1801. Case of winter cold causing an English colony in New England to fail occurred in 1607–1608 on the banks of the Kennebec River in present-day Maine. Surviving colonists returned to England in the autumn of 1608, before having to face the possibility of another severe winter in Maine.

1802. "Frost Fair" on the frozen Thames river in London occurred during the bitter winter of 1607–1608. The Great Frost of that year was the most severe in Britain's recorded history. Frost fairs, held to ease the monotony of bitter cold winters, became a London tradition for several generations.

1803. Weather records kept continuously in the United States were begun in 1644 by Rev. John Campanius of Swedes Ford DE, near Wilmington.

1804. Long-term failure of the cod fishery in and around the Norwegian Sea occurred from 1675 to 1700. The cause was colder sea temperatures associated with the Little Ice Age; the cods' livers failed when the water temperature fell below a certain point.

1805. Recorded long-term blockade of Iceland by sea ice occurred in 1695. Pack ice completely surrounded the island, and no ships could reach it for nearly six months.

CLIMATE AND WEATHER—*continued*

1806. Permafrost was discovered in 1735 in Siberia by Johann Gmelin, a German geographer and explorer. The information was published as *Riesen durch Sibirien* (Journey through Siberia) in four volumes from 1751 to 1752. His scientific expedition through Russia began in 1733 and lasted more than nine years.

1807. Weather reports kept continuously by a U.S. college were recorded from 1742 to 1778 at Harvard College in Cambridge, MA.

1808. Winter in which all the waters surrounding New York City froze solid occurred in 1779–1780, during the American Revolution. British soldiers crossed from New York City to Staten Island over several miles of harbor ice.

1809. Observer to propose a link between volcanic eruptions and the weather was Benjamin Franklin in 1784. Franklin, in Paris as the diplomatic representative of the new U.S. republic, attributed the severe winter of 1783–1784 to vast quantities of dust in the atmosphere from the eruptions of volcanoes in Iceland and Japan.

1810. Recorded instance of ice floes blocking the Mississippi River at New Orleans, LA, occurred in February 1784. It happened again in 1899. The floes disrupted and delayed commerce at the busy New Orleans port.

1811. Appearance of the "bosom friend," an article of warm clothing for women, was in the early years of the 19th century in England. The unusually cold winters of the Little Ice Age prompted the design.

1812. White Americans to systematically describe the weather of the western United States were Meriwether Lewis and William Clark, in 1804–1806. The journals of the Lewis and Clark expedition are packed with climatological details.

1813. Weather observations in Britain that are still being taken began in 1815 at the Radcliffe Observatory at Oxford University.

1814. Recorded "year without summer" in New England was in 1816. Average June temperatures in the region were 7 degreees below normal; frosts were experienced in every month of the year. The chill was attributed to the 1815 eruption of the Tambora volcano in Indonesia.

1815. Dew-point hygrometer was invented in 1820 by John Frederic Daniell, an English chemist and professor of chemistry at King's College, London. Known as the Daniell dew-point hygrometer, it measured humidity by recording the temperature at which condensation occurred when cooling devices were used to condense water vapor from the air.

1816. "Refreshment taverns" in the middle of the frozen Hudson River opened for business in January 1825. Thousands crossed on the ice daily between New York and New Jersey during the bitter winter of 1825. The refreshment taverns provided these people with warm drinks for a low price.

1817. Occasion of ice closing the Erie Canal for an entire month occurred in December 1831. Extreme cold gripped the northeastern United States in late 1831. New York City's mean temperature for December was only 22 degrees.

1818. Use of the term *acid rain* was in 1872, when Englishman Robert Angus Smith talked about its appearance in England during the Industrial Revolution. Smith said that acid rain—precipitation containing high amounts of acid-producing chemicals—was caused by high occurrences of industrial pollution. Despite this early mention, it would be nearly a century before the general public would become aware of the problem.

1819. El Niño known to have caused massive deaths occurred in 1877 when about 9 million people died in China and 8 million in India, mostly from famine. This catastrophe caused the first scientific effort to predict droughts and famines by atmospheric pressures. Henry Blanford, director of the India Meteorological Department, recorded atmospheric pressures above normal during the El Niño drought, and meteorologists in other countries confirmed this phenomenon.

1820. Ice jam to close the Ohio River for nearly two months formed in 1917. For 58 days, ice 30 feet high backed up the river for 100 miles along the Indiana, Kentucky, and Ohio banks.

1821. Report of wild temperature variations on simultaneous readings from nearby stations in South Dakota occurred on January 20, 1943. The temperature at Lead was 52 degrees while 1.5 miles away at Deadwood the thermometer had plunged to 16 degrees below zero. An elevation of 600 feet separated the two stations.

1822. World Meteorological Organization (WMO) Congress was held in Paris, France, in March 1951. At this first meeting, the executive council, technical commissions, and regional associations of the WMO were formed.

1823. Heat wave in the United States known to have lasted for two months occurred in July and August 1955 along the East Coast. Temperatures averaged 4 degrees above normal throughout the region. New York City, Philadelphia, Baltimore, and Washington reported their hottest Julys ever.

1824. Eight-day stretch of 100-plus degree heat in Los Angeles, CA occurred August 31 to September 7, 1955. The peak was 108 on September 1, the highest temperature since official records were started in 1877.

1825. Satellite to provide cloud-cover photographs was launched by a Thor-Able rocket on April 1, 1960, from Cape Canaveral, FL. Named *Tiros 1* (Television and Infra-Red Observatory Satellite), it took photographs of the earth's cloud cover from an altitude of 450 miles. The last transmission took place June 17, 1960.

1826. Complete freeze of Lake Constance (the Bodensee) in West Germany in more than two centuries occurred in 1963. The lake froze over completely during nine Little Ice Age winters from 1563 to 1699.

1827. Satellite to provide high resolution night photographs of cloud-cover was launched into polar orbit on August 28, 1964, from the Western Test Range, Point Arguello, CA, by Thor-Agena B rocket. The conical payload was 10 feet tall and 57 inches in diameter and weighed 830 pounds. On its first day in orbit, Nimbus 1 obtained pictures of Hurricane Cleo. It later sent back pictures of hurricanes Dora, Ethel, and Florence in the Atlantic, and typhoons Ruby, and Sally in the Pacific. It viewed Hurricane Dora in complete darkness. The satellite returned 27,000 photographs before ceasing operation on September 23, 1964.

1828. Climate hypothesis concerning animals and plants was the controversial Gaia hypothesis proposed in 1972 by James Lovelock, a British chemist. Gaia is the Greek name for Mother Earth. Lovelock said animals and plants sustain a balance of carbon dioxide and other substances in the atmosphere. Their interaction, he believed, makes the earth regulate itself and behave like a single living organism. He said the atmosphere is "an extension of a living system designed to maintain a chosen environment." He added that human disturbances could shift the earth into a new self-regulating state that might not support human life.

1829. World Climate Conference was held February 1979 in Geneva, Switzerland. It helped establish the World Climate Program by coordinating with other groups, such as the United Nations Environment Programme and the World Meteorological Organization. The Conference called upon national governments to "foresee and prevent potential man-made changes in climate that might be adverse to the well-being of humanity."

1830. Climate research agreement between the United States and Soviet Union was signed on December 8, 1987, in Washington, DC, by President Ronald Reagan and Soviet leader Mikhail Gorbachev. They agreed to pursue joint studies about "global climate and environmental change." In February 1988, a joint commission for the two nations agreed to study responses to climate change, identify measures to reduce greenhouse gas emissions, and to measure ozone changes in the Antarctic and Arctic.

1831. Joint United Nations organization to assess scientific information on climate change was the Intergovernmental Panel on Climate Change (IPCC), set up in 1988 by the World Meteorological Organization (WMO) and the United Nations Environment Programme (UNEP). More than 1,000 scientists participated. Their reports emphasized that the climate was warming faster during the late 20th century than at any other time over the last 10,000 years. The IPCC continues to assess and report signs of climate change.

1832. Meeting of the Intergovernmental Program on Climate Change (IPCC) was in Geneva, Switzerland, in November 1988. The IPCC was established by the United Nations Environment Programme and the World Meteorological Organization.

1833. Climate-impact forecasts by the U.S. government were given by the Climate and Global Change Program begun officially in 1989 by the National Oceanic and Atmospheric Administration. It is the only focused federal program to provide forecasts of climate events, such as El Niño, and their consequent impacts. The program is designed to complement the work of other federal agencies, such as the National Aeronautic and Space Administration (NASA) and the U.S. Department of Energy.

CLIMATE AND WEATHER—*continued*

1834. African-American president of the American Meteorological Society was Dr. Warren M Washington. When elected in 1993, Washington was a scientist and researcher at the National Center for Atmospheric Research in Boulder, CO.

CLIMATE AND WEATHER—GLOBAL WARMING

1835. Greenhouse effect description was made in 1827 by Baron Jean-Baptiste-Joseph Fourier, a French mathematician. Without using the term, he explained that the atmosphere acted like a transparent glass cover over a box.

1836. Warning about global warming was made in 1896 by the Swedish chemist Svante Arrhenius. He predicted that the Earth would become several degrees warmer if the level of carbon dioxide doubled. In 1908 he warned that the earth was already becoming warmer from the burning of coal and petroleum.

1837. Atmospheric carbon dioxide link to human activities was made in 1938 by British engineer G. D. Callendar. He also believed the increase in atmospheric carbon dioxide had created global warming, but postulated that it was good for agriculture and served as a defense against future glaciers.

1838. Global warming report to gain international scientific attention was issued in 1957 by the Scripps Institution of Oceanography at La Jolla, CA. The research team, led by Roger Revelle, concluded that half the carbon dioxide emissions of industry would remain in the atmosphere, contrary to the prevalent opinion of climatologists that they would be absorbed by the oceans.

1839. Daily measurements of atmospheric carbon dioxide were begun in 1958 by David Keeling of the Scripps Institution of Oceanography in La Jolla, CA, when he was a graduate student of Roger Revelle. To achieve purer samples, this was done at the Mauna Loa Observatory in Hawaii, and the measurements are still being taken.

1840. Global conference on environmental issues was the United Nations Conference on the Human Environment, convened in Stockholm in 1972. The conference, attended by 114 of the UN's 132 member countries, took up issues of world concern, including global warming, air and marine pollution, toxic waste, and population growth.

1841. Global warming report of significance made to the U.S. government was in June 1979 to the President's Council on Environmental Quality. It warned that man-made events "seem certain to cause a significant warming of world climate unless mitigating steps are taken immediately." The report was from David Keeling and Roger Revelle of the Scripps Institution of Oceanography, George Woodwell of the Woods Hole Biological Laboratory, and Gordon MacDonald of the Mitre Corporation.

1842. United Nations organization concerned with global warming was the World Commission on Environment and Development, established in 1983 by the General Assembly. It was called the Brundtland Commission for its chairman, Prime Minister Gro Harlem Brundtland of Norway. Among the proposals in its 1987 report were calls for international policies for reducing greenhouse gases, scientific monitoring of global warming and research into its effects, and international negotiations for a treaty on the climate.

1843. Global warming long-term policy of the government of the Soviet Union was included in a Supreme Soviet Directive of July 3, 1985. It required monitoring of the climate and research into human-induced changes in the climate and the effect of global warming on humans. The full title of the directive was the "General Long-term U.S.S.R. State Program of Environmental Protection and Rational use of Natural Resources."

1844. Climate-change international panel was the Intergovernmental Panel on Climate Change (IPCC), established in 1988 in Geneva, Switzerland, by the World Meteorological Organization (WMO) and the United Nations Environment Programme (UNEP). It meets to assess scientific, technical, and socioeconomic information concerning global warming and climate change. Its three working groups look at the climate system, impacts and response options, and the economic and social dimensions of climate change. The IPCC issues assessment reports and technical papers, and it conducts greenhouse gas inventories.

1845. "Greenhouse effect" report to the U.S. Congress was presented by a NASA scientist, James Hanson, in 1988. He said that global warming could lead to drought, melting of the polar ice cap, and rising sea levels.

1846. International climate conference held exclusively for developing countries was the Tata Conference, held February 21–23, 1989, in New Delhi, India. Co-sponsored by the United Nations Environment Programme and the World Resources Institute of the United States, the conference recommended a reduction in greenhouse gases, more efficient use of fossil fuels, and the phasing out of chlorofluorocarbons (CFCs) by the end of the century. It said the industrial nations had a special responsibility to "assist developing countries in finding and financing appropriate responses" to climate change, since richer countries were the major cause of the problem and had the resources to do something about it.

1847. Global-change research coordination program established by the U.S. government was the U.S. Global Change Research Program. The program was created as a Presidential Initiative in 1989 and formalized by the Global Change Research Act of 1990, passed by the U.S. Congress. The program, part of the National Science and Technology Council, coordinates and supports research by U.S. governmental agencies involved in global-change research. This includes research on global warming, ozone depletion, droughts, and floods. It also then provides the information to scientists, policymakers, educators, industries, and the public.

1848. Discussion of the greenhouse effect and other environmental issues at a "post-cold war summit" was at the Economic Summit of Industrialized Nations in Houston, TX, in July 1990. Despite support from European nations, no specific timetable was reached for efforts to limit emissions of greenhouse gases, but there was support for negotiation of a convention on climate change under the auspices of the United Nations Environment Program and the World Meteorological Organization.

1849. Daily forecasts of ultraviolet radiation issued by the Canadian government began in May 1992. The UV Index Program was launched by Environment Canada and Welfare Canada to measure the sunburning ultraviolet radiation. The forecast was issued for the next day and was reported on a scale of 0 to 10, with higher numbers indicating more ultraviolet radiation.

1850. Legally binding international document exclusively about global warming was the United Nations Framework Convention on Climate Change, opened for signing on June 4, 1992. It was signed by 154 countries, including those of the European Community, during the United Nations Conference on Environment and Development (UNCED), held June 3–14, 1992, in Rio de Janeiro, Brazil.

1851. Greenhouse-effect warning about recycling paper was made by the Royal Geographical Society at its annual meeting in London, England, on January 4, 1996. The organization said recycling of paper could lead to an increase in the carbon dioxides and gases that cause the greenhouse effect. The theory behind the warning was based on the belief that recycling leads to a reduction in reforestation and that the transportation of used paper increases vehicle emissions.

1852. Record-high global temperatures in consecutive years occurred in 1997 and 1998. These were the highest temperatures recorded since worldwide records began in the mid-19th century.

CLIMATE AND WEATHER—METEOROLOGY

1853. Recorded mention of a rain gauge occurred in an Indian manuscript of 400 B.C. The gauge was a bowl 18 inches in diameter; the writer suggested that the amount of water in the bowl should regulate the sowing of seeds.

1854. Weather vane was believed built in 48 B.C. by the Greek astronomer Andronicus of Cyrrhus for the Tower of the Winds in Athens, an octagonal tower he constructed. The weather vane, possibly as tall as 8 feet, honored the Greek god Triton, who was shown carrying a rod that would point in the direction of the quarter from which the wind blew. The figure had the head and body of a man and the tail of a fish. The tower also contained sundials and figures that symbolized the eight principal winds that the Greeks believed had divine powers.

1855. Person to demonstrate that air has weight was probably Hero of Alexandria, in the 1st century A.D. An engineer, Hero also deduced that air is a material substance, probably composed of particles.

1856. Weathercock wind vanes appeared atop Christian churches in Europe in the 9th century. The cock turned its head into the wind. The church chose the symbol as a reminder of the cock that crowed at Peter's denial of Christ.

CLIMATE AND WEATHER—METEOROLOGY—*continued*

1857. Known journal about the weather was kept between 1337 and 1344 by Walter Merle, a fellow at Merton College in Oxford, England. The entries, written in Latin, contain details about frosts and warm spells.

1858. Almanacs with yearlong weather forecasts appeared in Germany in the early 16th century. Compilers based their forecasts mainly on astrology.

1859. Recorded mention of the effects of El Niño dates from 1532–1533. El Niño is an ocean-atmospheric interaction in the equatorial Pacific and alters weather patterns in a large part of the world. Spanish conquistadors in South America were the first to observe the phenomenon.

1860. Open-air thermometer was devised by the Italian scientist Galileo Galilei in Padua in 1607.

1861. Observer to realize that the atmosphere becomes colder at the upper levels nearer the heat-giving sun was the Italian scientist Galileo Galilei, in 1640. He theorized, correctly, that sun rays are transformed into heat only when they encounter obstructions during their passage.

1862. Sealed thermometer was made by Ferdinand II, Grand Duke of Tuscany, in 1641. He used alcohol sealed in the tip of a glass tube.

1863. Instrument to measure atmospheric pressure was the mercury barometer, invented in 1643 by the Italian physicist and mathematician Evangelista Torricelli, who was a professor at the Florentine Academy. The instrument was first called the "Torricellian tube" or "Torricellian vacuum."

1864. Weather forecast using a barometer was in 1660 by Otto von Guericke, a German engineer and physicist. He also invented the vacuum pump.

1865. Angled barometer was invented in England in 1670 by Sir Samuel Morland. He bent the tube in an attempt to obtain more accurate measurements, but the device was more likely to break than a vertical barometer.

1866. Rain gauge was developed in England in 1676. Richard Towneley ran a lead pipe from the roof of his house to his bedroom so he could measure the rain in warm and dry comfort.

1867. Person to keep a continuous record of rainfall in England was Richard Towneley of Burnley, Lancashire. Towneley began his records of rainfall in 1677. Towneley continued to record English rainfall until 1703.

1868. Tide gauge was installed at Amsterdam in Holland in 1682. The gauge measured the level of the sea, which rises and falls with the tide, more or less sharply depending upon planetary alignments and local weather conditions.

1869. Mercury thermometer was invented by Daniel Gabriel Fahrenheit, a German physicist who lived most of his life in Holland, in 1714. He developed the Fahrenheit temperature scale now in widespread use in the United States and Canada.

1870. Weather vane maker in Colonial America was Deacon Shem Drowne. He created weather vanes for several of Boston's well-known buildings, including a copper Indian for Province House in 1716, a banner for the Old North Church in 1740, and a grasshopper for Faneuil Hall in 1742.

1871. Edition of Benjamin Franklin's *Poor Richard's Almanac* appeared in 1732. The almanac, with its yearly weather prognostications, sold an average of 10,000 copies annually for a quarter-century.

1872. Trade-wind explanation was provided in 1735 by George Hadley, an English philosopher who made many contributions to meteorology. He correctly reasoned that winds did not blow directly into the equatorial zone because the Earth rotates. Winds from the north thereby become northeast trade winds and those from the southern hemisphere become southeast trade winds. This theory was published in 1735 in the Royal Society's *Philosophical Transactions.*

1873. Continuous daily weather observations in the United States began in 1737 in Charleston, SC. Daily reports were compiled continuously in Cambridge, MA, beginning in 1742.

1874. Upper-air temperature measure of significance occurred in 1749 in Glasgow, Scotland, by Dr. Alexander Wilson. Wilson was the first to attach a thermometer to a kite in order to measure the temperature several yards above ground level.

1875. Simultaneous weather observations in the United States occurred in 1777 and 1778. Thomas Jefferson and James Madison kept daily records in Monticello, VA, and Williamsburg, VA, respectively.

1876. Meteorological society was the Societas Meteorologica Palatina established in Mannheim, Germany, in 1780. It published the influential *Ephemerides* and existed until 1792.

1877. Hygrometer utilizing human hair was devised in 1783 by Horace Bénédict de Saussure, a Swiss geologist and pioneer meteorologist who taught in Geneva. Hair (attached to a pointer) provided a way of measuring atmospheric moisture because hair lengthens with increasing humidity and shortens with decreasing humidity.

1878. *Farmer's Almanac* was published in 1792. Featuring a weather forecast for the year, it continues today as the *Old Farmer's Almanac*, compiled in Dublin, NH.

1879. Explanation of atmospheric condensation was made in 1800 by the British chemist and natural philosopher John Dalton. He explained the relationship between condensation and the expansion of air and also why water vapor varies in the atmosphere. He began a meteorological journal in 1787 and recorded more than 200,000 observations during his lifetime.

1880. Cloud classification of scientific significance was devised by English meteorologist Luke Howard, who published them in 1803 in his paper "On the Modifications of Clouds." His system, using Latin names, evolved into the international classifications used today.

1881. Meteorological research done by balloon was conducted September 16, 1804, by Joseph Gay-Lussac, a French chemist and physicist. He ascended to 23,012 feet from the Conservatoire des Arts in Paris to study the temperature and humidity and take samples of air at different heights. His research found no change in the air's composition at the different levels. This research made him one of the pioneers of meteorology. He and Jean-Baptiste Biot, a physicist and astronomer, had made an earlier flight to 13,120 feet on August 24 to observe the force of terrestrial magnetism at high altitudes. It was the first French balloon ascension for scientific research.

1882. Scale used internationally for wind forces was the Beaufort Wind Scale, devised in 1806 by British rear admiral Sir Francis Beaufort, a hydrographer with the Royal Navy. The British Admiralty adopted the scale in 1838, and it was adopted by the International Meteorological Organization in 1874. It is based on observations of the effects of the wind. The 12 Beaufort numbers of intensity are 0 for a calm wind up to 1 mph, 1 for light air of 1–3 mph, 2 for a light breeze of 4–7 mph, 3 for a gentle breeze of 8–12 mph, 4 for a moderate breeze of 13–18 mph, 5 for a fresh breeze of 19–24 mph, 6 for a strong breeze of 25–31 mph, 7 for a near gale of 32–38 mph, 8 for a gale of 39–46 mph, 9 for a strong gale of 47–54 mph, 10 for a storm of 55–63 mph, 11 for a violent storm of 64–72 mph, and 12 for a hurricane of 73 mph or more. Descriptive terms and wind speed are sometimes slightly altered.

1883. Weather observations by the U.S. military were in 1812, when the surgeon general of the U.S. Army ordered weather reports to be made at field hospitals.

1884. Weather map was believed to have been drawn in 1816 by the German scientist Heinrich W. Brandes. He used observations for a day in 1783.

1885. Map using isothermal lines to map temperatures was published in 1817 by German naturalist Alexander von Humboldt. An isothermal map displays the Earth's surface connecting points of equal temperature. Each point on the map generally reflects the average temperature of that area over a period of time. The spacing of the isothermal lines indicates the amount of temperature change over a certain distance.

1886. Meteorology society in Great Britain was the Meteorological Society of London, founded October 15, 1823. It existed for 20 years.

1887. Meteorological theory saying human activities affect the climate was made in 1827 by Jean-Baptiste-Joseph Fourier, a French mathematician and physicist who was the joint secretary of the Academy of Sciences of Paris.

1888. Scientist in the United States to conduct a detailed study of a single winter storm was the mathematician Elias Loomis, in 1836–1837. Loomis used data from 102 locations in the Ohio Valley and the Northeast to produce a map showing wind direction and the progress of the storm.

1889. Meteorologist to claim that lighting large forest fires would produce widespread rain was James Pollard Espy, in a series of lectures in 1837. The U.S. Congress declined, however, to appropriate funds to test Espy's theory.

1890. Meteorological records officially kept in Great Britain began in 1840 at the Royal Observatory at Greenwich. The records included temperatures and rainfall.

CLIMATE AND WEATHER—METEOROLOGY—*continued*

1891. Official meteorologist to work for the U.S. government was James Pollard Espy. Espy was appointed in 1842. He established himself as one of the leading meteorologists in the world after several years of studying energy sources of storms and describing frontal weather surfaces.

1892. Aneroid barometer was invented in France in 1843 by scientist Lucien Vidie. The aneroid barometer is easily transportable and accurate for most practical purposes.

1893. Cup anemometer was invented in England by John Robinson in 1846. The main function of Robinson's cup anemometer was to measure the speed of wind. It remains the most widely used instrument for measuring wind speed.

1894. Appeal for U.S. weather observers was issued in November 1848. Joseph Henry, director of the Smithsonian Institution, asked members of Congress to send notices to their constituents for volunteers for a weather network. At its peak, Henry's network numbered 600 observers.

1895. Maps of the trade winds were published in 1848 by Matthew Fontaine Maury, a hydrographer and U.S. naval officer. He also produced storm, rain, and thermal charts. In 1842, he was appointed superintendent of the navy's hydrographical office in Washington, DC. His studies of wind fields and ocean currents were turned into logbooks used by sailing ships. This demonstrated the need for international cooperation, and Maury coordinated the first international marine conference in 1853 in Brussels, Belgium.

1896. Coordinated weather observations in the United States began in 1849 when the Smithsonian Institution in Washington, DC, supplied instruments to telegraph companies. Their telegraphed reports were used by the Smithsonian to create weather maps. In 1853, regular observations were being made by nearly 100 stations, and by 1860 this had increased to 500.

1897. Use of the telegraph for reporting weather conditions was in the United States in 1849. Joseph Henry, the first director of the Smithsonian Institution, supplied weather instruments to telegraph stations in return for free use of the lines. By 1860, Henry's telegraphic system covered most of the eastern United States.

1898. Use of the word *forecast* relating to meteorology occurred in Great Britain in the 1850s. Admiral Robert Fitzroy, chief meteorologist of the British Board of Trade, coined the term to refer to weather predictions based on meteorological

1899. Meteorology society in Great Britain continuing through the 20th century was the British Meteorological Society, established on April 3, 1850, at Harwell House near Aylesbury, England. Its goal was to advance and extend meteorological science by "determining the laws of climate and of meteorological phenomena in general." The first president was Samuel Charles Whitbread. In 1883, Queen Victoria granted a royal charter that changed the name to the Royal Meteorological Society. It is now located in Reading, England.

1900. Weather maps published for general sale were offered to the London public at the Great Exhibition of 1851 at the Crystal Palace. The London Electric Telegraph Company sold the maps for a penny.

1901. Instrument to accurately measure the length of time the Earth's surface receives sunshine was the Campbell-Stokes sunshine recorder, devised in Great Britain in 1853 and improved in 1876. The inventor was John Francis Campbell; Irish physicist Sir George Stokes improved the device.

1902. First systematic weather forecasts were issued in 1854 by Britain's Meteorology Office, founded that year within the Board of Trade. It provided meteorological and sea current information to sailors. The first director was the meteorologist Robert Fitzroy, who had been captain of the HMS *Beagle* when it took Charles Darwin on his five-year journey. As director, Fitzroy issued storm warnings that evolved into daily weather reports. He also invented the "Fitzroy barometer."

1903. National forecasting map in the United States was displayed at the Smithsonian Institution in 1856. The map displayed in symbols information collected from a national network of weather observers and sent to Washington by telegraph.

1904. Newspaper weather forecast in the U.S. appeared in the Washington *Evening Star* on May 7, 1857. It was based on the national telegraphic reports collected by the Smithsonian Institution's corps of weather observers.

1905. Weather forecasts available to the public in Great Britain were issued in July 1861. The Board of Trade gave the newspapers forecasts based on reports from British and continental weather stations.

1906. Publication of modern weather maps was by the Paris observatory in 1863. The maps were created from the data of weather reports from various European sites.

1907. Weather maps in the modern version were based on the ideas of Sir Francis Galton, an English scientist and pioneer in meteorology, presented in 1863 in his book *Meteorographica.* It also introduced the term "anticyclone" for winds that rotate from a high-pressure center and usually bring about fair weather. Galton, who lived in London, was a cousin of Charles Darwin.

1908. Map showing movement of cyclonic depression was published in England by British meteorologist Alexander Buchan in 1868. This is often considered the first example of modern scientific meteorology.

1909. Weather charts in the United States were produced in 1869 by a telegraph service in Cincinnati, OH. It collected data from around the country.

1910. Weather bulletin in the United States was sent out September 1, 1869, by Cleveland Abbe, a meteorologist and the director of the observatory in Cincinnati, OH. In 1871, Abbe became the scientific assistant in the first national meteorological service established by the U.S. Army Signal Corps.

1911. Government weather service in the United States was established February 5, 1870, when President Ulysses S. Grant signed a bill creating a federal weather organization. The new weather service was made a part of the U.S. Army and placed under the direction of the Army's Signal Corps. It would remain as a part of the army until its direction was transferred to the U.S. Department of Agriculture in 1891 and officially named the U.S. Weather Bureau. In 1940 the U.S. Weather Bureau came under the jurisdiction of the Department of Commerce.

1912. "Blizzard" designation for a snowstorm was used March 14, 1870, by an Iowa newspaper for a storm of heavy snows and winds that swept over Minnesota and Iowa. It was commonly used by the 1880s. Originally in the United States, the word had referred to cannon shot and musket fire. It was derived from the German word for lightning.

1913. Weather observations atop Mount Washington in New Hampshire began November 13, 1870. The observatory, at 6262 feet above sea level, boasts "the worst weather in the world."

1914. Meteorological organization on an international scale was the International Meteorological Organization, established in 1873 in Vienna, Austria-Hungary, by directors of the world's main meteorological institutes. In 1951 it became a specialized agency of the United Nations, changing it name to the World Meteorological Organization.

1915. Weather bulletins for U.S. farmers were issued in 1873. These were among the earliest in a series of specialized weather reports from the U.S. Army Signal Service, forerunner of the National Weather Service.

1916. Collection of weather adages gathered under the aegis of the U.S. Army Signal Corps was published in 1883. The corps, a forerunner of the National Weather Service, *Weather Proverbs,* published sayings gathered from all over the country.

1917. Weather service of the U.S. government not run by the military was the U.S. Weather Bureau, which began operating October 1, 1890. An act by Congress established it as part of the Department of Agriculture.

1918. Postcard weather-forecasting service in the United States was launched in 1895. The U.S. Weather Bureau forecasts were stamped on the backs of cards and mailed to subscribers. At its peak, the service sent out 90,000 cards a day.

1919. Two-day U.S. Weather Bureau forecasts were issued in 1899. Weekly forecasts were added in 1910.

1920. Prediction of the existence of the ionosphere was made in 1901 by A. E. Kennelly of Harvard University. Oliver Heaviside, an English physicist, shortly afterwards independently predicted this electrically conducting region of the upper atmosphere. Between 1901 and 1930 it was commonly known as the Kennelly-Heaviside layer.

1921. Stratosphere and troposphere discoveries were made by Léon Teisserenc de Bort, a French meteorologist and physicist who also named the two atmospheric layers. His discoveries were the result of a nine-month research program in 1902 and 1903, during which he flew kites into the upper atmosphere to take samples of the air over Paris and Holland. Teisserenc de Bort became head of the Central Meteorological Bureau in Paris in 1880.

CLIMATE AND WEATHER—METEOROLOGY—*continued*

1922. Weather predictions using scientific methods were proposed in 1904 by the Norwegian meteorologist Vilhelm Bjerknes while teaching at the University of Stockholm. He was the first to have the idea of using physically defined units and equations to calculate the future state of the atmosphere. He suggested this could be done at any particular time by observations of temperature, humidity, pressure, and wind.

1923. Climatologist of repute was Helmut Landsberg, a German-born American. Born in Frankfurt in 1906, he began research at Pennsylvania State University in 1934 and remained in the United States. Called "the Father of Climatology," he helped develop it as a physical science in 1951 and was one of the founders of the National Weather Records Center (later named the National Climatic Center) at Asheville, NC. He was director of the Climatology Weather Bureau from 1954 to 1965 and president of the World Meteorological Organization Commission for Climatology from 1969 to 1978. Landsberg wrote an elementary textbook, *Physical Climatology*, and created an important data-processing system to facilitate climatology research.

1924. Rainfall predictions in Australia using atmospheric pressure were done in 1910 by Edward Quayle, a scientist with the Australian Bureau of Meteorology. He used tropical pressure fluctuations to predict the rainfall over east Australia.

1925. Cosmic radiation discovery was made in 1912 by Victor Hess, an Austrian physicist at Vienna University. During balloon ascensions, he realized that this radiation in the atmosphere must come from outer space because it increased with height. Hess immigrated to the United States in 1938 to become a professor of physics at Fordham University in New York City.

1926. Meteorological reports for British pilots were begun in 1912 by the Meteorological Office. The information and advice to pilots came from the first outstation at South Farnborough, England, established that year.

1927. Frontal air mass theory of forecasting dates from circa 1914. Norwegian physicist Vilhelm Bjerknes developed the theory from data obtained from a network of observation stations in Norway. The frontal theory remains in use to this day for forecasting.

1928. National meteorological organization in the United States was the American Meteorological Society, founded in St. Louis, MO, in December 1919.

1929. Indications of a jet stream emerged during the 1920s. A jet stream is a narrow belt of high wind near the tropopause. The tropopause is the region at the top of the troposphere. Jet streams were later confirmed by very high altitude flights during World War II.

1930. Long-range weather forecasting attempt using equations was in 1920 by the British meteorologist Lewis F. Richardson, who worked for the Meteorological Office. He used nonlinear differential equations. His calculation of numerical values for one sample day took six years and ended in failure. Richardson's concept, however, formed the basis for modern numerical weather predictions by computers.

1931. Outline of the mathematical concepts that would lead to numerical modeling for accurate weather forecasting dates from 1922, when British scientist Lewis Richardson published his book *Weather Prediction by Numerical Processes.*

1932. Weather reports on radio in Britain were first broadcast in 1922 by the British Broadcasting Company (BBC). Weather caption information first appeared on BBC television in 1936.

1933. Ionosphere direct measurements were made in 1925 by the English physicists Edward Appleton and M. A. F. Barnett, using wave interference. Appleton, a professor of physics at London University, also discovered that year a layer of electrically charged particles in the ionosphere. Named the Appleton Layer, its discovery was essential to the development of radio and radar. For his discovery, he received the 1947 Nobel Prize for Physics.

1934. Weather map to be telecast from a land sending station to a land receiving station was sent on August 18, 1926, from radio station NAA in Arlington, VA, and received at the U.S. Weather Bureau Office in Washington, DC. The demonstration was arranged by the Jenkins Laboratory in Washington, DC.

1935. Microclimatology research was done in 1927 by German scientist Rudolf Geiger, who established that field of study. It investigates the climatic conditions of limited or confined areas in the lowest part of the atmosphere, such as communities, houses, woods, and caves.

1936. Teletype communications of regularity between U.S. weather stations date from 1930. Forecasters collated information from telegraph and radio reports and aerial observation for speedy transmission.

1937. Weather map to be telecast to a transatlantic steamer was sent by the Radiomarine Corporation station of New York, NY, on June 30, 1930, to the S.S. *America*, nearly 3,000 miles away.

1938. Radar detection of air echoes from the lower atmosphere were made in 1935 by the British physicist Robert Watson-Watt, who headed the radio department of the National Physical Laboratory. Working with A. F. Wilkins and E. G. Bowen, he discovered the atmospheric echoes during tests to develop an aircraft-detection system.

1939. Television weather forecast in Great Britain was shown November 1, 1936, from Alexandra Palace. Televised forecasts in Great Britain were suspended during World War II; they resumed June 29, 1949.

1940. Meteorological sensors on balloons in Great Britain were used in 1939 as World War II began. The Meteorological Office received data at receiving sites on land from the upper air by the "radiosonde" sensors that recorded the temperature, pressure, and humidity.

1941. Wind chill calculations were published in 1939 by antarctic explorer Paul Siple in his doctoral dissertation, "Adaptation of the Explorer to the Climate of Antarctica." In this paper he created the term *wind chill*. He had defined the phenomenon during an exploration of antarctica when he measured how long it took for a pan of water to freeze and found that the heat loss could be calculated by air temperature and wind speed. In 1941, Siple and the explorer Charles Passel developed a wind chill chart.

1942. Detailed five-day U.S. Weather Bureau forecasts were issued in 1940. In 1941 the forecasts had an accuracy rate of only 48 percent for temperatures and 16 percent for precipitation.

1943. Use of high-speed electronic computers for U.S. weather forecasts dates from the 1940s. U.S. Weather Bureau meteorologists routinely began using computers in the mid-1950s.

1944. Proof that radar could be used to locate precipitation was established on February 20, 1941, in England. Using a 3,000-megahertz frequency, scientists identified a rain shower seven miles off the coast. Before World War II, it was assumed that radar was not sensitive enough to detect precipitation, but military radar operators often discovered images of rain and snow on their screens. In 1968, scientists with the U.S. Air Force found that radar could also detect wind shifts.

1945. Televised weather forecast in the United States was on October 14, 1941, over WBNT in New York City, the forerunner of WNBC. It featured the animated cartoon character Wooly Lamb.

1946. Meteorological data reported from airplanes in England was issued in 1942 during World War II. The Meteorological Research Flight was established in 1942 by the Meteorological Office to provide information about atmospheric parameters.

1947. Artificial snowstorm was created November 13, 1946, by the American physicist Vincent Joseph Schaefer. He seeded a cloud with dry ice (carbon dioxide) dropped from an aircraft. Four months earlier, while working for the General Electric Company in Schenectady, NY, on the problem of icing on airplane wings, Schaefer had accidentally discovered that dry ice transformed water vapor in a cold box into snow. In 1947, his assistant at General Electric, the physicist Bernard Vonnegut, improved cloud-seeding by using silver iodide instead.

1948. Joint arctic weather stations were established by the World Meteorological Organization in 1947. At one, Eureka in Canada's Northwest Territories, the mean annual temperature was recorded at -2.9 F.

1949. Weather prediction system using modern numerical systems was devised in 1949 on a computer at Princeton University in New Jersey. The system was devised by American meteorologist Jule Charney and the Norwegian meteorologist Ragner Fjortoft. A 24-hour forecast on this elementary computer, however, took 24 hours to calculate.

1950. Numerical weather forecast on an electronic computer was run in April 1950 at the Institute for Advanced Study in Princeton, NJ. The program took nearly 50 minutes to analyze data from 768 weather stations.

CLIMATE AND WEATHER—METEOROLOGY—*continued*

1951. Upper-atmosphere study with a rocket launched from a balloon was in 1952 by the American physicist James Alfred Van Allen, head of the physics department at the University of Iowa. He developed the "rockoon" to investigate the physics of the atmosphere. In 1958 he added a geiger counter to the first U.S. satellite, *Explorer 1*, and it discovered the radiation belt that is named for him.

1952. Live television reports of weather in England began in 1954 on the British Broadcasting Company from London. The regular weather reports were five minutes long. The information for the reports was provided by the Meteorology Office. The television weatherpersons continue to be employees of the office.

1953. Long-range numerical weather forecasts by computer were achieved in 1954 by Jule Charney, a meteorologist and mathematician at the Institute for Advanced Study in Princeton. He used equations he had created to predict the development of depressions. Charney increased his five-day forecasts to 30-day ones, including 30 days' warning of hurricanes. He had previously helped to establish the long-range forecasting branch of the U.S. Weather Bureau.

1954. National Hurricane Center in the United States was established in Florida in 1955. The U.S. Weather Bureau established the center after hurricanes Connie and Diane killed nearly 200 people earlier in the year. The main purpose of the National Hurricane Center was to provide people with a better warning system about approaching hurricanes.

1955. Index for temperature-humidity was established in 1959 by the U.S. Weather Bureau. It was created by Earl C. Thom and first called the "discomfort index." An index level of 70 indicates that 10 percent of the population of a specific area would feel discomfort, while a reading of 80 would affect 100 percent. At 85 or above, the danger of heat exhaustion and heat stroke exists, so precautions should be taken.

1956. Satellite to offer useful weather data was the U.S. *Explorer 7*, launched in October 1959 from Cape Canaveral, FL. *Tiros 1*, later launched in 1960, was the first specialized weather satellite.

1957. Satellite to transmit weather information to the earth was *Vanguard 2*. It was launched by the United States from Cape Canaveral on February 17, 1959. The satellite, only twenty inches in diameter, was launched at a perigee of 350 miles and an apogee of 2,065 miles. It carried equipment to measure cloud cover over the earth.

1958. Attempts at climate modeling date to the early 1960s at the Geophysical Fluid Dynamics Laboratory in Princeton, NJ. Other early climate models were developed by the National Aeronautics and Space Administration (NASA) and the British Meteorological Office.

1959. Modern analysis of significance into the chemistry of rain was carried out in the 1960s by Christian Junge, a German scientist working at the U.S. Air Force Cambridge Research Laboratory in Massachusetts. Junge detected chloride, sulfate, and nitrate ions in rainwater.

1960. All-weather satellite was launched by the U.S. from Cape Canaveral, FL, on April 1, 1960. Called *Tiros 1*, the polar-orbiting satellite would send nearly 23,000 photos back to earth during its first three months in orbit.

1961. Meteorological satellite funded by the U.S. Weather Bureau was *Tiros 10*, launched by the National Aeronautics and Space Administration on July 2, 1965, at Cape Kennedy (Cape Canaveral), FL. It carried two television cameras to send back global weather pictures. The satellite weighted 290 pounds in orbit.

1962. Weather satellite monitoring station north of the Arctic Circle was the Esrange Satellite Station established in 1966 about 25 miles from Kiruna, Sweden, by the European Space Research Organization. It tracks and controls polar orbiting satellites, recording and processing the data. It also launches sounding rockets and balloons. In 1972, the Swedish government formed the Swedish Space Corporation to become the owner and operator of the station. Esrange was originally an abbreviation of European Space Research Organization Sounding Rocket Launching Range but is now the official name of the station.

1963. Doppler radar system to study storms was used by U.S. meteorologists in 1971. Meteorologists were able to examine rainfall and other weather events by using Doppler radar. Doppler radar employs a high-powered antenna that rotates and sends out pulses of radio waves. The pulses bounce off the falling rain and return to the radar source. By measuring the time between pulses and the time it takes the radio echoes to return, Doppler radar can calculate the distance and direction of the rain.

1964. Meteorology satellite launched by Europeans was Meteosat in 1977 by the European Space Agency (ESA), whose headquarters are in Paris. It carried an imaging radiometer and a special telescope to measure radiance from the Earth and its cloud cover in the visible, thermal infrared and water-vapor spectral bands. It provided weather images every 30 minutes. Several other Meteosat versions have been launched since, and Meteosat Second Generation (MSG) satellites are planned. The system has reduced weather forecasting errors by 80 percent. Its operation was transferred on December 1, 1995, to the European Organization for the Exploitation of Meteorological Satellites (EUMETSAT) in Darmstadt, Germany.

1965. Geostationary Operational Environmental Satellite (GOES) was launched by the U.S. in 1980. A second GOES was launched in 1981. The satellites are used to photograph weather patterns from orbit. Each has the ability to photograph 7,000 milcs of weather patterns in 18 minutes.

1966. Television network to broadcast weather programming exclusively was the Weather Channel, established in 1982. The founder was John Coleman, a former ABC television network forecaster.

1967. Meteorological organization in Europe organized specifically to control weather satellites was the European Organization for the Exploitation of Meteorological Satellites (EUMETSAT), established in 1986. It consists of 17 European nations and works to establish, maintain, and exploit European systems of operational meteorological satellites. The headquarters are in Darmstadt, Germany. On December 1, 1995, it took over the Meteosat satellite program from the European Space Agency.

1968. International declaration on El Niño was issued on November 13, 1998, at the First International Meeting of Experts in Quayaquil, Ecuador. The declaration by the 450 delegates called for urgent action to increase intergovernmental programs, early warning systems, and the monitoring of the climate. The conference, which began November 9, was called for by the United Nations General Assembly under the International Decade for International Disasters Reduction. It was sponsored by the United Nations Task Force on El Niño and the Permanent Commission on the South Pacific. It noted that 27 nations reported 21,000 deaths related to El Niño with 117 million people affected, 4.9 million displaced or homeless, and worldwide economic losses estimated at up to $34 billion.

CLIMATE AND WEATHER—OZONE LAYER

1969. Discovery of the ozone layer was in 1913 by Charles Fabry, a French physicist and professor at Marseilles University. He was assisted by M. Buisson.

1970. Chlorofluorocarbons (CFCs) were developed in 1928 as a coolant for automobile air conditioners by Thomas Midgley of General Motors. It was later discovered that released CFCs rise to the upper atmosphere and produce chlorine monoxide that destroys the ozone layer. Midgley had previously invented leaded gasoline, which would prove to be another pollutant.

1971. Measurements of chlorofluorocarbons (CFCs) in the atmosphere were made in the early 1970s by James Lovelock, a British chemist. He used the electron capture detector he invented in 1958 to measure pesticide residues in the environment. Lovelock first regarded CFCs as harmless and helpful for tracing air movements, but he later warned that they should be eliminated to reduce the greenhouse effect.

1972. Nobel Prize was awarded to ozone-layer researchers in 1995—to a Dutch professor, Paul Crutzen, at the Max Planck Institute for Chemistry in Mainz, Germany, and to two American professors, Mario Molina at the Massachusetts Institute of Technology (MIT) in Cambridge, MA, and F. Sherwood Rowland at the University of California–Irvine. They all did pioneer research into the formation and decomposition of ozone, explaining the chemical mechanisms that affect the layer's thickness. The award said they had "contributed to our salvation from a global environmental problem that could have catastrophic consequences."

1973. Ozone-layer warning was given in 1971 by Harold S. Johnston at the University of California–Berkeley. He warned that the layer could be threatened by a planned fleet of supersonic aircraft because the planes would release destructive nitrogen oxides into the middle of the ozone layer.

CLIMATE AND WEATHER—OZONE LAYER—*continued*

1974. Ozone-layer data concerning damage by chlorofluorocarbons (CFCs) was published in the June 1974 issue of the British journal *Nature* by Sherry Rowland and Mario Molina, two researchers at the University of California–Irvine. They calculated that between 20 and 40 percent of the ozone shield would be depleted within 30 years if the 1972 rate of CFC release continued.

1975. Ozone depletion warning about Freon was in 1976 by the National Academy of Sciences in Washington, DC. It said the Freon used in spray cans depletes the ozone layer and increases ultraviolet radiation on the Earth.

1976. United Nations Environment Programme Governing Council meeting on the ozone layer was held in Washington, DC, in 1977, to draw up a plan of action for the study of the natural ozone layer.

1977. Ozone-layer committee of the United Nations was the Coordinating Committee for the Ozone Layer (CCOL), established in 1978 within the United Nations Environment Programme (UNEP) in Geneva, Switzerland. It has played a major role in deciding on scientific ozone research priorities. In 1980, the committee warned that the continuing release of chlorofluorocarbons (CFCs) would eventually deplete the ozone layer and have serious impacts on the health of people and the biosphere. The work of the committee formed the basis for the important 1987 Montreal Protocol on Substances That Deplete the Ozone Layer.

1978. Ban of chlorofluorocarbons (CFCs) by the U.S. government took effect on October 15, 1978. Although the ban covered only nonessential uses of CFCs, it reduced their production by 98 percent within two months. The ban had been jointly announced in March 1978 by the U.S. Environmental Protection Agency, the U.S. Food and Drug Administration, and the Consumer Product Safety Commission.

1979. Ozone-layer measurements from space were by the Total Ozone Mapping Spectrometer (TOMS), launched on October 24, 1978, in the NASA spacecraft *Nimbus 7*. TOMS began taking measurements in December of that year and operated until May 1993. It produced the longest continual data for international total-column ozone.

1980. Ozone depletion of major significance over Antarctica began in 1979. The early 1980s saw the first depletion of the ozone layer on a worldwide level.

1981. Ozone-layer vertical profile from space was done by NASA's Stratospheric Aerosol and Gas Experiment (SAGE), launched in 1979 at Cape Canaveral, FL, aboard the *Applications Explorer Mission-B* spacecraft. It continued vertical ozone measurements until 1981, using the solar occultation technique. In 1983, SAGE II was launched to study the Arctic ozone hole and the depletion of ozone over the Earth's mid-latitudes.

1982. Total ban of chlorofluorocarbons (CFCs) for aerosols by a country took effect in Sweden on January 1, 1979. The Swedish Ministry of Agriculture had announced the ban in December 1977 on CFCs as an aerosol-can propellant.

1983. Ban on chlorofluorocarbons (CFCs) on shipments between U.S. states took effect on April 15, 1979, when the federal government made the interstate conveyance of CFCs illegal.

1984. Hole in the earth's ozone layer was discovered in 1985 over Antarctica. Scientists had known since the mid-1970s that chlorofluorocarbons (CFCs) destroy high-atmosphere ozone molecules. In the early 1980s, scientists gathered evidence proving that ozone in the upper atmosphere was decreasing. In 1985, scientists discovered that the tiny ice particles in the atmosphere over Antarctica, combined with the dangerous gasses, increased the rate of ozone breakdown. This combination produced an ozone hole during the Antarctic summer, when there is virtually continuous sunlight.

1985. Chlorofluorocarbons (CFCs) limitation on an international level was the United Nations' Vienna Convention for the Protection of the Ozone Layer, signed by 20 countries in March 1985 in Vienna, Austria. It stated that nations had an obligation to control emissions that damaged, or were likely to damage, the ozone layer. Put into effect in September 1987 in Montreal, Canada, it created the first practical international cooperation to deal with the problem.

1986. Workshop hosted by the European Community on the ozone layer was held in Rome, Italy, in May 1986, as a preliminary planning session to the negotiations that resulted in the Protocol on Substances That Deplete the Ozone Layer (Montreal Protocol) in September 1987.

1987. Ozone pollutant agreement on an international level was the Montreal Protocol on Substances That Deplete the Ozone Layer, which was initially signed by 27 nations on September 16, 1987, in Montreal, Canada. They agreed to reduce the use of five chlorofluorocarbons (CFCs) and three halon gases by half within ten years. The agreement was prompted by the discovery three years previously of the ozone hole over Antarctica.

1988. Scientific evidence that humans caused the Antarctic ozone hole was collected by the international Airborne Antarctic Ozone Experiment, conducted from August 23 to September 16, 1987. It was chiefly organized by NASA, with major contributions by the National Oceanic and Atmospheric Administration, the National Science Foundation, and the U.S. Chemical Manufacturers Association. More than 100 scientists, engineers, and technicians traveled to Punta Arenas on the southern tip of Chile, and two aircraft flew dangerous missions over Antarctica to gather information about the stratosphere. The information was coordinated with ground-based observations at the McMurdo Station in the Antarctic, and in April 1988 a report was issued stating the evidence that chlorine from chlorofluorocarbons (CFCs) definitely caused the ozone hole.

1989. Ozone fund for developing nations was the Multilateral Fund established on September 16, 1987, in Montreal, Canada, by the Montreal Protocol on Substances That Deplete the Ozone Layer. The 27 nations agreed to contribute $160 million over three years to help developing nations phase out ozone-depleting substances. The Multilateral Fund, based in Montreal, went into effect in January 1991 and was raised to $240 million when China and India signed the Protocol. The fund was made permanent in November 1992 at a further meeting in Copenhagen, Denmark.

1990. Conference held after the signing of the Protocol on Substances That Deplete the Ozone Layer (Montreal Protocol) was the International Ozone Conference, convened by British Prime Minister Margaret Thatcher in London, England, in March 1989, with more than two-thirds of the participants at the ministerial level. This conference was held one month before the Helsinki Conference, at which the contracting parties of the Montreal Protocol were to meet.

1991. Ozone research coordination program of the European Community (EC) was the European Ozone Research Coordinating Unit, which began work on April 18, 1989, in the Department of Chemistry at the University of Cambridge, England. Also called the Ozone Secretariat, it was formed to review and coordinate existing European Union (EU) and European national programs on stratospheric ozone, prepare a research plan, and provide information and advice to the EC, national programs, and research groups. The unit had been formally established in 1988 at a meeting in the Hague by the EU and the European Free Trade Association (EFTA). The unit was jointly funded by the European Community and the Global Atmospheres division of the United Kingdom Department of the Environment.

1992. Ozone measurements of chlorine monoxide from space was by NASA's *Upper Atmosphere Research Satellite (UARS)* launched September 15, 1991, from the space shuttle *Discovery*. Chlorine monoxide is the main component of ozone depletion. After three years, *UARS's* 10 instruments had provided conclusive proof from space that human-made chlorine in the stratosphere had caused the Antarctic ozone hole. NASA announced this on December 19, 1994. *UARS* was the first satellite launched as part of NASA's Mission to Planet Earth, a comprehensive study of how the global environment changes and how human activities contribute to this. *UARS* is managed by the Goddard Space Flight Center at Greenbelt, MD.

1993. Ozone weekly summary by the Canadian government began in 1992. *Ozone Watch* gave ozone levels recorded for the previous two-week period at 10 of Canada's 11 ozone-monitoring stations. These figures and the long-term averages were issued each week by Environment Canada.

1994. Ozone-protection plan to phase out methyl chloroform took effect in August 1992. This London Amendment to the 1987 Montreal Protocol on Substances That Deplete the Ozone Layer was passed at a meeting in June 1990 in London, England. The signatories' commitment was to phase out methyl chloroform by 2005.

1995. Ozone-protecting household refrigerator was Greenfreeze sold in 1993 in Germany. Greenpeace International brought together scientists and DKK Scharfenstein, a German company, to develop the refrigerator that uses hydrocarbons for the refrigerant. These have zero ozone depletion, and the product uses no chlorofluorocarbons (CFCs), hydrofluorocarbons

CLIMATE AND WEATHER—OZONE LAYER—*continued*

(HFCs), or hydrochlorofluorocarbons (HCFCs). Major companies, such as Bosch and Siemens, sell the refrigerators in Germany, Great Britain, Austria, Denmark, France, Italy, the Netherlands, Swizerland, Argentina, India, and China.

1996. Ozone-depletion Executive Order by a U.S. President was President Bill Clinton's Executive Order 12843 of April 21, 1993. It set down procurement requirements for federal agencies, requiring them to maximize safe alternatives to ozone-depletion substances, to modify contracts that required ozone-depletion substances, to substitute non-ozone-depletion substances to the extent it was economically practicable, and to recycle ozone-depleting substances.

1997. Ozone depletion recorded over the Dead Sea was published on January 20, 1999, in the journal *Science*, by German and Israeli meteorologists. They found lower levels of ozone and higher levels of bromide oxide over the Dead Sea between Israel and Jordan. Bromide, known to be one cause of ozone depletion, is given off mostly by polar ice.

1998. Year the normal ozone level will return to the stratosphere will be 2060, by scientific estimate, if the present global schedule is maintained to eliminate substances that destroy the ozone. U.S. meteorologists also believe the Antarctic ozone hole could be totally repaired by that year.

CLIMATE AND WEATHER—PALEOCLIMATOLOGY

1999. Glacial epoch that is verifiable occurred 2.7 to 2.3 gigayears before the present. (A gigayear is a thousand million years.) The glaciation appears to have been extensive, though its cause is uncertain.

2000. Ice age to cover tropical areas occurred during the Permian Period of the Paleozoic Era. This was about 250 million years ago, and the ice covered parts of India and the continents of Africa and South America.

2001. Climate change to lead to the peopling of a continent occurred circa 100,000 B.C. Dropping sea levels during the Ice Age exposed a land bridge between Asia and North America that allowed the eastward migration of Homo sapiens.

2002. Reversal of the post-glacial warming trend in China occurred beginning circa 1100 B.C. Cooler winter temperatures forced the southward retreat of the bamboo, an extremely valuable plant in ancient China.

2003. Stabilization of climates of the present-day United States to their current levels occurred circa 8000 B.C. There have been, however, significant regional climate fluctuations over the millennia.

2004. Ice Age calculations done scientifically were by Baron Gerhard de Geer, a Swedish geologist who taught at the University of Stockholm and established the Geochronological Institute there in 1924. He devised a system of comparing sediment patterns (varves) at the bottoms of lakes that had been fed by melting ice from glaciers. The deposits differed by seasons and years, allowing him to compare patterns made by the receding ice. He did this research in Sweden, Iceland, Newfoundland, New Zealand, and other countries. From these studies, he estimated that the last ice age ended circa 6740 B.C.

2005. Evidence of long-term cooling and drying of Australia dates from circa 4000 to 2500 B.C. Before that period, the continent was considerably wetter than it is today.

2006. Post-glacial cold spell of significance in the Northern Hemisphere occurred from circa 3500 to 3000 B.C. Alpine glaciers advanced and the forest retreated in Europe; in Europe and North America, warmth-demanding trees such as elms and linden declined.

2007. Surviving year-by-year record of flood levels in the Nile Valley in Egypt begin circa 3100 B.C., during the first dynasty of the pharaohs. Flood levels usually depended on the summer monsoon over Ethiopia.

2008. Retreat of human settlement from the North African and Arabian deserts began circa 3000 B.C. and continued for several centuries. A long-term climate change—hotter and drier—caused the retraction.

2009. Natural catastrophe theory about animal extinctions was set forth in 1825 by the French naturalist Georges Cuvier, secretary of the National Institute in Paris. He believed great floods and other natural disasters had caused the extinction of animal species in different areas and had altered the face of the Earth. Then, he believed, those areas had been repopulated by animals from other regions.

2010. Scientific evidence to support the theory of an ice age was provided in 1840 by Louis Agassiz, a Swiss glaciologist and naturalist, professor at Neuchâtel. He proved that glaciers move (because rocks are transported by them), a discovery that convinced the scientific community that an ice age had come and gone. Agassiz published his discovery in 1840 in his *Études sur les glaciers* (*Studies of Glaciers*). The theory of an ice age had first been argued in detail by Johann von Charpentier, a German-Swiss geologist. Agassiz later became a professor at Harvard University and a U.S. citizen.

2011. Use of pollen analysis and vegetation history to determine successive climate patterns was by Norwegian botanist Axel Blytt in 1876. He presented his findings in his *Essay on the Immigration of the Norwegian Flora.*

2012. Geologist to discovery that several ice ages had existed was the American Thomas Chrowder Chamberlin in 1899 while he was professor of geology and director of the Walker Museum at the University of Chicago. His research into the origin of the atmosphere and glaciation led him to challenge the theory that only one ice age had occurred and to dispute the theory that the Earth was only 100 million years old, an idea held by several scientists, including the British natural philosopher William Thomson, 1st Baron Kelvin. Chamberlin also was one of the first scientists to believe that the climate varied according to the amount of carbon dioxide in the atmosphere.

2013. Use of dendroclimatology as a means of determining historic yearly rainfall and temperatures by tree rings dates from 1929, when Harvard astronomer A. E. Douglass found he could date the timbers of prehistoric Indian pueblos by matching their rings with cores from 1,000-year-old living conifers. The discovery grew out of Douglass's research into the relationship of sunspots and climatic change.

2014. Detailed model for determining prevailing temperatures in prehistoric northern Europe was developed in 1944 by J. Iversen of the Danish Geological Survey. Iversen deduced temperature values from the presence of holly and mistletoe.

2015. Paleoclimatology research in the United States using a global network was by the Paleoclimatology Program, established in 1992 by the National Oceanic and Atmospheric Administration. By 1997, it had 10,000 data sites in its global network. The program brings together this paleoclimatology data to improve the understanding of Earth's climate variability and abrupt changes, to improve predictive climate models, and to distinguish between climate changes caused by humans and those that are natural.

COAL AND NATURAL GAS

2016. Commercial use of coal was in China circa 1000 B.C. Historians believe the Fu-shan mine in northern China yielded coal for the smelting of copper.

2017. Use of natural gas as a fuel occurred in China as early as 1000 B.C. Bamboo tubes were used as pipes to carry the gas.

2018. Authenticated use of coal by the Romans in England occurred before A.D. 400. Coal cinders found in Roman ruins suggest the colonizers burned coal for fuel.

2019. Use of natural gas for illumination was in the Sichuan province of China beginning circa 900. The gas existed from 1,500 to 1,600 feet deep under rock salt, and wells were sunk to extract it through bamboo pipes. The gas was used to light homes and salt mines.

2020. Coal use known in North America was in circa 1000 by the Hopi people in mesa pueblos located in what is now part of northeast Arizona. They used the coal as fuel to bake their clay pottery.

2021. Coal use documented in Europe was recorded circa 1200 in the chronicles of a monk, Reinier of Liège,in what is now Belgium. He described "black earth very similar to charcoal" used in the furnaces of metalworkers.

2022. Report to Europe of vast coal deposits in China was by Marco Polo in 1275. The Venetian traveler wrote that in China "there was a black stone existing in beds in the mountains which they dig out and burn like firewood."

2023. Substantial waterborne export trade in British coal developed in the 14th century. Merchants shipped coal into Europe from ports on the rivers Forth and Tyne.

2024. Law binding Scottish coal miners to their employers was enacted in 1606. The measure, approved by the Scottish parliament, provided harsh penalties for miners who left the pit without permission.

2025. Natural gas sighting by Europeans in North America was in 1626 by French missionaries. They recorded seeing Indians ignite gas in the shallows of Lake Erie and streams flowing into it.

2026. Coal discovery recorded in North America was during the 1673–1674 expedition of two French explorers, Louis Joliet and the Jesuit missionary Father Jacques Marquette. Their accounts of the journey mention coal deposits near the Illinois River.

COAL AND NATURAL GAS—*continued*

2027. Suction pump to draw water from coal mine shafts was patented in England in 1699. Thomas Savery called his invention, a forerunner of the steam engine, the "miner's friend."

2028. Use of the process of smelting by coke came in 1709. English iron master Abraham Darby used the process of producing coke from coal in his works at Coalbrookdale in Shropshire and it was in widespread use by the middle of the 18th century.

2029. Organized strike against a colliery in England occurred in 1765. Miners struck over issues of pay and working conditions.

2030. Coal-burning steam engine was invented by the Scottish engineer James Watt in 1769. The invention ultimately led to a vast expansion of the coal industry.

2031. Coal discovery in Australia occurred in 1791 in the Newcastle area on the southeast coast. The coal deposits were found by escaped convicts.

2032. Gas used to light a house was coal gas in 1792 for the office of William Murdock, a Scottish steam engineer, then living at Redruth in Cornwall, England. Murdock distilled the gas from coal in large iron containers. It was conveyed to the house through 70-foot metal pipes. He did further experiments for lighting in 1796 and then in 1803 installed coal-gas lighting for the engineering factory of his employees, Matthew Boulton and James Watt, at Soho, near Birmingham, England.

2033. Coal exported from Australia was in 1799 when coal from Newcastle was shipped to India. The coal industry had begun the year before when ship owners transported surface coal from Newcastle to Sydney.

2034. Widespread use of canaries to detect deadly carbon monoxide gas in coal mines occurred in England and other European countries in the early 19th century. The death of a canary indicated that the level of carbon monoxide could be dangerous. If a canary died, workers increased ventilation in the mine.

2035. Gas lamps used to fully light a public thoroughfare was in 1807 for Pall Mall, a street in London, England, noted for its private clubs. The gas lamps replaced the feeble oil lamps that had lighted the streets for years.

2036. Anthracite coal burned experimentally was used on February 11, 1808, by Judge Jesse Fell in his home in Wilkes-Barre, PA, much to the surprise of his neighbors, who regarded coal as valueless.

2037. Gas supplied by a private company was in 1813 in London, England. The Gas Light and Coke Company, previously named London and Westminster Gas Company, produced as well as supplied the gas. Other similar companies were soon established throughout the city.

2038. Use of natural gas for commercial lighting in the United States was in 1816 in the New Theater in Philadelphia, PA.

2039. Use of natural gas for street lamps in a U.S. city was in 1816 in Baltimore, MD. The gas was produced from coal and had more impurities and less energy output than modern natural gas.

2040. Natural gas well in the United States was in 1821 in Fredonia, NY. The well was dug by William Hart to a depth of about 27 feet, next to a creek on the town's outskirts. Gas bubbles had earlier been seen rising from the creek. Hart became known as "America's Father of Natural Gas."

2041. Natural gas used to light a U.S. house was on June 4, 1825, in Fredonia, NY. The house was used for a reception honoring General Lafayette, French-born hero of the American Revolutionary War. Thirty burners were lit by natural gas piped from a well.

2042. Hot-blast iron furnace fueled by anthracite coal in the United States was in operation in Allentown, PA, in 1840. The use of hard anthracite as a metallurgical fuel made possible the mass production of cheap, high-quality iron.

2043. Use of natural gas for cooking was in 1841 in London, England. Alexis Soyer cooked the first meal in the famous private Reform Club.

2044. Law in England banning women from working underground was enacted in 1842. The legislation barred women from working in coal mines—for many, the only means of livelihood.

2045. Natural gas invention for heating was the Bunsen burner, developed in 1855 by the German chemist Robert Wilhelm Bunsen while he was a professor of chemistry at the University of Heidelberg. He discovered that a hot, nonluminous flame could be produced by mixing air with natural gas. This encouraged a heating industry that used the mixture to heat rooms and water, and it also was used for cooking.

2046. Anthracite coal mine disaster of major significance in the United States occurred at the Avondale Mine in Pennsylvania on September 6, 1869. A fire broke out in the shaft and the entire work force, 179 men and boys, perished.

2047. Gas conversion law by a state was enacted by Indiana on March 2, 1891, making it illegal to convert "what are known as flambeau lights" to burn natural gas. Violators were subject to a fine of up to $25 and second offenders to a fine not exceeding $200.

2048. Time that natural gas was overtaken by electricity as the primary power source for lighting in the United States was circa 1900. Natural gas had been virtually the only source of lighting in the United States for most of the 19th century.

2049. Government study on U.S. coal industry conditions appeared in 1907. The U.S. Geological Survey reported that death rates in American mines were substantially higher than elsewhere and recommended mine safety legislation. It was several years before any action was taken to improve conditions of coal mines.

2050. Interstate gas pipelines in the United States were built in the 1930s. A massive pipeline construction program was launched after World War II to deliver gas from the Southwest to the energy-deficient Northeast.

2051. Natural energy directly regulated by the U.S. government was natural gas. The Natural Gas Act of 1938 gave the Federal Power Commission (FPC) the authority to regulate rates and to approve certificates for companies to build interstate pipelines and ship gas. The FPC, especially after 1954, set prices, allocated supplies, and established other policies.

2052. Natural gas piped ashore from the North Sea was in 1967 to Scotland by the British Gas Corporation, now named British Gas. It was mainly used in the British domestic market.

2053. Natural gas well in Sudan was sunk in 1975 on the continental shelf of the Red Sea by Chevron Oil Company.

2054. Plant in the United States to recover methane gas from garbage opened in Palos Verdes, CA, in 1975. In the following years, methane plants went on-line in many localities. Many saw this as a way to lessen the burdens on overflowing landfills.

2055. Offshore natural gas plant to separate liquids from the gas began operating in 1976 in the Java Sea about 90 miles northeast of Jakarta, Indonesia.

2056. Natural gas emergency legislation passed by the U.S. Congress was the Emergency Gas Act of 1977. Shortages had been caused by the most severe winter of the century. The act authorized the Federal Power Commission to transfer natural gas between pipelines when needed.

2057. Natural gas processing plant in the United Arab Emirates began operating in 1977 on Das Island at Abu Dhabi. It pumps gas from an offshore field, and most is exported to Japan.

2058. Law in the United States deregulating the price of natural gas extracted from below 15,000 feet was the Natural Gas Policy Act of 1978. The measure provided a range of incentives for deep drilling for gas.

2059. Natural gas extraction in Ireland began in 1978 from an offshore field south of Kinsale Harbour in the southern part of the country.

2060. Synthetic fuel plant in the United States built to operate on a commercial scale was constructed beginning in 1980 at Beulah, ND. It was designed to convert lignite into natural gas. When the operators defaulted on $1.5 billion in federally guaranteed loans in 1985, the plant's title passed to the Department of Energy, which was unable to find a buyer.

2061. Use of compressed natural gas to power a ship was in 1981 in Australia. An Adelaide cement company used the gas-fueled ship for daily trips across the St. Vincent Gulf.

2062. Natural gas field in Thailand was the Erawan field, which began production in August 1981. The Erawan field in the Gulf of Thailand was the first of several to be developed.

2063. Natural gas price deregulation by the U.S. government was the National Gas Policy Act (NGPA), which became effective January 1, 1985. Passed by the U.S. Congress in 1978, it deregulated gas prices for new wells to stimulate competition. This helped initiate later deregulation by the Federal Energy Regulatory Commission (FERC), which is in charge of interstate prices for natural gas, and also instigated deregulation by state public utility commissions. In 1989, the National Gas Wellhead Decontrol Act lifted all the remaining price controls at old wells, effective January 1, 1993.

2064. Clean-coal joint program by the U.S. government and industry was the Clean Coal Technology Demonstration Program, begun in 1986. Its goal was to expand and develop pollution-control devices to curb the release of pollutants and to process coal into cleaner fuels. About $6.7 billion has been invested in

COAL AND NATURAL GAS—*continued*

the program, which is developing and improving such technology as cost-effective environmental retrofit technologies, low-pollution coal burners, sulfur-removing devices, and better coal-to-electricity processes to reduce carbon dioxide releases.

2065. Natural gas fuel cell to produce electricity commercially was introduced May 22, 1992, by the Southern California Gas Company. The cells were used to produce power for the headquarters of the South Coast Air Quality Management District in Diamond Bar, CA.

2066. Mine gas extracted from coal fields in France began in November 1994 at the GIE Methamine in the northern part of the country. It is a joint venture by Gaz de France, who operate the process, and Charbonnages de France (CDF).

2067. Natural gas production in Vietnam began in 1995 with gas pumped from the offshore field of Bach Ho. In the first 100 days of operation, 108 million cubic yards were pumped to the Ba Ria power station near Vung Tau, the center of the country's oil industry.

D

DAMS

2068. Dam organization of international significance was the International Committee of Large Dams (ICOLD), established in 1928. Located in Paris, France, the nongovernmental organization was originally formed to encourage advances in the planning, design, construction, operation, and maintenance of large dams by gathering and disseminating information and by research into technical problems. Since the late 1960s, it has concentrated on dam safety and the environmental impact of dams. In 1999, ICOLD had national committees from 81 nations and about 6,000 individual members, such as engineers, geologists, and scientists from governmental organizations, universities, laboratories, and construction companies.

2069. Hydroelectric dams on Russia's Volga river were built in the late 1930s. Eventually, eight large dam and reservoir complexes were developed along the Volga, Europe's longest river.

2070. Multi-purpose dam development of significance in the United States was the network of dams along the Tennessee River built by the Tennessee Valley Authority, established in 1933. The TVA operates 50 dams that supply electric power, water, and recreational opportunities that control flooding.

2071. Major hydroelectric power-producing dam on the Zambesi River in Africa was built between 1955 and 1959. The Kariba Dam, a concrete arch straddling the Zambia-Zimbabwe border, created a lake 175 miles long.

2072. Systematic dam inspection by the U.S. government began with passage of the National Dam Inspection Act by the U.S. Congress on August 8, 1972. It required the Secretary of the Army to have safety regulations drawn up and inspections of all dams done through the U.S. Chief of Engineers. In May 1975, the U.S. Army Corps of Engineers published its National Program of Inspection of Dams, which included inspection guidelines and an inventory of dams.

2073. Dam safety guidelines issued by the U.S. government were published in 1979 as the Federal Guidelines for Dam Safety. These were drawn up because of the failure of the Teton Dam in Idaho on June 5, 1976.

2074. Dam safety organization among U.S. states was the Association of Dam Safety Officials (ADSO), established in 1983. The organization was begun in response to several massive failures of dams in the late 1970s. The ADSO, located in Lexington, KY, was founded to exchange ideas and experiences about dam safety through interstate cooperation. It provides information and assistance to improve dam safety programs, lobbies state legislatures and the U.S. Congress, and provides information to foster public awareness of dam safety.

2075. International dam opposition group was the International Rivers Network (IRN), established in 1985 in Berkeley, CA. An organization of volunteers supported by grants from foundations, the IRN acquired its first permanent staff in 1989. It works to stop the construction of dams and other projects that harm and destroy rivers, and it promotes sound river management. IRN has international programs and supports such groups as the European Rivers Network, established in 1994. It also has a research unit and publishes a quarterly newsletter, *World Rivers Review*.

2076. Dam safety organization in Canada was the Canadian Dam Safety Association (CDSA), established in 1989. It was formed to help implement the safe operation of dams in Canada. In 1997, the association was amalgamated with the Canadian National Committee on Large Dams (CANCOLD) to form the Canadian Dam Association (CDA). It exchanges ideas and experiences in dam safety by fostering interprovincial cooperation, promoting the adoption of regulatory policies and safety guidelines for dams and reservoirs throughout Canada, and sharing information with Canadian and international groups interested in dam safety.

2077. Dam safety code issued by the U.S. government was in the Dam Safety Act, passed by the U.S. Congress and signed by President Bill Clinton on October 12, 1996. The act codified a national dam safety program and authorized funds for grants to state dam safety programs, research into dam safety, training for state dam safety inspectors, and a National Inventory of Dams. The 1999 funding was $3.8 million.

2078. Convention of people affected by the construction of dams was the First International Meeting of People Affected by Dams, held from March 11 to 14, 1997, in Curitiba, Brazil. Delegates from 200 countries attended. They issued a declaration, "Affirming the Right to Life and Livelihood of People Affected by Dams," which called for an immediate moratorium on building large dams until environments damaged by dams were restored, even if the dams had to be removed, and until provisions of reparations were made for millions of people whose livelihoods were adversely affected by dams. The declaration also called for the establishment of an international, independent commission to review all dams financed by international aid and credit agencies.

2079. Dam cost analysis on a global scale was begun by the World Commission on Dams (WDC), launched on February 16, 1998 in Cape Town, South Africa. It was supported by the World Bank and the IUCN-World Conservation Union. The cost analysis compared the cost of droughts and floods with that of building dams and their environmental and social impact. The commission was created also to develop internationally accepted standards, guidelines, and criteria for decision-making in the planning, design, construction, monitoring, operation, and decommissioning of dams. The idea for such an organization was formulated at the First International Meeting of People Affected by Dams, held in March 1997 in Curitiba, Brazil.

2080. Government demolition of a U.S. dam without the owner's permission began on July 1, 1999, on the Kennebec River near Augusta, ME. The 915-foot-long Edwards Dam had for 162 years prevented salmon and other fish from swimming upstream to spawn. By demolishing the dam, the federal government hoped that the fish population would again flourish in the Kennebec River.

2081. Dam opposition day worldwide was the International Day of Action Against Dams: For Rivers, Water and Life, held on March 14, 1999. It consisted of events designed to unite groups that oppose dams and protect rivers, and included programs to discover the best ways to manage rivers and other waterways. The date was chosen to coincide with Brazilian action groups' Day of Action Against Large Dams, and the event was organized by delegates who had attended the First International Meeting of People Affected by Dams, held in March 1997 in Curitiba, Brazil.

DAMS—CONSTRUCTION

2082. Dams built whose remains have been discovered were constructed circa 3,000 B.C. near the town of Jawa in what is now Jordan. The small dams were part of a system that supplied water to the town via a canal leading into 10 reservoirs. The largest dam was more than 13 feet high and more than 260 feet wide.

2083. Dam constructed in Ancient Egypt was built between 2950 and 2750 B.C. in the Wadi el-Garawi valley south of Cairo. Its purpose was to collect rainwater for use at nearby quarries. The dam was 348 feet long, 276 feet thick at its base, and constructed of nearly 91,000 tons of earth covered with masonry. It collapsed after a few years from the water pressure.

2084. Dam known to have a curved shape was the Daras Dam constructed from A.D. 527 to 565 in Persia under the Byzantine emperor Justinian I. It used the principal of the arch turned horizontal with the curve upstream to increase its strength. Details of the construction were described in 560 by Procopius, the Byzantine historian.

2085. Masonry dam built that is still in use was constructed circa 1500 at Almanza, Spain. It has a facing of cut stone over an interior of rubble masonry. Its width is only 10 feet at the top and 34 feet at the bottom.

DAMS—CONSTRUCTION—*continued*

2086. British dam of significance built entirely of concrete was in 1876 at Woodhead, Scotland, about 25 miles northeast of Aberdeen.

2087. Dam built in the United States to run a hydroelectric plant was constructed in 1889 on the Willamette River at Oregon City, OR.

2088. Dam built using the hydraulic fill process was the Tyler Dam, completed in 1894. The Tyler Dam was built by Julius M. Howells and spans Indian Creek in Tyler, TX. The hydraulic fill process uses water to move earth, which is then drained at the site.

2089. Steel dam in the United States was the Ash Fork Dam, built in 1898 in Johnson Canyon, four miles east of Ashfork, AZ, by the Atchison, Topeka and Santa Fe Railway Company. The steel portion was 184 feet long with masonry abutments and the height of the spillway 30 feet above the bottom of the reservoir. The width of the canyon at the stream bed was 40 feet. The canyon drained about 30 square miles.

2090. Dam project to provoke major opposition from environmentalists was a proposal in 1901 to build the Hetch Hetchy dam and reservoir on the Tuolumne River in Yosemite National Park. Rapidly growing San Francisco was in need of an independent water supply. The Tuolumne River was the purest and most abundant source, and a dam built on it would also generate hydroelectric power and provide irrigation water for farmers. Naturalists, among them John Muir, opposed the flooding of part of Yosemite National Park and the possible destruction of its ecology and diverse plant species. The dam would not be built until 1923.

2091. Reinforced concrete buttress dam was built in Theresa, NY, in 1903.

2092. Dam to have a maximum height of 295 feet was the New Croton Dam, spanning the Croton River in New York and completed in 1907. The New Croton Dam was designed by Alphonse Fteley and became a standard for gravity

2093. Major dam on the Mississippi River below St. Paul, MN was the Keokuk Dam, completed in 1913. The Keokuk Dam was built by Hugh L. Cooper on the rapids between Keokuk, IA, and Hamilton, IL.

2094. Dams on the Tennessee River near Muscle Shoals, AL were the Wheeler Dam and the Wilson Dam, begun in 1916. Wheeler Dam and Wilson Dam were acquired by the Tennessee Valley Authority in the 1930s.

2095. Hydroelectric power-plant licenses in the United States were created by the Federal Water Power Act, passed by the U.S. Congress on June 10, 1920. It created the Federal Power Commission (FPC), which was required to license approved nonfederal hydroelectric power projects that affected navigable rivers, occupied federal lands, used water or water power at a government dam, or affected interstate commerce. The FPC was instructed to license only those projects that best improved or developed one or more waterways. Prior to the act, proposed plants on navigable streams or federal lands required a congressional act.

2096. Hydroelectric dam of significance in the United States financed by private capital was the Conowingo Dam on the Susquehanna River in Pennsylvania, completed in 1928. At the time, the 4,700 feet long and 105 feet high dam's generating power was second only to the Niagara Falls plant.

2097. Hydroelectric dams on the Columbia River in the U.S. Pacific Northwest were built during the 1930s. Grand Coulee, the largest, took form between 1933 and 1942.

2098. Dam built as part of the Tennessee Valley Authority (TVA) was the Wheeler Dam, completed in 1936. The Wheeler Dam spans the Tennessee River south of Elgin, TN.

2099. Dam to be constructed with a height greater than 152 meters (500 feet) was the Hoover Dam, constructed on the Colorado River between Nevada and Arizona in 1936. Lake Mead, created by the dam, was the world's largest reservoir at the time it was built and remains the largest in the United States.

2100. Dam with a height greater than 214 meters (800 feet) was the Alvaro Obregon Dam, constructed in Mexico. The Alvaro Obregon Dam, also known as the El Oviachic Dam, was built on the Yaqui River and completed in 1946.

2101. Reservoir to cover four towns and affect seven others in the United States was the Quabbin Reservoir, filled to capacity in 1946. The Quabbin Reservoir was created by the Winsdor Dam, which spans the Swift River in Ware, MA. Four towns in the swift river valley were destroyed to make the dam. Members of the communities of Dana, Enfield, Greenwich, and Prescott fought the proposed dam, but they were unsuccessful. The Massachusetts legislature passed the Swift River Act of 1927. The act allowed the state to buy out the towns.

2102. Dam built by Canada's Prairie Farm Rehabilitation Administration was the St. Mary Dam, completed in 1951 on the St. Mary River about 25 miles from Lethbridge in southern Alberta. The earth-filled dam was the main structure of the St. Mary Irrigation Project, which irrigated 410,000 acres of previously arid grasslands. In December 1992, the St. Mary Hydroelectric Plant was completed at the base of the dam.

2103. Dam with a height greater than 273 meters (900 feet) was the Grande Dixence Dam, constructed on the Dixence River in Switzerland in 1962. The Grande Dixence Dam is still the world's highest concrete dam.

2104. Dam to harness tidal power was completed in 1967 on the estuary of the Rance river near St. Malo in Brittany, northwest France. The dam, about 2,500 feet long, converts power from tides that rise 44 feet high. Its power station has 24 turbines. Each can create 10 megawatts of electrical power.

2105. Dam built to totally control the annual flooding of Egypt's Nile River was the Aswan High Dam, opened in 1968 in the southern part of the country at Aswan. Begun in 1960, it is 364 feet high with a length of 12,562 feet. It created from the desert some 100,000 new acres of arable land and also Lake Nasser, which is 1,930 square miles in extent. It generates electricity from its 12 hydroelectric generators. However, the dam held back 40 million tons of enriching silt that were deposited annually on adjacent soils. It also required the relocation of about 90,000 people and the ancient temples of Abu Simbel, whose move cost $35 million.

2106. Dam with a reservoir that could hold more than 100,000 cubic meters of water was the Daniel Johnson Dam on the Manicouagan River in Quebec, Canada, completed in 1968. The Daniel Johnson Dam has the world's largest reservoir, with a capacity of 141,852 cubic meters.

2107. Dam in India with a double curvature and parabolic arch was the Idikki Dam on the Periyar River, completed in 1974. It is 555 feet high and 1,201 feet long, with a width of 75 feet at its base and 25 feet at its top. Two other dams under construction to create the Idikki hydroelectric system are the Kulamavu masonry dam and the Cheruthoni concrete gravity dam.

2108. Dam to reach 300 meters (984 feet) in height that is still in use was the Norek Dam in Vakhsh, Tajikistan, completed in 1980.

2109. Dam constructed totally with roller-compacted concrete (RCC) was the Willow Creek Dam in Oregon, completed in July 1983 on the North Fork River. The $37 million project marked improvement in the construction of concrete gravity dams by using a drier concrete mixture spread in one-foot layers and compacted with large roller equipment. The structure was harder but one-third cheaper than the conventional concrete construction with buckets. The dam, begun in September 1980, was built to protect Heppner, OR, which in 1903 had suffered one of the worst U.S. floods, which killed 247 people.

2110. Phase of construction on the gigantic Three Gorges Dam on the Yangtze River in China began in 1997. The dam, projected to be the world's largest, will be 600 feet high and 1.2 miles wide when completed.

DAMS—DISASTERS

2111. Dam failure of major significance in the United States occurred when the earthen dam on Mill Creek near Williamsburg, MA, was destroyed in a landslide on May 16, 1873. The dam failure killed 144 people and caused $1 million in damages.

2112. Warnings of structural problems at the South Fork Dam near Johnstown, PA, were presented to the owners in the early 1880s. Daniel Morrell, a Pennsylvania businessman, hired a firm of structural engineers to inspect the dam. Their report concluded that the dam was in danger of collapsing. Morrell's report was ignored and in 1889 the dam broke, causing the great Johnstown Flood.

2113. Dam disaster in the United States with a high death toll occurred at Johnstown, PA, on May 31, 1889, when the South Fork Dam burst. A wall of water, one-half mile wide and reaching 75 feet high, swept down upon the city, causing more than 2,200 deaths and property damage of about $10 million. The dam had held back the waters of Conemaugh Lake, which was about 2.5 miles long and 1.5 miles wide, with an average depth of 50 feet.

2114. Dam burst to cause a disaster by chemical contamination occurred on January 18, 1921, in the mining center of Pachuca de Doto, on the Moctezuma River north of Mexico City. Two dams, built by the Rosario Fresnillo Company to contain waters used in the chemical treatment of ores mined in the mountains above Pachuca, burst, killing 50, injuring 200, and rendering more than 1,000 homeless. Most of the deaths came not from drowning but from poisioning through the chemically contaminated waters.

DAMS—DISASTERS—*continued*

2115. Dam in the United States to fail during an earthquake was the Sheffield Dam in Santa Barbara, CA, on June 29, 1925. It is the only U.S. dam to collapse under seismic forces. Up to 13 people died when the dam, holding 30 million gallons of water, lost its central section and flooded the city. The dam was built in 1917 on sandy soil at the base of the Santa Ynez Mountains at the northern end of Santa Barbara. It was 720 feet long and 25 feet wide.

2116. Dam of significance in the United States to collapse when completely filled was the St. Francis Dam in California on March 12, 1928. It caused 450 deaths. The tragedy has been called "the greatest American civil engineering failure in the 20th century." An investigation revealed that the foundation rock had lost strength when saturated. The curved concrete gravity dam was built by the city of Los Angeles in 1925–1926. It was located in San Francisquito Canyon, and the collapse flooded towns in the canyon and the Santa Clara Valley.

2117. Dam disaster of significance in Spain in the 20th century occurred on January 9, 1959, when the Tera River Dam collapsed. The town of Rivaldelago was destroyed and at least 130 villagers died.

2118. Modern disaster blamed on faulty geologic analysis occurred at Vaiont Dam in the Italian Alps on October 9, 1963. Four thousand people perished when a rock slide behind the dam sent an enormous wave over the barrier, flooding the valley below.

2119. Televised dam collapse was broadcast when the Baldwin Dam in Los Angles burst on December 14, 1963. The dam began sprouting small leaks about two hours before it burst, which gave the Los Angeles authorities time to warn people of its impending collapse. Newspeople also had enough time to cover the event. Thousands in Los Angeles watched film footage, shot from a helicopter, of the dam collapsing.

DROUGHTS AND WATER SHORTAGES

2120. Drought recorded known to have caused a famine occurred circa 3500 B.C. in Egypt. Several thousand people died of starvation.

2121. Evidence of extended drought causing the decline and collapse of a great civilization dates from circa 1900 B.C. in the Indus Valley of the Indian subcontinent. The cause of the drought may have been a permanent southward shift of summer monsoon rainfall.

2122. Recorded drought that affected a Roman military campaign in Britain occurred in 55 B.C. Julius Caesar reported that short autumn harvests after a dry summer compounded his supply problems and contributed to his inability to completely conquer the island.

2123. Drought of major significance in Central America occurred between A.D. 800 and 1000 in the Yucatan peninsula of present-day Mexico. With the water supply nearly gone for some 5 million people, peasants cleared the forests to plant more crops, such as corn and beans. The devastation of these forests led to severe soil erosion and ecological disaster. Some scientists now believe the drought instigated the decline of the Mayan civilization.

2124. Drought in England known to have created slavery occurred in 1069 in the area of Durham in the northeast region of the country. The long drought destroyed crops and led to a famine that killed about 50,000 people. Thousands more became slaves in exchange for food to survive.

2125. North American civilization known to end due to a drought was that of the Mesa Verde Anasazi, ancestors of the Pueblo peoples. After the Great Southwest Drought of 1276–1299, the Anasazi left the area for better farm lands and dispersed among other tribes.

2126. Drought of significance in colonial America occurred from 1587 to 1589. Researchers believe it contributed to the mysterious disappearance of the "Lost Colony" on Roanoke Island, then in the colony of Virginia and now in North Carolina. The drought was identified by reading rings of ancient trees in the Tidewater area, a project funded by the National Park Service and carried out by research teams from the College of William and Mary and the University of Arkansas.

2127. Abandonment of a city in India because of a failure of the water supply occurred in 1588. The city of Fatepur Sikri was emptied only 16 years after its construction.

2128. Drought affecting English colonists in Massachusetts occurred in 1621, the first summer of Pilgrim settlement at Plymouth. No rain fell in Plymouth for 24 days.

2129. Drought in India to cause massive deaths by starvation took place from 1669 to 1670 when about 3 million deaths were estimated. The famine was especially severe in the west central area of the country, including the port city of Surat.

2130. Drought of significance recorded in the American colonies occurred in 1727 in New England, where there was virtually no rainfall from the second week of April through the middle of July.

2131. Great famine due to a drought recorded in detail occurred in India in 1769–1770. An 18-month drought caused a famine that killed one-third of the population in Hindustan.

2132. Drought known to have caused cannibalism in India occurred from 1790 to 1791. Several thousand people died from the drought, and numerous survivors ate human remains in what became known as the Poij Bara (Skull Famine).

2133. Drought recorded in Australia was mentioned in a letter of March 4, 1791, by the governor of New South Wales, Arthur Phillip. The severe drought had lasted since June 1790, and Phillip said crops of corn had been ruined and rivers had been dry for several months. During this period, known as "the hungry years," the colonists had to import food.

2134. Drought to cause severe agricultural damage in Canada occurred in 1806 in the Red River area of Manitoba. Potato crops were scorched and ruined throughout the region. In 1846, another drought in the Red River area caused a total crop failure.

2135. Explorer to describe the High Plains of the western United States as the Great American Desert was Zebulon Pike, in his expedition journal published in 1810. Pike's label discouraged several generations of settlers from migrating to the region.

2136. Drought to create the conditions for a devastating fire in a major U.S. city occurred on October 8, 1871, in Chicago. Although not the cause of the Great Chicago Fire, the dry conditions in the city due to the drought helped spread the fire across the city, killing 250 persons and causing nearly $200 million in damages.

2137. Drought known to have caused cannibalism in China occurred in Manchuria from 1876 to 1878. Between 9.5 and 13 million people died during the accompanying famine, and some of those victims were eaten by survivors. The Western world did not become aware of the severe drought in China until January 28, 1878.

2138. Geologist/explorer to state flatly that much of the western United States was too dry to farm was John Wesley Powell, in his *Report on the Arid Regions of the United States*, published in 1878. Powell challenged a powerful body of hopeful thinkers who believed that "rain would follow the plow."

2139. Drought patterns discovered in the Southern hemisphere were identified in 1888 by Charles Todd, a meteorologist for the South Australia state government. He drew "teleconnections" between drought patterns and between climate fluctuations of the "Southern Oscillation."

2140. Drought to depopulate the Great Plains state of Nebraska by half occurred from 1888–1890. During the same prolonged dry spell, Kansas lost one-quarter of its population.

2141. Widespread and catastrophic drought to afflict farmers in the U.S. Great Plains occurred in 1894–1895. Crop failures were complete in many areas. Normal rainfall returned only during the last years of the decade.

2142. Reported period of more than 900 days without measurable precipitation in a southern California town occurred at Bagdad in San Bernardino County from August 18, 1909, to May 6, 1912. This 993-day drought was followed by another 767 consecutive days without rain, from October 1912 to November 1914.

2143. Person to compare tree ring width to climate and droughts was astronomer Andrew Ellicott Douglass of the Lowell Observatory in Flagstaff, AZ. In 1914 he established a quantitative relationship between ring width and amount of rainfall. This method gave scientists a tool to study droughts of the past.

2144. Extensive peacetime drought relief project in the United States was operated by the Red Cross in 1930 during the Midwestern drought that would lead to the Dust Bowl conditions of the 1930s. The Red Cross raised more than $10 million and aided more than 2,765,000 drought victims.

2145. Drought used for political purposes was a drought in the Soviet Union that lasted from 1932 to 1933. Soviet leader Joseph Stalin used the disaster to compel the people of the Ukraine into industrialization, which they bitterly opposed. About 5 million peasants died as Stalin refused international relief aid. In the course of the 1930s, about 25 million peasants were forced off farms into factories.

DROUGHTS AND WATER SHORTAGES—*continued*

2146. Dust storm known to carry dust from Montana all the way to the U.S. eastern seaboard occurred on November 12–13, 1933. "Black rain" fell in New York State while "brown snow" fell in Vermont. The dust storms were caused by severe droughts in Montana.

2147. Dust storm known to drop massive quantities of Great Plains topsoil on Chicago occurred in May 1934. The storm dumped 12 million tons of dust onto the city and dimmed the sun at midday.

2148. Severe drought year affecting nearly one-half of the United States was 1934. This was during the worst period of the Dust Bowl in the Midwest.

2149. Major windstorm of the Dust Bowl drought in the U.S. Midwest took place in May 1934. It picked up approximately 350 million tons of topsoil and deposited it over the eastern United States. Dust was even detected on ships 300 miles away in the Atlantic Ocean. Between 1934 and 1938, 10 million acres of land in west Kansas, southeast Colorado, northwest Oklahoma, north Texas, northeast New Mexico, and parts of Nebraska and the Dakotas had lost the top five inches of soil. Another 13.5 million acres lost the top 2.5 inches. The Dust Bowl drought, which was the product of unwise farming methods colliding with a succession of windstorms, drove 3.5 million people from their lands, often into poverty in other states. Respiratory diseases rose 25 percent in four years, and infant mortality increased by a third.

2150. Use of the term *Dust Bowl* to describe the drought-devastated Southern Plains came in April 1935. Journalist Robert Geiger used the term in an Associated Press dispatch on the notorious "Black Sunday" of April 14, 1935.

2151. Drought rehabilitation program by the Canadian government was the Prairie Farm Rehabilitation Administration, created by the Prairie Farm Rehabilitation Act, which was passed by Parliament and approved by royal assent on April 17, 1935. Prompted by widespread drought, farm abandonment, and land degradation in the early 1930s, the act provided technical and financial assistance to the prairie provinces of Manitoba, Saskatchewan, and Alberta. It supported research and demonstrations of "drought-proofing" technologies, such as techniques in irrigation and water conservation.

2152. Return of normal rainfall to the area of the central United States hit by the Dust Bowl occurred in November 1940. From 1930 to the end of the decade, drought affected at least some part of the region every year. The Dust Bowl, coupled with the coinciding economic depression of these years, severely crippled the central United States.

2153. Drought in modern United States to affect 28 percent of the population lasted five years, between 1962–1967. The drought hit all or part of 14 eastern states. During those five years, rain fell mostly over the ocean. The Northeast drought was the longest in modern U.S. history, but compared to the Dust Bowl it caused relatively little hardship. This was due, in part, to the Northeast's abundant supply of water resources.

2154. Index to scientifically measure droughts was the Palmer Drought Severity Index (PDSI), devised in 1965 by an American meteorologist, W. C. Palmer. It identifies prolonged periods of abnormal dryness (or wetness), usually by months or years. The measurements are based on precipitation, temperature, and local available water content in the soil. The index is calibrated roughly between -6 and +6, with -4 or less indicating extreme drought and +4 or more indicating extreme wetness. The index can predict disaster areas of drought or wetness and the long-term water supplies in reservoirs and streams, which helps with water resource management. It is used by such organizations as the National Oceanic and Atmospheric Administration, the National Climate Data Center, and the Climate Prediction Center.

2155. Drought forcing the Soviet Union to buy U.S. grain occurred in 1972. On July 8, 1972, the Nixon administration announced that the Soviets were purchasing 18 million tons of corn, wheat, and other grain for $750 million. Continuing dry weather produced an agreement announced by the Ford administration on October 20, 1975, to sell the Soviets 6 million to 8 million tons annually. Prior to this, the cold war hampered much possible trade between the two countries.

2156. Water shortage in China causing the Yellow River to run dry was in 1972. Increased volumes of water pumped into industries and fast-growing cities made the river's water level fall severely, and the river bed became dry before it reached the sea at Bo Hai Gulf. This has now happened each year since 1985.

2157. Drought to cause damages of $1 billion or more in the United States occurred with a heat wave from June to September 1980. The total estimated damage was about $20 billion. About 10,000 deaths were estimated to be related to the extreme heat and drought in the eastern and central United States

2158. Drought to draw worldwide aid through television was the great African drought from 1981 to 1985. Twenty countries were affected. The victims' plight was televised in October 1984 by the BBC. The pictures of starving refugees prompted an unprecedented international response of financial and food relief, including money raised by the rock festival Live Aid. Crop failures had caused millions to starve, and more than 20,000 children died each month during the height of the famine. The country hit hardest was Ethiopia, where a civil war hindered famine relief.

2159. Drought aid from a global rock music festival came from Live Aid, televised throughout the world on July 13, 1985, from London, England, and Philadelphia, PA. About 1.5 billion people worldwide saw the 16-hour event held to aid victims of the 1984–1985 African drought. It was organized by the Irish singer Bob Geldof, and many famous rock stars performed without pay.

2160. Drought education and training center for developing nations was the International Drought Information Center, established in September 1988 at the University of Nebraska in Lincoln. It provides information and seminars to help developing countries prepare for droughts, as well as seminars on water resources. The staff have published the *Drought Network News* three times yearly since 1989 and have written a guidebook sponsored by the United Nations Environmental Program. The Center is located in the university's Department of Agricultural Meteorology.

2161. Drought in the United States to strand boats on the Mississippi River occurred in 1988. In the Memphis, TN, area alone, about 1,000 barges could not move as the river's level fell to the lowest in recorded history.

2162. Drought year producing worldwide environmental responses was 1988, the hottest year on record. There were droughts in Africa and several other places, including the Soviet Union, China, and India. Directors of United Nations agencies met that summer in Oslo, Norway, to consider many environmental factors, including global warming and how to protect the atmosphere and climate. From June 27 to 30, 1988, more than 300 scientists and officials from 48 countries met in Toronto, Canada, for an international conference, called "The Changing Atmosphere: Implications for Global Security." The European Ozone Research Coordinating Unit was also created in 1988. Numerous other national meetings were held to discuss the causes of the worldwide drought and what measures should be taken for the future.

2163. Year in which half of the agricultural counties in the United States were designated drought disaster areas by the federal government was 1988. By the end of June, much of the country was suffering through the worst drought in five decades. It caused erosion on about 13 million acres and was compared to the Dust Bowl drought of 1934.

2164. Drought policy by Australia's national, state, and regional governments was formalized in 1992. The ministers of the Commonwealth (national), states, and territories agreed to the National Drought Policy (NDP) with provisions for maintaining needed resources during droughts, encouraging farmers to assume greater responsibility for managing the risks of drought, and carrying out research and development, especially concerning drought prediction, monitoring, and management. The ministers also agreed to provide financial assistance to farmers harmed by droughts.

2165. Drought research and development program nationally established in the United States was the National Drought Mitigation Center, established in 1995 at the University of Nebraska in Lincoln. The Center analyzes different state plans for droughts, conducts research about droughts, and publishes reports. Housed in the university's Department of Agricultural Meteorology, it is funded by the Cooperative State Research, Education and Extension Service of the U.S. Department of Agriculture and by the Climate Prediction Center of the National Oceanic and Atmospheric Administration (U.S. Department of Commerce).

2166. Drought predictions for the 21st century by a U.S. government agency were made on December 15, 1998, by the National Oceanic and Atmospheric Administration (NOAA). Based on research by its Paleoclimatology Program, it predicted droughts in the next century to be worse than the Dust Bowl of the 1930s. NOAA said its investigations revealed that the Dust Bowl was moderate compared to earlier severe droughts, especially in the last quarter of the 13th century and the last quarter of the 16th century. The information came from tree rings, archaeological sites, various sediments, and historical documents.

DROUGHTS AND WATER SHORTAGES—PLUVICULTURE

2167. Christian rainmaking ceremony in New England occurred at Plymouth Plantation during the drought summer of 1623. Colonists set aside a day of fervent prayer and were rewarded that night with a series of gentle life-restoring showers.

2168. Attempts to cause rain with gun blasts into vapor-laden clouds occurred in Brisbane, Australia, in September 1901. The experiment was abandoned when the Stiger gun, specially designed for the purpose, fractured; no rain was produced.

2169. "Cloud compeller" to claim success in bringing rain to California with ill-smelling chemical cocktails set in pans atop high wooden towers was Charles M. Hatfield, in the first years of the 20th century. Hatfield won a $1,000 bet when at least 18 inches of rain fell in Los Angeles between December 1904 and April 1905.

2170. Cloud-seeding demonstration was made by an American female meteorologist, Florence W. van Straten, in 1958. Van Straten's method could increase or decrease the amount of clouds, and it was hoped that this would provide a way of producing rain in drought-stricken areas. However, although her method did increase cloudiness, it did not always produce rainfall.

E

EARTHQUAKES

2171. People to keep regular records and collect data on earthquakes were the ancient Chinese, beginning with an earthquake of 1177 B.C. Accurate quake records in the Fen and Wei river valleys of north China have been kept since 466 B.C.

2172. Biblical prophet to use a famous earthquake to date his prognostications was Amos, circa 750 B.C. The quake happened in the time of Uzziah, king of Judah. Amos foretold the destruction of Israel, an event that shortly came to pass.

2173. Earthquake known to cause massive deaths throughout Greece occurred in 464 B.C. it killed more than 20,000 people. The city of Sparta and surrounding Laconia suffered heavy destruction, and the earthquake also inspired a revolt by Sparta's Messenian subjects.

2174. Recorded scientific theories explaining earthquakes were proposed by Democritus and Aristotle in the 5th and 4th centuries B.C. Democritus thought rainwater seeping into the earth caused them. Aristotle believed they were caused by pockets of gas under the Earth's surface.

2175. Precise description of an earthquake was recorded in Strabo's *Geography*. There he describes in great detail how an earthquake destroyed the city Helice in Greece in 373 B.C.

2176. Earthquake recorded in Great Britain was in A.D. 103 in the county of Somerset. It is thought to have totally destroyed a town (whose name has never been recovered).

2177. Seismograph was invented by Chinese mathematician Zhang Heng circa A.D. 132. While others had written about and studied earthquakes, Zhang was the first to build an instrument that indicated the direction of the first motion of an earthquake.

2178. Earthquake in which a massive loss of life was recorded was in 526, in Antioch, Syria. More than 250,000 people perished.

2179. Earthquake to kill more than 1 million people occurred in 1201 in the Near East and eastern Mediterranean area. Cities and towns were destroyed throughout the region by severe earthquakes and aftershocks.

2180. Earthquake in Japan causing massive deaths was on May 20, 1293, at Kamakura on the coast of Honshu Island. About 30,000 people were killed while most of the city's buildings were devastated.

2181. Earthquake causing thousands of deaths in Portugal was on January 26, 1531, in Lisbon. The earthquake killed an estimated 30,000 people and destroyed about 1,500 houses.

2182. List of earthquake deaths was ordered by the Ming emperor of China after an earthquake hit the Shaanxi province on January 24, 1556. The list eventually named about 820,000 of the estimated 830,000 people who had perished in the quake.

2183. Death in England known to have been caused by an earthquake occurred on April 6, 1580, in London. A quake dislodged masonry at Christ's Hospital Church near Newgate, and it fell on an apprentice, Thomas Grey. His death and the subsequent death of a young girl, Mable Everet, in this event are the only two earthquake deaths ever recorded in Great Britain.

2184. Earthquake to cause massive loss of life in Jamaica occurred in 1692 when about 2,000 lives were lost. Most of the town of Port Royal (then the colonial capital) disappeared when the earth beneath it opened up and a tsunami (giant wave) swept over it. The disaster led to the establishment of Kingston as the capital.

2185. Archaeology site preserved by an earthquake occurred as a result of the earthquake that sank Port Royal, Jamaica, into the Caribbean at 11:53 A.M. on June 7, 1692. A watch stopped at 11:53 and many eyewitness accounts give us the exact time of the slide into the sea. More than 300 years later most of the area has remained intact, and many organic items were preserved by the water and silt. Study of this undisturbed site has given archaeologists insights into English colonies in the Americas.

2186. Earthquake and tsunami (giant wave) to destroy Japan's capital occurred in 1703 in Edo (present-day Tokyo), killing about 200,000 residents.

2187. Earthquake to cause massive loss of life in Peru occurred on October 28, 1746. The quake killed about 18,000 people and devastated the city of Lima and the seaport of Callao. After the tremor subsided, only 21 of the area's 3,000 homes remained.

2188. Earthquake causing massive loss of life in Portugal occurred on November 1, 1755, in Lisbon. It struck at 9:40 A.M., killing people worshiping in cathedrals and churches on All Saints' Day. The area's estimated deaths were 60,000, some dying in fires that burned for six days. The epicenter of the quake, now estimated at 8.75 on the Richter scale, was thought to be about 200 miles offshore in the Atlantic Ocean. Shock waves were felt as far away as Scotland and Africa. The six-minute shaking also began a tsunami (giant wave) that was 60 feet high when it reached Cadiz, Spain.

2189. Scientific investigation of an earthquake was ordered by the Marquês de Pombal of Portugal after the November 1, 1755, earthquake in Lisbon, Portugal. He told priests in Portugal to record the time, direction of the wave, and any effects they observed.

2190. Description of an earthquake or aftershock as a wave was given by John Winthrop IV, a Harvard professor, in 1755. After the November 18, 1755, earthquake in New England, Winthrop was sitting in his Boston home by his fireplace. His feet were on the brickwork around the fireplace. When an aftershock hit he felt the bricks move and described it as "one small wave of earth rolling along."

2191. Description of two types of earthquake motion was given by England's John Mitchell around 1756. After his study of the reports of the 1750 earthquake in England and the 1755 Lisbon earthquake, he said there were two types of motions—the first vibration or tremor being followed by a wavelike motion.

2192. Theory on how to locate the epicenter of an earthquake was postulated by John Mitchell of Cambridge University in England around 1756. After reviewing accounts of England's earthquake in 1750 and the Lisbon earthquake in 1755, Mitchell proposed that the direction of the seismic wave be mapped in various places and all lines extended until they meet. He did not use his method to locate the epicenter of the Lisbon earthquake, but later others did successfully use his theory to locate its epicenter.

2193. Earthquake causing thousands of deaths in Chile was in 1757. The death toll was about 5,000, and most people died in the city of Concepción from the tsunami (huge wave) activated by the earthquake. Another 10,000 were injured.

2194. Earthquake to devastate Guatemala struck in 1773, causing many deaths and devastation in the capital of Antigua in the central highlands. Three years after the disaster, the Spanish crown moved the capital 15 miles away to the site of present-day Guatemala City.

2195. Earthquake intensity scale was created by Italian physician Domenico Pignataro around 1786. After reviewing accounts of the 1,181 earthquakes in Italy between January 1, 1783, and October 1, 1786, he devised a four-part scale measuring earthquakes as slight, moderate, strong, and very strong.

2196. Earthquake causing massive loss of life in Ecuador occurred on February 4, 1797, in Quito, with most of its population of around 41,000 killed. Several rivers changed course after the violent earthquake.

EARTHQUAKES—*continued*

2197. Earthquake known to have changed the course of the Mississippi River occurred on December 16, 1811, in Missouri. The quake, now estimated to have been between 7 and 7.5 on the Richter scale, shifted the river and destroyed the town of New Madrid near the Kentucky border, but few deaths were reported in the sparsely populated area.

2198. National Disaster Relief Act in the United States was passed in 1815 as a result of the December 16, 1811, earthquake centered in New Madrid, MO. The initial shock was felt over two-thirds of the United States, and a series of lesser shocks continued until February 7, 1812. Some 150,000 acres of timberland and farmland were destroyed. In the relief act of 1815, Congress authorized the exchange of ruined lands for equal amounts of government lands.

2199. Known earthquake on the historical record occurred in China in 1831 B.C. The Chinese were the first to keep regular records of quakes.

2200. Earthquake to be described in detail by Charles Darwin occurred at Concepción, Chile, on February 20, 1835. Darwin, a passenger aboard the HMS *Beagle*, published scientific speculations that later influenced theories about the global distribution of earthquakes and volcanoes.

2201. Systematic catalog of earthquakes was started by Robert Mallet in 1837 and ultimately listed more than 6,831 quakes worldwide. Mallet used any printed material he could find that contained the record of an earthquake. His listings included date, location, number of shocks, probable direction, and duration for each quake as well as any effects. The study took him 20 years and was finished in 1857.

2202. Theory stating that earthquakes come from the Earth's crust and not from outside forces working on the crust was presented by Dublin engineer Robert Mallet in 1857 after he studied the 1857 Naples earthquake. Mallet transferred his knowledge of how internal pressure acts on metals to the Earth's crust and said that it could cause the crust to warp and bend.

2203. World seismic map of significance was plotted by Dublin engineer Robert Mallet in 1857. In his map, Mallet plotted more than 6,500 historic earthquakes and showed clear belts of seismic activity for the first time.

2204. Seismograph in its modern form was invented in 1880 by John Milne, an English seismologist and mining engineer. He later set up a worldwide network of seismological stations using this instrument. Milne, who was professor of geology and mining at the Imperial Engineering College in Tokyo for 20 years, is credited with pioneering modern seismology.

2205. Seismology association was the Seismological Society of Japan formed by Englishman John Milne. After the 1880 earthquake in Yokohama, Japan, Milne gathered Japanese and foreign scientists to form the society.

2206. Severe earthquake to strike the U.S. east coast in modern times occurred in Charleston, SC, on August 31, 1886. Around 100 buildings were destroyed in the quake, whose initial shock lasted 35–40 seconds. More than 20 people were killed directly; others died later of injuries or exposure. Shock waves were felt as far away as New York and Nebraska.

2207. Seismograph in the United States was installed in 1888 at the Lick Observatory on Mount Hamilton in California.

2208. Scientific formulation of the "fault theory" as we know it was presented by Japanese scientist Bunjiro Koto. Koto interviewed survivors of the October 28, 1891, earthquake in Mino-Owari, Japan, and found that the land shifted as the earthquake began. After studying these interviews Koto said that the "sudden formation of the great fault of Neo was the actual cause of the earthquake." The previous theory was that the earthquakes caused the faults.

2209. Recorded incidence of an earthquake reversing the paths of two glaciers occurred on September 3 and 10, 1899, in Yakuta Bay, AK. As a result of the second shock the Yakuta Glacier and the Muir Glacier reversed direction.

2210. Vertical surface displacement of more than 45 feet caused by an earthquake occurred after the second shock of the earthquake at Yakuta Bay, AK, on September 10, 1899. Disenchantment Bay was raised 47 feet and 4 inches by the shock.

2211. Scientist to describe the type of earthquake surface wave that acts like waves caused by a pebble tossed into a pond was the Brtish physicist Lord Rayleigh, in 1900. It was named the Rayleigh wave in his honor.

2212. Scale widely used to measure earthquake intensity was devised in 1902 by the Italian seismologist Giuseppe Mercalli. The Modified Mercalli Scale is still used to describe effects at the epicenter, not the energy released. The scale runs from Intensity 0 (felt only by seismographs) to Intensity 12 (nearly total destruction).

2213. International seismographic recording network was centered at John Milne's home on the Isle of Wight in England in 1902. Between 1902 and 1913, 40 stations were established to send data back to the Seismological Committees of the British Association for the Advancement of Science, for John Milne to analysis. His reports, called the Shide Circulars, were sent to seismologists throughout the world.

2214. Earthquake to register an 8.9 on the Richter scale occurred 200 miles off the Colombian coast in 1906. This earthquake was one of two to reach 8.9. Because it was so far from land, it caused no loss of life.

2215. Earthquake in the United States believed to have surpassed 8.0 on the Richter scale occurred on April 18, 1906, in San Francisco, CA, with an intensity now estimated at 8.3. More than 500 people were reported dead or missing, and more than 500,000 were made homeless. The quake hit at 5:15 A.M. for 40 seconds and, after 10 calm seconds, returned for another 25 seconds. The combination of the earthquake and the fires that burned for three days afterward destroyed more than 28,000 buildings in four square miles. Property loss was estimated at more than $300 million.

2216. Earthquake organization formed because of a U.S. earthquake was the Seismological Society of America (SSA) in 1906. It was established in San Francisco, CA, after the earthquake of April 18, 1906, and the society had its first meeting on November 20 that year. Its original goal was "for the acquisition and diffusion of knowledge covering earthquakes and allied phenomena, and to enlist the support of the people and the government in the attainment of these ends." The future U.S. president Herbert Hoover was a member from March 1911 to 1959, as was his wife, Lou Henry Hoover, from November 1911 to 1944. The SSA now has members throughout the world, such as seismologists, geophysicists, geologists, engineers, and insurers.

2217. Theory of continental drift was conceived by the German meteorologist and Arctic explorer Alfred Wegener circa 1910. The theory sought to explain, among other things, the occurrence of earthquakes and volcanoes.

2218. Seismologist to advance the elastic rebound theory to explain earthquakes in tectonic was Harry Fielding Reid of Johns Hopkins University in 1911. Reid developed the theory after studying the San Francisco earthquake of 1906.

2219. Scientist to use records of seismic waves to demonstrate the existence of the earth's core and measure its distance from the surface was German-born Beno Gutenberg, in 1914. Gutenberg later migrated to the United States and in 1935 developed with Charles Richter a scale of earthquake magnitude named for Richter but known more accurately as the Richter-Gutenberg scale.

2220. Earthquake of significance in New Zealand occurred on February 3, 1931, on North Island. The largest earthquake in the country's history, it registered a 7.9 on the Richter scale and killed 258 people, 162 of them in Napier. Hastings had 93 deaths and Wairoa had 3. The area was raised by six feet and rivers changed courses.

2221. Statewide construction and design standards for earthquakes in the United States was mandated by the Field Act of 1933, passed by the California legislature. The act came about as a result of the destruction caused by the Long Beach, CA, earthquake on May 10, 1933. The Field Act regulated construction at public schools.

2222. Evidence of water reservoir-induced earthquakes began to emerge in 1936. With the filling of Lake Mead behind Hoover Dam on the Nevada-Arizona line, seismologists registered many shallow focus quakes in an area where they previously had been rare.

2223. Scale to accurately measure earthquake strength was designed by the American seismologist Charles F. Richter. Assisted by Beno Gutenberg, he devised the Richter scale at the Carnegie Institution in Washington, DC, in 1935. The scale measures the energy released by earthquakes, using a magnitude of 1 to no limit, although 10 is often considered the upper level. An increase of one unit represents about 30 times an increase in energy. Measurements are made on a seismograph and adjusted for the distance to the earthquake's epicenter. An earthquake of 3.5 can cause damage, and major quakes usually occur at 5 or above on the scale.

EARTHQUAKES—*continued*

2224. Seismic Tidal Wave Warning System was established at Honolulu, HI, in 1948. A devastating quake-caused tidal wave (or tsunami) that claimed 173 lives at Hilo, HI, in 1946 spurred development of the international warning network.

2225. Earthquake to register a magnitude of 9 or more on the Richter scale occurred from May 21 to 30, 1960, on the southern coast of Chile. The quake registered a 9.5 on the Richter scale, killing some 5,000 people in the sparsely populated area. Seismic waves from the quake reached the Pacific, causing 138 deaths in Japan and 56 in Hawaii.

2226. Use of elevation benchmarks to measure settling or uplifting of the earth occurred after the March 27, 1964, earthquake near Valdez, AK. The benchmarks had been set at Lake Kenai for other purposes by the United States Coast and Geological Survey. While studying the effects of the earthquake in the area, David McCulloch observed the benchmarks on opposite ends of Lake Kenai and discovered the lake had tilted. Benchmarks have now been set around all Alaskan lakes.

2227. Formal earthquake prediction research program in quake-prone Japan was launched in 1965. The research surveys extended to a range of 12,000 miles.

2228. Earthquake data center established by the U.S. government was the National Earthquake Information Center in 1966 in Rockville, MD. It was established as part of the National Ocean Survey of the U.S. Department of Commerce to mitigate the risks of earthquakes. It moved to Boulder, CO, in 1972, and the next year became part of the U.S. Geological Survey before moving in 1974 to Golden, CO. The center determines the location and size of earthquakes worldwide, runs a national seismic database for scientists and the public, and provides 24-hour information to federal and state agencies that provide emergency services.

2229. Recorded evidence of quakes on the Moon was gathered between 1969 and 1976. The Passive Seismic Experiment, part of the Apollo Space Program, used seismographic stations at five lunar sites to record as many as 3,000 moonquakes a year.

2230. Seismographs on Mars were installed in 1976 by *Viking 2*, a U.S. spacecraft. Only one instrument on the *Viking 2* worked. It detected only one wave motion over a 12-month period.

2231. Earthquake in China with its epicenter directly under a city of 1 million people occurred July 28, 1976, at Tangshan in the country's northeast. Registering 8.3 on the Richter scale, it killed about 242,000 people in the area. Damage was also reported in Beijing and the port city of Tianjin. More than 100,000 members of the People's Liberation Army helped to search for survivors and build temporary shelters.

2232. Earthquake in the Soviet Union allowed open media coverage struck on December 7, 1988, in Armenia. News of earthquakes and other natural disasters had been strictly censored by the state until Soviet leader Mikhail Gorbachev introduced *glasnost*, or openness. The Armenian earthquake, recorded at 6.9 on the Richter scale, killed about 25,000 people. Damage extended over 400 square miles of Armenia and part of Turkey. Because of the worldwide coverage, rescue teams were sent from several countries and the survivors received international aid.

2233. Earthquake to be broadcast on live television nationwide was on October 17, 1989, in San Francisco, CA. Some 50 million Americans were watching a pregame telecast of a World Series game in Candlestick Park when the earthquake struck the Bay area, killing 62 people and injuring about 3,000.

2234. Earthquake to occur directly under a large U.S. metropolitan area was on January 17, 1994, in Los Angeles, CA. The quake measured 6.8 on the Richter scale. A total of 61 people died and more than 8,000 were injured; 11 highways and about 11,000 buildings were destroyed. The damage was estimated at up to $25 billion. The mountains surrounding the San Fernando Valley rose by one foot. About 14,000 aftershocks followed the original quake.

2235. Earthquake workshop canceled by an earthquake occurred on January 17, 1995, in Kobe, Japan. Scientists and engineers had gathered in the city for the Urban Earthquake Hazard Reduction Workshop scheduled for the day the surprise quake hit. It registered 7.2 on the Richter scale. More than 6,300 people in the area were killed; about 56,000 buildings were destroyed by the tremor or fire; and an expressway collapsed. The estimated damage was $147 billion.

2236. Major earthquake in the Izmit region of western Turkey, an area that holds nearly half the nation's population, struck on August 17, 1999. The death toll from the quake, at 7.4 on the Richter scale and one of the most powerful of the 20th century, exceeded 14,000.

ECOTOURISM

2237. Tourists in Yosemite National Park arrived in 1855.

2238. Tourists to visit the Glen Canyon–Lake Powell area were George Flavell and Ramón Montez, who in 1896 floated through Glen Canyon simply for the fun of it.

2239. Commercial whale-watching ventures date from 1955 along the California coast. Whale-watching has since become a big business, in more than 40 countries and Antarctica.

2240. Tourists to visit the continent of Antarctica arrived in 1958. The industry grew slowly but steadily, and by the late 1990s some 10,000 people a year were visiting the ice-covered continent.

2241. Ecotourism project started by the Indian government was Project Tiger, begun on April 1, 1973, with 9 tiger reserves; by 1999, there were 23. Many of these reserves have accommodations for tourists or are near national parks. Although the Indian government did not involve the local people at the beginning of the project, by 1991 it saw the advantages of doing so and began to address the problem.

2242. Laws in Kenya against commercial trade in wildlife trophies or products were passed in 1978. This began the Kenyan ecotourism business. Hunters without work became guides and coined the slogan "Come shooting in Kenya with your camera."

2243. Wilderness expeditions operator in Costa Rica was American Michael Kaye. In 1978 he came to Costa Rica to open a river rafting travel agency. Soon after he began to get requests for additional excursions. His Costa Rica Expeditions travel agency is still in business today.

2244. Environmentally conscious codes of ethics for whale-watching operators and participants in North America were developed in the 1980s. Such voluntary codes regulated the number of vessels and the time spent observing a group of whales.

2245. Hospitality ranch in Wyoming was the Masters Ranch started by Dick and Jean Masters in the mid-1980s. Located in Ranchester, WY, it offers tourists an authentic western experience.

2246. Costa Rican ecotourism organization whose purpose was to save an endangered part of the environment was Rara Avis created by Amos Bien, a tropical biologist, and Rainforest, Inc., in 1983. The success of Rara Avis is attributed to the involvement of the local community; today it is a major source of income for the people of Horquetas, Costa Rica.

2247. Use of the term *ecotourism* was in 1983 by Hector Ceballos-Lascurain, a Mexican architect and environmentalist. He coined the word to describe educational travel to relatively undisturbed nature areas. Ceballos-Lascurain later became director of Mexico's Program of International Consultancy on Ecotourism and director of the Ecotourism Conservation Program of the World Conservation Union.

2248. Environmental program to limit ecotourism damage in Nepal was the creation of the Annapurna Conservation Area Project in 1986. Thousands of tourists and trekkers through the fragile mountain environment in this north-central part of the country had ventured off paths, destroying vegetation and littering the landscape. Deforestation was also occurring because of the large amounts of wood needed to heat mountain lodges. Nepal's environmentally responsible program provided seedlings for volunteers to plant, educated lodge managers on alternate heating methods, and encouraged villagers to protect their environment. Ecotourists were also required to use kerosene stoves instead of burning wood.

2249. Year that tourism was Kenya's leading source of foreign currency was 1989. Ecotourism activities such as photo safaris into Kenyan game preserves have flourished since the early 1960s.

2250. Efforts to develop ecotourism in the Lake Kenyir region of Malaysia began in the 1990s. The man-made lake is surrounded by dense jungle and borders a national park.

2251. Substantial growth of ecotourism in Hawaii occurred during the 1990s. Ecotourism focuses on nature study and other outdoor activities such as whale and bird watching with minimal impact on the environment.

2252. Tourism restrictions on Ecuador's Galápagos Islands were imposed in the early 1990s. Travel to the islands had become so heavy that their natural habitats were being disturbed.

ECOTOURISM—*continued*

2253. Virtual whale-watching tours were developed on the World Wide Web in the late 1990s. "The Prince of Wales Whale Watching" site of Victoria, B.C., promised browsers "just a taste of the real thing."

2254. Ecotourism research center in Australia was the International Centre for Ecotourism Research (ICER) opened on August 8, 1993, on Griffith University's Gold Coast campus in Southport, Queensland. Ten other Australian universities participate in the ecotourism studies, which are funded by the Australian government.

2255. Ecotourism appeal coordinated by governments was in March 1997 in Berlin, at the International Tourismus Borse (ITB), known in English as the International Tourism Exchange Berlin. Environmental officials from 20 countries made a joint appeal to the tourism industry to limit ecotourist and other visits to ecologically sensitive areas, to restrict motor-vehicle safaris, and to protect small islands, beaches, and coral reefs.

EDUCATION AND RESEARCH

2256. Professorship of botany in Europe was held by Dr. Francesco Bonafede at the Botanical Garden of Padua, Italy. In 1561, the University of Padua appointed Bonafede to conduct a "Showing of the Simples," a post that authorized him to teach the medicinal properties of the plants being cultivated at the garden.

2257. Botany professor at a college was Adam Kahn, appointed in January 1768 to Philadelphia College in Philadelphia, PA. He taught there for 21 years.

2258. College instruction course in mineralogy was offered at Rhode Island College in Providence, RI, by Dr. Benjamin Waterhouse in 1786.

2259. Use of the term *biology* appeared in Gottfried Reinhold Treviranus's *Biologie oder Philosophie der Lebenden Natur* in 1802. The German naturalist coined the term to describe the study of all living plants and animals.

2260. Continuously operating natural sciences institution in North America was the Academy of Natural Sciences in Philadelphia, PA, opened in 1812. The nonprofit organization includes a museum, a library, and two research divisions in environmental research.

2261. Degree program in natural science in the United States was offered by the Rensselaer School, now called Rensselaer Polytechnic Institute, in Troy, NY, in 1825. The program was designed by botanist and geologist Amos Eaton. Classes were given in biology, chemistry, geology, land surveying, mineralogy, philosophy, and zoology. The first class graduated in May 1826.

2262. Summer research term at a U.S. college was conducted in the summer of 1826 by botanist and geologist Amos Eaton of the Rensselaer School, now called Rensselaer Polytechnic Institute. Eaton hired and equipped canal boats as laboratories and living quarters for a floating field research course in natural sciences along the Erie Canal in New York.

2263. Botany professorship in the United States was established at the University of Michigan in 1840. Botanist Asa Gray was hired to fill the post, but the university's financial problems doomed the program before Gray began teaching. In 1842, he accepted a professorship of natural history at Harvard University, where he became renowned as the leading U.S. plant taxonomist of the 19th century.

2264. Female college professor with the same rights and privileges as her male colleagues was Rebecca Mann Pennell. She began teaching physical geography and natural history at Antioch College in Yellow Springs, OH, on October 5, 1853.

2265. Entomology professor at a U.S. college was Hermann August Hagen, who taught at Harvard University in Cambridge, MA, from 1870 to 1893.

2266. Female awarded the Sage Fellowship in Entomology and Botany was Elizabeth Grotecloss on June 19, 1884. Cornell University in Ithaca, NY, awarded her the $400-per-year fellowship. This award also made her the first woman to receive a graduate fellowship.

2267. Biology course offered at a U.S. college was at Bryn Mawr College in Bryn Mawr, PA, beginning on September 23, 1885, by Edmund Beecher Wilson. There were five weekly lectures plus eight hours of laboratory practice. The course began by examining the structure of typical animals and plants, first of familiar species, then of unicellular organisms. From there, students worked their way upward through the higher animals and plants and concluded the course by studying the embryological development of the chick. An advanced class studied animal morphology. Lectures on specific phases of biology had, however, been given earlier.

2268. Woman to win a Nobel Prize was the Polish-born French scientist Marie Curie. In 1903, she and her husband Pierre shared the Nobel Prize in Physics with French Physicist Antoine Henri Becquerel for the discovery and research of radioactivity. In 1911 Madame Curie won a second Nobel Prize, for chemistry, for isolating radium.

2269. U.S. institute for study and research on oceans was the Scripps Institution of Oceanography of the University of California, in La Jolla. From its small beginnnings as the San Diego Marine Biological Station, established by William E. Ritter in 1903, it became the Scripps Institution for Biological Research in 1912 and assumed its present name in 1925.

2270. Research conducted in the United States by a woman in the field of occupational health was by Dr. Alice Hamilton and was published in *Charities and the Commons* in 1908.

2271. Climatology professor at a U.S. college was Robert DeCourcy Ward, appointed in 1910 by Harvard University in Cambridge, MA.

2272. Statewide survey of industrial poisons was conducted by Dr. Alice Hamilton for the state of Illinois in 1910.

2273. Society of professional ecologists was the British Ecological Society. The organization was founded by English botanist Arthur George Tansley in 1913. Since then the London-based group has become an international independent network of educators, researchers, and environmental specialists. The society promotes nature conservation, sound environmental management, and sustainable development through ecological research, publications, and conferences.

2274. Society of professional ecologists in the United States was the Ecological Society of America, founded in 1915 in Washington, DC, to promote communication among ecologists and influence public policy. Since 1991, the organization's programs have focused on the ecological aspects of global change, the ecology and conservation of biodiversity, and strategies for a sustainable biosphere.

2275. Female professor at Harvard was appointed to the medical school in 1919. Dr. Alice Hamilton was a pioneer in occupational health and the study of industrial poisons.

2276. Textbook on industrial poisons in the United States was *Industrial Poisons in the United States* (1925) by Dr. Alice Hamilton. Dr. Hamilton used the book in her classes at Harvard University.

2277. School of conservation was established at the University of Michigan in Ann Arbor, in 1927.

2278. School devoted exclusively to training wildlife conservation officers was the Game Conservation Training School, which opened near Brockway, PA, on July 7, 1932. Its graduates, who studied subjects including tree identification and predator control, were expected to become state wildlife officers. In 1948 the school was renamed and continues to be known as the Ross Leffler School of Conservation, after a prominent Pennsylvania conservationist. The school was moved from Brockway to Harrisburg in 1986.

2279. Funding for multisector wildlife research began in 1935 with the Cooperative Wildlife Research Unit Program in which state wildlife agencies, the federal government, and the private sector provided funding to state land-grant colleges and universities to undertake wildlife research.

2280. Zoologist to win a Nobel Prize was Hans Spemann, who won the 1935 prize for Physiology or Medicine while he was Professor of Zoology at the University of Freiburg-im-Breisgau in Germany. Spemann, whose research concentrated on experimental embryology, won for his discovery of the organizer effect in embryonic development. He also performed the first nuclear transfer cloning and conducted pioneering work in microsurgery.

2281. Wildlife dioramas exhibited in a museum that were accurate were conceived by American naturalist, sculptor, photographer, and taxidermist Carl Ethan Akeley. His methods of mounting zoological specimens brought new anatomical verisimilitude to taxidermy, while his expeditions in Africa helped him to create dioramas that displayed the animals in environmentally correct settings. The Akeley Hall of African Mammals, which opened in 1936 in New York's American Museum of Natural History, is named for him.

2282. United Nations agency to develop environmental education and research programs was the United Nations Educational, Scientific, and Cultural Organization (UNESCO), created in London, on November 16, 1945. Its natural sciences sector includes programs in earth sciences, sustainable development, coastal ecology, hydrology, oceanography, and biodiversity protection. UNESCO's Education Information Service also began maintaining an International Directory of Environmental Education Institutions in 1971.

EDUCATION AND RESEARCH—*continued*

2283. U.S. National Science Foundation (NSF) was created as an independent government agency by the National Science Foundation Act, signed by President Harry S. Truman on May 10, 1950. The NSF is responsible for awarding thousands of government grants to fund research and education in scientific and technological fields. Programs of particular environmental relevance include biology, geosciences, and polar research. The NSF is based in Arlington, VA.

2284. Female pioneer in the field of rubber recycling was Austro-Hungarian chemist Desirée Le Beau. In 1958, when working for Midwest Rubber Recycling Company in Illinois, Le Beau patented a railroad tie pad made from recycled rubber.

2285. Use of the term *bionics* was by U.S. scientist J. E. Steel in 1960. He coined the word to describe the study of living organisms as a model for man-made devices.

2286. Science center originally designed for a world's fair in the United States was the Pacific Science Center, opened in 1963. It occupies part of the building complex originally designed as the United States Science Pavilion for the 1962 Seattle World's Fair.

2287. Regulations on animal research in the United States came into being when President Lyndon Johnson signed the Animal Welfare Act (Public Law 89-544) on August 24, 1966. The legislation regulated the sale, transport, and handling of live dogs, cats, monkeys (nonhuman primate mammals), guinea pigs, hamsters, and rabbits intended to be used for research or for experimentation, and established humane treatment and transportation of these animals. However, the act set no specific standards and did not include enforcement provisions.

2288. Research grant program administered by the U.S. Department of the Interior was authorized on October 15, 1966, by the Research Grants Act. The law allowed the Secretary of the Interior to enter into contracts with educational institutions, public or private organizations, and individuals to conduct scientific or technological research.

2289. U.S. National Sea Grant Program was signed into law in 1966 by President Lyndon Johnson. The program, which became part of the National Oceanic and Atmospheric Administration in 1970, dispenses grants to qualified colleges and universities for scientific, technological, commercial, and public education research involving coastal, ocean, and Great Lakes resources.

2290. Conservation legal organization was the Natural Resources Defense Council (NRDC), incorporated in February 1970 in New York City. The main purpose of the NRDC is to utilize and research American laws and the legal system to protect the nation's environment. NRDC monitors the activities of federal regulatory agencies charged with environmental duties.

2291. Conference on Third World environmental problems was held in Fournex, Switzerland, in 1971. A panel of experts from both developed and developing countries agreed that many serious environmental problems of Third World countries were often the result of extreme poverty and a lack of economic and social development.

2292. International registry of data on chemicals in the environment was proposed at the United Nations Conference on the Human Environment, held in Stockholm, in 1972. The request resulted in the establishment of the International Register of Potentially Toxic Chemicals (IRPTC), a computerized data bank for the exchange of information on production and consumption of these substances, in 1976.

2293. *Landsat* satellite was launched by the United States from Cape Kennedy (Cape Canaveral), FL, in 1972. Three more *Landsats* were launched in 1975, 1978, and 1982. The satellites orbit the Earth and transmit spectral images to receiving stations on the ground. The images are used to study mineral, agricultural, and other resources.

2294. Proposed concept of entitlement assistance for environmental conservation was laid out at the United Nations Conference on the Human Environment in Stockholm in 1972. At this conference, the groundwork was developed for Third World countries to be compensated if they declined development options to preserve environmental resources that are of special interest to the world.

2295. Plant to use the pollution-to-homes process was Cholla I Station, which began operation in Arizona in 1973. The process uses wastes from scrubbers to create drywall or sheetrock.

2296. Ice meteor to be subject to intensive study dropped to the ground in Manchester, England, on April 2, 1973. A lightning observer for the Electrical Research Association picked up a large lump of the meteor and stored it in a freezer, preserving it for study.

2297. Ocean Thermal Energy Conversion (OTEC) laboratory established by a U.S. state was the Natural Energy Laboratory of Hawaii (NELH). The laboratory was funded by the state in 1974 for research and development of OTEC and other natural energy sources. NELH was built at Keahole Point on the island of Hawaii. The establishment of an adjacent "eco-industrial park" whose businesses use energy created at NELH resulted in the facility changing its name to the Natural Energy Laboratory of Hawaii Authority (NELHA) in 1984.

2298. Solar energy research funded by the U.S. Congress came about when it passed the Solar Energy Research, Development, and Demonstration Act in 1974 to accelerate development of renewable energy sources that would not pollute the environment. President Gerald R. Ford signed the legislation on October 26, 1974. It established the Solar Energy Coordination and Management Project and the Solar Energy Research Institute.

2299. Industrial-scale selective catalytic reduction system was installed at the Chiba works of Kawasaki Steel in Japan in 1977. This technology reduced air pollution at the steel mill and has been sold to industries and power plants worldwide.

2300. Intergovernmental conference on environmental education was held in Tbilisi, Georgia, in the Soviet Union from October 14 to 27, 1977. The conference was organized by the United Nations Educational, Scientific, and Cultural Organization (UNESCO) in cooperation with the UN Environment Programme (UNEP). The meeting unanimously approved the Tbilisi Declaration, which set forth goals recognizing education as necessary for global environmental preservation and sustainable development.

2301. Cool water or integrated gasification combined-cycle utility began operation in the mid-1980s in Dagget, CA. The advent of cheaper forms of energy production, such as natural gas, caused this plant to close.

2302. Long-term ecological research network in the United States was established in 1980 by the National Science Foundation to support ongoing research programs. Areas of scientific inquiry have included forestry, prairies, deserts, urban and rural ecosystems, and Antarctica. The network's main office is located at the University of New Mexico's Biology Department in Albuquerque.

2303. Woman to graduate from the Pennsylvania School of Conservation was Cheryl A. Stauffer, on March 13, 1982.

2304. Office of Environmental Education in the United States was established as part of the Environmental Protection Agency (EPA) after President George Bush signed the new office into law on November 16, 1990. It was created to oversee grants for environmental education and training programs. The legislation also authorized private individuals to donate money to a nonprofit foundation furthering EPA educational programs.

2305. Education center devoted exclusively to genetics and biotechnology was the DNA Learning Center, established by the Cold Spring Harbor Laboratory on Long Island, NY, in 1992. The learning center conducts public programs in molecular genetics education for students ranging from elementary school to college faculty.

2306. Supercomputer used solely for environmental research was at the U.S. Environmental Protection Agency's National Environmental Supercomputing Center (NESC) in Research Triangle Park, NC. Since beginning operations in October 1992, NESC computers have been used for global environmental research and educational programs.

2307. License plate in Massachusetts designed to benefit environmental education programs was introduced by the state's Registry of Motor Vehicles in 1994. Use of the plates, which picture the endangered right whale and roseate terns, raises millions of dollars annually for the Massachusetts Environmental Trust, a philanthropic body created by the state legislature in 1988. The trust disperses the proceeds through grants to educational and conservation programs that benefit the state's environment, particularly its waterways.

2308. Annual report on the environment and American foreign policy was *Environmental Diplomacy: The Environment and U.S. Foreign Policy*, issued by the U.S. Department of State in 1997.

2309. Doctorate program in ecological economics in the United States was established in 1997 at Rensselaer Polytechnic Institute in Troy, NY. The increasingly popular interdisciplinary field examines the interrelationships of economic and environmental issues.

EDUCATION AND RESEARCH—EXPOSITIONS

2310. World's fair with an environmental theme was Expo '74, The International Exposition on the Environment, held in Spokane, WA, from May 4 to November 3, 1974.

EDUCATION AND RESEARCH—EXPOSITIONS—*continued*

2311. World's fair with an ocean theme to be sanctioned by the Bureau of International Expositions (BIE) was the International Ocean Exposition, held on Motobu Peninsula in Okinawa, Japan between July 20, 1975, and January 18, 1976.

2312. World's fair in the American Southeast to be sanctioned by the Bureau of International Expositions (BIE) was the Knoxville International Energy Exposition in Knoxville, TN, between May 1 and October 31, 1982.

2313. World's fair to select energy as its theme was the Knoxville International Energy Exposition in Knoxville, TN, between May 1 and October 31, 1982.

2314. World's fair to select rivers as its theme was the Louisiana World Exposition held in New Orleans, LA, between May 12 and November 11, 1984. The theme was "The World of Rivers: Fresh Water as the Source of Life."

2315. International horticultural show in Asia was the International Garden and Greenery Exposition held in Osaka, Japan, from April 1 to September 30, 1990.

2316. Officially sanctioned international horticultural fair in the United States was AmeriFlora '92, held in Columbus, OH, from April 3 to October 12, 1992, having been approved by the Association of International Horticulture Producers (AIPH) in 1987. AIPH is recognized by the Bureau of International Expositions (BIE) as the sanctioning body for international horticultural exhibitions.

2317. Exposition with an environmental theme to be held in Korea was Expo '93, the Taejon International Exposition, held from August 7 to November 7, 1993, in Taejon, South Korea. Its central theme was "The Challenge of a New Road to Development." One of its subthemes was "Towards an Improved Use and Recycling of Resource," which the U.S. Pavilion incorporated into its own exhibition design on recycling and the environment.

EXTINCT AND ENDANGERED SPECIES

2318. Laws protecting the vicuña were instituted during the Middle Ages by the rulers of the Incan Empire, whose clothing was woven from the animal's fine wool. Ordinary Incans were prohibited from hunting or killing vicuña. Although the animal has been in danger of extinction at times since the collapse of the Incan Empire, it is now protected by the Peruvian government.

2319. Endangered species laws in the Western Hemisphere were issued by the government of Bermuda in the early 1600s. Despite acts in 1616 and 1621 protecting the cahow or Bermuda petrel, the birds were considered to be extinct for nearly 300 years until a few were discovered alive in the 20th century. Today the cahow is still listed as an endangered species.

2320. State in the United States to protect an endangered plant species by law was Connecticut. On July 8, 1869, the state legislature passed a law prohibiting the taking of any creeping fern from another person's property. Possible penalties included a seven-dollar fine or a jail sentence.

2321. Law in Vermont to protect endangered plants was passed by the state legislature in 1921. By restricting the taking of listed plants to a single plant or cutting annually on personal property, the law intended to prevent commercial trafficking in endangered plants. Violators faced an individual fine for each plant taken.

2322. Law in Massachusetts to protect an endangered plant species was a law prohibiting the picking or injuring of mayflowers. Passed in 1925, it made violators liable to a $50 fine.

2323. Species of bird to be resuscitated from a single live specimen was the Laysan duck. In 1930, the population of Laysan ducks had been reduced to a single pregnant female. In 16 days of searching tiny Laysan Island, a biologist found a single pair of the ducks, whose nest contained only eggs that had been punctured by a predator. The drake subsequently disappeared, but the female had enough remaining semen in her oviduct to lay another clutch of fertilized eggs. By the 1980s, the wild population of Laysan ducks had reached approximately 500, and a captive population was also well established.

2324. Program to stop extinction of trumpeter swans was established in the United States in 1935. Swans were transplanted to wildlife refuges across the country. By 1972 breeding populations were established in Oregon, Nevada, and South Dakota.

2325. U.S. National Wildlife Reserve created specifically for Kirtland's warbler was established in 1935 in the Upper Peninsula of Michigan. The 6,500-acre reserve set aside to protect the endangered bird now lies within the much larger woodland and lake ecosystem of the Seney National Wildlife Reserve.

2326. Law in Rhode Island to protect endangered plants was passed by the state legislature in 1939 to prohibit the taking of mayflowers, mountain laurel, flowering dogwood, and other dwindling species. Violators faced fines of up to $50.

2327. U.S. National Wildlife Reserve created specifically for Key deer was established in 1957 to stop overhunting and protect habitat critical to the survival of the endangered small deer. The islands and pineland of the National Key Deer Refuge in Big Pine Key, FL, are now surrounded by the Great White Heron National Wildlife Refuge.

2328. U.S. National Wildlife Refuge created specifically for the dusky seaside sparrow was St. Johns National Wildlife Refuge, established in 1963 on Merritt's Island, FL. Despite the refuge and a captive breeding program, however, the bird was extinct by 1987; habitat losses and pesticides were the principal causes. The dusky seaside sparrow was removed from the Endangered Species List in 1990.

2329. U.S. National Wildlife Refuge created specifically for Attwater prairie chickens was established in 1972 near Eagle Lake, TX. The prairie grassland habitat of the Attwater Prairie Chicken National Wildlife Reserve in southeastern Texas is critical to the survival of the endangered birds, who have suffered due to human development and overgrazing.

2330. U.S. National Wildlife Refuge created specifically for Columbian white-tailed deer was established in 1972 on the Columbia River flood plain straddling the border between Washington and Oregon. The flood plain and islands are a protected habitat for the species, which was listed as endangered on March 11, 1967. The Columbian White-Tailed Deer National Wildlife Refuge was renamed in honor of Washington Congresswoman Julia Butler Hansen in 1988.

2331. U.S. National Wildlife Refuge created specifically for Mississippi sandhill cranes was established in 1975 in southeastern Mississippi to provide a secure habitat for the endangered birds. The protected area of Mississippi Sandhill Crane National Wildlife Refuge counters population losses due to human development, commercial deforestation, and pesticides. The cranes had been listed as endangered on June 4, 1973.

2332. U.S. National Wildlife Refuge created specifically for wintering bald eagles was Bear Valley National Wildlife Refuge. The forested reserve near Worden, OR, was established in 1978 to provide an undisturbed winter roosting habitat for the endangered birds. Although bald eagles were later relisted as "threatened" rather than "endangered," the area remains closed to the public except for limited deer hunting.

2333. Lawsuit on behalf of an endangered species to be decided by the U.S. Supreme Court was the case of *Tennessee Valley Authority (TVA)* v. *Hill.* In its landmark decision on June 15, 1978, the court halted completion of the multi-million dollar Tellico Dam on the Little Tennessee River, citing the Endangered Species Act and holding that the dam would deprive a listed fish, the snail darter, of habitat necessary for its survival. The TVA eventually won the right to complete the dam through a controversial law passed by Congress. The snail darter survived after being successfully introduced into a different habitat.

2334. U.S. National Wildlife Refuge created specifically for watercress darters was established in 1980 near Bessemer, AL. Industrial and residential development in the Birmingham area had already diminished the tiny fish's critical habitat of freshwater ponds sufficiently for it to be listed as "endangered" on October 13, 1970. Watercress Darter National Wildlife Refuge consists of a few protected acres and a small pond where the last known examples of the species live.

2335. Species Survival Plan (SSP) was formulated by the American Zoo and Aquarium Association (AZA) in Wheeling, WV, in 1981 to help prevent the extinction of threatened and endangered species. The cooperative program combines the resources of nearly 200 North American zoos and aquariums in efforts to breed and study selected species, increase public awareness of conservation issues, and implement wildlife reintroductions when appropriate.

2336. Recovery plan approved by the U.S. Fish and Wildlife Service for a plant focused on the roundleaf birch, a very rare tree. The species had been discovered in southwestern Virginia by W. W. Ashe in 1914, but as no one was able to locate the stand of trees he had described it was subsequently listed as extinct. However, in 1975 Douglas Ogle and Peter Mazzeo found 41 of the birches growing along Cressy Creek, VA. Depredations by vandals

EXTINCT AND ENDANGERED SPECIES—*continued*

and collectors soon reduced this population by half, and in 1978 the birch was declared endangered. A recovery plan approved by the Fish and Wildlife Service in 1982 resulted in new plantations of the tree, which have flourished.

2337. Habitat Conservation Plan (HCP) in the United States was implemented in 1983 for land on San Bruno Mountain near San Francisco, CA. Prior to the use of HCPs, privately-owned land could not be developed if any possible harm or disturbance to species listed under the Endangered Species Act resulted. HCPs established the use of "incidental take" permits, which allowed development of such land on the condition that the developers preserve similar habitats for the listed species nearby. Although the first HCP was considered a success in preserving the rights of landowners and the survival of threatened mission blue and San Francisco silverspot butterflies at San Bruno Mountain, growing use of HCPs has made the program increasingly controversial.

2338. U.S. National Wildlife Refuge created specifically for West Indian manatees opened at Crystal River, FL, on August 1, 1983. The numerous islands of the Crystal River National Wildlife Refuge provide a secure winter habitat for much of Florida's population of endangered manatees, a popular tourist and educational attraction.

2339. U.S. National Wildlife Refuge created specifically for endangered Nevada fish species was established in June 1984 in Amargosa Valley in southern Nevada. The desert and springs of Ash Meadows National Wildlife Refuge protect 24 fish and plant species unique to Nevada, including listed endangered varieties of pupfish, dace, milk vetch, and other endemic wildlife.

2340. African elephant protection law in the United States was passed by the U.S. Congress on October 7, 1988, in response to the rapid decline of African elephant populations. The African Elephant Conservation Act sought to monitor conservation programs abroad, and effectively banned the importation of ivory from any country that failed to maintain an adequate elephant conservation program. The act also established an African Elephant Conservation Fund to provide financial aid to support the protection, conservation, research, and management of elephants in African countries.

2341. U.S. National Wildlife Refuge created specifically for Iowa pleistocene snails was established in 1989, protecting the endangered snails on separate parcels of land scattered across three northeastern Iowa counties. The Driftless Area National Wildlife Refuge is also one of the few remaining habitats of the northern monkshood flower, which is listed as threatened.

2342. Law in the United States to specifically ban importation of endangered exotic birds was signed by President George Bush on October 23, 1992. Its intent was to stop the depletion of endangered wild exotic bird populations abroad and encourage U.S. pet suppliers to sell only exotic birds bred in captivity.

2343. Canadian law to protect all endangered wildlife and flora was the Wild Animal and Plant Protection and Regulation of International and Interprovincial Trade Act (WAPPRIITA). Before the act was passed, Canada protected endangered species as a signatory of the Convention on International Trade in Endangered Species of Wild Fauna and Flora (CITES). WAPPRIITA extended protection to game species and all wild animals and plants not already listed as threatened or endangered by CITES. The new law set substantial fines and prison terms for transporting listed species across provincial or territorial lines. WAPPRIITA was enacted December 17, 1992, and came into force on May 4, 1996.

2344. Fly to be designated as an endangered species in the United States was the Delhi Sands fly in 1993. The small orange-brown insect is found only in the Delhi Sands dunes 60 miles outside of Los Angeles. It is the only fly listed as an endangered species.

2345. Trade sanctions imposed by the United States to protect endangered wildlife were announced by President Bill Clinton on April 4, 1994. Responding to Taiwan's refusal to stop the sale of tiger bones, rhino horns, and other products made from endangered species, the United States banned the importation of any wildlife products from Taiwan.

2346. Rhinoceros and tiger protection law was passed by Congress on October 22, 1994. The Rhinoceros and Tiger Conservation Act provided financial aid for conservation programs in countries whose activities affected endangered rhino and tiger populations. When the law was reauthorized by President Bill Clinton on October 30, 1998, it included the Rhinoceros and Tiger Product Labeling Act, banning the sale, import, and export of products containing or claiming to contain substances derived from any species of either animal.

2347. Genetic recovery program for wildlife in the United States was initiated in 1995, when the Fish and Wildlife Service approved a plan for introducing Texas cougars into southern Florida, the habitat of a closely related cat, the endangered Florida panther. That spring eight female cougars were released in panther territory, and by 1997 at least nine kits had been borne. The Florida panther, one of America's rarest mammals, is threatened not only by urban sprawl but by the genetic consequences of prolonged inbreeding, which have impaired its resistance to disease and its ability to reproduce. Although "genetic enhancement" is a controversial strategy, many biologists supported it in this case as the best hope for the animal's survival. The National Parks Conservation Association helped finance the project.

2348. Environmental lawsuit to challenge the United States under the North American Free Trade Agreement (NAFTA) was decided in U.S. District Court in Phoenix, AZ on May 13, 1995. The nonprofit environmental law group Earthlaw charged that the U.S. Fish and Wildlife Service (FWS) was prevented from enforcing the Endangered Species Act of 1973 by the so-called "Hutchison Rider." The rider, which had been inserted in an unrelated defense appropriations bill by Senator Kay Bailey Hutchison of Texas, suspended the FWS's power to list species as endangered or threatened and reduced funding for the FWS endangered species program budget. Earthlaw successfully argued that the FWS should retain the power to designate endangered wildlife habitats, under NAFTA provisions pledging to protect the environment along the U.S.–Mexico border.

2349. Jaguar Conservation Team in Arizona was established in April 1997. The group is composed of state, federal, and local governments, private individuals, and other groups in the United States and Mexico that promote jaguar conservation through research, species data gathering, and various conservation programs. The jaguar was listed as endangered under the U.S. Endangered Species Act on August 21, 1997.

2350. National Threatened Species Day in Tasmania was held on September 7, 1998. The date was selected because it was the anniversary of the death of the last known Tasmanian tiger or thylacine, which died in the Hobart Zoo in 1937.

2351. Building developments halted to protect an endangered species were several multimillion-dollar projects planned for the dunes in Southern California in 1999. The dunes are the only known breeding ground for the Delhi Sands fly, a orange-brown insect designated an endangered species in 1993. The U.S. Fish and Wildlife Service halted all development in the dunes until it can be determined how many flies are left in the area and how to best protect them. The cities of Fontana and Colton each had to halt building projects with budgets of nearly $500 million. Smaller building projects in the cities of Rancho Cucamonga, Ontario, Rialto, and Hemet were also halted to protect the flies.

2352. Regional endangered-species plan to protect an aquatic ecosystem was announced on May 18, 1999, by Secretary of the Interior Bruce Babbitt. The plan, designed to protect a wide variety of species by restoring much of the Florida Everglades, would engage local land developers and governments with federal agencies in a shared process of designating protected areas within the Everglades as reserved for endangered species.

2353. Systematic survey of the endangered Asiatic cheetah commenced in the fall of 2001 when biologists George B. Schaller and Timothy O'Brien journeyed to eastern Iran to report on the remnant population of this animal and the conditions threatening its survival, in a joint project of the World Conservation Union, the Wildlife Conservation Society, and Iran's Department of Environment. The biologists did not see any Asiatic cheetahs—the worldwide population is estimated at fewer than 60 individuals—but did capture the image of one in a photograph taken by an unmanned, heat-and-motion-sensitive camera. Threats to the species's survival in Iran include an expanding human population, increased truck traffic, drought, and the general disruption resulting from the war in neighboring Afghanistan.

EXTINCT AND ENDANGERED SPECIES—EXTINCTIONS

2354. Mass extinction of species is known as the Ordovician extinction. It is thought to have occurred 440 million years ago and destroyed most of the planet's aquatic organisms. Geological evidence reveals that four subsequent mass extinctions have occurred—the Devonian, Permian, Triassic, and Cretaceous. During the 1990s, many biologists pointed to profound species losses due to pollution, global warming, habitat reduction, and other factors as evidence that a sixth mass extinction is occurring, the first to be caused by humans.

EXTINCT AND ENDANGERED SPECIES—EXTINCTIONS—*continued*

2355. Extinctions of North American animals caused by humans occurred circa 9000 B.C. when settlers from Asia moved south out of Alaska. Two-thirds of the large mammals present when humans first arrived were driven to extinction, partly by being pushed out of their accustomed climate. Three species of elephants, six of giant edentates (the family that includes sloths and anteaters), fifteen of ungulates (hoofed mammals), and a large number of giant rodents and carnivores were extinguished.

2356. Dodos encountered by Europeans were discovered by Portuguese sailors in 1518. The dodos inhabited Mauritius, in the Mascarene Islands east of Madagascar. The birds, which had no natural fear of humans or other predators, were slaughtered by later Dutch settlers for meat and amusement. Dogs and other species introduced by humans helped kill the remaining dodos, which became extinct by 1681.

2357. Spectacled cormorant description was recorded by German naturalist Georg Steller during the disastrous Russian-sponsored Bering expedition of 1741. Like his other discovery, Steller's sea cow, the spectacled cormorant was quickly extirpated. The large Arctic seabirds were hunted to extinction for food by the mid-19th century.

2358. Animal hunted to extinction in North America was Steller's sea cow. It was named for Georg Steller, a German naturalist whose memoirs described how the starving Bering expedition of 1741 relied on sea cow meat to survive. Subsequent fur hunters also used the large marine mammal for meat and had hunted Steller's sea cow to extinction by 1768, only 27 years after its discovery by Europeans.

2359. Scientist to establish the fact of extinction was French biologist Georges Cuvier, whose comparative studies of vertebrate fossils in 1796 proved that some life forms were no longer to be found alive on Earth. Cuvier also founded the sciences of comparative anatomy and vertebrate paleontology.

2360. Report of the extinction of the Bonin Islands thrush was in 1828, when the last four specimens were collected on the small Japanese island of Chichi Shima. The bird was probably eradicated by rats that had escaped from whaling ships.

2361. Report of the death of the last great auk in the British Isles was in 1834, when an auk captured alive in Waterford harbor in southeast Ireland was beaten to death by local people who believed it to be a witch. Another report indicates that this last specimen was immediately collected by a museum. A subsequent, similar account came from St. Kilda in the Western Isles in 1840, where an islander reportedly caught the last specimen and, after keeping it alive for three days, killed it when he suspected it of being a witch responsible for a recent storm.

2362. Scientific description of a moa was written by British anatomist Richard Owen and published in the journals of the Zoological Society of London in 1839. From a single bone, Owen correctly theorized that the extinct New Zealand creature was a large flightless bird. Subsequent discoveries of moa bones revealed that the birds were capable of growing as tall as seven feet.

2363. Use of the word *dinosaur* appeared in the work of British anatomist Richard Owen in 1841. Owen created the term by combining Greek words meaning "terrible lizard" to describe extinct Mesozoic Saurian reptiles.

2364. Report of extinction of the great auk was made by three fishermen—Jon Brandsson, Sigourer Isleffson, and Ketil Ketilsson—on either June 3 or June 4, 1844, after they clubbed the last incubating pair to death and smashed the pair's single egg on Eldey Rock, an island near Iceland. They killed the auks to sell their carcasses to a bird collector. Over the years the species had been virtually extirpated by hunters for food, feathers, and oil, and a major breeding site had been destroyed in 1830 by a volcanic eruption. In 1852, however, Colonel Drummond-Hay, later to become the first president of the British Ornithologists' Union, claimed to have seen a great auk off the Grand Banks of Newfoundland, but this sighting was never confirmed.

2365. Report of the extinction of the quagga was on August 12, 1883. The last surviving example of the once plentiful South African horse died at the Amsterdam Zoo in the Netherlands.

2366. Species of bird to be eliminated by a single cat was the Stephen Island wren of New Zealand. The seminocturnal wrens were observed only two times, in 1894. Within the year, the entire species had been eliminated by the island's lighthouse keeper's cat, who brought home the remains of twenty wrens.

2367. Carolina parakeet extinction report came in September 1914, when the reported last survivor died at the Cincinnati Zoo. Another account has it that the last two birds, a pair named Lady Jane and Incas, who had lived at the zoo for more than 30 years, expired in 1917 and 1918, respectively. Once plentiful in the southeastern states, Carolina parakeets became extinct due to habitat loss and extirpation as an agricultural pest.

2368. Extinction caused by humans of an animal endemic to North America in modern times was that of the passenger pigeon (*Ectopistes migratorius*). The last known specimen, named Martha, died in the Cincinnati Zoo on September 1, 1914. Until the mid-19th century the passenger pigeon was the most numerous bird on Earth. It was hunted extensively for meat and had vanished from the wild by 1890.

2369. Report of the extinction of the paradise bird of Australia was in November 1927, when the last documented sighting occurred in southern Queensland.

2370. Heath hen extinction report was made after the last sighting of the bird on March 11, 1932. Once common throughout the eastern United States, the heath hen began to dwindle in colonial times due to overhunting and habitat loss. By the late 19th century, the only surviving population lived on the Massachusetts island of Martha's Vineyard, where natural disasters and predatory birds contributed to the heath hen's eventual disappearance.

2371. Amphibian to become extinct in U.S. history was the Vegas Valley leopard frog. The frog existed only in Clark County, NV, where the growth of the Las Vegas area hastened the species' extinction. Freshwater springs were capped or diverted for the city's use, destroying the frogs' habitat. They were extinct by 1942.

2372. Caribbean monk seal extinction report occurred after the last seal was seen in 1952. Attempts to find surviving populations in the Gulf of Mexico and the Caribbean Sea have been unsuccessful, and the species appears to have been hunted to extinction. The Caribbean monk seal remains on the U.S. Endangered Species List.

2373. Species of bird to be eliminated by volcanic eruption was the San Benedicto rock wren. The entire population was vaporized when a volcano on the island on which the birds lived, Isla San Benedicto in Mexico, erupted in 1952.

2374. Reports of the virtual disappearance of the mayfly in Lake Erie were in 1953. Pollution killed off the mayfly, a chief source of food for fish.

2375. Mounted bird specimen to sell for more than £9,000 sterling was a well-preserved great auk, which the Natural History Museum of Iceland purchased on March 4, 1971.

2376. Theory that an asteroid caused the extinction of dinosaurs was developed in 1980 by American geologists Walter and Louis Alvarez and their colleagues Chris McKee, Frank Asaro, and Helen Michel. The Alvarezes researched an unusually thick layer of iridium found globally between sediments, separating the Mesozoic and Cenozoic periods. The theory proposed that debris in the atmosphere resulting from the impact of an asteroid plunged the Earth into darkness long enough to kill vegetation, thus disrupting the food chain and hastening mass extinctions. The scenario was published in the June 6, 1980, issue of the journal *Science* in Berkeley, CA.

2377. Reports of possible extinction of the dusky seaside sparrow in Florida came in the early 1980s when only a few males were listed living in captivity. The last known specimen died in 1987. Losses of habitat and possible toxins from pesticides are thought to be the cause.

2378. Fish removed from the U.S. Endangered and Threatened Species List was the Tecopa pupfish. It was delisted on January 15, 1982, due to extinction.

2379. Bird removed from the U.S. Endangered and Threatened Species List due to extinction was the Santa Barbara song sparrow. It was declared extinct and removed from the list on October 12, 1983.

2380. Mollusk removed from the U.S. Endangered and Threatened Species List was the Sampson's pearly mussel. It was removed from the list on January 9, 1984, due to extinction.

2381. Species of bird to fall victim to space exploration was the dusky seaside sparrow, a highly specialized bird, which in 1987 finally succumbed as a result of the space race. The construction of the NASA center in Cape Canaveral, FL, entailed extensive diking of Merritt Island to control the mosquito population. This riparian redistribution ended tidal flooding of the island marshes, depriving the sparrow of its habitat.

EXTINCT AND ENDANGERED SPECIES—EXTINCTIONS—*continued*

2382. Extant species on the U.S. Endangered and Threatened Species List to become extinct was the Amistad gambusia. The Texas fish was the sixth listed species to be declared extinct. After the initial listing, however, the first five extinctions were declared to have occurred before the Endangered Species Act was passed in 1973. The Amistad gambusia was delisted on December 4, 1987.

EXTINCT AND ENDANGERED SPECIES—LISTS

2383. Short-tailed albatross protection law was instituted by Japan in 1933. The species, which breeds on the volcanic island of Torishima, was once common in the Pacific. In the forty years before the hunting ban was instituted, however, the birds were hunted almost to extinction for their plumes; they remain endangered.

2384. Federal law to protect the bald eagle was the Bald Eagle Protection Act of 1940. The law made it illegal for any person within the United States to "possess, sell, purchase, barter, offer to sell, transport, export, or import, at any time or in any manner, any bald eagle or any golden eagle, alive or dead, or any part, nest, or egg." The Secretary of the Interior could permit exceptions for scientific, exhibition, or religious purposes.

2385. Endangered species list by the U.S. government was issued by the Department of the Interior in 1966. The list included 78 species of rare and endangered plants, reptiles, birds, and mammals.

2386. Bat listing under the U.S. Endangered Species Act was the Indiana bat. The small species, which can be found throughout the eastern United States, was the only bat included on the first official U.S. endangered species list on March 11, 1967. Despite this protection, the bats' numbers have declined severely since their initial listing, due primarily to vandalism and other disturbances by humans.

2387. Bear listing under the U.S. Endangered Species Act was the grizzly bear. Grizzlies in the lower 48 states were the only bears included on the first official U.S. Endangered Species List on March 11, 1967. Grizzly populations stabilized sufficiently by the 1990s, when their proposed delisting from "threatened" status became a controversial political issue.

2388. Darter listed under the U.S. Endangered Species Act was the Maryland darter, which appeared on the first Endangered Species Act (ESA) list on March 11, 1967. Man-made dams and pollution are considered the primary threats to the extremely rare fish, which is thought to now survive only in Maryland's Harford County. Less well known than the snail darter, whose protection under the ESA delayed the completion of Tennessee's Tellico Dam in the 1970s, the Maryland darter was only one of 17 types of darters listed as endangered or threatened by the late 1990s.

2389. Deer listings under the U.S. Endangered Species Act were the Columbian white-tailed deer and the Key deer. Both the Columbian deer, whose range is in the Pacific Northwest, and Florida's Key deer were included on the first official U.S. endangered species list on March 11, 1967. Despite federal protection and recovery efforts, neither species has recovered significantly since.

2390. Fox listing under the U.S. Endangered Species Act was the San Joaquin kit fox, which was included on the first official U.S. endangered species list on March 11, 1967. The small nocturnal species is confined to southern California, where it lives in arid portions of Kern County and on U.S. military bases.

2391. Parrot listed under the U.S. Endangered Species Act was the Puerto Rican parrot, which was included on the first ESA list on March 11, 1967. Habitat loss due to human development and other factors caused the once common green parrots to be reduced to a total population of about a dozen by the 1970s.

2392. Snake listed under the U.S. Endangered Species Act was the San Francisco garter snake. Human development of the snake's native California habitats sufficiently reduced its numbers for it to be listed on the first ESA list on March 11, 1967. It survives in scattered hospitable habitats in counties adjoining the San Francisco area.

2393. Peregrine falcons listed under the U.S. Endangered Species Act were the Arctic and American subspecies. Both were placed on federal endangered species lists in 1970 after it was determined only a few hundred nesting pairs remained, due largely to pesticide contamination that made the birds' eggs too brittle to survive until hatching. The banning of DDT and recovery efforts allowed peregrine falcon populations to recover sufficiently for the U.S. Fish and Wildlife Service to propose delisting in 1998.

2394. Crocodilians listed under the U.S. Endangered Species Act were the Yacare caiman and four crocodile subspecies—Cuban, Morelet's, Nile, and Orinoco—on June 2, 1970. Despite being native to diverse areas located in two hemispheres, all five reptiles suffered from the effects of overhunting, with the Central and South American subspecies contending with habitat loss. Such problems were so common to crocodilians that the number of listed subspecies quadrupled in the following two decades.

2395. Gorilla listing under the U.S. Endangered Species Act became effective June 2, 1970. Despite this law and numerous other protective measures implemented by the World Conservation Union (IUCN), African governments, and conservation programs, the gorillas of central and west Africa remain endangered, mainly because of poaching and the loss of habitat due to human settlement.

2396. Lion listing under the U.S. Endangered Species Act became effective on June 2, 1970. The Asiatic lion, whose native range once extended from Turkey to India, became rare by the 1900s due to overhunting and habitat displacement by human settlement and agriculture.

2397. Prairie dog listed under the U.S. Endangered Species Act was the Mexican prairie dog. The declining species, which lives in the plains and valleys of northern Mexico, was listed on June 2, 1970. The controversial listing of the Utah prairie dog as "threatened" followed later, on June 4, 1973.

2398. Rhinoceros listings under the U.S. Endangered Species Act included three distinct species and became effective on June 2, 1970. The northern white rhinoceros, which is native to central Africa, was found to be endangered. So were smaller Javan and Sumatran rhinos, whose ranges extend from Bangladesh to Indonesia. All three subspecies suffered from habitat displacement by humans and poaching for their horns.

2399. Snail listed under the U.S. Endangered Species Act was the Manus Island tree snail, a native of Manus or Admiralty Island, north of New Guinea. Listed on June 2, 1970, the snail is endangered by humans who collect it for its attractive shell and by logging, which is destroying its habitat. Of nearly twenty snails listed under the ESA, only the Manus Island tree snail and Hawaii's Oahu tree snail are not native to the continental United States.

2400. Tortoise listed under the U.S. Endangered Species Act was the Galápagos tortoise, a native of the Galápagos Islands off the coast of Ecuador. Listed on June 2, 1970, the giant tortoises became endangered due to competition with other species for food and to alien species introduced by humans, such as domestic animals and rats, which preyed on the tortoises' eggs.

2401. Turtles listed under the U.S. Endangered Species Act were the hawksbill and leatherback sea turtles and the western swamp turtle. All were listed on June 2, 1970. Both sea turtles are threatened by consumption of hatchlings by both natural and introduced predators and by humans, who overharvest their eggs. Hawksbill populations have also suffered from human demand for their supply of "tortoise shell" material. Apart from captive populations, Australia's western swamp turtle is nearly extinct.

2402. Sea turtle listing to include all species occurred in the U.S. Endangered Species Act of 1973. Green, hawksbill, Kemp's Ridley, leatherback, loggerhead and olive Ridley sea turtles were all listed as endangered or threatened, placing them under dual protection, by the Fish and Wildlife Service and the National Marine Fisheries Services.

2403. Endangered plants list issued by the United States resulted from the Endangered Species Act of 1973. The act, passed by Congress on December 20, directed the Smithsonian Institution to research and compile a list of endangered and threatened plant species. The study began on July 1, 1975, and resulted in more than 3,000 species being recommended for listing. Further research by the U.S. Fish and Wildlife Service pared the list to fewer than 1,800 species. The first listing, which appeared in 1977, included only 21 species.

2404. Horse listed under the U.S. Endangered Species Act was the Przewalski horse, effective June 14, 1976. Some examples of the wild species survive along the border between Mongolia and China. A larger population exists in captivity.

2405. Hawaiian monk seal listing under the U.S. Endangered Species Act became effective November 23, 1976. The species, which is found mostly in the northwestern Hawaiian Islands, has declined since the 1950s due to disturbance by humans, entanglement in fishing nets, competition with fisheries for food supplies, shark attacks, and fatal mobbing attacks of females by male seals.

EXTINCT AND ENDANGERED SPECIES—LISTS—*continued*

2406. Elephant listed under the U.S. Endangered Species Act was the comparatively small and often tuskless Asian elephant, whose native habitats are found in south central and southwestern Asia. The four subspecies of Asian elephant were listed as "endangered" on June 24, 1976. The larger African elephant was not listed under the ESA until May 12, 1978.

2407. Butterfly listing under the U.S. Endangered Species Act was the Schaus swallowtail butterfly, which was classified as "threatened" on April 28, 1976. The species further declined due to pesticides and butterfly poachers, causing it to be reclassified as "endangered" on August 31, 1984. Poison controls, captive breeding, and reintroduction to its native Florida habitat allowed the Schaus swallowtail to recover considerably by the late 1990s.

2408. Species removed from the U.S. Endangered and Threatened Species List was the Mexican duck. It was removed from the list on July 25, 1978, because the taxonomic data by which it had been classified required revision.

2409. Amphibian removed from the U.S. Endangered and Threatened Species List was the Pine Barrens tree frog. It was removed from the list on November 22, 1983, because new populations were discovered.

2410. Reptile removed from the U.S. Endangered and Threatened Species List was the Indian flap-shelled turtle. It was removed from the list on February 29, 1984, when more accurate data revealed that the species was not endangered.

2411. Insect removed from the U.S. Endangered and Threatened Species List was the threatened Bahama swallowtail butterfly. It was removed from the list on August 31, 1984, due to amendment of the Endangered Species Act.

2412. Species removed from the U.S. Endangered and Threatened Species List due to recovery was the brown pelican. Although it remained in danger elsewhere, the Atlantic coast and eastern Gulf populations of the species had increased sufficiently for it to be delisted on February 4, 1985.

2413. Reptile removed from the U.S. Endangered and Threatened Species List due to recovery was the American alligator. Populations of the southeastern U.S. species had increased sufficiently for it to be delisted on June 4, 1987.

2414. Dolphin listing under the U.S. Endangered Species Act became effective May 30, 1989. The Chinese whitefin dolphin, whose declining population is found mostly in the Yangtze River, was classified as "endangered." Pakistan's Indus River dolphin was added to the endangered list on January 14, 1991.

2415. Plant removed from the U.S. Endangered and Threatened Species List was the Rydberg milk vetch. It was declared to have recovered and removed from the list on September 14, 1989.

2416. Northern spotted owl listing under the U.S. Endangered Species Act became effective on June 26, 1990, when the bird was classified as "threatened." Efforts to preserve the owl's habitat, which was found in the woodlands owned by lumber companies in the American Northwest, led to contentious lawsuits involving conservationists, loggers, and the federal government.

2417. Sockeye salmon listing under the U.S. Endangered Species Act became effective January 3, 1992. The listing protected sockeyes found in Idaho's Snake River. The species spawns and nurses in freshwater lakes and rivers in the Pacific Northwest, where local populations are under review for possible listing as threatened or endangered.

2418. Mexican spotted owl listing under the U.S. Endangered Species Act occurred on March 16, 1993, when the species was granted "threatened" status. Found in the southwestern United States and northwestern Mexico, it was threatened by habitat loss caused by timber harvests, fires, and increased predation by hawks and other owls displaced from their native habitats.

2419. Mammal removed from the U.S. Endangered and Threatened Species List was the gray whale. Its eastern North Atlantic populations recovered sufficiently for it to be removed from the list on June 16, 1994.

2420. Land mammals removed from the U.S. Endangered and Threatened Species List were the eastern, red, and western gray kangaroos. Although found in Australia, the marsupials were listed in the United States because of international wildlife treaties. Populations of all three species recovered sufficiently for them to be simultaneously delisted on March 9, 1995.

2421. Global list of endangered plants was the 1997 IUCN Red List of Threatened Plants, published on April 8 of that year by the World Conservation Union (IUCN). Years of IUCN data gathering concluded that nearly 34,000 or 12.5 percent of the planet's plants were in danger of extinction. The report was simultaneously made available in Australia, England, South Africa, and the United States.

2422. Sturgeon listing by the Convention on International Trade in Endangered Species of Wild Fauna and Flora (CITES) to include all species became effective on April 1, 1998. The decline of sturgeon and paddlefish populations due to overfishing and habitat loss prompted CITES to declare the fish in danger of extinction in June 1997. The CITES listing made international trade in caviar and sturgeon meat subject to permit requirements that were intended to reduce poaching.

2423. Steelhead trout listing under the U.S. Endangered Species Act became effective on June 17, 1998. The listing protects spawning populations of steelheads in California between the Santa Maria River and Malibu Creek. The species spawns and nurses in coastal rivers and streams. Several other local populations were subsequently listed as threatened or endangered.

2424. Australian indigenous species habitat listed as endangered was the Cumberland Plains Woodland, whose shale soil was vital to the survival of disappearing gray box, forest red gum, and narrow-leaved ironbark trees. The area west of Sydney was declared endangered on September 23, 1998, under the Commonwealth Endangered Species Act by Australia's Ministry for the Environment.

2425. Proposed removal of bald eagles from the U.S. Threatened and Endangered Species List was announced by President Bill Clinton on July 2, 1999. The banning of DDT in 1972 and conservation efforts allowed eagle populations to recover sufficiently for the birds to be reclassified from "endangered" to "threatened" on August 12, 1995. The proposal for complete delisting was published in the Federal Register on July 6, 1999, initiating a customary six-month period for public comment before a final decision.

EXTINCT AND ENDANGERED SPECIES—REINTRODUCTIONS

2426. Nene (Hawaiian goose) breeding program was begun in 1918 by Herbert C. Shipman at Hilo on the island of Hawaii. The birds became nearly extinct in the wild during the 1930s, leaving Shipman's successful captive breeding effort as the best hope for the species' survival. A state breeding effort was begun in 1948 with Shipman's help. Various reintroduction programs began in the early 1990s. The nene was declared the state bird of Hawaii in 1957.

2427. European bison reintroduction took place in Bialowieza Forest in eastern Poland in 1952. The bison or wisent had been extirpated in the wild since 1919.

2428. Elk reintroduced in Europe were released in Poland's Kampinos Forest. At the time of their release in 1958, the elk had been extinct in the wild since the 1700s.

2429. Wolves reintroduced to Soviet Georgia were released in 1974. Wolf populations in the region had been severely reduced by years of previous Soviet government policy, which favored killing the animals on the false assumption that they were competitors for human resources.

2430. Condors reintroduced to Arizona were released on December 12, 1996. Prior to the release of six captive-bred birds at Vermillion Cliffs, north of the Grand Canyon, condors had been absent in the wild in Arizona since the 1920s.

2431. Oryx reintroduction program was Operation Oryx. Although the last of the northern African antelopes were extirpated in the wild in 1972, captive oryxes surviving in facilities around the world formed the basis of a population that was imported to the Jiddat al Harisis desert region in Oman. After a reintroduction period, the first oryxes to be released into their native habitat were freed on January 31, 1982.

2432. Pére David's deer reintroduced to China arrived at a new deer park in Beijing's Nanhaizi Zoo in 1985. The deer, which had been extinct in China since 1900, were bred in English and European zoos. In January 1995, the species was reintroduced to the wild at Tianzhou Natural Reserve.

2433. Bali starling reintroductions took place in Indonesia's Bali Barat National Park in 1987. The Bali starling, also known as Rothschild's mynah, remains critically endangered because of habitat loss and illegal poaching for bird collectors. It is Bali's only endemic bird.

EXTINCT AND ENDANGERED SPECIES—REINTRODUCTIONS—*continued*

2434. Wolf to be reintroduced into the wild was the red wolf (*Canis rufus*), the only species believed to have evolved completely in North America. In 1987, four breeding pairs and their pups were released in the Alligator River National Wildlife Refuge, NC, by the U.S. Fish and Wildlife Service. Once common in the southwestern United States, the red wolf had been extinct in the wild since the 1970s. The only survivors were maintained in captive breeding programs.

2435. Reintroductions of black-footed ferrets in the United States took place in Shirley Basin, north of Medicine Bow, WY, in autumn 1991. Ferret mortality due to disease made the effort unsuccessful.

2436. Raven reintroduction in Italy was begun by the Italian State Forest Service in September 1991 in the Monte Velino nature reserve in northern Italy. Although raven populations existed in Italy's southern regions, human harvesting of eggs and the aftereffects of poisoning local carnivores had caused the birds to disappear from the central Apennine Mountains by the 1960s.

2437. European lynx reintroduced in Poland were released in Kampinoski National Park in 1992. The European lynx had virtually disappeared from the wild due to overhunting and habitat loss caused by agricultural development. It was hoped that the reintroduction would help check an overpopulation of herbivores.

2438. Ostriches reintroduced to Arabia were released in June 1994 into a Saudi Arabian reserve, the Mahazat as-Sayd Protected Area. Ostriches had been considered extinct in Arabia since the 1950s due to overhunting.

2439. Griffon vulture reintroduction in Italy was begun in July 1994 in the Monte Velino nature reserve in northern Italy. The birds had been considered extirpated on the Italian mainland since the 17th century.

2440. Legal challenge to wolf reintroduction to Yellowstone National Park occurred on December 21, 1994, in the U.S. District Court in Cheyenne, WY. An injunction against the reintroduction was requested by the American Farm Bureau and the Mountain States Legal Foundation on behalf of ranchers who feared that the program would imperil their livestock. On January 3, 1995, the injunction was denied on grounds that such fears were insufficient proof of harm.

2441. Elk reintroduced to Wisconsin were released in Chequamegon National Forest on May 17, 1995. They had been extirpated due to overhunting for over a century.

2442. Intergovernmental carnivore reintroduction strategy in Europe was started in 1996 by the World Wildlife Fund (WWF). The WWF's Large Carnivore Initiative for Europe enlisted the agreement of governments to aid endangered carnivores like wolves, lynxes, and bears through public education, reduction of subsidies for agricultural development of wildlife habitats, and compensation for farmers losing livestock to increased carnivore populations.

2443. Black-footed ferret reintroduction program to use acclimatization pens in the United States was the reintroduction site in the Aubrey Valley near Seligman, AZ. The pens are used to protect the ferrets from predators while conditioning them to the wild prior to their release. Before their reintroduction to Arizona on March 27, 1996, no black-footed ferrets had been seen there since 1931.

2444. Mexican wolf reintroduction took place in Arizona's Apache National Forest on January 27, 1997. "Lobos," or Mexican wolves, had been extinct in the wild since the 1960s.

2445. Lemurs reintroduced to Madagascar were released at Betampona Nature Reserve on November 10, 1997, by scientists from the Duke University Primate Center and the Madagascar Fauna Group. Lemurs had been hunted nearly to extinction in Madagascar.

2446. Reintroduction of Aplomado falcons to Mexico took place on July 29, 1998, when seven of the endangered birds were sent to Tamaulipas, Mexico from the Peregrine Fund's World Center for Birds of Prey in Boise, ID. Prior to their breeding in captivity and reintroduction to the southwest, Aplomado falcon populations had been decimated by pesticides.

2447. Lynx reintroduction in Colorado occurred on February 3, 1999. The first of 50 of the rare cats scheduled for release in 1999 was freed in the Weminuche Wilderness Area, located in the San Juan Mountains of southwestern Colorado. Prior to the restoration, the last sighting of a lynx in Colorado had been in 1973.

F

FISH AND FISHING

2448. Written description of fly fishing appeared circa A.D. 200 in *De naturà animalum* (On the nature of animals), a zoological treatise by Roman writer and scholar Claudius Aelianus. He described how Macedonians used feathers to bait their fishhooks.

2449. Official "fish days" in England were established by a 1563 act of Parliament during the reign of Queen Elizabeth I. The law declared that fish was to be eaten on Wednesdays and Saturdays to ensure the strength of commercial fishing.

2450. Reference work to identify Brazilian fish was Georg Marcgrav's *Historiae Rerum Naturalium Brasiliae* (Natural History of Brazil), published in 1648. Marcgrav's descriptions of unique Brazilian fish species were the first read by many Europeans. The German naturalist's accurately illustrated work was published in the Netherlands several years after his death.

2451. Description of the giant squid was a hypothetical portrait constructed from sailors' tales of sea monsters by Danish zoologist Johann Japetus Steenstrup, who named the genus *Architeuthis* in 1857. The first example of the rare, enormous squid to be observed at sea was seen in 1861 near the Canary Islands by the crew of the French ship *Alectron*, who managed to retrieve the squid's tail and deliver it to zoologists.

2452. Professional association of fisheries scientists in the United States was the American Fisheries Society (AFS). The organization was established in 1870 and is based in Bethesda, MD. The AFS promotes scientific research, education, and training programs to further the conservation, development, restoration, and wise utilization of fisheries.

2453. Vessel constructed solely for fisheries research was the *Albatross*, which was built and launched in 1883. It operated from the U.S. Commission of Fish and Fisheries headquarters in Woods Hole, MA, until 1921, collecting marine organisms, taking soundings, and doing other research in oceans around the world.

2454. Fish research laboratory operated by the U.S. government on a permanent basis was built at Woods Hole, MA, in 1885. The facility served as headquarters of the Commission of Fish and Fisheries.

2455. Texas state agency to regulate fishing was the Fish and Oyster Commission, established by the Texas legislature in 1895. The agency was expanded in 1907 to regulate all wildlife. Its name was changed in 1951 to the Game and Fish Commisson. A merger with the state's park system in 1963 created the Texas Parks and Wildlife Department, which administers both inland fisheries and Texans' commercial and sport fishing activities in the Gulf of Mexico.

2456. Fisheries management in U.S. national parks was mandated by the legislation that produced the National Park Service (NPS), the National Park Service Organic Act, which became law on August 25, 1916. It directed the Secretary of the Interior and the National Park Service to manage and conserve wildlife resources. The law made the NPS responsible for regulating all sport and commercial fishing within the park system.

2457. Fishing legally allowed on a U.S. government wildlife refuge was on the Upper Mississippi River Wildlife and Fish Refuge. Created by an act of Congress on June 7, 1924, the refuge allowed fishing throughout its wetlands and rivers, which follow the Mississippi River from western Wisconsin's Chippewa River to Rock Island, IL.

2458. Law passed to protect black bass in the United States was the Black Bass Act on May 20, 1926. It extended to fish the kind of protection the Lacey Act of 1900 gave game by prohibiting the interstate shipment of illegally taken wildlife. The Black Bass Act and the Lacey Act were re-authorized periodically until November 16, 1981, when both were repealed and their provisions consolidated in the Lacey Act Amendments of 1981.

2459. U.S. Fish and Wildlife Service jurisdiction over sockeye salmon conservation resulted from Executive Order 9892 signed by President Harry S. Truman on September 20, 1947. The law made the U.S. Fish and Wildlife Service responsible for enforcing the Sockeye Salmon Fishery Act of 1947. The order further directed the service to enforce the May 26, 1930, convention between the United States and Canada for the protection, preservation, and extension of the sockeye fishery of the Fraser River system and to uphold the regulations of the International Pacific Salmon Fisheries Commission.

FISH AND FISHING—*continued*

2460. Inter-American Tropical Tuna Commission was established by Costa Rica and the United States in Washington, DC, on May 31, 1949. The two nations agreed to manage fisheries in the eastern Pacific Ocean by monitoring the ecology and populations of yellowfin and skipjack tuna, as well as those fishes used for bait by the tuna industry.

2461. Control method used on sea lampreys in the Great Lakes was mechanical and electrical weirs, deployed in the 1950s to prevent the alien predatory fish from spawning and worsening its severe depredations on native species. TFM, a chemical lethal to larval sea lamprey but not to other fish, was tested successfully in Lake Superior in 1958 and became commonly used as a control method in the other Great Lakes. Sterilization and new barrier designs were later developed to lessen dependence on chemicals.

2462. National fisheries center approved by Congress was enacted through the National Fisheries Center and Aquarium Act on October 9, 1962. The act authorized the Secretary of the Interior to plan, construct, and maintain a National Fisheries Center and Aquarium in the District of Columbia. The act also established an advisory board to assist in management decisions and authorized the purchase, operation, and maintenance of vessels to collect marine specimens. Entrance fees were authorized to pay for the center's construction, annual operation, and maintenance.

2463. Anadromous fish conservation law in the United States was the Anadromous Fish Conservation Act, enacted on October 30, 1965. The law authorized the Secretaries of the Interior and Commerce to enter into cooperative agreements with states and other nonfederal interests for conservation of fish that depend on rivers to spawn. An October 17, 1991, amendment to the act shifted its original purpose of investigating declines in anadromous fish populations to restoration and management information programs. The act authorizes engineering and biological surveys, stream clearance, construction projects, and maintenance of hatcheries. It also authorizes the Fish and Wildlife Service to make recommendations to the Environmental Protection Agency regarding the reduction or elimination of pollutants substances detrimental to fish and wildlife in interstate or navigable waters or their tributaries.

2464. Turtle excluder device was developed in 1980 at the Mississippi Laboratories Harvesting Systems Division of the National Marine Fisheries Services. Turtle excluders help endangered sea turtles escape from shrimp trawls.

2465. Law in the United States to protect Atlantic striped bass was the Atlantic Striped Bass Conservation Act enacted on October 31, 1984, to conserve the fish for its commercial and recreational value. The law required the Secretary of Commerce to impose a moratorium on fishing for striped bass in any state not in compliance with the Atlantic States Marine Fisheries Commission's Plan for Striped Bass. Subsequent reauthorizations of the act have modified its conservation provisions and geographical scope.

2466. New England Fishery Resources Restoration Act was passed on November 16, 1990. It was intended to ensure implementation of programs concerning New England waterways, including restoration of Atlantic salmon and other fishery resources and studies of fish passage impediments like dams. It also required the U.S. Fish and Wildlife Service to inventory fish and wildlife habitat and other natural areas in New England river basins.

2467. Use of the Atlantic Coastal Fisheries Cooperative Management Act of 1993 against a state occurred on December 5, 1994. The National Marine Fisheries Service (NMFS) cited New Jersey for noncompliance with the Atlantic States Marine Fisheries Commission's Interstate Coastal Fishery Plans for Atlantic sturgeon, bluefish, and weakfish. The U.S. Department of Commerce threatened the state with a moratorium on fishing in New Jersey State waters if it did not come into compliance with the commission's plans by April 1, 1995.

2468. DNA test to detect whirling disease in trout and other salmonids was developed at the University of California-Davis School of Veterinary Medicine. The test was hoped to be useful in discovering and tracking the parasite that causes whirling disease, a major cause of declining fish populations in the western United States. Development of the DNA test was announced in January 1997 by the U.S. Fish and Wildlife Service.

FISH AND FISHING—COMMERCIAL

2469. English law to support North American fisheries was a 1542 act of Parliament that forbade English fishermen and fishmongers from buying their wares from foreign sources. The act was ordered by King Henry VIII to strengthen the English cod fishing industry in the North Atlantic, particularly off the coast of Labrador, against rival nations.

2470. Commercial fishery in an American colony is believed to have been established at Medford, MA. The colonists were given instructions on April 17, 1629, to let the fish "be well saved with the said salt, and packed up in hogsheads; and send it home by the Talbot or Lion's Whelpe." The industry flourished and on May 28, 1639, received "salt, lime hooks, knives, boots, etc., for the fisherman."

2471. Cannery in Alaska was built in 1878 in Klawock, on Prince of Wales Island. Like most of the canneries that sprang up in the Alaskan territory in the subsequent decade, it was owned by a firm located in the United States and depended on contract labor.

2472. U.S. Supreme Court decision regarding a sponge harvesting law was *The Vessel "Abby Dodge"* v. *United States* in 1912. The *Abby Dodge*'s owner, Anthony Kalimeris of Tarpon Springs, FL, was fined $100 for violating a 1906 federal law banning the use of diving equipment in sponge fishing. Kalimeris appealed the fine on grounds that the law was unconstitutional. On February 19, 1912, the Supreme Court overturned the fine, stating that the law improperly tried to regulate the taking of sponges within the exclusive territorial jurisdiction of a state.

2473. Commercially marketed frozen fish were sold by Birdseye Seafoods in New York City in September 1922. The growing availability of quick-frozen seafood throughout the United States helped to increase the popularity of fish as a staple food, eventually creating unprecedented pressures on U.S. fisheries.

2474. General Fisheries Council for the Mediterranean was created by France, Greece, Italy, Lebanon, Turkey, Great Britain, and Yugoslavia in Rome, Italy, on May 22, 1963. The contracting nations formed the new body within the Food and Agriculture Organization of the United Nations, which approved the new group on December 3, 1963. The council agreed to coordinate and share information about fisheries research, oceanography, and the occupational health of fishermen.

2475. International organization to regulate tuna fisheries was the International Commission for the Conservation of Atlantic Tunas (ICCAT). The group's 25 member countries meet annually in Madrid, Spain to share fisheries research, determine stock status, and set quotas for the take of tuna species, swordfish, marlin, and other billfish. ICCAT was opened for signature on May 14, 1966, in Rio de Janeiro, Brazil and entered into force on March 21, 1969.

2476. Catfish farms in the United States went into business in southern states around 1970. Aquaculture, which is the propagation of marine animal or plant life in a controlled environment, took hold in "Deep South" states like Alabama, Arkansas, Louisiana, and Mississippi, where inactive cotton fields were converted into artificial ponds suitable for raising catfish, crawfish, trout, salmon, and tilapia. Aquaculture has become a major industry, accounts for a significant percentage of fish sold in the United States, and has increased the culinary popularity of the species it produces.

2477. Involvement of the U.S. Department of Commerce with commercial fisheries resulted from Reorganization Plan No. 4 of 1970. The plan transferred duties formerly handled by the Bureau of Commercial Fisheries to the Secretary of Commerce, with several exceptions. Great Lakes fishery research, Missouri River Reservoir research, and trans-Alaska pipeline investigations were instead assigned to the Bureau of Sport Fisheries and Wildlife. Functions vested in the Secretary of the Interior relating to migratory marine species of game fish also were transferred to the Commerce Department.

2478. Fishing vessel buyback program in the United States was funded by the federal government and administered by the state of Washington in 1976. The program provided economic relief and retraining for non–American Indian salmon fishermen, whose businesses were suffering because of declining salmon fishery and the 1975 allocation of 50 percent of Washington's salmon catch to American Indian tribes by treaty. The program also attempted to preserve the salmon fishery by reducing the size of Washington's fishing fleet.

2479. "Dolphin-safe" labels for tuna packaging in the United States appeared in 1990, after major American tuna companies announced that they would no longer purchase any tuna caught by fishing enterprises that used purse-seine nets, which trap and drown dolphins. Congress later enacted the Dolphin Protection Consumer Information Act to set standards for labeling tuna as "dolphin-safe."

FISH AND FISHING—COMMERCIAL—*continued*

2480. United Nations moratorium on drift net fishing in international waters was passed by the General Assembly on December 20, 1991. Although the non-binding resolution became effective on December 31, 1992, noncompliance was widespread. Because of their ability to ensnare all marine animals within their target area, not just food fish, drift nets were identified as a major cause of overfishing and a threat to fisheries survival.

2481. Embargo in the United States against tuna that was not "dolphin-safe" resulted from the International Dolphin Conservation Act of 1992, which banned the sale, purchase, transport, or shipment in the United States of any tuna not caught by "dolphin-safe" methods. The law was passed by Congress on October 8, 1992, and resulted in stronger efforts to minimize dolphin mortality by countries hoping to regain access to the U.S. market. In 1995, in the Declaration of Panama, the United States and other countries agreed that the restrictions would be modified in exchange for binding dolphin protection agreements, a proposal that ran into some trouble in Congress. Eventually, however, despite controversy and claims that the new guidelines weakened the 1992 Act, they were agreed upon by Congress and signed into law by President Bill Clinton on August 15, 1997.

2482. Federal limits on shark fishing enacted by the United States were established by the National Marine Fisheries Services (NMFS). To sustain severely declining shark populations, the NMFS's April 1993 Fisheries Management Plan (FMP) for Sharks of the Atlantic Ocean set fishing regulations for 39 shark species. The conservation program imposed quotas and weight regulations for commercial shark fishing, set recreational bag limits, and instituted mandatory permit and reporting requirements. For certain species, the FMP banned the practice of finning, in which sharks were stripped of their valuable fins, then thrown back into the ocean to die. The plan was found to be insufficient to protect sharks and was expanded upon by more restrictive quotas on harvesting.

2483. Fishing vessel buyback program for New England was begun when the historically rich fishery was declared to be in a state of disaster in August 1995. The pilot program was called the Fishing Capacity Reduction Initiative (FCRI) and retired both vessels and licenses from active use in the Atlantic fishery. Vessels bought by the National Marine Fisheries Service under the plan were initially destroyed, although a modification to the program allowed subsequent buyouts to be used for alternative, nonfishery purposes such as research.

2484. Permit system enacted in the United States for fishing vessels on the high seas was created through the High Seas Compliance Act of 1995. The law was designed to prevent U.S. commercial fishing vessels from avoiding regulations by flying the flags of countries that were not parties to international conservation treaties. The law was enacted on November 3 and is administered by the Department of Commerce. The law brought the United States into compliance with the Agreement to Promote Compliance with International Conservation and Management Measures by Fishing Vessels on the High Seas, which was adopted by the UN Conference of the Food and Agriculture Organization of the United Nations on November 24, 1993.

2485. Regulations enacted in the United States to limit swordfish harvesting according to size became effective on November 20, 1997. The Commerce Department and National Marine Fisheries Service prohibited the possession aboard U.S. commercial fishing vessels of any swordfish shorter than 29 inches or less than 33 pounds in dressed weight. The rule also applied to swordfish imported into the United States.

2486. Mandatory international Atlantic bluefin tuna conservation plan was implemented by the International Commission for the Conservation of Atlantic Tunas (ICCAT) in Santiago de Compostela, Spain on November 24, 1998. The 20-year plan limited the 25 ICCAT member states to lower catch quotas in the eastern Atlantic, while raising quotas in the western Atlantic. The plan also provided incentives to minimize the discard of undersized bluefin and required all countries to monitor and report all sources of fishing mortality. Conservation groups praised the measures but were critical of ICCAT's decision to raise quotas in the western Atlantic.

FISH AND FISHING—HATCHERIES AND FISH FARMING

2487. Fish farms are thought to have originated in China circa 1000 B.C. Early writing suggests that carp was the freshwater species most commonly produced by the aquaculturalists of ancient China.

2488. Oyster farms were constructed near Naples circa 95 B.C. by Romans. Pliny the Elder wrote of heated artificial oyster beds created by Sergius Orata, the inventor of central heating.

2489. Oyster farming under state auspices began in Rhode Island, which in June 1779 set aside some state-owned waters for the cultivation and propagation of oysters.

2490. Fish hatchery to breed salmon in the United States was an experimental laboratory established in 1864 in New York by James B. Johnson. He imported salmon eggs from Europe and hatched them in his laboratory.

2491. Known successful introduction of a fish species within the United States occurred in 1871 when the California Fish Commission transported 10,000 shad fry from New York's Hudson River to the Sacramento River in California by train. Most of the fish survived and became a permanent part of West Coast fisheries.

2492. Fish hatchery run by the U.S. government was established for the propagation of Atlantic salmon in 1872 at Bucksport, ME. It was a joint activity, with the cooperation of state agencies of Maine, Massachusetts, and Connecticut, under the supervision of Charles Grandison Atkins. It continued experiments initiated by these agencies in 1871 and was later moved to East Orland, ME.

2493. Floating hatchery run by the U.S. government in ocean waters was the steamer *Fish Hawk* in 1880. The hatchery ship operated mostly in coastal waters, promoting shad, herring, and striped bass reproduction.

2494. Brown trout in the United States were imported in 1883 from Germany for breeding at the New York State–operated hatchery at Cold Spring Harbor on Long Island. The present-day Cold Spring Harbor Hatchery still operates as an educational aquarium and houses the largest exhibit of native freshwater fish species in New York.

2495. Federal law in the United States to develop salmon fisheries in the Columbia River Basin was the Mitchell Act, enacted on May 11, 1938. It directed the Secretary of the Interior to establish salmon hatcheries, conduct engineering and biological surveys and experiments, and install fish protective devices. It also authorized agreements with state fishery agencies and construction of facilities on state-owned lands in Oregon and Washington.

2496. Federal fish-agriculture rotation program in the United States was authorized by the Fish-Rice Rotation Farming Program Act. Congress approved the act on March 15, 1958. Under the direction of the Secretary of the Interior, the program included research to determine which species of fish were most suitable for culture on a commercial basis in shallow reservoirs and flooded rice lands. The law also directed the U.S. Department of Agriculture to study the effect of fish-rice rotations on both the fish and the crops.

2497. Aquaculture Development Act in the United States was enacted on September 26, 1980. The act directed the Secretary of Commerce to develop a National Aquaculture Development Plan to identify aquatic species that could be cultured on a commercial or other basis. Research and development, technical assistance, education, and training activities were also initiated. A November 8, 1984, amendment to the act established the Office of Aquaculture Coordination and Development and the National Aquaculture Board.

2498. Permits issued by the U.S. Fish and Wildlife Service to kill migratory birds at aquaculture sites were issued following an April 18, 1990, FWS policy directive. The emergency measure was to be allowed only as a last resort in cases where all other efforts by Animal Damage Control agents to prevent birds from preying on fish farm populations had failed. The permits did not allow the killing of endangered species and were a short-term solution to allow fish farmers time to construct protective enclosures or other deterrents to save their businesses.

2499. International Fisheries Gene Bank (IFGB) was established in 1994 in Victoria, British Columbia. The gene bank was affiliated with the Vancouver Aquarium and aimed to preserve species depleted by overfishing, pollution, dams, and other development.

2500. Genetic manipulation of hatchery fish stocks was pioneered in 1996 by the Tulalip Indian Tribe at their coastal reservation near Marysville, WA. By selectively breeding and monitoring stocks of chum salmon, the program was able to steer harvesting toward runs of salmon produced in the hatchery, thus protecting decreasing populations of wild salmon using the same waterways.

FISH AND FISHING—LEGISLATION AND REGULATION

2501. American colony to restrict oyster fishing was New Jersey. On March 27, 1719, a law was passed prohibiting nonresidents from taking oysters or placing them in any boat belonging to a nonresident.

2502. Fish protection law enacted by an American city was passed by the city of New York on May 28, 1734. It prohibited fishing in freshwater ponds by nets or by any means other than by angling with angle-rod, hook, and line. Violators were subject to a fine of 20 shillings.

2503. State fishing commission in the United States was California's Board of Fish Commissioners. The three-member panel convened in 1870 to oversee the protection and restoration of fish in California waters.

2504. Law in the United States banning tin-based paint on boats was signed by President Ronald Reagan on June 16, 1988. The ban was proposed after tin-based paints, which had been used on hulls as a barrier against barnacles, were shown to harm commercially harvested shallow-water organisms like salmon, oysters, and mussels.

FISH AND FISHING—SPORT FISHING

2505. Sport fishing guide written in English is attributed to Dame Juliana Berners, a Benedictine nun. Her "Treatyse of Fysshynge wyth an Angle" was printed as part of the *Boke of St. Albans* in 1496.

2506. Fishing club in the United States of any duration was the Schuylkill Fishing Company, founded in 1732 in Philadelphia, with a limited membership of 25.

2507. Tax on fishing equipment in the United States to specifically benefit wildlife habitats resulted from the Federal Aid in Sport Fish Restoration Act, which became law on August 9, 1950. The law was also known as the Dingell-Johnson Act. A 10 percent tax on fishing tackle provided federal funds disbursed to states specifically for acquiring and managing fish habitats, fish stocking, and research programs.

2508. Survey of recreational fishing on a nationwide basis in the United States was conducted in 1955 as part of the National Survey of Fishing, Hunting, and Wildlife-Associated Recreation. The survey is conducted every five years. The Fish and Wildlife Service maintains collected data on the number of salt- and freshwater anglers, as well as the amount of money spent annually on equipment, licenses, tags, membership fees, transportation, literature, and other fishing-related expenses.

2509. Salmon caught by rod and line in the River Thames was recorded in 1983. The reintroduction of salmon came after efforts to clean up industrial and sewage pollution in the Greater London region.

2510. Tax on boats and motors in the United States to benefit wildlife habitats resulted from 1984 amendments to the Federal Aid in Sport Fish Restoration Act of 1950. The new laws, which were known as the Wallop-Breaux amendments, expanded the terms of the earlier law by enlarging the list of fishing equipment taxed at 10 percent to provide funds for fisheries restoration. New 3 percent taxes were also imposed on fish finders and electric trolling motors. The Wallop-Breaux amendments were approved by Congress on July 18 and became effective on October 1, 1984.

2511. National Fishing Week in the United States took place June 1–7, 1987. A proclamation by President Ronald Reagan on April 10, 1987, established the first week of June as an annual occasion sponsored by government wildlife agencies and sport fishing enthusiasts as a means of educating anglers and promoting recreational fishing.

2512. U.S. Sport Fishing and Boating Partnership Council was established to advise the secretary of the Interior and the director of the U.S. Fish and Wildlife Service on conservation issues beneficial to recreational fisheries and recreational boating. Partnerships between industry, the public, and government were reflected by the council's membership, which included the director of the U.S. Fish and Wildlife Service, and president of the International Association of Fish and Wildlife Agencies, and 16 appointees from saltwater and freshwater recreational fishing and boating industries, conservationists, and tourism experts. The first council was chartered on October 16, 1997, with a provision to be re-chartered every two years.

2513. California waters in the state Heritage Trout Program (HTP) were designated on April 19, 1999. The freshwater areas stocked and opened for trout fishing included Golden Trout Creek, Clavey River, Eagle Lake, Heenan Lake, Upper Kern River, and the Upper Truckee River.

2514. State in the United States to ban lead sinkers and jigs was New Hampshire. The ban, which took effect January 1, 2000, prohibited the use of lead sinkers weighing 1 ounce or less and lead jigs less than 1 inch long in the state's freshwater ponds and lakes. The lead was found to result in the fatal poisoning of loons, which swallowed the sinkers, mistaking them for the grit that birds use to grind food.

FLOODS AND FLOOD CONTROL

2515. Glacial floods of major significance known in North America occurred in 8000 B.C. and covered the scabland (barren volcanic land) of vast areas in the present-day states of Montana and Washington. The waters from a glacial lake more than 3,200 feet deep, located at present-day Missoula, MT, were released when an ice barrier weakened and broke.

2516. Flood known in Asia Minor occurred circa 7000 B.C. when ice melted from the last ice age and the Mediterranean Sea rose about 400 feet. It burst through a thin strip of land that spanned the Bosphorus and flooded northeastward to create the Black Sea.

2517. Recorded flood on China's Yellow River occurred in 2297 B.C.; tens of thousands of villagers were killed. Subsequent records suggest that the river flooded fairly regularly, although not often to such a lethal degree. The Yellow River, which is 3,000 miles long, gets its distinctive color from the silt its waters carry, which is deposited in flood times on the alluvial Great North China Plain. This is some of the most fertile soil in China and is intensively cultivated. When a great flood occurs, the loss of life and shelter is compounded by the loss to the whole country of huge crops of corn, winter wheat, vegetables, and cotton.

2518. Surviving account of a great flood covering the Earth was the Gilgamesh epic of 2000 B.C. in Babylonia. The details are similar to those of the great deluge of Noah's time.

2519. Recorded flood in Europe occurred on the Tiber river in Rome in 413 B.C.

2520. Flood known to cause massive loss of life in Europe was in 1228 in the Friesland region of Holland. An estimated 100,000 people died.

2521. Flood to cause massive deaths in what is now Germany occurred in 1362. About 30,000 people died when the Baltic Sea flooded the low region around the city of Schleswig, then part of the Holy Roman Empire.

2522. Flood to cause massive deaths in the Dordrecht region of Holland occurred in 1421. Water overwhelmed the dike system where four rivers met, and 72 of the region's villages were inundated, causing the deaths of more than 100,000 residents.

2523. Tsunami (giant wave) with great loss of life recorded in Japan occurred in 1489 after an earthquake. It killed about 1,000 people in Kii on Honshu island.

2524. Flood recorded by Europeans in North America was along the Mississippi River in 1543, observed by the Spanish explorer Hernando de Soto. He observed the river overflowing its banks on March 18. He also witnessed the river's crest on April 20. The river had returned to its banks by the end of May.

2525. Great flood at Tombouctou on the River Niger in present-day Mali occurred December 16–17, 1592. The flood, greater than all previously known inundations, resulted from unusually heavy summer rains in the Niger's upper basin.

2526. Flood deaths of major significance deliberately caused by humans occurred in 1642 in Kaifeng, China. Some 300,000 people died after invaders destroyed the city's defenses along the Yellow River, causing the river to flood the region.

2527. Floods known to have lasted 15 years in China occurred between 1851 and 1866. Waters constantly overflowed from the Yellow River and Yangtze River onto the triangle of land between them, and an estimated 50 million people died during this period as hundreds of communities were flooded continually.

2528. Connecticut River flood crest at Hartford to reach 28 feet, 10.5 inches occurred on May 1, 1854. The record crest followed 66 hours of incessant rain over all of Connecticut and other parts of southern New England.

2529. Heavy rainstorm and local flooding to cause the complete failure of an offensive operation in the American Civil War occurred near Fredericksburg, VA, on January 21, 1863. Union commanders called off the campaign, soon derisively dubbed the "Mud March."

FLOODS AND FLOOD CONTROL—*continued*

2530. American song written about a flood warning was "The Ride of Collins Graves" written by John Boyle O'Reilly, a Boston songwriter. The song tells how Collins Graves rushed ahead of the flood on horseback and warned the town of Williamsburg, MA, about the Mill Creek Flood on May 16, 1874.

2531. Flood victims to receive aid from the hand of Clara Barton were those in the Ohio and Mississippi River floods of 1884. In her role as president of the Red Cross, Barton led the organization in its first flood relief effort.

2532. Flood death toll to exceed 1.5 million occurred during the Great China Flood of 1887. More than 2,000 villages were inundated when the Yellow River flooded.

2533. Flood covered with burning oil in the United States occurred in 1892 in Oil City, PA. The flood broke oil pipelines that emptied a thick coat of oil over the surface. This caught fire and sent a blazing river through town, burning buildings in its path. The disaster killed 130 people, many of them burned to death in the water.

2534. Flash flood to cause $100 million in damages occurred in Heppner, OR, on June 14, 1903. Surging Willow Creek, fed by a mountain cloudburst, claimed 236 lives in what became known as the Heppner Disaster.

2535. Large-scale flood control project by a single U.S. state was completed in the Miami River Valley of Ohio in 1922. The Miami Conservancy District included a system of reservoirs.

2536. Legislation of significance in the United States to promote reforestation and soil conservation as flood control measures was the Clarke-McNary Act passed by Congress in 1924.

2537. Mississippi River flood control project of significance was initiated on May 15, 1928, when the U.S. Congress passed the Flood Control Act. The project, which took a decade to complete, received $325 million from the federal government.

2538. Catastrophic flood caused by a hurricane in the United States happened on September 13, 1928, at Lake Okeechobee, FL, when more than 2,000 people died from the rushing waters. The lake's mud dikes had disintegrated under the force of the storm, which lashed southern Florida from September 6 to 20, killing an additional 1,836 people.

2539. Flood legislation by the U.S. government that called floods a national problem was the Flood Control Act signed by President Franklin D. Roosevelt on June 22, 1936. It said federal funds should be used to address the problem by preventive efforts including dams, levees, and channel improvements. This emphasis directed U.S. flood policies for three decades.

2540. Great flood to maroon the entire city of Louisville, KY, occurred in January 1937. In a sudden thaw with heavy rains, the Ohio River flooded along its entire course. Water stood 80 feet deep in places in Cincinnati, OH. More than 100 people perished, and 1.5 million people were forced from their homes.

2541. Flood in the United States to claim more than 600 lives occurred in September 1938 in Connecticut, Rhode Island, Massachusetts, New Hampshire, and Maine. Tidal surges and river flooding associated with the great hurricane of September 21, 1938, caused the deaths.

2542. Flood control legislation of significance in the United States was the Flood Control Act of 1946. Congress authorized 123 projects for river regulation, flood control, and power development.

2543. Tsunami of significance to hit a U.S. territory occurred on April 1, 1946 in Hawaii. The giant wave, 20 to 30 feet high, killed 159 people, most of them in Hilo where the downtown area was flooded. Several children and teachers were killed in the Laupahoehoe School 25 miles north of the city. Hawaii is the highest-risk area in the world for tsunamis.

2544. Flood leaving 1 million people homeless in the Netherlands was the North Sea Flood of February 1, 1953. The sea rose 18 feet above normal level and broke through hundreds of dikes to sweep 37 miles inland. Flooding had begun on January 31 and also ravaged Great Britain and Belgium, with the final death toll for the North Sea coastal areas passing 2,000.

2545. Time the Yangtze and Hwai rivers flooded to over 96 feet was August 1954. The flood turned an area about twice the size of Texas into an inland sea. More than 40,000 people drowned as a result of the flood in the Tungting Lake region.

2546. Renovation of significance to the dike system in the Netherlands commenced in 1958. Known as the Delta Project, the renovation and construction became necessary after the devastating floods of 1953. In addition to reinforcing and renovating the system of dikes, the Delta Project resulted in a series of new dams linking islands in the Rhine, Maas, and Schelde rivers to the mainland. The project was completed in 1985.

2547. Widely publicized version of the Christian fundamentalist theory of flood geology was *The Genesis Flood* in 1961. In the book, John C. Whitcomb Jr. and Henry M. Morris argued that violent natural events such as the biblical flood may explain large-scale changes in the Earth's crust.

2548. Flood of significance in Spain during the 20th century occurred on September 26, 1962, in Barcelona. Nearly 450 people were killed and more than 10,000 left homeless. Salvador Dali and Pablo Picasso later auctioned some of their paintings to help the homeless.

2549. Tsunami to devastate the U.S. West Coast occurred immediately after an Alaskan earthquake on March 27, 1964. The giant wave 10 to 20 feet high hit parts of the coasts of Alaska, Washington, Oregon, and California. The tsunami caused 123 deaths and about $84 million in damages, while the earthquake killed 66 people and caused about $500 million in damages.

2550. Flood legislation by the U.S. government to emphasize the environment was the Water Resources Planning Act passed in 1965 by the U.S. Congress. It reversed a 30-year policy of building dams and levees and emphasized environmental protection through prevention. The act created the U.S. Water Resources Council, an independent executive agency that supports conservation, and created the Unified National Program for Flood Plain Management, to provide guidelines for federal and state agencies.

2551. Flood crest on the Mississippi River at St. Paul, MN, to exceed the previous record by 4 feet occurred on April 17, 1965. Flood damage there and downstream was estimated at close to $100 million.

2552. Flood-danger evaluations required by a U.S. presidential executive order were required in 1966 by President Lyndon Johnson. Special research had to be done before a federal agency could take any action that might icrease the risk of flood, according to Presidential Executive Order 11296.

2553. Flood-plain protection advocated by a U.S. presidential executive order was in 1977 by President Jimmy Carter. Flood plains should be protected as natural phenomena, according to Presidential Executive Order 11988, which also gave support for nonstructural measures to control floods whenever possible.

2554. Use of wrecked cars from a junkyard to make a barrier against a flood occurred on the Santiago Creek in Santa Ana, CA, in 1969. During the subtropical storm from June 18 to 26, marine helicopters flew cars from a junkyard to form the barrier on the creek.

2555. Substantial rainfall after more than 400 years at Calama, Chile, occurred on February 10, 1972. Torrential storms in Calama, known as the driest place in the world, caused catastrophic floods and landslides.

2556. Disaster settlement to include provisions to address the long-term psychological needs of the victims occurred after the Buffalo Creek, WV, flood. A dam made from mining wastes collapsed on February 26, 1972, and after the flood the Buffalo Mining Company paid an out-of-court settlement to the 645 survivors from Man and Logan City, WV.

2557. Flood limitation program of significance by the Canadian government was the Flood Damage Reduction Program passed by Parliament in 1975. It required the creation of maps to define the probability of a flood in a flood-risk zone. Under the program, the Canadian government also signed agreements with provincial governments in which they agreed not to engage in, or provide assistance to, undertakings that would be vulnerable to flood damage in any designated flood-risk area.

2558. Flash flood national program in the United States was the National Flash Flood Program Development Plan established in 1978 within the Hydrological Service Branch of the National Weather Service. In 1979 it created the Integrated Flood Observing and Warning Systems (IFLOWS) to reduce loss of life, property damage, and disruption from flash floods. The IFLOWS prototype program was in 12 counties in Kentucky, Virginia, and West Virginia, and in 1985 the U.S. Congress increased funding to include Tennessee, North Carolina, New York, and Pennsylvania. It uses the Automated Flood Warning Systems (AFWS) network to share flash flood information.

2559. Flood in the United States to cause $1 billion in damages occurred along the Pearl River in Mississippi in April 1978. Fifteen fatalities were reported.

FLOODS AND FLOOD CONTROL—*continued*

2560. Catastrophic flood of the 20th century in Wales occurred December 26–27, 1979. Ceaseless rain caused inundations that killed 4 people and damaged more than 2,000 dwellings.

2561. Flood barrier in Britain rotated from beneath the water was the Thames Barrier completed in 1982 in London. Located near the mouth of the river at Woolwich, it has curved steel floodgates that move 90 degrees to rise from concrete housings on the riverbed to form a barrier across the river's 569-yard width. Their movement is controlled by hydraulic machinery in a series of island casings.

2562. Artificial flood of significance on the Colorado River in the western United States occurred in March 1986 when the federal government released 100 billion gallons of water from the Glen Canyon Dam in Arizona. The flood added to beaches downstream and cleared fish spawning grounds of sediment.

2563. Floods to cause damages of more than $1 billion in the United States occurred in May 1990. Torrential rains flooded the Red, Arkansas, and Trinity Rivers, causing disasters in Arkansas, Louisiana, Oklahoma, and Texas.

2564. Flood to affect one-fifth of China's population occurred in the summer of 1991 on the Yangtze and other rivers. A long monsoon season beginning in May in central China caused the deluge that reached about 220 million people and had killed about 1,800 by the middle of July. The worst hit provinces were Jiangsu and Anhui. An estimated 10 million people were evacuated before the waters receded in September. The floods also destroyed at least 20 percent of the area's summer harvest.

2565. Floods known to cover 10 U.S. Midwestern states occurred from June to August 1993 in Illinois, Wisconsin, Missouri, Kentucky, Iowa, Minnesota, Kansas, Nebraska, South Dakota, and North Dakota. With 31,000 square miles covered with water, this was the worst flooding ever recorded in the Midwest. After 49 straight days of rain, more than 100 rivers flooded and many reached record levels; St. Louis recorded water 20 feet above flood level. Civilians and military units laid more than 26 million sandbags to stem the torrents. A total of 48 people died, and about 70,000 were left homeless. The estimated damage to property and agriculture was $12 billion, and President Bill Clinton signed a $6.2 billion flood relief bill in August.

FLOODS AND FLOOD CONTROL—LEVEES

2566. Diversion of the Yellow River was attempted circa 3,000 B.C. The 2,700-mile-long river, which rises near the border between Tibet and China and runs through China's Great Plains, carries yellow water that is at times 40 percent mud by weight. Although its silt provided rich agricultural land, the river had a tendency to flood to disastrous effect. It was diverted into several parallel channels, but these efforts to tame the river did not succeed.

2567. Levees along the Mississippi River were built at New Orleans, LA, in 1724. Compared with the enormous levees of today, they were only rudimentary dikes. Sieur Le Blond de la Tour, a knight of St. Louis and chief engineer of the French-held colony, began construction of the levees in 1718. They were completed in 1727 and extended 18 miles north and 18 miles south of New Orleans.

2568. Federal government program to finance construction of flood control levees in the United States dates from 1879. The levees were built under the pretex of improving river navigation, a power granted the federal government under the Constitution.

2569. Levees destroyed in China to halt a foreign invasion were leveled in 1938 by General Chiang Kai-shek along the Yellow River to stop the advance of Japanese forces that had captured the ancient city of Kaifeng. The invaders were halted and had to take a more southerly route. The levee destruction, however, killed more than 1 million Chinese as floods covered 21,000 square miles, and it also changed the course of the river in that region of Henan province.

FLOODS AND FLOOD CONTROL—MUDSLIDES

2570. Mudslide catastrophe in Guatemala caused by a volcano occurred on September 11, 1541, burying the town of Ciudad Vieja and killing more than 1,000 people. Three days of pouring rain had filled the crater of the volcano, later named Aqua (water), and transformed the ash into mud. The sudden tremors at 2 A.M. on September 11 shook the mud from the volcano and sent it down the slopes to engulf the town.

2571. Landslide known to kill more than 2,400 people happened in the Chiavenna Valley in Italy on September 4, 1618. The massive landslide destroyed everything in its path. The only three survivors dug their way out of the mud with their bare hands.

2572. "Moss flood" of significance in England occurred in November 1771 in Yorkshire. With heavy rains, Solway Moss, a hill 3 miles long and a mile wide, swelled and breached the earth shell that contained it. Moss and mud spilled into the valley, killing animals and destroying 14 farms. Moss covered 900 acres of farmland to depths of up to 20 feet.

2573. Landslide in Mexico on record to kill more than 2,000 people happened on the night of October 29, 1959. Minatitlan was the first town hit, and more than 800 people were trapped in their beds. More than a dozen towns along the Pacific coast were destroyed.

2574. Major landslide of a slag heap in Wales happened on October 21, 1966, in Aberfan. The 800-foot-high pile of mining debris was weakened by a natural spring beneath it and released more than 2 million tons of rock, coal, and mud when it burst. The first building it ran into was a school, and 116 children were killed.

2575. Mudslides to cause property damage of over $135 million were landslides of January 18–26, 1969, in Los Angeles, CA. Exclusive homes of the San Gabriel and Santa Monica Hills were destroyed during the eight days of landslides.

2576. Mudslide in the 20th century to bury a city in Colombia occurred on November 13, 1985, killing 21,000 of the 23,000 residents of Armero. The site of the buried city has not been occupied since. The disaster was caused by the eruption that day of the Nevado del Ruiz volcano. At 11 P.M., a 130-foot wall of mud buried Armero, 30 miles away. Another 1,000 or so people were killed in other villages. During a previous eruption in 1845, a mudslide in Armero had caused about 1,000 deaths.

FORESTS AND TREES

2577. Conifer forests appeared some 230 million years ago. For around 100 million years, conifer forests covered most land areas of the globe.

2578. Broadleaf deciduous forests appeared some 50 million years ago. At one time, deciduous forests grew in the polar regions; with global cooling, they shifted south to the midlatitudes.

2579. Clearing of the natural forests of northern Europe occurred from 5000 to 2000 B.C. with the introduction and spread of settled agriculture. Oak, beech, elm, and lime forests in Germany, France, and parts of the Netherlands were hewn down to open land for food crops.

2580. Official designation of the "Mendocino Tree" as the world's tallest living thing occurred in 1999. The California redwood near Ukiah, CA, is estimated to be 600 to 800 years old and stands 376.5 feet above the forest floor.

FORESTS AND TREES—DEFORESTATION

2581. Rapid deforestation of Italy began circa 300 B.C. Eroded soil washed into rivers and carried to the sea silted up the ports of Paestum, Ravenna, and other towns.

2582. Acute timber shortages in north China occurred in the early 13th century. Demand for wood and charcoal had resulted in widespread deforestation of China.

2583. Signs of a timber shortage caused by the deforestation of western Europe occurred in the 15th century in the shipbuilding industry of Venice. By late in the 16th century, Venetian shipyards had exhausted local supplies and were importing completed hulls for their vessels.

2584. Widespread teak harvesting by British companies in Burma followed the British conquest of the country in 1826. Within 20 years, the entire province of Tenasserim had been stripped of teak.

2585. Serious warnings about a timber shortage in the United States were issued in the 1850s. Widespread deforestation, a series of devastating fires, and extensive homesteading alarmed conservationists.

2586. Widespread commercial logging in the Philippines began under American direction in 1904. By the 1980s, less than one-third of the islands' virgin forests were standing.

2587. Widespread, widely reported blowdown in the United States occurred in 1921 on the Olympic Peninsula of Washington State. The gale cut a swath 30 miles wide and destroyed 5 million board feet of timber.

2588. Diebacks of significance to western white pine forests in the United States were in the 1930s. The cause was long-term drought.

FORESTS AND TREES—DISEASES AND PESTS

2589. Observation of the disease known as damping off was in Europe circa 1795. The disease, probably caused by fungi, causes seedlings to topple over and die just after they poke above the soil.

2590. Gypsy moth infestation in a U.S town occurred in Medford, MA, near Boston, in the 1880s. The leaf-attacking pest had been accidentally introduced from Europe in 1869.

FORESTS AND TREES—DISEASES AND PESTS—*continued*

2591. National and state efforts to eradicate the foliage-destroying gypsy moth in the United States were begun in 1890. Despite these and later efforts, the pest's range has continued to expand.

2592. Chestnut blight in the United States appeared in 1903. The disease virtually wiped out the country's native chestnut trees.

2593. Bark beetle in the United States was identified near Boston in 1904. The beetle (*Scolytus multistriatus*) would later be the leading vector of Dutch elm disease, which destroyed millions of trees in the United States.

2594. Introduction of gypsy moth-attacking parasites into the United States from Europe and Asia began in 1905. The parasites had only limited effect in controlling the leaf-eating moths, which were then infesting Massachusetts and Connecticut.

2595. Bounty for tree-damaging porcupines was offered in New York State in 1908. Porcupines feed off the bark of northern hardwoods in the east and off the bark of ponderosa and lodgepole pine in the west.

2596. Tree-killing white pine blister rust in the United States was found on imported German white pine seedlings in 1909. The fungus was well established in the eastern pine belt by 1915.

2597. Appearance of white pine blister rust on the Pacific coast of North America was in 1910, when the fungus turned up on nursery stock sent from France to Vancouver, British Columbia. The disease eventually spread south to California and east to Montana.

2598. Large-scale widely reported attack of the larch sawfly occurred in Minnesota between 1910 and 1926. The pest destroyed 1 million board feet of tamarack.

2599. Large-scale widely reported outbreak of spruce budworm occurred in Quebec, Canada, between 1910 and 1920. The epidemic destroyed 200 million cords of balsam fir.

2600. Dutch elm disease description was published in the Netherlands by scientist Marie Beatrice Schwartz in 1921. The disease was named for the country of first appearance; there is no "Dutch elm" tree. By 1930 the disease had reached the United States, where it spread rapidly, killing most American elms.

2601. Confirmation that Dutch elm disease-carrying bark beetles deposit fungi spores on feeding wounds of trees was by William Middleton in 1934. The spores germinated and grew into the conductive tissues of elms; this soon led to the illness and death of the tree.

2602. Appearance of the Asiatic chestnut blight in Europe was in 1938, in Switzerland and Italy. The tree-destroying blight spread rapidly to many parts of Europe.

2603. Widespread spraying of the pesticide DDT to eradicate gyspy moths in the United States occurred in the late 1940s. DDT proved an efficient killer, but brought only short-term relief. Gypsy moth infestations dramatically increased in the 1950s.

2604. Significant federal legislation attacking forest pests in the United States was the Forest Pest Control Act, passed in 1947. The measure made federal funds available to control diseases and pests on state and private land.

2605. Widespread use of DDT to spray for the tree-killing spruce budworm in Canada occurred in 1952, in New Brunswick. The treatment seemed to work. By the following summer, though, the budworm had spread over a four times greater area of forest.

2606. Signs of damage to trees far from automobile pollution sources were seen in the San Bernardino National Forest 80 miles downwind of Los Angeles, CA, in the early 1960s. Excess ozone reduced growth rates there by 80 percent.

2607. Release of imported gypsy moth–killing nematodes in North America occurred in New Jersey in 1974. Nematodes (roundworms) had infected and killed leaf-destroying gypsy moths in Germany and Austria.

2608. Asian long-horned beetle seen in North America was discovered in 1992 in a warehouse in Ohio that contained raw wood packing materials from China. In August 1996, the beetle, which bores holes in deciduous trees and kills them, was found infesting some Norway maples in Brooklyn, NY. By 2000 it had appeared in adjoining boroughs and the Long Island suburbs, and a separate population had been identified in Chicago. More than 5,000 trees were affected, and either died or had to be destroyed. Experts believe the beetle has been introduced into port cities repeatedly, through wooden crates and pallets, and consider it a major threat to American parks and forests. Early in 1999 a ban was imposed on all imports involving untreated wood from China.

The U.S. Department of Agriculture's Animal and Plant Health Inspection Service, working with state and city agencies, began an aggressive eradication program, seeking to identify and remove infested trees and to inject healthy ones with imidaclopride, a systemic insecticide.

2609. Appearance of an oak tree blight previously unknown in the United States occurred in California in 1995. The blight, which came to be called "sudden oak death," affected the roots and then the bark of trees, causing their death within a few years. It spread rapidly along the West Coast, seeming to prefer oaks but sometimes attacking other species as well. Within five years it had killed about 100,000 trees. On July 31, 2000, plant pathologists at the University of California at Davis announced that they had discovered the cause of this plague: a new kind of parasitic fungus, *Phytophthora ramorum*. The *Phytophthora* genus contains a number of notorious plant-killers, including the one that caused the potato famine of the 1840s.

FORESTS AND TREES—FIRES AND FIRE PREVENTION

2610. Firm evidence dating a major U.S. forest fire was burn scars on California giant sequoia trees. The scars dated the fire to A.D. 245.

2611. Forest fire ordinances were enacted in Germany toward the close of the 16th century.

2612. Forest fire to cause a large number of deaths in the United States began on October 8, 1871, at Peshtigo, WI, a few miles north of Green Bay, and claimed 2,682 lives while burning some 400 square miles of timberland. The forests were dry, since there had been no rain for three months, and the fires may have been started by burning railroad debris.

2613. Forest fire lookout tower was erected by the M. G. Shaw Lumber Company of Greenville, ME, on Squaw Mountain, southwest of Moosehead Lake. The first watchman to occupy the tower—a log cabin with a flat roof—was William Hilton of Bangor, ME, whose service started on June 10, 1895.

2614. Legislation in the United States to provide federal funds to states for cooperation in forest fire control was the Weeks Act. Congress enacted the measure in 1911.

2615. Forest fire air patrol was established on June 1, 1919, by the Forest Service of the U.S. Department of Agriculture and discontinued on October 31, 1919. The U.S. Army covered most of the costs. Five routes were covered twice a day out of March Field in Riverside, CA. The five airplanes flew 2,457 hours and covered 202,009 miles.

2616. Provision for federal cooperation in fire control on private timber or forest-producing lands came with the passage of the Clarke-McNary Act of 1924. The measure provided up to $20 million annually and also called for a study of the effects of tax laws on forest preservation.

2617. Large-scale fire prevention publicity campaign in the United States was launched in Washington State in 1940. The Keep Green movement rapidly spread throughout the country.

2618. Environmental public service symbol of the U.S. government was Smokey Bear. Adopted by the U.S. Forest Service during World War II to raise public awareness about the danger of forest fires, Smokey's image was used on posters and in comic books distributed through schools and later in television commercials. Albert Staehle, an illustrator, originated the image of the firefighting bear on August 9, 1944. In 1947, a Los Angeles ad agency coined the slogan "Only you can prevent forest fires." The Forest Service named a cub rescued in 1950 during a New Mexico forest fire "Smokey Bear." The bear was exhibited in the National Zoo in Washington, DC, until he died in 1976.

2619. Use by Smokey Bear of the fire prevention slogan "Remember—only you can prevent forest fires" occurred in 1947. Smokey himself had been introduced three years earlier.

2620. Forest fire battled using artificial rain was attacked on October 29, 1947, near Concord, NH, by seeding cumulus clouds with dry ice. Seeders from the General Electric Company of Schenectady, NY, flew over the burning area in "rain-making" planes and rain followed. However, it was impossible to determine the extent of artificial rainfall since rain caused by natural conditions followed. The experiment, dubbed Project Cirrus, was a joint weather research program of the Army Signal Corps and the Office of Naval Research.

FORESTS AND TREES—FIRES AND FIRE PREVENTION—*continued*

2621. Interstate forest-fire protection agreement in the United States was adopted in 1949. Widespread fires in Maine in 1947 prompted the Northeastern Interstate Forest Fire Protection Compact, involving the six New England states and New York. State firefighting equipment and manpower were pooled for emergency use.

2622. Interstate forest-fire protection agreement in the southern United States was the South Central Interstate Forest Fire Protection Compact of 1954. In it the states of Arkansas, Louisiana, Mississippi, Oklahoma, and Texas agreed to pool their resources for battling forest fires.

2623. $100 million forest fire–fighting campaign in U.S. history was carried out during the hot, dry summer of 1988 against seven major fires in Yellowstone National Park in Wyoming. Wildfires burned 36 percent of the park's area that summer. More than 25,000 firefighters were mobilized to contain the blazes.

2624. Legislation in California to require fire-resistant roofing in that fire-prone state was adopted in 1991 in the wake of the October 1991 fire in the hills above Oakland and Berkeley. The fire killed 25 people and destroyed 2,810 homes.

FORESTS AND TREES—LEGISLATION AND REGULATION

2625. Forestry law enacted by a British colony was passed by Plymouth Plantation on March 29, 1626. The law required the approval of the governor and the council to export lumber.

2626. Significant legislation to manage French forests to assure a continuous yield of timber was the Ordinance of Waters and Forests of 1669. The goal was to achieve self-sufficiency in timber for France.

2627. Environmental law in the American colonies is said to be a British government edict of 1691 reserving New England pine trees for naval ships' masts. Three cuts of an axe, known as the "King's broad arrow," marked reserved trees.

2628. Legislation in an American colony regulating timber cutting to protect the soil was enacted in Massachusetts in 1709. The law curbed logging in Truro on Cape Cod to protect the village from encroaching sand.

2629. Timber trespass laws in French Canada were adopted in 1720. Local timber shortages led to the adoption of the new laws.

2630. Legislation allowing settlers and miners to cut timber on U.S. public lands for their own use was the Timber Cutting Act of 1878.

2631. Forest conservation legislation in the United States was approved in 1891. With the Forest Reserve Act, the U.S. Congress authorized the president to set aside forest lands in the public domain and close timber areas to settlers. Two later measures, the Carey Land Act (1894) and the Newlands Reclamation Act (1902), supplemented the forest reserve legislation.

2632. National legislation mandating protection of timber and water resources in U.S. national forests was the Organic Act of 1897. It also provided for the administration and use of existing and new U.S. forest reserves. The measure allowed for the tapping of forest reserves when private timber supplies were depleted, as in wartime or other periods of high demand.

2633. Legislation allowing exchanges of U.S. public domain forest land for other lands of equal value was approved in 1922. The General Exchange Act stipulated that swaps could only happen when the public interest would benefit.

2634. Substantial federal legislation to provide funds for research into reforestation in the United States was the McSweeney-McNary Research Act of 1928. The measure also called for a comprehensive survey of U.S. forests.

2635. Federal Lumber Code Authority in the United States was adopted in 1934 as part of the New Deal's National Recovery Act (NRA). Private industry groups took over the code when the U.S. Supreme Court declared the NRA unconstitutional in 1935.

2636. Legislation authorizing federal-state cooperation in U.S. farm forestry was the Norris-Doxey Farm Forestry Act. Congress approved the measure in 1937.

2637. Important federal legislation to provide technical forestry services to private nonindustrial landowners was the Cooperative Forest Management Act of 1950. The measure also provided technical services to local governments and private agencies.

2638. Legislation preserving a significant extent of coastal redwood forest in the United States was the Redwoods National Park Act of 1968. The act created a reserve of 58,000 acres in California.

2639. Federal court ruling declaring clear-cutting in national forests a violation of U.S. law came in 1973. The U.S. District Court for the Northern District of West Virginia ruled the lumbering practice illegal.

2640. Significant forestry legislation in the United States in the second half of the 20th century was the Forest and Rangeland Renewable Resources Planning Act, also known as the Humphrey-Rarick Act. Approved in August 1974, the measure mandated long-range planning to assure future forest resources.

FORESTS AND TREES—LOGGING

2641. Commercial logger on the East Coast of North America was probably the Viking Thorfinn Karlsefni, who in 1012 "caused the trees to be felled, and hewed into timbers, wherewith to load his ship."

2642. Recorded imports of timber into England were in 1230, from Norway. By the 16th century, deforestation had forced England into large-scale imports of fir and oak from Scandinavia and Russia.

2643. Spanish ships built from pine trees felled along the southern reaches of the Mississippi were launched in 1543. Soldiers of Hernando de Soto's exploring expedition carried out the logging and shipbuilding.

2644. Pacifist ecological protests in India occurred in the 17th century when Amrita Devi clung to a tree marked to be cut for the Maharajah of Jodhpur's new palace. She clung to the tree and cried, "A chopped head is cheaper than a felled tree."

2645. Tree species in North America to be widely harvested commercially was the eastern white pine, a conifer, in the early 17th century. It grew naturally in a broad band from Newfoundland to Manitoba in Canada and southward from the Great Lakes region to New England.

2646. Sawmill established in America was in Jamestown Colony, in Virginia, in 1607. Historians regard it as America's first commercial enterprise.

2647. Exports of New England white pine to England were shipped in 1631. Riven or hewn boards were sent across the Atlantic from the colony of New Hampshire.

2648. New England pines for Royal Navy masts were felled in 1652. By the end of the century, timber dominated the economy of the New Hampshire colony.

2649. Steam-powered sawmill was established in New Orleans in 1803.

2650. Great "lumber town" in the United States was Bangor, ME, a 1769 settlement incorporated as a city in 1834. During the 1820s and 1830s, lumber harvested in Maine's vast white pine forests and shipped from Bangor made the city one of the world's busiest ports. By the 1850s, the lumber industry had begun moving west.

2651. Logging crews to form the advance guard of what would become known as the Big Cut in the U.S. Great Lakes region reached the Saginaw Bay area of Lake Huron about 1835. Between 1835 and 1900, the great primeval forests of the region would be substantially exhausted.

2652. Appearance of the Scribner Log Rule for determining board-foot measurements of timber was in the United States in the 1840s. J. W. Scribner's is the oldest measuring rule in general use in the country.

2653. Company forests in the United States date back before the Civil War, to the 1840s. Large firms bought up vast areas of virgin timber throughout the United States and hired logging crews to work them.

2654. Use of a locomotive in logging operations occurred in the United States in 1852.

2655. Use of the "Big Wheel" in logging dates from the early 1870s. A wheel-and-axle arrangement with a five-foot clearance at the hub, the Big Wheel could raise logs high enough to clear stumps, meaning that logging could be carried on in the absence of deep snow cover. It was first used in Michigan.

2656. Appearance of the two-man crosscut saw for felling timber dates from the early 1880s. The so-called "misery whip" substantially increased loggers' productivity.

2657. Skyline lead for logging was invented in 1886 in Michigan. Along with the "Big Wheel," it substantially extended the range of summer logging, allowing timber companies to exhaust the Great Lakes forests all the sooner.

2658. International Log Rule for measuring timber yield was published in 1906. In modified form, it became an official rule of the U.S. Forest Service, in use mainly in eastern national forests.

2659. Use of tractors in timber harvesting occurred during World War I. The use of tractors became widespread during the 1930s.

2660. Widespread replacement of railroad logging by trucks in the United States occurred in the late 1930s. Logging railroad lines continued to be used in some parts of the country as late as the 1960s.

FORESTS AND TREES—LOGGING—*continued*

2661. Emergence of Oregon as the leading U.S. timber-producing state was in 1938. Oregon supplanted Washington and remained at the top of the list for many years.

2662. Introduction into the United States of the method of timber cruising called "variable plot cruising" occurred in 1952. Developed in Germany, the method required a cruiser to stand at a spot in the forest and, turning a circle, count all the trees whose diameters seemed to be greater than that of an angle gauge held at eye level.

2663. Balloon logging experiments occurred in Sweden in 1956. The method now is used on steep hillsides in the western United States and Canada.

2664. Helicopter logging experiments in the United States occurred in the 1960s. Two helicopters were used; one flew the logs while the other hovered near on standby.

2665. Attack by loggers on a children's book occurred in Vancouver, British Columbia, in February 1992. Loggers asked a local school board to pull *Maxine's Tree*, about a girl who adopts a tree as part of her campaign to save a forest from clear-cutting, from an elementary school library.

2666. Logging company in California history to have its timber license suspended for environmental violations was the Pacific Lumber Company, in November 1998. In early 1999, the company sold 10,000 acres of ancient redwood groves to the United States for $480 million. In a deal that environmentalists questioned, the government agreed to let Pacific Lumber harvest trees on the rest of its 211,000 acres in northern California.

FORESTS AND TREES—MANAGEMENT

2667. Evidence of widespread forest clearing by felling and the use of fire dates from circa 30,000 B.C. in New Guinea. The purpose was to open the forest cover so wild food plants such as yams and taro could flourish.

2668. Controlled forest management began in Germany in the 16th century. Foresters tried to balance harvesting and regeneration to assure a sustainable yield of firewood and wood products.

2669. Development of the science of forestry began during the 16th century in England and the Netherlands. It grew out of a concern over the scarcity of quality timber for naval construction.

2670. Edict setting aside for Royal Navy use all Massachusetts trees more than 24 inches in diameter was included in the provincial charter for the colony issued in 1691. The decree soon spread to the rest of New England, Nova Scotia, New York, and New Jersey.

2671. Post–American Revolution "broad arrow" policy marking trees for naval use was adopted in Massachusetts in 1783. In modified form, it later became U.S. policy.

2672. Use of fertilizers in forests was in western Europe and in India in the mid–19th century. The practice did not, however, become widespread for many decades.

2673. Forest policy for British India was laid down in 1855. Lord Dalhousie, the Governor General of India, established the Indian Forest Service in response to a threat to the supply of teak ship timbers in Burma and India.

2674. Division of Forestry in the United States was established in 1876. A forerunner of the U.S. Forest Service, its first head was Franklin B. Hough.

2675. Forest management on a professional scale was begun in 1891 in Asheville, NC, on George Washington Vanderbilt's Biltmore estate.

2676. Law of significance to permit timber cutting in U.S. forest preserves was the Organic Act of 1897. The Act opened federal preserves to logging in the event of a timber shortage.

2677. Substantial federal effort to promote industrial forestry on private land in the United States dates from 1898. The Forestry Division of the Department of Agriculture launched a cooperative program of technical assistance and advice to owners of large private timber tracts.

2678. Forestry school at a U.S. college was established as the New York State College of Forestry on September 19, 1898, at Cornell University, Ithaca, NY. The law establishing the institution, signed by Governor Frank Swett Black on April 8, 1898, made New York the first state to establish a forestry course. Bernhard Eduard Fernow was the first director and dean. Lectures on forestry and tree culture were given at Yale University in 1873 and at Cornell in 1874. The operation of the school was suspended in 1903.

2679. Authorization of the U.S. Forest Service to make arrests for violations of its regulations came in 1905. Conservation-minded President Theodore Roosevelt gave the Forest Service arrest powers.

2680. Establishment of U.S. Forest Service Field Districts came in 1908. Six districts were set up, with headquarters at Denver, CO; Albuquerque, NM; Ogden, UT; Missoula, MT; Portland, OR; and San Francisco, CA.

2681. Forestry school privately operated in the United States to give scientific training in the care and preservation of shade trees was a department of the Davey Tree Expert Company, Kent, OH, incorporated on February 9, 1909. The first president was John Davey.

2682. U.S. Forest Service study of recreational opportunities and values in the national forests was completed in 1917. The study led to the designation of the world's first wilderness area, the Gila National Forest, in New Mexico in 1924.

2683. Employment program of significance for forestry and other natural resource projects was the New Deal's Civilian Conservation Corps, established in 1933. Congress authorized the corps with the Reforestation and Relief Bill, approved on March 29, 1933. The measure led to the establishment of a chain of camps where unemployed young men were put to work protecting and preserving forest land. The program was dismantled in 1942, after the start of World War II.

2684. State to require the registration of foresters was Georgia, in 1950. More than a dozen states followed Georgia's lead over the next two decades.

2685. Law stating that U.S. forests would be exploited for a variety of recreational, commercial, and conservation purposes was the Multiple-Use–Sustained Yield Act of 1960. The measure mandated that national forests be administered for recreation; range, timber, and watershed development; and fish and wildlife preservation. President Dwight D. Eisenhower signed it into law on June 12, 1960.

2686. Establishment of a federal Bureau of Outdoor Recreation in the United States occurred on April 2, 1962. The agency oversees recreational activities in national forests and on other federal land.

2687. World tropical timber marketing organization was the International Tropical Timber Association, established in 1985 in Yokohama, Japan. More than 60 countries agreed to take the need for conservation into account in the production and marketing of timber.

2688. "Ecosystems management plan" for U.S. national forests was announced on June 4, 1992. The U.S. Forest Service estimated its plan could reduce clear-cutting by as much as 70 percent.

2689. Claim by the U.S. House of Representatives that overestimates of reforestation efforts in the Pacific Northwest had led to dangerously high logging quotas was made on June 15, 1992. A House committee reported that the exaggerated quotas could not be sustained.

2690. Forest Summit in the United States convened in Portland, OR, in 1993. President Bill Clinton met with loggers and environmentalists concerned with the survival of the northern spotted owl, an inhabitant of old-growth forests in the Pacific Northwest.

FORESTS AND TREES—ORGANIZATIONS

2691. National forestry association was founded on September 10, 1875, as the American Forestry Association in Chicago, IL. It merged with the American Forestry Congress in 1882.

2692. State forestry association was the Minnesota Forestry Association, founded in St. Paul, MN, on January 12, 1876, to promote the planting of forest trees. On March 2, 1876, the state appropriated $2,500 for the association to use to carry out this work. E. R. Drake was the first president.

2693. American Forest Congress was convened in 1882. The American Forestry Association called the session to address fears of a timber famine caused by over-harvesting and fires.

2694. International forestry research agency was the International Union of Forestry Research Organizations, established in 1892 with headquarters in Vienna, Austria.

2695. "Master school" for forestry training in the United States was established at Biltmore, the Vanderbilt estate in North Carolina, in 1898. The organizer was Carl Alwin Schenck, who patterned the school on German models.

2696. Office of chief forester in the United States was established in 1905, when the U.S. Forest Service was organized on its present basis.

2697. National professional organization for foresters in Canada was the Canadian Society of Forest Engineers, established in 1908. It became the Canadian Institute of Forestry in 1950.

2698. Private organization for protecting original redwood forests was the Save-the-Redwoods League, founded in California in 1918. The League played an important role in the establishment of Redwoods National Forest.

FORESTS AND TREES—ORGANIZATIONS—*continued*

2699. Organization to gather international forestry statistics was the International Institute of Agriculture, beginning in 1921. The program's headquarters were in Rome, Italy.

2700. National trade group to promote the U.S. timber industry's positions on environmental issues was founded in 1926, forerunner of today's American Forest Council. The council sponsors the American Tree Farm System, which includes 40,000 tree farms in the United States.

2701. World Forestry Congress convened in Rome, Italy, in 1926. Some 1,200 delegates from 58 countries attended the congress, sponsored by the International Institute of Agriculture.

2702. International organization to promote the timber trade was the Comité International du Bois, established in Vienna, Austria, in 1932. Fifteen European nations, the United States, and Canada were members.

2703. League of Nations conference of world timber experts took place in 1932 in Geneva, Switzerland. The experts discussed potential and actual shortages of timber.

2704. National timber industry trade group in the United States was American Forest Products Industries Inc., established in 1941. Renamed the American Forest Institute in 1968, the organization represents lumber, logging, and other forest industry interests.

2705. Large-scale United Nations program to fund forestry projects was the United Nations Development Fund, established in New York City in 1958. Later renamed the United Nations Development Program, it financed projects to help developing countries strengthen their forestry, agricultural, and nutritional programs.

2706. International professional forestry society was established in the United States in 1969. The International Union of Societies of Foresters promotes cooperation in the field.

2707. Forestry volunteers for the U.S. Peace Corps were recruited in 1971. About 75 volunteers worked in forestry programs in Asia, Africa, and Latin America.

2708. Administrator named to head the U.S. Forest Service who was not a career bureaucrat or a timber planter was wildlife biologist Jack Ward Thomas, appointed in 1993. The appointment signaled a shift in Forest Service policy from mainly timber production to a broad ecological range of public forest issues.

FORESTS AND TREES—PRESERVATION AND RESERVES

2709. Community forest was established in Newington, NH, in 1710. The "town forest" set aside an area of 110 acres of pine trees.

2710. American naturalist to suggest that every town preserve a "primitive forest" was Henry David Thoreau, in an essay of the 1840s titled "Huckleberries."

2711. Forest reserve set aside by a state was the New York State Forest Preserve in the Adirondack Mountains, so designated on May 15, 1885. Logging and other commercial forms of exploitation were prohibited in the forest preserve, which functioned as a state park. Over the years, additional public and private lands were added, to make up what is now called Adirondack Park, a complex patchwork of 6 million acres, with 2.6 million directly owned and managed by the state. Adirondack Park is the largest park in the lower 48 states.

2712. National forest reserve in the United States was the Yellowstone Timberland Reserve in Yellowstone National Park, Wyoming. The reserve was established under an act of Congress signed by President Benjamin Harrison on March 30, 1891. The Land Office of the Department of the Interior administered it. Its territory was extended later the same year and extended again in 1902, at which time it was separated into four divisions. One of these, the Shoshone Division, was officially designated the Shoshone National Forest on July 1, 1908.

2713. Federal program of significance to purchase forestland for watershed protection in the United States came about as a result of congressional passage of the Weeks Act in 1911. The American Forestry Association, a private conservation organization founded in 1875, sponsored the legislation.

2714. Use of the term *primitive area* by the U.S. Forest Service to describe wilderness came in the 1930s.

2715. Voluntary international principles to conserve the world's forests were agreed upon at the United Nations Earth Summit in Rio de Janeiro in June 1992. The document asserted the right of countries to exploit forests economically so long as this was done on a sustainable basis.

2716. Pencil made of recycled newspaper and cardboard fiber was introduced on April 20, 1992. Faber-Castell called its forest-friendly pencil the EcoWriter.

FORESTS AND TREES—RAIN FORESTS

2717. Announcement that Brazil would transfer its capital from Rio de Janeiro to a new city to be built in the undeveloped interior came in 1957. The building of the new capital, Brasilia, spurred development of the interior and put severe pressure on Brazil's fragile rain forests, in turn threatening the global environment.

2718. Decision to build a highway into Brazil's Amazon wilderness was announced on June 16, 1970. Brazilian President Emilio Garrastazu's initiative to open the Amazon rain forest to economic development led to the rapid destruction of habitat and other environmental damage.

2719. Introduction of the concept of a "debt for nature" swap came in 1984. Thomas E. Lovejoy III, president of the World Wildlife Fund in the United States, advanced the notion that developing countries could ease their national indebtedness by agreeing to protect fragile environments within their borders. In 1991, Brazil, where rain forests covered 40 percent of the total land area, agreed to a $100 million-a-year debt for nature package.

2720. Large-scale boycott of a restaurant chain in an effort to halt imports of tropical beef began in the U.S. on April 14, 1984. The Rainforest Action Network organized a boycott of the Burger King chain, charging that the restaurant's imports of Latin American beef encouraged the clearing of tropical rain forests for cattle pasture. The group called off the boycott in 1987 after Burger King dropped many of its Latin American suppliers.

2721. Formal promise from Brazil to take steps to protect its rain forests came on October 12, 1988, in a televised speech by President José Sarney. International pressure forced Brazil to act; the World Bank, fearing Brazilian development projects would endanger the global environment, threatened to block release of funds promised to Brazil if officials there failed to act to stop the despoliation of the forests.

2722. Environment minister in the ecologically critical country of Brazil was named in 1990. Fernando Collor de Mello, the first directly elected president since 1960, named José Lutzenburger, an outspoken environmentalist, to the post.

FORESTS AND TREES—REFORESTATION AND PLANTING

2723. Forest replanting schemes promoted by the British Admiralty date from the late 17th century. British naval constructors had faced shortages of locally grown timber from at least 1650.

2724. Arbor Day celebration took place on April 10, 1872, in Nebraska. The holiday's founder was Julius Sterling Morton, who, along with the Nebraska State Board of Agriculture, offered $25 worth of farming books to the person who planted the most trees. The winner of the prize, J. D. Smith, planted 35,000 trees in one day. Arbor Day is now generally celebrated on the last Friday of April in all 50 states and more than 70 countries.

2725. Legislation of significance by the U.S. government promoting tree farming was the Timber Culture Act of 1873. The measure allowed an individual who kept 40 acres of timber in healthy condition for 10 years, with trees planted not more than 12 feet apart, to acquire title to the land. The act was repealed in 1891.

2726. Forest planted by the U.S. government was begun in 1891 in the sand hills four miles west of Swan, NE. For the purpose of forming shelter belts to hold the sand in place, a small planting of jack and Norway pines was established in cooperation with private individuals. The land was acquired under authority of the act of March 3, 1891, "an act to repeal timber culture laws, and for other purposes."

2727. Plantings of ornamental Japanese cherry trees along the Potomac River in Washington, DC were in 1912. The trees were a gift of the emperor of Japan.

2728. U.S. legislation providing federal aid for up to 50 percent of the cost of reforestation of farmlands on the Great Plains was the Norris-Doxey Act. Congress approved the measure on May 18, 1937.

2729. Tree farm movement in the United States dates to 1940 in Washington State. Timber trade associations adopted the idea and promoted it in the Pacific Northwest.

2730. Tree farm in the United States was dedicated in 1941 in Grays Harbor County, WA. This first farm covered 120,000 acres.

FORESTS AND TREES—REFORESTATION AND PLANTING—*continued*

2731. Large-scale international private reforestation campaign aimed at individuals and communities was launched on October 12, 1988. American Forests (formerly the American Forestry Association) initiated the program, dubbed ReLeaf, in response to global warming. It involved public education, forest management, and the planting of millions of trees.

FORESTS AND TREES—RESEARCH

2732. School of forest practice was probably that of Hans Dietrich von Zanthier, established at Ilsenburg, Germany, in 1768.

2733. Forestry school in France was established at Nancy in 1825. Its first director, J. B. Lorentz, had been trained in the German forestry school at Tharandt.

2734. Forestry experiment stations were established in Germany in 1863. These research stations were the origin of the modern science of forestry.

2735. Forest experiment station to systematically study the effects of the removal of litter from the forest floor began operations in Prussia in 1865. Forest cover (litter) holds soil in place and is an important component in the health of forests.

2736. Issue of *Forestry Quarterly* was published in 1902. Started by Dr. Bernhard E. Fernow, chief of the U.S. Division of Forestry, it carried technical articles and news.

2737. Forest experiment station in the United States was established in the Coconino National Forest in Arizona in 1908. Scientists there researched timber preservation, forest nursery, and other subjects.

2738. Forest products laboratory of the U.S. Forest Service was established in Madison, WI, in 1910.

2739. Truly international forestry literature was the proceedings of the first World Forestry Congress, published in Rome, Italy, in 1926.

2740. Symptoms of nutrient deficiencies in exotic pine plantations in Australia appeared in the 1930s. This led to expanded research in forest fertilization.

2741. Issue of *Forestry Abstracts* was published in Oxford, England, in 1939. A quarterly of the Commonwealth Forestry Bureau, it publishes English-language abstracts of the world's forestry literature.

2742. Multilingual forest terminology was developed beginning in 1949. A United Nations committee of French, German, and English foresters began compiling a standard glossary of technical forest terms.

2743. Use of the term *desertification* to describe deforestation of tropical and subtropical Africa was by the French forester André M. A. Aubreville in 1949. The term came into general use in the 1970s.

2744. Government program in the United States to screen plant and animal species for natural products with anticancer properties began in 1960. Between 1960 and 1981, the National Cancer Institute tested 114,045 plant and 16,196 animal extracts. One compound, dubbed taxol, showed significant cancer-fighting potential, especially for ovarian cancer.

2745. United Nations conference on desertification took place in 1977. The conference attempted to mobilize global resources to address the issue of environmental degradation through deforestation.

2746. Inventory and assessment of the world's forest resources was undertaken by the United Nations Food and Agriculture Organization in 1980. A 1989 update showed that tropical deforestation had accelerated during the decade in consequence of forest fires, harvesting wood for fuel, and conversion to agricultural uses.

2747. Study that defined the factors making old-growth Douglas fir forests unique was published in 1981 by a team of U.S. Forest Service researchers under the direction of Jerry Franklin. The study contradicted the conventional view that old-growth forests consisted of mainly dead and dying trees that should be harvested.

2748. Systematic research documenting the slow death of the sabal palm along Florida's Gulf of Mexico coastline was released on March 15, 1992. Scientists believed rising salt water levels were drowning the root systems of Flordia's state tree.

2749. Forest canopy crane used in North America was erected in 1995 in the Wind River Experimental Forest, part of the Gifford Pinchot National Forest in the Cascade Range in southern Washington State. The 285-foot-tall crane—the largest of three such cranes in the world—raised researchers to the canopy of an old-growth conifer forest that was impossible to reach by climbing. This was the first time such a crane was used to study a temperate forest canopy.

G

GENETIC ENGINEERING AND BIOTECHNOLOGY

2750. Cell description and use of the term *cell* appeared in *Micrographia*, by the English experimental philosopher Robert Hooke. The work was published in 1665, when Hooke was a professor of geometry at Oxford University. He observed plant tissues (such as cork), protozoa, and other cellular matter through a microscope, which was still a relatively new invention.

2751. Nucleic acids were discovered in 1874 at the University of Tübingen in Germany by Swiss scientist Friedrich Miescher. Miescher's early identification of nucleic acids ultimately laid the basis for understanding the genetic code.

2752. Use of the word *gene* to describe units of heredity appeared in University of Copenhagen plant geneticist Wilhelm Johannsen's *Elements of an Exact Theory of Heredity* (1909).

2753. Accurate DNA structure description was published in 1953 in the journal *Nature* by English biophysicist Francis Crick and American biochemist James Watson, who were working together at Cambridge University. With the help of data and an X-ray crystallographic image created by chemist Rosalind Franklin, Crick and Watson developed a hypothetical model of the double helix structure of the dioxyribonucleicacid (DNA) molecule. The model successfully revealed DNA as the source of hereditary information in all organisms. The discovery ultimately made genetic manipulation and cloning possible. Watson and Crick shared the 1962 Nobel Prize in Physiology or Medicine with Maurice Wilkins, on whose research their theory was partially based (Franklin had died in 1958 and was thus ineligible, despite her contribution).

2754. Artificial gene was synthesized by Indian-born U.S. scientist Har Gobin Khorana in 1970 at the University of Wisconsin. Scientists had previously synthesized genes using natural genes as templates. However, analine-transfer RNA, the gene synthesized by Khorana and his team, was assembled directly from its chemical components.

2755. Genetic engineering was accomplished by U.S. scientists Stanley N. Cohen and Herbert W. Boyer in 1973. The two successfully implanted foreign genetic material into a bacterium.

2756. Biotechnology company in the United States was Genentech, Inc., formed on April 7, 1976. Originally based in San Francisco and cofounded by prominent geneticist Herbert W. Boyer, the corporation developed and marketed drugs produced through recombinant DNA technology.

2757. U.S. Supreme Court decision to rule that living things are patentable material was decided on June 16, 1980, in the case of *Diamond v. Chakrabarty*. The court's decision required that the U.S. Commissioner of Patents and Trademarks award biotechnician Ananda Chakrabarty a patent for creating a new man-made, genetically engineered bacterium capable of breaking down crude oil.

2758. State-sponsored biotechnology initiative in the United States was the North Carolina Biotechnology Center in Research Triangle Park. The center was created by the state legislature in 1981 and became a private, nonprofit corporation in 1984. It supports North Carolina-based biotechnology research, business, and educational programs through grants, information sharing, and a public biotechnology library.

2759. Genetically engineered product to become commercially available was an insulin produced from human material called Humulin, marketed by the Eli Lilly Company of Indianapolis, IN, beginning on May 14, 1982. Previously, all insulin had been produced from animal sources.

2760. Genetically altered plants were approved for outdoor testing in 1986 by the U.S. Department of Agriculture. The cultivars were high-yield tobacco plants.

2761. Genetically altered bacterium tested outside of a laboratory in the environment was called "Frostban." The frost-resistant bacterium was produced by Advanced Genetic Services to minimize crop damage due to cold weather. Frostban was first tested outdoors on April 24, 1987, on strawberries in Contra Costa County, CA.

2762. Genetically engineered mammal to be patented was the "oncomouse" or "Harvard mouse." Invented by Harvard Medical School molecular geneticists Philip Leder and Timothy Stewart, such mice were genetically engineered to possess cancer genes and were bred to aid medical research. The patent was awarded in 1988, a year after the U.S. Patent Office ruled that nonhuman animals were patentable material.

GENETIC ENGINEERING AND BIO-TECHNOLOGY—*continued*

2763. Human Genome Project was begun in the 1980s by the U.S. Department of Energy and National Institutes of Health (NIH), who officially consolidated their research efforts in 1990. The project planned to map and analyze the genetic content of the entire human body in order to further health-related research. The project was accompanied by ongoing analyses of the social, ethical, and legal implications of genetic research. The international program is coordinated by the NIH's National Human Genome Research Institute in Bethesda, MD.

2764. Boycott of genetically engineered food products in the United States was launched in 1992. Under the direction of Jeremy Rifkin, the Pure Food Campaign urged chefs, independent grocers, and some large food chains not to use genetically engineered products after the federal Food and Drug Administration said it wouldn't regulate or label them.

2765. National policy on marketing genetically engineered food products in the United States was issued by the Food and Drug Administration (FDA) in May 1992. Commissioner David A. Kessler announced the FDA would neither regulate such products nor require they be labeled.

2766. Genetic privacy legislation proposal in the United States was drafted in 1994 by George J. Annas, Leonard H. Glantz, and Patricia A. Roche of the Health Law Department at Boston University's School of Public Health. Their report was approved by the U.S. Human Genome Project's Ethical, Legal, and Social Issues Group. The draft suggested laws to regulate the collection, analysis, storage, and use of DNA samples and genetic information. Despite numerous bills introduced in Congress, such as the Genetic Confidentiality and Nondiscrimination Act of 1996, the Senate was unable to agree on any definitive federal legislation regarding genetic privacy.

2767. Genetically altered food was the Flavr Savr tomato, approved for sale by the U.S. Food and Drug Administration on May 18, 1994. It began appearing in stores about two weeks later. Developed by Calgene of Davis, CA, the tomato contained a gene that caused it to ripen quickly on the vine. The company also claimed an improved flavor.

2768. Genetically engineered corn was developed and marketed by Ciba Seeds, a Greensboro, NC, division of the Swiss corporation Ciba-Geigy. The corn's DNA was modified with a gene from the bacterium *Bacillus thuringiensis* or Bt, which is fatal to the European corn borer. The U.S. Environmental Protection Agency agreed with Ciba's claim that the "Maximizer" brand of transgenic corn seed was harmless to other organisms and approved its sale on August 9, 1995.

2769. Mammal to be cloned from an adult cell was Dolly, a Dorset lamb born on July 5, 1996, who was the exact genetic copy of her mother. Dolly was created by embryologist Ian Wilmut, biologist Keith Campbell, and their colleagues at the Roslin Institute near Edinburgh, Scotland, by resetting the biological clock of an adult cell from a sheep. Previous cloning by other researchers had relied on embryonic cells, a technique of much more limited application.

2770. Bioprospecting agreement in the United States was signed August 17, 1997, between Diversa Corporation of San Diego, CA, and Yellowstone National Park. The five-year contract allowed Diversa rights to collect microorganisms sampled from the park's hot springs, in return for $100,000 and royalty payments on any commercially useful discoveries that might benefit the company's enzyme research. The money was earmarked for park conservation and public education programs. After a legal challenge by environmentalist groups, the agreement was suspended on March 24, 1998, by a federal judge, who ordered the Park Service to complete an environmental impact assessment before proceeding with the plan.

2771. State in the United States to ban human cloning was California. On October 4, 1997, a bill passed by the state legislature and signed into law prohibited any person from cloning a human being and from purchasing or selling ovum, zygote, embryo, or fetus for cloning purposes. Possible penalties included fines of $1 million for a corporation or $250,000 for an individual, as well as the loss of professional licenses.

2772. Genetically engineered papayas were developed for Hawaiian fruit growers by researchers from the New York State Agricultural Experiment Station at Cornell University, the University of Hawaii, and the U.S. Department of Agriculture. The plants, called "Rainbows," resisted a virus that had been destroying Hawaii's papaya crop. The first Rainbow seeds were distributed free on May 1, 1998.

2773. State in the United States to ban state funding for human cloning research was Michigan. On June 3, 1998, a bill passed by Michigan's state legislature was signed into law, banning the use of state funds for cloning of human beings or for any research into such cloning.

2774. Human embryonic stem cells to be artificially cultivated were produced by separate groups of researchers at the University of Wisconsin and Johns Hopkins University School of Medicine. Their achievements were announced simultaneously on November 5, 1998. All cells in the human body can be grown from such stem cells. The potential for growing healthy organs was praised by medical researchers but criticized by those who saw ethical problems in using embryonic material to develop the cells.

GEOLOGY AND GEOPHYSICS

2775. Mineralogy book was written by the Greek scholar Theophrastus (374–287 B.C.), a student of Aristotle's. Mineralogists and geologists used Theophratus's work *De lapidibus (Concerning stones)* as a sourcebook for centuries. Its influence lasted through the Middle Ages.

2776. Statement of the theory of groundwater was made by Leonardo da Vinci in 1508. Da Vinci postulated that rivers were fed by rain and melting snow that also sank into the ground. According to him this underground water traveled through the porous strata beneath the surface of the earth.

2777. Terrestrial magnetism theory was proposed by English physicist and physician William Gilbert, who was also a pioneer in the study of electricity. In *De Magnete, Magneticisque Corporibus*, in which he attempted to analyze the Earth's magnetic nature, Gilbert was the first to use the term *magnetic poles*. Gilbert's work, which was published in 1600, was the first significant scientific book written in England.

2778. Dating of the origin of Earth through biblical analysis was proposed in 1654 by the Irish churchman and scholar James Ussher. Using the Book of Genesis, Archbishop Ussher calculated that the planet was created October 23, 4004 B.C. Most Western philosophies about the origin of Earth were grounded in, or at least accommodated, the assumption that the planet had been created by God as described in the Bible. Consequently, Ussher's calculations were widely accepted until the 19th century, when scientific analyses revealed the planet to be considerably older.

2779. Formulation of the Principle of Original Horizontality was proposed by Danish physician Nicolaus Steno (Niels Stensen). In 1669 in Italy he published *Prodromus to a Dissertation Concerning a Solid Naturally Enclosed within a Solid.* He proposed that sediments accumulated in layers.

2780. Proposal to create a geological map was presented to the Royal Society in 1683 by Englishman Martin Lister. Lister called it a soil or mineral map and suggested colored lines be used to show the various soils and rocks in the land.

2781. Stratigraphy theory was proposed in Tuscany in 1669 by Danish physician Nicolaus Steno (Niels Stensen), who realized that geographical strata can be interpreted as a chronological record of the Earth's history. Steno's discovery that rock strata lie in order of their age, with each lower strata being progressively older, is known as the law of superposition.

2782. Geological map was published in 1743 by Christopher Pack. The *Philosophico-Choreographical Chart of East Kent* shows topography and broad categories of minerals or soils such as "clay hills."

2783. Statement that rocks and minerals were not randomly scattered but deposited in bands was made by French geologist Jean Étienne Guettard in 1752. As a result of his study of France he made maps, showing the distribution of minerals, rocks, and fossils, that were published by the Academy of Sciences in 1752. From his maps he could see the bands of minerals.

2784. Geological survey of Russia was ordered by Empress Catherine II. Pierre Simon Pallas led the expedition that began in June 1768 and ended six years later in July 1774.

2785. Use of the term *geology* appeared in Swiss naturalist Horace Bénédict de Saussure's multivolume work about his expeditions in the Alps, *Voyages dans les Alpes*, published between 1779 and 1796. Saussure was a professor of philosophy at Geneva and sponsored a 1786 contest which resulted in the first successful ascent of Mont Blanc.

2786. Systematic work on American geology was published by Johann David Schopf in Germany in 1787. He worked as a surgeon with Hessian troops in the Napoleonic wars. After the war he toured the eastern states and gathered material for his geology book.

GEOLOGY AND GEOPHYSICS—*continued*

2787. Extensive geological study of the Alps was conducted by Swiss geologist Horace Bénédict de Saussure in the 1790s. Saussure believed the study of mountains could reveal much about the history of the earth. The publication of his 1779–1796 work *Voyages dans les Alpes* helped ease the common dread of mountains.

2788. Faunal succession theory was conceived between 1794 and 1799 by William Smith, an English surveyor and geologist whose work in canal-building led him to observe that similar types of fossils lay in the same geologic strata and sequences. This "theory of faunal succession" allowed geologists to date strata at disparate locations.

2789. Publication of the geological principle that the present is the key to the past was written by Scottish geologist James Hutton in *Theory of the Earth* in 1795. Hutton's belief that the "study of the earth's present features and changes explain its past" became the guiding principle of Earth science.

2790. Scientist to establish "uniformitarianism" as a principle of geology was James Hutton of Scotland, who published his theories in 1795. Hutton proposed that the geophysical processes at work in the Earth do not substantially change. His work was ignored until the early 1830s. Uniformitarianism has helped to explain the planet's geological history, as well as natural processes like erosion and volcanism.

2791. Calculation of the Earth's mean density was determined by English philosopher and chemist Henry Cavendish in 1798. By using a torsion balance, a gravitational instrument invented by English geologist and astronomer John Michell, the "Cavendish Experiment" also allowed Cavendish to approximate the Earth's gravitational constant.

2792. Geological maps of the United States were published in 1809 and 1817 by Philadelphia geologist William Maclure. His maps represented the geology of the eastern United States as far west as the Mississippi River.

2793. Geological publication in the United States was the *American Mineralogist Journal*, founded in 1810. Professor Archibald Bruce established the journal. It ceased publication in 1814.

2794. Classification of minerals by their chemical composition was developed by Swedish chemist Jöns Jakob Berzelius. In his book *A New System of Minerology* (1814), Berzelius presented the classification. He discovered and isolated numerous elements, and also introduced the system of writing chemical symbols and formulas still in use today.

2795. Modern geological map was published in England by William Smith in 1815. Smith used his theories of stratigraphical and faunal succession to produce a geological map of England and Wales.

2796. Geological map of the eastern United States was presented to the American Philosophical Society of Philadelphia in 1817. William Maclure traveled extensively from the Canadian border to the Gulf of Mexico and recorded his observations on the distribution of rocks.

2797. Classification of the layers of the Earth's crust in North America were found in Amos Eaton's *Index to the Geology of the Northern States* (1818). It was published as a textbook for his class at Williams College in Williamstown, MA.

2798. Geological periodical still published today was the *American Journal of Science*, founded in July 1818 by Benjamin Silliman. The journal was often referred to as "Silliman's Journal."

2799. Geological organization in the United States was the American Geological Society founded in 1819 at Yale University. The society's lifespan was short; the last meeting was in 1828. Many prominent geologists of the day were members, including William Maclure and Benjamin Silliman.

2800. Geothermal heat exchange theory was proposed by French engineer Nicholas Carnot in 1824. Although Carnot did not put his theory into practice, his ideas were the basis for the later manufacture of geothermal heat pumps.

2801. Scientist to popularize uniformitarianism was Scottish geologist Charles Lyell. James Hutton's theory that geological history can be understood in terms of processes still occurring in the Earth was largely ignored until Lyell's *Principles of Geology* was published in 1830. Uniformitarism replaced earlier theories that geological changes were caused by violent events (catastrophism) or that the earth's geology was the sedimentary remnant of a universal ocean (neptunism).

2802. Complete state geological survey in the United States was authorized by the Massachusetts State Legislature on June 5, 1831. Professor Edward Hitchcock of Amherst College conducted the survey. It was completed in early 1833.

2803. Paper describing dinosaur footprints was published by Amherst College professor Edward Hitchcock in 1836. In *A Report on the Sandstone of the Connecticut Valley Especially Its Fossil Foot Marks*, he describes tracks he believed to be made by giant birdlike creatures. Though he did not name dinosaurs, he was the world's first ichnologist.

2804. Geologist hired by the U.S. government was David Dale Owen, official geologist of the state of Indiana. In 1839, Congress hired Owen to survey 11,000 square miles in the Upper Mississippi Valley, so the government could determine prices for selling mineral assets in the territory.

2805. Statement of the theory of a world wide ice age was written by Louis Agassiz in *Studies of Glaciers* in 1840. Agassiz studied glaciers in his native Switzerland and proposed that the Earth was once covered with ice. Over the next decade he added more data to the theory with studies in Europe and North America.

2806. Geological Survey of Canada was founded in 1842 by Sir William Edmond Logan. The agency was Canada's first scientific institution. Under Logan's direction, *Geology of Canada*, the first comprehensive book about Canada's geological and mineralogical resources, was published in 1863.

2807. Geological survey of Niagara Falls was published in 1842 by American geologist James Hall. Like Scottish geologist Charles Lyell, who visited the area the previous year, Hall concluded that the massive erosive power of the Niagara River causes the rim of the falls to recede. Hall was the first to calculate the rate of the falls's recession.

2808. Geologist to classify rocks from Precambrian times was Canadian geologist Sir William Edmond Logan. Precambrian rocks most closely represent the geological character of Earth's formation and early existence. Logan's work was the first attempt to classify the rocks according to their ages. Logan was also the first to realize that coal resulted from earlier plant life. He was the first director of Canada's Geological Survey, established in 1842.

2809. Director of the U.S. Geological Survey (USGS) was geologist Clarence King, who led the agency from its founding on March 3, 1879. King earned his reputation as leader of a ten-year government-sponsored geological survey along the 40th parallel from eastern Colorado to California begun in 1867. King was succeeded as USGS Director in 1881 by his friend John Wesley Powell, the Grand Canyon explorer, with whom he had urged the government to create the agency.

2810. Geological observations of the Grand Canyon were made by American explorer John Wesley Powell in 1869. After his initial exploration, Powell returned to the canyon in 1871 with a government grant to map the Colorado Plateau. His detailed geological descriptions appeared in Powell's *Exploration of the Colorado River*, which was published in 1875 by the Smithsonian Institution.

2811. Agency created by U.S. government specifically for geological issues was the United States Geological Survey, which was established by Congress and President Rutherford B. Hayes on March 3, 1879. The new agency was part of the U.S. Department of the Interior. It was responsible for "classification of the public lands, and examination of the geological structure, mineral resources, and products of the national domain."

2812. Geothermal district heating system was established for the community of Boise, ID, in 1892. The system is still in use, providing heat by piping water from hot springs into local homes and businesses.

2813. Female scientist employed by the U.S. Geological Survey (USGS) was American geologist Florence Bascom in 1896. She was also a respected researcher and educator who taught at Bryn Mawr College in Bryn Mawr, PA.

2814. Female college graduate in geology was Lou Henry, later Mrs. Herbert Hoover. She graduated from Stanford University in Palo Alto, CA, in 1898.

2815. Geothermal power plant began to produce electricity at Lardarello, Italy, in 1904. As geothermal technology progressed, the small generators initially used to utilize steam from the Lardarello wells were replaced by larger ones capable of providing a significant amount of power to the region.

GEOLOGY AND GEOPHYSICS—*continued*

2816. X-ray analysis of minerals was developed by German physicist Max von Laue at the University of Munich in 1912. By focusing X-rays on a crystal and recording the resulting pattern on a photographic plate, Laue created an image that allowed study of the crystal's atomic structure. For this discovery he received the 1914 Nobel Prize for Physics.

2817. Cohesive theory of the concept of "continental drift" was proposed by German astronomy professor Alfred Wegener on January 6, 1912, at a scientific conference in Frankfurt. Wegener cited oceanographic and geological evidence that suggested the hemispheres were once joined in one large continent, which he called Pangaea. He published his theories in *The Origin of Continents and Oceans* in 1915.

2818. Theory that seafloor spreading was caused by convective heat in the Earth's mantle was proposed by English geology professor Arthur Holmes in 1928. Unlike earlier continental drift theorists, Holmes proposed that cooling molten magma beneath the oceans created new crust capable of pushing the continents apart. The theory gradually gained acceptance after the 1962 publication of American geologist Harry Hess's *History of Ocean Basins*. Hess's findings were confirmed as the study of plate tectonics developed during the 1960s.

2819. Biological evidence that continents were once joined was developed by South African geologist Alexander du Toit in *Our Wandering Continents*, published in 1937. Inspired by German continental drift proponent Alfred Wegener, du Toit discovered that certain types of earthworms were found only in southern Africa, South America, Australia, and India. The geographical distribution of the animals corresponded to Wegener's hypothesis that common land masses had separated over time. Du Toit also based his theory on similarly distributed geological strata and fossils.

2820. Radiocarbon or C-14 dating was developed in 1947 by Willard F. Libby, a chemist at the University of Chicago and Institute for Nuclear Studies. Professor Libby discovered that by comparing the relative carbon-14 content of new organisms to the amount found in older objects like rocks, petrified vegetation, or even ancient human bones, an object's age can be determined. The process is a vital tool in measuring how time relates to geological events. For his discovery Libby won the 1960 Nobel Prize for Chemistry.

2821. Plate tectonics theory verification came with the discovery, by Drummond Matthews and Frederick Vine, of a series of magnetic anomalies found in new crust below the Indian Ocean. In 1963 the British geophysicists published their theory that alternating magnetic fields in new and old crust confirmed the concept of seafloor spreading, in which the ocean floor is in a constant state of change. The discovery provided confirmation for theories of plate tectonics. They revealed that the Earth's surface is a series of plates of oceanic and continental crust that move at rates relative to their thickness.

2822. Orbiting geophysical observatory was launched at 9:23 P.M. on September 4, 1964, from Cape Canaveral, FL, by an Atlas-Agena B rocket. The payload weighed 1,073 pounds and was 6 feet long and 3 feet wide. Dubbed *OGO 1*, the satellite had 32,250 solar cells and two 28-volt nickel cadmium batteries. It had an apogee of 92,827 miles and a perigee of 175 miles.

2823. Satellite to transmit color photographs of the full Earth face was named *Dodge* and was launched on July 1, 1967, from Cape Canaveral, FL, by a Titan 3C rocket. It had an apogee of 20,925 miles and a perigee of 20,685 miles and weighed 430 pounds.

2824. Marine geology vessel was the research vessel *Glomar Challenger*, named for the pioneering 19th century oceanic studies ship HMS *Challenger* and the Glomar Challenger's manufacturer, Global Marine Inc. The ship's construction was commissioned by American universities participating in the Joint Oceanographic Institutions for Deep Earth Sampling (JOIDES) and the National Science Foundation. The drill ship was designed to collect core samples from the ocean floor and began operating above the Mid-Atlantic Ridge in 1968.

2825. Geologist to go to the Moon was U.S. astronaut Dr. Harrison Hogan Schmitt. Dr. Schimitt left for the Moon in *Apollo 17* on Dec 7, 1972. Four days later, on Dec 11, 1972, he landed on the Moon and spent three days gathering geological data.

2826. Geothermal building in Manhattan was Foundation House, built in 1997 on East 64th Street. The building's energy system was maintained by water pipes heated by the bedrock below New York City.

H

HAZARDOUS WASTE

2827. Observer known to write about the toxicity of lead was the Roman author Pliny the Elder, in the 1st century. One of the most poisonous substances on earth, lead is known to harm the brain, nerves, kidneys, and reproductive system of humans.

2828. Law in the United States regulating explosives and flammable materials was enacted by Congress in 1866.

2829. Arachlor to be successfully synthesized was in 1881 in Germany. Arachlors are organic chemical compounds, byproducts of coal tar. The highly toxic arachlors were eventually used as lubricants and hydraulic fluids in industry and renamed polychlorinated biphenyls—PCBs. By the 1920s, they were widely available for coatings, insulators, coolants and other industrial used. In the United States, the Monsanto Company was a leading producer of highly toxic, hard-to-dispose-of PCBs. They have entered into the environment primarily through equipment leaks, aging and weathering of materials containing PCBs, and interaction with food products. Many countries have banned the production of PCBs.

2830. Federal effort to regulate hazardous materials on railroads in the United States dates from 1908. The Explosives and Combustibles Act addressed safety issues on railroads.

2831. Year all the states in the United States had at least one law regulating hazardous wastes was 1924.

2832. Statement of the theory that constant exposure over time to small doses of toxins causes poisoning was published in Alice Hamilton's *Industrial Poisons in the United States* (1925). Hamilton was an expert in lead toxicology.

2833. State government concerns about public landfills began in the 1930s when the Illinois Geological Survey and the California Department of Public Works both conducted hydrological reviews of proposed landfill sites. These reviews showed that landfills could lead to contamination.

2834. Law in Great Britain in the modern era addressing hazardous waste was passed as the Public Health Act of 1936. Among many public health issues, the act was the first to address the growing concern of hazardous wastes.

2835. Warnings against tapping groundwater near industrial developments were published by the Committee on Groundwater Supplies in 1937. The report said "offensive, poisonous or dangerous liquid wastes" could reach the nearby groundwater.

2836. Laws in the United States to include a cost-benefit analysis of industrial wastes were the Restatement of Torts laws passed in 1939. The Restatement of Torts laws allowed this test: "An intentional invasion of another's interest in the use and enjoyment of land is unreasonable unless the utility of the actor's conduct outweighs the gravity of the invasion."

2837. Maximum Allowable Concentrations (MACs) for hazardous substances in the United States were published in the *Journal of Industrial Hygiene and Toxicology* in 1940.

2838. Conference on industrial wastes was held in the fall of 1944 at Purdue University. Professor Don E Bloodgood, an associate professor at Purdue, planned the Purdue Industrial Wastes Conference, which continues to the present day.

2839. U.S. government manual with guidelines for placement of public wells and instructions about possible contamination from industrial sites was published by the U.S. Public Health Service in 1944. *The Sanitation Manual for Public Groundwater Supplies* establishes guidelines for how close to the source of contamination a well should be placed.

2840. Legislation in Canada to establish procedures for handling radioactive waste was the Atomic Energy Act of 1946. The measure created the Atomic Energy Control Board, responsible for, waste among other things, waste storage and disposal.

2841. Accidental distribution in Turkey of feed grain treated with toxic hexacholorobenzene occurred in the 1950s. Some 5,000 people were poisoned after eating the contaminated grains. Liver deterioration and skin blistering were reported; many people died.

2842. Evidence of toxic chemicals in groundwater near the U.S. government's Rocky Mountain arsenal in Colorado emerged in 1951. Beets, alfalfa, corn, and barley irrigated from shallow wells near the arsenal turned yellow and brown; livestock sickened and died. Defoliants, pesticides, and chemical warfare agents were to blame.

HAZARDOUS WASTE—*continued*

2843. Instrument that could detect hazardous chemicals such as PCBs (polychlorinated biphenyls) was introduced in 1952 in England. Scientists Tony James and Archer Martin developed the gas chromatograph at the National Institute for Medical Research in London.

2844. Groundwater contamination inventory in the United States was conducted by the American Water Works Association in 1957. The inventory revealed that 25 states had some groundwater contamination.

2845. Commercial land disposal of low level nuclear waste in the United States began in the early 1960s. Three of the six facilities built to handle waste from nuclear power plants, biomedical and industrial research, and nonmilitary government projects were closed in the 1970s as a consequence of leakage and other contamination problems; a fourth shut down in 1993.

2846. PCB (polychlorinated biphenyl) testing to show that the chemical was highly toxic and dangerous was conducted in Sweden in 1966. Two years later, PCB contamination killed five people in Japan and sickened many others.

2847. Research showing that PCBs (polychlorinated biphenyls) were present in human and wildlife tissues was made public in December 1966. Swedish chemist Soren Jensen's research alerted the world to the presence of highly toxic PCBs in the food chain.

2848. Hazardous waste exchange program in Europe was established by the Netherlands in 1969.

2849. Company to voluntarily limit production of PCBs (polychlorinated biphenyls) was the Monsanto Corporation in 1970. Monsanto agreed to sell the highly toxic compound only to manufacturers of high-load electrical equipment.

2850. Destruction of Michigan cattle, chickens, and eggs contaminated with the chemical compound PBB (polybrominated biphenyl) occurred in 1974. Between 1974 and 1979, 30,000 cattle, 150,000 chickens, and millions of eggs were destroyed to purge PBBs from the food supply.

2851. Evidence of hazardous waste poisoning in the Michigan town of Hemlock emerged in 1974, when 10 Holstein cows sickened and died. Later, a large number of calves and wild animals died, and many people became mysteriously ill—poisoned, so it turned out, from groundwater contaminated with waste brine laced with PCBs (polychlorinated biphenyls) and other toxins from industrial chemical operations in the region.

2852. Cease-and-desist order to the U.S. Army to stop toxic contamination from the Rocky Mountain National Arsenal at Denver, CO, was issued in April 1975. The Colorado Department of Health sought the order after confirming contamination of groundwater near the arsenal.

2853. Discovery of radioactive contamination in the town of Port Hope, Ontario, came in 1975. Radioactive waste buried at sites in Port Hope from 1933 to 1948 poisoned the harbor and the soil in some areas.

2854. U.S. government moves to limit permissible amounts of arsenic in drinking water came in 1975. Between 1975 and 1980, the government reduced allowable levels in water, limited air emissions, and regulated the disposal of arsenic.

2855. Hazardous Materials Transportation Act in the United States was adopted on January 3, 1975. The measure regulated rail, air, water, highway, and pipeline transport of toxic products.

2856. PCB contamination to attract widespread national attention in the United States came to light in 1976. In the early 1970s, two General Electric (GE) plants north of Albany, NY, dumped 440,000 pounds of PCBs into the Hudson River in New York State. By late 1974, researchers were reporting high PCB levels in Hudson River fish. When GE was ordered to stop discharging PCBs, the company threatened to close the plants and to put 1,200 people out of work. A public outcry forced GE to back down; the company eventually agreed to a $4 million cleanup of the Hudson River.

2857. Comprehensive legislation in the United States banning PCBs and regulating their disposal was enacted in 1976. The Toxic Substance Control Act also regulated the use and management of a range of hazardous materials.

2858. Comprehensive legislation in the United States to regulate the generation, transport, and management of a wide range of hazardous wastes was the Resource Conservation and Recovery Act of 1976. Congress approved important amendments to the act in 1984.

2859. Canadian town to be partially demolished because of radioactive waste was Port Hope, Ontario, in 1977. By 1979, some 5,000 truckloads of waste were shipped out of the town for storage at the Chalk River Nuclear Laboratories in central Ontario.

2860. Hazardous waste treatment site to accept waste from 48 U.S. states opened at Emelle, AL, in 1978. The landfill, operated by WMX Technologies, was the largest such facility in the United States.

2861. Announcement that New York State would no longer accept out-of-state hazardous waste came in 1979. The state acted to ban imports at the urging of the Department of Environmental Conservation.

2862. Widely publicized scare in hazardous materials transport in Canada occurred in 1979 when a train carrying chlorine gas derailed near Toronto. Some 240,000 people were evacuated from the area. The accident sparked public concern over potential exposure to toxic materials.

2863. Federal trust fund for the cleanup of hazardous waste sites in the United States was set up as part of the Comprehensive Environmental Response, Compensation, and Liability Act of 1980. The act has come to be better known as the Superfund.

2864. Significant legislation in Massachusetts to address the shortage of waste treatment and storage capacity there was the Massachusetts Hazardous Waste Facility Siting Act, approved in 1980. The measure was designed to reduce the state's reliance on out-of-state exports of hazardous waste.

2865. Legislation by the United States to give considerable authority for disposal of low-level radioactive waste to the states was the Low Level Radioactive Waste Policy Act of 1980. With its 1985 amendments, the measure established a system of incentives and timetables for states to take responsibility for waste produced within their borders.

2866. Comprehensive legislation in New Jersey to deal with hazardous waste disposal was enacted in 1981. The Major Hazardous Waste Facilities Siting Act, an effort to alleviate one of the nation's most serious hazardous waste problems, provided for the planning, siting, and licensing of treatment facilities.

2867. Bilateral U.S.–Mexican agreement requiring American companies in Mexico to return hazardous waste to the United States was reached in 1983. Some 1,800 U.S. manufacturing plants were operating just over the border that year.

2868. National Priorities List (NPL) of hazardous waste sites in the United States was created on September 8, 1983. The NPL is used under the Hazard Ranking system of the U.S. Environmental Protection Agency to guide the agency in setting "Superfund" cleanup priorities. The NPL established listing and delisting procedure for sites, as well as a process by which the public can comment on controversial environmental risk listings.

2869. Recommendation from a California state waste management agency against siting hazardous waste incinerators in middle- to upper-income communities came in 1984. The California Waste Management Agency noted that inhabitants of low-income rural areas were less likely to object to such facilities.

2870. Successful effort to block a large hazardous waste disposal plant under the Massachusetts Hazardous Waste Facility Siting Act occurred in 1984. The IT Corporation, in the face of intense opposition from the townspeople of Warren, MA, withdrew its plan for what would have been the largest such plant in the United States.

2871. Town plebiscite to approve a large hazardous waste treatment facility in Alberta, Canada, took place in 1984. A majority of voters approved the Swan Hills Special Waste Treatment Center, which opened in 1987.

2872. Government agency in the United States to regulate leaking underground storage tanks was the Office of Underground Storage Tanks (OUST) within the Environmental Protection Agency. OUST was created through the 1984 Hazardous and Solid Waste Amendments (HSWA) to the Resource Conservation and Recovery Act of 1976 (RCRA) because previous legislation did not address the problem of gasoline and oil leaking into groundwater from underground storage tank systems.

HAZARDOUS WASTE—*continued*

2873. Bilateral U.S.-Canadian agreement to waive "consent notification" for hazardous waste shipments from the United States into Canada dates from 1986. Waste shipments between the two nations moved without the formal consent of the importing government.

2874. Known U.S. waste entrepreneurs to be convicted for fraudulent export of hazardous waste overseas were Jack and Charlie Colbert. The Colbert brothers, with headquarters in Mount Vernon, NY, and warehouses in several states, went to federal prison in July 1986 in connection with a shipment of toxic material to Zimbabwe.

2875. Legislation to provide comprehensive planning for hazardous waste treatment in California was the Tanner Act, passed in 1986. The act failed to achieve its goal of reducing conflict over the siting of treatment facilities, and California continued to export hazardous waste to other U.S. states and to foreign countries.

2876. Toxic waste–carrying ship to draw international media attention was the *Khian Sea*, which wandered the oceans for 27 months in 1986–1988, seeking a place to offload 10,000 tons of Philadelphia incinerator ash. Turned away by a number of countries, the ship, renamed the *Pelicano*, arrived mysteriously empty in Singapore in November 1988.

2877. Comprehensive hazardous waste treatment facility in the Canadian province of Alberta opened at Swan Hills in September 1987. Before that, hazardous waste generated in the province was shipped to the U.S. for disposal.

2878. Greenpeace campaign to stop exports of hazardous waste was launched in 1987. Greenpeace became involved in the issue after its researchers learned about substantial overseas shipments of toxic materials from the United States.

2879. Reported profit of Concord Resources, a U.S. company, for its hazardous waste disposal facility in Blainville, Quebec came in 1987. The plant relied heavily on increased business through imports of toxic wastes from other Canadian provinces and the United States.

2880. International core list of hazardous substances was accepted by the Organization for Economic Cooperation and Development member countries in 1988. The organization's headquarters are in the Netherlands.

2881. Reported disposal of hazardous waste by a "poison ship" from the United States occurred early in 1988 on the island of Kassa, part of the West African nation of Guinea-Bissau. The ship *Bark* offloaded 15,000 tons of toxic ash from Philadelphia's garbage incinerators.

2882. Significant expansion of federal presence in hazardous waste policy in Canada began with the passage of the Canadian Environmental Protection Act in 1988. The act established uniform definitions and procedures for handling hazardous waste.

2883. U.S. legislation to regulate the disposal of hazardous medical materials was the Medical Waste Tracking Act. Congress approved the measure in 1988.

2884. Widely publicized epidemic of medical waste washing up on U.S. beaches occurred in July 1988. Various materials, including syringes and AIDS-tainted vials of blood, washed up on beaches from Maine to the Gulf of Mexico.

2885. Negotiating session to prepare for a global convention on the control of transboundary movements of hazardous wastes was held in Geneva, Switzerland, in February 1988. The convention was to draw upon the Cairo Guidelines and Principles for the Environmentally Sound Management of Hazardous Wastes, which were adopted by the United Nations Environment Programme Governing Council in June 1987. The Geneva session was the first of six that in March 1989 resulted in the Basel Convention on the Control of Transboundary Movements of Hazardous Wastes and Their Disposal.

2886. African-American organization to focus on pollution and hazardous waste in minority and poor communities was the Black Environmental Scientific Trust, founded in 1989 by African-American environmental scientist Dr. Warren M. Washington.

2887. International agreement on hazardous waste was the Basel Convention on the Control of Transboundary Movements of Hazardous Wastes and Their Disposal, approved in 1989. As of 1996, 97 countries had ratified the convention.

2888. Government ban on the import of discarded batteries into Taiwan came in 1990. Taiwan had become a major processor of batteries; lead and other toxic substances released during recycling were causing health problems for smelter workers.

2889. Felony indictments under a U.S. environmental law for smuggling hazardous waste from California into Mexico were issued on May 9, 1990. Two men, Raymond Franco and David Torres, were charged under the Resource Conservation and Recovery Act of 1976.

2890. U.S. Supreme Court ruling striking down a federal requirement that states "take title" to all radioactive wastes after specified deadlines had passed was issued in 1992. Deadlines were implemented to force states to take responsibility for wastes generated within their borders.

2891. Meeting of the contracting parties to the Basel Convention on the Control of Transboundary Movements of Hazardous Wastes and Their Disposal was held in Periapolis, Uruguay, in December 1992.

2892. Law in Minnesota to treat as hazardous waste commonly used products such as motor oil, antifreeze, and fluorescent lamps dates from 1993. The measure required that such products, if not recyclable, be taken to hazardous waste treatment facilities.

2893. U.S. moratorium on building hazardous waste incinerators was imposed in May 1993. Growing environmental concerns prompted the U.S. Environmental Protection Agency to order the moratorium and put new controls on existing incinerators.

2894. U.S. Supreme Court decision that municipal incinerator ash be treated as hazardous waste dates from May 1994. Ash traditionally had been handled as solid waste.

HUNTING AND TRAPPING

2895. Bow and arrow was invented between 50,000 B.C. and 30,000 B.C. Used primarily for hunting, the bow and arrow was most likely the first composite mechanism created by man.

2896. Evidence of humans using birds to hunt appears in an Assyrian bas-relief dated from the early 7th century B.C.

2897. English falconer of record was Ethelbert II, the Saxon king of Kent in the 8th century A.D.

2898. Falconry book was *Art of Falconry*, written in the 13th century by Frederick II of Hohenstaufen, the Holy Roman Emperor. It took him 30 years to write and was the first scientific work on the anatomy of birds.

2899. Female falconer of note was Philippa of Hainault in the mid-14th century. She was the wife of Edward III of England.

2900. Game laws in England restricting hunting to the nobility were first enforced in 1390. The laws existed in one form or another until about 1830.

2901. Written reference to a mousetrap appears in 1450 in Germany and Italy. A German steel trap and an Italian cage trap are described and illustrated in *Mashal la Kadmoni*.

2902. Furs exported from an American colony were shipped to England on December 13, 1621, aboard the *Fortune* under the charge of Robert Cushman, a Plymouth colonist. The ship was captured by the French and the cargo seized.

2903. Fur trading post in the American colonies was established in Augusta, ME, in 1628 by the Pilgrims of Plymouth Colony to trade with the Norridgewock tribe of Native Americans. Most pelts were exported to England.

2904. Hunting law enacted by an American colony was passed by Virginia on March 24, 1629. It provided that "no . . . hides of skins whatever be sent or carried out of this colony upon forfeiture of thrice the value, whereof the half to the informer and the other half to public use."

2905. Closed season on deer hunting in the United States took place in Newport, RI, in 1639. The ban on hunting was established to control the slaughter of valuable wild population.

2906. Hunting privileges in a North American colony were granted in 1647 by the West India Company to potential settlers of New Amsterdam, the Dutch colony that later became New York. Hunting rights and participation in the fur trade were among the economic enticements for colonists who accepted the company's offer of free passage to the New World.

2907. Hunting hounds in North America were imported by Robert Brooke of Maryland on June 30, 1650. The dogs were used to retrieve the dead animals.

2908. Game law in Pennsylvania was enacted on August 26, 1721. Provincial governor Sir William Keith set deer hunting season from July 1 to January 1. A 20-shilling fine was charged for breaking the law. Indians were exempt from the law.

2909. Colonial American game warden system was established in Massachusetts in 1739. New York followed with a similar law in 1741.

HUNTING AND TRAPPING—*continued*

2910. Fox hunting pack in North America was established by Sir William Gooch, lieutenant governor of Virginia, in 1742 on his colonial estate.

2911. English law forbidding night poaching was enacted around 1770. The poacher received 6 months in prison for the first offense.

2912. Metal trap mass-produced in the U.S. was designed by Sewell Newhouse of Oneida, NY, in 1823. Previous to Newhouse's design, metal traps had been individually crafted by blacksmiths.

2913. Hunting license required by a state was mandated in a New York law passed on April 30, 1864, requiring Suffolk County deer hunters to pay a license fee of $10, the money to be "paid over to the overseers of the poor of such town for the benefit of the poor thereof."

2914. Year in which salaried game wardens were appointed in the United States was 1887. A Michigan law approved on March 15 led to the appointment of William Alden Smith of Grand Rapids to a four-year term as warden at an annual salary of $1,200, plus expenses. On April 12 Wisconsin passed a similar law, authorizing the appointment of four game wardens for two-year terms at annual salaries of $600, plus up to $250 for expenses.

2915. Game laws in Oklahoma were passed in 1895 by the territorial government, prohibiting the killing of wild game and insectivorous birds. The regulations also set seasons for hunting quail, prairie chicken, turkey, doves, and plover. Local authorities were responsible for enforcing the laws, which grew increasingly restrictive in the succeeding decade as wildlife became severely depleted. Oklahoma hired its first game and fish warden in 1909.

2916. State agency in Texas to regulate game was the Texas Game Department, which became part of the state's Fish and Oyster Commission in 1907. The agency's name was changed in 1951 to the Game and Fish Commisson. A merger with the state's park system in 1963 created the Texas Parks and Wildlife Department, whose divisions regulate seasons and bag limits, enforce game laws, and conduct species inventories.

2917. State in the United States to ban use of automatic weapons in hunting was Pennsylvania in 1907.

2918. Pennsylvania state law prohibiting aliens from owning firearms was passed in 1909. The law was not repealed until 1967.

2919. Hunting legally allowed on a U.S. government wildlife refuge took place on the Upper Mississippi River Wildlife and Fish Refuge. Created by an act of Congress on June 7, 1924, the refuge allowed hunting on its nearly 200,000 acres of wetlands, which lie in the states of Minnesota, Wisconsin, Iowa, and Illinois.

2920. Position of director of game research was created at the University of Iowa in Ames in 1932.

2921. Survey of recreational hunting on a nationwide basis in the United States was conducted in 1955 as part of the National Survey of Fishing, Hunting, and Wildlife-Associated Recreation. The survey is conducted every five years. The U.S. Fish and Wildlife Service maintains collected data on the number of sport hunters as well as on the amount of money spent annually on equipment, licenses, tags, land leasing, transportation, literature, and other hunting-related expenses. The statistics also reflect the percentages of hunters pursuing big and small game, migratory birds, waterfowl, and agricultural pests.

2922. Game animals on U.S. postage stamps in 1956 were a wild turkey, antelope, and a king salmon.

2923. State to own one million acres of game lands was Pennsylvania. On June 19, 1965, Pennsylvania dedicated its millionth acre of state game lands.

2924. Workshops on hunter safety in the United States were started by the National Rifle Association in 1966 with their sponsorship of a Hunter Safety Education Coordinators Workshop. The workshops were held in cities throughout the United States and continued as an annual event.

2925. Federal aid to hunter education programs in the United States resulted from an October 23, 1970, amendment to the Federal Aid in Wildlife Restoration Act of 1937, also known as the Pittman-Robertson Act. The 1937 law levied an excise tax on guns (and later archery equipment) and uses the resulting revenue to fund wildlife restoration projects. The 1970 amendment made aid available for hunter training programs and the development, operation, and maintenance of public shooting ranges. The U.S. Fish and Wildlife Service's Division of Federal Aid distributed the funds on the basis of state population.

2926. Tax on bows and arrows in the United States was approved on October 25, 1972. The 11 percent excise tax amended the Federal Aid in Wildlife Restoration Act of 1937, or Pittman-Robertson Act. Like the earlier legislation, the tax on bows, arrows, and related equipment was levied to benefit wildlife and hunter safety programs.

2927. National Hunting and Fishing Day in the United States was declared by Congress on April 20, 1973, to be the fourth Saturday of September 1973. After 1975, the observance was designated by individual presidential proclamations, not by Congress. In 1979, President Jimmy Carter permanently designated the third Saturday of every October as National Hunting and Fishing Day.

2928. Tax on crossbow arrows in the United States was approved on July 18, 1984. Like a 1972 tax on conventional arrows, the tax on crossbow arrows, which are commonly used in game hunting, was levied to benefit wildlife and hunter safety programs.

2929. Agreement on International Humane Trapping Standards was signed on December 15, 1997, in Brussels, Belgium, by government representatives of the 15 European Union nations and Canada. The parties agreed to ban the use of steel-jawed leghold traps and to establish binding standards for other mechanical traps used for hunting or pest control.

HUNTING AND TRAPPING—GAME MANAGEMENT

2930. Partridge propagation in the United States was encouraged in 1790 when Richard Bache, son-in-law of Benjamin Franklin, stocked his land at Beverly, NJ, with the game birds. Four years previously, the Marquis de Lafayette had sent a few partridges to General George Washington from France.

2931. Game preserve in the United States was established circa 1860 by Judge John Dean Caton on his Ottawa, IL, estate. He stocked the hunting preserve with many species of American native game.

2932. Game management chair at a U.S. college was established at the University of Wisconsin in August 1933, when Aldo Leopold was appointed to fill the post. The principal objective was to establish an information clearinghouse that would promote graduate research and game production as a new use for Wisconsin land. The University of Michigan had established a school of conservation in 1927, and the University of Iowa had created the position of director of game research in 1932.

2933. Annual license required of all U.S. waterfowl hunters 16 years of age or older resulted from the Migratory Bird Hunting Stamp Act. The law, popularly known as the "Duck Stamp Act," was passed by Congress and signed by President Franklin D. Roosevelt on March 16, 1934. Millions of dollars in revenue generated annually by the law are used by the Department of the Interior and the U.S. Fish and Wildlife Service to buy or lease waterfowl sanctuaries.

HUNTING AND TRAPPING—LEGISLATION AND REGULATION

2934. Closed hunting season in North America was proclaimed by the colony of Massachusetts in 1694 to reverse the effects of overhunting deer.

2935. Closed season on bird hunting in North America was instituted in several counties of the colony of New York in 1708 to reduce overhunting of heath hen, grouse, quail, and turkey.

2936. Prohibition against using camouflaged hunting boats in North America was established in 1710 by the Massachusetts colony. The law forbade the use of camouflaged boats, boats with sails, or camouflaged canoes in hunting waterfowl.

2937. Closed hunting term in North America was a three-year moratorium on deer hunting imposed by the Massachusetts colony in 1718 in response to dwindling deer populations.

2938. Federal game law in the United States was passed by the Continental Congress in 1776. It established closed season regulations on deer hunting for all of the first 13 states except Georgia.

2939. Protection of American Indian hunting grounds by the U.S. Congress was authorized on May 19, 1796, in an act "to regulate the trade and intercourse with the Native American tribes and to preserve peace on the frontiers." Hunting or destroying game within Indian territory was punishable by a fine of $100 and six months in jail. A later treaty signed in 1832 with the American Indians is generally regarded as the first national hunting law.

2940. Hunting law to protect non-game birds in the United States was an 1818 Massachusetts law that imposed a closed-season regulation pertaining to robins and larks.

HUNTING AND TRAPPING—LEGISLATION AND REGULATION—*continued*

2941. Closed season on moose hunting in the United States was established by the Maine legislature in 1830, restricting the hunting of moose to a September–December season. In one version or another, the law remained on the books until 1935, when declining populations led the state to ban moose hunting altogether. It was not allowed again until 1982, and then only by state permit.

2942. Prohibition against the use of gun batteries in hunting in the United States was passed by New York State, in 1838. The law, which outlawed the use of multiple guns in shooting waterfowl, was later repealed.

2943. Law in the United States to outlaw spring shooting was passed by Rhode Island in 1846 to protect wood duck, black duck, snipe, and woodcock from overhunting. The law was later repealed.

2944. Closed season on antelope and elk hunting in the United States was imposed by California in 1852.

2945. Hunting law in California was passed in 1852. It prohibited the hunting of elk, antelope, deer, quail, mallard, and wood ducks for six months of each year.

2946. Closed season on hunting mountain goats and mountain sheep in the United States was imposed by Nevada in 1861.

2947. State in the United States to protect bison was Idaho. In 1864, Idaho imposed an annual five-month closed season on hunting bison, deer, elk, antelope, goat, and sheep.

2948. Closed season on caribou hunting in the United States was imposed in the state of Maine in 1870.

2949. Hunting law establishing "rest days" for waterfowl in the United States was imposed by Maryland in 1872.

2950. Nonresident hunting license in the United States was required in New Jersey. Under the Acts of 1873, which incorporated the West Jersey Game Protective Association and set rules for the taking of game in six New Jersey counties, nonresidents were required to purchase a membership which amounted to a hunting license.

2951. State in the United States to restrict market hunting was Arkansas. In a move to curtail commercial or "market" hunters, who devastated wildlife populations in the 19th century, Arkansas on March 6, 1875, levied a $10 license fee on all nonresident hunters, trappers, and fishermen using nets. In 1903, hunting by nonresidents was completely prohibited.

2952. Bag limit on game bird hunting in the United States was imposed by Iowa in 1878 as a response to commercial or market hunting.

2953. Game commission in the United States was formed in California in 1878. The state's Board of Fish Commissioners, which had overseen the restoration and preservation of fish in California waters since 1870, was expanded to regulate game laws. New Hampshire also formed its state game commission in 1878.

2954. General hunting license in North America originated in Canada. Unlike U.S. states, where the depredations of market hunters led to complete state prohibitions on hunting by nonresidents, New Brunswick instituted a general license system on all game hunting in 1878. Other Canadian provinces and U.S. states soon followed this model.

2955. Bird hunting regulation enacted by the U.S. Congress was the Weeks-McLean Act, also known as the Migratory Bird Act, signed into law on March 4, 1913. The bill was sponsored by Representative John Wingate Weeks of Massachusetts and Senator George Payne McLean of Connecticut. It placed certain migratory species under federal protection, authorizing the Department of Agriculture to limit hunting, and reinforced state laws by prohibiting the interstate transportation of birds, feathers, and other bird parts.

2956. Soviet law to ban polar bear hunting was issued in 1956. This conservation law forbade the shooting of polar bears in the Soviet Arctic and prohibited the capture of cubs without a license.

2957. Law in China to outlaw hunting giant panda was passed by China's State Council in 1962. The law also banned the export of panda skins and provided for the establishment of reserves for panda conservation.

2958. Canadian ban on hunting baby seals was passed on October 15, 1969. The law established a minimum weight at which harp seals could be killed. Because the white coats of baby seals desirable to furriers turned brown by the age at which the seals reached the minimum weight, profits in seal hunting collapsed. The law also banned killing of seals with clubs, while allowing hunting with firearms and exempting from the rules indigenous peoples who depended on seal harvesting.

2959. Banning of aircraft for hunting in the United States occurred on November 18, 1971, with the passage of Public Law 92-159. Known as the Airborne Hunting Act, the law prohibited "shooting or attempting to shoot or harass any bird, fish, or other animal from aircraft except for certain specified reasons, including protection of wildlife, livestock, and human life as authorized by a Federal or State issued license or permit."

2960. Mandatory hunter education law in Idaho was enacted by the state legislature in 1979. It required any person born after January 1, 1975, to produce evidence that he or she had successfully completed a hunter education course accredited by Idaho or another U.S. state. The program was designed to reduce the fatality rate from accidental shootings and increase awareness of wildlife, game management, and hunting law issues.

2961. Penalties enacted by the United States for harassing hunters were established by the Recreational Hunting Safety and Preservation Act. The law was enacted on September 13, 1994, setting fines for anyone convicted of obstructing a lawful hunt. The act was added to the Violent Crime Control and Law Enforcement Act of 1994 and was implicitly aimed at aggressive antihunting protestors. Funds derived from fines were to be directed to federal wildlife conservation programs.

2962. Ban of lead shot in Canada took effect September 1, 1999. An earlier 1996 ban mandated fines and loss of hunting privileges only for people found possessing lead shot in Canada's National Wildlife Areas. A 1997 amendment banned the use of lead shot within 200 meters or 220 yards of any body of water. The 1999 ban applied to all of Canada and made it illegal to use toxic shot in hunting all but three species of migratory game birds.

2963. Ban on fox hunting in Scotland was the Protection of Wild Mammals (Scotland) Bill, passed on February 13, 2002, by the Scottish Parliament in Edinburgh. The bill made it illegal to hunt foxes with packs of dogs, a traditional equestrian sport in Britain.

HUNTING AND TRAPPING—ORGANIZATIONS

2964. Fox hunting club in the United States was the Gloucester Fox Hunting Club, organized in Philadelphia, PA, in 1766.

2965. Fox hunting club in Canada was the Montreal Hunt Club, founded in Canada in 1826.

2966. Archery club in the United States was the United Bowmen of Philadelphia, PA, established in 1828.

2967. Game protection society in the United States was founded on May 20, 1844, in New York City as the New York Sportsmen's Club. On March 10, 1873, it was renamed the New York Association for the Protection of Game. The first president was B. J. Meserole.

I

INDUSTRIAL POLLUTION

2968. Area in Europe to become industrialized was Holland beginning in the 16th century. By 1622, half of Holland's population lived in cities and towns.

2969. Recognition of cancer as an occupational hazard was made by English surgeon Percival Pott in 1775. Pott discovered that chimney sweeps in London suffered an abnormally high rate of scrotal cancer.

2970. Electrostatic device to remove particles from industrial smoke was invented by American scientist and educator Frederick Gardner Cottrell. The device gave unburned particles in industrial emissions an electrical charge. The particles could then be removed by attracting them to an opposite charge. Cottrell's "electrostatic precipitator" could effectively reduce industrial emissions, recover unburned material for reuse, and purify the air in research facilities. Cottrell demonstrated the device in 1906 and applied for a patent in 1907.

2971. Polyvinyl chloride (PVC) was invented in 1926 by Waldo L. Semon, an American researcher at the B. F. Goodrich Company in Akron, OH. Once manufactured, products made from the popular and versatile plastic were apparently harmless. During the 1970s, however, it was learned that production workers exposed to vinyl chloride gas were dying of liver cancer, prompting pollution prevention rules.

INDUSTRIAL POLLUTION—*continued*

2972. Minamata Disease or "cats dancing disease" cases were reported in 1953 in Japan. The disease, caused by methyl mercury waste, affects the cental nervous system. A chemical and plastics factory had discharged such waste into Minamata Bay, contaminating the fish eaten by the local people. Cats also ate the fish, and their gyrations were early signs of trouble. Michiko Ishimori documented the disease in her book *Paradise in a Sea of Sorrows* in 1969.

2973. Industrial waste treatment program at an airport was established in 1962 at Washington's Seattle-Tacoma airport. The airport's treatment plant separates rainwater runoff from jet fuel, engine oils, and other contaminants spilled during maintenance, fueling, and deicing in the terminal areas. The treated water is discharged offshore to prevent pollution of local freshwater sources.

2974. Industrial pollution lawsuit in Japan that was successful was decided in favor of plaintiffs who charged that cadmium runoff from the Kamioka Mining Station into the Jinzu River was the cause of a degenerative bone affliction called Itai Itai disease. In June 1971, Mitsui Mining and Smelting, owners of the mine, were forced to pay damages and install pollution control measures.

2975. United Nations agency for management of hazardous chemicals was UNEP Chemicals, a section of the United Nations Environment Programme, which was founded in 1972. UNEP Chemicals is located in Geneva, Switzerland, where it maintains the International Register of Potentially Toxic Chemicals (IRPTC). UNEP Chemicals dispenses free information and advice on chemical safety, production, use, and disposal. Because it accepts that use of chemicals is necessary to world economies, the agency promotes the use of safe chemical technologies as an element of sustainable development.

2976. National pollution control agency in India was the Central Pollution Control Board (CPCB). The CPCB was established through the Water Act of 1974. The national board is responsible for coordinating activities of state boards, providing technical assistance, and conducting research, training, enforcement, and public education, and investigative work to eliminate pollution.

2977. Federal law to restrict the dumping of hazardous waste in the United States was the Toxic Substances Control Act. The law also instituted a gradual ban on cancer-causing polychlorinated biphenyls (PCBs) and gave the U.S. Environmental Protection Agency the power to ban the sale of chemical products suspected of being hazardous. The bill was signed into law on October 12, 1976.

2978. National amendment to restrict industrial polluters was India's 42nd Amendment, which became effective on January 3, 1977. The amendment's Directive Principles of State Policy require the national government to seek the improvement of polluted environments by allowing legislative restrictions on industries that produce pollution.

2979. Warsaw Pact government to recognize that industrial wastes cause health problems was Poland. New laws were enacted in Krakow, Poland, during the early 1980s to cut the flow of industrial wastes into the air of the city. Maria Guminska and the Polish Ecological Club's activism in the early 1980s caused the Polish government to recycle aluminum wastes at the aluminum smelter in Skawina, downwind from Krakow.

2980. "Superfund" environmental cleanup law in the United States was the Comprehensive Environmental Response, Compensation, and Liability Act (CERCLA). The act, commonly known as Superfund, was enacted by Congress on December 11, 1980. Its provisions established financial liability for businesses responsible for the release or threatened release of hazardous substances causing a threat to public health or the environment. The program also established a trust fund to pay for cleanup of sites where no responsible party could be identified. The trust fund was funded by a tax on the oil and chemical industries, whose companies were responsible for a majority of Superfund cases.

2981. National Contingency Plan (NCP) and Hazard Ranking System (HRS) for hazardous wastes in the United States were authorized under a July 16, 1982, reauthorization of the Comprehensive Environmental Response, Compensation, and Liability Act, also known as the "Superfund" act. It enlarged the mandate of the NCP by providing for emergency responses to hazardous waste releases. The HRS provided a mechanism for the U.S. Environmental Protection Agency to evaluate the character and possible risk of hazardous waste escaping from specific sites through ground or surface water, soil, or air.

2982. Law in the United States to establish asbestos as a health hazard in schools was the Asbestos School Hazard Abatement Act, enacted on August 11, 1984. The act and subsequent amendments stated that asbestos was a proven carcinogen to which children are particularly susceptible. The administrator of the U.S. Environmental Protection Agency was directed to assist states and local educational agencies in evaluating the extent of the danger, to the health of schoolchildren and employees, from asbestos materials in schools. The act also provided funding and scientific and technical assistance to asbestos abatement programs and prohibited the disciplining of any school system employee for reporting asbestos hazards.

2983. Leaking Underground Storage Tank (LUST) trust fund in the United States was created in 1986 by a Congressional amendment to the Resource Conservation and Recovery Act of 1976 (RCRA). The LUST trust fund provided money for overseeing corrective action taken by owners or operators of leaking underground storage tank (UST) systems. The trust fund also provided money for emergency situations and cleanups at LUST sites where the owner or operator was unknown, unwilling, or unable to respond to the problem. The fund was financed by a national 0.1-cent tax on each gallon of motor fuel.

2984. Law in the United States requiring public disclosure of pollutants through the Toxic Waste Inventory was the Emergency Planning and Community Right-To-Know Act (EPCRA) of 1986. Through EPCRA, Congress mandated that a computerized Toxics Release Inventory (TRI) be compiled annually. The TRI is available to the public so that communities can be informed about toxic chemicals released into their environment and the pollution prevention practices of local companies whose activities fall under EPRCA guidlines. EPCRA also required the establishment of state and regional commissions to deal with environmental emergencies. The act is also known as Title III of the Superfund Amendments and Reauthorization Act of 1986 (SARA).

2985. Reauthorization of the "Superfund" program was the Superfund Amendments and Reauthorization Act (SARA). SARA amended the Comprehensive Environmental Response, Compensation, and Liability Act (CERCLA), or "Superfund" act, on October 17, 1986. SARA expanded state and public participation in hazardous waste cleanup decisions and increased the size of Superfund's trust fund.

2986. "Right to Know" law in the United States covering hazardous substances was the Emergency Planning and Community Right-To-Know Act (EPCRA), which was signed as part of a Superfund reauthorization bill by President Ronald Reagan on October 17, 1986. The law increased the monitoring powers of the U.S. Environmental Protection Agency in order to provide the public with information about local use of hazardous chemicals, particularly in instances when toxic substances were released into the environment.

2987. Toxics Release Inventory (TRI) in the United States was mandated by Congress and signed into law as part of the Emergency Planning and Community Right-To-Know Act (EPCRA) on October 17, 1986.

2988. Financial responsibility requirements for underground storage tanks (UST) were published on October 26, 1988, by the U.S. Environmental Protection Agency. The rules established minimum levels of insurance that UST owners and operators need to respond to any leaks, and to compensate anyone who is harmed by a release of UST contents, which commonly include gasoline, solvents, methanol, or antifreeze.

2989. National database in the United States for pollution source reduction was mandated by Congress on November 5, 1990, through the Pollution Prevention Act. The act required the U.S. Environmental Protection Agency to establish a public Source Reduction Clearinghouse that would compile information and maintain a computer database containing information on management, technical, and operational approaches to source reduction.

2990. U.S. federal source reduction policy was contained in the Pollution Prevention Act (PPA), enacted on November 5, 1990. Passed by Congress, the act stated that pollution should be prevented or treated at its source whenever possible. Barring effective source reduction, recycling or environmentally safe disposal of unrecyclable waste would be required. The act shifted emphasis to source reduction rather than waste management or control as the most effective means of reducing pollution.

2991. Environmental accident insurance guarantee in India was created by the Public Liability Insurance Act. The law was enacted on January 22, 1991. It allowed India's government to establish a national environmental insurance fund to provide immediate relief to persons affected by accidents occurring while han-

INDUSTRIAL POLLUTION—*continued*

dling hazardous substances. The law was created under the no-fault principle, so that victims would not have to sue those responsible for the hazardous materials in order to receive damage payments.

2992. Brownfields Economic Redevelopment Initiatives program in the United States was created by the U.S. Environmental Protection Agency (EPA) in June 1993. Brownfields are former or unused industrial facilities where pollution may have contaminated the environment. The EPA's brownfields program sought to make reuse of such sites possible by offering grants for cleanup.

2993. Brownfields grants dispensed by the U.S. Environmental Protection Agency (EPA) were awarded in 1994. The EPA gave $200,000 pilot-program grants to cities to help clean up the former sites of an armory in Richmond, VA; an automobile plant in Cleveland, OH; and a valve factory in Bridgeport, CT. In each case, the grants were combined with private funds from businesses that helped redevelop the formerly contaminated sites.

2994. Cabinet-level environmental protection agency in Mexico was the Secretariat of the Environment, Natural Resources, and Fisheries (SEMARNAP). It was established in 1994 to coordinate the strategies of the government's major environmental departments. SEMARNAP oversees the hazardous waste and environmental policy work of the National Ecology Institute, the water quality regulatory work of the National Water Commission, and the enforcement activities of the Office of the Attorney General for Environmental Protection.

2995. "Environmental justice" action by the U.S. government was Executive Order 12898. The order, also known as "Federal Actions to Address Environmental Justice in Minority Populations and Low-Income Populations," was signed by President Bill Clinton on February 11, 1994. The order directed all federal agencies to devise strategies by which their policies might improve the environment in low-income communities, where living conditions suffer disproportionately because of proximity to polluted and pollution-producing industrial areas.

2996. Natural resource damage settlement paid to New Jersey under the "Superfund" law was made under an agreement reached on March 30, 1994, in the District Court of New Jersey. In the case of *U.S.* v. *Carborundum Company et al.*, 19 companies were sued for the cost of cleaning up waste dumped into Fairfield Township lagoons by the Caldwell Trucking Company between the 1950s and the 1970s. Nine of the defendants agreed to pay $2.46 million for the EPA's costs and to perform an estimated $32 million worth of remedial and natural resource restoration work at the site. The state of New Jersey was paid $984,000 in natural resource damages for injury to the area's aquifer.

2997. Enforcement action by the U.S. Environmental Protection Agency (EPA) against a company for failure to comply with radionuclide emissions standards was an order issued on September 26,1994, by EPA-New England. Syncor International Corporation, a radiopharmecutical company in Woburn, MA, was ordered to comply with the National Emission Standards for Hazardous Air Pollutants standard. The company was also ordered to submit monthly reports and a compliance plan. Man-made radioactivity can result from the manufacture of radiopharmecuticals and medical x-ray use.

2998. Lawsuit over industrial sewers initiated by the U.S. Resource Conservation and Recovery Act (RCRA) was settled in U.S. District Court for the Northern District of New York on October 7, 1994. The case concerned the failure of the Eastman Kodak Corporation to report hazardous waste leaking from miles of sewers at its plant in Rochester, NY. Kodak agreed to pay $8 million in damages and to spend millions more to repair the sewers and to implement pollution control and monitoring programs.

2999. Governmental tribunal in India for litigation relating to environmental accidents was created through the National Environment Tribunal Act (NETA). The act was passed on June 17, 1995. It brought India into agreement with the 1992 United Nations Conference on Environment and Development at Rio de Janeiro, Brazil, which called for nations to establish laws and legal structures to deal with pollution accidents. NETA established a national board of environmental experts to assess compensation for damage to persons, property and the environment caused by mishandling of hazardous substances.

3000. Nuclear byproduct technology capable of cleaning smokestack pollutants was discovered at the University of Liverpool in England in 1996. Researchers discovered that exposing industrial pollutants to a uranium oxide catalyst broke the material down into carbon dioxide, hydrochloric acid, and trace amounts of carbon monoxide. The technique was significant for the promise it held in chemically dismantling organic compounds containing chlorine before they could escape into the environment and enter the food chain.

3001. Scottish Environment Protection Agency (SEPA) began operating on April 1, 1996. SEPA was established specifically to address Scotland's industrial pollution problems, as well as the special environmental needs of the region's many rivers and islands. SEPA's headquarters are in Stirling.

3002. United Nations emergency response system for industrial pollution disasters was established by the UN Department of Humanitarian Affairs (DHA) and its Relief Coordination Branch in Geneva, Switzerland. The DHA typically handles disaster response logistics and communications. In cooperation with the UN Environment Program (UNEP), the DHA emergency response system decided in June 1996 to cover environmental aspects of industrial accidents when international assistance was not otherwise provided for under existing conventions or programs.

3003. Environmental justice complaint filed with the U.S. Environmental Protection Agency pitted environmentalists against Shintech Inc., which planned to build a large polyvinyl chloride plant in Convent, LA, in 1997. Citizens groups represented by the Tulane Environmental Law Clinic protested that the facility would worsen the environment in the already severely polluted area, causing disproportionate suffering in minority communities along the Mississippi River. A test case of the environmental justice policy was avoided when Shintech chose an alternate site.

3004. Modern industrial pollution crime penalties in China were contained in Chapter Six of China's Criminal Code. It became effective October 1, 1997. Revisions of the code set fines and prison terms for anyone responsible for major environmental pollution accidents. Anyone convicted of causing losses of public or private property or creating serious health hazards is liable to fines, imprisonment, or both. A conviction also renders the superiors of the responsible party open to prosecution.

3005. Brightfields initiative by the U.S. Department of Energy was inaugurated in August 1999. The program used brownfields—reclaimed contaminated industrial sites—as locations for solar energy-related businesses. Chicago, IL, launched the program by finding a site for a solar panel manufacturer, who brought jobs to the community and sold alternative energy equipment that was helpful in reducing pollution in the city.

INTERNATIONAL LAW

3006. Recorded statement of the law of the sea was written by Byzantine Emperor Justinian I, probably early in the sixth century. In *The Digest of Justice* he said that no country had jurisdiction beyond the high water mark and that the seas and their fish were available to all.

3007. Papal declaration regarding world exploration was issued in 1481. In the papal bull *Aeterni Regis*, Pope Sixtus IV declared that all lands south of the Canary Islands belonged to Portugal. King John II of Portugal was given the mission to Christianize these lands.

3008. East-west division of the world for exploration was declared in 1493 in the papal bull *Inter Caetera*. On May 4, 1493, Pope Alexander VI gave to King John II of Portugal all new lands discovered east of the 38° meridian and to King Ferdinand and Queen Isabella of Spain all lands west of that line. By the Treaty of Tordesillas the following year, the Atlantic line was moved further west, which allowed the Portugese to lay claim to Brazil. However, the exact location of the dividing line in the Pacific—for the meridian ran around the globe—remained unclear.

3009. Statement of the idea that nations should be held responsible for their actions just as individuals were was published in 1625 in *On the Law of War and Peace* by Dutch scholar Hugo Grotius. Acceptance of this viewpoint led to the mid-20th-century agreements to punish nations who pollute and abuse the Earth's natural resources.

3010. Trilateral agreement to protect the Wadden Sea was the Joint Declaration on the Protection of the Wadden Sea. The document was signed December 9, 1982, in Copenhagen, Denmark, by representatives of the governments of Denmark, Germany, and the Netherlands. The three countries, which adjoin the sea and share its coastal ecosystem, subsequently agreed to coordinate efforts to control pollution, manage fisheries, maintain the area's natural beauty for tourism, and protect Wadden Sea fauna, particularly seals and waterfowl.

INTERNATIONAL LAW—*continued*

3011. Environmental lawsuit decided by the International Court of Justice occurred on September 25, 1997. The case concerned the course of the Danube River, which divides Slovakia and Hungary. In 1992 Slovakia diverted the Danube onto its own territory to provide hydroelectric power and canal transportation, severely endangering wetlands in Hungary. In 1995 a coalition of environmental and human rights groups joined Hungary's lawsuit protesting Slovakia's action. The International Court of Justice ruled that diversion of the river was illegal.

3012. Superpower to effectively withdraw from the Kyoto Protocol was the United States, on March 14, 2001, when President George W. Bush announced that his administration would not require utility companies to control carbon dioxide emissions. The Kyoto Protocol, an international attempt to slow the pace of global warming, had not been ratified by any of its signatory nations, but environmentalists considered it an important first step and the President's action was widely criticized. On February 14, 2002, he presented his own plan, featuring incentives rather than requirements.

INTERNATIONAL LAW—CONVENTIONS, AGREEMENTS, AND TREATIES

3013. Treaty between nations dividing the world for exploration was the Treaty of Tordesillas, signed on June 7, 1494, in Tordesillas, Spain. King John II of Portugal was unhappy with the papal bull Inter Caetera, which in 1493 had divided the world according to Pope Alexander VI's plan, so he went directly to King Ferdinand and Queen Isabella of Spain and renegotiated the terms.

3014. Treaty in which England recognized the fishing rights of the United States was signed by Benjamin Franklin and John Jay on September 3, 1783. The Paris Peace Treaty of 1783, article 3, gave the United States fishing rights to areas off its coast, which had been fished by both countries.

3015. Treaty to recognize the three-mile limit for fishing off the North American coast was signed on October 20, 1818. Albert Gallatin and Richard Rush from the United States and Frederick John Robinson and Hanry Goulburn from Britain negotiated the treaty.

3016. List of wetlands of international importance was created at the Convention on Wetlands of International Importance Especially as Waterfowl Habitat, held in Ramsar, Iran, on February 2, 1971. Each treaty signer agreed to set aside and protect at least one wetland of international importance.

3017. Bird protection treaty was the Migratory Bird Treaty, signed on August 16, 1916, by the United States and Great Britain at Washington, DC, to protect migratory birds in the United States and Canada. It was signed by President Woodrow Wilson on September 1, 1916, and ratified by Great Britain on October 20, and proclaimed on December 8.

3018. International wildlife treaty upheld by the U.S. Supreme Court was the Migratory Bird Treaty of 1916. In *Missouri* v. *Holland*, the state of Missouri attempted to invalidate the treaty by arguing that it abridged sovereign states' rights as set forth in the Tenth Amendment. In its decision of April 19, 1920, the court upheld the federal government's right to make treaties, thus validating the landmark wildlife protection law.

3019. International treaty to propose African wildlife preserves was the Convention Relative to the Preservation of Fauna and Flora in Their Natural State, signed in London, England, on November 8, 1933. The treaty did not enter into force until January 14, 1936. The convention, aimed primarily at nature in Africa, advocated establishing national parks and other reserves "within which the hunting, killing or capturing of fauna, and the collection or destruction of flora" would be limited or prohibited. The agreement advocated regulating the traffic in hunting trophies and prohibiting certain methods of hunting or capturing fauna. It was signed by South Africa, Belgium, the United Kingdom, Egypt, Spain, France, Italy, Portugal, and the Anglo-Egyptian Sudan.

3020. International agreement on whaling regulations was signed on June 8, 1937, in London, England. The agreement called for the conservation and development of the world whaling industry.

3021. Treaty obliging the United States to maintain a list of endangered species was the 1940 Convention on Nature Protection and Wildlife Preservation in the Western Hemisphere. The treaty was signed October 12, 1940, by numerous Western Hemisphere nations and ratified by the U.S. Congress on May 1, 1942. The convention obliged signatories to maintain an "Annex of Wild Fauna and Flora" in danger of depletion or extinction and to designate "Nature Monuments" or refuges for species protection. This convention was an outgrowth of the Migratory Bird Treaty Act of 1918 and a precursor to the Endangered Species Act of 1973.

3022. Treaty to establish national parks and preserves and to identify protected flora and fauna in the Western Hemisphere was the Convention on Nature Protection and Wildlife Preservation in the Western Hemisphere, adopted on December 10, 1940, in Washington, DC. A provision of the convention called for national parks to have educational felicities.

3023. Controls on the import and export of plants and produce were adopted on June 12, 1951, at the International Plant Protection Convention in Rome, Italy. The purpose of the treaty was to control the spread of plant diseases and pests across international boundaries.

3024. International convention to address oil pollution of the oceans was the International Convention for the Prevention of Pollution of the Sea by Oil (OILPOL), signed on May 12, 1954, in London, England. The convention became effective on July 26, 1958. OILPOL set rules for the discharge of oil by sea vessels and for the cleaning of ballast tanks. OILPOL was superseded in 1973 and 1978 by provisions of the International Convention for the Prevention of Pollution from Ships (MARPOL), which included rules for sea disposal of sewage, garbage, and harmful substances. MARPOL is overseen by the International Maritime Organization (IMO).

3025. International treaty regarding the use of Antarctica was the Antarctic Treaty. Although it was signed by more nations in succeeding years, the original treaty was signed by 12 nations in Washington, DC, on December 1, 1959, and entered into force June 23, 1961. The signatories agreed that Antarctica would forever be used only for peaceful scientific purposes and agreed to share research information about the continent.

3026. Agreement governing treatment of Antarctic wildlife and plants was contained in the treaty known as the Agreed Measures for the Conservation of Antarctic Fauna and Flora. The treaty expanded the terms of the Antarctic Treaty, signed in 1959. The new treaty was signed at Brussels, Belgium, on June 2, 1964, and entered into force November 1, 1982. It established strict guidelines and a permit system for the killing or taking of wildlife. It also banned the importation of non-indigenous species to the continent.

3027. Treaty to ban the introduction of nuclear weapons into orbit around the Earth was the Outer Space Treaty, signed of January 27, 1967, by Great Britain, the United States, and the Soviet Union. The treaty prohibited the installation of nuclear weapons or other weapons of mass destruction on the Moon, other planets, and any future space stations.

3028. International convention protecting wetlands was signed February 2, 1971, in Ramsar, Iran. Officially titled the Convention on Wetlands of International Importance Especially as Waterfowl Habitat, the convention entered into force in 1975 and maintains a secretariat with the World Conservation Union (IUCN) in Gland, Switzerland. Member nations promote conservation through wetlands listing, research, training, and management programs.

3029. U.S.–Canadian effort of significance to improve the water quality of the Great Lakes was the International Great Lakes Water Quality Agreement, signed on April 15, 1972, by U.S. President Richard Nixon and Canadian Prime Minister Pierre Trudeau. The agreement emphasized a joint effort to solve the environmental problems of the Great Lakes. The International Joint Commission was charged with assisting both nations in implementing the goals of the agreement

3030. United Nations conference devoted to the global environment was the Conference on the Human Environment, held in Stockholm, Sweden, in June 1972. The landmark meeting resulted in the establishment of the UN Environment Programme (UNEP), whose secretariat is based in Nairobi, Kenya.

3031. International convention to specifically protect Antarctic seals was the Convention for the Conservation of Antarctic Seals. The treaty was signed in London, England, on June 1, 1972. It expanded the Agreed Measures for the Conservation of Antarctic Fauna and Flora (1964), which prohibited the disturbance of seals by persons on foot or by use of aircraft, vehicles, guns, or dynamite. The new convention allowed the killing of seals only for scientific research, to provide food necessary for the survival of humans or dogs, or to provide specimens for museums or educational or cultural institutions. It also imposed regulations, including catch limits, open and closed seasons, preserve areas, and a permit system.

3032. Laws regulating space debris were agreed upon on September 1, 1972, at the Convention on International Liability for Damage Caused by Space Objects. The treaty was signed in triplicate in London, England; Moscow, USSR; and Washington, DC.

INTERNATIONAL LAW—CONVENTIONS, AGREEMENTS, AND TREATIES—*continued*

3033. Treaty to combine preservation of natural and cultural heritage sites was the Convention Concerning the Protection of World Cultural and Natural Heritage, adopted by the United Nations Educational, Scientific, and Cultural Organization (UNESCO) on November 16, 1972. Growing out of international concern about the decision to build the Aswan High Dam in Egypt, which would result in flooding the valley containing the Abu Simbel temples, the convention encourages international cooperation in the identification, protection, and preservation of natural and cultural heritage sites such as geological formations and monuments with historical, scientific, aesthetic, anthropological, or conservation value.

3034. Treaty protecting the polar bear and its natural habitat was the Agreement on Conservation of Polar Bears, signed in Oslo, Norway, on November 15, 1973, and in force May 23, 1976. The agreement stipulated that in most instances polar bears should not be removed from their natural habitat. Signers also agreed to protect and preserve the polar bear's ecosystem.

3035. Laws stipulating the registration of objects sent into space were agreed on at the 1974 Convention on Registration of Objects Launched into Outer Space, held in New York City. The regular marking and centralized registry, as a result of this treaty, made it easier to enforce the Convention on International Liability for Damage Caused by Space Objects signed in 1972.

3036. Global agreement to protect plant and animal species from unregulated international trade was formulated by the Convention on International Trade in Endangered Species of Wild Fauna and Flora (CITES). The first CITES agreement to monitor and regulate a biennially updated list of endangered species was signed by 21 countries in Washington, DC, in 1973 and became effective July 1, 1975. By 1998, CITES membership had grown to 144 countries. Its secretariat is based in Geneva, Switzerland.

3037. Treaty to insure humane slaughter of animals was the European Convention for the Protection of Animals for Slaughter, adopted in Strasbourg, France, on May 10, 1979. The convention stipulated immediate slaughter when possible and minimal suffering during the slaughtering process, which was to be conducted by trained workers.

3038. Bilateral negotiations between the United States and Canada on acid rain began in July 1979. The negotiations came a year after a small group of congressmen, whose states bordered on Canada, requested them. Shortly after this, the U.S. Congress passed a resolution calling on the U.S. president to negotiate with Canada an agreement aimed at preserving the quality of air above both nations.

3039. Binding international treaty to address regional air pollution was the Convention on Long-Range Transboundary Air Pollution. The convention was signed in Geneva, Switzerland, on November 13, 1979, and entered into force in 1983. Concerned about deteriorating air quality in Europe, its signatories agreed to share research and prevention information to combat the harmful effects of air pollutants, particularly sulphur compounds, on humans and the environment.

3040. Joint antipollution pact in the Mediterranean region occurred in 1980. Fifteen Mediterranean nations signed a pollution control agreement that was backed by the United Nations Environment Programme. The pact aided Mediterranean seacoast countries in their efforts to fight marine pollution and other environmental problems. The pact built upon earlier agreements: the 1975 Mediterranean Action Plan and the 1976 Barcelona antipollution treaty.

3041. Agreement among nations to work together to create guidelines, policies, and procedures for protecting the ozone layer occurred March 22, 1985, at the Vienna Convention for the Protection of the Ozone Layer. Members also agreed to exchange research information when the agreement went into effect on September 22, 1988.

3042. U.S.–Canadian bilateral report on acid rain was issued in January 1986. The *Joint Report of the Special Envoys on Acid Rain* specifically addressed the transport of acid rain across the U.S.–Canadian border. The report was a continuation of the attempt to formulate a bilateral agreement on transboundary air pollution that culminated in the 1991 Air Quality Accord between the United States and Canada. The two envoys, Drew Lewis of the United States and William Davis of Canada, were appointed by their respective leaders, President Ronald Reagan and Prime Minister Brian Mulroney, to study specific environmental issues and to then report their findings.

3043. International treaty to protect the Earth's atmosphere was the Montreal Protocol. The 24 countries signing the treaty on September 16, 1987, agreed to halt the production of chlorofluorocarbons (CFCs), which were found to be damaging the planet's ozone layer.

3044. International convention attempting to regulate Antarctic mineral exploitation was the Convention on the Regulation of Antarctic Mineral Resource Activities. The convention was set forth in Wellington, New Zealand, on June 2, 1988, but it was not ratified by the minimum of 16 countries required for it to come into force. It was replaced by the Protocol on Environmental Protection to the Antarctic Treaty (1991), which prohibited any mineral resource activities unrelated to scientific research.

3045. International convention on transporting hazardous waste opened on March 22, 1989, and was called the Basel Convention after the Swiss city where it was opened for signatures. Parties to the Basel Convention on the Control of Transboundary Movements of Hazardous Wastes and Their Disposal agreed that environmentally hazardous waste should be treated in its country of origin, thus minimizing the dangers inherent in dumping waste in countries technologically unprepared for its disposal. The convention entered into force May 5, 1992, and is administered through the United Nations Environment Programme.

3046. Specific treaty guidelines for Antarctic ecosystem protection were contained in the Protocol on Environmental Protection to the Antarctic Treaty. The protocol was adopted October 4, 1991, in Madrid, Spain. It reaffirmed the Antarctic Treaty of 1959, an agreement that the continent should be reserved for peaceful scientific purposes. The new treaty instituted procedures for environmental impact assessments, monitoring, emergency response, and handling of disputes. It also set forth guidelines for waste management and species protection. Marine pollution and any mineral activities unrelated to scientific research were banned.

3047. Accord on transboundary effects of industrial accidents on the environment is the Convention on the Transboundary Effects of Industrial Accidents. Its first parties signed the convention on March 17, 1992, in Helsinki, Finland. Signatory nations agreed to share information and technology for the prevention of, preparedness for, responses to, and mutual assistance in cases of industrial accidents whose hazardous effects are capable of crossing borders.

3048. Global agreement on all aspects of biodiversity was the United Nations Convention on Biological Diversity signed on June 5, 1992, in Rio de Janeiro, Brazil, and entered into force December 29, 1993. The convention encourages conservation of biological diversity and wise use of genetic resources.

3049. International convention on biodiversity and sustainable development collected its first signatures on June 5, 1992, at the United Nations Conference on Environment and Development (UNCED), the Rio de Janeiro "Earth Summit." The convention's parties agree to share scientific, technical, and technological expertise in reconciling global competition between development and conservation. The agreement entered into force December 29, 1993.

3050. Treaty setting the framework to address the production of greenhouse gases was signed May 9, 1992, and entered into force March 21, 1994. The United Nations Framework Convention on Climate Change created a process identifying specific problems and recommending specific actions related to greenhouse gases.

3051. International body to recognize the obligations of countries to fight and prevent marine pollution was founded by the United Nations in Montego Bay, Jamaica, through the Convention on the Law of the Sea. The convention, which went into effect November 16, 1994, states that countries are "bound to prevent and control marine pollution and are liable for damages caused."

3052. Treaty establishing an international 200-nautical-mile Exclusive Economic Zone (EEZ) went into effect November 16, 1994. The treaty gives countries freedom of navigation, the right to regulate fishing, use, and exploration of natural resources. It was the end result of the United Nations Conference on the Law of the Sea.

3053. International body to govern use of the ocean was the United Nations Convention on the Law of the Sea, which began enforcing regulations on November 16, 1994. Representatives of more than 150 countries signed the convention on December 10, 1982, at Montego Bay, Jamaica.

INTERNATIONAL LAW—CONVENTIONS, AGREEMENTS, AND TREATIES—*continued*

3054. International treaty to specifically protect the sea turtles of the Western Hemisphere was the Inter-American Convention for the Protection and Conservation of Sea Turtles. The international agreement was formalized in Caracas, Venezuela, on December 1, 1996. Upon ratification, signatory nations agreed to establish national sea turtle conservation programs, ban the intentional capture, killing, or sale of sea turtles, and require the use of turtle excluder devices (TEDs) on most shrimp trawl vessels.

3055. International treaty to ban the reproductive cloning of human beings was the Universal Declaration on the Human Genome and Human Rights. Meeting in Paris, France, the United Nations Education Science and Culture Organization (UNESCO) acted on November 11, 1997, on recommendations of UNESCO's International Bioethics Committee setting universal ethical standards on human genetic research. The declaration attempted to balance the freedom of scientists to pursue their research with the need to safeguard human rights and protect humanity from potential abuses.

3056. International treaty to set binding limits on carbon dioxide emissions was the 1997 Kyoto Protocol. The signatories also set limits on five other greenhouse gases.

INTERNATIONAL LAW—ORGANIZATIONS

3057. International whaling commission was established in the United Kingdom to regulate whaling worldwide. More than 65 countries agreed to abide by the commission's recommendations when they signed the convention on December 2, 1946, in Washington, DC. The commission's power went into effect on November 10, 1948.

3058. Version of the World Conservation Union (IUCN) was established as the International Union for the Protection of Nature (IUPN) at Fontainebleau, France, on October 5, 1948. Today the IUCN is one of the world's oldest and most influential conservation organizations, in matters ranging from international environmental law to sustainable growth programs. It is also responsible for compiling and maintaining the international "Red List" of threatened and endangered species.

3059. Organization for the control of desert locusts was created in Rome, Italy, on December 3, 1963. The Commission for the Control of Desert Locusts included members from Afghanistan, India, Iran, and Pakistan. They agreed to maintain reporting, control services, and control equipment and to do research on the desert locusts. A second commission for the Near East was created in 1965 and a third for northwest Africa was created in 1970.

3060. International body to regulate conservation of Atlantic tunas was created on May 14, 1966, in Rio de Janeiro, Brazil. The International Commission for the Conservation of Atlantic Tunas conducts research on tuna and makes recommendations to the members.

3061. Organization to regulate dumping of wastes in the high seas was created at the London Dumping Convention on December 12, 1972.

3062. Intergovernmental organization devoted exclusively to protection of migratory species worldwide was the Convention on Migratory Species (CMS). The organization represents parties to the treaty also known as the Convention on Migratory Species, or the Bonn Convention, which was signed in Bonn, West Germany, on June 23, 1979, and entered into force November 1, 1983. CMS is devoted to conservation of migratory species through programs including intergovernmental sharing of information and sustainable use strategies.

3063. Organization to study the North Pacific was the North Pacific Marine Science Organization (PICES), established at a convention in Ottawa, Canada, on December 12, 1990, and in force March 12, 1992. The organization provides a forum for international cooperation in the studies of the North Pacific.

3064. Inter-American institute for research on global change was agreed to in a conference in Montevideo, Uruguay, on May 13, 1992, and in force March 12, 1994. The Inter-American Institute for Global Change Research was established to promote inter-regional cooperation in research on environmental issues and their global impact.

3065. International organization to regulate use of the ocean floor was the International Seabed Authority (ISA), created by the United Nations on November 16, 1994. The ISA is an autonomous agency that sets rules for nations adhering to the UN Convention on the Law of the Sea. The ISA has jurisdiction over the exploration, exploitation, protection, and preservation of the ocean floor and its subsoil beyond the national jurisdictions of territorial waters.

3066. U.S. government body devoted solely to mediating environmental disputes was the Institute for Environmental Conflict Resolution (IECR). The Institute was created by Congress through the Environmental Policy and Conflict Resolution Act of 1997. By offering compromise solutions, the IECR was intended to provide an alternative to lengthy, expensive, and emotionally charged legal battles clogging U.S. district courts.

L

LAND USE AND DEVELOPMENT

3067. Major land reclamations in low-country Flanders and Zeeland occurred circa 900. The need for farmland to support a growing population drove the reclamation projects.

3068. Soo Lock, to carry shipping over the rapids at Sault Ste. Marie from Lake Huron to Lake Superior opened in 1797. Successive longer locks were built to exploit vast copper and iron deposits in the Lake Superior mining region. The passage became the most heavily used waterway in the world.

3069. Civil works project undertaken by the U.S. Army Corps of Engineers dates from 1824. The corps today is mainly a civilian agency that manages navigation on inland waterways and flood control projects. It also regulates some pollutant discharges and has other environmental regulatory functions.

3070. Canal connecting Lake Ontario with the other four of the Great Lakes was the Welland Canal, opened in 1829. The canal, which has been rebuilt and enlarged three times, opened the Great Lakes to heavy shipping, logging, settlement, and industry. It also became a major factor in the rise of pollutants in the Great Lakes.

3071. Proposal to dig a canal across the Florida peninsula dates from 1829, when a U.S. Army engineer presented surveys of possible routes to Congress. President Franklin D. Roosevelt authorized a sea level canal across Florida in 1935 and work on the canal commenced in 1964. However, with environmental groups citing water pollution concerns, Congress in 1990 passed a bill deauthorizing the Cross-Florida Barge Canal. President George Bush signed the measure on November 28, 1990.

3072. Street railway in Boston opened in 1852. This form of fast transit touched off a development boom; by 1910 the Boston area's population had increased fivefold, to more than one million.

3073. Published presentation of a more holistic view of U.S. public land management was *Man and Nature* (1864) by George Perkins Marsh. In the later part of the 19th century, U.S. government scientists stopped investigating separate issues like irrigation and grazing and focused on land management as the sum of these parts.

3074. Tract housing development in the United States was Concrest, a development of 100 identical six-room houses near East Rochester, NY. The development was built by Kate Gleason, a local machine-shop saleswoman turned contractor, beginning in 1913. Gleason had seen car engines mass-produced on an assembly line and applied the method to the construction of low-cost housing.

3075. Modern-day large-scale planned U.S. suburban development to use assembly-line methods of construction was Levittown, NY, built between 1947 and 1951 by William Jaird Levitt on former potato fields in Long Island. The development consisted of 17,447 nearly identical four-room houses, 7 village greens and shopping centers, 14 playgrounds, 9 swimming pools, 2 bowling alleys, and a town hall. The houses initially sold for $7,990 each. Levitt had been a Navy Seabee during World War II and adapted the Seabees' rapid construction techniques in his development. Levittown and similar suburban developments built in the late 1940s and 1950s were intended to meet the demand for housing by veterans who returned home to marry and start families. The mortgages on such homes were often financed under the benefits of the GI Bill.

3076. Restoration to prairie of U.S. farmland previously reserved as a site for a nuclear power station was begun in Iowa in November 1991. The U.S. Fish and Wildlife Service undertook to clear the site for plants that grew there in the mid-19th century.

LAND USE AND DEVELOPMENT—LEGISLATION AND REGULATION

3077. National law regulating the surveying and sale of U.S. public lands was the Land Ordinance of 1785. The statute made public land available in 640-acre sections at $1 an acre and established the procedures for parceling out western lands until 1862, when the Homestead Act was passed.

LAND USE AND DEVELOPMENT—LEGISLATION AND REGULATION—*continued*

3078. Legislation by the United States barring private acquisition of American Indian lands was the Indian Nonintercourse Act, passed in 1790. The act required congressional approval of any sale of Indian lands.

3079. Cabinet-level agency concerned primarily with natural resource development and conservation in the United States was the Department of the Interior, established in 1849. Thomas Ewing of Ohio served as the first interior secretary, in the administration of Zachary Taylor. Land is considered a natural resource.

3080. Law to open vast areas of the Great Plains in the United States to systematic settlement was the Homestead Act of 1862. The measure promoted large-scale wheat farming that consequently disrupted the ecosystems of the Great Plains.

3081. Public Land Review Commission in the United States was established in 1879, during the presidential administration of Rutherford B. Hayes. Congress, however, ignored the commission's recommendations on managing public lands.

3082. Protected historic district in the United States was established in Charleston, SC, in 1931. A second historic district was declared in the Vieux Carré of New Orleans, LA, in 1937.

3083. Green belt legislation for London England, was approved by Parliament in 1938. The London Green Belt Act recognized publicly owned open spaces as a buffer around the city. London's green belt has since been expanded.

3084. Housing program for U.S. military veterans subsidized by the federal government was approved in 1944. The Servicemen's Readjustment Act guaranteed long-term loans at low interest to World War II veterans, spurring the postwar development of suburban housing tracts.

3085. Legislation of significance in the United States promoting and funding mass public housing was the Housing and Slum Clearance Act of 1949. The act spurred construction of high-rise housing projects in many large U.S. cities.

3086. Law in the United States to require an environmental impact statement for all major federal projects that could significantly affect the environment was the National Environmental Policy Act. Congress passed the measure in 1969.

3087. State-level legislation in the United States mandating review of any development for impact on natural beauty, aesthetics, and historic sites was adopted in Vermont in 1970. The law set up a district environmental commission, appointed by the governor, to review each proposed development project.

3088. National land-planning measure to receive U.S. congressional approval was the Coastal Zone Management Act of 1972. The act mandated federal control over large-scale development along the shorelines of 30 coastal and Great Lakes states.

3089. Land-use plan for a public desert area in the United States was developed under the terms of the Federal Land Policy and Management Act of 1976. The act allocated $4 million for a plan for 37,000 square miles of California desert.

3090. Wetlands protection and restoration law to provide funds for protection was the North American Wetlands Conservation Act. Approved by Congress, the measure became law on December 13, 1989.

3091. Bill in California forcing developers to show that there would be an adequate supply of water for any proposed large subdivision was signed into law by Governor Gray Davis on October 9, 2001. The bill affects developments of 500 houses or more, saying they cannot be built unless local water agencies agree that supplies will be adequate for the next 20 years. Davis also signed a companion bill requiring cities and counties to consult with water agencies at the early stages of all development projects. Sheila Kuehl and Jim Costa, state senators, were the authors of the two bills, which reflected increasing concern over the pace of development in California, concern intensified by the 2001 energy crisis and water shortages in the northern counties.

LAND USE AND DEVELOPMENT—PLANNING

3092. New technology to modify city design in late medieval Europe was gunpowder, introduced in the late 14th century. Straight stone walls were found to be vulnerable to cannon fire; sloping ramparts and, eventually, triangular bastions replaced the old city walls.

3093. City plan in America was William Penn's for Philadelphia in 1692. Penn laid out a two-square-mile tract that reserved space for parks and public buildings. Philadelphia became a chartered city in 1701.

3094. Modern "artificial" capital built in Europe was St. Petersburg, founded by Peter the Great in 1703. With Moscow, it was one of two Russian capitals.

3095. City planned and built as a seat of national government was Washington, DC, the U.S. capital since 1800. The basic plan for the city was submitted by Pierre L'Enfant in 1791 and approved by Congress that same year. L'Enfant oversaw the construction of the city from the outset, but his overzealous approach and use of funds without the approval of Congress led to his dismissal in 1792. As the city grew over the next century, L'Enfant's plans were largely ignored and forgotten. However, by the 1890s, the haphazard planning of the growing city led many to call for an organized plan to rebuild the central region of Washington. In 1901, L'Enfant's original plans were used as the basis for Washington's system of gridiron streets and avenues and its elaborate system of parks.

3096. Introduction of the famous "grid pattern" of New York City came in 1811. The New York City Board of Commissioners adopted the plan that determined the development of America's largest city.

3097. Plan of significance for a suburban community in the United States was for Riverside, IL, in 1868. Frederick Law Olmsted's plan for the Chicago suburb included winding streets, houses set back at least 30 feet from the streets, and mass plantings of shrubs and trees.

3098. National urban beautification movement in the United States was the City Beautiful Movement, launched in the aftermath of the World's Columbian Exposition in Chicago in 1893. It produced park and boulevard plans for many American cities.

3099. Introduction of the "green belt" concept came from British social reformer Ebenezer Howard in 1898. He proposed "garden cities" surrounded by a country farming belt to separate urban areas and particularly to control sprawl from fast-growing London.

3100. Use of the term *green belt* came around 1900, when British architect and planner Raymond Unwin coined it. He was a contemporary of the British reformer Ebenezer Howard, who introduced the green belt concept in 1898.

3101. Planned green belt town was Letchworth, England, developed in 1903.

3102. City planning commission in the United States was established in Hartford, CT, in 1907.

3103. National association for city planning in the United States was founded in 1910. The organization took the name of the National City Planning Association.

3104. Interstate planning and development agency in the United States was the Port Authority of New York and New Jersey, established on April 30, 1921, by the neighboring states after a long history of conflict over shipping and docking in the Hudson River harbor. (At least one dispute led to an exchange of gunfire in midstream by rival police.) The agency was based on a clause in the Constitution allowing states to enter into compacts with one another and was modeled upon Britain's Port of London Authority, the oldest such body in the world. The Port Authority of New York and New Jersey soon expanded its role from coordinating harbor traffic to planning and constructing bridges, tunnels, airports, and other regional facilities, as well as high-rise buildings, notably the World Trade Center. A self-supporting public corporation, governed by appointees from both states, it was called the Port of New York Authority until 1972, when it adopted its present name.

3105. Large shopping center built on the outskirts of an urban area specifically to serve people with cars opened in Kansas City, MO, in 1922.

3106. U.S. city to decide to widen roads rather than build a subway system was Detroit, MI, in 1926. Detroit was, and remains, the center of the American automobile industry.

3107. Comprehensive reform of land-use planning laws in the United States occurred in Oregon under conservation-minded Governor Thomas McCall in 1969 and 1973. The two-stage reforms led to a Land Conservation and Development Commission that regulated development in Oregon based on statewide goals.

LAND USE AND DEVELOPMENT—PRESERVATION

3108. Restoration of a colonial-era town to museum status began in Williamsburg, VA, in 1927. John D. Rockefeller Jr. funded the large-scale restoration of the city. To preserve its colonial appearance, more than 700 buildings were removed, 83 were renovated, and 413 were rebuilt on their original sites. Today the town is a very popular tourist attraction.

3109. Charter for the National Trust for Historic Preservation in the United States was granted by Congress in 1949. Within 20 years, the trust would count 140,000 members.

LAND USE AND DEVELOPMENT—PRESERVATION—*continued*

3110. Statewide statutes authorizing preferential tax assessment of farmland in the United States were enacted in Maryland in 1957. Virtually every state now has some form of farmland preservation program.

3111. National Historic Preservation Act was passed in the United States in 1966. The measure directed the secretary of the interior to maintain and expand the national register of places of historic, architectural, or cultural significance.

3112. Tax breaks for owners of historically significant commercial property in the United States were included in the Tax Reform Act of 1976. Owners were offered tax deductions for improving their historic properties.

3113. Federal government interagency study to evaluate methods by which states and localities in the United States could slow farmland losses was authorized in 1979. The study group included representatives from the Department of Agriculture, the Council on Environmental Quality, the Environmental Protection Agency, and other cabinet departments.

LAND USE AND DEVELOPMENT—PROPERTY RIGHTS

3114. Indications of private landowning emerged around 2500 B.C. in the cities of Sumer in Mesopotamia. Previously, cities and temples had owned the land.

3115. Court ruling upholding Boston's right to regulate the boundary line for wharves was issued in 1853. In what amounted to an early ruling on zoning ordinances, the court declared an overlong wharf "injurious to the general public."

3116. Ruling by the U.S. Supreme Court that historic preservation was a form of community welfare was issued in 1978. The court upheld lower court rulings denying Penn Central's bid to lease the air rights above Grand Central Station in New York City to a developer who planned to build an office tower above the historic building. The ruling also denied compensation to Penn Central.

3117. Rejections of state-mandated shoreland-zoning ordinances in coastal Maine were reported in 11 towns in April 1992. Defenders of property rights accused the environmental movement of a massive "land grab" conspiracy in pushing for forest and shoreline preservation.

LAND USE AND DEVELOPMENT—SETTLEMENT

3118. Semi-permanent human settlements developed in southwest Asia, particularly Mesopotamia, circa 8500 B.C. The domestication of some animals and grains made a sedentary life possible.

3119. Large-scale human settlement developed at Tell-es-Sultan in Syria circa 8000 B.C. Domestication of wild grains made this community of about 2,000 people possible.

3120. Farming communities in China developed circa 6000 B.C., following the domestication of millet, a cereal grass. They were small villages of 200 or so people on the North China plain.

3121. Agricultural settlements in the fertile Nile Valley of Egypt appeared circa 5500 B.C. Gathering and hunting groups had occupied the valley for some 20,000 years before the first permanent settlements.

3122. Settlement of the Indus Valley of the Indian subcontinent occurred circa 3500 B.C. Farmers migrating eastward from southwest Asia cultivated wheat and barley and herded sheep and goats in the valley.

3123. Settled communities in Mesoamerica emerged circa 2000 B.C. The domestication of maize made settled life possible.

3124. Colonization of Iceland from Norway occurred in A.D. 874. Viking settlers reached Greenland farther to the north and west in 896.

3125. Permanent Spanish settlement in what is now the United States was at St. Augustine, FL, in 1565. The founding of St. Augustine inaugurated Spanish colonization of the region.

3126. Permanent English settlement in America was Jamestown, in Virginia, founded in 1607. The settlement survived until the end of the 17th century, when it fell into decay. The area today is a national historic shrine.

3127. Permanent French settlement in North America was Quebec, founded by Samuel de Champlain in 1608. French explorers had established claims to the St. Lawrence region in the 1530s.

3128. Land grants of 100 acres to each Virginia colony settler were issued in 1616. Successful development of the tract entitled the colonist to a second 100-acre allotment.

3129. Court ruling in colonial New England barring settlers from direct purchase of American Indian lands was handed down in 1633. The Massachusetts General Court issued the ruling, which barred colonists from personally buying land from Indians.

3130. Large-scale urban middle-class movement from townhouses to "suburbs of privilege" commenced in the London, England, metropolitan area in the late 18th century. Such communities would become the predominant form of suburban settlement in 19th-century England and the United States.

3131. British government attempt to restrict American colonial settlement to the area east of the Appalachian Mountains was made in 1763. By decree, the British established a boundary line along the mountain chain and prohibited settlement west of the line. Settlers widely ignored the boundary.

3132. Introduction of the rectangular system of land surveys in the United States was in 1785, under the Land Ordinance of that year. The system used meridians and baselines as points of reference to lay out townships divided into 36 sections of 640 acres apiece. Land could be sold by the section for not less than $1 per acre; one section in each town would be used to fund a local school, and four sections would be reserved for government. With modifications, the Land Ordinance of 1785 provided the basis for settlement on public lands until the Homestead Act of 1862.

3133. U.S. congressional "treaty" acquiring American Indian lands for the government was ratified in 1789. Between 1789 and 1850, Congress ratified 245 such treaties covering more than 450 million acres. The treaty sales price worked out to 20 cents per acre.

3134. Local land-sales offices in the United States were established in 1800. These bureaus were charged with regulating the sale of government lands.

3135. Acquisition of land of significance by the United States from a foreign government was the Louisiana Purchase of 1803. The administration of Thomas Jefferson bought 828,000 square miles of western land from France for $15 million, or about four cents an acre. The lands were eventually opened to settlement; 13 states were created out of the purchase.

3136. Legislation permitting settlers to live on U.S. public land until they could pay for it or until someone else bought it was the Intrusion Act, adopted in 1807.

3137. Government agency in the United States to keep track of the sale of federal land was the General Land Office, established in 1812. The bureau was a branch of the Treasury Department.

3138. Pre-emption act in the United States involving lands settled before the government had purchased and surveyed them was passed in 1830. Settlers routinely squatted on empty land, then lobbied Congress for legal title to the property. The Pre-emption Act of 1830 was the first of several affecting the sale and distribution of public lands.

3139. Government land grant to a railroad by the United States was issued to the Illinois Central in 1850. Between 1850 and 1871, 300 million acres of public lands were given to railroad companies.

3140. Declaration that the American frontier had closed came in 1893, in an address by historian Frederick Jackson Turner before the Amercan Historical Association. Turner's thesis was that cheap lands on a receding frontier were the main factors in forging American democracy and national character.

3141. Land boom of significance in Florida began in 1896 when developers drained a mangrove swamp that became the site of Miami. Railroad magnate Henry M. Flagler built the first resort hotel in Miami; by 1930 the city's population exceeded 110,000.

3142. Large-scale peripheral suburban growth in the United States occurred in the 1920s with the mass production of cheap automobiles and the development of the road network. Thousands of subdivisions were laid out for urban working-class and lower-middle-class homebuyers.

3143. Formal identification of the line of continuous development along the U.S. eastern corridor from Washington to Boston came in 1961 in geographer Jean Gottman's book *Megalopolis*. Airline pilots flying eastern coastal routes were the first to call attention to the phenomenon.

LAND USE AND DEVELOPMENT—WETLANDS

3144. Federal government action granting the U.S. states sovereignty over swamplands occurred in 1850.

3145. Draining and filling of the marshy area of the Charles River near Boston, MA began in 1858. By 1886 the residential district of Back Bay had been developed on the reclaimed land.

LAND USE AND DEVELOPMENT—WETLANDS—*continued*

3146. Acquisition of fragile wetlands on a large scale by the United States was the Everglades in south Florida. The area became the 1.3-million-acre Everglades National Park in 1947.

3147. National survey of wetlands in the United States was published by the U.S. Fish and Wildlife Service in 1956. "Wetlands of the United States" provided a benchmark for assessing losses of wetlands to development.

3148. Legislation aimed at protecting U.S. wetlands from drainage and development was the Water Bank Act of 1970. The measure authorized the federal government to pay landowners to leave their wet areas in a natural state.

3149. Regulations by the U.S. Army Corps of Engineers requiring applicants for permits to fill wetlands to prove the work would be consistent with ecological considerations were adopted in May 1970.

3150. Catalogue of critical world wetlands was produced as a result of the Convention on Wetlands of International Importance Especially as Waterfowl Habitat, adopted in 1971 at Ramsar, Iran. The convention, which went into effect in 1975, promotes prudent use of wetlands. Within two decades, the catalogue would list more than 500 wetlands, and 50 countries had joined the convention.

3151. Emergence of wetlands science as a distinct discipline occurred in the early 1980s. Loss of wetlands has caused soil erosion, degraded water quality, and reduced populations of certain plants and animals.

LAND USE AND DEVELOPMENT—ZONING

3152. District zoning in the United States was implemented in Los Angeles, CA, in 1909. New York City adopted the country's first comprehensive zoning ordinance in 1916.

3153. Zoning ordinance in the U.S. was put into effect in New York City on July 25, 1916, limiting the total floor area of skyscrapers to no more than 12 times the area of the building site and requiring setbacks, which were based on the building's height and the size of the building site. The regulation was prompted by the unprecedented size of the 39-story Equitable Building, erected in lower Manhattan the previous year.

3154. National guidelines for local zoning ordinances in the United States were issued in 1922 by the Department of Commerce. By the mid-1920s, 300 communities had used either the Department of Commerce Standard Act or New York City's 1916 zoning statute as a model for their own zoning ordinances.

3155. U.S. Supreme Court ruling affirming the legality of zoning came in 1926. New York City had enacted a zoning law in 1916. Private landowners challenged zoning as unjustified public control over private property.

3156. Zoning law in the United States to encourage a mix of shops, offices, and residences was enacted in New York City in 1971. The mixed-use zoning law applied to Fifth Avenue between 38th and 57th streets.

3157. Zoning scheme to take time into account was the "Petaluma Plan," implemented in 1972. The plan, designed to limit population growth, came after three years of explosive suburbanization of the San Francisco–area community that followed the expansion of Highway 101. It limited new housing construction to 500 units a year over five years.

3158. Limits on the height of skyscrapers in New York City were proposed on April 20, 1999. The limits were part of the first overhaul of New York City zoning laws in 40 years.

LITERATURE AND ARTS—ART AND ARTISTS

3159. Artistic representation of a bird or bird parts was of a prehistoric bird-man dating from 15,000 to 10,000 B.C. and painted on the walls of Lascaux Cave in France.

3160. Bird pictured in John James Audubon's *The Birds of America* was the wild turkey, which appeared on plate one of Audubon's masterpiece, engraved by Robert Havell Jr. and published in an "elephant folio" edition in London, England, between 1827 and 1838.

3161. Postage stamps depicting the American bald eagle were the one-cent carrier's stamps in blue, issued on November 17, 1851.

3162. Ledger-art drawings of the Plains Indians were made in the 1870s, after they were confined to reservations, and featured mostly scenes of hunting and warfare. The drawings were made with pencil, then colored in with crayon, colored pencil, or paints, on pages literally torn out of the ledger books of agency trading posts in the Dakotas. The artwork conveys a clear sense of the end of wilderness.

3163. Monument to a bird in the United States was unveiled October 1, 1913, at Salt Lake City, UT, to honor the sea gulls from the Great Salt Lake that attacked a horde of black crickets, or grasshoppers, that were devouring the wheat fields of the Mormon settlers in May 1848. The monument was designed by Mahonri Mackintosh Young, a grandson of Brigham Young, the leader of the settlers.

3164. Monument to an insect in the United States was dedicated December 11, 1919, by the citizens of Enterprise, AL, "in profound appreciation of the boll weevil and what it has done as the herald of prosperity." The weevil's devastation of cotton crops had prompted local farmers to diversify their crops, which soon increased their income to triple what it had been in the best cotton years.

3165. Federal duck stamp in the United States was issued in July 1934, after passage of the Migratory Bird Hunting Stamp Act (popularly known as the Duck Stamp Act) on March 16 that year. Proceeds from sales of the stamps, which must be purchased by all individuals wishing to hunt waterfowl on federal lands, are used to acquire wetlands that have become part of the National Wildlife Refuge System. Sales thus far have generated more than $500 million.

3166. Federal duck stamp in the United States to feature the mallard was the first duck stamp, issued in July 1934. Designed by Jay N. "Ding" Darling—who as chief of the Bureau of Biological Survey was responsible for instituting the duck stamp program—this first issue consisted of 635,001 stamps and was priced at one dollar.

3167. Federal duck stamp in the United States to feature the canvasback was designed by F. W. Benson and appeared in 1935. This issue consisted of 448,204 stamps.

3168. Sales of federal duck stamps to collectors in the United States were permitted on July 15, 1935. Since then certain issues—especially those dating from before 1941—have become very valuable.

3169. Federal duck stamp in the United States to feature the Canada goose was designed by R. E. Bishop and appeared in 1936. This issue consisted of 603,623 stamps.

3170. Federal duck stamp in the United States to feature the greater scaup was designed by J. D. Knapp and appeared in 1937. This issue consisted of 603,623 stamps.

3171. Federal duck stamp in the United States to feature the pintail was designed by Roland Clark and appeared in July 1938. This issue consisted of 1,002,715 stamps.

3172. Federal duck stamp in the United States to feature the green-winged teal was designed by Lynn B. Hunt and appeared in 1939. This issue consisted of 1,111,562 stamps.

3173. Federal duck stamp in the United States to feature the black duck was designed by Francis Lee Jaques and appeared in 1940. This issue consisted of 1,111,561 stamps.

3174. Federal duck stamp in the United States to feature the ruddy duck was designed by Edwin Richard Kalmbach and appeared in 1941. This issue consisted of 1,439,967 stamps.

3175. Mountain range to become a work of art was Mt. Rushmore in South Dakota, whose presidential sculptures were completed in 1941.

3176. Federal duck stamp in the United States to feature the widgeon was designed by A. L. Ripley and appeared in 1942. This issue consisted of 1,383,629 stamps.

3177. Federal duck stamp in the United States to feature the wood duck was designed by Walter E. Bohl and appeared in 1943. This issue consisted of 1,169,352 stamps.

3178. Federal duck stamp in the United States to feature the white-fronted goose was designed by Walter A. Weber and appeared in 1944. This issue consisted of 1,487,029 stamps.

3179. Public service symbol created for the U.S. government was Smokey Bear, who was created in August 1944 by illustrator Albert Staehle. Staehle created Smokey for the U.S. Forest Service, which then used the illustrations to educate the public about the dangers of forest fires.

3180. Two-time winner of the U.S. federal duck stamp contest was Walter A. Weber, who won first in 1944 for his picture of white-fronted geese and then a second time in 1950 for his picture of trumpeter swans.

3181. Federal duck stamp in the United States to feature a shoveler was designed by Owen Gromme and appeared in 1945. This issue consisted of 1,725,505 stamps.

3182. Federal duck stamp in the United States to feature the redhead was designed by Robert W. Hines and appeared in 1946. This issue consisted of 2,016,841 stamps.

LITERATURE AND ARTS—ART AND ARTISTS—*continued*

3183. Federal duck stamp in the United States to feature a snow goose was designed by Jack Murray and appeared in 1947. This issue consisted of 1,722,677 stamps.

3184. Federal duck stamp in the United States to feature a bufflehead was designed by Maynard Reece and appeared in 1948. This issue consisted of 2,127,603 stamps.

3185. Federal duck stamp in the United States to be priced at two dollars appeared in 1949. Designed by Roger E. Preuss, it pictured a pair of goldeneye ducks. Previously, federal duck stamps had sold for one dollar apiece.

3186. Federal duck stamp in the United States to feature a goldeneye was designed by Roger E. Preuss and appeared in 1949. This issue consisted of 1,954,734 stamps.

3187. Federal duck stamp in the United States to feature a trumpeter swan was designed by Walter A. Weber and appeared in 1950. This issue consisted of 1,903,644 stamps.

3188. Federal duck stamp in the United States to feature a gadwall was designed by Maynard Reece and appeared in 1951. This issue consisted of 2,167,767 stamps.

3189. Federal duck stamp in the United States to feature a harlequin duck was designed by John H. Dick and appeared in 1952. This issue consisted of 2,296,628 stamps.

3190. Federal duck stamp in the United State to feature a blue-winged teal was designed by C. B. Seagers and appeared in 1953. This issue consisted of 2,268,446 stamps.

3191. Federal duck stamp in the United States to feature a ring-necked duck was designed by H. D. Sandstrom and appeared in 1954. This issue consisted of 2,184,550 stamps.

3192. Federal duck stamp in the United States to feature a blue goose was designed by Stanley Stearns and appeared in 1955. This issue consisted of 2,369,940 stamps.

3193. Federal duck stamp in the United States to feature a merganser was designed by E. J. Bierly and appeared in 1956. This issue consisted of 2,332,014 stamps.

3194. Federal duck stamp in the United States to feature an American eider was designed by J. M. Abbott and appeared in 1957. This issue consisted of 2,355.353 stamps.

3195. Waterfowl species to appear twice on U.S. federal duck stamps was the Canada goose, which appeared on the winning design for the second time in 1958. The Canada goose first appeared on a federal duck stamp in 1936.

3196. Reproduction of U.S. federal duck stamps was made legal in September of 1958, when the United States Code was amended to permit black-and-white reproductions of the stamps to be published for certain purposes. In 1984, the law was further amended to permit color reproductions.

3197. Federal duck stamp in the United States to feature an image other than waterfowl alone was designed by Maynard Reece and pictured a Labrador retriever holding a mallard. This issue, which appeared in 1959, consisted of 1,628,365 stamps.

3198. Federal duck stamp in the United States to sell for $3.00 appeared in 1959. Duck stamps had previously been priced at two dollars each.

3199. Artist to win the U.S. federal duck stamp contest three times was Maynard Reece, who won the competition for a third time in July 1959. Reece would win again in 1969 and 1971.

3200. Federal duck stamp in the United States to feature a Pacific brant was designed by E. A. Morris and appeared in 1963. This issue consisted of 1,455,486 stamps.

3201. Monument dedicated to a songbird was a 4-foot-high stone monument to Kirtland's warbler. It was located on the lawn of the county courthouse in Mio, MI. The monument was dedicated on July 27, 1963.

3202. Federal duck stamp in the United States to feature a Hawaiian nene goose was designed by Stanley Stearns and appeared in 1964. This issue consisted of 1,573,155 stamps.

3203. Federal duck stamp in the United States to feature a whistling swan was designed by Stanley Stearns and appeared in 1966. This issue consisted of 1,805,341 stamps.

3204. Federal duck stamp in the United States to feature an old squaw duck was designed by Leslie C. Kouba and appeared in 1967. This issue consisted of 1,934,697 stamps.

3205. Federal duck stamp in the United States to feature a hooded merganser was designed by C. G. Pritchard and appeared in 1968. This issue consisted of 1,837,139 stamps.

3206. Federal duck stamp in the United States to feature a white-winged scoter was designed by Maynard Reece and appeared in 1969. This design by Reece was his fourth winning duck-stamp design. This issue consisted of 2,087,115 stamps.

3207. Federal duck stamp in the United States to feature a Ross's goose was designed by E. J. Bierly and appeared in 1970. This issue consisted of 2,420,244 stamps.

3208. Federal duck stamp in the United States to feature a cinnamon teal was designed by Maynard Reece and appeared in 1971. This design by Reece was his fifth winning duck stamp design. This issue consisted of 2,428,647 stamps.

3209. Federal duck stamp in the United States to feature an empress goose was designed by Arthur M. Cook and appeared in 1972. This issue consisted of 2,183,981 stamps.

3210. Federal duck stamp in the United States to be priced at five dollars appeared in July 1972. The stamps had previously sold for three dollars.

3211. Federal duck stamp in the United States to feature a Steller's eider was designed by Lee LeBlanc and appeared in 1973. This issue consisted of 2,113,594 stamps.

3212. Federal duck stamp in the United States to feature a waterfowl decoy, rather than a live bird was designed by James L. Fisher and consisted of a picture of a canvasback decoy. Appearing in 1975, this issue consisted of 2,218,589 stamps.

3213. Federal duck stamp in the United States to be priced at $7.50 appeared in July 1980. The stamps had previously been priced at five dollars.

3214. Federal duck stamp in the United States to feature a fulvous duck was designed by Burton E. Moore Jr. and appeared in 1986. This issue consisted of 1,794,484 stamps.

3215. Federal duck stamp in the United States to be priced at ten dollars appeared in July 1987. Previously the stamps had been priced at $7.50.

3216. Federal duck stamp in the United States to be priced at $12.50 was issued in July 1989. Previously the stamps had been priced at ten dollars.

3217. Federal duck stamp in the United States to feature a lesser scaup was designed by Neal Anderson and appeared in July 1989. This issue consisted of 1,415,882 stamps.

3218. Federal duck stamp in the United States to feature a black-bellied whistling duck was designed by Jim Hautman and appeared in 1990. This issue consisted of 1,409,121 stamps.

3219. Federal duck stamp in the United States to be priced at $15 was issued in 1991. Federal duck stamps had previously been priced at $12.50.

3220. Federal duck stamp in the United States to feature a king eider was designed by Nancy Howe and appeared in 1991. This issue consisted of 1,423,374 stamps.

3221. Woman to win the U.S. federal duck stamp contest was Nancy Howe in 1991 for her picture of a pair of king eiders.

3222. Federal duck stamp in the United States to feature a spectacled eider was designed by Jim Hautman and appeared in 1992. This issue consisted of 1,342,588 stamps.

3223. Federal duck stamp in the United States to feature a red-breasted merganser was designed by Neal Anderson and appeared in 1994.

3224. Federal junior duck stamp in the U.S. was issued in July 1994, after a national contest was held for all public and private school students from kindergarten to grade 12.

3225. Female winner of the federal junior duck stamp contest in the United States was Jie Huang, whose picture of a pintail in a pond was featured on the 1995 issue.

3226. Federal duck stamp in the United States to feature a surf scoter was designed by Wilhelm Goebel and appeared in 1996.

LITERATURE AND ARTS—BOOKS AND WRITERS

3227. True book devoted to birds was *De Artis Venandi cum Avibus*, a book about hunting with birds written by the Holy Roman Emperor Frederick II circa 1240, although it remained unpublished until 1596. Its illustrations of hawking and bird life constituted the first known serious paintings of birds.

3228. Book devoted entirely to birds was *Praecipuarum Quarum apud Plinim et Aristotelem Mentio est Brevis & Succincta Historia*. The book was written in 1544 by William Turner, an Englishman living in Germany who had sought refuge there after he was persecuted for his Protestant beliefs. A few copies of this volume still exist.

LITERATURE AND ARTS—BOOKS AND WRITERS—*continued*

3229. Scholarly text on the international law of the sea was *Mare Liberum* (Freedom of the Seas) by the Dutch jurist Hugo Grotius. It was published in 1633. The dissertation was subtitled *The right which belongs to the Dutch to take part in the East Indian trade.*

3230. Book devoted to birds and written in English was *The Ornithology of Francis Willughby*, edited by John Ray and published in London in 1678.

3231. Agricultural book published in the United States was *The Husbandman's Guide: In Four Parts. Part First: Containing Many Excellent Rules for Setting and Planting of Orchards, Gardens and Woods, the Times to Sow Corn, and All Other Sorts of Seeds. Part Second: Choice Physical Receipts for Divers Dangerous Distempers in Men, Women and Children. Part Third: The Experienc'd Farrier, Containing Many Excellent and Profitable Receipts for the Curing of All Diseases in Horses, Sheeps, Cow[sic]es, Oxen, & Hogs. Part Fourth: Certain Rare Receipts to Make Cordial Waters, Conserves, Preserves, with Many Useful Writings, &c.* The book was a 107-page reprint of an English work and was published in 1710 in Boston, MA, by John Allen for Eleazar Phillips.

3232. Distinctively American agricultural book was *Essays upon Field-Husbandry in New England* by Jared Eliot, printed and sold in Boston, MA, in 1760 by Edes and Gill. The book consisted of six essays which had originally appeared separately.

3233. Textbook on ecology was *Natural History and Antiquities of Selborne* by English clergyman and naturalist Gilbert White (1789). White's book is still considered a classic in scientific writing.

3234. Almanac with a continuous existence in the United States was *The Farmer's Almanack*, created by Robert Bailey Thomas of Sterling, MA, and printed at the Apollo Press in Boston, MA, by Belknap and Hall in 1792. In 1832, the title was changed to *The Old Farmer's Almanac.*

3235. Coast survey book written and published in the United States was Captain Lawrence Furlong's *The American Coast Pilot, Containing the Courses and Distance from Boston to All the Principal Harbors, Capes and Headlands Included between Passamaquady and the Capes of Virginia with Directions for Sailing Into, and Out of, All the Principal Ports and Harbours, with the Sounding on the Coast.* It was printed by Edward March Blunt and Angier March in Newburyport, MA, in March 1796.

3236. Gardener's manual written and published in the United States was the *Young Gardener's Assistant* by Thomas Bridgeman, published in 1835 in New York City.

3237. Text of modern oceanography detailing the trade winds and ocean currents was *A New Theoretical and Practical Treatise on Navigation*, by Matthew Fontaine Maury, an American hydrologist and oceanographer. The book was published in 1836.

3238. Book on natural history written by Charles Darwin was *Journal of Researches into the Geology and Natural History of the Various Countries Visited by H.M.S. Beagle*, published in 1839. Darwin based the book on observations he made from 1831 to 1836 during his round-the-world voyage aboard the *Beagle.*

3239. Account of the wonders of the Rocky Mountains to reach the East Coast of the United States was *Scenes in the Rocky Mountains.* It was by Rufus Sage, an explorer from Connecticut. It was published in 1843.

3240. Herd book for livestock compiled and published in the United States was *The American Herd Book* by Lewis Falley Allen, published in 1846 in Buffalo, NY.

3241. Cranberry treatise was *A Complete Manual for the Cultivation of the Cranberry: with a Description of the Best Varieties*, by B. Eastwood. It was published in 1856 by C. M. Saxton of New York City.

3242. Significant statement about the value of wilderness that appeared in the United States was Samuel Hammond's *Wild Northern Scenes; or Sporting Adventures with Rifle and Rod*, published in 1857. Hammond was an attorney who lived in Albany, NY, and also an avid sportsman who advocated setting aside a portion of the Adirondack region as a permanent wilderness preserve.

3243. Agriculture Bureau scientific publication was *A Report on the Chemical Analysis of Grapes*, by Charles Mayer Wetherill, dated October 15, 1862, and printed by the Government Printing Office in Washington, DC.

3244. Comprehensive study of human impact on the environment was George Marsh's *Man and Nature; or Physical Geography as Modified by Human Action.* The book was published in 1864.

3245. Written record of the discovery of Glen Canyon was by explorer John Wesley Powell. After leading the first scientific expedition through the Colorado River gorges in 1869 and making a second trip there in 1871, Powell published his account of his discovery in 1875 in *Report on the Exploration of the Colorado River and Its Tributaries.*

3246. Air pollution book to dramatize London's exploding 19th-century air pollution problem was *London Fogs* by F. A. R. Russell. Published in 1880, it attempted to calculate England's economic losses from air pollution.

3247. Forestry book written and published in the United States was Franklin Benjamin Hough's *The Elements of Forestry: Designed to Afford Information Concerning the Planting and Care of Forest Trees for Ornament or Profit and Giving Suggestions upon the Creation and Care of Woodlands with the View of Securing the Greatest Benefit for the Longest Time, Particularly Adapted to the Wants and Conditions of the United States.* It was published in 1882 by R. Clarke of Cincinnati, OH.

3248. Person to write a book promoting the idea of a U.S. national parks system was John Muir, whose book *The Mountains of California*, published in 1894, helped return control of the Yosemite Valley to federal management and brought him to the attention of President Theodore Roosevelt, whose conservation efforts eventually resulted in the formation of the National Park Service.

3249. Serialization of Upton Sinclair's *The Jungle* was in the socialist weekly *Appeal to Reason* in 1905. Published in book form the following year, it exposed unsanitary conditions in the meat-packing industry. The book is considered instrumental in the passage of the Pure Food and Drug Act and the Meat Inspection Act in 1906.

3250. Appearance of "The Great American Fraud," Samuel Hopkins Adams's article on patent medicines was in *Collier's* magazine on October 7, 1905. The article created a sensation and was a key in the passage of the Pure Food and Drug Act of 1906

3251. Major attack on the U.S. consumer products industry was *100,000,000 Guinea Pigs* by Arthur Kallet and Frederick J. Schlink, published in 1933. The book shocked Americans with its accusations that well-known food, drug, and cosmetics companies displayed gross irresponsibility in producing and marketing their products.

3252. Textbook on wildlife management was *Game Management* by Aldo Leopold, published in 1933.

3253. Field guide to birds was Roger Tory Peterson's *Field Guide to the Birds*, first published in 1934. Peterson's book, a pocket-sized volume that used illustrations and text to aid bird observation in the wild, was the first true field guide unencumbered by scientific apparatus. It was an immediate success, introducing millions to bird-watching and ultimately selling more than 3 million copies.

3254. Edition of *Empire of Dust* by Lawrence Svobida was published in 1940. Svobida was a young wheat farmer in Kansas who wrote by kerosene lantern and told his story of the Dust Bowl in *Empire of Dust.* He left the plains in 1939 believing the area had gone to desert for good.

3255. Population control book published after World War II was *Road to Survival* (1948)by William Vogt, an ecologist who predicted worldwide misery if population growth was not brought under control. Vogt's book heightened concern over the limited availability of natural resources to meet the needs of an ever-expanding population; he advocated birth control and family planning.

3256. Popular book to warn of pesticide threat was *Silent Spring*, published in 1962 and written by Rachel Carson, who became the literary midwife of the U.S. environmental movement. In the book Carson wrote about the environmental damage and potential dangers of unwise and widespread use of pesticides and herbicides. *Silent Spring* struck a nerve in the public consciousness, and made it aware of the often disastrous effects modern technology had upon the natural environment. This in turn gave rise to the first organized citizens' groups dedicated to environmental preservation.

3257. Presentation of the essay "The Historical Roots of Our Ecological Crisis" by historian Lynn Townsend White Jr. was in December 1966 before the American Association for the Advancement of Science. The essay, which was later published in *Science* on March 10, 1967, explored the roots of modern environmental problems in the Judeo-Christian tradition.

LITERATURE AND ARTS—BOOKS AND WRITERS—*continued*

3258. Publication that offered "access to tools" information for the energy-efficient and promoted ecologically aware lifestyles was *The Whole Earth Catalog*, founded and edited by Stewart Brand. It first appeared in November 1968. The book was a "truck store" and catalog of resources that was very popular in the late 1960s and 1970s.

3259. Publication inspired by *The Whole Earth Catalog* was a magazine, *Whole Earth News*, whose first issue appeared in January 1970. It was edited by John Shuttleworth. The magazine emphasized rural lifestyles.

3260. Example of a whole-systems approach to the environment was found in *The Closing Circle* (1971), a highly influential book by Barry Commoner. The book outlined the magnitude of the environmental crisis facing the United States. Working from the premise that all occurrences are interrelated, Commoner showed how natural cycles could be interrupted by technology and become one-way, linear events.

3261. Nontechnical text on population growth to be widely distributed was *The Limits of Growth*, issued by the Club of Rome in 1972. The book described how the computer could be used to model the global future of humankind. In 1992, *Beyond the Limits* was published as the 20th-anniversary companion to *The Limits of Growth*

3262. Comprehensive overview of the global environmental problem was *Our Common Future*, also called the Brundtland Report, released on April 27, 1987, by the World Commission on Environment and Development (WCED), a special independent commission of the United Nations. The commission, established in Geneva, Switzerland, was chaired by Gro Harlem Brundtland. The report was considered a major step in the evolution of global responsibility and sustainable development concerns.

3263. Book on birds to sell for $3.96 million was a set of, *The Birds of America* by John James Audubon. It sold for a record price in July of 1989 at Sotheby's in New York City. It was one of only about 134 complete sets remaining in the world.

3264. Papal document to deal with ecology was Pope John Paul II's, *Peace with God the Creator, Peace with All of Creation*, first issued on December 5, 1989. The document was actually prepared for the World Day of Peace on January 1, 1990. It confirmed the pope's teaching of the moral duty of Christians to protect the environment and heralded the beginning of "ecotheology"

LITERATURE AND ARTS—PHOTOGRAPHY AND TELEVISION

3265. Photographer to capture successfully the beauties of Yellowstone on film was William Henry Jackson when he accompanied the geological survey of the area led by Ferdinand Hayden in 1871. Jackson's photographs were instrumental in convincing Congress to grant Yellowstone protected status as a national park in 1872.

3266. Phase-contrast cinemicrography film to be shown on television was a presentation of "The Birth of a Plant," about plant reproduction, shown February 28, 1954, on KPIX-TV in San Francisco, CA.

LITTERING

3267. Littering law in New York City became effective when the city was New Amsterdam, a Dutch colony owned by the West India Company. In 1648, colonial director-general Peter Stuyvesant instituted ordinances forbidding the disposal of rubbish or dead animals in city streets, requiring residents to clean the roads in front of their homes, and regulating the condition of privies to prevent the discharge of sewage.

3268. State in the United States to control littering on levees was Mississippi in 1906. An anti-dumping provision is included in a state law prohibiting damage to and ensuring the right of way on riverfront levees. The law was amended in 1948 and 1968. The law prohibits the depositing of refuse or garbage on levees or otherwise obstructing the right of way with dead animals or commodities like wood, cotton, or bricks.

3269. State in the United States to forbid the posting of advertising flyers on telephone poles was Connecticut in 1949. The first version of the law was passed under the Parks, Forests, and Public Shade Trees section of the state's statutes. The law prohibits the posting of advertising on trees, rocks, and telephone or power poles on public ways. A violator is liable to a fine of $50 for each offense.

3270. National antilitter organization in the United States was Keep America Beautiful. The nonprofit educational organization was created in 1953. It has expanded its original antilitter campaign to support recycling, beautification efforts, and neighborhood revitalization. Cleanup programs by local chapters in more than 40 states range from litter collection to special programs like an initiative aimed at recycling building materials discarded at construction sites. Headquarters for Keep America Beautiful are in Stamford,

3271. National antilitter organization in Great Britain was the Keep Britain Tidy Group, a volunteer group organized in 1955 by the Women's Institute. In 1961 the group became an independent organization and started to receive government funding. Now known as the Tidy Britain Group, the organization is a coalition of locally based campaigns that promote litter prevention, solid waste disposal, recycling, marine pollution, and cleanup efforts in England, Scotland, and Northern Ireland.

3272. State law to revoke camping permits for littering convictions was enacted in Montana in 1971. The law voids for one year the permit privileges of anyone convicted of littering public or private lands, streams, or lakes while camping. The same statute voids hunting, fishing, and trapping licenses for the offense for one year from the date of a conviction.

3273. State law to revoke fishing, hunting, and trapping licenses for littering convictions was enacted in New Hampshire in 1955. The law voids the license of anyone convicted of placing refuse into or on the ice over any public water, streams or watercourse, or bordering land. The loss of license lasts for the current year in which the conviction is obtained.

3274. Antilitter campaign in Canada was Pitch-In Canada in 1967. The effort was launched by a small group of volunteers and grew to become a national non-profit organization supported by public, municipal, and corporate groups. Pitch-In Canada's priorities include preventing the harmful effects of litter on land and marine environments, recycling and source reduction, and promoting civic pride. Its headquarters are in White Rock, British Columbia.

3275. State in the United States to mandate citizen participation in refuse collection systems was Alabama in 1969. To ensure that refuse is collected rather than improperly discarded, the state requires every person, household, business, industry, or property generating solid wastes, garbage, or ash to subscribe to a waste collection service. Only citizens who prove that Social Security payments are their sole source of income are exempt. The law was amended in 1971, 1982, 1989, and 1997 to further curtail littering.

3276. State in the United States to suspend vehicle registrations of litterers was Rhode Island. The penalty provisions of the state's Litter and Recycling General Laws, which went into effect during the 1970s, 1980s, and1990s, allow a court to hold the registration of any vehicle owned by a litter violator and used in an act of littering until fines have been paid for the offense. Rhode Island is the only state to have imposed this judicial remedy.

3277. Adopt-A-Highway program in the United States was created in 1985 by the Texas Department of Transportation and the Tyler Civitan Club, which agreed to adopt two miles of Highway 69 in Tyler, TX, and pick up litter there at least four times a year for two years. On March 9, 1985, signs were erected at each end of the adopted highway section recognizing the civic group's effort. The program provided a model for successful Adopt-A-Highway programs in Texas, other U.S. states, Canada, Australia, and New Zealand.

3278. State in the United States to pass a beautification and antilitter act was New Mexico with the Litter Control and Beautification Act of 1985. The statute allowed New Mexico's State Highway Department to award grants to aid local litter-control and beautification programs.

3279. Law in New York City outlawing the sale of spray paint to minors was Title 10-117 of the New York Administrative Code. Effective July 1985, the graffiti prevention law forbade the sale of aerosol spray paint cans and broad-tipped ink markers to anyone under 18 years of age. The law also required vendors to keep spray paint cans behind their counters and locked up. Violators were liable to a $350 fine for each infraction.

LITTERING—*continued*

3280. State in the United States requiring retailers to obtain an annual litter-control permit was Rhode Island in 1990. The regulation applies to all state sales-tax permit holders whose sales relate in whole or in part to the sale of food or beverages. The permits are issued in five different classes and are issued on the basis of the holder's gross receipts for the previous year. The "Litter Control Participation Permit" law became effective December 31, 1990.

3281. Litter and recycling strategy funded by fines and a statewide retail tax was implemented by Nebraska, effective July 1, 1991. A tax on all retail businesses began to raise funds for the state's Waste Reduction and Recycling Incentive Fund. The antilitter portion of the state strategy, the Litter Reduction and Recycling Fund, was supported by fines.

3282. State in the United States to set a time limit on accumulating litter near a public highway was Texas in 1991. The state allows county commissioners to prohibit the accumulation of litter for more than 30 days on a person's property within 50 feet of a public highway, remove the refuse, and bill the property owner for the removal. The law became effective September 1, 1991.

3283. State in the United States to tax litter-producing industries was Washington in 1992. The state law levied a tax on the value of selected packaged products considered to contribute to Washington's litter problems. The taxable items included food, cigarettes, newspapers, and drinks sold in glass, metal, or plastic containers. The tax was levied on industries that produced the items and applied only to items used or consumed in the state.

3284. City in the United States to ban the sale of spray paint to reduce graffiti was Chicago, IL, in 1992. Many U.S. municipalities passed laws in the last three decades of the twentieth century to eliminate graffiti, which was considered a visual pollutant akin to litter and a violation of private property rights. Effective in May 1992, the Chicago City Council prohibited the sale of spray paint within city limits and instituted a $100 fine for every infraction. The law was challenged by the paint industry but was upheld on appeal.

3285. State in the United States to impose a "hard to dispose" materials tax was Rhode Island in 1993. The statute became effective January 1, 1993. It classified a variety of liquids that are environmentally troublesome when improperly discarded under the state's litter control laws. It imposed a tax on sellers of listed "hard to dispose" items, which include vehicle lubricating oils, recycled and re-refined oils, antifreeze, and organic solvents like paint thinners.

3286. State in the United States to enact a "conspiracy to dump" law was Mississippi in 1994. Under the state's "Penalties for unauthorized dumping of solid wastes" laws, anyone who conspires to dump illegally is liable to the same penalties as someone who has illegally disposed of solid waste. The statute covers a complete range of waste, from common litter to unauthorized commercial dumping. The law became effective on July 1, 1994.

3287. State in the United States to enact a "strict liability" antilitter law was Ohio in 1994. Under Ohio's nuisance statutes, anyone who deposits litter on public property without authorization has violated the law, regardless of their intent. The statute was designed to streamline prosecutions by eliminating any need to prove that the person charged had intentionally littered. The law also prohibits illegal dumping by prohibiting the disposal of trash on another person's property without permission. Such disposal is likewise prohibited in public receptacles, with the exceptions of beverage containers, food wrappers, and automobile ashtray contents. The law became effective October 20, 1994.

3288. State commission in the United States funded by taxes on litter-producing corporations was the Ohio Division of Recycling and Litter Prevention. Taxes that became effective September 27, 1997, were levied both as part of the state's corporate tax rate and on manufacturers of cigarettes, beer, liquor, and soft drinks, and other products that contributed to Ohio's "litter stream."

3289. Satellite station built to track "space litter" was Globus II, which was built at Vardö in northern Norway in 1998. The facility is operated by Norway's military intelligence service and the U.S. Air Force Space Command. It was built specifically for monitoring thousands of pieces of orbiting man-made debris in order to minimize the danger they present to new spacecraft and to life on earth in the event of their re-entry into the atmosphere.

3290. International Adopt-a-Highway Day was March 3, 1999. The event coordinated cleanup and educational public-awareness efforts by citizens, governments, and business groups involved in Adopt-a-Highway antilitter campaigns. The event was observed in Australia, Canada, and 48 of the 50 United States.

3291. Tax in Ireland on shoppers' plastic bags began to be levied in March 2002. The tax was collected at the store cash register, if the customer wanted a bag to carry purchases (meats and tender vegetables excepted). It was hoped that customers would resort to reuseable totes and so reduce the number of plastic bags littering the landscape. Money raised by the tax went to environmental projects.

M

MINES AND MINING

3292. Metal mines believed to be Egyptian copper mines in the Sinai peninsula operated circa 5,000 B.C.

3293. Mines known in Great Britain were discovered at what is now Church Hill in Findon, West Sussex. They were flint mines that had been worked in circa 3300 B.C.

3294. Tin mines in ancient Britain were described in 320 B.C. by the Greek explorer Pytheas, a navigator from the Greek colony of Massilia, now Marseilles, France. The mining operations described in his *On the Ocean* were located in present-day Cornwall, England.

3295. Monastic and manorial records of the coal mining industry in England date from the 13th century A.D. Coal has the longest continuously documented history of any British industry.

3296. Treatise on mining that was scholarly, detailed, and systematic was *De Re Metallica* in 1556 by Georgius Agricola, the Latinized name of George Bauer, Germany's first major mineralogist and metallurgist. His illustrated text contained mining procedures similar to 20th-century methods. He spent more than 25 years compiling the information, and the work was translated into English in 1912 by Herbert Hoover, future U.S. president and a mining engineer, and his wife, Lou Hoover.

3297. Mining book detailing methods used by miners was *Beschreibung Allerfurnemisten mineralischen ertzt und bergwerksarten (Description of Leading Ore Processing and Mining Methods)* published in 1574 in Prague by Lazarus Ercker, a Bohemian metallurgist. The book was still widely read into the middle of the 18th century.

3298. Iron mines in the American colonies were developed in New England during the 1640s. By 1775, the colonies had pushed ahead of England in iron manufacture.

3299. Coal mined in Canada was in 1685 from "crop openings" at Baie des Espands, or Sydney Harbour, Nova Scotia. It was mined by the French military. Coal was first recorded there in 1672 by Nicholas Denys, a French explorer and governor of Acadia, who called the discovery "a mountain of very good coal."

3300. Steam engine of high pressure used in a mine was "The Miner's Friend," patented in 1698 in England. It was invented by Thomas Savery, a military engineer, to pump water from coal mines. It had no safety valve to prevent possible explosions in the mine. In 1712, he and Thomas Newcomen worked together to produce an improved version.

3301. Coal deposits recorded in the American colonies were found by Huguenot settlers in 1701 in Manakin on the James River near present-day Richmond, VA.

3302. Coal mine used on a commercial basis in Canada was opened in 1720 at Cow Bay, or Port Morien, in Nova Scotia. It was used to supply the fortress at Louisbourg and also exported coal to the New England colonies.

3303. Mining school was the Schemnitz Mining Academy established in 1733 at Schemnitz, Hungary. It was also the world's first technical college.

3304. Bituminous coal commercially mined in the American colonies was in 1745 on the James River near present-day Richmond, VA. The coal was later used to fuel the manufacture of weapons for the American Revolution.

3305. Mining by European settlers in Australia occurred in 1788 on the eastern seaboard. Sandstone near the Hawkesbury River was quarried and shaped for the buildings of Sydney Cove.

3306. Gold discovery of significance in the United States occurred in North Carolina in 1799.

3307. Gold discovery in Georgia was in 1828. The find touched off the first large-scale U.S. gold rush.

MINES AND MINING—*continued*

3308. Coal mines in England to use steam engines began operation in 1830. The machines were used to lift coal in containers up shafts and to pump out water that had seeped into the mines.

3309. Pure copper deposits discovery in Michigan was made by Lewis Cass in his exploration of the region in 1830. The United States eventually became the world's largest supplier of copper.

3310. Lead mine in Australia was opened in 1841 in the Glen Osmond Hills on the outskirts of Adelaide

3311. Gold deposit discovery of significance in California occurred in 1848. The find near Sutter's Mill on the Sacramento River touched off the massive California Gold Rush of 1849.

3312. Comstock Lode exploration in Nevada began in 1859. Prospector Henry Comstock sold his mining rights to the lode, the largest silver deposit ever found in the United States, for $11,000.

3313. Diamond in South Africa was found in 1866. A "pebble" picked up on the banks of the Orange River turned out to be a 21-carat diamond. South Africa's diamond fields soon became the greatest in the world.

3314. Coal-mine safety laws in a U.S. state were passed in Pennsylvania in 1870. The protective legislature followed a fire in a coal mine there that suffocated 179 miners.

3315. Mining law in Pennsylvania was enacted in 1870. The legislation established minimum regulation of the industry in Pennsylvania, then the leading coal producer in the United States.

3316. Mining law that freely released federal land for public exploitation was the Mining Law passed by the U.S. Congress in 1872. It allowed individuals and companies to purchase the mining rights to federal land generally without a fee, lease, or contract, although sometimes five dollars (or less) an acre was charged. The impact on the environment was not questioned, so prospectors could use bulldozers or other equipment to recover the minerals.

3317. Copper mine in Australia was opened in January 1843 at Kupunda in South Australia. The copper deposits had been discovered the previous year separately by Francis Dutton and Charles Bagot. The mine closed in 1877.

3318. Mining of significance in the iron-rich Mesabi Iron Range in northeast Minnesota began in 1890. Leonidas Merritt and his brothers organized the Mountain Iron Company to mine the ore from the range. For half a century, the Mesabi was the chief supplier of iron ore for the U.S. steel industry.

3319. Coal-mining safety law in the United States was passed in 1891 by Congress. It applied only to mines in U.S. territories, creating minimum ventilation regulations and prohibiting the employment of children under the age of 12.

3320. Gold diggings discovery on Klondike Creek in the Yukon Territory of Canada was in 1896. By 1898, 18,000 prospectors had poured in with the Klondike gold rush.

3321. Coal mining accident of significance in the United States occurred in Scofield in central Utah, when 200 miners were killed on May 1, 1900. The accident was caused by the explosion of blasting powder.

3322. Coal mining accident in Europe to cause catastrophic deaths was in France. An explosion on March 10, 1906, in a mine at Courrieres resulted in 1,060 deaths.

3323. Coal-mine explosion with massive loss of life in the United States occurred on December 6, 1907, in the Fairmont Coal Company's mine at Monongah, WV. The explosion killed 362 people. The large explosion was caused when a small methane gas explosion triggered a larger one of coal dust.

3324. Miners' health protection agency established by the U.S. government was the Bureau of Mines in 1909. It promoted safety in mines and the general welfare of the miners. That same year, it began a joint study with the U.S. Public Health Service into lung disease and mining. In 1912, the bureau did research into smoke control in mines.

3325. Coal-mining agency in the U.S. government was the Bureau of Mines established in 1910 by the U.S. Congress. The bureau was given the responsibility of conducting research and reducing accidents in the coal mining industry. It was not allowed to conduct inspections, however, until 1941 when the U.S. Congress empowered federal inspectors to enter mines.

3326. Coal-mine emergency legislation of significance passed by the British Parliament was the Coal Mines Act of 1911. It required coal companies to provide and maintain equipment for rescue and first-aid work. A certain number of miners had to be trained in this work and an ambulance service was stipulated.

3327. Coal-mine inspection law by the U.S. government was the Coal Mine Inspection Act of 1941. It required the Secretary of the Interior to establish federal inspectors through the Bureau of Mines. They inspected mines to investigate the workers' health and safety conditions, although the federal act gave the government no power to enforce standards.

3328. Coal mining accident in the Far East causing catastrophic deaths occurred on April 25, 1942, in the Honkeiko colliery in Manchuria. An underground explosion killed 1,549 miners. Manchuria was then independent but controlled by the Japanese, who named it Manchukuo.

3329. Coal mines to be fully automated in the United States were opened in 1950, following an agreement between the industry and the United Mine Workers of America, led by John L. Lewis. Before this agreement, only 1.2 percent of underground coal was mined by machine.

3330. Coffinite mineral was discovered in the United States, in Colorado in 1955. The mineral is a high-grade ore more than 60 percent uranium in content.

3331. Sulfur mine offshore was the Grand Isle offshore mine, which produced its first sulfur on March 14, 1960. Located about seven miles off the Louisiana coast 2,000 feet beneath the bottom of the Gulf of Mexico, the mine was operated using a steel drilling rig equipped with boilers and generators standing in 50 feet of water. The deposit was discovered by the Humble Oil and Refining Company; the mine was built and operated by the Freeport Sulphur Company.

3332. Coal-mine fire to depopulate a U.S. city began in May 1962 under Centralia, PA, and is still burning. The population has fallen from 1,100 to less than 20 and almost all buildings in the city's center have been removed as efforts to put out the fire and smoke have failed. The fire began accidentally from a trash dump fire.

3333. Mine safety law in the United States applying to non-coal mines was the Federal Metal and Nonmetallic Mine Safety Act, passed in 1966 by the U.S. Congress. It gave the Bureau of Mines the power to set safety standards, many of which were advisory, and to conduct inspections and investigations. The bureau, however, had little enforcement authority.

3334. Coal-mining law with penalty fines imposed by the U.S. government was the Federal Coal Mine Health and Safety Act of 1969, signed by President Richard M. Nixon on December 30, 1969. It was prompted by a 1968 disaster that killed 78 miners at the Consol Coal Company's No. 9 mine in Farmington, WV. The act required four annual inspections for underground mines and two for surface ones. It provided the first fines for all violations and also criminal penalties for known willful violations. Inspectors could close mines that were in imminent danger. The act also provided benefits to miners who were disabled by black lung disease.

3335. Productive mine above the Arctic Circle in Canada was the Nanisivik lead-zinc mine that began operations in October 1976 on Baffin Island. The lode had been discovered by Texasgulf, Inc., in the 1950s.

3336. Strip mining controls by the U.S. government began with the signing of the Surface Mining Control and Reclamation Act (SMCRA) in August 1977. The new law required companies to restore stripped land to its original function before mining. It established a $4.1 billion fund to restore stripped land never restored, using funds from a tax on coal. Previous controls of strip mining had been established by the states, but this law required them to implement these regulations within 36 months or forfeit jurisdiction over surface mines to the U.S. Office of Surface Mining Reclamation and Enforcement.

3337. Law in the United States to require mine spoil waste land to be reclaimed was the Surface Mining Control and Reclamation Act of 1978. Under the act, when mining activity ceases the waste produced by the mine must be buried and covered with vegetated topsoil. This helps prevent the generation of acid mine drainage by limiting the exposure of pyrite to oxygen and water. In addition, mining companies are required to monitor the effectiveness of their compliance with the regulation and post bonds against possible abatement efforts.

3338. Diamond mine in Canada opened on October 14, 1998, in the Northwest Territories about 190 miles northeast of Yellowknife. Prior to this, an environmental impact statement was issued after two years of research and a cost of $14 million (Canadian dollars). The Ekati Diamond Mine, owned by BHP Minerals and the Dia Met Mineral Ltd., was expected to produce 4 million carats of gems annually. With a second mine due to open 22 miles away in 2002, Canada expects to produce 10 percent of the world's diamonds.

N

NATURAL RESOURCES

3339. Natural resources exploited in Indonesia were spices like nutmeg and mace, which were traded with Indian and South Asian merchants as early as the 10th century. Rubber and oil resources were discovered in the late 19th century and became economically significant by the early 20th century, as did timber, which led Indonesia to become one of the major global providers of plywood.

3340. Natural resource exploited in Tanzania was gold found in the Zambezi River area and traded extensively in the 13th century by the region's powerful city-states. Colonial exploitation of the land began in earnest under German administrators after 1887, mostly in the form of rubber and coffee plantations.

3341. Natural resource exploited in Ghana was gold. The first Portuguese colony in the area was established in 1482. Mineral resources made the West African region well-known as the "Gold Coast," even before it achieved notoriety in the 16th century as a source of slave labor, but significant gold mining began with the introduction of mechanized extraction equipment by European companies in the 1890s. The region now known as the country of Ghana also became an important supplier of cacao during the 19th century.

3342. Natural resource exploited in Argentina was land. The region's colonization evolved slowly during the late 16th century. Because it lacked the more profitable mineral resources of other Spanish colonies like Peru, Argentina's economy developed around the use of its grassy interior for livestock production. Agricultural development gradually made the country a major producer of grains and oilseeds.

3343. Natural resource exploited in Bolivia was silver in the 16th century. As a Spanish colonial department of Peru, the region became a major global silver producer during the 16th century. The southern city of Potosı was one of the largest cities in the Western hemisphere until the decline and collapse of the silver market in the early 19th century.

3344. Natural resource exploited in Brazil was brazilwood, which was used to produce red and purple dyes. The wood was exported to Europe by Portuguese traders after 1500. As settlement of the colony named after the wood progressed, agricultural land was developed for sugar and tobacco in the mid-16th century and for coffee plantations in the 19th century. The region's mineral exploitation began in the 18th century after the discovery of gold and diamonds.

3345. Natural resource exploited in Cuba was land. The island's mineral resources were quickly depleted after its discovery by Christopher Columbus in 1492. Cattle ranching developed during the 16th century, when Cuba was still considered a stop on colonial trade routes. Agricultural production of sugar and tobacco also begun and increased significantly in importance after the Haitian slave revolt of 1796 significantly diminished Haiti's power as an economic competitor.

3346. Natural resource exploited in Panama was land. In 1511, the first Spanish governor, Vasco Núñez de Balboa, advised colonists to develop agricultural produce, like corn, cacao, cotton, and vegetables. The region remained important primarily as a trade route, centuries before the Panama Canal opened in 1914.

3347. Natural resource exploited in El Salvador was indigo used in dye-making. Indigo exports increased after the Spanish conquest in 1528. Development of land for agricultural production of cacao followed in the mid-16th century.

3348. Natural resource exploited in Peru was silver in the 16th century. Mining remained the center of Peru's economy for several centuries after the founding of Lima by Spanish conquistadors in 1535. Cotton production and petroleum development were the country's main economic activities by the twentieth century.

3349. Natural resource exploited in Honduras was gold, which was exported by the Spanish after 1536. Silver mining, cattle ranching, and agricultural development followed. The significant Honduran banana industry grew after the first exports to New Orleans in the 1890s.

3350. Natural resources exploited in Colombia were precious metals. After colonization by the Spanish commenced in 1539, the mining and export of gold and copper to Spain dominated the region's economy until it was surpassed by agricultural production in the 18th century. The colony's first major export crops were sugar and tobacco in the late 18th century, followed by coffee in the 19th century.

3351. Natural resource exploited in Chile was land developed for agriculture in the 16th century. Apart from some gold mining after the first Spanish governorship was established in 1540 and later extractions of copper and nitrates in the mid-19th century, Chile's early economy was dominated by production of wheat for other Spanish colonies.

3352. Natural resource exploited in Mexico was silver, which was discovered in Zacatecas in 1546. Exports of gold from the Spanish colony during its early years were produced mainly by melting Aztec treasures, not from mining. Silver dominated Mexico's economy until the mid-17th century, when mercury needed for refining was diverted to Bolivian silver producers, causing a collapse of the Mexican mining industry.

3353. Natural resource exploited in Costa Rica was land. Although the first permanent European settlement was established in 1564, development proceeded slowly in comparison to mineral-rich colonies in Central and South America. The Costa Rican coffee industry did not develop significantly until the mid-19th century.

3354. Natural resource exploited in North America was wildlife. After the use of beaver in felt hats became popular in Europe around 1600, exports of pelts by North American trading companies spurred exploration of the continent and created a major industry which reached its peak between 1800 and 1850. By the time the industry collapsed in the mid-19th century, populations of beaver everywhere but in the most remote areas of Canada and the United States had been decimated.

3355. Natural resource exploited in South Africa was land. Expanding development of wheat farming, livestock, and vineyards in the late 17th and early 18th century by Dutch and French settlers resulted in competition for available land with indigenous farming and herding societies. Development of South Africa's significant mineral resources did not begin until the nineteenth century, after the discoveries of diamonds (1867) and gold (1869). The first Dutch settlement at Cape Town was established in 1652.

3356. Natural resource exploited in Nigeria was the supply of palm trees in the coastal areas in the late 18th century. Exports of palm oil used as a lubricant increased significantly after abolition of the slave trade in 1807, as did later agricultural production of peanuts, cotton, and cacao. The Nigerian economy and environment were significantly transformed after the discovery of major oil supplies in 1956.

3357. U.S. census to collect information on a natural resource was the census of 1840, which attempted to assess the nation's fisheries. Previous censuses had counted people (1790) and manufacturing assets (1810). From the early 19th century onward, data collected by the census increasingly included human demographics and resource-related information useful in projecting the country's economic and social health.

3358. Department of the Interior in the United States was originally named the Home Department, created by Congress on March 3, 1849. The Home Department was created to administer miscellaneous agencies whose duties were unrelated to the established executive State or Foreign Affairs, Treasury, and War departments. Over the next 150 years, the Department of the Interior evolved to oversee a wide variety of domestic affairs agencies, including the Geological Survey, National Park Service, Indian Affairs, Land Management, and the Fish and Wildlife Service.

3359. Electric generating station began operation in 1880 in London, England. The plant was designed by Thomas Edison and provided power for London's streetlights.

3360. Electric generating plant in the United States began operation in New York City in 1882. The plant provided power for the first system of electric lighting in the United States.

3361. U.S. government study into the effects of wildlife on natural resources was ordered by Congress on June 3, 1886. The study appropriated funds for the Department of Agriculture's Division of Economic Ornithology and Mammalogy to study the food, habitats, distribution, and migrations of North American birds and mammals in relation to agriculture, horticulture, and forestry. The studies were renewed with a similar appropriation and directive given to the Division of Biological Survey on April 25, 1896. In 1905 the Division became the Bureau of Biological Survey.

3362. Natural resources exploited in Madagascar were land and minerals. During the French colonial period, which lasted from 1896 to 1959, Madagascar's export economy developed around agriculture and mining. The island's mines produced graphite, chomites, and uranium. Crop exports included indigo, sugar, and palm

NATURAL RESOURCES—*continued*

3363. Natural resource exploited in Kenya was land, whose agricultural use was heavily promoted and regulated by the British colonial government during the first three decades of the twentieth century. By the 1920s, Kenya was a major exporter of coffee and tea. The country also became the major global source of pyrethrum, a natural biodegradable insecticide made from the dried flowers of the daisy-like *Chrysanthemum cinerariaefolium* plant.

3364. Use of the expression *wise use* in discussing U.S. resource issues occurred during the first decade of the 20th century, when the phrase was coined by conservationist Gifford Pinchot while he was serving as U.S. Forest Service director. Pinchot believed that natural resources should be conserved to ensure their sustained use. Ironically, the phrase was later appropriated by militant anti-environmentalists in the western United States to describe their own cause.

3365. U.S. government conference devoted to conservation issues was the Conference on the Conservation of Natural Resources. Also called the Governors Conference, the three-day meeting was convened on May 13, 1908, by President Theodore Roosevelt. It was organized by prominent conservationist and Roosevelt advisor Gifford Pinchot, who brought citizens, state governors, Congressional representatives, and other government officials together at the White House to discuss protection and management of the nation's natural resources. The meeting was a landmark in attracting public interest in conservation issues.

3366. Canadian–U.S. International Boundary Treaties Act was signed in Washington, DC, on January 11, 1909, by the governments of the United States and Great Britain, on behalf of Canada. The law officially established the waterways dividing the two countries as a common resource to be shared for irrigation, power, navigation, and "domestic and sanitary purposes." The law also established the International Joint Commission to oversee boundary water relations between the two countries.

3367. Intergovernmental conservation conference in North America was the North American Conservation conference. Sponsored by President Theodore Roosevelt, representatives of Canada, Mexico, the British colony of Newfoundland, and the United States met in Washington, DC, on February 18, 1909, to discuss natural resource-related issues.

3368. Environment and Natural Resources Division of the U.S. Department of Justice was established on November 16, 1909. It was originally called the Public Lands Division and was directed to handle legal matters concerning federal lands, water and American Indian disputes. Its responsibilities have evolved to include handling litigation concerned with the protection, use, and development of natural resources and public lands, wildlife protection, Indian rights and claims, cleanup of hazardous waste sites, the acquisition of private property for federal use, and defense against challenges to government environmental programs and activities.

3369. National natural-resource protection law in India was the Indian Forest Act of 1927. The legislation gave the national government complete rights over natural resources. Licensing of the timber industry and forest protection programs were relegated to control by individual states.

3370. Natural Resources Conservation Service of the U.S. Department of Agriculture was created in 1933 and originally called the Soil Erosion Service. It was later known as the Soil Conservation Service. The agency's initial role was creating conservation districts to aid agriculture and prevent erosion. Its work has evolved to include preservation and incentive programs in watershed, wildlife habitat, farmland, and forestry conservation.

3371. National natural resource inventory document in the United States was the "1934 National Erosion Reconnaissance Survey." The study was conducted by the Soil Erosion Service, an early version of the Natural Resources Conservation Service of the U.S. Department of Agriculture. The report inventoried and analyzed erosion damage caused by wind or water throughout the United States.

3372. Natural resources inventory to employ statistical sampling in the United States was the Conservation Needs Inventory of 1958. The inventory was ordered by the Secretary of Agriculture, who required the Natural Resources Conservation Service to collect data on all privately owned lands, with the exception of forest land.

3373. Wetlands Loan Act in the United States was approved by Congress on October 4, 1961. The act allowed federal agencies to borrow funds from the U.S. Treasury that were anticipated from the sale of "duck stamp" hunting fees as a means of accelerating the purchase of migratory wildfowl habitat for conservation. The loan program was expanded in 1969 and 1976. Loans for wetlands purchases were forgiven on November 10, 1986.

3374. International conference on wetlands was the "MAR Conference" held in November 1962 at Saintes Maries de la Mer, France. The conference was sponsored by the World Conservation Union (IUCN), the International Council for Bird Protection (ICBP), and the International Wildfowl Research Bureau (IWRB). Experts from 16 nations discussed economic, scientific and moral aspects of wetlands conservation and management. Their recommendations were distributed by UNESCO (the United Nations Educational, Scientific, and Cultural Organization) in 1964.

3375. Wetlands conservation book published by the U.S. government for a general audience was *Waterfowl Tomorrow*. The book was produced by the U.S. Department of the Interior in 1964. Aimed at hunters and conservationists, it examined the decline of waterfowl populations and their relationship to wetlands lost through overhunting, pollution, drainage, and other habitat encroachment.

3376. Iowa State Preserves System was created by the state's legislature in 1965. The program allows designation of lands for protection and preservation under criteria in five categories: natural, geological, archaeological, historical, or scenic. Iowa's preserves, on both public and private lands, are established and overseen by the State Preserves Advisory Board.

3377. Regional wetlands development commission in New Jersey was the Hackensack Meadowlands Development Commission (HMDC). The commission was established by the New Jersey Legislature in 1968 by the Hackensack Meadowlands Reclamation and Development Act, which became law in Janury 1969. Years of pollution controls and cleanup, wildlife protection, and sustainable development programs have helped restore the district's wetlands and wildlife while increasing its economic value through ventures like the Meadowlands Sports Complex.

3378. United Nations convention on African natural resources was the African Convention on the Conservation of Nature and Natural Resources. The convention was adopted in Algiers, Algeria, on September 15, 1968, and entered into force on October 9, 1969. The issues addressed by the convention were dealt with separately in many later UN resolutions and programs: soil, water, flora, and fauna conservation; forest management; hunting regulation; and sustainable development plans.

3379. Geothermal Steam Act in the United States was enacted December 24, 1970. The act governed geothermal steam and related resources on public lands, including national forests. It prohibited geothermal leases on lands in the National Park System, federally administered fish hatcheries, wildlife refuges and management areas, wildlife or game ranges, waterfowl production areas, lands reserved for the protection of endangered fish and wildlife, and tribally or individually owned Indian trust lands. The act was amended in 1977, 1988, and 1993.

3380. Water Bank Act in the United States was approved December 19, 1970. It authorized the secretary of agriculture to enter into ten-year contracts with landowners to preserve wetlands and retire adjoining agricultural lands. The program authorized annual payments to and conservation cost-sharing with landowners. A total annual limit of $10 million in payments to participating owners was increased to $30 million on January 2, 1980. At that time, coverage under the act was extended to include wetlands classified as shrub or wooded swamps and artificially created inland freshwater lakes.

3381. Satellite composite map of the U.S. was assembled in November 1974 by the Department of Agriculture's Soil Cartographic Unit in Hyattsville, MD. The photomosaic of the contiguous 48 states was constructed from 595 photographs taken from a 560-mile altitude by the Landsat satellite launched on July 23, 1972. This Earth Resources Technology composite map measured 10 feet by 16 feet.

3382. American Indian organization to collectively negotiate over control of natural resources in modern times was the Council of Energy Resource Tribes (CERT). The group formed during the 1975 energy crisis, when anxiety about U.S. energy reserves provoked new interest in developing domestically-produced energy. CERT officials used their experience in the private sector to negotiate tribal participation in managing rich energy resources

NATURAL RESOURCES—*continued*

on or near Indian lands. Resulting income helped build stable tribal economies, increased energy self-sufficiency on reservations, funded scholarships, and placed affiliated tribes in a better bargaining position in lease negotiations.

3383. National Wetlands Resource Center (NWRC) in the United States began in 1975 as the National Coastal Ecosystems Team, part of the U.S. Fish and Wildlife Service's Office of Biological Services. Headquartered at Stennis, MI, the group researched and disseminated reports on coastal, estuarine, and species conditions. When wetland issues became more prominent nationwide, the team was renamed the National Wetlands Research Center in 1986. The NWRC is now based at the University of Southwestern Louisiana in Lafayette, where it provides technical assistance and information about wetland losses, restoration, and management.

3384. Natural resource profits fund to benefit all Alaskans was the Alaska Permanent Fund, which was created by voters in 1976 through an amendment to the state constitution. The fund is a public savings account that receives 25 percent of all state oil and other mineral lease royalties. The Alaska Permanent Fund Corporation, established in 1980, invests the income and distributes dividend checks to Alaska citizens.

3385. National Resources Inventory conducted by U.S. Natural Resources Conservation Service (NRCS) was completed in 1977. The report was more comprehensive than earlier surveys done by the Department of Agriculture's NRCS, which were called Conservation Needs Inventories. Systematic collection of data on wind and water erosion in the inventory prepared the way for analyses of resource depletion in future NRCS inventories, which are done at five-year intervals.

3386. Nation to pass a constitutional amendment to protect its public health, forests and wildlife was India. The 42nd Amendment, which became effective January 3, 1977, required the national government to protect health and natural resources in formulating legislation. The amendment also gave the government the power to regulate industries capable of producing pollution.

3387. Survey of U.S. wetlands and wetlands losses that was comprehensive and statistically valid was produced by the National Wetlands Inventory (NWI) in 1982. The NWI is a division of the U.S. Fish and Wildlife Service. The agency is responsible for monitoring the status of U.S. wetlands and deepwater habitats. Starting in 1990, it was required to report its findings to Congress every ten years.

3388. "Subsistence rights" legislation for Alaska was Title VIII of the Alaska National Interest Lands Conservation Act. The federal act was signed into law December 2, 1982. It required the State of Alaska to manage resources in a way in which rural Alaskans who depended on wildlife for food, clothing, and shelter would receive priority in hunting and fishing rights. The state managed a subsistence rights policy until 1989, when it was overturned by Alaska's Supreme Court, which ruled it unconstitutional in the case of *McDowell* v. *Alaska*. In 1999 the federal government threatened to assume management of the program unless the state instituted a program of its own.

3389. Federal law in the United States to withhold development funds for construction on coastal barrier islands was the Coastal Barrier Resources Act (CBRA), enacted October 18, 1982. The statute prohibited direct or indirect federal financial assistance that might support development on barriers designated in the Coastal Barrier Resources System. The law expanded a provision of the Omnibus Reconciliation Act of 1981, which sought to limit such development by ending the availability of federal flood insurance after October 1, 1983, for all new construction and improvements to structures on undeveloped coastal barriers designated by the secretary of the interior.

3390. African country to produce a National Conservation Strategy was Zambia. In 1984, the Zambian government devised a management policy for the country's natural resources, including wildlife, farmland, minerals, and timber. The strategy addressed national ecological issues like soil erosion, agricultural production, the illegal ivory trade, and planning for sustainable development.

3391. U.S. government agency to adopt a formal environmental policy with American Indian tribes was the Environmental Protection Agency (EPA). On November 8, 1984, EPA administrator William D. Ruckelshaus implemented a policy that recognized the importance of tribal governments in environmental regulation. The policy included declarations that the

EPA would work directly with tribal nations to protect public health and the environment, as well as a promise to include tribes' policy goals in the EPA's overall decision-making process.

3392. United Nations educational center for African natural resource management was the United Nations University Institute for Natural Resources in Africa (UNU/INRA). The research and training program was established in 1986 to strengthen African national economies and sustainable development through educational programs in resource management and conservation. Its main facility is based in Accra, Ghana. A Mineral Resources Unit (MRU) is located at Lusaka, Zambia.

3393. Debt-for-nature swap was arranged between the non-profit organization Conservation International (CI) and the government of Bolivia in 1987. In the increasingly common practice known as a "debt-for-nature swap," conservation groups purchase and eliminate debt obligations of underdeveloped countries in exchange for agreements that payments on such debts will be used in projects fostering the conservation of natural resources. The agreement between CI and Bolivia benefited funding for and legal protection of the Beni Biosphere Reserve.

3394. Federal law in the United States to withhold development funds for construction on coastal barrier islands in the Great Lakes was the Great Lakes Coastal Barrier Act. The law was enacted November 28, 1988. It required that the secretary of the interior prepare maps and recommend to Congress undeveloped barriers along the Great Lakes that were appropriate for inclusion in the national Coastal Barrier Resources System. Highways in Michigan and coastal barriers that were publicly owned or protected by nonprofit organizations were exempt from the statute.

3395. National Environmental Action Plan (NEAP) in Africa was a 15-year development plan begun in Madagascar in 1988. The plan designated the biologically rich Masoala Peninsula and its surrounding marine environment as a conservation zone, although it was not protected by law until 1997, when a new government instituted a variety of sustainable use programs.

3396. Hybrid geopressure-geothermal power plant began operating at Pleasant Bayou, TX, in 1989. The plant used an organic Rankine gas engine. The U.S. Department of Energy test well used both heat and methane gas recovered from beneath the earth's surface.

3397. Modern U.S. National Park Service (NPS) strategy for natural resource management was the "Vail Agenda." Published in 1991, the NPS document agreed that park managers need comprehensive and reliable scientific information about the nature, condition, and evolving status of the major biotic and abiotic natural resources placed under their stewardship. To accomplish this, the agenda prompted a variety of natural resource inventory and monitoring programs.

3398. Arctic-subarctic biome surveyed by a U.S. National Park Service Inventory and Monitoring Program (I&M) was Denali National Park and Preserve in Alaska. In 1992 Denali was selected for prototype monitoring of watersheds in large Alaskan parks. Initial research at Rock Creek was centered on the characteristics of water and soil at the site, as well as the structures and dynamics of vegetative and aquatic communities in the watershed area. Although the initial research was limited to the Rock Creek area, its results were expected to be useful in resource and ecosystem management throughout Alaska.

3399. Biogeographic areas of the U.S. National Park Service Inventory and Monitoring Program (I&M) included ten types of ecosystems. Pilot Inventory & Monitoring sites were chosen in 1992 and subsequently launched at Atlantic/Gulf Coast, Pacific Coast, arctic/subarctic, prairie and grassland, deciduous forest, and tropical/subtropical sites. Insufficient federal funding delayed the start of I&M data gathering at sites representing the remaining biome types, which included caves (Mammoth Cave National Park, Kentucky), coniferous forests (Olympic National Park, Washington), rivers and lakes (North Cascades National Park, Washington), and arid lands (Northern Colorado Plateau Cluster, Colorado and Utah).

3400. Eastern deciduous forest biomes surveyed by a U.S. National Park Service Inventory and Monitoring Program (I&M) were the Great Smokey Mountains and Shenandoah National Parks in 1992. Monitoring in the Great Smokey Mountains of Tennessee and North Carolina focused on the effects of air pollution and climatic change on spruce-fir forests and the population dynamics of black bears and white-tailed deer. Other monitored resources included water quality, rare plants, exotic plants and animals, brook trout populations, and the effects of habitat losses to neigh-

NATURAL RESOURCES—*continued*

boring human development. Monitoring in Virginia's Shenandoah Park focused on selected species, forest and aquatic communities, and hydro-geochemical processes, including the effects of acid rain.

3401. Pacific coast biome surveyed by a U.S. National Park Service Inventory and Monitoring Program (I&M) was Channel Islands National Park, a system of five islands off the California coast. Monitoring began in 1992 to assess the status of the park's natural resources and wildlife populations, particularly the effects of air and water quality on kelp forests, rocky intertidal communities, sandy beaches and lagoons, terrestrial vegetation, seabirds, pinnipeds, and land birds.

3402. U.S. National Park Service (NPS) Inventory and Monitoring Program (I&M) began surveying natural resources in 1992. An initial ten-year program was to create a baseline inventory of basic biological and geophysical resources in all the natural resource parks in the United States. A subsequent phase would use this information to establish ongoing programs to monitor ecosystems, integrate NPS management policies with other government agencies, and inform natural resource management decisions. The program's initial progress was hampered by low federal funding.

3403. Guidelines issued by the U.S. Fish and Wildlife Service (FWS) on foregoing lawsuits in natural resource damage cases were enacted as FWS policy on March 27, 1992. Before the FWS would agree not to sue in cases involving oil spills and hazardous waste sites, legally enforceable environmental restoration to damaged sites would be ensured. Legal settlements would be reached only with the coordination of other federal agencies, such as the U.S. Environmental Protection Agency or Office of the Solicitor. Lack of data would not be accepted as a reason to settle or forego legal action to redress environmental damage.

3404. Coniferous forest biome chosen for a U.S. National Park Service Inventory and Monitoring Program (I&M) was Olympic National Park in Washington. Despite being designated in 1993 an Inventory & Monitoring site, lack of federal funding resulted in pilot monitoring protocols being developed outside the Inventory & Monitoring program by the Olympic Field Station and the Forest and Rangeland Ecosystem Science Center of the U.S. Geological Survey. Initial monitoring concentrated on ecological indicators, particularly in areas concerned with resource management, species status, and ecosystem health.

3405. U.S. Natural Resource Inventory and Monitoring (I&M) mapmaking program was the Base Cartographic Data program. Collection of data was begun in 1993 to provide necessary maps for monitoring of resources within the National Park system. Cost of the program was shared by the National Park Service and the U.S. Geological Survey. Complete or partial digital maps of nearly every natural resource park in the United States were produced by 1999.

3406. U.S. Natural Resource Inventory and Monitoring (I&M) Program on water quality was the Baseline Water Quality Status and Trends Project. The study was a national inventory and analysis of all water resources in the National Park system, initiated by National Park Service (NPS) Water Resources Division in 1993. By late 1999, complete inventories of annual and seasonal water quality data collected in and near every national park was filed with the STORET national water-quality database of the U.S. Environmental Protection Agency (EPA). Individual reports were also available from the National Technical Information Service of the Department of Commerce and from the NPS Technical Information Center.

3407. Comprehensive database of natural resource information about U.S. national park lands was organized during the 1990s as the National Resource Bibliography (NRBIB). The project compiled natural resource studies and historical scientific material, including rare event records, maps, photographs, manuscripts, and specimen collections. Federal funding for the project was obtained in 1994 and 1995 through the National Park Inventory & Monitoring Program. The material contributed to the Automated Natural Resource Bibliographic Database, whose adaptation to the Internet began in 1997. The NRBIB Coordinator's office is based at the Columbia Cascades Library in Seattle, WA.

3408. Comprehensive inventory of vegetation in U.S. national parks began in 1994 under the direction of the Biological Resources Division of the U.S. Geological Survey (USGS). The USGS enlisted the help of other government agencies in field sampling and compiling aerial photographs of park vegetation for systematic analysis and mapping. In addition to providing accurate information to individual parks to aid their management of vegetation resources, the study was intended to support natural resource inventories of soil, geology, and species. The program was slowed by gradual funding and did not include Alaskan parks.

3409. Inventory of vegetation in U.S. national parks in Alaska began in 1994 under the direction of the Biological Resources Division of the U.S. Geological Survey (USGS) and National Park Service. The inventory used satellite images and was undertaken separately from vegetation mapping in the lower 48 states because of the size of Alaska's national parks, which comprise 54 million acres. The resource inventory is coordinated from the Park Service's Alaska regional office in Anchorage.

3410. Prairies and grasslands biome surveyed by a U.S. National Park Service Inventory and Monitoring Program (I&M) was the Great Plains Prairie Cluster, a group of six small prairie parks in Iowa, Minnesota, Missouri, and Nebraska. Long-term monitoring began at the parks in 1994 to ensure the sustainability of small remnant and restored prairie ecosystems, determine the effects of external land use and watersheds on small-prairie preserves, and evaluate the effects of fragmentation on the biological diversity of small-prairie parks. The cluster of parks is administered from Wilson's Creek National Battlefield in Missouri.

3411. U.S. Environmental Protection Agency (EPA) office for American Indian environmental affairs was the American Indian Environmental Office (AIEO). Established in 1994, the AIEO institutionalized the EPA's goals of improving and properly regulating environmental conditions on the lands of 560 tribal nations through better government-to-government communication and technical assistance.

3412. Federal environmental justice grants in the United States were authorized by Executive Order 12898, signed by President Bill Clinton on February 11, 1994. The order addressed air, land, and water quality problems whose impact was felt acutely in minority and low-income communities. Grants were authorized for three programs: Small Community Groups (including American Indian tribes), Community/University Partnerships, and Environmental Justice through Pollution Prevention. The latter category, in particular, helped to fund state resource pollution prevention programs.

3413. Standardized national vegetation classification system in the United States was developed by the Nature Conservancy, based in Arlington, VA. The system consolidated data collection approaches used by agencies like the National Biological Survey, the U.S. Fish and Wildlife Service, the U.S. Forest Service, the National Park Service, the U.S. Environmental Protection Agency, and various academic groups. The system was presented to the National Biological Service and National Park Service in November 1994 to provide a consistent national framework for mapping vegetation, conservation planning, and biodiversity protection.

3414. Ecosystem management plan upheld in a U.S. court was the Forest Plan, which was challenged in the 1994 case of *Seattle Audubon Society* v. *Lyons*. The case concerned federal lands in Washington, Oregon, and Northern California. Because the lands were within the range of the northern spotted owl, the Seattle Audubon Society sued the departments of Agriculture and the Interior, charging that the Forest Plan ran counter to provisions of the National Environmental Protection Act (NEPA) and did not properly consider the plan's environmental impact. In a decision of December 21, 1994, a Washington court found the plaintiff's charges to be without merit.

3415. Natural Resources Inventory & Analysis Institute (NRIAI) was chartered in 1995 by the Natural Resources Conservation Service (NRCS) of the U.S. Department of Agriculture. The NRIAI is not a centralized facility but a network of research institutions sharing technological tools and research information beneficial to the NRCS. The program's statistical library is hosted by the Department of Statistics of Iowa State University in Ames, IA.

3416. Atlantic–Gulf Coast biome surveyed by a U.S. National Park Service Inventory & Monitoring Program (I&M) was Cape Cod National Seashore in Massachusetts. The Inventory & Monitoring survey of the park's natural resources began in 1996. The study concentrated on coastal ecosystem components, including shoreline margins, estuaries, maritime forests, kettle ponds and freshwater habitats, and barriers like islands, sand spits, and dunes.

3417. Tropical-subtropical biome surveyed by a U.S. National Park Service Inventory & Monitoring Program (I&M) was the Virgin Islands–Southern Florida Cluster, comprised of Buck Island Reef National Monument, Dry Tortugas National Park, and Virgin Islands National Park. Monitoring of the three areas as a cluster began in 1996 and is headquartered at Virgin Islands National Park. Inventory & Monitoring activities in the region focus on the condition of coral reefs, marine fish communities, water quality, and the now rare tropical dry forests found on St. John and Buck islands.

NATURAL RESOURCES—*continued*

3418. World Wetlands Day was February 2, 1996. Since then, February 2 of each year is designated World Wetlands Day, to commemorate the signing at Ramsar, Iran, in 1971, of the Convention on Wetlands of International Importance Especially as Waterfowl Habitat. Member nations of the convention use the occasion to alert the public to the values and benefits of wetlands.

3419. Environmental Quality Incentives Program (EQIP) in the United States was signed into law by President Bill Clinton on April 4, 1996. The program was designed to aid farmers and ranchers in addressing natural resource concerns with soil erosion, water quality and quantity, wildlife habitat, wetlands, or forest and grazing lands. The program offers five- to ten-year contracts that provide incentive payments and cost sharing for implementing conservation plans. EQUIP was authorized under the Federal Agriculture Improvement and Reform Act of 1996, also known as the 1996 Farm Bill. The program was administered by the Department of Agriculture's Commodity Credit Corporation.

3420. Wetlands Reserve Program (WRP) in the United States was authorized by the Food Security Act of 1985, but states were not allowed to institute a continuous application process until October 1, 1996. WRP dispenses financial incentives like easements to landowners who restore and act as stewards of wetlands created by retiring unprofitable farmland. WRP is administered by the Natural Resources Conservation Service of the U.S. Department of Agriculture.

3421. Executive Order issued to protect the Lake Tahoe region was Executive Order 13057, signed by President Bill Clinton on July 26, 1997. The order created the Tahoe Federal Interagency Partnership, which ordered federal agencies with jurisdiction over the natural resources of the Lake Tahoe area to consolidate their efforts with the states of California and Nevada, tribal nations, and local governments. The order attempted to provide coordinated policies that would consistently manage air, soil, and water quality, fish and wildlife habitats, scenery, noise, recreation, transportation, research, and restoration projects.

3422. Modern natural resource crime laws in China are contained in Chapter Six of China's criminal code. It became effective October 1, 1997. Revisions of the code made the destruction of forestry or mineral resources punishable by prison terms of three to five years, fines, or both. Conviction for illegal mining or logging results in identical penalties, although longer prison sentences are acceptable if the damage is significant.

3423. American Heritage Rivers program was created by a presisential executive order signed by President Bill Clinton on September 11, 1997. The program added federal support to community-based initiatives for natural resource and environmental protection, economic revitalization, and historic and cultural preservation of rivers and riverfronts designated by the program. President Clinton designated the first 14 American Heritage Rivers on July 30, 1998.

3424. Comprehensive inventory of geologic resources in U.S. national parks commenced in 1998 as a cooperative venture involving the National Park Service's Geologic Resources Division, the U.S. Geological Survey, and state geological surveys. The inventory was designed to produce digital maps, a bibliography of geologic literature and maps called GeoBib, an evaluation of geologic maps of parks, resources, and issues, and a report with geologic baseline information.

3425. Modern environmental management policy in South Africa resulted from the National Environmental Management Act (NEMA). The law was passed by the National Assembly on November 5, 1998. NEMA was the first act to standardize environmental management policies in a post-apartheid government and entitled South Africans to seek redress for environmental damage. National policy acts for water and forests were also passed in 1998.

3426. National assessment of biological resources in the United States was released by the U.S. Geological Survey on September 17, 1999. The two-volume report was entitled "Status and Trends of the Nation's Biological Resources." It assessed and detailed changes to the health of land and waterways caused by human activities like urbanization, agriculture, draining of wetlands, and the development of navigation, irrigation, and hydroelectric power projects. The report also identified invasive non-native species as a major threat to aquatic natural resources.

NOISE POLLUTION

3427. Noise control commission in the United States was New York City's Noise Abatement Commission, formed in 1930. The group was organized to assess metropolitan noise problems by the New York City Health Department.

3428. Municipal office for noise control in the United States was the New York City Bureau of Noise Abatement created in 1967. The office, a division of the city's Environmental Protection Administration, treated noise as a health hazard.

3429. Noise control agency in the United States was the Office of Noise Abatement and Control (ONAC), an agency of the U.S. Environmental Protection Agency (EPA) created by authority of the Noise Control Act of 1972. ONAC duties included identifying noise problem sources, setting emission standards, public education programs, and research. The office was abolished in 1982, when President Ronald Reagan ordered the Office of Budget and Management to end funding for ONAC.

3430. Study of aircraft noise effecting U.S. national parks and forest wilderness was mandated by Congress on August 18, 1987. Congress directed the Federal Aviation Administration, the National Park Service, and related state agencies to conduct the study in the interests of air safety and reduction of noise over the parks.

3431. World Health Organization (WHO) working group to address noise problems was PACE, the Prevention and Control Exchange. The group was formed in Geneva, Switzerland, on September 21, 1994. It acts on information sharing and public education initiative to address occupational health problems, including the damage done by unprotected exposure to noise in the workplace.

3432. Town in the United States relocated due to airport noise was Minor Lane Heights, KY. Instead of accepting individual government buyouts to escape the noise from nearby Louisville International Airport, the community negotiated a deal in which state and federal funding supported construction of and relocation to a new town called Heritage Creek. Relocation to the new homes began in 1999.

NOISE POLLUTION—LEGISLATION AND REGULATION

3433. Nation to adopt building codes with noise standards was Germany, in 1938. Under the new building codes, construction businesses were required to conform to national noise insulation standards in building projects in the same manner by which new construction was subject to safety and fire regulations.

3434. New York highway-noise emissions limits became effective October 1, 1965. Section 386 of New York's Vehicle and Traffic Laws set maximum limits for the amount of motor-vehicle noise emissions permissible on the New York State Thruway.

3435. Law in California to limit motor-vehicle noise emissions on highways became effective January 1, 1968. In Section 23130 of the Vehicle Code, the California legislature set maximum allowable decibel levels for motor vehicles on highways, calculated and classified by vehicle weight and speed.

3436. Law in California to limit motor-vehicle noise emissions through sales restrictions became effective January 1, 1968. In Section 27160 of the Vehicle Code, the California legislature set maximum allowable decibel-production levels for all new motor vehicles sold in the state.

3437. City in the U.S. to adopt a building code with internal noise suppression standards was New York, NY, in October 1968. Under the new code, New York began to require the use of acoustical materials to reduce the amount of noise that can be transmitted within a building.

3438. Aircraft noise certification standards in the U.S. were instituted in 1969 by the Federal Aviation Administration (FAA) after passage of an amendment to the Federal Aviation Act of 1958. The new rules, known as Part 36, became effective December 1, 1969. They set a limit on noise emissions of large aircraft of new design under certification standards known as Stage 2.

3439. Regulation of toy guns as a noise hazard in the United States was made possible by the Toy Safety and Child Protection Act, which became effective November 6, 1969. On December 19, 1970, the Food and Drug Administration banned the sale of toy guns or toy gun ammunition capable of producing 138 decibels or more.

NOISE POLLUTION—LEGISLATION AND REGULATION—*continued*

3440. State in the United States to pass a comprehensive noise-control law was New Jersey. The Noise-Control Act of 1971 identified noise as a potential health hazard and assigned its control to the New Jersey Department of Environmental Protection.

3441. Noise pollution legislation of significance in the United States was the Noise Control Act of 1972. The law, which Congress approved October 18, 1972, directed the U.S. Environmental Protection Agency (EPA) to set standards for maximum noise levels produced by commercial products like motors. It also required the EPA to study and set standards for noise levels capable of affecting human health.

3442. Legislation in the United States to allow noise pollution lawsuits was contained in the Noise Control Act, passed October 27, 1972. A legal provision was made for any U.S. citizen wishing to begin a civil action against any person or governmental agency illegally violating noise control regulations.

3443. Noise emission standards for motor vehicles involved in U.S. interstate commerce resulted from the Noise Control Act. The Secretary of Transportation was ordered to set limits on such commercial motor vehicle noise within nine months of the law's passage on October 27, 1972.

3444. Noise standards set by the United States Occupational Safety and Health Administration (OSHA) were implemented June 27, 1974. OSHA acted to prevent occupational hearing loss and other noise-related health problems by publishing noise exposure guidelines and requiring employers to provide protection equipment to workers subjected to long periods of loud noise.

3445. Law controlling aircraft noise over the Grand Canyon went into effect January 3, 1975. Federal law directed the Secretary of the Interior to advise appropriate government agencies of any aircraft activities which threatened visitors' health, safety, welfare, or enjoyment of the natural quiet of Grand Canyon National Park. Complaints about aircraft noise were to be directed to the U.S. Environmental Protection Agency for resolution under the Noise Control Act of 1972.

3446. Quiet Communities Act in the United States was enacted November 8, 1978. The law provided federal funding for state and local governments for noise control programs. It was the first of several identically titled congressional attempts to support noise reduction programs.

3447. Sonic boom regulations for aircraft, deemed a noise pollutant and public health issue in the United States were ordered by Congress July 5, 1994. The head of the Federal Aviation Administration was required to set standards and limits for such aircraft noise.

3448. Noise control law in England with seizure provisions was the Noise Act, which became law on July 18, 1996. The law allowed local authorities to impose a £100 fine and confiscate equipment in complaint cases where a warning to cease making noise between 11 P.M. and 7 A.M. was ignored.

3449. Leaf blower ban of significance in Los Angeles, CA went into effect July 1, 1997. The ordinance was passed by the Los Angeles City Council and prohibited the use of gas-powered leaf blowers within 500 feet of a residence. Unlike quieter, electric-powered blowers, which were unaffected by the law, gas-powered blowers were accused of worsening the California city's already serious noise and air pollution problems. The law superceded earlier rules limiting the blowers to daytime use.

3450. Helicopter noise bill to reach a committee in the U.S. Congress was the Helicopter Noise Control and Safety Bill. The legislation would have required the Federal Aviation Administration to develop a noise reduction plan for any county or municipality of more than 500,000 inhabitants where helicopter noise and safety were at issue. Upon being introduced as H.R. 2957 in the House of Representatives on November 8, 1997, no vote was taken. The bill was instead referred to the Committee on Transportation and Infrastructure, which in turn referred it to the Subcommittee on Aviation on November 21. The bill never made it to the House floor for a vote.

3451. Congressional attempt to re-establish the Office of Noise Abatement and Control (ONAC) was the Quiet Communities Bill (QCA) of 1997, introduced as H.R. 536 and S. 951. President Ronald Reagan's administration ended federal noise control programs by abolishing ONAC in 1982, leaving the U.S. Environmental Protection Agency without an office to address noise pollution, as it was still required to do under the Noise Control Act of 1972. The 1997 QCA proposed restoration ONAC to the EPA, and was introduced to the Senate on June 4. A Senate vote to table the bill effectively killed it on September 2, 1998.

NUCLEAR POWER

3452. "Atomic city" in the United States was Oak Ridge, TN, founded in 1942 as the site of the U.S. government's Manhattan Project to develop an atomic bomb. Oak Ridge was incorporated as a city in 1959.

3453. Test of an atomic weapon occurred at the Alamogordo Bombing Range in New Mexico, on July 16, 1945. The Trinity Test was part of the U.S. military's Manhattan Project.

3454. Use of an atomic weapon in war occurred August 6, 1945, when a U.S. B-29 dropped a primitive uranium bomb on Hiroshima, Japan. Casualties totaled 136,000; 45,000 people died on the first day.

3455. Nuclear energy development agency of the U.S. government was the U.S. Atomic Energy Commission (AEC), established by the Atomic Energy Act of 1946. It was formed to regulate the development and operation of nuclear energy for peaceful purposes. In 1975 it was replaced by the Nuclear Regulatory Commission, which also set standards of safety for the construction and operation of nuclear energy plants.

3456. Use of nuclear power to produce electricity in Great Britain was at the Harwell laboratory, established in January 1946 and located 14 miles south of Oxford. The electricity was produced by a gas-cooled reactor that used uranium rods encased in a magnesium alloy called magnox.

3457. Research of significance by the U.S. government on the peaceful uses of nuclear energy was begun on October 6, 1947, by the U.S. Atomic Energy Commission. Its suggested program was issued as a report in 1948.

3458. Museum dedicated to explaining the peaceful uses of atomic energy to the public was the American Museum of Science and Energy in Oak Ridge, TN. It began in 1949 as the American Museum of Atomic Energy.

3459. Nuclear generator accident of significance occurred December 13, 1950, at the NRX reactor at Chalk River, Canada, in Quebec Province about 150 miles northwest of Ottawa. An explosion of hydrogen killed one worker and contaminated five others, damaged the reactor core, and flooded the reactor building with millions of gallons of radioactive water.

3460. Plutonium-producing nuclear reactor in England began operating in October 1950 at Windscale, a site in Cumbria, northwest England, on the Irish Sea coast. Its Magnox reactor's Number 1 Pile produced plutonium for weapons, as did a second reactor that began operating in June 1951.

3461. Electric power from nuclear energy was produced during the summer of 1951 at the Experimental Breeder Reactor I at the national reactor testing station operated by the Atomic Energy Commission near Idaho Falls, ID. The core in use at the time enabled the fast-breeder reactor to produce 1,400 thermal kilowatts and 150 electrical kilowatts. On December 20, 1951, the reactor supplied steam to a turbogenerator that produced more than 100,000 watts of electricity to operate the pumps and other reactor equipment and provide light and electrical facilities for the building in which the reactor was housed. EBR-I was designed and operated by the Argonne National Laboratory, Argonne, IL, and was the first practical breeder reactor.

3462. Underground nuclear weapons test explosion at the Nevada Test Site (NTS) northwest of Las Vegas was in 1951. Over the following 35 years, the United States and Great Britain conducted more than 450 tests at the Nevada site.

3463. Nuclear power organization of the Australian government was the Australian Atomic Energy Commission, established by the Atomic Energy Act of 1953. Its recent work has included safeguards on nuclear materials in regard to nonproliferation of nuclear weapons.

3464. Nuclear energy proposal at an international level was made December 8, 1953, by U.S. President Dwight D. Eisenhower. Addressing the United Nations General Assembly, he suggested an "Atoms for peace" program involving international research and the control of nuclear stockpiles in order to produce energy for the world community.

3465. Legislation in the United States to specifically control the disposal of nuclear waste was the Atomic Energy Act of 1954. Despite this act, disposal of nuclear waste remains a controversial issue.

3466. Nuclear-powered submarine was the USS *Nautilus*, launched January 21, 1954, at Groton, CT, at the Electric Boat Company. This was the first use of atomic power for propulsion. The submarine's atomic reactor was controlled by inserting or removing rods into its core.

NUCLEAR POWER—*continued*

3467. Nuclear power station began operating on June 30, 1954, in Obninsk in the Soviet Union. It produced more than 5,000 kilowatts of electricty.

3468. Nuclear technology access of significance by U.S. corporations was granted by the Atomic Energy Act of 1954, passed by the U.S. Congress and then signed by President Eisenhower on August 30, 1954. It increased the amount of government information on detailed nuclear technology available to civilians. The act was the first major amendment of the original Atomic Energy Act.

3469. Development by the United States of small nuclear reactors for use in space was in 1955. The first nuclear-powered satellite was launched in 1961. The United States suspended the space reactor program in 1973; it resumed in the early 1980s.

3470. Nuclear energy nonmilitary joint projects of the U.S. government and industry were created by the Power Demonstration Reactor Program announced on January 10, 1955, by the U.S. Atomic Energy Commission. The programs involved cooperation in constructing and operating experimental nuclear power reactors.

3471. Electric power from nuclear energy used to illuminate a town was generated on July 17, 1955, by the Utah Power and Light Company's station at Arco, ID. At 11:28 P.M., the station released steam from a borax reactor into a turbine that drove a 3,500-kilowatt generator that supplied current for the town's 1,200 inhabitants. The power was the sole source of the town's light for an hour. The news was withheld until August 11, 1955, when it was announced at the Atoms for Peace Conference in Geneva, Switzerland.

3472. Electric power generated by nuclear energy to be sold commercially was delivered on July 18, 1955, by the Atomic Energy Commission at West Milton, NY, to the Niagara Mohawk Power Corporation. The power came from a prototype for the nuclear reactor to be used in the submarine *Seawolf*. The approximately 10,000 kilowatts generated by the reactor were integrated with current from traditional sources by Niagara Mohawk, which supplied power to homes and industry at three mills per kilowatt hour.

3473. United Nations meeting directly concerned with peaceful uses of nuclear energy was the United Nations International Conference on the Peaceful Uses of Atomic Energy held from August 8–20, 1955, in Geneva, Switzerland.

3474. Nuclear power agency established by the Turkish government was the Turkish Atomic Energy Commission (TAEK), established in 1956.

3475. Nuclear power plant in England opened in 1956 in Cumbria, northwest England, on the Irish Sea coast. It was then known as the Windscale Nuclear Reactor, from the name of its industrial site. The name was changed in 1981 to Sellafield, which had been the site's original name up to 1947.

3476. Nuclear power agency of the Japanese government was the Atomic Energy Commission, established in January 1956.

3477. Nuclear power from a civilian reactor in the United States was produced on July 12, 1957, by the Sodium Reactor Experiment at Santa Susana, CA, on the outskirts of Los Angeles, CA. This reactor produced power until 1966.

3478. Atomic energy organization of the United Nations for peaceful nuclear power was the International Atomic Energy Agency (IAEA), established on October 1, 1957, in Vienna, Austria. It had the dual role of promoting the peaceful and safe use of nuclear energy and preventing the spread of nuclear weapons. It is permitted to buy and sell nuclear fuels and materials.

3479. Nuclear power-plant accident of significance occurred on October 8, 1957, at the Windscale Nuclear Reactor in Cumbria, northwest England, on the Irish Sea coast. A physicist accidentally let the core temperature of Number 1 Pile rise until the fuel melted and the core caught fire. A large amount of radioactive iodine and polonium was released. By 1983, the Radiological Protection Board of Great Britain attributed 260 cases of thyroid cancer to the accident, and the area has had increases in other cancers, including leukemia. This was the world's worst nuclear power accident until Chernobyl in 1986. To improve its image, the British plant was renamed Sellafield in 1981.

3480. Accident aboard the USS *Nautilus*, the world's first nuclear-powered submarine, occurred at Groton, CT, in December 1957, before the vessel put to sea for the first time. A seamed pipe in the reactor compartment burst, leading to the replacement of ordinary piping with a seamless type.

3481. Nuclear power plant exclusively used for peaceful purposes was the Shippingport Atomic Power Station in Shippingport, PA, where the reactor attained criticality on December 2, 1957. The plant's full rated capacity of 60,000 net kilowatts, enough to provide for the residential needs of a city of 250,000 people, was reached December 23. The station consisted of a single pressurized water-type reactor, a radioactive waste disposal system, laboratory, shops, and administrative facilities. To allow for increased output from future nuclear fuel loads, the turbine generator was designed with a capacity of 1 million kilowatts. President Dwight D. Eisenhower broke ground for the station by remote control from Denver, CO, on September 6, 1954, and formally dedicated the plant by remote control from Washington, DC, on May 26, 1958.

3482. Nuclear accident of significance in the Soviet Union occurred in December 1957 or January 1958 at Chelyabinsk-40, a plutonium production plant in Russia's Ural Mountains. The accident contaminated thousands of square miles and may have caused hundreds of human deaths. Since the accident occurred at the height of the cold war, Western countries did not learn of it until years later.

3483. Harbor created by an atomic bomb was to have been located at Ogotoruk Creek, near the Inupiat Eskimo village of Point Hope in northwestern Alaska. The blast was planned in 1958 and scheduled to be detonated in 1962 by the U.S. Atomic Energy Commission (AEC), which called the plan Project Chariot. Despite the AEC's assurances that the resulting radiation would not effect the environment, local and national opposition forced it to abandon the project in 1962.

3484. International moratorium on above-ground nuclear weapons testing went into effect late in 1958. The moratorium held until 1961, when the Soviet Union conducted a series of tests.

3485. Nuclear energy produced by a fast-breeder reactor in Great Britain was in 1959 at the Dounreay nuclear power plant eight miles west of Thurso in the Highlands on the north coast. The Atomic Energy Authority operated the experimental reactors. The Dounreay Fast Reactor operated until 1977; the Prototype Fast Reactor operated from 1974 to 1994. The fast-breeder reactors did not use a moderator to reduce the speed of high-energy neutrons.

3486. Nuclear power agency of the South Korean government was the Office of Atomic Energy, established in 1959. In April 1967 it became the Atomic Energy Bureau under the newly established government Ministry of Science and Technology (MOST).

3487. U.S. nuclear-powered surface warship was the USS *Long Beach*, a cruiser, commissioned in 1959. The nuclear-powered aircraft carrier USS *Enterprise* went into service in 1960.

3488. Nuclear-powered merchant ship was the N.S. *Savannah* launched on July 21, 1959, at Camden, NJ. It was designed under the auspices of the U.S. Atomic Energy Commission and the U.S. Maritime Administration, with construction beginning May 22, 1958. It could travel 350,000 miles without a renewal of fuel, compared to a conventional ship of that size that could only go 13,000 miles before refueling.

3489. Nuclear power plant in the United States built without government funding was the Dresden-1 Nuclear Power Station in Illinois. It achieved a self-sustaining nuclear reaction on October 15, 1959.

3490. U.S. submarine to carry nuclear-armed ballistic missiles was the USS *George Washington*, built at the Electric Boat shipyard in Groton, CT, and commissioned on December 12, 1959. The ship was withdrawn from service in 1985.

3491. Law establishing comprehensive control of radioactive material by the British government was the Radioactive Substances Act passed by Parliament in 1960. It regulated by license the production, use, storage, and transportation of radioactive substances.

3492. Nuclear weapons test conducted by the French took place February 13, 1960, at a site in the desert of southern Algeria. At least 150 Algerian prisoners were exposed to radioactive contaminants from the atmospheric test.

NUCLEAR POWER—*continued*

3493. Nuclear reactor to power space vehicles was developed in the United States and announced November 18, 1960, by the Atomic Energy Commission. The reactor weighed 220 pounds.

3494. Nuclear power station to operate on a commercial basis was the Yankee Atomic Electric Company's $57-million plant on the Deerfield River at Rowe, MA. It began producing power for distribution on November 10, 1960. The pressurized light water reactor produced 135,000 kilowatts of electricity. The company, formed by 12 New England utility companies, signed a contract with the Westinghouse Electric Corporation as the principal contractor on June 4, 1956. The reactor became critical on August 19, 1960. It was permanently shut down for safety reasons in 1995.

3495. Deaths caused by an American nuclear reactor occurred January 3, 1961, when high levels of radiation from an experimental reactor killed three workers at a federal installation near Idaho Falls, ID.

3496. Satellite to use nuclear power for electricity was NASA's Transit 4A weather satellite, launched June 29, 1961, from Cape Canaveral, FL. It carried a nuclear auxiliary power system. Weighing 175 pounds in orbit, it had an apogee of 534 and a perigee of 623 miles.

3497. Nuclear power plant in Antarctica began operating March 4, 1962. It was put up by the United States at McMurdo Sound.

3498. Nuclear power barge was the converted USS *Sturgis*, a World War II liberty ship. The U.S. Army Nuclear Power Program had it modified, beginning in January 1963, into a pressurized nuclear steam plant producing 10,000 kilowatts. Its new designation was as a MH-1A (mobile high-powered, first of a kind, field installation). It produced power for the Panama Canal Zone from 1968 to 1975 but eventually became too expensive to maintain.

3499. Nuclear submarine from the U.S. Navy to be lost at sea was the USS *Thresher*, which plunged to the ocean floor during sea trials in the Atlantic on April 10, 1963. All 129 crewmen perished.

3500. Nuclear-powered lighthouse began operating on May 21, 1964, on Chesapeake Bay, MD. The Baltimore Light facility received electricity for a decade from a 60-watt radioisotope nuclear generator before refueling was required.

3501. Uranium fuel ownership by private U.S. companies was allowed by the Private Ownership of Special Nuclear Materials Act, passed by the U.S. Congress and then signed on August 26, 1964, by President Lyndon Johnson. After June 30, 1973, the companies were required to own their own uranium fuel.

3502. Titan II silo explosion occurred on August 9, 1965, at Searcy, AR. A fire and resulting explosion in the 174-foot-deep missile silo claimed 53 U.S. Air Force personnel.

3503. Recorded nuclear weapons accident involving a U.S. warship at sea occurred December 5, 1965, in the Pacific. An aircraft loaded with a nuclear weapon rolled off an elevator on the USS *Ticonderoga* and into the sea 70 miles from Japan's Ryukyu Islands. Pilot, plane, and weapon were lost.

3504. Reactor meltdown aboard a nuclear-powered ship occurred during the winter of 1966–1967 aboard the Soviet icebreaker *Lenin*. The vessel, too radioactive to tow, was left abandoned in the Arctic for a year.

3505. Nuclear power commercial reactor in Japan was Toka No. 1, which began operations in 1966. The first experimental reactor had been the JPDR that began in October 1963.

3506. Nuclear-powered ship launched by Japan was the *Mutsu*, in 1969. The cargo vessel, named for its homeport, was the subject of intense local protests. A series of technical problems plagued the *Mutsu*; redesign and repair took more than ten years.

3507. Nuclear power accident in Switzerland occurred on January 21, 1969, at Lucens Vad. An underground reactor suffered a problem with the coolant and released a vast amount of radiation into a cavern that workers then sealed.

3508. Nuclear power-plant requirement in the United States to assess overall environmental impact came from a ruling in September 1971 by the 5th Circuit Court of Appeals in New Orleans, LA. Environmental groups had filed a petition under the National Environmental Protection Act claiming the U.S. Atomic Energy Commission had not considered the total environmental factors when it approved construction of a nuclear power plant on Chesapeake Bay. This court decision eventually required more than 90 other uncompleted nuclear power plants to meet requirements for their total environmental impact.

3509. United Nations condemnation of French atmospheric nuclear weapons tests came in June 1972 at the UN Conference on the Human Environment in Stockholm, Sweden. In November, 106 UN member nations voted against the French weapons tests.

3510. Nuclear energy research museum was the National Atomic Museum at Kirtland Air Force Base in Albuquerque, NM. It opened in 1976 for the advancement of public knowledge regarding the history, development, and use of nuclear energy.

3511. Commercial nuclear reactor that was gas-cooled began operating in March 1977 about 45 miles north of Denver, CO. It used helium instead of water or sodium, to cool the reactor. The fuel was a mixture of uranium and thorium.

3512. Nuclear power plant in Taiwan was the Chinshan-1 plant which began operating in October 1978 in Taipei County. In March 1998, protests were held in front of it by environmental groups that organized an anti–nuclear power month. They included the Taiwan Environmental Protection Union and the Taiwan Green Party.

3513. Jury award of damages in the U.S. to a victim of radiation contamination from a nuclear facility came in May 1979. An Oklahoma jury ordered the Kerr-McGee Co. to pay $10.5 million to the estate of whistleblower Karen Silkwood. An appeals court later overturned the verdict; Silkwood's family eventually agreed to a $1.38 million settlement.

3514. Major accident in a U.S. nuclear power plant occurred March 28, 1979, at the Three Mile Island nuclear reactor complex near Harrisburg, PA, when a series of equipment failures and human errors caused a loss of coolant and a partial core meltdown. People in nearby neighborhoods were evacuated until the situation was stabilized.

3515. Nuclear energy program to disassemble a damaged reactor was begun by the U.S. Department of Energy (DOE) on March 26, 1980, to develop technology to disassemble and defuel the damaged Three Mile Island reactor. The program continued until April 19, 1990, when the last shipment of damaged fuel arrived at the DOE facility in Idaho to be investigated and temporarily stored. The ten-year program helped develop new technology for nuclear power plant safety.

3516. U.S. legislation making states responsible for their nuclear waste was the Low-Level Radioactive Waste Policy Act, passed by the U.S. Congress in December 1980. The new law forced states to dispose of their own low-level nuclear waste from reactors, as well as the contaminated equipment from hospital medical procedures and from industrial research and development projects.

3517. Official confirmation of a powerful hydrogen bomb dropped accidentally from an aircraft came in 1981. The bomb, released from a U.S. Air Force B-36 near Albuquerque, NM, on May 22, 1957, was unarmed, though the explosion of its TNT detonator caused some environmental contamination.

3518. Nuclear power plant accident of significance in Japan happened on February 11, 1981, at Tsuruga. About 100 workers making repairs were exposed to radioactive materials.

3519. Bombing of a nuclear reactor occurred at the nearly completed Osirak reactor near Baghdad, Iraq. On June 8, 1981, Israeli air force jets made the raid that destroyed the site. Iraq denied that the reactor was for military purposes, and on June 19 the United Nations condemned the attack.

3520. Nuclear power plant shelled by environmentalists was the Super-Phénix fast-breeder reactor being constructed in 1982 at Creys-Malville, France. A Soviet bazooka was fired from 2,000 feet away, and a "pacifist ecologist" group later claimed to have launched the attack. The damage was minor, although two shells were propelled into the reactor building and two others hit the steam generator.

3521. Nuclear-generated electricity exported from Canada was in April 1982. The Canadian National Energy Board gave approval for the electricity to be supplied from a plant at Point Lepreau, New Brunswick, to U.S. customers in Maine and Massachusetts.

3522. Spent nuclear fuel and high-level radioactive waste disposal law in the United States was the Nuclear Waste Policy Act of 1982 (NWPA). Enacted on January 7, 1983, the NWPA instituted a program to research and develop safe repositories for nuclear waste. The act established a nuclear waste fund, which would dispense user fees from nuclear-generated electricity to cover disposal costs. The act also established the Office of Civilian Radioactive Waste Management (OCRWM) within the Department of Energy. The OCRWM is responsible for the transportation, storage, and disposal of spent nuclear fuel from

NUCLEAR POWER—*continued*

commercial power plants and high-level radioactive waste from defense facilities. The provisions of the original NWPA legislation remained at the center of nuclear waste controversies for years.

3523. Nuclear energy agreement between the U.S. and China was the Agreement for Cooperation Concerning Peaceful Uses of Nuclear Energy, signed in 1985. In 1997 the two nations issued a joint statement saying they had taken steps to implement the agreement and that the U.S. Department of Energy and China's State Planning Commission had signed an agreement of intent to promote peaceful nuclear cooperation and research between the two countries. On January 12, 1998, President Bill Clinton signed a formal certification that China had met the conditions, set down by the U.S. Congress, assuring that imports of U.S. nuclear materials and equipment would be used solely for peaceful purposes.

3524. Accidental discharge of nuclear waste in British waters was in 1986 into the Irish Sea. Half a ton of uranium was discharged from the Sellafield nuclear power plant in Cumbria, northwest England. The plant, run by British Nuclear Fuels, normally pumps liquid waste labeled as safe through a pipe into the Irish Sea.

3525. Compensation payments by the Australian government to Aborigines whose tribal lands were used for British nuclear weapons testing were approved in August 1986. Cost estimates for decontaminating tribal lands ran upwards of $250 million.

3526. Nuclear agency established by the Chinese government for both civil and military activities was the China National Nuclear Corporation (CNNC), established in 1986. It is in charge of all steps in the processing of nuclear materials and has a research and development program.

3527. Year that nuclear power plants numbered 100 in the U.S. was in 1986, when the Perry Power Plant in Perry, OH, went into operation.

3528. Nuclear power plant disaster with worldwide fallout was on April 26, 1986, at the Chernobyl nuclear power plant near Kiev, Ukraine, in the Soviet Union. The graphite core of a reactor exploded and burned, spewing nearly 9 tons of radioactive material into the atmosphere. The accident caused 31 immediate deaths in the plant. Although some 135,000 residents in the area were evacuated, more than 25,000 people have since died prematurely. Birth defects in the area have also increased. Fallout from the disaster spread over the western Soviet Union and Europe, resulting in contamination of crops and meat and dairy products. Small amounts of radiation even reached the United States.

3529. Nuclear power plant in the United States to be totally decontaminated and decommissioned was the Shippingport Atomic Power Station in Shippingport, PA, in 1987. Parts of it could no longer be repaired because they were radioactive. The U.S. Congress decided in 1982 that the plant, built in 1957, should be decontaminated and decommissioned. This was done by the U.S. Department of Energy, which waited two years for the uranium to cool before beginning in 1984. The plant's seven-acre site was declared suitable for unrestricted use in November 1987. Its nuclear waste was buried in 1989 at the Hanford Nuclear Facility in Richland, WA.

3530. Nuclear waste site evaluation was ordered as part of the Nuclear Waste Policy Amendments Act of 1987. Signed by President Ronald Reagan on December 22, 1987, these amendments to the Nuclear Waste Policy Act of 1982 directed the Department of Energy to examine the suitability of Yucca Mountain, NV, as a repository for spent nuclear fuel and radioactive waste. The amendments also established the Office of the Nuclear Waste Negotiator in order to attempt to reach an agreement with any state or American Indian tribe willing to host a nuclear repository or monitored retrievable storage facility.

3531. Nuclear emergency plan by the Irish government came into effect in 1988. Although Ireland had no nuclear power plants, the Radiological Emergency Protection Plan was established to deal with any problems resulting from a nuclear power plant accident in Great Britain or Europe.

3532. Nuclear power plant to be closed by a popular state vote was the Rancho Seco plant near Sacramento, CA, in 1989.

3533. Nuclear waste site chosen by the U.S. Congress was near Carlsbad, NM. In October 1992, the Waste Isolation Pilot Plant (WIPP) Land Withdrawal Act reserved these public lands as a repository for transuranic nuclear waste disposal. This is part of the government's program for the environmentally responsible disposal of defense-generated radioactive waste.

The U.S. Geological Survey selected the Salado salt formation, which has little earthquake activity. The transuranic wastes, which have heavy radioactive elements like plutonium, are stored 1,250 feet deep in the salt bed.

3534. Nuclear power in Great Britain generated from a pressurized water reactor was produced in 1994 by the nuclear power plant at Sizewell, a village in Suffolk, England. The Sizewell B reactor, based on the type of pressurized water reactor commonly used in the United States, was the first of its kind in Great Britain.

3535. Uranium fuel bought from Russia by the United States began in 1994 with the "megatons to megawatts" program. Spread over 20 years, the purchases would cost the U.S. government $8 billion, and it would sell the uranium commercially. The Russian source was the high-enriched uranium recovered from warheads that had been dismantled. This was then processed into low-enriched uranium.

3536. Nuclear power plant in Romania was Cernavoda-1, which began operating April 16, 1996, in Cernavoda, about 93 miles east of Bucharest. This was nearly 20 years after the first agreements for its construction had been signed, and the plant is equipped with the technology of the 1960s. During its early construction, forced labor was used by the communist regime of President Nicolae Ceauşescu.

3537. Year that nuclear power was the main source of British electricity was 1997. It produced 35 percent of the country's electricity; coal (which had been the leading source) produced 33 percent, and gas contributed 29 percent.

3538. Nuclear accident reported falsely in Japan occurred March 11, 1997, at the Bituminization Demonstration Facility of the Tokai Reprocessing Plant, which handles low-level radioactive waste. It is located at Tokai in Ibaraki Prefecture, north of Tokyo. The facility suffered a fire and explosion that exposed 37 workers to radiation. When the operator, the Power Reactor and Nuclear Fuel Development Corporation (PNC), was found to have destroyed evidence and issued false reports about the incident, the government shut down the plant for one year.

3539. Nuclear energy law allowing expropriation of power plants in Sweden was passed in December 1997. The Swedish Parliament enacted legislation allowing the government to expropriate and close nuclear power plants without citing safety issues. The plant owners would receive financial compensation.

3540. Nuclear waste lawsuit of significance against the U.S. government was won on December 14, 1998, by 38 environmental groups led by the Natural Resources Defense Council. The suit, filed in 1988, charged the government with avoiding a federal law requiring an assessment of the environmental impact of the cleanup of radioactive and hazardous waste from plants and laboratories involved in nuclear arms. In the settlement, the Department of Energy agreed to create an Internet site so the public could follow its cleanup of more than one million tons of waste at the plants and laboratories involved. It also agreed to pay $6.25 million so those public interest groups can hire experts to monitor the cleanup.

3541. Deep underground nuclear depository was the Waste Isolation Pilot Plan (WIPP) facility near Carlsbad, New Mexico. WIPP received its first shipment of nuclear military waste on March 26, 1999. Operated by the U.S. Department of Energy, WIPP was the focus of intense controversy between government officials intending to use the remote site to reduce the amount of nuclear waste stored near heavily populated areas and environmental groups concerned with security and possible toxic leaks.

3542. Admission by the U.S. government that nuclear weapons production may have caused illnesses in thousands of workers came on July 15, 1999. The Clinton Administration proposed legislation to compensate victims for medical care and lost wages.

3543. Working prototype of a commercially viable pebble-bed nuclear reactor was scheduled for completion in 2002, in South Africa. The project was a joint venture by Eskom, a South African utility company, and Exelon, a Chicago-based group. Pebble-bed reactors, which had previously been used only in research, are proof against meltdown and are therefore safer than fuel-rod reactors. They can produce electricity more cheaply; however, they also create greater quantities of hazardous waste.

3544. Nuclear power plant in Turkey is scheduled to open by June 2006 after bids for construction were closed in October 1997. Turkey's national utility, TEAS, said it would be built on the country's Mediterranean coast.

O

OCEANS AND OCEANOGRAPHY

3545. Chronometer was invented by English carpenter John Harrison in 1735. Harrison's timekeeping device, which he continued to refine and improve, enabled navigators to determine longitude at sea. The chronometer became standard equipment on international oceanographic explorations and commercial voyages.

3546. Freedom-of-the-seas doctrine was proposed in Dutch jurist Hugo Grotius's 1609 treatise *Mare Liberum*. Grotius, who became respected as a pioneer of international law, advocated the right of the free passage and recommended that marine resources across the seas be open to all nations, with the exception of defined territorial zones along their shores.

3547. Law to assert control of offshore waters was proclaimed by England. In 1672, England declared control of waters extending one league or three nautical miles from its shores, and likewise from the shores of its colonies and possessions.

3548. Charts of the Gulf Stream were compiled by U.S. inventor Benjamin Franklin in 1769, when he was serving as colonial Postmaster General. Using data collected from whaling captains, Franklin produced surface current charts that enabled mailing ships to reduce the transit time between Europe and the American colonies.

3549. Survey of the coast of the United States was ordered by Congress in 1807. The Office of Coast Survey was established to accomplish the task. In 1878, the agency became the U.S. Coast and Geodetic Survey. After numerous subsequent name changes, the agency is now known as the Office of Coast Survey (OCS), an office within the National Oceanic and Atmospheric Administration (NOAA). The OCS is responsible for NOAA's mapping, charting, and tidal information.

3550. Evidence of life on the sea floor was collected by Scottish naturalist John Ross aboard HMS *Isabella* in Baffin Bay, Canada, in 1818. While searching for a northwest passage, Ross used a self-designed device he called a "deep sea clamm" to retrieve mud from the sea floor. The mud contained worms, proving that life existed over a mile below the ocean's surface.

3551. Wind and current analysis of significance was compiled by Matthew Fontaine Maury during his tenure as Superintendent of the U.S. Navy's Depot of Charts and Instruments, a precursor of the National Observatory. From his analyses of recorded data submitted by countless navigators, Maury published his *Wind and Current Charts* in the 1840s.

3552. Deep-sea soundings were made in the South Atlantic Ocean from the British ship HMS *Erebus* during an Antarctic exploratory expedition. On January 3, 1840, a lead weight attached to a hemp line was lowered to a depth of 2,425 fathoms.

3553. Theory that coral reefs are founded on sinking underwater land masses like subsiding volcanoes and continental shelves was correctly advanced by English naturalist Charles Darwin in 1842 in his *The Structure and Distribution of Coral Reefs.*

3554. Charts of the Atlantic Ocean winds and currents were compiled by Matthew Fontaine Maury, U.S. naval officer and government scientist, in 1847. His wind and current charts were a great aid in cutting the sailing time of most Atlantic Ocean trading ships.

3555. Significant study of North American marine algae was Irish botanist William Henry Harvey's *Nereis Boreali-Americana*, a monograph published by the Smithsonian Institution in 1852. At the time of its publication, Harvey was curator of the Herbarium at Trinity College, Dublin. It remains an important reference work on the subject, as does Harvey's work on the flora of Australia and South Africa.

3556. Bathymetric map of the North Atlantic Ocean was compiled by U.S. Naval officer and government scientist Matthew Fontaine Maury in 1854. Maury used depth soundings to create a map of the ocean floor, which he published in 1855 in *The Physical Geography of the Sea.*

3557. Oceanography textbook was Matthew Fontaine Maury's *The Physical Geography of the Sea*, published in 1855. A U.S. Naval officer and government scientist, Maury in his research and writing laid the groundwork for modern oceanographic studies.

3558. Underwater photograph was taken in 1856 by William Thompson, an English amateur naturalist. Thompson lowered his camera into the waters of Weymouth Bay to a depth of 3 fathoms, or 18 feet. The waterproof box enclosing the camera leaked, but the photographic plate nevertheless produced a weak image from beneath the sea.

3559. Marine biology station of significance was the Stazione Zoologica, located on the Bay of Naples in Italy. The institution, which was founded in 1872 by German zoologist Anton Dohrn, is still an important marine research facility.

3560. Oceanographic expedition was the voyage of the British schooner HMS *Challenger.* Between 1872 and 1874, the ship sailed around the world with a group of scientists who obtained soil and plant samples from the sea floor, collected species of fish, sounded depths, and studied currents, temperatures, salinity, and the weather.

3561. Marine biology station in the United States was established at the Anderson School of Natural History on Penikese Island in Buzzard's Bay, MA, in 1873. The research station's first director was renowned naturalist Louis Agassiz.

3562. U.S. oceanographic research ship was the *Albatross.* The steamship was built for the U.S. Commission of Fish and Fisheries and was launched in 1882. The ship was used by the renowned marine biologist Alexander Agassiz for his oceanographic expeditions in the Gulf of Mexico, Caribbean Sea, and Pacific Ocean.

3563. Scientist to state the law of relative proportion in seawater was William Dittmar, a German-born researcher at the University of Scotland. In 1884, after analyzing 77 samples collected around the globe by the oceanographic expedition of HMS *Challenger*, Dittmar realized that the mineral proportions of seawater are fairly constant even though its salinity may vary. This is known as Dittmar's Principle.

3564. Tide-predicting machine was invented in 1882 by American mathematician William Ferrel for the U.S. Coast and Geodetic survey. The Ferrel Maxima and Minima Tide Predictor was capable of analyzing a variety of complex interrelated tidal data.

3565. Intergovernmental agency devoted to marine and fisheries science was the International Council for the Exploration of the Sea (ICES), which was founded in Copenhagen, Denmark, in 1902. ICES provides information and advice to governments and international regulatory commissions for the protection of the marine environment and for fisheries conservation.

3566. International Council for the Exploration of the Sea (ICES) was established at Copenhagen, Denmark, in 1902 to create an international program for marine research. Decades later, members of the United Nations met in Copenhagen to update the council and promote new research of marine resources, primarily in the South Atlantic Ocean. The Convention for the International Council for the Exploration of the Sea was adopted on September 12, 1964, and entered into force July 22, 1968.

3567. Oceanography institution was the Scripps Institution of Oceanography of the University of California, located at La Jolla, CA. It was an offshoot of the Scripps Institution for Biological Research founded in 1912. The scope of its investigations encompassed the circulation of ocean waters, the interrelation between the oceans and the atmosphere, the chemistry of ocean water, sea floor sediments, and marine life.

3568. Underwater movie seen by the public was filmed by American newspaper reporter John E. Williamson in February and March 1914. After practicing in Chesapeake Bay, Williamson descended into the ocean near the Bahamas in a diving bell with his camera. His resulting documentary became an international sensation. He later provided the first underwater footage for a fictional feature, a 1915 movie version of *Twenty Thousand Leagues Under the Sea.*

3569. Oceanographic study of an entire ocean was undertaken aboard the German ship *Meteor.* While unsuccessfully attempting to refine a process to extract gold from ocean water, the *Meteor* expedition collected samples and conducted experiments in the South Atlantic Ocean from 1925 to 1927.

3570. Use of sonar (sound navigation and ranging) to measure ocean depths was carried out during the 1925–1927 oceanic expeditions of the German ship *Meteor.* The *Meteor*'s sonar findings were a major improvement over previous efforts to map the floor of the Atlantic Ocean.

3571. Multi-purpose oceanographic research vessel in the United States was the R.V. *Atlantis*, based at Woods Hole Oceanographic Institution in Massachusetts. Between its launching in 1931 and its decommission in 1964, the sail-powered ship was used for hydrographic research, charting the sea floor, and the first study of underwater acoustics.

OCEANS AND OCEANOGRAPHY—*continued*

3572. Bathysphere was designed by U.S. engineer Otis Barton and naturalist William Beebe. After four years of preliminary dives, the men used the cast iron sphere to descend to an unprecedented depth of 3,028 feet near Bermuda on August 15, 1934.

3573. Aqualung or diving tank was invented in France by Jacques-Yves Cousteau and Emil Gagnan in 1943. A valve ensured that the tank would dispense air whose pressure matched the water pressure at the depth to which a diver descended. The invention was the first of Cousteau's many contributions to oceanography.

3574. International agency to address pollution of the oceans by ships was the International Maritime Organization (IMO), a branch of the United Nations. Created at a UN conference in Geneva, Switzerland, on March 6, 1948, and convened for the first time ten years later, IMO was called the Inter-Governmental Maritime Consultative Organization (IMCO) until 1982, when its name was changed. The agency works to reduce oil pollution and increase maritime safety through international treaties, technical assistance, and training programs. IMO headquarters are in London, England.

3575. Coral reef field research station in Australia was the Heron Island Research Station. The facility was established by the Great Barrier Reef Committee, a precursor of the Australian Coral Reef Society, during the 1950s. The station is now affiliated with the University of Queensland and hosts ornithological, meteorological, and oceanographic studies of the Great Barrier Reef's ecosystem.

3576. Bathyscaph was designed by Swiss physicist Auguste Piccard and his son Jacques. The Piccards made their first descent in the vessel, which they named the *Trieste*, in August 1953 in the Mediterranean Ocean near Castellammare, Italy. The *Trieste* was sold to the U.S. Office of Naval Research in 1958.

3577. International convention on ocean pollution was the Pollution of the Sea by Oil Convention, convened in London, England, in 1953, to regulate oil commerce in order to contain destruction of the seas by oil spills. Twenty nations signed the agreement, but not the United States, whose oil companies convinced the State Department that technological developments, not regulation, would cure the problem. They did not, and the United States signed the agreement in 1961.

3578. Manned descent to the greatest depth on earth was made on January 23, 1960, in an improved version of the bathyscaph *Trieste*, designed by Swiss physicist Auguste Piccard and his son Jacques. The dive was made by Jacques Piccard and U.S. Navy Lieutenant Don Walsh, who descended approximately 5,966 fathoms or 35,800 feet to the bottom of the Mariana Trench in the Pacific Ocean.

3579. Major destruction by starfish of coral in the Great Barrier Reef was reported in 1963. Increasing populations of crown-of-thorns starfish (*Acanthaster planki*), which eat coral and thus cause reef ecosystems to collapse, were blamed for damage to marine environments across the South Pacific.

3580. Oceanographic research vessel to recover a lost hydrogen bomb was *Alvin*, a submersible owned by the Woods Hole Oceanographic Institution in Woods Hole, MA. *Alvin* was used to find and retrieve the bomb, which had fallen into the Mediterranean Ocean near Palomares, Spain, after a midair collision of U.S. aircraft on January 17, 1966. *Alvin* located the bomb on March 15, but the bomb was lost again when an attempt to retrieve it failed. After *Alvin* relocated the bomb, it was recovered on April 7.

3581. Law requiring the U.S. president to develop and conduct a comprehensive program of marine science activities was the Marine Resources and Engineering Development Act. The law was enacted on June 17, 1966, and has been amended numerous times. It requires the president to conduct policies that coordinate studies in a wide range of activities as they relate to marine science, including fisheries, technology, transportation, national security, education, and exploration. The president's advisory body in such policies was the National Council on Marine Resources and Engineering Development. The council, however, was terminated April 30, 1971.

3582. United Nations compensation policy on marine oil pollution was contained in the International Convention on Civil Liability for Oil Pollution Damage, which was originally adopted in Brussels, Belgium, on November 29, 1969. The convention entered into force on June 19, 1975, and was amended in 1976 and 1984. It held the owner of any ship causing oil pollution damage to be responsible for the damage, unless the incident was caused by an act of war, sabotage, or the navigational negligence of a third party. The convention also required ships carrying more than 2,000 tons of oil to maintain insurance.

3583. United Nations agreement on marine pollution from offshore industrial facilities was a regional agreement, the Agreement between Denmark, Finland, Norway and Sweden Concerning Cooperation in Measures to Deal with Pollution of the Sea by Oil. The agreement was opened for signature in Copenhagen, Denmark, on September 16, 1971, and entered into force October 16, 1971.

3584. Prohibition against nuclear weapons on the ocean floor was outlined in the Sea-Bed Arms Control Treaty. The treaty banned the emplacement of nuclear bombs and other weapons of mass destruction on the sea bed beyond 12 miles from territorial shores. On February 11, 1971, the treaty was opened for signature simultaneously in London, Moscow, and Washington, DC. and entered into force May 18, 1972.

3585. United Nations convention to define civil liability in nuclear accidents at sea was the Convention Relating to Civil Liability in the Field of Maritime Carriage of Nuclear Material. The convention was adopted December 17, 1971, in Brusssels, Belgium, and entered into force July 15, 1975. The convention holds the owner of a nuclear installation solely responsible for any accidents that might take place in the transportation of nuclear material by sea.

3586. United Nations compensation fund for marine oil pollution damage was established by the International Convention on the Establishment of an International Fund for Compensation. The convention was adopted in Brussels, Belgium, on December 12, 1971, and entered into force October 16, 1978. It created the International Oil Pollution Compensation Fund to adequately compensate anyone suffering from oil pollution damage who could not obtain compensation through the UN's 1969 International Convention on Civil Liability for Oil Pollution Damage. The fund's administration office is in London, England.

3587. United Nations agreement to regulate dumping of waste into the oceans was the Convention for the Prevention of Marine Pollution by Dumping from Ships and Aircraft. The convention was adopted in Oslo, Norway, on February 15, 1972, and entered into force April 7, 1974. It provided a list of substances whose disposal at sea was prohibited, created a permit system for other substances and disposal methods, and required signatories to enforce compliance by ships and aircraft registered in their territories.

3588. United Nations agreement on marine pollution from airborne sources addressed regional European pollution in the Convention for the Prevention of Marine Pollution from Land-Based Sources. The convention was opened for signature in Paris, France, on June 4, 1974, and entered into force May 6, 1978. It was later replaced by the 1992 Convention for the Protection of the Marine Environment of the North-East Atlantic.

3589. Coral reef off the coast of the United States protected by federal law was located three miles off the coast of Key Largo, FL. The largest living coral reef North America, it was designated a U.S. National Marine Sanctuary in 1975.

3590. Marine sanctuary in the United States was the Monitor U.S. National Marine Sanctuary located off Cape Hatteras, NC. The sanctuary was created in 1975. It was the first of a nationwide system of protected marine areas.

3591. United Nations attempt to define civil liability for oil pollution resulting from seabed drilling was the Convention on Civil Liability for Oil Pollution Damage Resulting from Exploration for and Exploitation of Sea Bed Mineral Resources. The convention called for the establishment of a compensation fund for such oil damage, insurance requirements, and civil liability rules. The document's emphasis on the North Sea, Baltic Sea, and North Atlantic Ocean was reflected by the countries that signed it—Germany, Ireland, the Netherlands, Norway, Sweden, and Great Britain. Twenty years after it opened for signature in London, England, on May 1, 1977, however, an insufficient number of countries had signed the convention for it to enter into force.

3592. Comprehensive United Nations agreement on use of the oceans was the UN Convention on the Law of the Sea. The convention was opened for signature at Montego Bay, Jamaica, on December 10, 1982, and entered into force November 16, 1994. The convention covers sovereignty issues relating to territorial waters, economic and scientific exploitation of marine resources, environmental protection, and settlement of international disputes. Later amendments addressed deep seabed mining (1994) and conservation of fish stocks (1995).

3593. Federal program to protect and restore an estuary in the United States was the Chesapeake Bay Program. On December 15, 1983, Maryland, Virginia, Pennsylvania, the District of Columbia, and the Environmental Protection Agency signed the Chesapeake Bay Agreement, committing their states and the Dis-

OCEANS AND OCEANOGRAPHY—*continued*

trict of Columbia to prepare plans for protecting and improving water quality and living resources in the Chesapeake Bay. The program's environmental management practices were a model for other estuarine reclamation and preservation programs. It is administered separately from estuaries designated under the National Estuary Program.

3594. Central American coral reef reserve was created by Janet Gibson in 1987. Gibson was concerned about the exploitation and pollution of the reef. With the support of the Belize Audubon Society and the Wildlife Conservation Society, she secured financial support for the five square miles of reserve that became known as the Hol Chan Marine Reserve.

3595. National Estuary Program (NEP) in the United States was established through a 1987 amendment to the Clean Water Act. Estuaries designated as NEPs are managed through a voluntary partnership of state, federal, business, environmental, and community groups. Information is shared in the interest of maintaining water quality, monitoring pollution, restoring wetlands and shoreline habitats, resisting invasive species, and identifying marine diseases in coastal waters.

3596. Intergovernmental agency devoted to marine and fisheries science of the Northern Pacific is the North Pacific Marine Science Organization (PICES). Modeled after the International Council for the Exploration of the Sea (ICES), PICES was formally organized on December 12, 1990, in Ottawa, Canada. The organization's committees study biological oceanography, fishery science, physical oceanography and climate, and marine environmental quality.

3597. Female oceanographer of distinction was marine biologist and environmentalist Sylvia A. Earle, who served in 1991 and 1992 as the first female chief scientist of the National Oceanographic and Atmospheric Administration (NOAA) in Washington, DC. Earle made the world's deepest untethered dive in 1979, when she walked along the ocean floor off Oahu, HI, for 2.5 hours at a depth of 1,250 feet.

3598. Federal law in the United States to reduce vessel sewage discharges was the Clean Vessel Act, which was enacted on November 4, 1992. The law required and funded state programs to provide pump-out stations and dump stations at marinas, reducing the amount of damage to fisheries, public beaches, and marine ecosystems caused by raw sewage being discharged in shallow waters by recreational boaters.

3599. Remote-controlled submersible to reach the oceans' greatest depth was *Kaiko*, a Japanese-designed deep-sea vehicle equipped with cameras and video equipment. On March 24, 1995, *Kaiko* descended to a depth of 35,798 feet in the Pacific Ocean's Mariana Trench.

3600. High-resolution maps of the ocean floor were declassified and released by the U.S. Navy in July 1995. Many previously unknown submarine features were revealed in the maps, which provided the first detailed view of the world's ocean basins. The data for the maps had been collected by Geosat, an American satellite launched in 1985, and by the European Space Agency's ERS-1 satellite.

3601. Executive Order protecting coral reefs in the United States was signed by President Bill Clinton on June 11, 1998. Executive Order 13089 directed federal agencies whose actions might affect U.S. coral reef ecosystems to ensure that such ecosystems be protected. The order also created the U.S. Coral Reef Task Force, which was charged with mapping, monitoring, researching, and conserving reefs inside U.S. waters.

3602. Deep-sea exploration of U.S. marine sanctuaries was Sustainable Seas Expedition, a partnership between the National Oceanic and Atmospheric Administration (NOAA) and the National Geographic Society. Using the NOAA ship *MacArthur* and *DeepWorker 2000*, a deep-sea submersible vehicle, the expedition set sail on April 22, 1999, for a series of deep-sea geological, biological, and oceanographic studies in marine sanctuaries off the coast of California.

3603. Report on the impact of human predation upon coastal marine ecosystems dating back to prehistoric times was published in the journal *Science* in 2001. In the extensively documented report, scientists presented evidence that overfishing and the hunting of sea turtles and marine mammals to near-extinction commenced very early in human history, with significant consequences for coastal ecosystems today.

OIL AND PETROLEUM INDUSTRY

3604. Discovery of "burning water" (petroleum) in Japan was circa 615 in the Echigo district.

3605. Recorded mention of oil in Persia was by Marco Polo in 1264 during his travels. He described "a fountain from which oil springs in great abundance" at Baku on the Caspian Sea. He estimated that 100 shiploads could be taken from the supply at one time.

3606. Street lamps lighted by using oil were in Krosno, Poland, circa 1500. The oil was hand-collected in the Carpathian Mountains from pits dug next to oil seeps. It was a thick, sticky liquid that when burned gave off an unpleasant smell along with much smoke and soot.

3607. Written mention of petroleum in what would become the United States was in 1627. A French missionary wrote of a "fountain of bitumen" he had seen issuing from a spot near Lake Ontario in New York State.

3608. Map showing a "fountain of bitumen" (petroleum) in what would become the U.S. appeared in 1650. The map placed the fountain near the present village of Cuba in New York State.

3609. Oil company in the United States was launched in 1854. The Pennsylvania Rock Oil Company owned 100 acres along Oil Creek in Venango County, PA.

3610. Oil production on a large scale in the United States was in 1860, when 500,000 barrels of crude oil were extracted. This was one year after the first commercial oil well at Titusville, PA.

3611. Oil-tank railroad car purposely built in the United States was the Densmore rotary oil car, first used in 1865 by Pennsylvania oil companies. It had two upright 150-gallon cylinder wooden containers on a flatbed. Three years later, the horizontal cylinder tank was used, having a dome to allow for expansion of the oil.

3612. Recorded oil seepages in Alaska were observed in 1882 at Oil Bay on the Iniskin Peninsula.

3613. Oil tanker built for long-distance conveyance was the *Murex*, owned by Marcus Samuel, who later founded the Shell Oil Company. The tanker's first trip began July 22, 1892, from West Hartlepool, England. It arrived at the Suez Canal on August 23, and proceeded on to Singapore and Bangkok.

3614. Oil pipeline in Russia was constructed from 1897 to 1907 between oil-producing Baku in Azerbaijan and Batumi on the Black Sea in Georgia.

3615. Crude oil production in Mexico began in 1901 at Ebano near Tampico.

3616. Oil pipeline of significance in Russia was completed in 1906 and ran 540 miles alongside the Trans-Caspian Railroad. It had 16 pumping stations and lost more than 3 percent of its oil because of joints that were fitted badly.

3617. Oil-drilling equipment to penetrate hard rock was developed in 1908 by the Hughes Tool Company in Houston, TX. The drilling mechanism, which had a steel-toothed bit, accelerated oil exploration throughout the entire industry.

3618. Oil and gas conservation law by a state was enacted on May 17, 1913, by Oklahoma. It authorized a chief deputy inspector of oil and gas wells and pipelines to supervise their use and operation.

3619. Gas station in the United States opened December 1, 1913, in East Liberty, PA. Owned by the Gulf Refining Company, it marked the beginning of easy access to fuel for the automobile, a major producer of air pollution in the United States. Soon after the opening of this first Gulf station, other gas stations rapidly began to open across the country.

3620. Oil pipeline of significance in Iraq was completed in 1934 and ran about 500 miles, connecting the Kirkuk oil field with the Mediterranean Sea. The twin-pipe system was constructed by the Iraq Petroleum Company and pumped about 11,000 tons of crude oil each day.

3621. Year oil became the major energy source in the United States was 1950. It replaced coal.

3622. International oil cartel was the Organization of Petroleum Exporting Countries (OPEC), established in 1960 to coordinate oil marketing policies and prices. The original members were Saudi Arabia, Iran, Iraq, Libya, Kuwait, and Venezuela. Counties that later joined were the United Arab Emirates, Qatar, Indonesia, Algeria, Nigeria, Gabon, and Ecuador. In 1973, OPEC created a world oil crisis by raising prices by 130 percent and imposing oil embargoes on the United States and the Netherlands. This was done because of Western support for Israel.

OIL AND PETROLEUM INDUSTRY—*continued*

3623. Petroleum in Vietnam was extracted in 1975 by the state-owned company. The wells are located in the north in an area also claimed by China.

3624. Flow of oil through the Alaska Pipeline from Prudhoe Bay to Valdez on Prince William Sound took place on June 27, 1977. On August 1, the first tanker laden with pipeline oil sailed out of Valdez.

3625. Pipeline giving access to Canada's northern oil was completed in 1985 in the Northwest Territories. The Norman Wells Pipeline brings oil to the south from that oil town on the northern shore of the Mackenzie River.

3626. Reformulated gasoline sold in the United States was EC-1 Regular in September 1989. The new gas was sold by ARCO Petroleum Company in Los Angeles, CA, but variants of the new gas under different names were soon sold by other companies.

3627. Legislation in the United States to mandate double hulls on oil tankers was the Oil Pollution Act of 1990. The March 1989 grounding of the *Exxon Valdez*, which spilled more than 10 million gallons of oil into Prince William Sound in Alaska, inspired the measure.

OIL AND PETROLEUM INDUSTRY—FIRES, SPILLS, AND BLOWOUTS

3628. Oil spill of significance recorded was in 1565 when the 330-ton *San Juan*, loaded with barrels of oil destined for Spain, sank as it readied to sail from present-day Red Bay Harbor in southern Labrador, Newfoundland, Canada. Divers have since recovered the damaged barrels from the wreckage on the seabed.

3629. Fire in a U.S. oil well occurred on April 17, 1861, when the Little and Merrick well on the Buchanan farm near Rouseville, PA, caught fire shortly after it gushed. The fire, which burned for three days, took 19 lives.

3630. Supertanker to founder and spill its cargo of oil was the *Torrey Canyon*, which ran aground at full speed on the shoals of Seven Stones, off the coast of Cornwall, England, on the morning of March 18, 1967. Some 119,000 tons of crude oil spilled into the Atlantic Ocean, fouling both the beaches of Cornwall and those of Brittany in France. More than 30,000 birds died and damages ran in the tens of millions of dollars.

3631. Oil spill of significance off South Africa happened on June 13, 1968, when the tanker *World Glory* suffered hull failure. The accident spilled more than 13.5 million gallons into the sea.

3632. Offshore oil spill to cause widespread damage occurred January 28, 1969, when offshore well A-21, owned by the Union Oil Company, burst. Leaking crude oil spread along a 200-mile stretch of coast centered on Santa Barbara, CA, fouling beaches and killing birds, fish, and marine mammals before the leak was plugged after 11 days.

3633. Oil spill of significance off Sweden occurred on March 20, 1970, when the tanker *Othello* collided and released 60,000 to 100,000 tons of oil into Tralhavet Bay.

3634. Offshore oil-well fires of significance in the Gulf of Mexico were finally extinguished in April 1971 off the coast of Louisiana. The fires had burned for 133 days, and the cost of bringing them under control was $26 million. It was then the most expensive offshore fire in history.

3635. Oil spill of significance off Japan happened on November 30, 1971, when a tanker broke in half and released 6,258,000 gallons into the sea.

3636. Oil spill of significance in the Gulf of Oman was on December 19, 1972. The tanker *Sea Star* collided with another vessel and lost 115,000 tons of oil.

3637. Oil spill of significance off Spain was on May 12, 1976. The *Urquiola* tanker ran aground off Corunna in northwest Spain and released 100,000 tons of oil into the Atlantic.

3638. Oil well blow-out in the North Sea occured on April 22, 1977, in the Ekofisk oil field. It released 8,200,000 gallons of oil into the sea near Scotland.

3639. Oil-tanker spill of significance off the coast of France was on March 16, 1978, when the *Amoco Cadiz* supertanker grounded near Portsall. It lost 223,000 tons of oil and caused an environmental catastrophe along more than 100 miles of the Britanny coast.

3640. Oil platform blowout occured at the Ixtoc I platform, off the Yucatan Peninsula, on June 3, 1979. During the nine months after its blowout, an oil slick that could have covered the entire Gulf of Mexico formed. Beaches in Mexico and Texas were grossly fouled.

3641. Offshore drilling-rig disaster near Canada occurred February 15, 1982, at the Grand Banks in the Atlantic Ocean 186 miles east of St. John's, Newfoundland. All 84 crew members of the Ocean Ranger rig died when it sank in a severe storm of hurricane-force winds, rain, and snow. Environmentalists continue to warn of the potential threat of oil drilling in the Grand Banks, one of the world's best fishing grounds, where millions of fish feed off the thriving plankton.

3642. Explosion of significance on an offshore oil rig occurred July 6, 1988, in the North Sea off Aberdeen, Scotland. In the world's worst offshore oil rig disaster, 167 men (including three rescuers) were killed when the Piper Alpha platform of the Occidental Petroleum Company suffered a series of explosions followed by a fire. A total of 64 were rescued. The British government later accepted all 106 suggestions for improved oil rig safety made in an official report on the disaster.

3643. Oil spill cleanup law in the United States was the Oil Pollution Act, passed by the U.S. Congress in 1990 in the wake of the tragedy of the *Exxon Valdez*. It established a network of regional oil spill response centers, promoted improved navigational safety, demanded crew licensing and new tanker construction standards, and made it easier to collect settlements with offending oil companies.

3644. Cleanup of an oil spill of significance in the United States with an oil-eating bacteria occurred June 8, 1990, in the Gulf of Mexico 50 miles off Galveston, TX. The Norwegian supertanker *Mega Borg* exploded and caught fire, spilling 4.3 million gallons of light crude oil that was eventually dissipated by the genetically engineered bacteria.

3645. Ecological disaster caused by a deliberate oil spill occurred in 1991 during the Gulf War. Iraq, occupying Kuwait, released about 130 million gallons of crude oil from January 23 to 25 into the Persian Gulf from tankers and a terminal 10 miles off Kuwait. This was the world's worst oil spillage.

3646. Intentional oil-well fires of significance were set in February 1991 in Kuwait by retreating Iraqi troops during the Gulf War. About 700 oil wells were sabotaged and black smoke blanketed Kuwait and spread south and west over neighboring countries and into southwest Asia. The last fires were not extinguished until November with international help.

3647. Oil-tanker spill of major significance off Scotland occurred January 5, 1993, off the Shetland Islands. The *Braer*, a tanker traveling from Norway to Canada, had engine failure in a gale and hit the rocks of Fitful Head, spilling about 26 million gallons of light crude oil into the sea. The ecological damage was limited somewhat because the heavy winds helped disperse the oil.

3648. Oil-pipeline spill of significance in Russia publicly reported occurred on August 12, 1994, in the Arctic region near Ursinsk. A dam hurriedly built to contain the spill broke on September 8 and more than 4 million gallons of crude oil flowed into the Pechova and Kolva rivers.

3649. Oil-tanker spill of significance off Wales occurred February 15, 1996. The *Sea Empress* supertanker grounded off Milford Haven in southwestern Wales and lost 79,000 tons of crude oil that covered the Irish Sea with an oil slick 25 miles long.

OIL AND PETROLEUM INDUSTRY—OFFSHORE DRILLING AND PRODUCTION

3650. Oil drill offshore rig was patented on May 4, 1869, as a "submarine drilling apparatus" by Thomas F. Rowland of Greenpoint, NY.

3651. Oil wells in the ocean floor to be successfully drilled were off the shore of Summerland near Santa Barbara County, CA, in 1896.

3652. Offshore oil well was drilled by Gulf Oil Company in 1911. The well was drilled in the floor of a lake on the Texas and Louisiana border.

3653. Offshore oil to be accessed from a modern platform out of sight of land was found in 1947 in the Gulf of Mexico. The well was drilled by the Kerr-McGee Co. using a relatively small platform in conjunction with a tender vessel. Earlier in 1947, the Magnolia Oil Company had drilled a drut Superior failed to hit oil until the next year. The Superior type of platform was the ancestor of today's large rigs.

3654. Oil discovery in the Persian Gulf was made in 1951 by the Aramco Company. This underwater exploration led to the development of the extensive Safaniya oil deposits.

3655. Offshore oil wells in Trinidad were drilled in 1954 off the west coast of the island. In 1986, the first commercial oil and gas was discovered off the east coast.

OIL AND PETROLEUM INDUSTRY—OFFSHORE DRILLING AND PRODUCTION—*continued*

3656. Seagoing oil drilling rig for drilling in water more than 100 feet deep was placed in service March 24, 1955. The rig was built in the Beaumont Yard of the Bethlehem Steel Company in Bethlehem, PA, for the C. G. Glasscock Drilling Company and could drive piles with a force of 827 tons and could pull them with the force of 942 tons.

3657. Offshore oil drilling of significance in the Trucial States (now United Arab Emirates) was carried out in 1958 in the Persian Gulf by the Abu Dhabi Marine Areas company.

3658. Legislation in the United States to provide for use of fees from offshore oil drilling for land purchases and historic preservation was enacted in 1965. The measure created the Land and Water Conservation Fund.

3659. Exploratory offshore oil wells in Canada were drilled in 1966 off Newfoundland's coast. They were Tors Cove D-52 and Grand Falls H-09. A year later the first on the West Coast of Canada were drilled off British Columbia. No commercial quantities of oil or gas were found.

3660. Offshore oil discovered off Denmark was in 1966 in the North Sea by the Danish Underground Consortium, established by a shipping company, A. P. Möller, and several foreign oil companies. In 1972, the first oil came ashore.

3661. Oil discovery of significance in the North Sea was a strike by the Phillips Petroleum Company in February 1970 off the northern tip of Scotland. The British government had granted the first licenses to drill in the North Sea in 1964, and more than 100 companies had been searching for the deep deposits.

3662. Offshore oil produced in Norway was June 15, 1971, in the Ekofisk Field in the North Sea. It soon reached its capacity of 40,000 barrels a day. The oil was brought in by the Phillips Petroleum Company's Gulftide rig, which was officially opened June 9. The oil had been discovered 18 months earlier.

3663. Offshore U.S. oil leases in the outer continental shelf of the Atlantic were offered in August 1976 by the U.S. Bureau of Land Management. The high bids on 101 of 154 Atlantic tracts totaled $1.1 billion.

3664. Year that North Sea oil production exceeded home demand in Great Britain was 1981. The first viable oilfields had been discovered off Scotland only 10 years before.

3665. Extension of the offshore drilling ban occurred in 1990. It included 84 million acres off the coasts of California, Alaska and the eastern states. The main purpose of the ban was to prevent pollution due to offshore drilling.

OIL AND PETROLEUM INDUSTRY—WAR AND DIPLOMACY

3666. Use of fuel oil to power warships occurred in 1903. Admiral John A. Fisher's experiments with two Royal Navy battleships showed that oil as a fuel was economical without impairing efficiency.

3667. Attack on the German-controlled oilfields at Ploesti, Romania, by the U.S. Air Force occurred August 1, 1943. The mission, flown from bases in North Africa, sought to destroy the last source in Europe of oil for Germany's war machine.

3668. Major oil producer to fully nationalize the oil industry was Iran in 1951. The measure forced British Petroleum to quit Iran after more than 50 years of operations there.

3669. Embargo on oil supplies by the oil producers of the Middle East was imposed in October 1973 in retaliation for Western political and military support of Israel. In effect until March 1974, the embargo caused an energy crisis in the industrialized world and led to a short-lived interest in energy conservation.

OIL AND PETROLEUM INDUSTRY—WELLS AND REFINERIES

3670. Description of an oil production industry was recorded by the Greek historian Herodotus circa 450 B.C. He wrote of an oil pit in Susiana, a southern province of ancient Persia.

3671. Oil wells were drilled in A.D. 347 in China using bits attached to the ends of bamboo poles. The wells reached a depth of about 800 feet.

3672. Oil wells in Persia were hand dug in 1594 at Baku, to a depth of about 115 feet. This series of wells is often considered to be the first real oil field.

3673. Reported use of a refined oil for lighting occurred in 1810 in Prague, in the Czech Republic, then a part of the Austrian Empire.

3674. Oil well in the United States was drilled unintentionally in 1818 at the mouth of Troublesome Creek, on the Big South Fork of the Cumberland River, 28 miles southeast of Monticello, KY. Seeking salt-bearing brine, the drillers, led by Martin Beatty, bored a five-inch hole with pole and augur to a depth of 536 feet, where they struck oil. Since oil had no known value at the time, sand was thrown down the well to plug it up. The "devil's tar," as Beatty called the oil, was allowed to flow into the Cumberland River and covered its surface for a distance of 35 miles. The oil caught fire and the ensuing conflagration destroyed trees along the banks of the river, as well as the salt works.

3675. Industrial distillation of oil from coal and shale began in France in 1832. The industry quickly spread across the Channel to England.

3676. Exploitable petroleum "seepage" in England was discovered in 1847 in a Derbyshire coal mine. It yielded 300 gallons a day and required refining.

3677. Oil well with a modern design was drilled in 1848 by Russian engineer F. N. Semyehov, on the Aspheron Peninsula of Russia at the Caspian Sea. The nearby city of Baku later became a worldwide oil center with refineries and the production of oil-drilling equipment and chemicals.

3678. Modern oil refining processes were devised in 1850 by James Young, a Scottish industrial chemist. His work in distilling oil from shale and cannel coal led to the large-scale production of paraffin oil and helped develop the shale oil industry in Scotland that existed until 1962.

3679. Patent for distilling kerosene from crude oil was granted in 1852 to Abraham Gesner, a Canadian physicist and geologist. Kerosene was marketed as a clean-burning lamp fuel.

3680. Oil wells in Europe were hand-drilled in 1854 at Bobraka, Poland, by Ignacy Lukasiewicz, Titus Trzecieski, and Mikolaj Klobassa. Called "oil mines," the wells reached a depth of about 100 to 160 feet.

3681. Oil refinery in the world was established in 1855 by Samuel M. Kier, a druggist of Pittsburgh, PA. His small refinery produced the petroleum, which he called Kier's Rock Oil, for medicinal purposes. The bottled product was sold for 50 cents a half-pint. Kier discovered in his laboratory that the light fractions from the crude oil would burn and that the heavy fractions could be used to clean wool.

3682. Published research suggesting a wide range of uses and products for distilled petroleum was authored by Yale University chemistry professor Benjamin Silliman, Jr., in 1855. He is credited with launching the modern petroleum industry.

3683. Crude oil wells in Germany were dug in 1857. The first well to catch the world's attention, however, was drilled near Titusville, PA, in 1859.

3684. Oil well drilled intentionally and successfully in North America was in 1858 by James Miller Willkins near Sarnia in Ontario, Canada. He drilled to a depth of about 50 feet.

3685. Oil refinery was established in 1859 in Ulaszowice (city district) of Jaslo, southeast Poland. It was built by Ignacy Lukasiewicz.

3686. Oil well that was commercially productive was created at Titusville, PA, on August 27, 1859, when E. B. Bowditch and Edwin Drake of the Seneca Oil Company bored through rock in an area already known as Oil Creek. From a depth of 69.5 feet, the well produced about 400 gallons a day. Drake was the first to tap petroleum at its source and prove oil could be found in reservoirs beneath the earth's surface.

3687. Commercial oil refinery was built in Oil Creek Valley, PA, in June 1860 by William Barnsdall and William Hawkins Abbott. The refinery's sole purpose was to manufacture kerosene, which was sold in competition with whale oil and rock oil as fuel to burn in lamps. The small amount of gasoline produced was run into Oil Creek. The refinery and some property were sold in 1864 for $50,000 to six men who organized an oil company.

3688. Oil deposits in Trinidad were discovered in 1866. The first crude oil production began in 1908, and the first oil refinery was established in 1912.

3689. Oil well in India was drilled in 1886 near Jaipur in the state of Rajasthan. The McKillop Stewart Company struck oil, but not enough for commercial production.

3690. Oil well in India that was commercially successful was drilled in 1889 at Digboi in Assam by the Assam Railway and Trading Company (ARTC). It began India's oil production, and the well was later sold to the Burma Oil Company.

3691. Petroleum discovery in Sumatra was in 1890. The Royal Dutch Shell Co. was founded that year to exploit the oil find.

3692. Recorded wildcat oil drilling team in Alaska showed up in 1901. Between that year and the early 1950s, more than 150 wells were drilled, though none proved commercially productive.

OIL AND PETROLEUM INDUSTRY—WELLS AND REFINERIES—*continued*

3693. Oil well in the United States with a massive petroleum deposit was Spindletop in southeast Texas near Beaumont. Oil exploded from the well on Spindletop Hill on January 10, 1901, and spewed 200 feet into the air. It was drilled by the engineer, Anthony F. Lucas, with financial backing from Andrew Mellon of Pittsburgh, PA. Spindletop was called "the greatest oil well ever discovered in the United States." It led to the establishment of the Gulf and Texaco oil companies.

3694. Oil discovery in the Middle East was on May 26, 1908, at Mas jid-i-Suleiman in Persia.

3695. Oil discovery of significance in Persia was on May 26, 1908, by Englishman William Knox D'Arcy at Mas jid-i-Suleiman in the country's southern region. After several other wells in the area proved successful, D'Arcy formed the Anglo-Persian Oil Company on April 14, 1909. He had signed a 60-year oil concession agreement with the Persian government on May 28, 1901.

3696. Oil well in Malaysia was drilled in 1910 on Canada Hill at the city of Miri in northeast Sarawak bordering Brunei.

3697. Use of the term *wildcatter* in oil exploration dates from the 1920s in Texas. Prospectors encountered many wildcats as they cleared land for exploratory wells.

3698. Oil well that was productive in Iraq was drilled in 1927 at Baba Gurgur, about 140 miles north of Bagdad.

3699. Oil wells seized from foreign companies were those nationalized in Mexico in 1938. The Mexican government established a national oil company, Petroleos Mexicanos, to take control of the oil fields from the United States and Great Britain. In April 1942, Mexico agreed to pay a total of $23.9 million to the two countries as compensation for the seizures.

3700. Oil discovery at Prudhoe Bay on Alaska's North Slope was in 1944. The strike spurred the oil companies' interest in Alaskan oil exploration.

3701. Oil refinery to achieve complete conversion of waste gasses into useful power with a carbon monoxide boiler was the Sinclair Oil Corporation's refinery in Houston, TX. The boiler went into operation in November 1953. By injecting a stream of air into the waste gases from the generator, carbon monoxide was converted into carbon dioxide. The catalyst was regenerated in the catalytic oil cracking process.

3702. Oil discovered in Nigeria was in 1956 at Oloibiri by a Shell-BP exploration. This find in the Niger Delta came after a 50-year search for oil in the country. The first well began to produce two years later, bringing in 5,100 barrels a day.

3703. Oil discovery of major significance in Alaska was made on the Swanson River by the Richfield Oil Company on July 15, 1957. At 900 barrels a day, it was the first well in Alaska to produce oil in commercial quantities.

3704. Oil refinery in Singapore was the Bukom refinery set up by the Shell Singapore Company in 1961 in Singapore City. It is the largest of Shell's worldwide refineries in terms of crude distillation capacity.

3705. Oil refinery construction halted by environmental protests in Japan occurred in Shizuoka Prefecture in 1964. Residents from the towns of Numazu, Mishima, and Shimizu organized protests against the building of the petroleum complex in what is regarded as Japan's first environmental protection movement. Public opinion and the news media were swayed by information distributed by the residents.

3706. Oil refinery in South Korea built with private funds was the Kukdong Oil Company, established in 1964. After four name changes, it is now the Hyundai Oil Refining Company. It built a large, modern refinery in 1989 at Chungnam.

3707. Oil exploration and production in a U.S. National Forest was in 1965 by the Phillips Petroleum Company. This was a Bridger Lake operation in Wasatch National Forest in Utah.

3708. Pipeline to carry natural gas from the Soviet Union to a European country was opened in February 1984. The pipeline, opposed by the U.S. Government, transported oil 2,800 miles from Siberian gas fields to Czechoslovakia and onward to other European countries. The first Western European country to receive the gas was France.

3709. Oil production in Papua New Guinea was in June 1992 at the Kubutu Oil Well. It brings in 140,000 barrels a day.

P

PARKS

3710. European mission of significance to the interior of North America was begun in the vicinity of the De Soto National Memorial in Bradenton, FL. In 1539 the Spanish explorer Hernando De Soto landed on the southwest Florida coast with an army of 600 soldiers and began a 4-year, 4,000-mile journey inland.

3711. Historic trail designated in the Western Region of the U.S. National Park Service was the Juan Bautista de Anza National Historic Trail, which runs from Nogales, AZ, to San Francisco, CA. Authorized by Congress on August 15, 1990, the trail commemorates the 1775 trek of Spanish colonists, under the leadership of Col. Juan Bautista de Anza, searching for an overland route from Mexico to California.

3712. Jointure of the Gulf of Saint Lawrence and the Gulf of Mexico occurred when the Illinois and Michigan Canal was built, between 1836 and 1848, connecting Lake Michigan with the Illinois River at Lockport, IL. The Illinois River flows into the Mississippi, which continues to the Gulf, while the Great Lakes ultimately connect to the Saint Lawrence River, which flows into the North Atlantic. The Illinois and Michigan National Heritage Corridor, commemorating the event, was established as a joint effort of federal, state, and local governments in 1984.

3713. Public preserve in the United States was established in 1864, when the federal government granted the Yosemite Valley and Mariposa Grove to the state of California.

3714. National forest in the United States was the Shoshone National Forest in Wyoming, part of the Yellowstone Timberland Reserve established on March 30, 1891. The Shoshone received its present official name on July 1, 1908, in a proclamation by President Theodore Roosevelt. It contains approximately 2.4 million acres.

3715. Lands in the United States to have constitutional protection were in the Adirondack Forest Preserve in upstate New York. In 1894, the people of New York added a clause to their state constitution specifically decreeing that this area be kept "forever wild." This action was the culmination of a long struggle to protect the Adirondack wilderness from despoliation by logging, mining, tanning, and papermaking industries. Nine years earlier, New York had led the way in establishing a forest reserve in the Adirondacks—the New York State Forest Preserve—but by 1894 additional protection was deemed necessary.

3716. National forest established east of the Mississippi was Chippewa National Forest, established in 1908 and originally known as the Minnesota National Forest. The name was changed in 1928 to honor the original inhabitants of this northern Minnesota region.

3717. Full-time landscape architect hired by the U.S. Forest Service was Arthur H. Carhart, who began his service in 1919.

3718. Private organization to support national parks in the United States was the National Parks Association, founded in Washington, D.C., in May 1919 by Stephen T. Mather, the first director of the National Park Service, and Robert Sterling Yard. The main function of the organization was to give support to the idea of national parks in the United States. In 1970 the organization changed its name to the National Parks Conservation Association

3719. Application of the wilderness concept to U.S. Forest Service areas was in 1920, when the Trappers Lake area in White River National Forest in Colorado was designated as an area to be kept roadless and undeveloped.

3720. Wilderness area in New Mexico was established by the U.S. Forest Service in 1924.

3721. Citation for heroism ever bestowed by the U.S. Department of the Interior was given to Ranger Charlie Browne in 1929. Browne had led an effort to save injured climbers and recover the bodies of a six-member climbing party that fell into a deep crevasse in Mt. Rainier National Park.

3722. National monument to commemorate the world's largest gypsum dune field was the White Sands National Monument near Alamagordo, NM, designated on January 18, 1933, by President Herbert Hoover. The monument preserves a major portion of the 275 square miles of desert and plants associated with the dunes.

3723. National Trails system unit opened was the Appalachian National Scenic Trail, completed by volunteers on August 14, 1937. The trail runs through the states of New Hampshire, Vermont, Massachusetts, Connecticut, New York, New Jersey, Pennsylvania, Maryland, West Virginia, Virginia, Tennessee, North Carolina, and Georgia. More than 98´of the trail is on public land.

PARKS—*continued*

3724. Research park in the United States was established at Stanford University in Palo Alto, CA, in 1951 at the request of the school's engineering dean.

3725. Corporation to join the newly created Research Triangle Institute in North Carolina was Chemstrand Corporation in 1959. Later arrivals established operations in such fields as pollution control and biotechnology

3726. Research park inside an "education triangle" was the Research Triangle Institute in North Carolina, founded in January 1959 inside the academic corridor of Duke University in Durham, the University of North Carolina at Chapel Hill, and North Carolina State University in Raleigh.

3727. National park in Thailand was the Khao Yai National Park established in 1962. The park is known for it herds of wild elephants and various tropical birds.

3728. Underwater state park in the United States was John Pennekamp Coral Reef State Park opened to the public in 1963. The park is located in Key Largo, FL, and contains more than 52,000 acres underwater

3729. National lakeshore in the U.S. was Pictured Rocks National Lakeshore, in Michigan, authorized in 1966. The area consists of 73,000 acres of cliffs, beaches, dunes, waterfalls, and lakes located along the shore of Lake Superior.

3730. National Wilderness Area in New Jersey was Great Swamp, NJ, established on September 28, 1968.

3731. Classification of rivers and federally-owned adjoining lands to determine their use by the public was initiated under the Wild and Scenic Rivers Act, October 2, 1968. The categories were wild, scenic, and recreational.

3732. National reserve in the United States was the New Jersey Pinelands, designated by Congress in 1978 as an affiliate of the National Park System. As a national reserve, it differs from national parks in that its primary mission is to preserve and protect areas of natural and cultural significance through state and local management.

3733. Program of the National Park Service intended to help local groups undertake river and trail conservation projects was the Rivers, Trails and Conservation Assistance Program, begun in 1988 as an offshoot of the Wild and Scenic Rivers Act of 1968.

PARKS—LEGISLATION AND REGULATION

3734. "Forest preserves" legislation in the United States was the Forest Reserve Act, passed in 1891. This act created the foundation for what would become the National Forest System.

3735. Legislation in Canada that preserved parks for future generations was the Canadian National Parks Act of 1930 that removed the parks from authority of the Dominion Forest Reserve and Parks Act, which allowed resource exploitation, and stated that the parks should be used and preserved for the enjoyment of future generations

3736. Wilderness bill introduced in the U.S. Congress was the Wilderness Bill introduced by Hubert Humphrey in 1957. The bill was not passed and signed into law until 1964. President Lyndon B. Johnson signed the bill on September 3.

3737. Environmental law case in which citizens were granted standing for a noneconomic interest was the "Scenic Hudson Case" of 1965, in which the Sierra Club brought suit to prevent a power project from being established in Storm King State Park in New York State.

PARKS—NATIONAL MONUMENTS

3738. Human resident in what is now Russell Cave National Monument in Alabama has been traced back as far as 9,000 years ago.

3739. European to set foot on what would later become the west coast of the United States. was Juan Rodriguez Cabrillo, who on September 28, 1542, sailed into San Diego Bay as part of the Spanish Empire's North American colonization. His efforts are commemorated by the Cabrillo National Monument in San Diego, CA.

3740. Non–Native Americans to enter what is now Dinosaur National Monument were William H. Ashley and his band of fur trappers in 1825. The area was named Echo Park by John Wesley Powell, who explored it during his 1869 scientific expedition into the Colorado Plateau. In 1915 it was renamed Dinosaur National Monument after a quarry of Jurassic fossils was discovered there. The monument is located on the Utah-Colorado border.

3741. White men to see the great chasm at what later became Black Canyon of the Gunnison National Monument in Colorado were members of the Ferdinand Hayden expedition in 1873–1874.

3742. Discovery by non–American Indians of the Horsecollar Ruins in what would become Natural Bridges National Monument in Utah was in the late 1880s. The ruins consist of cliff dwellings once occupied by ancestral Pueblos.

3743. Formal ascent of Devils Tower was on July 4, 1893, when William Rogers, a local rancher, accomplished the feat to great fanfare. At least a thousand people came to see him climb what would later be protected as Devils Tower National Monument in Wyoming.

3744. Scientific expedition to what would later be designated the John Day Fossil Beds National Monument was in June 1899, when students from the University of California at Berkeley, led by Professor John C. Merriam, collected specimens that can still be found in the university archives.

3745. Monument erected to commemorate the first flight by an airplane was the Wright Brothers National Monument in Kitty Hawk, NC. The first flight took place in December 17, 1903, when Wilber and Orville Wright briefly lifted off in their airplane at Kitty Hawk.

3746. Natural national monument in the United States was Devils Tower in Wyoming. President Theodore Roosevelt, invoking the newly passed Antiquities Act, bestowed federal protection on the prehistoric natural formation on September 24, 1906.

3747. National monument in the United States created from land donated by a private individual was Muir Woods National Monument, designated January 9, 1908. The land had been purchased three years earlier by Congressman William Kent, who donated 295 acres. President Theodore Roosevelt suggested naming the monument in Kent's honor, but Kent insisted on honoring the conservationist John Muir.

3748. Monument to commemorate the first major exploration of the American Southwest was the Coronado National Monument, established on May 24, 1911. It is located along the U.S.-Mexico border near Hereford, AZ. The monument honors the 1540–1541 expedition led by Francisco Vázquez de Coronado in search of the fabled Seven Cities of Cibola.

3749. Prairie restoration in the U.S. National Parks System was Homestead National Monument in Beatrice, NE, authorized on March 19, 1936, to commemorate the signing of the Homestead Act in 1863.

PARKS—NATIONAL PARKS

3750. Humans to cross the Bering Land Bridge from Asia to North America did so more than 13,000 years ago near the site of what is now Bering Land Bridge National Preserve in Alaska. The land bridge is now overlain by the Chukchi and Bering seas.

3751. Human beings to witness volcanic eruptions of Mount Mazama in what is now Crater Lake National Park probably did so 7,700 years ago.

3752. Human occupation of the area that later became Zion National Park occurred between A.D. 285 and 1200, when the Anasazi lived there.

3753. European to see the Grand Canyon was Spanish explorer García López de Cárdenas in 1540. Cárdenas was a member of Vázquez de Coronado's expedition through the North American Southwest. He was selected to lead a party from Cibola (in present-day New Mexico) in search of a river of which the Hopi had spoken. During the search, he became the first European to see the Grand Canyon.

3754. Written account of a journey through the area of the Blue Ridge Mountains that later became Shenandoah National Park was penned in 1669 by German physician and scholar John Lederer, who possibly was the first European explorer to visit the region.

3755. Documented journey through the Glen Canyon–Lake Powell area was in 1776, when a ten-man party led by two Spanish priests, Dominquez and Escalante from Santa Fe returned home this way after an unsuccessful attempt to find an overland trail to California.

3756. Person to write about what later became Waterton-Glacier International Peace Park was British trapper David Thompson, who recorded his impressions of the area in the 1780s.

3757. Discovery by Europeans of Mount Baker in Washington was in 1790, when Ensign Manuel Quimper of the Spanish navy explored the Strait of Juan de Fuca.

3758. European settlement in the area that later became Olympic National Park was built by the Spanish in 1792, but it was abandoned after only 5 months.

3759. Non-Hawaiian to climb Mauna Loa in the area that later became Volcanoes National Park was Scottish botanist Archibald Menzies, who did so in 1794 as part of the George Vancouver expedition.

PARKS—NATIONAL PARKS—*continued*

3760. European to lead an expedition into the area that later became Sequoia and Kings Canyon National Park was the Spaniard Gabriel Moraga, who discovered the Kings River on January 6, 1806, the day of Epiphany. He named the major tributary accordingly, calling it El Rio de Los Santos Reyes, or the River of Holy Kings—later shortened simply to Kings River.

3761. Non-Native American to see the geysers of Yellowstone National Park was probably John Colter in 1807. Colter was a member of an expedition to the West led by Manuel Lisa, who sent him forth alone to make contact with the Crow. In the course of this mission, Colter is believed to have crossed the Wind River Mountains and the Teton range, traversing areas that today are parts of Yellowstone National Park.

3762. Recorded description of the area that would become the Fort Laramie National Historic Site at the mouth of the Laramie River was made in 1812 by Robert Stuart, who was returning with others to the fur-trading port of Astoria in the Oregon Territory. What he described would later become the Oregon Trail, commemorated by the Fort Laramie National Historic Site in Wyoming.

3763. Non-American Indian to visit the area that later became Mount Rainier National Park was Dr. William Tolmie, a Scottish physician employed at nearby Fort Nisqually, who, aided by native guides, was in the area in 1833 as part of a foraging expedition searching for medicinal herbs.

3764. European to discover what is now Canyonlands National Park was Denis Julien, a fur trader who explored the area in 1836.

3765. Women to pass along the Oregon Trail were Elizabeth Spauling and Narcissa Whitman, who visited Fort Laramie—now Fort Laramie National Historic Site—in 1836, thus also becoming the first known white women to enter the future state of Wyoming.

3766. Railroad traffic in the United States occurred in 1842, when the Baltimore and Ohio Railroad steamed into Cumberland, MD, an event commemorated by the Chesapeake and Ohio Canal National Historic Park, which extends along the Potomac River from Washington, DC, to Cumberland, MD.

3767. Non-American Indian to see Crater Lake was prospector John Wesley Hillman in 1853, when he was searching for the Lost Cabin Gold Mine.

3768. Non-American Indian settler in what is now Rocky Mountain National Park was Joel Estes, a Kentuckian who in 1860 founded the town now known as Estes Park, CO.

3769. Scientific exploration of the area that later became Sequoia and Kings Canyon National was in 1864, when Harvard Geology professor Josiah Dwight Whitney, the director of the newly formed California Geological Survey, sent a team headed by William Brewer to map the terrain. Brewer named the highest peak in the area Mount Whitney in honor of Professor Whitney.

3770. Detailed geologic and topographic information about the area that later became Canyonlands National Park was recorded by explorer John Wesley Powell during two trips to the area in 1869 and 1871.

3771. Documented ascent of Mount Rainier was achieved in 1870 by General Hazard Stevens and Philemon Van Trump.

3772. National park in the United States was the Yellowstone National Park in Wyoming, authorized March 1, 1872, by an act of Congress, which set aside 2,142,720 acres near the head waters of the Yellowstone River "as a public park." Nathaniel Pitt Langford served as the first superintendent. Harry Yount, a Civil War veteran, was hired in 1880 as the first ranger. Later Wyoming, Montana, and Idaho made additional grants of land to Yellowstone Park. Hot Springs National Park in Arkansas, consisting of 911 acres with 46 hot springs, was established as a reservation by an act of Congress on April 20, 1832, but was not designated as a national park until March 4, 1921.

3773. Superintendent of Yellowstone National Park was Nathaniel Pitt Langford, who was appointed after Yellowstone became the first national park on March 1, 1872.

3774. Woman to climb Long's Peak in what is now Rocky Mountain National Park was Anna Dickinson, who accomplished this feat in 1873.

3775. Official exploration of the area that later became Mount Rushmore National Park was led by Lieutenant George Armstrong Custer, who in the course of his 1874 expedition discovered gold in the Black Hills.

3776. Non-American Indian to discover Glacier Bay and the Muir Glacier was John Muir, who in 1879 traveled to the area that would later become Glacier Bay National Park and Preserve.

3777. Ranger hired in Yellowstone National Park was Harry Yount, who was hired in 1880.

3778. Person to enter into the cave at what would become Wind Cave National Park was Charlie Crary in the fall of 1881. Crary went far enough into the cave to see its extensive box work formations composed of thin calcite forms resembling honeycombs. He left a twine trail that others were able to follow later.

3779. Recorded discovery of what would become Wind Cave National Park in South Dakota was in 1881, when brothers Jesse and Tom Bingham were attracted by the sound of whistling. As they approached the mouth of the cave, the whistling and rushing wind blew Tom's hat off.

3780. Hotel in Mt. Rainier National Park was the Mineral Springs Resort, built in 1883 by James Longman at the site of a spring on the mountain's south side.

3781. Cave discovered in American Forks Canyon was found in the autumn of 1887 by Martin Hansell, a Mormon settler who was following the tracks of a mountain lion. American Forks Canyon is located in what would later become Timpanogos National Monument in Utah. The cave Hansell discovered still bears his name.

3782. National park in Canada was established in 1887 in Banff, British Columbia, as Rocky Mountains Park. Between 1887 and 1930, when the National Parks Act was passed, fourteen national parks were created in Canada.

3783. Woman to climb Mount Rainier was Fay Fuller, a schoolteacher from a small town near Olympia, WA, who climbed the peak in 1890.

3784. Ascent of Mount Elias in what is now Wrangell–Mt. Elias National Park and Preserve in Alaska took place between April and August 1897, when a party led by Luigi Amedeo di Savoia-Aosta climbed the arctic peak.

3785. Automobile in the area of Mt. Rainier National Park was carrying future President William Howard Taft, who in 1899 traveled from Washington to see the nation's newest national park. The car had to be towed by horses the last few miles.

3786. Sierra Club outing to a national park was in 1901 to the Tuolomne Meadows in Yosemite.

3787. National park established to protect a cave was Wind Cave National Park in the Black Hills of South Dakota. President Theodore Roosevelt signed enabling legislation on January 3, 1903.

3788. Cultural park set aside in the U.S. National Park System was Mesa Verde, established June 29, 1906. The first ancestral Puebloan people, known as "Basketmakers" because of their artistry in basketmaking, settled there circa A.D. 550.

3789. Director of the U.S. National Park Service was Stephen T. Mather, who served in that role from 1916 to 1929.

3790. Federal legislation to establish the national parks system was the National Park Service Organic Act of 1916, according to which the purpose of national parks is "to conserve scenery and the natural and historic objects and the wildlife therein and to provide for the enjoyment of the same in such manner and by such means as will leave them unimpaired for the enjoyment of future generations."

3791. National park in the United States made up entirely of lands donated by private citizens was the present-day Acadia National Park in Maine. It was established on July 8, 1916, after Charles Eliot Norton, president of Harvard University and a summer resident of Mount Desert Island, ME, formed a public land trust for the purpose in 1910.

3792. National park east of the Mississippi River and the first located on an island was established by President Woodrow Wilson on July 8, 1916, as the Sieur de Monts National Monument on Mount Desert Island, in the Atlantic Ocean about a mile south of Bar Harbor, ME. Its name was subsequently changed twice: first to Lafayette National Park on February 26, 1919, and then to Acadia National Park on January 19, 1929. It contains 27,871 acres, including most of Mount Desert Island and parts of Isle au Haut and the mainland Schoodic Peninsula.

3793. National park in the United States to contain an active volcano was Lassen Volcanic National Park in the Sierra Nevada mountains of California, established by an act of Congress on August 9, 1916. There are 104,526 acres in the park. Lassen Peak, 10,457 feet high, at the southern end of the Cascade mountain range, last erupted in 1916.

PARKS—NATIONAL PARKS—*continued*

3794. National parks management agency in the United States was created August 25, 1916. The National Park Service, a division of the Department of the Interior, was formed to manage national parks and monuments and to insure that these places are preserved for "the enjoyment of future generations."

3795. National park created after the establishment of the National Park Service was Mount McKinley National Park, in the Territory of Alaska, on February 26, 1917. It was created to protect from extinction the white Alaskan mountain sheep, the caribou, the Alaska moose, and the grizzly bear.

3796. Nature reserve in the Soviet Union was the Astrakhan Preserve. The creation of the preserve was approved by Soviet leader Vladimir Ilich Lenin in January 1919. The legislation was signed on February 1, 1919; it set aside a portion of the Volga River delta as a *zapovednik* (nature reserve) and established a nationwide system of nature reserves in the Soviet Union.

3797. National wilderness area in the world was established by the U.S. Forest Service on June 3, 1924, in what became the Gila National Forest in New Mexico, which was preserved in its natural state.

3798. Park to be designated an international peace park was Waterton-Glacier International Peace Park, created in 1932 by joining together the Canadian national park Waterton Lakes, located in Alberta, and nearby Glacier National Park, located in Montana.

3799. National Park Service unit including coral reef resources was Fort Jefferson National Monument in Key West, FL, first designated on January 4, 1935. It was redesignated on October 26, 1992, as Dry Tortugas National Park.

3800. National seashore in the United States was Cape Hatteras National Seashore along the coast of North Carolina. Set aside by Congress in 1937, it was one of the longest stretches of undeveloped shoreline on the Atlantic seaboard with an area of 30,319 acres.

3801. National historic site in the United States National Park System was the Salem (MA) Maritime National Historic Site, designated in 1938 to commemorate the area's whaling and fishing culture.

3802. National park dedicated to the first transcontinental railroad was the Golden Spike National Historic Site, designated on April 2, 1957. The first U.S. transcontinental railroad was completed on May 10, 1869, when the Central Pacific and the Union Pacific were joined near Promontory, in Box Elder County, UT.

3803. Undersea park established by the U.S. government was the Key Largo Coral Reef Preserve, established on March 15, 1960, by President Dwight D. Eisenhower's presidential proclamation. This wildlife refuge, 21 miles long and 3.5 miles wide in the Atlantic Ocean off Key Largo, FL, contains 40 of the 52 known coral species. Previously, it had been the John Pennekamp Coral Reef State Park, established by the Florida Board of Parks and Historic Monuments on December 3, 1959.

3804. National parkland acquired with congressionally authorized funds was Cape Cod National Seashore, for which Congress appropriated funds in 1961. All previous parklands had been either federally owned or donated to the U.S. government.

3805. River to be declared a national Scenic Riverway in the United States was the Ozark National Scenic Riverway in the highlands of southeast Missouri, so designated by an act of Congress on August 24, 1964. Scenic Riverways were a precursor of the Wild and Scenic Rivers Act passed four years later.

3806. National recreation area designated in the United States was Lake Mead in Arizona, which was authorized on October 8, 1964. National recreation areas consist of areas composed largely of lakes and reservoirs created by dams.

3807. Park to commemorate the cattle industry of the American West, from its inception in the 1850s was the Grant-Kohrs Ranch National Historic Site, enacted in 1972 at Deer Lodge, MT.

3808. World Heritage site established in the United States was Yellowstone National Park in 1978. A World Heritage site is designated as such by the United Nations Educational, Scientific and Cultural Organization (UNESCO) and must be of particular cultural or natural importance to humanity.

3809. Federal legislation in the United States to protect caves located on federal lands was the Federal Cave Resources Protection Act of 1988.

3810. International journal devoted to protected areas issues began publication in 1990. *Parks* magazine is published by the World Commission on Protected Areas (WCPA) three times a year and is dedicated to presenting material such as "Protected Areas and Post-Communist Reform." It was developed by the WCPA to reach a wider audience.

3811. Park in the Western Hemisphere to commemorate metal mining was Keweenaw National Historic Park in the Upper Peninsula of Michigan. Established October 27, 1992, the park is located on the site of the first large-scale hard rock industrial mining operation in the United States and the only site in the world where abundant quantities of commercially useful pure natural copper occur.

3812. National park to focus on conservation history and the evolving nature of land use was the Marsh-Billings-Rockefeller National Historic Park in Woodstock, VT. Created in 1992 when the Rockefeller family donated their estate's residential and forest lands to the people of the United States, the park officially opened in June 1998.

3813. National preserve in an American desert was created in October 1994, when Congress passed and President Bill Clinton signed the California Desert Protection Act to protect the diverse desert environment of the Mojave.

3814. Woman to reach the rank of major in the U.S. Park Police was Gretchen W. Merkle, who on July 19, 1998, became the commander of the San Francisco field office.

3815. Light rail system in a national park in the United States was begun in the summer of 1999, linking the scenic South Rim of Grand Canyon National Park with the nearby town of Tusayan, AZ. By providing tourists with a convenient alternative to using their cars inside the park, the rail system was designed to reduce air pollution, water loss, and other environmental stresses on the Grand Canyon caused by an estimated 4 to 5 million sightseers visiting annually.

PARKS—STATE AND URBAN PARKS

3816. Urban park in the United States was Boston Common, created in 1634 to provide a place for Bostonians to graze their livestock.

3817. Parkland purchased by a city in the United States was Elm Park, containing 27 acres, bought by the city of Worcester, MA, in March 1854 from Levi Lincoln and John Hammond.

3818. State park in the United States was in Yosemite Valley, CA. The park area included the valley itself and the Mariposa Grove of trees a several miles to the south. Although the land was granted to the state of California by an act of Congress on June 30, 1864, the state did not gain control of the area for some 10 years while the land claims of settlers in the area were resolved. Yosemite National Park was created in 1890, and in 1905 the California state legislature passed an act returning the valley and grove to the federal government for incorporation in the national park.

3819. Botanical garden in a city opened to the public in the United States was the Boston Public Garden, part of the "Emerald Necklace," a series of public parks, designed by Frederick Law Olmsted and begun in 1878, in Boston, MA.

3820. Instigation in the United States of the use of easements to preserve parklands from urban expansion was in Piscataway Park, located along six miles of the Maryland shore of the Potomac River, across from Mount Vernon. The park opened in 1952.

3821. Urban national park to exceed 75,000 acres was Golden Gate National Recreation Area, established in October 1972 in and around San Francisco, CA. The park covers 76,500 acres of land and sea.

3822. Federal legislation in the United States to grant federal funds for urban park rehabilitation and infrastructure maintenance was the Urban Park and Recreation Recovery Act of 1978.

POLAR REGIONS

3823. First recorded polar expedition was commanded by Greek astronomer Pytheas. Pytheas left Marseilles circa 330 B.C. in search of the true North Pole. He is believed to be the first person ever to have crossed the Arctic Circle.

3824. Europeans known to have wintered in the extreme north did so involuntarily, in a bay they named Ice Haven on the coast of Novaya Zemlya, about 15 degrees from the pole. There, on August 26, 1596, the Dutch ship *Gesandte* was caught in an ice floe and almost smashed. Captain Willem Barents and 17 men survived the winter in a cabin made from driftwood and wood from the ship.

3825. Person to reach latitude 80 degrees north in the arctic region was Englishman Henry Hudson in 1607 near Spitsbergen on the present-day Norwegian island of Svalbard.

POLAR REGIONS—*continued*

3826. Person to cross the Antarctic Circle was British Captain James Cook on January 17, 1773. Cook was the leader of an expedition of two ships, the *Resolution* and the *Adventure*. Cook and his crew crossed the Antarctic Circle, but huge ice packs prevented them from ever seeing or reaching the actual continent of Antarctica.

3827. Commercial sealing in the Antarctic region began around 1780 in South Georgia. Seal hunters from New England sailed to South Georgia, an island discovered by British Captain James Cook during his expedition in the region.

3828. Sighting of Antarctica is variously credited to the Russian Fabian von Bellingshausen, the Englishman Edward Bransfield, and the American Nathaniel Palmer. The three commanded separate expeditions, all of which sighted parts of the continent in 1820. Maori legend gives the credit to a Polynesian, Ui-te-Rangiora, who is said to have ventured into Antarctic waters circa 650.

3829. Ship to bring home Antarctic sealskins was the *San Juan Nepomúceno*, commanded by Carlos Timblon. The *San Juan Nepomuceno* returned to Buenos Aires on February 12, 1820, with 14,600 skins on board.

3830. Men to winter in the Antarctic were Englishman Captain Clark and his crew in 1821. Clark and the crew of *Lord Melville* were marooned on King George's Island after the ship was wrecked.

3831. Person to see the Antarctic territories of Peter I Island and Alexander I Island was Russian explorer Fabian von Bellingshausen in January 1821 during his two-year expedition in the Antarctic region. Bellingshausen was the first explorer to circumnavigate Antarctica since James Cook, nearly fifty years earlier.

3832. Explorer to land on Antarctica was American sea captain and sealer John Davis, who landed at Hughes Bay on the Antarctic Peninsula on February 7, 1821.

3833. American scientist to go to Antarctica was Dr. James Eights, who sailed on the Palmers-Pendleton expedition of 1829–1831. As a result of this trip, Eights published seven papers on the geology of Antarctica.

3834. Man to reach 74 degrees 15 minutes south latitude was Englishman James Weddell, who sailed there on February 20, 1833. The sea where he crossed the 74-degree mark was named after him.

3835. Antarctic expedition funded by the U.S. Congress was led by Navy Lieutenant Charles Wilkes. The 1838 Wilkes expedition, which was sponsored by the government in hopes finding new seal hunting grounds, reached Antarctica in January 1840. Although the trip was of limited scientific value, Wilkes's memoir of the voyage was popular among Victorian readers.

3836. Explorer to sail through the Northeast Passage from Norway to the Bering Sea was the Finnish-born Swedish explorer Nils Adolf Erik Nordenskjöld. He began his attempt to sail the Northeast Passage in 1878, but was stopped by ice at the entrance to the Bering Strait. He and the crew of his ship, the *Vega*, spent the winter frozen in the ice near the strait before they were able to successfully navigate the strait and sail into the Bering Sea in 1879.

3837. International Polar Year was observed in the winter of 1882–1883 by eleven countries. During this time, scientists conducted numerous experiments in the Arctic and Antarctic regions.

3838. Wood fossils found on the Antarctic were discovered by Swede Carl Larsen in November 1892. Larsen found the fossils on Seymour Island. This is also the first evidence of a warmer climate on Antarctica.

3839. Landing on Antarctica that initiated the "Heroic Age" of Antarctic exploration was made by a party from the Norwegian vessel *Antarctic*, commanded by Captain Leonard Kristensen. The Norwegians stepped ashore at Cape Adare on January 24, 1895, becoming the first men in almost fifty years to walk on the continent. Their landing set off a flurry of international activity that became known as the "Heroic Age" of Antarctic exploration, culminating in the expeditions of Roald Amundsen and Robert Falcon Scott to the South Pole nearly 20 years later.

3840. Expedition to survive a winter in the Antarctic was the crew of the Belgian ship *Belgica*, led by Captain Adrien de Gerlache. The *Belgica* became stranded in Antarctic ice during the winter of 1898. The ship's crew included renowned polar explorers Roald Amundsen and Frederick Cook.

3841. Significant theory that Antarctica was a continent rather than a series of islands was advanced by Canadian oceanographer John Murray on February 24, 1898, at a meeting of the Royal Society in London, England. Murray cited geological evidence from northward floating icebergs to suggest that Antarctica was a land mass.

3842. Scientific record of an Antarctic winter was compiled by members of an expedition led by Australian explorer Carsten Borchgrevnik in 1899. Borchgrevnik's group was the first to spend a winter on the Antarctic mainland.

3843. Long-distance sledge journey on the Antarctic continent was taken by Swedish geologist Nils Otto Gustaf Nordenskjöld and his five-man crew. In 1902 they traveled more than 400 miles across the continent.

3844. Scientific exploration of significance of Antarctica was led by English Navy Lt. Robert Falcon Scott. The *Discovery* expedition landed in Antarctica in January 1902. Members spent the next two years collecting data while exploring the new continent. Scott's goal of reaching the South Pole was not achieved.

3845. Aerial photograph in Antarctica was taken from a balloon over McMurdo Sound by Ernest Shackleton on February 2, 1902. Shackleton was part of the *Discovery* expedition led by Robert Falcon Scott.

3846. Polar explorer to ascend in a balloon in the Antarctic regions was British explorer Robert Falcon Scott on February 3, 1902. His ascent in a balloon made of cow gut lasted only a few minutes. Scott was the leader of the *Discovery* expedition, which failed in its attempt to reach the South Pole. Scot died in 1912 during another attempt to reach the South Pole.

3847. Whaling station in the Antarctic was built by Carl Larsen, a Swede, in 1904 at Grytoiken on South Georgia Island. Over the next 10 years, more than 20 stations and factory ships opened in the region.

3848. Motorized transportation in a polar region took place in Antarctica on September 22, 1908. British explorer Ernest Shackleton and four others drove a distance of eight miles in a car before the severe cold and wind disabled the car's engine. Shackleton was the leader of the *Nimrod* expedition, which failed in its attempt to reach the South Pole.

3849. Explorers credited with reaching the North Pole were Frederick A. Cook and Robert E. Peary. Admiral Peary claimed to have reached the earth's north geographic pole on April 6, 1909, accompanied by his assistant Matthew Henson and four Eskimos. By the time Peary's expedition returned, however, Cook, who had served as surgeon on an earlier Peary expedition, claimed to have reached the Pole on April 21, 1908, almost exactly a year earlier. Cook was initially lauded for the feat, but data supporting his claim was discredited, thus shifting credit to Peary. Later examination of Peary's data created doubts that either explorer had actually reached the Pole.

3850. Radio station in Antarctica was set up by Australian Sir Douglas Mawson. From 1911 to 1914 he had a station at his base in Commonwealth Bay. Radio contact could be made with MacQuarie Island, off Tasmania, and with passing ships, but only under favorable conditions.

3851. Japanese Antarctic expedition began in November 1911. The expedition, led by Lt. Nobu Shirase, landed at the Bay of Whales.

3852. Explorers to reach the South Pole were led by Norwegian explorer Roald Amundsen. After becoming the first to successfully navigate a Northwest Passage between the Atlantic and Pacific oceans, Amundsen mounted an Antarctic expedition, which reached the South Pole on December 14, 1911. Amundsen disappeared in 1928 while attempting to rescue survivors of a polar dirigible expedition led by Umberto Nobile

3853. Meteorite discovered in Antarctica was found in the winter of 1912 by the first Australian expedition on the continent, led by British explorer and geologist Douglas Mawson. Later scientists found Antarctica to be the best source of meteorites on earth.

3854. Aerial mapping photographs over Antarctica were taken by Colonel Ashley C. McKinley, the photographer for the Richard Byrd expedition of 1928–1930.

3855. Flight across the Arctic Ocean was made by Australian George Hubert Wilkins in 1928. Wilkins began his flight in Point Barrow, Alaska, flew over the North Pole, and landed in Spitsbergen on the Norwegian island of Svalbard.

3856. Airplane flight over the South Pole occurred when Richard E. Byrd and three others flew 10 hours from the Bay of Whales on November 28, 1929.

3857. Explosion seismology experiments in Antarctica were conducted by U.S. geophysicist Thomas C. Poulter during Admiral Richard E. Byrd's second expedition, between 1933 and 1935. Using explosive charges, Poulter measured sound waves to determine the thickness of ice between underlying rock and the ice's surface.

POLAR REGIONS—*continued*

3858. Woman to visit Antarctica was Caroline Mikkelson, the wife of Norwegian fishing captain Klarius Mikkelson. The couple briefly went ashore on February 20, 1935. Due to the policies of various governments, female researchers and explorers were a rarity on the continent until the late 1960s.

3859. Research on the interaction of wind and cold on the human body was conducted by Americans Dr. Paul Siple and Charles Passel on the U.S. Antarctic Service Expedition of 1939–1941. As a result of these studies they formulated the wind chill factor, which measures the combined effect of wind and cold on heat loss form the human body or other objects.

3860. Snow cruiser for the Antarctic research was developed under the direction of Thomas Charles Poulter by the staff of the Research Foundation of the Armour Institute of Technology in Chicago, IL. The cruiser was powered by an engine and could be driven like a truck. Containing living quarters, a combination galley and darkroom, a two-way radio station, an engine room, a scientific laboratory, a machine shop, and a control room, the vehicle was 55 feet 8 inches long and 19 feet, 10.5 inches wide. Built at a cost of $150,000, it moved for the first time under its own power on October 22, 1939, in Chicago. On October 24 it was driven to Boston, MA, where on November 15 it boarded the *North Star* for the Antarctic.

3861. Permanent American base in Antarctica was built in 1940 on Stonington Island, Antarctica, and used for exploration, mapping, and scientific research. The two expeditions that used the East Base camp were led by Richard Black and Finn Ronne. Before their final departure in 1948, the explorers had used the base to carry out dangerous airplane expeditions and epic dogsled journeys that mapped some of the last unknown areas of the continent. The first women in Antarctica—Edith "Jackie" Ronne, Finn Ronne's wife, and Jennie Darlington, the wife of pilot Harry Darlington—spent a year at East Base starting in 1947 as members of a private 23-person expedition.

3862. High altitude meteorological station in Antarctica was Mile-High Post, operated during November and December 1940 at 5370 feet on the Antarctic Peninsula. The station was set up by the United States Antarctic Service Expedition of 1939–1941.

3863. Women to spend the winter in the Antarctic were Edith Ronne and Jennie Darlington who traveled with the Ronne expedition of 1947–1948. Ronne and Darlington and the rest of those in the expedition spent the winter at Stonington Island.

3864. Woman to fly over the North Pole was Louise Arner Boyd in 1955. At the age of 68, Boyd rented an airplane and flew over the North Pole. After this historic flight, she retired to her home in San Francisco.

3865. International Antarctic meteorological station was Antarctic Weather Central, which began operating July 1, 1957, at Little America V on Kainan Bay. The station collected and disseminated meteorological information relayed from other weather reporting stations around the continent. Weather Central was relocated to Melbourne, Australia, in 1959.

3866. Evidence of prehistoric vertebrate life in Antarctica was an amphibian fossil discovered on December 28, 1967, by Peter J. Barrett. The fossil provided proof of early vertebrate life on the continent and added credence to the theory that Antarctica had once been joined to northern continents.

3867. Fossil found in Antarctica was the jaw of a Labyrinthodont, discovered on December 28, 1967, by New Zealand geologist Peter Barrett. The Labyrinthodont was an amphibious dinosaur with distinctively folded enamel in its teeth. Because Labyrinthodont remnants had earlier been discovered on other continents, discovery of the Antarctic fossil lent credence to the theory of continental drift.

3868. Woman to direct an Antarctic research station was American biologist Mary Alice McWhinnie. In 1974, Dr. McWhinnie and her colleagues conducted research on krill and other marine life in McMurdo Sound.

3869. Radiometric survey in Antarctica was begun in the Antarctic winter of 1976–77 by a team of radiation experts from the University of Kansas. Using a gamma ray spectrometer installed in a U.S. Navy helicopter, which traversed the Antarctic landscape, the scientists were able to detect and survey radioactive minerals from the air.

3870. Scientist to survive the eruption of an Antarctic volcano was geochemist Werner Giggenbach, an expert on volcanic gases from the New Zealand Department of Scientific and Industrial Research. In December 1978, Giggenbach's colleagues lowered him into Mt. Erebus to allow him to collect gas samples. His clothing was scorched when lava debris from a sudden eruption hit him, but Giggenbach emerged from the crater unharmed.

3871. Person to circumnavigate the North Pole on foot and solo was Helen Thayer in March and April 1988. Thayer circled the North Pole on her 29-day Arctic trek.

3872. Woman to travel solo to either pole was Helen Thayer, who walked to the North Pole in 1988 with her dog, Charlie. Without the aid of aircraft, dog team, or snowmobile, she began her trek to the Magnetic North Pole on March 30 and finished on April 27. She wrote of her journey in *Polar Dreams*.

3873. Arctic Environmental Protection Strategy (AEPS) resulted from the June 1991 First Arctic Ministerial Conference in Rovaniemi, Finland. Representatives of Canada, Denmark, Finland, Norway, Sweden, Russia, the United States, and indigenous peoples met to establish an ongoing cooperative environmental strategy for dealing with issues such as arctic biodiversity, resource management, ozone depletion, marine pollution, conservation, and research.

3874. Direct evidence that the West Antarctic Ice Sheet was once open sea was uncovered by geologists Reed Scherer and Slawek Tulaczyk and their colleagues from Uppsala University and the California Institute of Technology during a 1998 research project. In soil samples taken from the mud beneath the ice sheet, two-thirds of a mile down, the scientists found fossil remains of diatoms, tiny marine organisms that typically live on the ocean floor. A possible future meltdown of the West Antarctic Ice Sheet is the subject of much controversy in the scientific community.

POPULATION GROWTH

3875. U.S. national census took place in 1790. The census estimated the nation's population to be 3.9 million, which included free white men and women and slaves but not indigenous inhabitants. The 1790 national estimate was preceded by censuses that counted the populations of individual colonies or states.

3876. Proposed theory that human population growth might create a global catastrophe was advanced by English cleric Thomas Malthus in *An Essay on the Principle of Population* in 1798. Malthus believed that competition for resources like food could destroy humanity if population growth was not controlled. Malthus's essay was a major influence on Charles Darwin's theories of natural selection and on later population theorists.

3877. City to reach one million in population was Edo (Tokyo), Japan, around 1800.

3878. Mass-produced condoms were made of vulcanized rubber in the 1840s. The vulcanizing process was discovered by American inventor Charles Goodyear in 1839. Condoms made of linen were in use as early as 1000 B.C.

3879. Birth control clinic in the United States opened in Brooklyn, NY, in 1916. The clinic's operators—Margaret Sanger, Ethel Byrne, and Fania Mindell—were arrested, prosecuted, and imprisoned for violating the Comstock Law, which banned the circulation of birth control information as obscene material.

3880. Oral contraceptive was called Enovid. The birth control pill was developed by American research chemist Frank B. Colton at G. D. Searle & Co. in Skokie, IL during the 1950s. Enovid was approved by the U.S. Food and Drug Administration for distribution in 1960.

3881. United Nations Fund for Population Activities (UNFPA) was established in 1963 through the United Nations Declaration on Population, which declared family planning to be a basic human right. UNFPA provides assistance in reproductive health and family planning services to developing countries and nongovernmental agencies. UNFPA also helps formulate national population programs that support sustainable development.

3882. Extensive collection of quality data on fertility and family planning was by the World Fertility Survey (WFS), an international research program. The survey was carried out in 61 countries between January 1, 1972, and June 30, 1984 by the International Statistical Institute in The Hague in cooperation with the International Union for the Scientific Study of Population, the U.S. Agency for International Development, and the United Nations Fund for Population Activities. The main purpose of the WFS was to help plan for population growth and its effects. It interviewed 341,300 women and provided the most complete picture, to that

POPULATION GROWTH—*continued*

time, of population trends in the tropical Americas, Southeast Asia, Europe, and the United States. WFS was stopped when there were concerns that it would become a permanent institution.

POPULATION GROWTH—LEGISLATION AND REGULATION

3883. One-child-per-family policy in China was instituted by the Chinese government on August 11, 1979, in an effort to control the country's swiftly growing population. While not outlawing having more than one child, the policy applied financial incentives and penalties according to the number of children born per family.

3884. Government policy in the United States to deny aid to foreign family planning agencies that performed or promoted abortions as a method of population control was President Ronald Reagan's "Mexico City policy." Announced at the second UN International Conference on Population in Mexico City in December 1984, the new policy rejected the existence of a worldwide population crisis and resulted in an end to U.S. aid to the International Planned Parenthood Federation and the United Nations Fund for Population Activities. President Bill Clinton ended the controversial policy on January 22, 1993. President George W. Bush reinstated it on January 22, 2001.

3885. National population plan in Micronesia was adopted by the Federated States of Micronesia (FSM) in 1988, as one of their first acts as a federal government. The FSM addressed the relationships between population growth and the region's finite resources, poverty and crime rates, pollution, and available living space by pledging to lower the national birth rate. The policy included pledges to make family planning information comprehensively available, to provide incentives for women to delay their first pregnancy until 24 years of age or space their children at least 3 years apart, and to promote policies that allowed more women to enter the work force.

3886. National population program for Burundi was adopted in 1990. The government's strategy for lowering a high and potentially unsustainable birth rate included increasing education for women and rural youth, so that information about family planning, quality of life, and sustainable development could be assimilated by the population. The policy also included a call for urbanization to relieve demand for living space in rural areas.

3887. National population plan for Sudan was adopted on September 16, 1990. The policy's main goals were slowing the fertility rate without resorting to coercive measures and reversing the population flow to urban areas by strengthening development in Sudanese villages. In addition to expanding educational opportunities, promoting contraception, and improving the status of women, the plan's institutional goals included developing statistics and an annual report on the national population situation.

3888. National population guidelines for Bolivia were included in the National Development Strategy of 1992. The plan called for supporting programs devoted to reducing infant mortality and improving maternal health. The plan did not explicitly support contraception. Instead, it called for educational campaigns relating to reproductive health, sex education, and family planning. Without specifically mentioning the status of women in Bolivian society, the plan also called for a campaign to publicize the fact that individuals' freedom to decide the number and spacing of children is an internationally recognized right.

3889. National population policy for Indonesia was the Law on Population, Development, and Prosperous Family of 1992. The Indonesian parliament ratified the largely successful government policy, which included providing birth control aids and education, counseling on ideal marriage and delivery ages, and promoting the concept of a small family as being harmonious with sustainable national development.

3890. National population policy for Jamaica was adopted in 1992 to further decrease the national birth rate, which had declined by 50 percent due to programs instituted since the 1960s. Although the plan directed government agencies to continue specific policies like increasing contraceptive use nationally, the document was a broad statement of social goals intended to improve the quality of life in Jamaica and sustain life on the island by pursuing an eventual goal of zero population growth. The quality-of-life objectives in the report aimed at further increasing Jamaicans' life spans by preventing diseases, improving maternal and child health services, and reducing the homicide rate by strengthening the social environment.

3891. National population policy for Niger was declared by the national government on August 19, 1992. The document included numerous explicit goals to address an acknowledged ineffectiveness of the nation's past development policies. The plan's objectives included national promotion of nutrition, literacy, food production, health benefits, contraceptive use, women's rights, and data collection by government employees, particularly in the field of population research.

3892. National population plan for Vietnam was the Resolution of 1993 on Policy Concerning the Population and Family Planning Work. The plan institutionalized earlier family planning policies, which had been largely ineffective, and called for participation of all parts of society in family planning and education, promoting contraception, making imported birth control devices tax exempt, and improving social services for the young, illiterate, and elderly. The resolution, which blamed lackadaisical implementation of previous efforts for their ineffectiveness, also proposed holding all government officials accountable for the program's success or failure.

3893. National population policy for Ethiopia was adopted in April 1993 to address a high birth rate unsustainable in a region beset by ongoing natural and man-made disasters like drought and civil war. Specific goals included expanding a minimal percentage of contraceptive use, ensuring the economic and social rights of women, and increasing reproductive health services throughout the country. Educating both men and women about potential quality-of-life improvements in a society with a lower fertility rate was also a priority.

3894. National population plan for Sierra Leone was the National Population Policy for Development, Progress, and Welfare, adopted on April 2, 1993. The document institutionalized earlier government moves toward a family planning policy to relieve stress on the nation's precarious economy. In addition to setting goals for access to family planning services and education, the plan suggested that women have smaller families of 3 to 4 children, postpone pregnancy until 16 years of age, and maintain intervals of 24 months between births.

3895. Law to discourage population growth in Iran was enacted on May 23, 1993. The law made paid maternity leave a government benefit applicable to only the first three births in one family. Savings generated by the new law were to be used for funding public education in family planning and related health care matters, through school programs, films, and the news media.

POPULATION GROWTH—ORGANIZATIONS AND CONFERENCES

3896. International conference to address world population problems was the World Population Conference, held from August 31 to September 2, 1927, in Geneva, Switzerland. The meeting of scientists was organized by American birth control advocate Margaret Sanger.

3897. Worldwide organization to advocate family planning was the International Planned Parenthood Federation (IPPF). The first IPPF conference was held in Bombay, India in 1952 and was organized by family planning organizations from Germany, Hong Kong, India, the Netherlands, Singapore, Sweden, Great Britain, and the United States. The federation's headquarters is now located in London, England.

3898. Modern world conference on population was the United Nations World Population Conference, held in Bucharest, Romania, in 1974. The conference's World Population Plan was one of the first UN efforts to link the interrelated issues of population growth, resource management, environmental health, and sustainable development.

3899. Organization of African Unity (OAU) population agency was the African Population Commission. The OAU created the group in July 1994 in Addis Ababa, Ethiopia. It was composed of the leaders of national population commissions and similar institutions in OAU member states. The group's charter envisioned it as a continental board for coordinating its member nations' population and development policies, promoting partnerships between governments and non-governmental organizations, and presenting a common African position in population and development matters at international conferences.

3900. United Nations conference to link the rights of women to population growth issues was the International Conference on Population and Development, held September 5–13, 1994, in Cairo, Egypt. Responding to high pregnancy-related mortality rates in developing countries, the conference charged nations with creating reproductive health programs and ensuring certain women's rights as basic human rights. These included the freedom to decide the number and spacing of children, to have the information and means to do so, and to make decisions concerning reproduction free of coercion, discrimination, or violence.

POPULATION GROWTH—ORGANIZATIONS AND CONFERENCES—*continued*

3901. United Nations conference to address human habitation in a sustainable development context was the Second United Nations Conference for Human Settlements, also known as Habitat II. The conference was held June 3–14, 1996, in Istanbul, Turkey. Unlike the original 1976 U.N. Habitat conference, which called for governmental responses to urban overpopulation, Habitat II promoted nongovernmental and community-based initiatives to counter problems like pollution, waste disposal, resource depletion, and lack of shelter.

PUBLIC HEALTH AND SAFETY

3902. Smallpox epidemic of significance recorded broke out in the Roman Empire in 165. The epidemic raged for 15 years, reducing populations in some parts of the empire by 25 to 35 percent.

3903. Black Death (bubonic plague) outbreak was in about 1346 in Mongolia. Infected fleas infested millions of rats and the disease made its way swiftly across Asia. Millions died; China's population fell from 123 million in 1200 to 65 million in 1393, largely because of the plague and ensuing famine.

3904. Black Death (bubonic plague) outbreak in Europe was in the fall of 1347. Reaching Messina, Italy, via a ship from the Crimea, the epidemic raged until 1351, leaving an estimated 20 million to 30 million Europeans dead.

3905. Black Death epidemic in England struck in the summer of 1348. Within a year, this variant of the bubonic plague, more virulent than any previous outbreak, had reached virtually all parts of the country, killing between a third and half of the population.

3906. Cases of smallpox to reach the Caribbean island of Hispaniola were reported in 1518. Brought by Europeans, the disease decimated the island's population; only around 1,000 indigenes survived. Public health measures virtually wiped out smallpox in the late 20th century.

3907. Yellow fever epidemics in the New World occurred in the Yucatan, Mexico, and in Havana, Cuba, in 1648. African mosquitoes imported into the Caribbean in slave ships brought the disease.

3908. Systematic quantitative study of death in populations was by statistician John Graunt in England in 1662. His *Natural and Political Observations on the Bills of Mortality* noted deaths, ratios of births to deaths, differences in seasonal health rates, and other categories.

3909. Boards of health in American cities were established in the late 18th century. Boston, New York, Philadelphia, and Baltimore had the first public health agencies.

3910. Hospital in the United States for the sick, injured, and insane opened in Philadelphia, PA, in 1752. Maryland-born Thomas Bond founded the facility, called Pennsylvania Hospital.

3911. Yellow fever outbreak to force as much as half the population of Philadelphia, PA, to flee to the countryside occurred in 1793. The epidemic killed 15 percent of the city's population.

3912. Successful vaccination against smallpox was administered to an eight-year-old boy by Edward Jenner in England in 1796. Jenner announced his discovery in his *Inquiry into the Causes and Effects of the Variolae Vaccinae* in 1798.

3913. Federal health program in the United States was created in 1798 to provide medical care for merchant seamen. This act by the U.S. Congress helped develop federal-state cooperation to enforce quarantine laws and fight epidemics, such as yellow fever and cholera.

3914. National public health agency in the United States was established in July 1798 in Philadelphia, PA. A measure authorizing seamen's hospitals led eventually to the creation of the U.S. Public Health Service, now an agency of the U.S. Department of Health and Human Services.

3915. Public health advocate of significance in Great Britain was Sir Edwin Chadwick (1801–1890), who has been called the "founder of the modern public health system." In 1832, he became secretary of the Royal Commission on the Reform of the Poor Law and was later secretary of the Poor Law Commission (1834–1836). His *Report on an Inquiry into the Sanitary Conditions of the Labouring Population of Great Britain* (1842),contained advanced ideas about environmental sanitation, concentrating on populations rather than individuals and on prevention instead of cure. His campaigning work led to the passage of Britain's Public Health Act of 1848, establishing a Board of Health, and he served as its first commissioner (1848–54).

3916. Study in England of significance into the effects of industrialization on public health was Edwin Chadwick's *Report on an Inquiry into the Sanitary Conditions of the Labouring Population of Great Britain*, published in England in 1842. Chadwick noted that more than half of working-class children died before age 5 and that the average life span of laborers in England's industrial cities was 16 years, half that of the gentry.

3917. Public health legislation of significance was Britain's Public Health Act of 1848. Parliament created a three-man national Board of Health to establish local boards with elected members who concentrated on environmental public health, such as sewerage and water supply. Boards of health came into existence throughout the nation, but the only ones legally required by the act were in areas with a death rate above the national average. The boards could raise taxes and appoint a medical officer. The act was especially directed at public sanitary measures to stem a cholera epidemic and was based on the ideas of Edwin Chadwick, secretary of the Poor Law Commission (1834–1836).

3918. Public health housing law in a large U.S. city was enacted in New York City in 1867. Its mild protections for tenement dwellers were strengthened in 1879 with a provision prohibiting windowless rooms.

3919. Board of health established in the United States by a state was the Massachusetts State Board of Health in 1869. It began with the collection of thousands of samples of water and food by Ellen Swallow Richards, a crusader for public health.

3920. Public health department created by the U.S. government was the U.S. Public Health Service, established in 1870 by Congress. It centralized the federal program and was headed by a medical officer. It is now part of the U.S. Department of Health and Human Services (HHS).

3921. Burial of electric and other wires to be required by a state was required by a law enacted on June 14, 1884, in New York. It required that "all telegraph, telephone and electronic light wires and cables in any incorporated city having a population of 500,000 or over . . . be placed under the surface of the streets, lanes and avenues." It also specified that telegraph poles be removed by November 1, 1885. Overhead wires had killed hundreds of people in New York City alone.

3922. Municipal public health laboratory in the United States was established in New York City in 1894. William H. Park was the founder.

3923. Evidence conclusively showing that yellow fever was transmitted by a variety of mosquito emerged from U.S. research in Cuba in 1900. The work of the commission of U.S. Army physician Walter Reed soon after the Spanish-American War led to the virtual elimination of yellow fever in Cuba and the United States.

3924. Comprehensive public health housing law in New York City dates from 1901. The law, a model for other large U.S. cities, required improved ventilation, fire protection, and sanitation in tenements.

3925. International public health agency was the Pan American Sanitary Bureau, established in 1902 with headquarters in Washington, DC. It developed out of a series of conferences, each called an International Sanitary Congress, which began in 1851 and were held in the United States and Europe.

3926. Animal disease of American origin was recognized in 1910 in ground squirrels of Tulane County, CA, by George Walter McCoy. He and Charles Willard Chapin named the organism *Bacterium tularense*. The disease, which also affected rabbits, was epizootic, meaning that it was capable of spreading very rapidly in a wild animal population. Edward Francis of the U.S. Public Health Service named the disease tularemia and was awarded a gold medal by the American Medical Association for his research on it.

3927. Widespread deadly global influenza epidemic occurred in 1918, the last year of World War I. The airborne disease killed as many as 20 million people worldwide.

3928. Yellow fever vaccine was developed in 1927 in the United States. The first official global eradication campaign began that year.

3929. Public health grants from the U.S. government to states were created by the Social Security Act of 1935. Congress passed the act as part of Franklin D. Roosevelt's New Deal program. The money stimulated the establishment of state and local public health services.

3930. Hazardous substances transportation law in France was passed on February 5, 1942. It has been expanded over the years to include newer forms of transportation, and 15 new decrees were issued on June 7, 1988, to strengthen regulations concerning toxic and corrosive substances.

PUBLIC HEALTH AND SAFETY—*continued*

3931. Public health use of DDT occurred in Italy in 1943 during World War II when U.S. military forces used the compound to eliminate typhus-carrying lice and other pests. Within a few years, however, mosquitoes and flies in the treated areas had become immune to DDT.

3932. Use of penicillin in general clinical practice occurred in Great Britain in 1944. Discovered by the British scientist Alexander Fleming in 1928, this "miracle drug" proved widely effective and led to the quick discovery of many other antibiotics.

3933. Worldwide organization concerned with civil aviation safety was the International Civil Aviation Organization (ICAO), established on April 4, 1947. A technical agency of the United Nations, it has headquarters in Montreal, Canada. The ICAO consists of 180 member nations. It concerns itself with many aviation safety issues, including the environmental impact of airplane emissions.

3934. Public health agency of the United Nations was the World Health Organization (WHO) established on April 7, 1948, in Geneva, Switzerland. Among its many duties, WHO researches pollution; devises standards for sanitation, nutrition, drugs, and vaccines; promotes health education; helps nations to develop health services and combat epidemics; and keeps health statistics.

3935. National Health Service in Great Britain was established in 1948. The service initially was free to patients; charges for prescriptions, dental treatment, and eye care were imposed in 1951.

3936. Mass fatalities from eating poisoned fish and shellfish began occurring around Minamata Bay, Japan, in 1953. The deaths were traced to discharges into the bay, from 1932 to 1968, of methyl mercury chloride, which built up in the tissues of the fish and shellfish that constituted a large part of the local diet. Although the mercury appeared to be harmless to the shellfish, it affected the central nervous systems of other species; human deaths and disabilities from mercury poisoning continued for decades.

3937. Mass polio vaccination campaign began in 1955. Dr. Jonas Salk's vaccine proved so effective that polio cases in western Europe and the United States dropped from 76,000 in the first year of the campaign to fewer than 1,000 in 1967.

3938. Food additive regulation passed by the U.S. Congress was the 1958 "Delaney Amendment" to the Federal Food, Drug, and Cosmetic Act, originally passed in 1938. The new amendment stated that "no additive shall be deemed safe if it is found to induce cancer when ingested by man or animal." Such general language later posed problems for enforcers.

3939. International campaign to eradicate malaria began in 1958. More than $400 million over five years went to a series of unsuccessful attempts to eliminate the disease in Southeast Asia and Africa.

3940. Automobile safety warning having national impact was Ralph Nader's book, *Unsafe at Any Speed*, published in 1965. It particularly targeted General Motor's Chevrolet Corvair as dangerous. The automotive industry attempted to refute the charges of building unsafe cars, and GM hired a detective to examine Nader's background. Within a year, however, the U.S. Congress enacted the National Traffic and Motor Vehicle Safety Act, and soon thereafter production of the Corvair ceased.

3941. Claim by the Centers for Disease Control that malaria had been eradicated in the United States came in September 1966. The CDC also declared that typhoid, infantile paralysis, and diphtheria were close to eradication.

3942. Environmental health organization of significance started by the U.S. government was the Division of Environmental Health Sciences, established in 1966 in Durham, NC, as part of the National Institutes of Health. Its mission was to understand the impact of environmental exposures on human health and disease. In 1969, it was renamed the National Institute of Environmental Health Sciences (NIEHS). It conducts biomedical research programs, conducts prevention and intervention efforts, and provides community education, training, and technology.

3943. Occupational safety coverage for U.S. workers was introduced in 1970 with the passage of the Occupational Safety and Health Act. The act set up the Occupational Safety and Health Administration as an agency of the U.S. Department of Labor.

3944. Legislation in the United States limiting lead content in paint and providing funds for cleanup was enacted on December 31, 1971. The measure provided $30 million over two years to study the problem, to remove lead paint, and to treat people for lead poisoning.

3945. Chemical and oil waste law enacted by the Danish government was the Law on the Disposal of Oil and Chemical Wastes passed on May 27, 1972. Any person or organization dealing with chemical waste was required to ensure there would be no pollution of the air, soil, and water. The government had to be informed about the amount of waste chemicals or oil being produced by any person or firm. In 1983, Denmark passed the Chemical Waste Sites Act, which required waste sites to be cleaned up and regulated the disposal of chemical wastes.

3946. Public health hazard causing the U.S. government to buy a town was discovered in November 1972 in Times Beach, MO. Soil tests revealed that the town was being poisoned by dioxin in thousands of gallons of oil used in 1971 to spray roads for dust control. Soon afterward, animals had died and children had become ill. In February 1983, the U.S. government decided to purchase Times Beach for $33 million and relocate the residents. The town was officially closed in April 1985. Since then, former residents with cancers and other diseases have claimed the dioxin was the cause, but lawsuits against the company, Syntex, have had little success because of the lack of conclusive evidence.

3947. Pollution-related national health compensation law was passed in Japan in 1973. The Pollution-Related Health Injuries Compensation Law provided compensation of up to 80 percent of the salary of a worker whose health was damaged by pollution. Medical care was free and victims could participate in a rehabilitation program. Several Japanese cities had previously covered medical costs and living expenses for pollution victims.

3948. National government publication emphasizing risk factors and lifestyle in public health was the Lalonde Report in Canada in 1974. The report recognized the interaction among lifestyle, human biological, environmental, and health care factors as determinants of health.

3949. Nuclear power organization of the U.S. government to emphasize public health was the Nuclear Regulatory Commission (NRC) established by the Energy Reorganization Act of 1974. It is responsible for seeing that nuclear power plants and nuclear materials do not harm public health and safety and the environment. It also does research in such areas as radiation health risks and enforces regulations concerning high-level radioactive waste.

3950. Ebola virus outbreak was reported in August 1976 in Zaire (now Congo). The disease soon spread, and there was no known treatment; in one group of family and friends, 18 of 21 people infected with the virus died.

3951. Law in the United States to specifically target toxic substances was the Toxic Substances Control Act passed by Congress in 1976. It sought to eliminate substances newly recognized as dangerous, such as chlorofluorocarbons (CFCs). *Toxic* meant a substance posed a threat to human health or the environment. The law gave the Environmental Protection Agency the right to test dangerous chemicals and to limit or ban their manufacture. Previously, bans had only been allowed after damage had occurred.

3952. Large-scale reduction in levels of benzene permitted in U.S. factories was ordered by the Occupational Safety and Health Administration (OSHA) in 1977. Benzene is a toxic product of coal tar with many industrial applications. OSHA ordered a 90 percent reduction in allowable levels.

3953. Chemical accident prevention law in France was the Decree of September 21, 1977, controlling facilities that produce, use, or store chemicals. It required a technical report on all possible dangers of the work, details of dangers that might arise in an accident, measures taken to reduce accidents, and public emergency plans and rescue plans created by the applicant. In case of an accident, thc law required notification of the Regional Directorate of Industry and Research (DRIR).

3954. Dangerous wastes law passed by the European Economic Community (EEC) was Directive 319, issued in 1978, requiring member states to dispose of 27 different types of toxic or dangerous wastes without harming the environment. EEC nations were instructed to keep and publish their plans for the disposal of such wastes and to license their waste-disposal plants.

3955. Federal ban of lead paint in the United States was imposed in 1978. The U.S. Consumer Product Safety Commission prohibited the use of paints containing more than 0.06 percent lead in buildings and on products accessible to children.

3956. Comprehensive hazardous waste cleanup program by the U.S. government was the Comprehensive Environmental Response, Compensation, and Liability Act (CERCLA) passed by Congress in 1980. The act, administered by the U.S. Environmental Protection Agency, is better known as Superfund. It called for strate-

PUBLIC HEALTH AND SAFETY—*continued*

gies and technologies needed to identify and clean up sites of uncontrolled hazardous waste, such as landfills and nuclear power plants. It also provided for studies of the impacts of hazardous waste sites and emergency releases on the surrounding environment (communities, ecological systems, etc.), as well as ways to minimize the risk from exposure to those contaminants.

3957. Firm evidence of the outbreak of AIDS in the United States emerged in 1981. By 1993, the virus that causes AIDS affected 1.5 million people and the disease cost the federal government $12 billion a year for research, drug development, education, and treatment.

3958. Hazardous substances protection law in Germany was the Chemical Substance Law implemented on January 1, 1982, in the Federal Republic of Germany (West Germany). It was designed to protect the general public, workers, and the environment from the damaging effects of hazardous substances. The person or company dealing with the substance was required to notify the Federal Office for Occupational Health so it could be classified. The government was given the power to ban or limit the substance, as well as the power to issue fines and penalties.

3959. Toxic and hazardous wastes national policy of the Portuguese government was Decree Law No. 49/83, passed in 1983. It also covered public works necessary for utilizing wastes and for preserving the quality of the environment. The law placed these concerns under the General Directorate for Environmental Quality, part of the Environment Ministry created on April 7, 1981.

3960. Hazardous waste transportation law passed by the European Economic Community (EEC) was Directive 631 issued in 1984. It required a certificate to accompany any transportation of dangerous waste across national borders. Such a movement could only take place when it was explicitly agreed upon by the countries. The effective date of the directive was to be October 1, 1985, but did not occur until January 1989. A 1986 amendment to the directive regulated the export of dangerous wastes to third countries.

3961. Carbon tetrachloride ban in the United States took effect on December 31, 1985. The U.S. Environmental Protection Agency said the substance, sometimes used to remove grease and oil from machines, caused liver cancer and other health problems in humans.

3962. Asbestos legislation adopted by the U.S. government was the Asbestos Hazard Emergency Response Act passed in 1986 by Congress. It required school districts to locate and manage asbestos in their buildings, and it created strict regulations for asbestos removal. In 1979, the Environmental Protection Agency (EPA) had issued its "Orange Book" on the dangers of asbestos, saying the substance should be removed from buildings. However, the EPA's "Purple Book" of 1985 said asbestos should be managed rather than removed, and its "Green Book" of 1990 decided the asbestos risk to health was low.

3963. Asbestos mine rehabilitation in South Africa began in 1986 when the government assigned the task to Potchefstroom University. The country had 134 abandoned mines and 400 open asbestos dumps, and the work involved contouring the sites and planting shrubs over them. About $7.3 million was spent on the project during the first 12 years, and estimates of the final cost range to $25 million or more. The mines began closing in the 1960s when the dangers of asbestos became apparent. South Africa has the world's highest rate of mesothelioma, the fatal cancer caused by asbestos, and the government's National Center for Occupational Health estimates that one in every 500 South African men has a risk of getting the disease, and one in 100 in former mining areas.

3964. Hazardous chemical public disclosure law in the United States was the Emergency Planning and Community Right-to-Know Act that was part of the Superfund Amendments and Reauthorization Act (SARA) of 1986. It required chemical plants to provide information about their inventories of chemicals considered hazardous to community groups under the Occupational Safety and Health Administration's Hazard Communication Standard. Requirements included information for the community if any of those substances was released to the environment and an immediate public announcement of an accidental release of an extremely hazardous chemical.

3965. Hazardous waste law enacted by the Spanish government was the Hazardous and Dangerous Waste Law passed in 1986. It recognized the great risks such waste posed for public health and the environment. The law laid down preventive measures for identifying dangerous waste and for controlling its production, transportation, storage, treatment, recovery, and disposal.

3966. Nuclear power public safety organization of the South Korean government was the Korea Institute of Nuclear Safety (KINS) established by the KINS Act on February 4, 1990. It protects the public's health and safety, as well as the environment, by licensing and regulating nuclear energy.

3967. Environmental health organization of European women was Women in Europe for a Common Future established in 1992 at Utrecht in the Netherlands. It was founded by women from 15 countries during the Earth Summit in Rio de Janeiro, Brazil. The nongovernmental, nonprofit organization supports women and children in ecological disaster areas, such as the Chernobyl region of Ukraine and the Aral Sea region of Uzbekistan, in their efforts to reduce pollution and improve health.

3968. Death attributed to a "killer bee" attack in the United States occurred in 1993. Such bees killed hundreds of people in Latin America after 1957, when a Brazilian geneticist introduced aggressive African bees into the Western Hemisphere.

3969. Environmental health conference of U.S. physicians and environmentalists was "Physicians and the Environment" on February 23 and 24, 1993, in Washington, DC. Participants included more than 100 physicians, environmental leaders, federal officials, and health professionals. The conference was held under the auspices of the National Association of Physicians for the Environment (NAPE) and the American Academy of Otolaryngology–Head and Neck Surgery, Inc. It was funded by the National Institute of Environmental Health Sciences (NIEHS) and the National Institutes of Health (NIH).

3970. Designation by the U.S. government of cigarettes and smokeless tobacco as nicotine delivery systems was made in 1996. The U.S. Food and Drug Administration declaration treated nicotine as a drug.

3971. Environmental law challenged under the North American Free Trade Agreement (NAFTA) was the Canadian government's April 1997 ban on importing a U.S. gasoline additive. The law resulted from the government's decision that methylcyclopentadienyl manganese tricarbonyl (MMT) was a pollutant dangerous to public health. The Ethyl Corporation of Richmond, VA, who produced it, filed a suit that month under NAFTA's Chapter 11 asking for $350 million in compensation, declaring the additive, which improved car performance, was safe and even reduced nitrogen oxide emissions. In July, Canada agreed to end the ban, pay Ethyl $10 million, and issue a public statement that MMT was no health risk; Ethyl then dropped its suit.

3972. Appearance of the West Nile virus in the Western Hemisphere occurred in the summer of 1999 in New York City, where people were stricken with an illness that was at first diagnosed as St. Louis encephalitis. However, researchers at a Fort Collins laboratory and at the University of California at Irvine subsequently identified the pathogen from tissue samples as the West Nile virus, a type of encephalitis known in Africa and the Middle East but never before encountered in the Western Hemisphere. It is transmitted by mosquito bites and can infect people, birds, and horses, sometimes fatally; it has since spread to other states, leading to the reintroduction of pesticide spraying in affected areas. Scientists do not know how the virus crossed the Atlantic. Some have suggested that it was carried by a sick passenger or a stray mosquito aboard a commercial airliner.

S

SCENERY

3973. World Heritage site in the United States selected for environmental reasons was Yellowstone National Park, designated in 1978 for its scenic, wildlife, and recreational attributes as well as for its cultural importance as the world's first national park. The Anasazi pueblos of Mesa Verde National Park in Colorado were also listed in 1978, for their cultural importance.

3974. Natural scenery protection state law connected with advertising was passed by New York on March 28, 1865 in the form of an amendment to an 1853 law entitled "an act for the more effectual prevention of wanton and malicious mischief and to prevent the defacement of natural scenery." The amendment made painting and printing upon stones, rocks, or trees and the defacement of natural scenery in certain areas a misdemeanor punishable by a fine not exceeding $250, six months' imprisonment, or both.

SCENERY—*continued*

3975. Scenic highway in the United States was the Columbia River Highway in Oregon. The highway and its pedestrian trails were built along the cliffs above the Columbia River Gorge between 1913 and 1922 in an effort to combine complex engineering with preservation of the area's scenic beauty. The region encompassing the highway was designated a National Scenic Area by Congress in 1986. It was designated an All-American Road in June 1998.

3976. Law in the United States requiring preservation of scenery within national parks was the National Park Service Organic Act, which became law on August 25, 1916. The act directed the Secretary of the Interior and the National Park Service to manage and conserve the scenic beauty of park system lands, thus maintaining them for the enjoyment of future generations. The law made the National Park Service responsible for regulating all human interference with natural landscapes within the park system.

3977. Recommendations for protecting landscapes issued by the United Nations were agreed upon by the General Conference of the United Nations Educational, Scientific, and Cultural Organization (UNESCO) in Paris, France, on December 12, 1962. The conference recognized that natural landscapes and man-made cultural sites can directly influence the well-being of humans and wildlife. The agreement stated that such places should be protected from potential damage caused by road building, electrical lines, airports, gas stations, advertising signs, deforestation, noise, air and water pollution, mining refuse, industrial waste, and unregulated recreation. The conference recommended that damaged important landscapes be restored to their natural state wherever possible.

3978. National Scenic Riverway law in the United States was passed by Congress on August 27, 1964. The new law established the Ozark National Scenic Riverways in southeastern Missouri and preserved 134 miles of the Current and Jacks Fork rivers. The law was a precursor to the National Wild and Scenic Rivers Act of 1968.

3979. Lawsuit by a conservation group to overrule a decision by a U.S. federal regulatory agency was decided on December 29, 1965 in the case of *Scenic Hudson Preservation Conference* v. *Federal Power Commission (FPC)*. The conservation group successfully argued that the FPC had erred in approving plans by Consolidated Edison to build a huge hydroelectric plant on Storm King Mountain near Cornwall-on-Hudson, NY, without appropriately considering the scenic and ecological impact on the region.

3980. Federal law in the United States to restrict use of billboards in scenic areas resulted from the Highway Beautification Act, signed by President Lyndon Johnson on October 22, 1965. Although loopholes weakened the legislation, it restricted signs on federally funded interstate highways to designated commercial areas.

3981. National Scenic Trails in the United States were the Appalachian National Scenic Trail and the Pacific Crest Scenic Trail, both of which were authorized by Congress in 1968. The Appalachian Trail extends 2,155 miles from Mount Katahdin, ME to Springer Mountain, GA. It is administered by the National Park Service. The Pacific Crest Scenic Trail reaches from Canada to Mexico, covering 2,600 miles along the mountains of Washington, Oregon, and California. It is administered by the U.S. Forest Service.

3982. National Trails System in the United States resulted from authorization of the National Trails System Act on October 2, 1968. The act allowed Congress to designate National Scenic or Recreational Trails. The law was amended in 1978 to authorize the establishment of National Historic Trails. Individual National Trails are administered by the National Park Service, the Forest Service, or the Bureau of Land Management.

3983. Waterways in the U.S. National Wild and Scenic Rivers System were sections of the Clearwater and Salmon (both in Idaho), Feather (California), Eleven Point (Missouri), Rio Grande (New Mexico), Rogue (Oregon), Wolf (Wisconsin), and Saint Croix (Minnesota and Wisconsin) Rivers. The eight initial waterways were designated in the U.S. Wild and Scenic Rivers Act, signed by President Lyndon B. Johnson on October 2, 1968. President Johnson signed the act creating the National Trails System at the same ceremony.

3984. National Wild and Scenic Rivers System in the United States was established by Congress on October 2, 1968, through the Wild and Scenic Rivers Act. The law established a classification and usage system for designated rivers and adjoining lands under federal ownership to be administered by the Departments of Agriculture and the Interior. The system classifies rivers as wild, scenic, or recreational and permits hunting and fishing.

3985. Ohio State Scenic River was the Little Miami River in southwestern Ohio. Its botanical, geological, wildlife, and historical features led the state to designate it as a scenic river on April 23, 1969. When it was added to the U.S. Wild and Scenic Rivers System on January 28, 1980, the segment of the Little Miami from Clark County to the Ohio River became the state's first designated river in the national program.

3986. National Wild and Scenic River in New England was the Allagash Wilderness Waterway in northern Maine. The remote Allagash River and most of its contiguous ponds, rivers, and lakes were designated as part of the national system on July 12, 1970. Unlike many designated waterways whose multiple characteristics are divided into wild, scenic, or recreational segments, all 92.5 miles of the Allagash are classified as wilderness.

3987. International convention to preserve scenery, culture, and the environment was the Convention Concerning the Protection of World Cultural and Natural Heritage. The convention was adopted by the United Nations Educational, Scientific, and Cultural Organization (UNESCO) on November 23, 1972. UNESCO's resulting World Heritage program lists and works to protect natural, scenic, and cultural sites from badly managed development. Its Secretariat is located in Paris, France.

3988. National Wild and Scenic River in the southeastern United States was the Chattooga River, designated on May 10, 1974. Its 56 miles of protected river and whitewater rapids pass through three national forests (Chattahoochee, Nantahala, and Sumter) in the borderlands of North Carolina, South Carolina, and Georgia.

3989. National Scenic Trail in the Rocky Mountains was the Continental Divide National Scenic Trail, established in 1978. The rugged trail extends from Mexico to Canada and is administered by the U.S. Forest Service.

3990. World Heritage site in Africa selected for environmental reasons was Simien National Park in Ethiopia. The site was designated for the scenic beauty of its dramatic mountainous terrain and for its importance as a habitat vital to the survival of the endangered Walia ibex. The region was added to the World Heritage List in 1978 along with Ethiopia's rock-hewn churches of Lalibela, which were selected for cultural reasons.

3991. World Heritage site in North America selected for environmental reasons was Nahanni National Park in Canada. The park, which lies at the western edge of Canada's Northwest Territories, was listed in 1978 for its scenic beauty, limestone caves, and abundant wildlife in the forests surrounding the wild Nahanni River.

3992. National Wild and Scenic River in the Mid-Atlantic region was the Delaware River. Two segments of the river totaling more than 100 miles—the Upper Delaware on the New York–Pennsylvania border and the Delaware Water Gap in New Jersey and Pennsylvania—were designated on November 10, 1978 for their recreational, geological, and historical value.

3993. U.S. National Wild and Scenic River in the Chihuahuan Desert was a 196-mile portion of the Rio Grande that runs from Big Bend National Park to the Terrell–Val Verde county line. The river's spectacular canyons and the surrounding Chihuahuan Desert ecosystem make it a valuable scenic, recreational, scientific, and natural resource.

3994. World Heritage site in Asia was Sagarmatha National Park in central Nepal, home of Mount Everest, the world's highest mountain. Although the park provides habitats for endangered species like the snow leopard, it was listed in 1979 primarily for its unique mountain scenery.

3995. World Heritage site in South America chosen for environmental reasons was the Galápagos Islands, which are administered by Ecuador. The islands were listed in 1979 under all four criteria by which nature-related World Heritage sites may be designated: unique representations of a stage in the Earth's history, ongoing ecological or biological processes, scenic beauty, and vital habitat to endangered species.

SCENERY—*continued*

3996. World Heritage sites in Europe chosen for environmental reasons were designated in 1979. Croatia's Plitvice Lakes National Park was chosen for the beauty and unique character of the watershed created by its travertine barriers. Macedonia's Lake Ohrid was chosen as a scenic attraction, although its surrounding region was listed for cultural and historical reasons in 1980. Belovezhskaya Pushcha, the forest range along the border between Belarus and Poland, was also chosen for its scenery and natural attributes, which include endangered wildlife.

3997. National Scenic Trail in the United States devoted to glacial terrain was Ice Age National Scenic Trail, established in 1980, which features 1,000 miles of moraine hills left behind by retreating Ice Age glaciers in what is now Wisonsin. The trail is administered by the National Park Service.

3998. National Scenic Trail to link the eastern and western United States was the North Country National Scenic Trail, established in 1980. The trail extends from New York's Adirondack Mountains to the Missouri River in North Dakota, touching the shorelines of the Great Lakes and covering a wide variety of mountain, forest, agricultural, glacial, and prairie terrain and ecosystems. The trail is administered by the National Park Service.

3999. U.S. National Wild and Scenic Rivers in Alaska were designated on December 2, 1980, as part of the Alaska National Interest Lands Conservation Act. The landmark wilderness designation law entered 25 of the state's waterways into the Wild and Scenic Rivers System. Individual rivers are managed by the U.S. Forest Service, the National Park Service, and the Bureau of Land Management.

4000. World Heritage Sites in Australia were chosen in 1981 for nature-related reasons. The Great Barrier Reef was listed for its ecological vitality and importance as a habitat for endangered marine species. Kakadu National Park in the Northern Territory was listed for its varied ecosystems, as well as its legacy of aboriginal culture. Scenery was a consideration in choosing both sites. Distinctive geology, without any scenic considerations, resulted in the choice of the third site initially listed, the Willandra Lakes Region.

4001. Pennsylvania law creating a scenic rivers system was the Pennsylvania Scenic Rivers Act. The law was passed by the state's general assembly on May 7, 1982, in an effort to promote conservation policies and protect the recreational value of Pennsylvania's rivers. Within a decade, more than a dozen Pennsylvania rivers were included in the system.

4002. National Scenic Trail in the south central United States was the Natchez Trace National Scenic Trail. While the trail is notable for its role in Native American and U.S. history, it had an earlier role as an animal track, extending across central Mississippi from the present-day site of Natchez on the Mississippi River to Nashville, TN. The trail was established in 1983 and is administered by the National Park Service.

4003. Subtropical National Scenic Trail in the United States is the Florida National Scenic Trail, designated in 1983. When completed, it will extend 1,300 miles from Big Cypress National Preserve in South Florida through Florida's three national forests to Gulf Islands National Seashore in the state's western panhandle. The trail is administered by the U.S. Forest Service.

4004. National Scenic Trail in the eastern United States is the Potomac Heritage National Scenic Trail, established on March 28, 1983. Although its 700-mile route begins in Virginia and is rich in historical associations with early U.S. history, the trail also connects the tidewater environments of the Potomac River with the Laurel Highlands of Pennsylvania. The trail is administered by the National Park Service.

4005. Ohio highway beautification program to use wildflowers began in 1984, when the state Department of Transportation and the Garden Club of Ohio planted seeds in selected highway areas. In 1988 the Ohio Department of Transportation and the Dayton-Montgomery Park District established a nursery near Germantown for the production of indigenous seeds to supply the wildflower program.

4006. National Wild and Scenic River in the Sonoran Desert was the Verde River in central Arizona. It was designated on August 28, 1984, and winds through three national forests—Coconino, Prescott, and Tonto—all of which lie within the Mazatzal Wilderness Area. The river forms a rare riparian oasis in the arid highlands of the Sonoran Desert.

4007. Bayou in the U.S. National Wild and Scenic Rivers System was Saline Bayou, that connects Saline Lake to the northernmost of the four preserves that comprise Kisatchie National Forest in Louisiana. The 19 miles of slow-moving water are a preserve for wildlife observation and canoeing. Saline Bayou was designated a Wild and Scenic River on October 30, 1986, and is managed by the U.S. Forest Service.

4008. Illinois State Scenic River was the Middle Fork of the Vermilion River. The 17 miles of waterway first designated as a scenic river by the state of Illinois in 1986 were added to the U.S. National Wild and Scenic Rivers System on May 11, 1989. The river's use as a recreational and wildlife area is managed by the Illinois Department of Conservation.

4009. U.S. National Wild and Scenic River managed in partnership with local governments was Wildcat Brook in New Hampshire's White Mountains. The brook was designated as a National Wild and Scenic River on October 28, 1988, making it the first river in New Hampshire to be included. It was the first waterway in the national system to include both private and public lands. Unlike previous additions, rivers already on public lands that were put under control of federal agencies, Wildcat Brook was also the first river for which local towns participated in the designation process. The area is now managed by the Wildcat River Advisory Commission, the U.S. Forest Service, and the town of Jackson.

4010. Scenic Byways and All-American Roads program in the United States was created by the Intermodal Surface Transportation Efficiency Act of 1991 and structured through rules given by the Federal Highway Administration in the Federal Register on May 18, 1995. A designated National Scenic Byway must be a state-designated scenic byway and must meet specific criteria, including possession of at least one of six intrinsic qualities: scenic, cultural, historic, archaeological, recreational, and/or natural significance. All-American Roads must meet National Scenic Byways criteria and possess at least two of the six intrinsic qualities, making the route a destination unto itself.

4011. U.S. National Scenic Byway in the Black Hills of southwestern South Dakota was the Peter Norbeck Scenic Byway, designated in September 1996. The roads comprising the byway route climb through the forests and granite geology of the Black Hills, passing Mount Rushmore. The route was named after Governor Peter Norbeck, a businessman and wildlife conservationist who promoted preservation projects like Custer State Park during his three terms in the U.S. Senate in the early 1900s.

4012. All-American Road in the Appalachian Mountains was the North Carolina segment of the Blue Ridge Parkway, which connects Shenandoah and Great Smoky Mountains National Parks. When the North Carolina portion of the wooded, mountainous scenic route was designated an All-American Road in September 1996, the northern segment of the Blue Ridge Parkway remained classified as a Scenic Byway by the state of Virginia.

4013. All-American Road on the U.S. Pacific coast was California's Route 1, also known as the Big Sur Coast Highway. The road winds along the redwood forests and rocky cliffs of California's Pacific Coast Highway from Carmel to San Luis Obispo. The route was designated an All-American Road in September 1996.

4014. All-American Roads in the Rocky Mountains were two Colorado roads designated in September 1996. The scenic 85-mile Trail Ridge Road/Beaver Meadow Road climbs through the glacial valleys and high peaks of Rocky Mountain National Park. The 236-mile San Juan Skyway circles through the wooded mountains of San Juan National Forest in southwestern Colorado.

4015. Former hunting path designated as an All-American Road was the Natchez Trace Parkway, which covers more than 400 miles between Natchez, MS, and Boston, TN. Throughout the Natchez Trace's history, wild habitats along the route have provided important hunting areas, agricultural land, and a trade route for both its original Indian inhabitants and later settlers. In September 1996, the Mississippi, Alabama, and Tennessee segments of the parkway were designated as three individual All-American Road systems.

SCENERY—*continued*

4016. U.S. National Scenic Byway in the Great Lakes region was the Seaway Trail Scenic Byway designated in September 1996. It includes more than 450 miles of New York State roads along the southern shores of Lake Ontario, Lake Erie, and the St. Lawrence River. Its environmental landscape includes shoreline ecological systems, agricultural land, and Niagara Falls.

4017. U.S. National Scenic Byway on an American Indian reservation was the Pyramid Lake Scenic Byway in northwestern Nevada. The route adjoins the Truckee River and the massive, geologically rich lake at the center of the Paiute tribe's Pyramid Lake Reservation. The scenic byway across the southern end of the reservation was designated in September 1996 by the Federal Highway Administration.

4018. U.S. National Scenic Byway through Louisiana wetlands was the Creole Nature Trail, designated in September 1996. Sometimes within sight of the Gulf of Mexico, the narrow state highways comprising the byway traverse the southwesternmost of the state's parishes, whose lakes, marshes, and bayous are rich with migrating and indigenous wildlife, fish, and birds.

4019. State Scenic Byway in Ohio covered 110 miles between Cleveland and Dover, OH, along the path of the 19th-century Ohio and Erie Canal. The byway was designed by the Ohio Department of Transportation and inaugurated on October 1, 1996. Ohio's Scenic Byway program was established in April 1994 to preserve unique scenic, historic, and recreation areas for their environmental and economic value.

4020. All-American Road with glacial features was State Route 410 in central Washington, also known as the Stephen Mather Memorial Parkway. The road was among the first to provide access to Mount Rainier, the largest single-peak glacier system in the lower 48 United States. The parkway is named after the first director of the National Park Service, Stephen T. Mather. It was designated an All-American Road in June 1998.

4021. All-American Road with volcanic features was the Volcanic Legacy Scenic Byway, a 140-mile route in southern Oregon's Klamath Basin. The road passes through the collapsed volcanic landscape of Crater Lake National Park and wetlands containing six National Wildlife Refuges. The route was designated an All-American Road in June 1998.

4022. U.S. National Scenic Byway below sea level was California Route 190, which was designated as the Death Valley Scenic Byway in June 1998. Descending from the fault blocks of the Panamint Mountains across the desert floor of Death Valley National Monument, the scenic route crosses some of the lowest elevations in North America. The lowest spot on the continent lies approximately 15 miles south of the byway.

4023. U.S. National Scenic Byway in Alaska was the Seward Highway, which connects Anchorage with Seward on Blying Sound. The highway, which was designated in June 1998, crosses a variety of mountainous, oceanic, and glacial habitats as it runs over 100 miles between Kenai National Wildlife Refuge and the Portage Glacier Recreation Area.

4024. U.S. National Scenic Byway in an urban area was the Grand Rounds Scenic Byway in Minnesota. The natural and scenic features of the 50-mile network of roadways and pedestrian trails include the Mississippi River and the numerous lakes within the city limits of Minneapolis. The byway was designated in June 1998.

4025. U.S. National Scenic Byway in Grand Canyon National Park was the Kaibab Plateau–North Rim Parkway, designated in June 1998. The Arizona road, also known as Route 67, joins the town of Jacob Lake in Kaibab National Forest's northern tier with the North Rim of the canyon. Together the two mountainous ecosystems present spectacular scenic views as well as a variety of geological, forest, wildlife, and botanical features.

SOIL RESOURCES

4026. Federal recommendation that areas of the San Joaquin Valley in California be taken out of agricultural use because of naturally high selenium concentrations came from scientists of the U.S. Geological Survey in 1941. The first reports of horses and cattle dying after grazing in the valley dated from 1857; parts of the valley were later determined to have elevated levels of the heavy metal.

4027. International conference on the connection between soil health and human health took place in 1991. It was sponsored by the Rodale Institute of the United States, whose founder, Robert Rodale, promoted the concept of regenerative farming.

SOIL RESOURCES—CONSERVATION

4028. Widespread use of crop rotation to preserve soil health occurred in Europe in the early 1700s. It gradually replaced the old system of planting only one type of crop and letting the field lie fallow every third year.

4029. Soil conservation advocate of prominence in the United States was Thomas Jefferson in the early 19th century. Jefferson urged farmers to adopt soil conservation methods. As early as 1775, American rivers were described as dark with mud as a result of soil erosion.

4030. Planter to recognize that some plants could put acid in the soil was Edmund Ruffin, who presented his findings at the 1818 Isle of Wight County Agricultural Society Meeting in Virginia. Through soil tests at his plantation in Virginia, he discovered that certain vegetables add acids to the soil. He neutralized the acids by applying marl, earth rich in calcium carbonate from fossilized seashells. Ruffin would later earn notoriety as the man who fired the shot at Fort Sumter in 1861 that touched off the Civil War.

4031. Broadbase terrace for farming was built in North Carolina by P. H. Magnum in 1885. A broadbase terrace consists of a wide, shallow channel dug uphill from a wide, gradually sloping embankment.

4032. Government publication of importance issued by the United States telling farmers how to protect their soil was published in 1894. *Washed Soils: How to Prevent and Reclaim Them* argued that the government should assist in soil preservation efforts.

4033. Government Soil Survey in the United States was organized in 1899. The sponsor, the U.S. Bureau of Soils, was disappointed in the early results, but the survey did train a corps of capable soil science researchers.

4034. Test to identify critical elements missing in the soil was created by Cyril George Hopkins around 1900 while working for the University of Illinois. After he identified critical elements for plant nutrition, he devised the first test for these nutrients in the soil.

4035. Comprehensive scientific effort to conserve soil and water resources on a national scale began in the United States in 1933. The federal government program, called Emergency Conservation Work, evolved into the New Deal employment-creating agency known as the Civilian Conservation Corps.

4036. Law in the United States designed to stop overgrazing and its adverse effects on the soil was the Taylor Grazing Act of June 28, 1934. Congressman Ed Taylor authored the Act, which gave the Secretary of the Interior the power to cut down on grazing and to charge fees for grazing on public lands.

4037. Effort of significance by the United States government to develop soil conservation and rehabilitation programs for the southern Great Plains during the Dust Bowl era were launched in 1935. They included crop rotation, contour plowing, and the planting of shelter belts of trees for windbreaks.

4038. Legislation linking soil conservation and commodity policy in the United States was the Agricultural Conservation Program of 1936. The program urged farmers to voluntarily shift crop acreage from soil-depleting surplus crops to legumes, grasses, and other soil-conserving crops.

4039. Two states to adopt President Franklin Roosevelt's Standard State Soil Conservation District Laws were Arkansas and Oklahoma in 1937. They were the first of 23 states to adopt the laws that year as a result of the President's letter to all state governors.

4040. Soil conservation district in the United States was the Brown Creek Soil Conservation District created in Anson County, NC, in August 1937. By 1945, more than 1,300 districts had been created nationwide to manage areas of high erosion.

4041. Municipal composting plant in the United States was established in Oakland, CA, in 1950 by Dr. Ehrenfried Pfeiffer, a follower of the German social philosopher Rudolf Steiner.

4042. Issue of the periodical *Acres U.S.A.* appeared in 1971. The monthly, published by Charles Walters of Kansas, championed soil conservation through organic farming.

SOIL RESOURCES—DESERTIFICATION

4043. Use of the term *desertification* was in 1949 by André Aubreville, a French forester, to describe African deforestation.

4044. Emergence of desertification as an international issue occurred between 1968 and 1973 when the Great Drought in the Sahel region of North Africa called attention to the phenomenon. Desertification happens when the soil is no longer able to retain moisture, allowing the desert to encroach on croplands.

SOIL RESOURCES—DESERTIFICATION—*continued*

4045. International meeting about desertfication the United Nations Conference on Desertification, was held in Nairobi, Kenya from August 29 to September 9, 1977.

SOIL RESOURCES—EROSION

4046. Abandonment of villages because of deforestation and resulting soil erosion occurred circa 6000 B.C. in central Jordan. Damage to the soil led to declining crop yields and, eventually, famine, forcing villagers to relocate.

4047. Signs of large-scale erosion because of deforestation in Greece appeared circa 650 B.C. Rising population and expanded settlement stripped the Attica hills of trees.

4048. Terraced or horizontal plowing in the United States was introduced by Thomas Jefferson in the early 19th century. After noticing that rain often washed soil and crops down hillsides, Jefferson started plowing his fields with this new horizontal method at his Monticello, VA, farm. In this method, hillside fields are terraced slightly so that soils do not wash away with heavy rains.

4049. Effective promoter of tree planting to prevent soil erosion on the U.S. Great Plains was Julius Sterling Morton, a Nebraska newspaper editor. His campaign led to the first Arbor Day, celebrated on April 10, 1872.

4050. Introduction of kudzu into the United States for erosion control occurred in 1876. The plant flourished in the south and was also used as hay and forage for livestock.

4051. Substantial U.S. government funding for a study of causes of soil erosion and methods of erosion control was approved in 1929. Congress appropriated $160,000 for the study.

4052. Use of the term *Dust Bowl* to describe drought-seared and wind-eroded U.S. southern Great Plains occurred in the early 1930s. In some places, 3–4 inches of topsoil were blown away, destroying the farm economy and creating a serious air pollution problem.

4053. Watershed erosion control demonstration project was established in the Coon Creek Valley in Wisconsin in 1933. This New Deal project alleviated many of the erosion problems until the 1970s. It was the first of many projects completed by the Civilian Conservation Corps.

4054. National soil erosion survey in the United States was taken by the U.S. Soil Erosion Service in 1934. The survey began just as the Dust Bowl began to hit the prairie states.

4055. Federal agency charged with preventing soil erosion was the Soil Conservation Service established in 1935 by the administration of Franklin D. Roosevelt. The impetus for its founding was the Dust Bowl disaster. Set up under the Soil Conservation Act of 1935, the agency moved to check soil erosion and replenish the soil following the Dust Bowl years in the Great Plains.

4056. U.S. Soil Conservation Service chief was the eminent soil scientist Hugh H. Bennett, appointed on April 27, 1935. The nation's leading advocate of soil conservation, he served until November 1951.

4057. Federal legislation of significance to take erodable U.S. cropland out of production was approved in 1985. The Food Security Act, which Congress passed on December 23, 1985, contained a series of conservation measures for agricultural ecosystems. By 1991, erodable croplands had been reduced by more than one fifth.

SOIL RESOURCES—FERTILITY AND FERTILIZERS

4058. Widespread cultivation of legumes that fixed nitrogen and thus improved soil fertility occurred in about 1300 in Flanders (Belgium).

4059. Discovery that tricalcium phosphate could be converted into a soluble plant fertilizer by treatment with sulfuric acid was by Sir John Lawes in 1840. This product of rock and bones was labeled superphosphate.

4060. Scientist to be dubbed the "father of chemical agriculture" was Justus von Liebig, in the 1840s. He developed a new theory of soil science based on the use of potassium salts as fertilizer.

4061. Fertilizer factory was established in England in 1842 by Sir John Lawes. He was thus the founder of the artificial fertilizer industry.

4062. Law in the United States requiring ingredient labels on fertilizers was passed in the 1850s in Connecticut. Samuel W. Johnson lobbied for the law after tests showed that fertilizers costing $4 contained only 35 cents' worth of ingredients.

4063. Discovery of a laboratory process for extracting liquid ammonia from free nitrogen in the air was by Fritz Haber, a German chemist, in 1905. Ammonia is an important constituent of chemical fertilizers.

4064. Major opposition to chemical fertilizers came from Sir Albert Howard, a British council officer in India in 1916. His research demonstrated that animal and vegetable wastes maintained healthy soil, so there was no need for chemical fertilization. Sir Albert was the founder of the modern organic movement.

4065. Woman to advocate organic gardening was Lady Evelyn Balfour, who published *The Living Soil* in 1943. Intrigued by the ideas of Sir Albert Howard on composting and Sir Robert McGarrison on health and the soil, she began her own research on her farm in Haughley, England, which led to the publication of her book.

4066. Published warnings against using chemical fertilizers appeared in Sir Albert Howard's book *The Soil and Health*, published in England in 1947. Howard warned that chemical fertilizers lead to many diseases in plants, animals, and humans.

4067. Standards for organic farming in Great Britain were printed in the October 1967 issue of *Mother Earth, Journal of the Soil Association* and included acceptable organic fertilizers like pig bristles, sheep's trotters, manure, and fish meal.

SOIL RESOURCES—RESEARCH

4068. Scientist to explain the reasons for the value of adding manure and ashes to the soil was Sir Humphry Davy, of England, in 1813. He called for chemical analyses to determine the causes of soil sterility.

4069. Establishment of the concept that soils are individual and natural and have their own form and structure occurred during the 1870s as a result of work by a school of Russian soil scientists. The notion finally reached the West in 1914, through a Russian scientific text by K. D. Glinka, published in German.

4070. Renowned scientist to ennoble the earthworm as one of the most important animals in the history of the world was Charles Darwin. His classic *The Formation of Vegetable Mould through the Action of Worms with Observations on Their Habits* appeared in 1881.

4071. Soil survey of a state in the United States was conducted by Cyril George Hopkins of Illinois. This survey, which began in 1900, resulted in the "Illinois System" of soil analysis, which listed 10 plant foods and identified calcium, magnesium, potassium, nitrogen, and phosphorus as critical elements often lacking in soils. Other states soon adopted this system to survey soil.

4072. Aerial surveys of large parts of the United States to support soil conservation programs were conducted in the 1930s. Similar surveys were undertaken in support of forest management programs.

4073. Darwin Centenary Symposium on Earthworm Ecology was in 1981, 100 years after the publication of Charles Darwin's classic work on earthworms and the formation of vegetable mould. The symposium aimed to correct the gross neglect and mistreatment of the earthworm in modern farming practice.

SOLAR ENERGY

4074. Written record of the use of a burning mirror in China occurred in *Chou Li*, a book of ceremonies written circa A.D. 20 The author describes receiving brilliant fire from the sun through concave mirrors.

4075. Use of solar architecture by the Greeks occurred in the fourth century B.C. In response to the lack of wood at the time, the Greeks began building homes oriented to the southern horizon to take advantage of the sun's rays in the winter and avoid the sun's heat in the summer.

4076. Greek city to include solar designs in the original plans was Olynthyus on the north hills of Greece in the third century B.C. The city was built on a flat plateau. The street plan was optimized to take advantage of solar energy and all homes were built with a southern exposure. The northern walls were thick and most often without windows to protect from the cold winter winds.

4077. Western man to build a parabolic or "burning" mirror was Doisitheius, a Greek mathematician, in the third century B.C. He discovered that this type of mirror focused the sun's rays and created high temperatures at a fixed point.

4078. Geometric proof of the focal properties of parabolic and spherical mirrors was given by Greek mathematician Dicoles in his book *On the Burning Mirror* in the second century B.C.

4079. Greenhouses known in the West was built by the Romans in the first century Emperor Tiberius's desire for cucumbers year round prompted his kitchen gardeners to place transparent glass cold frames over the plants in winter.

4080. Public baths in the West to use solar energy for heat were built by the Romans in the first century. The Roman baths faced the winter sunset and often had large transparent windows looking south or southwest. They also used sand floors to absorb the sun's heat during the day and release it in the evening.

SOLAR ENERGY—*continued*

4081. Solar machine was created by Hero of Alexandria in the first century. Hero created a solar siphon to move water from one container to another. Heated air in one container expanded and pushed water to the next.

4082. Record of transparent glass windows in the West was found in a letter written in Rome by Seneca in A.D. 65. Transparent glass windows were a major innovation in solar heating because they allowed the sunlight in and trapped the heat.

4083. "Sun rights" laws to guarantee buildings access to solar warmth and light were contained in the Justinian Code, set forth by legal experts appointed by the Byzantine emperor Justinian I. The first statutes of the Justinian Code appeared in 534 and gave uniformity to the civil laws of the earlier Roman Empire. The laws were a model for later European laws.

4084. Evidence of solar architecture was found in the remains of the southwest American Anasazi Indian dwellings of the 11th and 12th centuries.

4085. Record of using solar energy to make perfume was written by Adam Lonicier in 1561. Lonicier describes European alchemists filling a vase with water and putting flowers in it. The vase was then placed in the focal point of a burning mirror. The heat caused the flower essences to diffuse into the water.

4086. Practical modern greenhouse was designed by Jules Charles, a French botanist, in 1599. Charles built the greenhouse in Leiden, in the Netherlands, and used it to grow medicinal tropical plants.

4087. Fruit walls were used by the French and English in the 17th century. Branches of fruit trees were nailed to sun-warmed walls so the fruit would ripen faster. At first walls were perpendicular to the ground and faced south. Later improvements included facing the walls southwest and tilting them at a 45-degree angle. Fadio de Doillier, a Frenchman, suggested the tilting in his 1699 article titled "Fruit Walls Improved."

4088. Experiments to measure the amount of heat trapped by glass were conducted in 1767 by Horace Bénédict de Saussure, a French-Swiss scientist working in Geneva. Saussure conducted tests with five nested glass boxes on a black tabletop. The results showed progressively higher temperatures inside each box, with the outermost box the coolest and the innermost the warmest.

4089. Solar collector was built by Swiss scientist Horace Bénédict de Saussure in 1767. He built a nest of glass boxes or "hot box," which effectively trapped solar-produced heat.

4090. Solar water heaters on record were metal water tanks painted black that became popular in the late 19th century. During the last part of the 19th century in the southwestern United States, these water tanks were placed outside to receive the most sun. Extreme temperatures and a loss of heat when the sun was not out were the two major problems with this method.

4091. Brandy distilled by the heat of the sun was made by Augustin Mouchot in Tours, France in 1861. Mouchot boiled two quarts of wine by means of a solar still and in a few hours had brandy.

4092. Combination of a glass heat trap and a burning mirror was created by Augustin Mouchot in 1861 in Tours, France. He improved upon de Saussure's glass hot box by making a glass-shaped bell that the sun could shine into all day. Even with this improvement, the small hot box would not create enough heat to run a machine. So Mouchot thought to aim a burning mirror into the hot box. This link of the two inventions led to the creation of the solar oven, solar still, and solar pump.

4093. Solar motor was invented in 1865 by French mathematics instructor Augustin Mouchot, who used reflected solar rays to generate enough heat to power a small steam engine.

4094. Solar-powered steam engine in the United States was created by American inventor John Ericsson in 1870, four years after Mouchot's, and used metal tubes to collect the heat instead of a hot box.

4095. Parabolic trough solar reflector was invented in 1884 by American engineer John Ericsson, inventor of the screw propeller and designer of the Civil War ironclad battleship *Monitor*. Parabolic trough reflectors were less expensive and could be aimed at sunlight more easily than round reflectors, but Ericsson died before they could be made commercially available. He is also credited with inventing the first solar hot-air engine in 1872.

4096. Nonreflecting solar motor was invented by French engineer and refrigeration pioneer Charles Tellier in 1885. Using a "flat-plate" solar collector, Tellier's device heated pressurized ammonia with sufficient force to power a water pump, without the accompanying high temperatures produced by earlier solar motors utilizing reflected sunlight.

4097. Commercially marketed solar water heater was the Climax, patented in 1891 by American inventor Clarence Kemp and first sold in Baltimore, MD. It was the first water heater capable of both producing and storing hot water with any degree of efficiency.

4098. Solar power company was the Solar Motor Company. The research venture was formed in Boston by American inventor Aubrey Eneas, who later moved to California to continue his work. In 1901 Eneas demonstrated an enormous, 33-foot reflective solar collector in Pasadena, using it to power an irrigation engine with solar-heated steam. After establishing the Solar Motor Company of California in 1903, Eneas had little success trying to sell similar solar-powered irrigation systems to farmers.

4099. Solar power plant using a low-boiling-point liquid to power an engine was developed by Henry E. Willsie and John Boyle of the Willsie Sun Power Company. During the spring of 1904 in their St. Louis power plant they used sun-warmed water to vaporize ammonia, which has a low boiling point. The ammonia vapor ran a six-horsepower engine.

4100. Solar-powered energy plant capable of operating at night was the Willsie Sun Power Company in Needles, CA, designed by American inventors Henry E. Willsie and John Boyle. The plant began operating at night in 1908. Using flat-plate solar collectors, their system stored enough solar-heated water to operate an engine after the sun had set.

4101. Solar water heater and hot water storage system to provide high-temperature water around the clock was patented by William J. Bailey in 1909. These heaters with their insulated storage tanks maintained water at a consistent temperature through the night, losing only one degree of temperature an hour. The Day and Night, as it was called, was first sold in a suburb of Monrovia, CA.

4102. Research results in the United States on the use of sunlight through windows to generate heat were published in William Atkinson's book, *Orientation of Buildings or Planning for Sunlight.* Atkinson, a Boston architect, studied the temperatures inside hot boxes facing different directions during 1910 and came to the common conclusion that the south-southwest side of buildings should be open to the sun with transparent glass. Though his conclusion was not new, the scientific evidence to support it was.

4103. Modern experiments on the effects of sunlight on buildings were conducted by French housing official Augustin Rey. In 1912 Rey determined that long apartment buildings should face south and be spaced apart two and one-half times their height to avoid shadows.

4104. Modern planned solar community in Switzerland was Neubuhl, built in the 1930s. The community used many principles of solar architecture including glass on the south walls, appropriate space between buildings, and a south/southwest orientation.

4105. Completely sun-oriented residential community in the United States was built by Chicago real estate developer Howard Sloan. In the early 1940s he built a 30-house development near Chicago called Solar Park.

4106. Solar-heated school was the Rose Elementary School in Tucson, AZ. The system, which used fans to circulate solar heat collected by the school's roof, was designed by solar architect Arthur Brown.

4107. House completely heated by solar energy in the United States was occupied on December 24, 1948, in Dover, MA. A unit consisting of a black sheet-metal collector behind two panes of glass trapped the sun's heat, which was stored in a "heat bin" containing an inexpensive sodium compound. Electric fans then blew the stored heat through vents as needed. The house was designed by Eleanor Raymond and the heating system was developed by Maria Telkes. The experimental house was sponsored by Amelia Peabody.

4108. Photovoltaic cell or solar battery to work effectively was announced on April 25, 1954. Made of specifically treated strips of silicon, the battery needed no fuel other than the light of the sun. Since it had no moving parts and nothing in it was consumed or destroyed, theoretically it was possible for it to last indefinitely. The battery was invented by Gerald Leondus Pearson, Calvin Souther Fuller, and Daryl M. Chapin at the Bell Telephone Laboratories, New York City.

4109. House with solar heating and radiation cooling in the United States was placed in operation on January 15, 1955, in Tucson, AZ. The system, built at a cost of nearly $4,000 for the labor and materials, consisted of a large slanting slab of steel and glass, which converted the sunlight into heat. Fans and ducts channeled the heat into the house. The same fan controls and ducts were used for summer cooling. The house was built by Raymond Whitcomb Bliss.

SOLAR ENERGY—*continued*

4110. Solar-powered car was demonstrated to the public on August 31, 1955, at the General Powerama in Chicago, IL. Built by William G. Cobb of the General Motors Corporation, it was propelled by 12 photoelectric cells made of selenium, which converted light into electric current. The current powered a tiny electric motor with a driveshaft connected by a pulley to the rear axle of the 15-inch "sunmobile."

4111. Solar-heated office building was built in Albuquerque, NM, in 1956 by the engineering firm of Bridgers & Paxton.

4112. Satellite instruments powered by solar energy were aboard the *Vanguard 1*, a test satellite launched from Cape Canaveral, FL, on March 17, 1958 as part of the Vanguard space program. The use of solar cells to power the instruments also gave scientists an opportunity to study the lifespan of solar-powered equipment in outer space.

4113. Flight of a solar-powered aircraft took place on November 4, 1974, at NASA's Dryden Flight Research Center in Edwards, CA. The unpiloted craft, *Sunrise II*, was launched from a catapult, then remote-controlled from the ground. *Sunrise II* was designed by Robert J. Boucher.

4114. International holiday to promote solar energy was Sun Day, celebrated on May 3, 1978 with events ranging from public demonstrations to professional conferences discussing solar power as a viable energy source.

4115. U.S. law exempting alternative energy producers from state and federal regulation was the Public Utility Regulatory Policies Act (PURPA) of 1978. Meant to encourage development of alternative energy sources like solar and wind power, the law also required larger existing utilities to buy power from smaller alternative energy producers at the same rate it would cost to produce the electricity themselves.

4116. Photovoltaic (PV) village was developed in 1979 by the National Aeronautics and Space Administration at Schuchuli, AZ. The site on the Papago Indian Reservation was the first of a growing number of worldwide PV village programs whose technology is especially appealing in rural and developing areas where supplies or pollution from conventional fossil fuels are problematic.

4117. Use of solar energy at the White House was a solar water heating system installed by President Jimmy Carter in June 1979 as part of his policy of promoting renewable energy resources. The system was powered by solar panels on the White House roof. It was removed by Carter's successor, Ronald Reagan.

4118. Solar-powered aircraft that could carry a pilot was designed by Paul MacCready in 1980, as part of a NASA-sponsored program.

4119. Manned solar-powered aircraft flight was a brief test flight by 13-year-old Marshall MacCready in the *Gossamer Penguin* on May 18, 1980. MacCready's father had designed the craft. The boy was chosen as a test pilot because of his light weight, which was also the criterion for charter pilot and schoolteacher Janice Brown, who made the first public demonstration of the aircraft on August 7, 1980, at NASA's Dryden Flight Research Center in Edwards, CA. She remained in the air for more than 14 minutes. The *Gossamer Penguin* received 541 watts of power from 3,920 solar cells.

4120. Power plant using solar cells was dedicated on June 7, 1980, at Natural Bridges National Monument, UT, by Governor Scott Matheson. The photovoltaic system, located 38 miles from the nearest power line, mounted 266,029 solar cells in 12 long rows and output 100 kilowatts to supply electricity for 6 staff residences, maintenance facilities, a water sanitation system, and a visitors' center. The $3 million plant was a joint venture of the National Park System, the Department of Energy, and the Massachusetts Institute of Technology's Lincoln Laboratory.

4121. Solar automobile race ran on November 1, 1987, in Australia. The Australian World Solar Challenge started in Darwin and finished 1,950 miles away in Adelaide. Paul MacCready's GM SunRaycer won the race in 5.25 days, averaging 41.6 mph.

4122. Feasibility studies of solar energy use in U.S. government buildings resulted from Executive Order 12902, signed by President Bill Clinton on March 8, 1994 and was entitled "Energy Efficiency and Water Conservation at Federal Facilities." It ordered audits that would produce recommendations for the acquisition and installation of energy conservation measures, including the use of solar power. A later Executive Order, 13101, called "Greening the Government Through Efficient Energy Management," which Clinton signed on June 3, 1999, promoted use of solar energy systems in federal buildings. Under its renewable energy provisions, the order set goals of installing 2,000 solar energy systems at federal facilities by the end of 2000 and 20,000 by 2010.

4123. Solar-powered vehicle to travel into deep space was *Deep Space 1* (DS-1), a U.S. spacecraft launched by the National Aeronautics and Space Administration (NASA) at Cape Canaveral, FL, on October 24, 1998. Solar energy powered the small spacecraft's ion propulsion engine, using a technology its designers hoped would be far more efficient than earlier rocket fuel systems. DS-1 was the first vehicle in NASA's New Millennium program and the first to carry an autonomous navigational system. Its mission was to observe comets and asteroids.

SOLID WASTE

4124. "Tin" can was patented by English merchant Peter Durand in 1810. Expanding on earlier research by French inventor Nicolas Appert, Durand developed an airtight, tinplated, wrought-iron container used to preserve food. Durand's cans became available in the United States in 1818.

4125. Commercially available plastic was celluloid, a nitrocellulose and camphor synthetic invented and manufactured by John Wesley Hyatt in Albany, NY, in 1868. While the invention of plastic revolutionized industry and the quality of life in industrialized nations, its non-biodegradable nature made it a potential source of troublesome waste.

4126. Rubbish-sorting facility in New York City began operation in 1898 under orders from Colonel George Edwin Waring, head of the city's Department of Street Cleaning. Colonel Waring's other innovations included new programs for recycling, street-sweeping, and the establishment of a uniformed municipal cleaning and collection force.

4127. Commercially marketed beer cans in the United States were sold in Richmond, VA, on January 24, 1935 by the Krueger Brewing Company, a Newark, NJ–based brewery. Disposable metal beer cans quickly became popular and were a major source of waste and litter until the advent of recycling.

4128. Styrofoam was invented in 1944 by Ray McIntyre, a Dow Chemical Company engineer, in Midland, MI. First marketed as insulation material, the polystyrene foam was later used by fast-food restaurants as packaging until it fell into public disfavor in the 1980s.

4129. Disposable diapers were marketed in 1961 by the Procter & Gamble Company of Cincinnati, OH. The convenience of the diapers made them a commercial success, but created a significant new source of non-biodegradable waste.

4130. Electric power in the United States from municipal garbage as a boiler fuel was generated on April 4, 1972, at the Union Electric Company's MeramecPlant in St. Louis, MO. In the first month, 200,000 kilowatt-hours of electricity were generated by shredding refuse and burning it with coal to heat the boiler.

4131. Agency in the United States to regulate solid waste was the Office of Solid Waste (OSW), created with the passage of the Resource Conservation and Recovery Act, enacted on October 21, 1976. The OSW is the Environmental Protection Agency's main office for national management of hazardous and nonhazardous waste through reduction, prevention, education, and cleanup programs.

4132. Comprehensive waste management legislation in the United States was the Resource Conservation and Recovery Act, enacted October 21, 1976. Issues covered by the act included open dumping, hazardous waste, regulation of landfills, and recycling. Provisions of the act addressed solid waste both as a harmful influence on the environment and as a potential source of alternative fuels.

4133. Successful solid waste program in the Philippines was the Integrated Solid Waste Collection System instituted in September 1989 by the city of Olangapo. The comprehensive program relied on an aggressive public relations campaign to establish community awareness, followed by volunteer cleanup drives, health education, and strict collection, antilitter, and recycling rules.

4134. Law in the United States to require federal government agencies to obey environmental laws was the Federal Facility Compliance Act of 1992 (FFCA). Passed by Congress on October 6, the law was designed to ensure compliance by federal and federally funded facilities run by agencies like the Department of Defense and Department of Energy, which are capable of producing solid waste, pollutants, radioactive matter, and other hazardous materials. The law allows states and the Environmental Protection Agency to sue and fine such facilities if they are not following environmental regulations.

SOLID WASTE—*continued*

4135. "Universal Waste Rule" in the United States was finalized by the Environmental Protection Agency with publication in the Federal Register on May 11, 1995. By identifying and less stringently regulating the processing of materials defined as Universal Waste, the EPA hoped to improve the collection, transportation, and disposal of common household or office items with hazardous content, such as nickel-cadmium batteries, pesticides, thermostats, and lamps. The rules were also intended to ease regulatory burdens on business and decrease the amount of Universal Waste items in community landfills by increasing recycling opportunities.

SOLID WASTE—DISPOSAL

4136. Municipal dump in the Western world was organized by the government of Athens, Greece, circa 400 B.C. Unwanted materials were required to be disposed of at least one mile from the city's walls.

4137. Municipal sanitation crews collected garbage and other refuse from the streets of Rome circa 200 A.D.

4138. Waste disposal law in England passed in 1388 by Parliament, outlawed dumping of waste in ditches and public waterways.

4139. Municipal garbage dumps in New York City were established by order of New Amsterdam director-general Peter Stuyvesant's Dutch colonial administration in 1657. Prior to the creation of the dumps and Stuyvesant's 1648 restrictions on roaming livestock, colonists had allowed hogs and goats to consume much of the city's garbage.

4140. Garbage incinerator that was effective was established in 1897 in St. Louis, MO, by a private contractor with a city contract for the collection and disposal of garbage. Water was drained off; cans, bottles, and rags were taken out; and grease was extracted by means of naphtha, using the Merz process.

4141. Law to prohibit the dumping of solid waste in U.S. waters was the Rivers and Harbors Appropriations Act of 1899. The law banned the unauthorized obstruction or alteration of any navigable waters of the United States, including the deposit of material in such waters.

4142. Proposed U.S. legislation regarding abandoned cars was the Junked Auto Disposal Bill of 1966, proposed by Senator Paul Douglas of Illinois on May 25, 1966. If it had become law, the bill would have added a one percent federal tax to the price of cars, which would be used for the disposal of abandoned vehicles. Opposition prevented it from being brought to a vote in Congress.

4143. Federal guidelines for disposing of hazardous waste were contained in the Resource Conservation and Recovery Act of 1976 (RCRA). The act required the Environmental Protection Agency (EPA) to regulate hazardous waste and set standards for its transportation and disposal, but placed responsibility for the control of solid waste with the states.

4144. Federal rules in the United States for closure of hazardous waste treatment, storage, and disposal facilities were included in the Resource Conservation and Recovery Act of 1976. The law was enacted on October 21, 1976. Its closure regulations specified rules for covering or capping landfills and for the disposal or decontamination of equipment, structures, and soils. The act also regulated postclosure plans for land disposal facilities and for facilities that could not decontaminate all equipment, structures, and soils. In such cases, site owners/operators were typically required to conduct monitoring and maintenance activities to preserve the integrity of the disposal system and continue to prevent or control releases of contaminants from the disposal units for a 30-year period.

4145. U.S. Supreme Court ruling on state bans on imported waste was given in the case of *City of Philadelphia* v. *New Jersey*. The case challenged a New Jersey law banning the importation of any solid or liquid waste that originated outside the state. On June 23, 1978, by a 7–2 vote, the court ruled that the state restrictions were an unlawful discrimination against interstate commerce, violating the commerce clause of the U.S. Constitution. Since New Jersey had not banned similar locally produced waste, the court's majority agreed that the disputed law was an illegal protectionist measure and not an environmental protection statute.

4146. U.S. Supreme Court ruling on plastic milk containers was *Minnesota* v. *Clover Leaf Creamery Co.* After the Minnesota legislature passed a law that banned the retail sale of milk in plastic nonreturnable, nonrefillable containers but permitted such sale in other nonreturnable, nonrefillable containers, such as paper cartons, the creamery sued the state. A Minnesota District Court agreed with the creamery that the rules were unconstitutional. On January 21, 1981, however, the Supreme Court reversed the District Court decision. In a 6–2 vote, the court upheld the ban because it bore a rational relation to the state's objectives of promoting resource conservation, easing solid waste disposal problems, and conserving energy.

4147. Federal policy in the United States on emergency removal of asbestos from schools was established through the Asbestos Hazard Emergency Response Act (AHERA) on October 2, 1986. New rules required inspection of schools for asbestos-containing material and mandated safe response and reinspection procedures. The law also required a study of the danger to human health posed by asbestos in public and commercial buildings and a report on possible responses.

4148. Law in the United States banning the disposal of plastics at sea was passed by Congress on December 19, 1987. Because plastics are not biodegradable, they are an enduring threat to marine life. The law forbade the disposal of plastics within 200 miles of the U.S. shoreline and applied the provisions of Annex V of the International Convention for the Prevention of Pollution from Ships (MARPOL), which pertains to the disposal of garbage at sea.

4149. Medical waste disposal rules set by the U.S. Environmental Protection Agency (EPA) resulted from PL104-42, which was signed into law by President Ronald Reagan on November 1, 1988. After widely publicized instances of syringes and other medical waste washing up on beaches in the northeastern United States in the summer of 1988, Congress quickly passed legislation requiring the EPA to monitor the disposal of such refuse. Its guidelines required the proper packaging and labeling of medical waste, ranging from surgical tools and drug containers to body parts.

4150. U.S. Supreme Court ruling on local bans on imported waste was given in the case of *Fort Gratiot Sanitary Landfill v. Michigan Department of Natural Resources*. The case resulted from St. Clair County's refusal to authorize the Fort Gratiot landfill to accept waste from out of state, citing the Waste Import Restrictions in Michigan's Solid Waste Management Act (SWMA). The landfill sued, charging that the restrictions violated the commerce clause of the U.S. Constitution. On June 1, 1992, by a 7–2 vote, the court agreed, declaring that the state restrictions were an unlawful discrimination against interstate commerce.

4151. General grants from the U.S. government to Indian tribes for solid waste disposal programs were provided under the Indian Environmental General Assistance Program Act, enacted on October 24, 1992. The act directed the Administrator of the Environmental Protection Agency (EPA) to establish an Indian Environmental General Assistance Program to provide grants to tribes and intertribal consortia for solid and hazardous waste management programs on Indian lands. The act was amended in 1993 and 1996.

4152. Law in the United States to require phasing out of mercury in batteries was signed by President Bill Clinton on May 13, 1996. Improper disposal of heavy-duty dry-cell batteries containing mercury was found to be capable of polluting the air and water supplies. The law also set new rules for labeling rechargeable batteries, which can contain potential pollutants like lead or nickel cadmium.

4153. Landfill tax in England was contained in the Finance Act of 1996. The tax applied to all waste disposed of by way of landfill or at a licensed landfill site on or after October 1, 1996, unless the waste was specifically exempt. Landfill owners were taxed £2 per ton of inert waste listed under the regulations, with a higher rate of £7 per ton for all other waste.

4154. Modern solid waste disposal crime laws in China are contained in Chapter Six of China's Criminal Code. Revisions of the code, which became effective October 1, 1997, made dumping, stockpiling, or disposing of solid waste from outside of the country crimes punishable by imprisonment and/or fines. Importation of solid waste for recycling without permission and leading to environmental pollution accidents leaves those responsible liable to similar penalties.

4155. U.S. Government Paperwork Elimination Act became law on October 21, 1998. The act directed the Office of Management and Budget to oversee the acquisition and use of alternative information technologies to provided for electronic submission and maintenance of information as a substitute for paper. The regulation was expected to massively reduce paper use by government agencies and citizens, who could now correspond with the federal government via computers.

SOLID WASTE—RECYCLING

4156. Recycled paper in the United States was produced in 1690 by the Rittenhouse Mill in Germantown, PA, near Philadelphia. Rags and discarded paper were used in the paper's manufacture.

SOLID WASTE—RECYCLING—*continued*

4157. Nonreturnable bottle and can law enacted by a U.S. state was enacted by Oregon on July 2, 1971. The bill outlawed pull-tab cans and nonreturnable bottles for soft drinks and beer.

4158. Executive Order requiring U.S. government agencies to use recycled paper was Executive Order 12873 on Federal Acquisition, Recycling, and Waste Reduction. It was signed by President Bill Clinton on October 20, 1993. Without increasing spending, the order required that all printing and writing paper bought by the federal government contain 20 percent postconsumer material by the end of 1994 and 30 percent by the end of 1998. By requiring the government to use recycled and environmentally preferable products and services, the order was intended to relieve stress on landfills and to promote new technologies and economic opportunities for environmentally friendly businesses. Each government agency was to set its own waste reduction goals and purchasing policies for buying recycled goods.

4159. U.S. Jobs Through Recycling grant to Arizona was awarded in 1994 to the Arizona Department of Commerce. The federal Recycling Economic Development Advocate grant was used to educate business and state government leaders about recycling. The grant also funded recycling-related efforts to create jobs, attract investment, and assist businesses with marketing recyclables. The program was sufficiently successful for the U.S. Environmental Protection Agency to award the state a second grant in 1996 to help create long-term forestry and timber industry waste-related job opportunities in rural and tribal areas of Arizona.

4160. U.S. Jobs Through Recycling grant to California was a 1994 federal Recycling Business Association Center award. The grant was awarded to help businesses comply with California's Integrated Waste Management Act, whose goal was diverting 50 percent of the state's solid waste toward recycling facilities instead of landfills. The grant was used to help establish new businesses and expand existing ones, particularly those dealing with plastics, construction, demolition debris, and discarded tires.

4161. U.S. Jobs Through Recycling grant to Delaware was awarded in 1994. A combination of state and federal grant funds bolstered Delaware's under-promoted recycling infrastructure by aiding the Green Industries Initiative, a state marketing effort devoted to recyclables. The project also sited a new facility for processing scrap tires and relieved stress on landfills. The problem wastes targeted included plastics, compost, and waste tires.

4162. U.S. Jobs Through Recycling grant to Maryland was a federal Recycling Economic Development Advocate grant, awarded in 1994 to the Maryland Department of Business and Economic Development. The resulting project created new jobs and attracted investments in recycling processors of construction-related wastes like wood, mixed paper, roofing shingles, and drywall material.

4163. U.S. Jobs Through Recycling grant to Minnesota was awarded in 1994. The grant aimed at helping new recycling businesses with start-up problems. The 1994 federal grant to Minnesota's Office of Environmental Assistance also helped the state's existing recyclables-related businesses to expand their markets in plastics, composite materials, and wood fiber.

4164. U.S. Jobs Through Recycling grant to Nebraska was a federal three-year Recycling Economic Development Advocate grant, awarded in 1994 to Nebraska's Department of Economic Development. The project promoted the use of recycled plastic feedstock, provided technological help, and acted as an advocate for recycling businesses seeking financial aid. The state assisted environmental trust, waste reduction, and scrap tire concerns by promoting tax incentives and loans.

4165. U.S. Jobs Through Recycling grant to New York was a 1994 federal Recycling Business Assistance Center grant to the state's Office of Recycling Marketing Development. The grant concentrated on helping businesses find uses for recycled materials. Its projects included creating facilities to manufacture pallets, furniture, and flooring from waste wood. Waste paper collection and quality control measures were pursued in community-industry partnerships, while marketing and technological assistance were offered to paper mills to help deal with sludge.

4166. U.S. Jobs Through Recycling grant to North Carolina was partly aimed at recycling waste created by the state's manufactured home industry. The 1994 federal Recycling Business Assistance Center grant to the Department of Environmental Health and Natural Resources tried to expand the state's well-developed recycling infrastructure through educational programs, public relations, and financial aid. Direct recycling programs focused on plastics, paper, organics, and gypsum waste created by manufactured home companies.

4167. U.S. Jobs Through Recycling grant to Ohio was a 1994 Recycling Economic Development Assistance grant to Ohio's Department of Development. State funds and the U.S. Environmental Protection Agency grant helped improve the recyclable market in Ohio by coordinating the Ohio Department of Natural Resources "Recycle Ohio!" government program with private, nonprofit efforts by the Association of Ohio Recyclers.

4168. U.S. Jobs Through Recycling grant to Oregon concentrated on creating recycling-related jobs. The project obtained industrial development bonding for a solid waste recovery plant, brought plastics and scrap tire businesses into the state, and aided processors of polystyrene, glass, and other recyclables. The 1994 Recycling Economic Development Assistance grant was awarded to Oregon's Economic Development Department.

4169. U.S. Jobs Through Recycling grant to New Hampshire was awarded in 1995 to New Hampshire's Office of State Planning. The federal Recycling Economic Development Advocate project promoted plans for a recycling trade association. The grant also established a network of service providers to assist recycling processors of materials like glass, demolition debris, food waste, and toner cartridges.

4170. U.S. Jobs Through Recycling grant to Virginia funded 1995 demonstration projects by the state's Department of Environmental Quality. The agency concentrated on promoting interest in and marketing options for plastics and mixed paper recycling in rural Virginia. The grant also helped update the Virginia Recycling Markets directory and funded a mobile baler, scrap pantyhose processing, and other business projects.

4171. Battery recycling program in Canada was called Charge Up To Recycle! The program was started in September 1997 by the Rechargeable Battery Recycling in Canada (RBRC) organization, with the Canadian Household Battery Association. The nationwide program placed collection boxes for dead nickel-cadmium rechargeable batteries in retail stores where batteries were sold.

4172. America Recycles Day was held on November 15, 1997. A coalition of federal, state, local, business, and nonprofit organizations sponsored the event to promote public awareness about recycling opportunities through education, media releases, and special events. The event quickly grew to include "America Recycles Day" efforts in nearly all U.S. states.

SOLID WASTE—STREET CLEANING

4173. Trash collection tax in Paris was levied in 1508 to finance garbage removal from the streets, where household waste was typically discarded. A 1539 ordinance by French King François I set strict rules for the storage of garbage, its transportation, and scheduling of municipal trash removal.

4174. Street-sweeping service in the United States was instituted in Philadelphia, PA, in 1757 by Benjamin Franklin, who reported: "After some inquiry, I found a poor industrious man who was willing to undertake keeping the pavement clean by sweeping it twice a week, carrying off the dirt from before the neighbors' doors, for the sum of six pence per month, to be paid by each house."

4175. Street-cleaning machine of importance in the United States was used on December 15, 1854, in Philadelphia, PA. A contemporary account described it as consisting of a "series of brooms on a cylinder about two feet wide, attached to two endless chains, running over an upper and lower set of pulleys, which are suspended on a light frame of wrought iron behind a cart, the body of which is near the ground. As the cart wheels revolve, a rotary motion is given to the pulleys conveying the endless chains, and series of brooms attached to them: which being made to bear on the ground successively sweep the surface and carry the soil up an incline or carrier plate, over the top of which it is dropped into the cart."

STORMS

4176. Invasion of England to be halted by gale-force winds occurred in August 1588 when ships of the Spanish Armada were sunk, mostly off the Shetland Islands of Scotland and off Ireland. More than 10,000 sailors were drowned and 51 ships lost.

STORMS—*continued*

4177. Tsunami to cause more than 100,000 deaths in Japan occurred in 1703 on Okinawa. The giant wall of water is estimated to have been the most destructive ever to engulf a Japanese island.

4178. Storm to defeat a naval invasion of Canada occurred in 1711 when an English fleet under Sir Hovenden Walker was sailing to attack French Quebec. Caught by the storm off Labrador, the armada lost 1,342 men in 8 ships (of 61) and foundered against the rocks of Belle Isle. The French held possession of Quebec until 1759 during the French and Indian War.

4179. Storm surge on record known to kill many thousands in India occurred at Calcutta in 1737. Fatalities were estimated at more than 300,000.

4180. Scientist to fly a kite in a thunderstorm in hopes of attracting a lightning bolt was Benjamin Franklin near Philadelphia in June 1752. The experiment succeeded, providing a scientific explanation for what had been a mysterious phenomenon.

4181. Storm that destroyed a lighthouse in the United States occurred on April 16, 1851, in Boston Harbor, MA. The Minot Light was crushed by towering waves and gales and its two lighthouse keepers died. The event became known as the "Lighthouse Storm."

4182. National storm warning service was started in France in 1856. The government launched the service as the direct result of a violent storm that caught a French fleet unawares off the Crimean coast in November 1854.

4183. Report of a Santa Ana wind to force temperatures to over 130 degrees came from Santa Barbara, CA, on June 17, 1859. The afternoon high temperature hit 133 degrees.

4184. U.S. Signal Corps storm warning was issued from Chicago for the Great Lakes region on November 8, 1870. Professor Increase Latham prepared the warning for the Signal Corps, the predecessor of the U.S. Weather Bureau.

4185. Radio gale warnings from Great Britain to ships in the eastern Atlantic were issued in 1911. These warnings, however, were not made on a regular basis until 1919.

4186. Severe storm forecasting service of significance in the United States was begun in 1952 by the U.S. Weather Bureau. The Severe Local Storms Forecasting Unit (SELS) had its first headquarters in Washington, DC, and moved in 1954 to Kansas City, MO, where it is now named the National Severe Storms Forecast Center.

4187. Tropical storm to hit the United States in February occurred on February 2, 1952, moving across southern Florida from the Gulf of Mexico. The only such storm ever recorded in this month, it had winds of 60 mph and brought 2 to 4 inches of rain to the state.

4188. Storm tide warning service in Great Britain was launched in 1953. The Royal Navy operated the service, which was established after a series of floods along the east coast earlier that year.

4189. Monsoon in Bangladesh to cause $1 billion in damage occurred in August and September 1988. More than 1,000 people died, 30 million were left homeless, and 160,000 later suffered from diseases related to the catastrophe. Two-thirds of the country was flooded in the monsoon.

STORMS—BLIZZARDS AND SNOWSTORMS

4190. Person known to have studied snow crystals was Olaus Magnus, the archbishop of Uppsala, Sweden, in the mid-16th century. His drawings were reproduced in a book published in Rome in 1555.

4191. Scientist to describe the characteristic six-sided symmetry of snow crystals was the German astronomer Johannes Kepler, in a pamphlet published in 1611.

4192. So-called "Winter of the Deep Snow" in New England was in 1748. The heaviest storm, on February 29, 1748, left 30 inches of snow on the ground at Salem, MA.

4193. Snowstorm known to have dropped 18 inches of snow on Savannah, GA occurred on January 10, 1800. Other portions of the southeastern United States received significant amounts of snowfall from the same storm. Charleston, SC, 50 miles north of Savannah, received 10 inches of snow.

4194. Summer blizzard recorded in the United States occurred in late June and early July 1816 in Connecticut. Snow also fell in Vermont and Massachusetts. As far south as Savannah, GA, the temperature did not exceed 46 degrees on July 4. The unusual weather was apparently linked to cloud cover caused by the eruption of the Tambora volcano in Java the previous year.

4195. "Luminous" snowstorm in Vermont and Massachusetts occurred on January 17, 1817. St. Elmo's fire appeared on roof peaks, fence posts, and people's hats and fingers.

4196. Disastrous snow avalanche in England occurred on December 27, 1836, at Lewes, Sussex. A snow cornice broke free, demolishing a row of houses and killing eight people.

4197. American migrants to be forced into cannibalism after being trapped in the High Sierra by heavy snow were the Donner party, in 1846–47. Forty-four of the 89 people perished. The survivors of the group resorted to cannibalism when the food supply ran out.

4198. Snow avalanche to destroy a mining camp near Alta, UT occurred in 1874. This mining boomtown of the 1860s and 1870s, was totally destroyed. However, Alta became the state's first ski resort in 1937.

4199. Blizzards to completely halt railroad traffic in the Great Plains for many weeks occurred in the winter of 1880–1881. Eleven feet of snow fell in the Dakota Territory, and no trains ran for 79 consecutive days.

4200. Snowfall recorded in excess of 3.5 inches in San Francisco occurred on February 5, 1887. A precisely measured 3.7 inches fell downtown; the city's western hills reported 7 inches.

4201. Blizzard to cause numerous deaths on the U.S. East Coast began on March 11, 1888, and continued until March 14, killing 400 people. New York City was the center of the storm on March 12. Snow fell there for 36 hours, halting transportation, closing businesses, and immobilizing and isolating the city for 48 hours.

4202. Snow avalanche to kill Austrian and Italian troops during a war occurred on December 13, 1916, along a World War I battlefront in Italy's Dolomite Alps. More than 10,000 Austrian and Italian soldiers were killed. Bodies of the dead soldiers were found as late as 1952.

4203. Meteorologist to photograph ice crystals with a camera-equipped microscope was Wilson A. Bentley of Vermont. His book *Snow Crystals*, with 2,000 of his best photographs, appeared shortly after his death in 1931.

4204. Officially reported measurable snowfall in downtown Los Angeles occurred on January 15, 1932. Dubbed the "Big Snow," the storm dropped 2 inches at the city's weather station.

4205. Meteorologist to reproduce natural snow crystals in plastic was Vincent J. Schaefer, in 1940. Schaefer was a researcher at a General Electric laboratory in Schenectady, NY.

4206. Winter in Great Britain in the 20th century during which snow fell somewhere in the country every day from January 22 to March 17 occurred in 1947. There were several daily recorded snowfalls of more than two feet in some parts of the country.

4207. Scientist to study snow with modern X-ray equipment was Ukichiro Nakaya of Hokkaido University in Japan in 1954. Through his experiments, Nakaya showed that the symmetry of snow crystals is due to molecular structure.

4208. Blizzard to cause more than $500 million damage in the United States occurred over the northeastern states on February 15 and 16, 1958, causing 171 deaths. The snowfall was especially heavy in Allentown, PA.

4209. Snowflakes reported to measure 8 by 12 inches were reported at Bratsk, Siberia, in the winter of 1971.

4210. Known trace of snow in the Miami, FL, area fell on January 19, 1977. There were no significant accumulations. Snowflakes were sighted in the air as far south as Homestead, FL.

4211. Official "concentrated heavy snowstorm," defined as 20 inches or more within a 48-hour period, in Boston, MA, struck on January 20, 1978. The second such storm in a century of record-keeping began only 18 days later, on February 6.

4212. Snowfall in living memory in the Kalahari Desert in Namibia fell on September 1, 1981. The temperature dropped to 23 degrees (F) in Namibia's capital, Windhoek. The light snow that fell in the desert took most people by surprise.

4213. Blizzard and winds to cause damages of $1 billion or more in the United States was the "Storm of the Century" from March 12 to 14, 1993, along the Atlantic coast. More than 200 people died, and the damages were estimated at $3 billion.

STORMS—BLIZZARDS AND SNOWSTORMS—*continued*

4214. Blizzard to close all major U.S. airports on the East Coast occurred from March 12 to 14, 1993, causing more than 200 deaths. Nicknamed the "Storm of the Century," it caused about $3 billion damages. Heavy snow fell as far south as Florida and 60 inches on the Appalachian Mountains. The snow was accompanied by winds up to hurricane force that destroyed many homes along the coast, including some 200 in North Carolina and 18 on New York's Long Island.

STORMS—HAILSTORMS

4215. Use of the Stiger gun as a defense against hail was in Steinmark, Austria, in 1896. Interest in Albert Stiger's gun, a variety of mortar gun, spread rapidly and by 1900, 7,000 firing stations had been established in Italy alone.

4216. Authenticated death by hail recorded by the U.S. Weather Bureau occurred northwest of Lubbock, TX on May 13, 1930. A man caught in an open field perished in a hailstorm.

4217. Hailstorm of significance known to have hit the western Hunan province of China occurred on June 19, 1932. At least 200 people were killed and several thousand injured by the storm.

4218. Rocket trials for hailstone protection were carried out near Kericho, Kenya in July 1963. Firing the rockets turned hail "mushy," limiting damage to the area's tea plantations.

4219. Single hailstone to measure 7.5 inches in diameter fell at Coffeyville, KS on September 3, 1970. At 1.67 pounds, it was the largest authenticated hailstone to hit the ground.

4220. Hailstorm blamed for the crash of a jetliner occurred in April 1977 near New Hope, GA. The DC-9 crashed while making a descent during a violent storm; 68 people perished.

4221. Hailstorm known to cause a famine in Syria hit on June 2, 1978, in the Jabalah coastal region. Crops were destroyed in 40 of the 53 villages affected; 60,000 people faced starvation.

STORMS—HURRICANES, CYCLONES, AND TYPHOONS

4222. Typhoon known to have prevented an invasion of Japan occurred in 1281, destroying the armada of Kublai Khan, the Mongol Emperor of China. Believing it had saved their country, the Japanese named the storm *kamikaze*, or "divine wind."

4223. European use of the term *hurricane* dates from the late 1400s. Spanish sailors in the Caribbean picked up an Indian word for "great wind" and approximated it as *hurricane.*

4224. Hurricane to destroy a European colony in the Western Hemisphere occurred in 1495, hitting Isabela, a colony that had been established on La Isla Espanola (Hispaniola) by Christopher Columbus.

4225. Hurricane reported in the Americas by a European explorer was in 1495 by Christopher Columbus while sailing to the New World. The storm was so terrifying, he later wrote that he would never again expose himself to such danger except in "the service of God and the extension of the monarchy."

4226. Known military expedition in North America sunk by a hurricane was in 1528 in the Gulf of Mexico. The ship of Spanish adventurers, led by Pánfilo de Narváez, went down in Apalachee Bay with only two of the 400 men surviving. They were sailing for Mexico after failing to find gold at the Native American village of Apalachee near present-day Tallahassee, FL.

4227. Hurricane to sink a fleet in the Gulf of Mexico occurred in 1553 when 18 Spanish ships went down off Veracruz, Mexico. Two ships survived, but their crews were murdered by natives after landing in Cuba.

4228. Hurricane to wreck a colonists' ship in North America hit a Spanish expedition in 1559. The group had intended to found a colony in north Florida, near today's city of Pensacola.

4229. Hurricane that caused islands to be settled occurred on July 28, 1609. Settlers sailing from England to the Jamestown colony in Virginia in the ship *Sea Venture*, damaged by the hurricane, landed on the Bermuda Islands, and the 150 men, women, and children settled there. They chose the name "Somers Islands" for the ship's captain, Sir George Somers.

4230. Storm on record to demolish an entire Dutch island occurred on October 21, 1634. The island of Nordstrand vanished in the disturbance.

4231. Hurricane recorded in an American colony ravaged Plymouth, MA, on August 15, 1635.

4232. Hurricane to kill thousands in the Lesser Antilles occurred in 1650 and destroyed the town of Basseterre on the island of St. Kitts. The storm sank 28 ships along the coast of Basseterre, which is now the capital of St. Kitts and Nevis.

4233. Hurricane report in colonial Virginia was a written account of one that struck Jamestown on August 27, 1667.

4234. Hurricane to sink a fleet off North America occurred in 1715 when 11 ships of a Spanish flotilla went down off present-day Miami. About 1,000 lives were lost along with 14 million pesos, gold, and gems. One ship escaped the destruction.

4235. Typhoon to destroy 20,000 ships in India hit the Bay of Bengal on October 7, 1737, and killed about 300,000 people. It hit at the mouth of the Hooghly River on the Ganges delta, propelling a storm surge 40 feet high along the river, devastating towns and villages, and sinking or wrecking ships in the vicinity.

4236. Hurricane known to be accurately tracked in colonial America was by Benjamin Franklin in September 1743. After his planned observation of a lunar eclipse was ruined by the tail of a hurricane, Franklin calculated it must have come from the Boston area because of the wind direction. When told that this had not happened, he made contacts that allowed him to determine its true direction up the Atlantic coast.

4237. Hurricane of significance recorded in North America swept over the colonies from September 2 to 9, 1775, and was thus called the "Hurricane of Independence." The death toll was estimated at more than 4,000 people, from North Carolina to Nova Scotia, Canada.

4238. Hurricane in the Atlantic causing massive deaths occurred in October 1780, killing some 22,000 people. This is the deadliest hurricane in that ocean's history. Some 9,000 died in Martinique, 4,000 to 5,000 in St. Eustatius, and 4,326 in Barbados.

4239. Observer to describe tropical storms as large vortices of air rotating around an "eye" of low pressure was engineer William Redfield, an American, in the 1820s. Redfield's close tracking of hurricanes supplied valuable data to early meteorologists.

4240. Hurricane to kill many slaves in the United States happened in 1824 on St. Simons Island off the central Georgia coast. The high winds and flood waters combined to kill 83 people, most of them African Americans still working in the plantation fields.

4241. "October Gale" to sink 40 Cape Cod fishing vessels off Nantucket Island, MA, occurred on October 3, 1841. Fifty-seven fishermen perished. The same gale dropped 18 inches of snow on Middletown, CT.

4242. Meteorologist to discover the processes of cyclones and tornadoes was the American James Pollard Espy. He was born in Westmoreland County in Pennsylvania and served as the chief of the Meteorological Bureau of the War Department from 1842 to 1857. After Espy published the influential *Philosophy of Storms* (1841), he was called "The Storm King." His analyses of cyclones and tornadoes led to several discoveries: the low pressure at their centers, the air movement toward the storm center, the latent heat of condensation that keeps warm air rising, and the upper wind that determines the movement of these systems.

4243. Appearance of the word *cyclone* was in 1848 in England. Captain H. Piddington coined the term, taken from the Greek word for *circle.*

4244. Hurricane warning was issued in 1873 by the Signal Corps of the U.S. Army. The hurricane, however, curved back into the open sea.

4245. Meteorologist to name hurricanes was Clement L. Wragge of the Queensland Weather Bureau in Australia. Wragge, at the bureau from 1887–1902, named every identifiable system, high or low, on his weather charts.

4246. Hurricane causing massive loss of life in the United States occurred on September 8, 1900, in Galveston, TX. The hurricane killed more than 8,000 people. The city was hit in the morning with little warning by winds of over 120 mph and a 12-foot-high tidal wave from the Gulf of Mexico. The city remained underwater the next day. Property damage was more than $20 million, including 3,000 homes destroyed. Bodies had to be removed by boats and dumped in the open sea. The tragedy resulted in a reorganization of Galveston's emergency procedures and the raising of streets, sidewalks, and telephone poles 17 feet above normal high tide.

4247. Typhoon to cause massive loss of life in Hong Kong occurred on September 18, 1906, killing about 10,000 people. Many died from the accompanying tsunami (giant wave) that hit the island.

4248. Cyclone theory was published in 1919 by the Norwegian meteorologist, Vilhelm Bjerknes, and his son, the Norwegian-American meteorologist, Jacob Bjerknes. Their paper was "On the Structure of Moving Cyclones," written at the Bergen Geophysical Institute in Norway. Their work helped establish modern weather forecasting. Jacob identified two lines of convergence within cyclones that were later called warm and cold fronts.

STORMS—HURRICANES, CYCLONES, AND TYPHOONS—*continued*

4249. Hurricane to sink an ocean liner off the United States occurred on September 2, 1919, in the Florida Straits when the *Valbanera* went down, killing its 488 passengers and crew. The hurricane continued on to Corpus Christi, TX, where 284 more victims perished. The total deaths were 488 at sea and 287 on land.

4250. Hurricane to cause massive deaths in the Dominican Republic hit the island on September 3, 1930. It killed about 2,000 people and injured about 6,000, especially in the Santo Domingo area.

4251. Hurricane in the United States rated Category 5 (most intense) hit the Florida Keys in 1935, killing 408 people. It was the most intense hurricane in U.S. history, with its pressure measured at 892 millibars and at 26.35 inches of mercury.

4252. Pilot to fly into a hurricane intentionally was Army Air Force Lt. Col. Joseph P. Duckworth on July 17, 1943, accompanied by a navigator, Ralph O'Hair. Duckworth piloted his single-engine AT-6 Texan trainer aircraft into the eye of a hurricane off the Texas coast. He penetrated the hurricane at 5,000 to 6,000 feet and flew into the eye. He was stationed at the instrument flight school at Bryan, TX, and was dared to make the flight after he said a Texan could withstand any weather. On his next flight into a hurricane, he was accompanied by the base weather officer, who analyzed the hurricane's makeup.

4253. Hurricane report that was censored in the United States concerned a storm that struck the Galveston and Houston areas of Texas on July 27, 1943. Because this was during World War II and German U-boats were expected in the Gulf of Mexico, the U.S. government banned all broadcasts by ships' radios, including weather reports and hurricane warnings. The surprise hurricane killed 19 people, injured hundreds more, and caused an estimated $17 million property damage. No official records were ever found in the areas' weather bureaus.

4254. Typhoon to sink U.S. warships occurred in 1944 off the Philippines during World War II. Nearly 800 sailors died as three destroyers, the USS *Hull*, the USS *Monahan*, and the USS *Spence*, went down in the South China Sea off Luzon Island.

4255. Hurricane entered intentionally with a large aircraft was on September 1945, entered by members of the "Hurricane Hunters" of the U.S. Army Air Force's 53rd Weather Reconnaissance Squadron. The plane was a B-17 "Flying Fortress."

4256. Hurricane experiment to reduce cloud intensity was carried out in 1947 by the U.S. Air Force's 53rd Weather Reconnaissance Squadron in conjunction with the U.S. Weather Bureau. The Air Force sprayed particles of dry ice into clouds associated with hurricanes to reduce their intensity, but the results were not conclusive.

4257. Hurricane warning service to operate around the clock was begun on June 16, 1947, by the U.S. Weather Bureau.

4258. Hurricanes named by the U.S. Weather Bureau were in 1950. They were named for the phonetic alphabet used by the U.S. military, so the first hurricanes were "Able," "Baker," and "Charlie." These names for hurricanes had been used internally since 1947 by the U.S. Air Force's "Hurricane Hunters."

4259. Hurricanes to be given feminine names occurred in 1953, when U.S. forecasters adopted the system.

4260. Hurricane of significance with a feminine name to hit the United States was Carol, hitting the northeastern coast from August 25 to 31, 1954, the second year such names were used. Carol caused 68 deaths and injured about 1,000 people in the Long Island, NY, and New England areas.

4261. Hurricane reported to occur during early summer in the United States was Audrey from June 25 to 28, 1957, causing 534 deaths in the Gulf states of Alabama, Mississippi, Louisiana, and Texas. It wiped out the town of Cameron, LA, killing 390. Warnings by the National Weather Service were ignored by many who thought the season too early for hurricanes.

4262. Tropical cyclone known to have lived for a month was Ginger, in the Atlantic from September 5 to October 5, 1971. On 20 of those days, Ginger was classified a hurricane.

4263. Scale widely used to measure hurricanes was the Saffir-Simpson Hurricane Damage Potential Scale devised in 1975 in the United States by an engineer, Herbert Saffir, and a hurricane expert, Robert Simpson. It uses a scale from 1 to 5 to predict damage and is based on wind speed and atmospheric pressure. Category 1 hurricanes cause minimal damage with winds of 74 to 95 mph; Category 2 cause moderate damage with 96 to 110 mph; Category 3 cause extensive damage with 111 to 130 mph; Category 4 cause extreme damage with 131 to 155 mph; and Category 5 cause catastrophic damage with winds over 155 mph. Only two Category 5 hurricanes have been recorded: the Labor Day hurricane of 1935 over the Florida Keys and Camille in 1969 on the Mississippi coast.

4264. Hurricanes to be given masculine names occurred in 1978 and were northeastern Pacific storms. Male names were first used the next year for Atlantic and Gulf of Mexico hurricanes.

4265. Hurricane in the Atlantic with a masculine name was David, which lasted from August 25 to September 7, 1979. It caused 1,200 deaths in the Dominican Republic and five in the United States, becoming weaker as it moved from Florida up the East Coast. This was the first year hurricanes received masculine names in the Atlantic and Gulf of Mexico.

4266. Hurricane in the Gulf of Mexico with a masculine name was Frederic on September 12, 1979, hitting Mobile, AL and causing 5 deaths and $3.5 billion damage in Alabama, Mississippi, and Florida. It was a Category 3 hurricane. This was the first year masculine names were used for hurricanes in the Atlantic and Gulf of Mexico.

4267. Hurricane to cause damages of $1 billion or more in the United States was Hurricane Alicia, which hit southern Texas on August 18, 1983. It killed 17 people and caused damage estimated at $3 billion.

4268. Cyclone to cause catastrophic loss of life in Bangladesh occurred on April 30, 1991, in the southeastern part of the country; about 139,000 people were killed and another 9 million left homeless. Many more thousands died from disease and hunger in the aftermath.

4269. Hurricane to cause more than $25 billion in damage was Andrew from August 24 to 26, 1992. Andrew hit Florida and Louisiana, causing 14 deaths and damages of around $26.5 billion. It was the most expensive natural disaster in U.S. history. In Florida alone, some 85,000 homes were destroyed and about 250,000 people made homeless.

STORMS—ICE STORMS

4270. Devastating ice storm in New England in the 20th century occurred on November 26–29, 1921. Ice up to 3 inches thick coated power lines. Trees were down, roads impassable, and towns in darkness along a line from northern Connecticut to Massachusetts to New Hampshire to southern Maine.

4271. Devastating storm of freezing rain to strike Illinois in the 20th century occurred on December 17–18, 1924. Sleet and ice covered the ground to a depth of two inches. Damages were pegged at more than $21 million.

4272. Ice storm to cause $100 million in damages in the United States occurred from January 28 to February 1, 1951. The storm left a coating of ice, in some cases as much as 4 inches thick, along a broad path from Texas to West Virginia.

STORMS—THUNDERSTORMS

4273. Attempt to explain the origin of thunder was made by Aristotle in the fourth century B.C. The Greek philosopher suggested thunder was the noise made when air expelled from one cloud struck another.

4274. Lightning conductor in the United States was installed at Benjamin Franklin's Philadelphia home in September 1752. Franklin used a pointed metal rod reaching nine feet above his roof to draw the charge.

4275. Brush charge of atmospheric electricity to destroy an airship occurred at Lakehurst, NJ, on May 6, 1937. Thirty-five passengers were killed or fatally injured when the *Hindenburg* blew up at its moorings. St. Elmo's fire was blamed for the explosion.

4276. Person known to survive being struck by lightning seven times was Roy C. Sullivan, a U.S. Park Service ranger. He was first hit by lightning in 1942. He was later struck in 1969, 1970, 1972, 1973, 1976, and 1977. In one instance, lightning set his hair on fire.

4277. Jetliner to be destroyed in flight by lightning exploded and crashed near Elkton, MD, on December 8, 1983. All 81 aboard were killed.

STORMS—TORNADOES

4278. Written record of a devastating tornado in London dates from October 17, 1091. The tornado was said to have lifted the roof off St. Mary le Bow Church and to have demolished 600 houses.

STORMS—TORNADOES—*continued*

4279. Recorded reference to a tornado as a "black horse" dates from 1195. A monk at Scarborough, England, wrote of the "footprints of this accursed horse," a reference to the suction marks, very like a horse's hoof in shape, a tornado makes when it touches ground.

4280. Tornado to cause a major death toll in a U.S. city hit Charleston, SC, in 1811. More than 500 people died.

4281. Tornado in the United States to leave a city in ruins struck Natchez, MS, on May 7, 1840. More than 300 people perished, mostly along the Mississippi River front. Most houses and businesses were destroyed.

4282. Tornado observation instructions in the United States were completed in 1862 by the director of the Smithsonian Institution and distributed to its network of volunteer weather observers around the nation. At the same time, instructions were also given on how to observe thunderstorms.

4283. Tornadoes in the United States reported to cause a great loss of life occurred on February 19, 1884, in seven states. An estimated 800 people were killed by multiple tornadoes in Alabama, Mississippi, South Carolina, North Carolina, Tennessee, Kentucky, and Indiana.

4284. Killer tornado to cause $3 million in damages in Chicago touched down on March 28, 1920. The death toll was 28.

4285. Tornado in the United States with a death toll in excess of 650 struck on March 18, 1925, in Missouri, Illinois, and Indiana. The official death toll was 695.

4286. Tornado to last 3 hours in the United States hit Missouri, Illinois, and Indiana on March 18, 1925, killing 695 people and injuring 2,027. Known as the "Great Tri-State Tornado," its life of 3.5 hours is the longest by far in the United States. It traveled 220 miles during that period at speed of up to 73 mph. After it passed over the mining town of West Frankfort, IL, some 800 miners surfaced to find their homes flattened and 127 people killed.

4287. Volunteer tornado spotters network established by the U.S. government began in 1942 as a joint program of the U.S. Weather Bureau and the military forces. Volunteers were organized originally to watch for lightning near ordnance plants but soon began to watch for tornadoes and other storms as well.

4288. Tornado forecast for a specific U.S. location was made on March 25, 1948, for the vicinity of Tinker Air Force Base outside Oklahoma City, OK. Air Force Captain Robert C. Miller and Major Ernest J. Fawbush correctly assessed the atmospheric conditions and predicted the tornado that struck and caused damage of $6 million.

4289. *Tornado* designation publicly used by the U.S. Weather Bureau was in 1952 forecasts. The term had been introduced by Sergeant John Park Finley in 1882 for tornado forecasts by the U.S. Army Signal Corps meteorological service, but the Weather Bureau felt the name would panic the public and banned its use from 1887. It was allowed after 1938, but only for tornado warnings to alert disaster officials.

4290. Tornado in the eastern United States to claim 90 lives struck Worcester, MA, on June 9, 1953. The twister also caused $53 million in damages.

4291. Set of tornadoes to cause $500 million in damages struck on April 11, 1965, in Iowa, Wisconsin, Illinois, Michigan, Indiana, and Ohio. The death toll reached 271; more than 3,000 people were hurt.

4292. Use of the phrase *tornado watch* by the U.S. Weather Bureau occurred in 1966. The new term replaced *tornado forecast*.

4293. Scale to measure tornado intensities was the Fujita-Pearson Tornado Scale devised in 1971 by Dr. Tetsuya Theodore Fujita and Dr. Allan Pearson. It was adopted in 1973 by the National Severe Storms Forecast Center of the National Weather Service. It rates a tornado by the damage caused and by measuring its approximate path length and width. The five classifications are F0 for winds of 40–72 mph causing light damage, F1 of 73–112 mph for moderate damage, F2 of 113–157 mph for considerable damage, F3 of 158–206 mph for severe damage, F4 of 207–260 mph for devastating damage, and F5 of 261 mph and over for incredible damage.

4294. Tornadoes tracked by Doppler radar occurred in 1971 while this technology was being tested at the National Severe Storms Laboratory in Norman, OK. The Doppler system can detect and measure the speed and direction of winds and severe turbulence accurately because the radar signals change frequencies when reflected from a moving object. This improves the accuracy in identifying areas threatened by a tornado and increases the tornado warning time.

4295. Authentication of 148 tornadoes over a two-day period in the United States came on April 3–4, 1974. Tornadoes whirled through 7 southern and midwestern states, killing more than 300 people and injuring 4,000. Half the town of Xenia, OH, population 25,000, was leveled.

4296. Tornado in Great Britain to cause $1 million damage occurred on January 3, 1978, in Newmarket, England.

4297. "Tornado chasers" systematically organized were members of the National Severe Storms Laboratory in Oklahoma. In 1994, they began a project called VORTEX (Verification of the Origin of Rotation in Tornadoes) in which they followed storms likely to develop into tornadoes. Their equipment included mobile Doppler radar units and high-altitude balloons.

SUSTAINABLE DEVELOPMENT

4298. World Bank meeting took place in Savannah, GA, on March 8, 1946. The organization resulting from the ten-day conference has become the world's largest single lender to countries requesting funds for education, infrastructure construction, hydroelectric power, population planning, pollution control, and other development programs.

4299. Program opening the Amazon region to economic development was "Operation Amazonia," formulated by Brazil's military government in 1966. Over the following decade, the unsustainable strategy facilitated major highway construction, cattle ranching, and mining in previously untouched Amazonian wilderness. The resulting erosion, pollution, rain forest deforestation, species extinctions, and destruction of native cultures are widely considered to have had catastrophic human and environmental consequences in the region.

4300. Eco-industrial park was developed during the 1970s in the Danish port of Kalundborg. Area businesses include a power company, a pharmaceutical plant, wallboard manufacturer, a fish farm, and an oil refinery. The companies minimize waste and expenses by trading waste energy and industrial byproducts, which are used to produce energy, gypsum, sulfur, fertilizer, and commercially farmed fish.

4301. Resort powered by renewable energy was Harmony Resort on St. John in the U.S. Virgin Islands. Recycled materials were extensively used in constructing the luxury resort's twelve buildings, which are powered exclusively by passive solar and wind energy. The resort's grounds and landscaping were carefully designed to minimize impact on the scenic environment from which it overlooks Maho Bay. Harmony is one of four resorts on St. John established since the 1970s by American engineer and ecotourism developer Stanley Selengut.

4302. Self-sustaining village experiment in Colombia was Gaviotas, founded in 1971 in a desolate region of eastern Colombia. The community uses solar, wind, and other power sources to maintain services like electricity and pumped water without emitting any pollutants.

4303. "New Urbanism" community in the United States was Seaside, FL. The town plan was created in 1981 by architects Andres Duany and Elizabeth Plater-Zyberk, who tried intentionally to create a sense of community by using 19th-century town planning ideals rather than modern urban planning. Transportation features include a network of pedestrian walkways and bike paths, low-speed roadways, and tree-lined streets. Seaside's homes all have porches close to the sidewalks and are within walking distance of the town center. Zoning is structured to be inclusive of all age and income demographics.

4304. Sustainable development organization for mountain regions was the International Centre for Integrated Mountain Development (ICIMOD), created in 1983. ICIMOD was established by the government of Nepal and the United Nations Educational, Scientific, and Cultural Organization (UNESCO). Its aims were to promote development of economically and environmentally sound mountain ecosystems and to improve the living standards of mountain populations, especially in the Hindu Kush-Himalayan area. ICIMOD training and problem-solving programs relate to mountain farming systems, population, employment, infrastructure, technology, and environmental management issues.

4305. Use of the term *sustainable development* was by the World Commission on Environment and Development (WCED), also known as the Brundtland Commission. The controversial term was coined in 1987 to describe a global strategy in which economic development, resource management mindful of the future, political freedoms, and human resources like literacy and health services grow together, not at the expense of one another.

SUSTAINABLE DEVELOPMENT—*continued*

4306. Eco-industrial park fueled by Ocean Thermal Energy Conversion (OTEC) was created in 1990. OTEC is administered by the Natural Energy Laboratory of Hawaii Authority (NELHA) at Keahole Point, HI. The state of Hawaii created the agency to combine NELHA's research activities with commercial development utilizing OTEC technology. Businesses use OTEC technology provided by NELHA for aquaculture, refrigeration, water desalinization, agriculture, electrical power, and the manufacture of pharmaceuticals.

4307. Eco-industrial park in the United States was the Cape Charles Sustainable Technology Park, located on Virginia's eastern shore peninsula. The facility was developed during the 1990s by the Sustainable Technology Park Authority, a managerial agency created by Northampton County and the town of Cape Charles. Its businesses include a seawater desalinization plant.

4308. U.S. program to use Christmas trees for wetlands restoration was the Jefferson Parish Christmas Tree/Marsh Restoration Program, with headquarters in Harahan, LA. It was established in 1991 and provided for collection of discarded Christmas trees, disposal of which is a perennial solid waste and litter problem. Hundreds of thousands of the trees are bundled and used as sediment traps to help lessen severe erosion problems that threaten the state's wetlands and fisheries.

4309. Sustainable development accord on the Alps was the Convention on the Protection of the Alps. Also known as the Alpine Convention, the document was opened for signatures on December 7, 1991, in Salzburg, Austria and entered into force on March 6, 1995. The Alpine nations who signed the convention agreed upon a broad program of environmental protection and prudent development strategies for the entire region, which later specifically included land and resource management, agricultural, recreational, waste, and pollution prevention policies.

4310. United Nations sustainable development strategy was adopted June 14, 1992, at the 1992 Conference on Environment and Development, the so-called Earth Summit held in Rio de Janeiro, Brazil. The program was entitled Agenda 21 and attempted to provide a comprehensive model for sustainable development policies. The 40 chapters of the Agenda 21 document stressed the interconnectedness of environmental management, social and political freedom, economic growth, protection of natural resources, financial planning, education, and poverty eradication measures.

4311. Building industry consensus coalition for "green" construction in the United States was the U.S. Green Building Council, created in 1993. The nonprofit coalition of building companies, utilities, architects, real estate concerns, financial institutions, and environmental groups is based in San Francisco. The national council was created to promote "Green Building" policies, programs, technologies, standards, and design practices.

4312. Environmental quality monitoring agency of the European Union (EU) was the European Environment Agency (EEA), established in 1993 by the 15 nations of the European Union. The agency is based in Copenhagen, Denmark, where it collects, coordinates, and disseminates information vital to environmental protection of the region. Its data is available to member countries weighing sustainable development policy decisions.

4313. Permanent United Nations sustainable development agency was the Commission on Sustainable Development (CSD). The organization was formed to promote and monitor the progress of Agenda 21, the sustainable development strategy adopted by the UN in 1992. The CSD held its first session at UN headquarters in New York City in June.

4314. U.S. President's Council on Sustainable Development was created by President Bill Clinton on June 29, 1993. After signing Executive Order 12852, Clinton appointed a national advisory council of 25 representatives from industrial, environmental, governmental, and not-for-profit organizations with experience in sustainable development matters. The new body replaced the earlier President's Council on Environmental Quality. Membership in the council was increased from 25 to 29 on November 20, 1995, by Executive Order 12980.

4315. International airport to be redeveloped as a community was Stapleton International Airport near Denver, CO. After the airport closed, Denver officials established the Stapleton Development Corporation in 1995 to oversee new development for the 4,700-acre site. The new plan mixed residential housing with easy access to new environmentally friendly commercial development, community recreation, and agricultural facilities, and open spaces. The first Stapleton housing units opened in 1999.

4316. Sustainable development handbook issued by the Canadian government was *A Guide to Green Government*, published in June 1995. Its purpose was to assist government agencies in carrying out their individual mandates while instituting sustainable development policies, as required by the Auditor General Act of 1995. The guide stated broad goals of integrating development management and policy decisions with environmental and ecological priorities. It also provided a bureaucratic guide by which agencies could prepare their policies under the new law. Appendices of the guide offer "best practices" in procurement, water and energy usage in federal buildings, motor vehicle use, and management of land, waste, and human resources.

4317. Sustainable development governmental post in Canada was that of Commissioner of the Environment and Sustainable Development (CESD). The position was established as part of Canada's Auditor General's Office by December 15, 1995, amendments to the Auditor General Act. The amendments also required other departments in the Canadian government to incorporate sustainable development principles and practices into their policies, programs, and operations. The annual sustainable development policies of these individual departments are overseen by the CESD.

4318. International Standards Organization (ISO) certification program for environmental management is called ISO 14000. The ISO, a nongovernmental organization based in Geneva that acts as a global standardization body, began producing guidelines in September 1996 for companies wishing to manage themselves in an environmentally responsible manner. Expected benefits range from lower production costs to good public relations. Impetus for ISO 14000 resulted from the 1992 United Nations Conference on Environment and Development in Rio de Janeiro.

4319. Permanent forum for coordinating sustainable development in the Arctic was the Arctic Council. The organization was established by Canada, Denmark, Finland, Iceland, Norway, the Russian Federation, Sweden, and the United States on September 19, 1996, in Ottawa, Canada. The Inuit Circumpolar Conference, Saami Council, and Association of Indigenous Minorities of North Siberia and the Far East of the Russian Federation are also permanent members. The council was established to promote intergovernmental cooperation on matters of Arctic environmental protection and sustainable development policy. Its members meet every two years.

4320. State in the United States to require utility companies to offer a minimum of electricity from renewable energy sources was Vermont on April 3, 1997. The state legislature passed a bill with a minimum renewable requirement, intending to increase competition in the power industry, reduce pollution, create jobs, and ensure the survival of sustainable energy businesses in the marketplace.

4321. U.S. state with a sustainable development agency was New Jersey. On April 22, 1997, the New Jersey Office of Sustainability opened in Trenton. It offers loans to help establish environmentally friendly businesses, promotes sustainable practices among existing businesses, and provides technical assistance to help state government agencies procure goods and services from local, environmentally sound firms.

4322. "Green" U.S. Post Office was the 8th Street Station in Fort Worth, TX, which opened on January 30, 1999. The variety of "environment-friendly" means used to construct and operate the station ranged from building materials to waste recycling, energy-efficient lighting, passive temperature controls, and rainwater runoff management.

SYNTHETIC FUELS

4323. Internal combustion engine in the United States ran on a mixture of ethyl alcohol and refined turpentine. Such fuel was commonly used for lighting before petroleum products and whale oil became popular. The experimental engine was patented on April 1, 1826, by American inventor Samuel Morey of Orford, NH. Morey was well known for his pioneering work in designing steamboats, which he claimed to have invented before Robert Fulton.

4324. Fuel cell was invented by Welsh-born British judge and physicist Sir William Robert Grove in London in 1839. Grove was able to create electrical voltage through an electrolytic chemical process, using platinum electrodes and sulfuric acid as an electrolyte bath. Despite Grove's discovery, no practical application of the technique was developed until over a century later.

4325. Diesel engine fuels were seed oils and peanut oil. German engineer Rudolph Diesel's experiments with the engine he patented in 1892 and continued to develop until his death in 1913 were conducted mainly with vegetable oils, which Diesel considered a reliable and renewable source of fuel.

SYNTHETIC FUELS—*continued*

4326. Mass-produced car to run on ethanol was the Model T, which Henry Ford designed in 1908. While ethanol and ethanol blends were widely used in succeeding decades, low petroleum prices led to ethanol being replaced by gasoline by the 1940s.

4327. Fuel alcohol plant in the United States was established in Atchinson, KS, by the Bailor Manufacturing Company, which began selling fuel alcohol on October 2, 1936. Rye, oats, sweet potatoes, barley, milo, kafir corn, molasses, and rice were used as the raw materials. Five percent of the total output was butyl alcohol and acetone, which were blended with ethyl alcohol, which in turn was blended with gasoline.

4328. Ethanol promotion program in Brazil was instituted in 1976 to reduce dependence on foreign oil and to utilize the country's abundance of sugarcane. The government's national promotion of ethanol resulted in a significant shift away from reliance upon gasoline as a vehicle fuel, a reduction in air pollution, and growth in Brazil's agricultural business.

4329. Biomass feedstock research funded by the U.S. government began in 1978 at the Oak Ridge National Laboratory in Tennessee. The Department of Energy's Biomass Feedstock Development Program (BFDP) at Oak Ridge was part of the laboratory's shift from solely nuclear research toward investigation of alternative energy sources. The BFDP develops environmentally acceptable crops capable of producing large quantities of low-cost, high-quality feedstocks for conversion into biomass energy.

4330. Tax credit in the United States for ethanol was included in the Energy Tax Act, which became effective on December 1, 1978. The act exempted fuel blended to contain at least 10 percent ethanol from the 4-cent federal excise tax imposed on motor fuels. The legislation was passed in response to the energy crisis of the late 1970s and was among the laws discarded during the 1980s.

4331. Alternative fuel measure adopted by the U.S. Postal Service was the conversion of 500 Jeeps to Compressed Natural Gas (CNG), which began in 1979. Twenty years later, the Postal Service's CNG- and ethanol-powered vehicles comprised the largest fleet of alternative-fueled vehicles in the United States.

4332. Federal plan in the United States for developing biomass energy regionally was the Regional Biomass Energy Program (RBEP). In 1983, Congress directed the Department of Energy (DOE) to support biomass technologies based on regionally available resources, like wastes from agriculture, forestry, food processing, and municipal trash collection. DOE established four regional programs: Pacific Northwest and Alaska, Southeast, Great Lakes, and Northeast. A Western region was created in 1987. The decentralized, regionally managed federal programs cooperate with industries, trade associations, universities, farmers, and state agencies in efforts to develop biomass conversion technologies.

4333. Law in the United States requiring government agencies to purchase alternative-fueled vehicles was the Alternative Motor Fuels Act of 1988. The law encouraged the development of methanol, ethanol, and natural gas as transportation fuels and called for the federal government to purchase as many alternative fuel-powered cars and trucks as was practical.

4334. Transatlantic flight powered by ethanol took place on October 21, 1989. After flying a single-engine Velocity aircraft from Waco, TX, Baylor University aviation scientists Maxwell Shauck and Grazia Zannin flew the plane from Newfoundland to the Azores Islands. They continued on to Lisbon and Paris. The flight was intended to demonstrate the viability of ethanol as an aircraft fuel.

4335. Buses to use liquid natural gas (LNG) and diesel were placed in service in Houston, TX, in 1991. The Metropolitan Transit Authority of Houston (Metro) ran a pilot program of 14 buses that used diesel to ignite vaporized LNG. The success of the program led to Metro utilizing the largest fleet of LNG-fueled buses in the world. Local availability of LNG, convenience of refueling, and increased passenger loads because of the lighter fuel weight helped make it an attractive option.

4336. Car fueled by methanol was introduced in June 1992 by General Motors in Detroit, MI. The Chevrolet Lumina Variable Fuel Vehicle could be powered by unleaded gas, by methanol (a fuel derived from plant fermentation that is similar to alcohol), or by any combination of the two. The first production model was delivered to Governor Tommy G. Thompson of Wisconsin.

4337. Tax deduction in the United States for purchasers of Hybrid Electric Vehicles (HEVs) was created by the Energy Policy Act, which became law on October 24, 1992. Although it was primarily aimed at electric-powered vehicles, the law also applied to qualified HEVs and electric cars powered by fuel cells. The Electric Vehicle Tax Credit allowed a one-time 10 percent or $4,000 maximum deduction for qualified vehicles.

4338. Tax deductions in the United States for purchasers of Clean Fuel Vehicles (CFVs) were created through the Energy Policy Act, which became law on October 24, 1992. The Internal Revenue Service allowed deductions ranging from $2,000 for small CFVs and $5,000 for vans to $50,000 for new purchases of larger vehicle like buses.

4339. Survey of U.S. biomass resources was undertaken in 1995 and was called the Biomass Resource Assessment Task. The program, funded by the U.S. Department of Energy's Resource Assessment Program, was established to develop biomass resources and make information about them available to federal energy agencies, environmental groups, public utilities, and state economic development, energy, conservation, and regulatory agencies.

4340. Biodiesel plant in Illinois was opened in 1996 by the Columbus Food Company. In a building donated by the city of Chicago, the oil-packaging firm converted used cooking oil and fresh soybean oil into biodegradable fuel suitable for operating buses.

4341. Commercial fuel cell powered by landfill gas began operating in June 1996 in Groton, CT. A 200-kilowatt fuel cell system maintained by the town's Public Works Department converted methane produced by waste at the dump into electricity. The project was a partnership between the town, the Environmental Protection Agency, Connecticut Light & Power, and International Fuel Cell Inc.

4342. Use of biodiesel fuel at a U.S. national political convention occurred at the 1996 Democratic Convention in Chicago. Between August 26 and 29, delegates and media representives were transported by Chicago Transit Authority and American Sightseeing Tours buses, whose fuel was provided by a local biodiesel plant.

4343. Fuel cell passenger car to use methane was the NECAR 3, built by Daimler-Benz and first publicly displayed in 1997. The car used methanol to create hydrogen that powered fuel cells through a chemical process. By using hydrogen, the car's electrical engine didn't cause the pollution problems associated with combustion engines. The methanol process eliminated the need for hydrogen tanks that were a fixture on earlier fuel cell vehicles. The car had its debut at the Frankfurt International Motor Show.

4344. Plant to produce methanol from coal was a liquid-phase methanol demonstration project in Kingsport, TN. The project was the result of a commercial research and development program involving the Eastman Chemical Company, the Federal Energy Technology Center, and Air Products Liquid Phase Conversion Company. At the time the plant was dedicated on July 25, 1997, it had begun synthesizing storable methanol from coal, with possible applications in producing electric power, transportation fuels, and chemical manufacturing.

4345. Music festival powered by biodiesel fuel was the Green Mountain Know Your Power Festival, held on September 26, 1998, at the Mann Center For the Performing Arts in Philadelphia, PA. More than 50,000 people attended the concert, whose electrical generators and vehicles were powered by biodiesel and solar energy. The concert was sponsored by Green Mountain Energy, an independent power company specializing in renewable power sources.

4346. Biomass-to-ethanol conversion plant was built in Jennings, LA, by the Massachusetts-based BC International Corporation. Construction began on October 20, 1998. The plant intended to use a patented, genetically engineered bacterium to convert organic waste from rural Louisiana's agriculture into industrial chemicals and fuel.

4347. U.S. Interagency Council on Biobased Products and Bioenergy was created by Executive Order 13134, "Developing and Promoting Biobased Products and Bioenergy", signed by President Bill Clinton on August 12, 1999. It created a council composed of the Secretaries of Agriculture, Commerce, Energy, and the Interior; the Administrator of the Environmental Protection Agency; the Director of the National Science Foundation; and other officials. The council was charged with creating a national strategy by which the federal government could assist in the promotion of biomass-produced energy as well as biobased commercial and industrial products.

V

VOLCANOES

4348. Volcano eruption known in North America happened circa 2 million B.C. in what is now Yellowstone National Park in Wyoming. The ash darkened the atmosphere worldwide, and some scientists believe this began the Great Ice Age, or Pleistocene Epoch, which lasted until 8000 B.C.

4349. Volcano eruption of significance in Japan was in 8540 B.C. by Mount Fuji. Located on the island of Honshu, it is still active and last erupted in 1707.

4350. Volcano eruption of significance in Canada was in 8055 B.C. by Mount Garibaldi in British Columbia.

4351. Volcano eruption of significance in Chile was in 6880 B.C. by Mount Llaima. It is still active and last erupted in 1995.

4352. Volcano eruption to create a lake in North America was circa 5000 B.C. at Mount Mazama in present-day southwestern Oregon. The explosion created a caldera (volcanic crater) 6 miles across and 1,932 feet deep; the blue lake that has formed is preserved as Crater Lake National Park. Mazama is believed to have originally been about 10,000 feet high. The ash from the eruption covered all of the northwestern United States and drifted as far as Saskatchewan, Canada. Later small eruptions occurred—one creating Wizard Island in the lake—and some volcanologists believe Mazama is only temporarily dormant.

4353. Volcano eruption known on a Mediterranean island was at Santorini (now called Thíra), probably between 1645 and 1628 B.C. The eruption caused the center of the island to sink and fill with sea water. Some believe that this was the basis for the myth of a lost Atlantis, or that the eruption caused earthquakes that severely damaged the Minoan civilization on Crete. The volcano is still active and last erupted in 1950.

4354. Volcano eruption that was documented was by Mount Etna in 1500 B.C. Located in northeast Sicily, it has continued to have an active history of eruptions. One of the world's most famous volcanoes, Mount Etna is 10,750 feet high, with a circumference of 90 miles.

4355. Volcano eruption known to occur in Mexico was circa 400 B.C. from Popocatepetl, 45 miles southeast of Mexico City. Volcanologists believe the explosion and mudflow devastated an area 24 miles in diameter.

4356. Volcano eruption whose devastation was fully documented was from Mount Vesuvius on August 24, 79, in Italy. An estimated 16,000 to 20,000 people were killed when hot ash and boiling mud covered the cities of Pompeii, Herculaneum, and Stabiae, all near present-day Naples. Pliny the Younger, 20 miles away, recorded the eruption in detail, and archaeologists have since excavated the cities, recovering many preserved bodies and household items.

4357. Volcano eruption in New Zealand occurred circa 150 in the center of the North Island, where the city of Taupo now lies. The only remaining sign of the volcano, which turned the island into a desert of ash, is Lake Taupo in the craterlike basin.

4358. Volcano eruption in Central America known to have moved a civilization occurred circa 265. Mount Ilopango exploded in what is now El Salvador, and the surviving Mayan people resettled hundreds of miles away from the destruction. It is thought that their culture did not fully recover until two centuries later.

4359. Volcano eruption of significance in Iceland was in 1362 by Oraefajokull, the largest since settlement began in the nineth century. Floods were created by the blast, and about 200 people were killed as well as livestock.

4360. Volcano eruption known to cause massive deaths in Indonesia was by Mount Kelut on Java in 1586. The eruption killed some 10,000 people. Kelut erupted again on May 19, 1919, causing another 5,000 deaths.

4361. Volcano eruption in North America for which a date can be estimated occurred about 1694 at Cinder Cone, in the Lassen Peak district of California.

4362. Volcano eruption of significance recorded in today's Ecuador occurred in 1744 by Mount Cotopaxi. It is located in the Eastern Cordillera of the Andes in the central part of the country between Ambato and Quito. One of the world's highest active volcanoes at 19,347 feet, Cotopaxi last erupted in 1940.

4363. Paper on the extinct volcanoes of Europe was presented by Jean Étienne Guettard on May 10, 1752, to the French Academy of Sciences. On his geological travels in France he found volcanic rock, and he proposed that there were extinct volcanoes in central France.

4364. Volcano eruption of significance in Java was in 1772 by Papandayan. It collapsed about 4,000 feet and killed about 3,000 people, destroying 40 communities. Papandayan is still active.

4365. Volcano known to have caused significant deaths by starvation was on June 8, 1783, in Iceland. About 10,000 people starved when crops and livestock were killed by the lava flow and poisonous gases expelled from the 20-mile Laki fissure southwest of Vatnajokull. Even fishing was halted because of the toxic gas. The flow covered 218 square miles.

4366. Volcano to cause deaths by famine in Indonesia was Mount Tambora, which erupted in 1815 on the island of Sumbawa in the Lesser Sundas. About 12,000 people died in the ash and fumes that covered farms and fishing areas to create a famine that killed more than 90,000.

4367. Non-Hawaiian to observe Kilauea Volcano erupting in the area that later became Volcanoes National Park was Reverend William Ellis, a missionary from New England, who toured the Big Island in 1823.

4368. Volcano eruption mistaken for a war was in 1835 when Cosiguina exploded in northwest Nicaragua. The rumbling was heard some 800 miles away in neighboring Belize, which thought it was a large-scale military invasion and mobilized its army.

4369. Volcanoes discovered in Antarctica were seen in 1841 by Scottish explorer and naval officer Sir James Clark Ross on his expedition for the British Admiralty. He named them Mount Erebus and Mount Terror for his two boats. Erebus is the only active volcano at the Antarctic, having a crater of molten lava and hurling balls of cooling lava through the air two or three times a day.

4370. European to "discover" the volcanic Crater Lake in Oregon was John Wesley Hillman in 1853. Hillman rode a mule up the outer slope of the volcano until he suddenly reached the rim of the caldera.

4371. Volcano eruption to be audible from 3,000 miles away was that of Krakatoa, Indonesia, in 1883. The explosion was heard on the distant Indian Ocean island of Rodriguez.

4372. Volcano eruption known to cause catastrophic tsunamis (giant sea waves) occurred from August 26 to 28, 1883, on the volcanic island of Krakatoain Indonesia. The island was uninhabited, but some 36,000 people living on the nearby islands of Java and Sumatra were killed by the waves, which reached all the way to Cape Horn in South America and as far north, some scientists believe, as England. Two-thirds of Krakatoa was destroyed in this, the Earth's greatest recorded explosion.

4373. Eruption in modern times of the Mexican volcano El Chichon in the southeastern state of Chiapas occurred on March 28, 1982. The eruption killed 167 people, left 60,000 homeless, and caused a substantial temporary rise in air temperature in the northern hemisphere.

4374. Volcano eruption recorded as a major disaster in the Western Hemisphere was on April 24, 1902, in Guatemala. Some 1,000 people were killed by the eruption of Santa Maria. Another 3,000 people died from an outbreak of malaria that followed.

4375. Volcano causing massive deaths in the West Indies was Mount Pelee on the island of Martinique in the Windward group, killing some 30,000 to 40,000 people during an eruption on May 8, 1902. Its former capital of St. Pierre was totally destroyed.

4376. Significant use of the natural heat of volcanic systems for electrical power in Italy dates from 1904. Geothermal power production continues today in the Tuscan region of Italy.

4377. U.S. national park in a volcanic range was the Hawaii Volcanoes National Park, established on August 1, 1916. The park consists of the active volcanoes Kilauea and Mauna Loa.

4378. Volcano information network of significance set up internationally was the Volcanology Section established in 1922 by the International Union of Geodesy and Geophysics (IUGG) at their first General Assembly held in Rome. The section organized its Central Bureau of Volcanology, which first met on May 22, 1922, in Naples. At the IUGG's fourth General Assembly in 1930 in Stockholm, the Volcanology Section became the International Association of Volcanology (IAV). It informed the IUGG national committees by telegraph of the onset of volcanic eruptions, established a central international volcanological library, and published the *Bulletin Volcanologique*.

4379. Volcano eruption to end a World War II battle happened in March 1944 when Mount Vesuvius exploded. Fighting between tens of thousands of Allied and German forces in the Naples area was halted and soldiers helped in rescue efforts. The death toll was fewer than 100. This eruption raised Vesuvius's height by about 500 feet and widened its top threefold. The volcano's vent was also obstructed, turning its previous bellowing smoke into trails of steam.

VOLCANOES—*continued*

4380. Scientific explanation for the distribution of volcanoes around the world was the theory of plate tectonics, formulated in the early 1960s. Most volcanoes are found along or near the boundaries of the dozen or so large shifting plates that make up the Earth's crust.

4381. Volcanic island scientifically studied during its creation was Surtsey, a new island south of Iceland created from 1963 until 1967. It is 11 miles southwest of Iceland's island of Heimaey. The volcanic explosion began November 14 on the ocean's floor about 426 feet in depth. The interaction of magma and sea water produced volcanic ash that immediately formed a crater above the surface. By February 1964, the crater was complete, enclosing the magma so it then became a lava flow that formed rock. Fully developed in 1967, Surtsey rises 492 feet above the sea level and covers two square miles. Visits are now restricted and scientific research teams continue to study how the world's youngest island evolves with flora and fauna.

4382. Large-scale astronomical observatory on a volcano was the Mauna Kea Observatory, located on Mauna Kea, an extinct volcano on the island of Hawaii. The volcano, between Hilo and Honolulu, is the highest island-mountain in the world. The initial optical telescopes were built in 1968 and today number eleven working telescopes.

4383. Close-up evidence of volcanoes on Mars dates from 1971, when the U.S. *Mariner 9* spacecraft sent back images of the planet. The largest volcano measured 75,000 feet high and covered an area the size of Ohio.

4384. Volcano lava flow to be fought with sea water happened on Heimaey, an island off Iceland, after a fissure opened up on January 23, 1973. One third of Vestmannaeyjar, a harbor town, was destroyed. After the island was evacuated, volunteers returned to successfully spray 6 million tons of sea water to harden the lava before it could block their harbor. The fissure closed again on June 26.

4385. Steam well in a Hawaiian volcano was drilled in 1976 in Mt. Kilauea. In its first test, the geothermal well produced steam for four hours.

4386. Lava flow known to have attained a speed of 40 mph occurred in Zaire (now the Democratic Republic of Congo) in 1977, when a lake of molten rock that had collected in the crater of Nyiragongo burst through a fissure and poured down the volcano's side.

4387. Photographs of erupting volcanoes on the Jovian moon Io were taken by the U.S. spacecraft *Voyager 2* in 1979. *Voyager 2*, launched in September 1977, encountered Jupiter on July 9, 1979. Io's were the first volcanoes found on another body in the solar system.

4388. Volcano eruption in the United States photographed by weather satellites was Mount St. Helens in the Cascade Range of Washington State on May 18, 1980. The pictures taken hourly allowed meteorologists to warn pilots of the eastward drift of ash and dust clouds and to calculate the effects of the eruption on the worldwide climate.

4389. Volcano known to claim human life in the contiguous 48 states was Mount St. Helens, located in the Cascade Range of Washington State. Its massive eruption began on May 18, 1980. The rain of hot ash, pyroclastic flows of superheated gas and mud, and flash floods took the lives of 61 people.

4390. Volcano in Hawaii tapped to supply a geothermal power plant was Mount Kilauea in 1981. The three-megawatt plant was located at Puna on the island of Hawaii and supplied energy for the local grid system.

4391. Volcano eruption causing massive deaths in Colombia was from Nevada del Ruiz on November 13, 1985, when an estimated 25,000 people died. Mudslides caused by the eruption buried the town of Armero. The volcano is 85 miles north of Bogotà.

4392. Volcanic gas from a lake to cause massive deaths in Cameroon was in 1986 from Lake Nyos, killing 1,750 people. Investigating scientists said the cause was carbon dioxide from the volcano's eruption underwater. Two years earlier, a similar tragedy had occurred in Cameroon at Lake Monoun, killing 37 people.

4393. Documented case of severe damage to a jetliner flying through a volcanic ash cloud dates from December 1989. All four engines of a Boeing 747-400 shut down after the aircraft entered the ash cloud from Redoubt Volcano southwest of Anchorage, AK, at 26,000 feet. The crew managed to restart the engines at 13,000 feet and land safely at Anchorage. The December 1989 explosion of the Redoubt volcano sent ash columns rising to altitudes exceeding 40,000 feet.

4394. Evidence of volcanic activity on Triton, a moon of the planet Neptune, came from images sent by the U.S. spacecraft *Voyager 2* in 1989. Stereo images of a gas plume rising five miles above Triton's surface suggested volcanic activity.

4395. Evidence that Martian volcanic rocks are in some respects similar to Earth's came from the U.S. *Mars Pathfinder* spacecraft, which landed on Mars on July 4, 1997.

W

WATER POLLUTION

4396. Major pollution damage to West Coast oyster beds occurred during the first decade of the 20th century. Between 1899 and 1925, California oyster production fell from 3.4 million pounds to 700,000 pounds a year, in part because of heavy pollution in San Francisco and San Diego bays.

4397. Synthetic detergents for clothing were developed in Germany during World War I. Widely available after 1945, synthetic detergents created a major water pollution problem in the 1950s and 1960s. The development of biodegradable detergents largely corrected the problem.

4398. Government declaration of large stretches of the Hudson River as dangerous for recreation was issued in 1915, due to wastewater and industrial pollution.

4399. Long-standing survey of water quality conditions in the Western Approaches to the English Channel dates from the 1920s.

4400. Potomac River pollution abatement conference met on August 22, 1957. Conferees found the river unfit for recreation, commercial use, or game habitation. Despite these findings, conference activities did not lead to effective cleanup action until 1969.

4401. Reports of decreases in commercial fishermen's catches in Lake Superior date from the early 1960s. In 1963, U.S. Senator Gaylord Nelson asked for federal action against a growing pollution problem in the largest of the Great Lakes.

4402. Federally sponsored water resource research in the United States began in 1964 with congressional passage of the Water Resources Research Act.

4403. Public statement by a U.S. president that all of the country's major river systems were polluted came in February 1965. In his message to Congress on natural beauty, President Lyndon B. Johnson said U.S. rivers were so fouled they were no longer sources of beauty and pleasure.

4404. Declaration of phosphates as a major source of water pollution issued jointly by two countries was made by a U.S.–Canadian international commission in 1969. Later that year, Congress conducted hearings on phosphate pollution that spurred the development of phosphate-free laundry detergents.

4405. Phosphate-free laundry detergent was developed in the United States in the summer of 1970 by Sears, Roebuck and Co. Discharges of phosphorous compounds contributed substantially to the rapid eutrophication of U.S. lakes and rivers. Widespread use of the new product significantly reduced the amount of phosphates entering these waters.

4406. U.S. government listing of the nation's ten most polluted rivers appeared in the spring of 1970. The list included the Ohio River, the Houston Ship Canal, the Cuyahoga River (Ohio), the Arthur Kill (New York–New Jersey) and the Androscoggin River (Maine–New Hampshire).

4407. Ban on the use of phosphate-containing laundry detergents in the United States was by Suffolk County, NY, in 1979. Other states and localities soon followed suit. By 1988, more than a quarter of all city and state agencies had banned detergents with phosphates.

4408. International meeting of American Indians and environmentalists was held in 1979 to protect the Columbia River from more dams. Organizer Hazel Wolf met with 26 U.S. and Canadian tribal leaders, conservationists, and farmers to discuss the American government's new irrigation project along the Columbia River. Their activism caused the government to drop the project.

4409. Eutrophication monitoring in the Baltic Sea began in 1980 and showed evidence of increasing oxygen concentration and levels of nutrients. Eutrophication is the aging of bodies of water, usually lakes, by biological enrichment of the water.

4410. Time-constrained goals to be adopted by a group of nations to curb water pollution were set at a meeting in Genoa, Italy in 1985. The three goals, part of the ten priorities that were set for the next ten years, included the establishment of sewage treatment plants in all cities with more than 100,000 inhabitants; identification and protection of at least 100 sites of common interest; and identification and protection of at least 50 new marine and coastal sites and reserves.

WATER POLLUTION—*continued*

4411. Laundromat with an environmental theme opened in New York City on April 1, 1992. Called Ecowash, it used biodegradable laundry detergents and machines that required less water than conventional washers.

WATER POLLUTION—DESTRUCTION OF SPECIES

4412. Evidence that human pollution had reduced fish spawns in the Saginaw River and Saginaw Bay of Lake Huron emerged about 1845. Generally, rivers of the Great Lakes region were being described as "turbid" by the end of the 1840s.

4413. Fish-stocking program failures in acidic lakes in upstate New York occurred during the 1920s in at least a dozen lakes, where it was believed the acid content caused the death of the fish.

4414. Reports of sharp decreases in dissolved oxygen levels in Lake Erie were in 1929. Pollution-caused falling oxygen levels made it increasingly difficult for aquatic life to survive in the lake.

4415. Census of fish kills by pollution in the United States was taken in 1960. The U.S. Public Health Service conducted the census in cooperation with leading conservation groups.

4416. Report that pollutants discharged into the Detroit River may have killed bald eagle hatchlings along Lake Erie was issued March 2, 1992, by the National Wildlife Federation.

4417. Evidence that a natural occurrence rather than toxins was causing deformities in frogs around the United States emerged in April 1999. Scientists in New York and Oregon reported that, at least in some localities, parasitic worms appeared to be responsible for the deformities.

WATER POLLUTION—HEALTH

4418. Paralytic shellfish poisoning (PSP) cases were recorded in Canada in 1793. By the late 20th century, cases of PSP, caused by a lethal toxin stored in the bodies of clams, oysters, mussels and other shellfish, were common throughout the world.

4419. Water purity tables were created by Ellen Swallow at the end of the 1800s. These tables, a result of a water survey done by the Massachusetts Institute of Technology for the Massachusetts Board of Health, gave governments the first set of guidelines to set water quality standards.

4420. Researcher to prove the transmission of cholera via a contaminated water supply was British epidemiologist John Snow in the 1850s. Snow demonstrated the connection after the London cholera epidemic of 1849 and strengthened his findings with research in the wake of a second outbreak in the British capital in 1855.

4421. Standards for water quality in the United States were set in Massachusetts in 1900. They were based on results of a water survey done by Ellen Swallow for the Massachusetts Board of Health at the end of the 19th century.

4422. DES hormone ban by the U.S. government was issued in 1960. The U.S. Food and Drug Administration banned the use of the synthetic hormone in chicken feed. The FDA found that DES, (diethylstilbestrol) an additive used to fatten chickens, polluted rivers and streams in agricultural runoff and was carcinogenic.

4423. Significant conference on the risks of chemical contaminants to future generations to be sponsored by the U.S. government was convened under National Institutes of Health sponsorship in 1966. The conference warned of the dangers of contaminants in water from industrial and other sources.

4424. Study of a major U.S. water supply linking pollutants to high cancer rates was conducted by the Environmental Defense Fund in New Orleans. The findings led to the passing of the Safe Drinking Water Act in 1974.

4425. Aquifer in the United States granted sole source protection from pollution was the Edwards underground reservoir near San Antonio, TX, so designated on December 16, 1975, one year after the federal Safe Drinking Water Act became law.

4426. Evacuation of an entire chemically contaminated neighborhood in the United States occurred at Love Canal in Niagara Falls, NY, from 1978 to 1980. From 1942 to 1953, the Olin Corporation and the Hooker Chemicals and Plastics Corporation had buried more than 20,000 tons of highly toxic chemical waste in Love Canal. In the 1970s, a rise in the water table caused cellar flooding and a release of chemical odors into the air. Children and animals suffered chemical burns; trees and plants died; residents soon developed serious illnesses, including cancer. The grassroots Love Canal

Homeowners Association successfully lobbied the state government for a complete evacuation and resettlement, and by 1980 nearly everyone had gone. A massive and expensive cleanup followed.

4427. Town evacuated by the Environmental Protection Agency for dioxin levels 100 times emergency standards was Times Beach, MO. In 1986 the EPA evacuated and bought out the town so it could be decontaminated.

4428. Beach closings in the United States due to medical wastes washing ashore occurred in 1988. Beaches on the East Coast, Lake Michigan, and Lake Erie were closed during the summer of 1988 to protect beachgoers from possible health risks.

WATER POLLUTION—INDUSTRIAL DISCHARGE

4429. Corporate effort at pollution control at Du Pont de Nemours chemical plants in the United States began in the mid-1930s. Over a 30-year period to the mid-1960s, the company claimed to have spent or authorized spending of $91 million on pollution abatement.

4430. Dumping of highly toxic PCBs (polychlorinated biphenyls) in waterways occurred in the late 1930s. PCBs were used as electrical insulators and as additives in paints; when discarded, they contaminated harbors, rivers and water supplies.

4431. Federal program of significance to seal up abandoned mines in the United States to prevent water entry and polluting drainage was launched in the 1930s as part of the Works Progress Administration (WPA). Initial WPA efforts were not sustained, however.

4432. Widespread attempts to grow yeast on pulp wastes from paper manufacturing in the United States occurred as early as 1943. The Sulfite Pulp Manufacturers League set up yeast plants as part of an effort to curb polluting discharges of pulp waste into rivers and streams.

4433. Significant pollution abatement at the giant Du Pont de Nemours chemical plant on the Kanawha River in Belle, WV began in 1945. The plant was one of a dozen large industrial polluters in the Kanawha Valley in the Charleston, WV, region.

4434. Widespread incidence of poisoning on account of mercury pollution occurred in the 1950s along Minamata Bay, Kyushu, Japan. People and animals became ill from eating mercury-contaminated fish, and at least 11 people died or were severely disabled between 1953 and 1960; ultimately, more than 3,000 were affected. A local chemical and plastics factory was blamed for the discharge of mercury-laced effluent into the bay.

4435. Public hearing on meatpackers' discharge of offal into the Missouri River sponsored by the U.S. Public Health Service was in 1958 in Sioux City, Iowa. The dumping contaminated the river water of four states.

4436. Strict standards for maintaining stream temperatures in a U.S. state were set in Pennsylvania in 1960. The state decreed that warm effluent discharges from industrial sources could not raise water temperatures above 93 degrees F.

4437. Generally known instance of a river catching fire because of pollution occurred on the Iset near Sverdlovsk in Russia in 1965.

4438. Water pollution national conference of significance sponsored by the U.S. Chamber of Commerce to address the issue of industrial waste met in Washington, DC, in December 1965. Major industrial and municipal organizations were represented. One speaker, Oklahoma governor Henry Bellmon, told the convention that a principal purpose of a waterway was to serve as a "repository of waste."

4439. Federal government move to relax water pollution controls for U.S. companies that discharged nontoxic substances came in August 1979. The Environmental Protection Agency estimated the policy shift would save business $200 million a year.

4440. Large-scale mercury pollution of streams in northern Brazil occurred in the late 1980s. Gold prospectors used mercury to try to find gold; in 1988, 100,000 were at work in northern Brazil.

4441. Agreement among the Big Three U.S. automakers to reduce the dumping of toxins into the Great Lakes was announced on October 1, 1991. The reductions were voluntary, however, and the car makers noted it would be up to them to find the most cost-effective ways of making them.

4442. Joint commitment to "zero discharge" of dangerous chemicals into Lake Superior by the United States and Canada was announced on October 2, 1991. Lake Superior is one of four Great Lakes (out of five) with a common U.S.-Canadian border.

WATER POLLUTION—INDUSTRIAL DISCHARGE—*continued*

4443. Significant hog waste leakage into the New River of North Carolina occurred in 1995. A holding lagoon leaked 25 million gallons into the river, causing an algal bloom, oxygen depletion, and fish kills.

WATER POLLUTION—LEGISLATION AND REGULATION

4444. Order to Dutch linen bleachers barring the dumping of wastes in canals was issued in 1582. The bleachers were instructed to use separate channels known as "stinkerds."

4445. Proclamations against pollution caused by London starchmakers were issued by King James I in 1607.

4446. Law making water closet connections to surface streams compulsory in London was enacted in 1847. Waste was thus flushed through, turning watercourses into open sewers.

4447. Criminal statutes against water pollution in Great Britain were enacted in 1876. They required local governments to prosecute polluters.

4448. Water pollution legislation passed by Congress was the Refuse Act of 1886. The measure barred dumping of refuse that obstructed navigation in New York Harbor.

4449. Law making it illegal to dump waste in U.S. navigable waters except by special permission was the Rivers and Harbors Appropriations Act, passed by Congress in 1899.

4450. Rivers and harbors antipollution law in the United States was the Rivers and Harbors Appropriations Act, passed in 1899 to control the pollution of navigable rivers and river and ocean harbors.

4451. Formal complaint about Missouri River pollution from stockyards and municipal sewers in St. Joseph, MO, was made to the U.S. Public Health Service in 1912. The government did not, however, order Armour, Swift, and the city of St. Joseph to take corrective action until 1957.

4452. Legislation on health issues involving water pollution passed by the U.S. government was the Public Health Service Act of 1912. The measure led to studies that demonstrated a link between polluted water and human health.

4453. Federal law in Germany to clean up the Ruhr river dates from 1913. The measure set up two public corporations, one to assure adequate water supplies and the other to keep the water clean. The Ruhr, which flows through a heavily industrialized region between Winterberg and Duisburg, has been regarded as Germany's cleanest major river.

4454. State law in the United States to protect wild and scenic rivers from development was the Wild and Scenic Rivers Act passed by the Wisconsin legislature in 1915.

4455. Law to regulate the discharge of petroleum products into coastal waters was the Oil Pollution Control Act. Congress approved the measure in 1924.

4456. Anglers' Co-operative Association request for an injunction against a water polluter under English Common Law was made in 1948. The association was formed that year to challenge three major corporations that were polluting the River Derwent and the River Trent.

4457. Multistate clean streams campaign for the Ohio Valley was launched in 1948. Eight states agreed to act jointly to encourage industries and municipalities to build sewage and waste control facilities.

4458. Water pollution law enacted by the U.S. Congress took effect on June 30, 1948. The Water Pollution Control Act authorized the Justice Department to file suit against polluters and provided funds for sewage treatment systems and pollution research.

4459. Large-scale federal legislation aimed at purifying salty waters in the arid western United States was the Saline Water Act of 1952. The measure called for the conversion of saline water to water suitable for agricultural, industrial, and other "beneficial" uses.

4460. Agency in the United States created to deal exclusively with the problem of water pollution was the Federal Water Pollution Control Administration, established in 1956 under the provisions of the Water Pollution Control Act of that year.

4461. Federal grants-in-aid to U.S. municipalities for pollution cleanup and control were issued in 1956. Federal aid supplemented local levies, which provided 90 percent of the funding for stream cleanup.

4462. Government-built saline water conversion demonstration plants in the United States were authorized in 1958. Congress extended the saline water conversion program, which mainly benefited the arid West, in 1965 and 1967.

4463. List of protected rivers in the United States was prepared by the National Park Service in 1964.

4464. Proposals for a federal tax on water polluters in the United States date from the autumn of 1964. The proposals called for an "effluent charge" to be levied on municipalities as well as businesses found to be polluting water sources.

4465. "Save Lake Erie" petition was turned over to Ohio governor James A. Rhodes in April 1965. It contained more than one million signatures and demanded action to clean up the terribly polluted lake.

4466. Water quality standards established at the U.S. federal level were in the Water Quality Act of 1965, which set standards for stream water quality by taking into account the use and value of a stream for public water supplies, propagation of fish and wildlife, and recreational, agricultural, and industrial uses. These water quality standards were enforceable by law and applied to all navigable waters in the United States.

4467. "Shellfish clause" in the U.S. Clean Water Amendments was approved in 1966. The measure gave the federal government jurisdiction over waters where pollution had caused substantial injury to the shellfishing industry.

4468. Ban on phenylmercury as a slimicide in the pulp and paper industry in Sweden was imposed in 1968. Because of the ban, and with dredging and treatment of spoils, mercury levels in fish dropped in some areas along the Swedish Baltic coast, enabling fishing bans to be lifted.

4469. Federal law in the United States protecting and preserving rivers was the National Wild and Scenic Rivers Act passed by Congress in 1968. The act created three classes of protected rivers—wild, scenic, and recreational.

4470. Estuary protection funding in the United States was appropriated under the Estuary Protection Act of August 3, 1968. It provided $500,000 to study and inventory estuaries in the United States.

4471. Ban on swimming in Indiana's Brandywine Creek because of bacterial contamination was issued in 1969. The creek, which flows through Greenfield, inspired James Whitcomb Riley's poem "The Old Swimmin' Hole."

4472. Announcement that the U.S. Army Corps of Engineers would require mandatory disclosure of industrial effluent came in December 1970. The corps said it had the authority to establish a national permit system under the Refuse Act of 1899.

4473. Legal proceedings against Alabama polluters by the Bass Anglers Sportsmen Society were initiated in July 1970. The environmental group resurrected the Refuse Act of 1899 to challenge 214 polluters of Alabama rivers.

4474. Pollution-related ban on commercial fishing in Lake St. Clair on the Canada-U.S. border took effect in March 1970, when Ontario provincial officials acted on reports of mercury contamination of walleye and perch stocks. The Ontario ban also covered commercial fishing in western Lake Erie.

4475. U.S. Justice Department lawsuit against U.S. Steel for failing to follow government recommendations to abate pollution from its Gary, IN, works was initiated in February 1971. The government had originally given the company a December 1968 deadline to begin corrective actions.

4476. Joint effort of major significance to solve pollution problems in the Great Lakes was the International Great Lakes Water Quality Agreement between the United States and Canada, signed on April 15, 1972, in Ottawa, Canada. The pact addressed eutrophication, high levels of toxins, the dumping of raw human wastes, and acid rain.

4477. Law giving the U.S. Environmental Protection Agency authority to set water quality standards and monitor water supplies was the Safe Drinking Water Act, approved by congress in 1974.

4478. Comprehensive U.S. legislation establishing minimum safety standards for chemical contaminants in water was the Safe Drinking Water Act, which became law on December 16, 1974. The act gave the Environmental Protection Agency authority to monitor levels of arsenic, barium, fluoride, lead, mercury, and other contaminants.

4479. Significant federal legislation to ban the discharge of sewage sludge in U.S. waters was the Ocean Dumping Act, approved by Congress in 1977.

WATER POLLUTION—LEGISLATION AND REGULATION—*continued*

4480. Significant multinational agreement to curb pollution of the Mediterranean Sea was reached in 1980. More than 15 Mediterranean nations signed the antipollution pact, which called for extensive studies and a blueprint for cleaning up the sea, in Geneva in 1980.

4481. Congressional action on "nonpoint" source pollution in the U.S. came in 1987. Congress required the states to develop programs to fight pollution from multiple or hard-to-detect sources.

4482. Coordinated assault on "nonpoint" sources of pollution in the United States began with a 1987 congressional amendment to the Clean Water Act requiring the states to develop programs to fight pollution from multiple or hard-to-detect sources. Nonpoint sources are pollution sources that can't be traced to a single discharge, such as runoff from streets and erosion from hillsides.

4483. Systematic legislation to clean up and restore Florida's polluted lakes, rivers, streams and bays was approved on June 29, 1987, by the Florida legislature. Among other things, the Surface Water Improvement and Management Act (SWIM) ended close to 170 years of ditching, diking, and other swamp reclamation efforts in Florida.

4484. National Wetlands Policy Forum to call for a program of no net loss of wetlands was issued in 1988. Under the program, new or restored wetlands were to balance the loss of wetlands to development.

4485. International compact barring the dumping of wastes into the oceans was the London Dumping Compact of 1990. Sixty-five nations signed the agreement, which prohibits the discharge of radioactive and toxic wastes.

4486. Revision of the U.S. National Wetlands Policy to counter the "no net loss" theory occurred in 1991. The revision by the Bush administration reduced the amount of standing water necessary to classify an area as a wetland and led to the opening up of about ten percent more wetlands to development.

4487. Massive cleanup plan for the Florida Everglades was approved on March 13, 1992. The state of Florida agreed to spend $395 million over 10 years to reduce agricultural pollution flowing into the environmentally fragile Everglades.

WATER POLLUTION—MARINE POLLUTION

4488. Use of the term *mare liberum* or "freedom of the seas" was in Dutch scholar Hugo Grotius's work of that title, published in 1608. The idea that the seemingly limitless opportunities offered by the seas should be available to all, without any significant restrictions or regulations, prevailed until the United Nations Conference on the Law of the Sea in the mid-20th century.

4489. Use of detergents and first-generation dispersants to clean up a beach after an oil spill occurred off the coast of Wales in 1967. The *Torrey Canyon* supertanker ran aground on Seven Stones Rocks off Wales, spilling about 35 million gallons of oil into the sea. The massive use of detergents and dispersants, however, caused unexpected ecological damage and led to a more judicious use of them in subsequent oil spill cleanups.

4490. Incineration of liquid organohalogen compounds at sea occurred in 1969. More than 100,000 tons a year were incinerated during the 1980s, mostly in the North Sea.

4491. Public disclosure of a U.S. Army proposal to dump nerve gas and live explosives 200 miles off the New Jersey coast came in May 1969. Public outcry forced the army to abandon the plan.

4492. Notice of the effects of tributylin (TBT) on marine life came in France in the mid-1970s in Arcachon Bay. Researchers traced serious malformations of oyster shells to the leaching of TBT from the antifouling paint on boats.

4493. Reports of high concentrations of toxic metals in bivalves from the U.S. Mussel Watch Program were issued in the mid-1970s. The program found "hot spots" of contamination in shellfish from Raritan Bay, NJ; Tampa Bay, FL; Delaware Bay, DE; and Chesapeake Bay, MD, among other sites.

4494. International compact on marine pollution from discharges of ships and aircraft was signed in Oslo, Norway, in 1972. The Convention for the Prevention of Marine Pollution by Dumping from Ships and Aircraft became known as the Oslo Convention.

4495. Report on the health of the oceans by the United Nations–sponsored Group of Experts on the Scientific Aspects of Marine Pollution (GESAMP) was issued in 1982. The report addressed chemical pollution, coastal development, dredging, fisheries, and other issues concerning marine pollution.

4496. TBT (tributylin) restrictions were enacted in France in 1982. Great Britain followed in 1985 and 1987 with bans on most uses of TBT.

4497. Regional antipollution pact covering the South Pacific region was the Noumea Convention of 1986. The compact sought to control marine pollution from waste disposal at sea.

4498. Large fish kill in Chesapeake Bay in the 1990s occurred in late June and early July 1999. More than 200,000 fish died in two tributaries of the Chesapeake. The kills were attributed to a buildup of oxygen-depleting nutrients such as nitrogen and phosphorus from fertilizers.

WATER POLLUTION—ORGANIC POLLUTION

4499. Use of the name Green Bay for an arm of Lake Michigan along the Wisconsin shore dates from 1815. The bay derived its English name from the frequent algal blooms that colored it.

4500. Appearance of alligatorweed in the United States was in Florida and Alabama in the 1890s. The weed spread rapidly up the Atlantic coast as far as North Carolina and along the Gulf Coast west to Texas, clogging waterways and polluting streams.

4501. Introduction of the water hyacinth in the United States occurred in the 1890s in New Orleans, LA. The South America native spread rapidly, clogging waterways and fouling water supplies.

4502. Sea lamprey in Lake Erie was discovered in 1921. The anadromous fish arrived via the Welland Canal and soon devastated Great Lakes commercial fisheries.

4503. Introduction of alligatorweed-eating adult flea beetles in Florida occurred in 1965. Within a year or so, this method of biological control cleared the test site of the weed.

WATER POLLUTION—RESEARCH AND TECHNOLOGY

4504. Use of rapid sand filtration for treating industrial wastewater in the United States dates from the 1890s. Sanitary engineers later applied the method to municipal sewage treatment.

4505. Commercial use of the ion exchange method for treating industrial wastewater in the United States dates from 1905. The method was later tested for municipal wastewater treatment.

4506. Trophic-level system application to classify lakes was used in 1919 by German limnologist Einar Naumann. Trophic levels describe the amount of nutrients in the water of a lake or stream. A eutrophic lake is well nourished—too well, for unwanted plant blooms crowd out other forms of life.

4507. Detergent industry research for an answer to the problem of foam in tap water dates to 1951. The Soap & Detergent Association, a trade group, decided that the foaming ingredient in detergent, which produced foaming tap water in many localities, presented a sales image problem for the industry.

4508. Classification of rivers is credited to wildlife biologist John Craighead, of Montana. Craighead believed that if people realized how few true "wild rivers" there were, they would see the need to protect them. Craighead's system, detailed in the June 1957 issue of *Montana Wildlife*, included "wild," "semiwild," and "harnessed" types of rivers.

4509. Research and development program for water pollution control sponsored by the U.S. government dates from 1960. The office was part of the Federal Water Quality Administration.

4510. Scholarly article to show that Lake Erie was dying a slow death from pollution was "Environmental Changes in Lake Erie" by Alfred M. Beeton, published in *Transactions of the American Fisheries Society* in 1961.

4511. International Joint Commission on Boundary Waters study of Lake Erie, Lake Ontario, and the upper St. Lawrence River came in October 1964. The formation of the international commission was requested by the U.S. State Department. It coordinates U.S. and Canadian policies on shared water resources, including the Great Lakes.

4512. Documentation of industrial pollutants in a lake on the Lake Superior wilderness preserve Isle Royale dates from the mid-1970s. A U.S. Environmental Protection Agency team found high levels of contaminants, including PCBs, in fish taken from Lake Siskiwit. The pollutants fell from the sky with rain in what scientists dubbed "toxic precipitation."

4513. Reports of mercury contamination in tuna and swordfish came in December 1970. Dr. Bruce McDuffie, a State University of New York chemistry professor, made the discovery.

WATER POLLUTION—RESEARCH AND TECHNOLOGY—*continued*

4514. Research of significance pointing to a problem in the U.S. drinking water supply was published in 1974 by the Environmental Protection Agency. In 1974 the EPA tested drinking water in New Orleans and found 66 different organic chemicals. This report, along with the Environmental Defense Fund's cancer study in New Orleans, led to the passing of the Safe Drinking Water Act of 1974. Before these two studies, the act had languished in Congress for years.

4515. Tests of leeches and mussels to monitor levels of freshwater contaminants were conducted by Claude B. Renaud and others in the Saint Lawrence River Basin in Quebec in the early 1990s. They reported in a 1995 article in the *Canadian Journal of Fish and Aquatic Sciences* that concentrations of nine pesticides, including DDT, were found in the animals.

WATER POLLUTION—SEWERS AND HUMAN WASTE

4516. Sanitary sewers may date from the prehistoric cities of Crete. Evidence of sewers has been found at Aegean Civilization sites on Crete circa 12,000–3000 B.C.

4517. Drainage systems that ran into brick-lined sewers in Indus Valley cities were developed as early as 2500 B.C.

4518. Sewers of Rome were laid out during the reign of Lucius Tarquinius Priscus, by tradition the fifth king of Rome. They date from circa 600 B.C.

4519. Aedile office in Rome was established in 493 B.C. Aediles supervised the maintenance of sewers, aqueducts, streets, and public buildings, among other duties.

4520. Recorded evidence of serious pollution of the River Thames in London dates from early in the 13th century. By 1236, pure water had to be piped into the city from Tyburn Spring.

4521. Use of carbolic acids to treat foul-smelling sewers dates from the mid-19th century in Great Britain.

4522. Decision of the Paris authorities to ban the taking of water from the polluted Seine came in 1852. The city tapped a large artesian well at Grenelle for cleaner water.

4523. Underground city sewer system in the United States was constructed beginning in 1856 in Chicago, IL. The main sewer lines were circular, ranging from 3 to 6 feet in diameter with brick walls 8.5 inches thick, while branch sewers were 2 feet in diameter. Manholes were provided every 100 feet, and the typical gradient was 1 foot in 500. By June 30, 1860, about 46 miles had been completed in a grid pattern. Single uncoordinated sewer lines had been used earlier.

4524. Dual sewer system in the United States for sewage and storm water was built under the direction of Colonel Julius Adams in Brooklyn, NY, in 1857. The capacity was calculated to accommodate a rainfall equal to one inch per hour.

4525. Cancellation of the British House of Commons sittings because of the stench of human waste in the River Thames occurred in 1858. The period of most intense odor from decaying sewage was dubbed "the Great Stink."

4526. Sewage-disposal system in the United States separate from the city water system was built in Memphis, TN, beginning on January 21, 1880. Within four months, a system comprising 18 miles of pipe, with 152 flush tanks and 4-inch connecting drains, had been constructed under the supervision of George Edwin Waring. The pipes were for sewage only and were constantly washed clean with water. The total cost of 20 miles for the two main sewers was about $137,000. A similar system was adopted at about the same time by Pullman, IL (now part of Chicago).

4527. Sewer district established by a city in the United States was the Sanitary District of Chicago, IL, authorized by a referendum on November 5, 1889, to construct and operate the sewage system that would protect the public water supply. Murray Nelson was the first president.

4528. Sewage disposal system in the United States by chemical precipitation was installed in 1890 in Worcester, MA. The raw sewage, screened and then treated with milk of lime, was passed through a mixing channel into six chemical precipitation settling basins, each 66.67 by 100 by 7 feet. After settling for a few hours, the top water was drawn off and the sludge run off through a 6-inch centrifugal pump into lagoons.

4529. Sewer in Moscow was built in 1898. Seven years later, only 6,000 houses in Russia's capital had been connected to the system.

4530. Reversal of course of the Chicago River to keep raw human waste from washing up on beaches in front of Chicago Gold Coast apartment buildings occurred in 1900. Engineers built the Chicago Sanitary and Ship Canal to reverse the river's flow and allow water from Lake Michigan to flush the city's waste westward into the Illinois River.

4531. Lawsuit against the state of Illinois and the city of Chicago claiming the diversion of Lake Michigan water to flush sewage down the Chicago Sanitary and Ship Canal was lowering water levels in the Great Lakes system was filed by New York State in 1927. The suit was settled in 1930 but was resurrected in 1959 by other Great Lakes states.

4532. Reports of widespread fouling of Great Lakes beaches by ill-smelling seaweed date from the early 1930s. Massive discharges of sewage into the lakes helped feed the explosion of seaweed growth.

4533. Storm sewer building project of significance in the District of Columbia was started in 1937. The system, meant to keep raw sewage from discharging into the Potomac every time it rained, was only 40 percent complete in 1969.

4534. Federation of Sewage Works Association meetings took place in 1940. Sanitation professionals discussed issues involving sewage pollution of rivers and streams and variations in water quality standards from place to place.

4535. Agency to administer national water pollution legislation in the United States was the Division of Water Pollution Control. Part of the U.S. Public Health Service, the division was created under the terms of the Water Pollution Control Act of 1948. Initially, ten field units were established around the United States.

4536. Secondary sewage treatment program in the District of Columbia was begun in 1949. The program aimed to treat raw sewage then flowing into the Potomac River.

4537. Disposable diapers mass-marketed in the United States came in 1961, when Procter & Gamble launched its Pampers line. By 1979, disposable diapers were a $1 billion-a-year business—and a major environmental concern, for although they somewhat reduced the amount of water-borne sewage, they also created a massive amount of non-biodegradable solid waste.

4538. Reports that latrines at the U.S. Military Academy at West Point were seriously polluting the Hudson River surfaced in 1965. The academy was dumping 1.2 million gallons of partially treated sewage into the river every day.

4539. Effort in the United States to address severe pollution in Lake Erie was launched in 1965. At a conference in Cleveland, OH, on August 11 and 12, the federal government and the five states in the Lake Erie drainage area agreed to clean up waste discharges, reduce agricultural runoff, and carry out other measures to revive the dying lake. The first signs that Lake Erie was beginning to recover appeared in the mid-1970s.

4540. Artificial wetlands treatment of wastewater in the United States dates from the 1970s. The 154-acre wetlands park in Arcata, CA, acted as a living filter, purifying sewage and pouring clean water into Humboldt Bay.

4541. Large-scale natural waste disposal system in the United States was approved in June 1970 for Muskegon County, MI. The new system, costly to develop but inexpensive to operate, incorporated a natural earth filtration system.

4542. Municipality to adapt the Federal Water Quality Administration's tertiary treatment system to raw sewage rather than effluent that had undergone secondary treatment was Rocky River, OH, in the early 1970s. The system used lime in a chemical process and a physical process called carbon adsorption.

4543. Pledge to halt all raw sewage flow from Victoria and Vancouver Island, British Columbia, into the Strait of Juan de Fuca was made by British Columbia Premier Mike Harcourt on May 8, 1992. Vancouver Island communities had been pumping 15 million gallons of sewage a day into the strait.

WATER POWER

4544. Water frame or water-powered cotton spinning machine was patented by English inventor Richard Arkwright in 1769. The water-powered spinning mill Arkwright built revolutionized the production of textiles, which had previously been made by individual workers.

WATER POWER—*continued*

4545. Water-powered cotton mill in an American colony was built by Samuel Slater in Pawtucket, RI, in 1793. Slater's design was based on English inventor Richard Arkwright's water frame. As Arkwright's invention had done in England, Slater's mill revolutionized industry in North America.

4546. Water-powered turbine was built in 1827 by French engineer Benoît Fourneyron in Le Creuset. Expanding upon the theories of his teacher Claude Burdin, who coined the term *turbine*, Fourneyron built a water-powered generator capable of producing six horsepower of energy.

4547. Electricity generated by Niagara Falls waterpower was made in 1879, when a small dynamo powered by the falls lit 16 arc lights in Prospect Park, Niagara Falls, NY. The first large-scale use was made by the Niagara River Hydraulic Tunnel Power and Sewer Company (later the Niagara Falls Power Company), which broke ground for a power plant on October 4, 1890. On October 24, 1893, a contract was executed with the Westinghouse Electric and Manufacturing Company of Pittsburgh, PA, for three 5,000-horsepower generators delivering two-phase currents at 2,200 volts, 25 cycles. Power was first transmitted commercially on August 26, 1895, to the Pittsburgh Reduction Company, which used the current in the reduction of aluminum ore. The city of Buffalo, NY, received its first power for commercial purposes on November 15, 1896.

4548. Hydroelectric power plant to furnish arc lighting service in the United States went into operation on July 23, 1880, in Grand Rapids, MI. Organized on March 22, 1880, the Grand Rapids Electric Light and Power Company installed a 16-arc-light Brush generator, driven by a water wheel, which supplied power to the factory of the Wolverine Chair Company. The first president and organizer was William T. Powers. On August 1, 1881, a new building was occupied from which current was generated to supply street lighting.

4549. Theory of Ocean Thermal Energy Conversion (OTEC) was advanced in 1881 by Jacques d'Arsonval. The French engineer proposed that the flow of warm solar-heated ocean water toward colder water could be used to power a turbine and create electrical energy.

4550. Hydroelectric power plant in the United States to furnish incandescent lighting was opened in Appleton, WI, on September 30, 1882. It consisted of a single dynamo of 180 lights, each of 10 candlepower.

4551. Turbogenerator to produce hydroelectric power was built in England by Charles A. Parsons in 1884.

4552. Hydroelectric power plant in the United States transmitting alternating current over a long distance supplied current on June 2, 1889, to Portland, OR, a distance of 13 miles from the plant operated by the Willamette Falls Electric Company at Willamette Falls, Oregon City, OR. The plant operated two 300-horsepower Stilwell and Bierce water wheels belted to a single-phase generator rated at 720 kilowatts.

4553. Hydropower agency of the U.S. government was the Reclamation Service, whose name was later changed to the Bureau of Reclamation. The agency was founded in 1902 by authority of the Reclamation Act. It was responsible for managing water resources and aiding development in the arid western states by creating irrigation and hydropower projects.

4554. Hydroelectric power plant built by the U.S. government was located at the Minidoka Dam on the Snake River in Idaho. Built by the Bureau of Reclamation, Department of the Interior, the power plant started its first unit on May 1, 1909 with a capacity of 1,400 kilovolt amperes.

4555. Canadian–U.S. treaty to regulate use of the Niagara River for water power was the International Boundary Treaties Act. The law set specific limits on the amount of water above Niagara Falls that could be diverted by electricty-producing power plants in New York and Ontario. The act was signed in Washington, DC, on January 11, 1909, by the governments of the United States and Great Britain, on behalf of Canada. The article governing water diversion lasted until October 10, 1950, when it was replaced by a new agreement between the United States and the government of Canada.

4556. Hydropower facilities operated by the U.S. Bureau of Reclamation were generators used to build the Roosevelt Dam and power plant on the Salt River in Arizona. The dam itself was officially opened by former President Theodore Roosevelt on March 18, 1911, eight years after construction began. The plant produced 4,500 kilowatts of power for the city of Phoenix, AZ, 76 miles to the west, while the dam stored water for regional irrigation projects. The dam was raised 77 feet in 1996 to increase safety and provide better recreational usage.

4557. Law in the United States to license construction of commercial hydropower plants was the Federal Power Act of 1920 (FPA). Prior to its passage on June 10, hydropower developers were required to obtain permission to build dams on public land or in navigable waters by an act of Congress. The FPA established the Federal Power Commission (FPC), which gave joint responsibility for approving such projects to the Secretaries of War, Agriculture, and the Interior. The FPC was replaced by the Federal Energy Regulatory Commission in 1977.

4558. Pumped storage hydroelectric plant of significance in the United States was the Rocky River hydroelectric station in New Milford, CT. The facility, which began generating power in 1929, uses water from Candlewood Lake, the largest man-made lake in the state.

4559. Ocean Thermal Energy Conversion (OTEC) power plant was constructed by French engineer Georges Claude in 1930 at Matanzas Bay, Cuba. Claude's experimental open-cycle OTEC system used the evaporation of warm seawater to run a steam-powered turbine capable of producing approximately 22 kilowatts of electricity. The facility was destroyed by a storm before it could be improved to produce a more significant amount of power.

4560. Hydroelectric power plant in the United States to produce a million kilowatts was the Boulder Dam, Boulder City, NV, which reached this production peak in June 1943. The concrete arch-gravity dam was contracted on March 11, 1931, by the Bureau of Reclamation of the Department of the Interior. The first of its four generators went into operation on October 26, 1936, providing electricity to the Los Angeles area. Its name was changed to the Hoover Dam in 1947.

4561. U.S. law allowing consideration of recreation, fish, and wildlife enhancement as purposes of federal water development projects was the Federal Water Project Recreation Act. The law was enacted on July 9, 1965, and was amended in 1974 and 1976. Such purposes were to be given full consideration on the condition that nongovernmental conservation bodies share and in some instances support the costs involved in maintaining reservoirs, wildfowl refuges, and other nonutility related usage.

4562. Electric power project to harness tidal motion was the Brittany Dam, completed in 1967 in France. Built in the estuary of the Rance River in central Brittany, the dam remains a technical prototype for generating electrical energy from the ebb and flow of tides.

4563. Tidal power plant north of the Arctic Circle began operation at the Kislaya Guba facility near Murmansk in the Soviet Union in 1968, utilizing tides from the Barents Sea.

4564. Successful offshore closed-cycle Ocean Thermal Energy Conversion (OTEC) operation took place in 1979 aboard the *Mini-OTEC*, a converted U.S. Navy barge operating 2 kilometers off Keahole Point, HI, home of the Natural Energy Laboratory of Hawaii (NELH). The plant used cold ocean water to power a small plant capable of producing approximately 50 kilowatts of electricity.

4565. U.S. legislation to promote Ocean Thermal Energy Conversion (OTEC) development for commercial use was the Ocean Thermal Energy Conversion Act, Public Law (PL) 96-320. The act became law on August 3, 1980. It authorized and regulated the construction, location, ownership, and operation of ocean thermal energy conversion facilities in a manner consistent with U.S. and international laws pertaining to the sea, coastlines, and protection of marine life.

4566. Tidal power plant in North America began operation at Annapolis Royal in western Nova Scotia, Canada, in 1984. It utilized the tides of the Bay of Fundy.

4567. Law in the United States requiring environmental considerations to be assessed in the licensing of hydropower projects was the Electric Consumers Protection Act (ECPA), passed by Congress on October 16, 1986. Among its provisions, which amended the Federal Power Act, the ECPA required the Federal Energy Regulatory Commission (FERC) to give the same level of consideration to the environment, recreation, fish and wildlife, and other nonpower values that it gave to power and development objectives in making licensing decisions. It also ordered the FERC to negotiate with agencies like the U.S. Fish and Wildlife Service in disputed cases.

WATER SUPPLY AND PURIFICATION

4568. Aqueduct built to carry water into Rome was constructed in 312 B.C. Romans decided the Tiber River was too polluted for drinking.

4569. Irrigation ditches in what is now the United States were dug by the Hohokam Indians near Arizona's Salt and Gila Rivers around 300 B.C. Their canals irrigated thousands of acres and provided water for the tribe's farmland.

WATER SUPPLY AND PURIFICATION—*continued*

4570. Water supply system built for a U.S. city was built for Boston, MA, in 1652 by Water Works Company. Wooden pipes were used to convey the water from a nearby spring to a 12-foot-square central reservoir.

4571. Public water supply in New York City was a public well dug in 1677 opposite Fort James at the southern end of Manhattan, near the future sites of Bowling Green and the financial district. Previously, New York residents had relied on private wells.

4572. Water pumping plant in the United States was installed on May 27, 1755, by Hans Christopher Christiansen in Bethlehem, PA. Spring water was pumped through wooden pipes into a 70-foot-high water tower.

4573. Waterworks system in the United States to use cast-iron pipes was installed in 1817 in Philadelphia, PA. The 4.5-inch diameter pipeline was 400 feet long and constructed using pipes imported from England. They proved so superior to the old wooden water pipes that the city's Watering Committee decided in 1818 to make all future installations with cast-iron pipe.

4574. Water filtration system was invented in 1828 by English engineer James Simpson. The system used sand to filter impurities from drinking water pumped from the River Thames.

4575. Aqueduct to supply New York City was completed in 1842. It routed water from the Croton River and Old Croton Reservoir in Westchester County to reservoirs at 42nd Street and Central Park near 86th Street in Manhattan.

4576. Drinking water conduit in the United States to be built underwater was built in 1848 by the Water Department of Boston, MA. The wooden tunnels were approximately 4 feet 8.5 inches in diameter and some 50 feet or more in length. They contained cast-iron pipes 20 inches in diameter that carried drinking water from central Boston to the South Boston, Charlestown, and Chelsea sections of the city. The conduits were constructed on the shore, floated into place, and sunk into a prepared trench below the surface of the channel under the Dover Street Bridge. The entire system was in operation by 1852.

4577. Aqueduct to supply Washington, DC, was constructed between 1852 and 1862 under the direction of U.S. Army Corps of Engineers Captain Montgomery Cunningham Meigs. The aqueduct routed water from the Great Falls of the Potomac River to Washington and Georgetown through a structure that also served as a roadway and railway line. Engineer Meigs rose to the rank of general and participated in the creation of other Washington-area landmarks, including the Capitol dome and Arlington National Cemetery.

4578. National water quality laws in India were contained in the Penal Code of 1860. The laws made the intentional defilement of water of a public spring or reservoir an imprisonable or fineable offense. The laws did not, however, restrict the pollution of rivers.

4579. Water tunnel to supply Chicago was designed by engineer Ellis S. Chesbrough. Completed in 1867, the tunnel delivered water from Lake Michigan to a pumping station in the city. It was two miles long, extending far enough from shore to obtain fresh water unpolluted by the city's considerable discharges of sewage and waste from meat-packing plants.

4580. Commercially bottled water in the United States was sold by the Saratoga Spring Water Company of Saratoga Springs, NY. The company began bottling naturally carbonated water from local sources in 1872 and marketing it for the professed health benefits of its high mineral content.

4581. Legal basis for U.S. federal drinking water standards was the Interstate Quarantine Act of 1893. The law empowered the Surgeon General of the U.S. Public Health Service to make and enforce regulations to prevent the introduction, transmission, or spread of communicable diseases from foreign countries or between the states.

4582. Water filtration system for bacterial purification of a U.S. city water supply was completed in September 1893 in Lawrence, MA. Designed by Hiram Francis Mills, the installation was an open filter of 2.75 acres that purified water from the Merrimack River by slow sand filtration. The idea dates from 1873, when an English-type slow sand filter was built at Poughkeepsie, NY, from plans prepared by James Pugh Kirkwood.

4583. Drinking water supply in the United States to be chemically treated with chlorine compounds on a large scale was put into operation in Jersey City, NJ, in 1908, under the supervision of George Arthur Johnson.

4584. Federal drinking water safety regulations in the United States were adopted in 1912. The U.S. Public Health Service banned the use of common drinking water cups on trains and other carriers of interstate commerce.

4585. Aqueduct to supply Los Angeles, CA, began operating on November 5, 1913. Diversion of water into the Los Angeles Aqueduct created an ecological disaster in California's Owens Valley, revolutionized agricultural production in the San Fernando Valley, and accelerated Los Angeles's growth into a modern metropolis.

4586. Federal drinking water standards in the United States were instituted by the U.S. Public Health Service (USPHS) in 1914. The USPHS set bacteriological standards for water available to the public on interstate carriers like trains. The standards were widely adopted by other government agencies.

4587. Water tunnel to supply New York City began operating in 1917. City Tunnel No. 1 supplies water propelled by gravity to Manhattan, Queens, Brooklyn, and the Bronx. Both City Tunnel No. 1 and a second tunnel, City Tunnel No. 2, convey water from Hillview Reservoir in Yonkers, which receives its water from the Catskill and Delaware supply systems. Construction on a third tunnel began in 1970; it is scheduled to begin operation in 2020.

4588. Federal chemical standards for drinking water in the United States were set by the U.S. Public Health Service in 1925 and included guidelines for the amounts of lead, zinc, and other mineral and chemical contents allowable in public water supplies.

4589. Tunnel in the Massachusetts water system was the Ware-Colebrook Tunnel, which increased the Wachusett Reservoir's ability to supply the Boston area by diverting some of the flow of the Ware River. Construction on the tunnel began in 1926. In 1946 it linked the Wachusett supply with the massive man-made Quabbin Reservoir, the state's main source of fresh drinking water.

4590. Federal requirement in the United States for bacteriological analysis of drinking water systems was made mandatory by the U.S. Public Health Service in 1942. Maximum allowable limits on arsenic and lead were also set, while the new rules made salts of barium, hexavalent chromium, and heavy metals unacceptable in public water systems.

4591. Water supply in the United States to be fluoridated in order to reduce tooth decay was that of Grand Rapids, MI. Fluoridation started on January 25, 1945, with the addition of one part of fluoride ion to each million parts of water passing through the water treatment plant.

4592. Aqueduct to supply San Diego, CA, was opened on November 26, 1947. The First San Diego Aqueduct, which routed water from the Colorado River to the San Vicente Reservoir, took the U.S. Navy two years to build. Its construction was prompted by San Diego's population boom during World War II, when the city served as a major naval seaport.

4593. Law in the United States for watershed protection was the Watershed Protection and Flood Prevention Act. The law brought the federal government into cooperation with state and local efforts to control erosion and sediments capable of destroying watersheds and causing floods. The law was enacted on August 4, 1954. Subsequent amendments further strengthened federal support for water quality improvements in rural areas, conservation of fish and wildlife, and watershed restoration.

4594. International organization to promote water quality monitoring was the World Health Organization (WHO), which released its first International Drinking Water Standards in 1958, a year after its European office released recommendations on water quality to governmental organizations. The reports were the first WHO guidelines on global water supplies, sanitation, and quality.

4595. Seawater conversion plant in the United States on a practical scale was opened at Freeport, TX, on May 8, 1961, by the Office of Saline Water, Department of the Interior. The plant was designed to produce about a million gallons of water a day at a cost of $1 to $1.25 per 1,000 gallons. President John Fitzgerald Kennedy dedicated the plant on June 21 by pressing a switch installed in his office at Washington, DC.

4596. United Nations hydrological agency was the International Hydrological Program (IHP). In 1965, the program was originally established as the International Hydrological Decade in Paris, France, by the United Nations Educational, Scientific, and Cultural Organization (UNESCO). In 1975, IHP became a coordinating body for scientific research of global water cycles, supplies, and management issues.

4597. Modern drinking water improvement legislation of significance in India was the Water Act of 1974. It established national and state pollution control boards, set penalties for violations, and created an appeals process for disputes over water samples taken in monitoring programs.

WATER SUPPLY AND PURIFICATION—*continued*

4598. Clean drinking water standards in the United States that were legally binding resulted from the Safe Drinking Water Act of 1974. The act authorized the U.S. Environmental Protection Agency (EPA) to set national drinking water regulations, monitor both chemical and microbial contaminants, and oversee implementation of the act. State governments were required to administer and enforce the act through their health departments and environmental agencies. President Gerald Ford signed Public Law 93-523 on December 16, 1974, establishing the first binding standards to apply to all public water supplies in the United States.

4599. Federal Sole Source Aquifer (SSA) Protection Program in the United States was authorized by Section 1424(e) of the Safe Drinking Water Act, which was signed into law on December 16, 1974. The program requires the Administrator of the Environmental Protection Agency (EPA) to identify aquifers that provide more than 50 percent of the drinking water source for an area and that, if contaminated, would create a significant hazard to public health. Publication of such an EPA notice in the Federal Register prohibits any federal funds from being used in activities that might contaminate the aquifer.

4600. Aquifer in the United States to be designated as a "sole source" under the Clean Water Act of 1974 was the Edwards Aquifer in central Texas. It provides most of the fresh water for drinking and agriculture in the San Antonio area and waas designated a sole source aquifer by the Environmental Protection Agency on December 16, 1975.

4601. Worldwide freshwater monitoring program was the Global Environment Monitoring System Freshwater Quality Program (GEMS/WATER), created in 1976 by the United Nations Environment Program and the World Health Organization. The international network monitors the world's freshwater resources and assists national water quality agencies in improving monitoring and assessment programs, especially in developing countries. It also provides global reports on pollution trends. GEMS/WATER is based at the National Water Research Institute in Burlington, Ontario, Canada.

4602. Federal agency in the United States to monitor water use was the U.S. Geological Survey (USGS). The agency's National Water-Use Information Program was established in 1978 to act as a central source of information about demands on the national water supply. In cooperation with state and local governments, the program collects, stores, analyzes, and disseminates information on the use and status of water in the United States. The information is available to the public and to agencies involved in water planning, management, and regulation.

4603. Government program in the United States established to monitor waterborne toxic materials was the U.S. Geological Survey (USGS) Toxic Substances Hydrology Program in 1982. The USGS created the program to study and provide scientific information about the behavior of toxic substances like petroleum products, metals, pesticides, and chemicals in hydrologic environments. The information is used to develop prevention and cleanup strategies for dealing with contaminants in the U.S. water supply.

4604. National water use survey of the United States by county was completed in 1985 by the U.S. Geological Survey Water Use Information Program through the program's partnerships of federal, state, and local conservation commissions and monitoring efforts in all 50 states and Puerto Rico. Uses of the data included the calculation of pollution cleanup grants to state and local governments, based partially on the amount of water used in a given locality.

4605. National Water Quality Laboratory in the United States opened in Denver, CO, in 1986. The facility combined the resources of three regional laboratories operated by the U.S. Geological Survey since 1972, which had previously consolidated 22 regional laboratories dating back to 1918. The Denver laboratory analyzes organic and inorganic constituents in samples of ground and surface water, river and lake sediment, aquatic plant and animal material, and precipitation collected in the United States and its territories.

4606. Ban on the use of lead pipes in public water systems in the United States was established through an amendment of the Safe Drinking Water Act. The amendment was signed into law by President Ronald Reagan on June 19, 1986. After that date, the use of any pipe, fitting, fixture, solder, or flux containing lead was prohibited in the installation or repair of any public water system or facility providing drinking water.

4607. Law in the United States to require the removal of lead from water coolers was the Lead Contamination Control Act, enacted on October 31, 1988. It created a program to eliminate lead-containing drinking water coolers from schools and required the Environmental Protection Agency to provide guidance to states and localities to test for and remedy lead contamination in schools and day care centers. It also established civil and criminal penalties for the manufacture and sale of water coolers containing lead.

4608. National water use survey of U.S. aquifer systems was completed in 1990 by the U.S. Geological Survey Water-Use Information Program. The nationwide survey collected water use data to aid the management of major aquifers that serve as important local or regional sources of water.

4609. National Water Quality Assessment Program in the United States became a permanent U.S. Geological Survey (USGS) program in 1991. The program studies the water quality of 60 aquifers and river basins that represent the majority of water resources in the nation. The resulting data is used to assess how natural and human activities relate to historical, current, and future water quality conditions.

4610. Federal standards in the United States for radon in drinking water were proposed by the U.S. Environmental Protection Agency on July 18, 1991. While the EPA sought to set a maximum contamination level for the carcinogenic radioactive gas in public drinking water supplies, the standard was removed on August 8, 1997, by an amendment to the Safe Drinking Water Reauthorization Act of 1996. Critics were concerned about the economic effects of its implementation. The EPA and National Academy of Sciences were ordered to conduct a health risk and cost analysis of the issue.

4611. Government facility in the United States established to analyze chlorofluorocarbons (CFCs) in water and air was the Reston Chlorofluorocarbon Laboratory. Their analyses of CFCs are used to trace seepage from rivers into groundwater systems, provide diagnostic tools for detection and early warning of leakage from landfills and septic tanks, and assess susceptibility of deep water-supply wells to contamination from shallow sources. The laboratory is in Reston, VA, and has been operated by the U.S. Geological Survey since 1994.

4612. Nationwide survey of drinking water system infrastructures in the United States was the Drinking Water Infrastructure Needs Survey, conducted by the Environmental Protection Agency (EPA) during 1996. It was published in January 1997 and presented to Congress the following month. The study surveyed communities nationwide to forecast the expense of complying with current and future federal regulations, replacing aging infrastructure to protect public health, and consolidating or acquiring neighboring systems without safe supplies of drinking water through the year 2014. The survey estimated that $138.4 billion would need to be spent to maintain safe drinking water supplies. Subsequent surveys were to be made every 3 years over a 20-year period.

4613. United Nations training center for water supply issues was the International Network on Water, Environment, and Health (INWEH), a branch of the UN University system. INWEH was created by the UN University Governing Council in 1996 with core funding provided by the government of Canada. Its mission is to strengthen water management capacity and provide direct expert support for water programs, particularly in developing countries where lack of safe drinking water is a major cause of child mortality and other health problems. INWEH is based at McMaster University in Hamilton, Ontario, Canada.

4614. Federal Drinking Water State Revolving Fund (DWSRF) in the United States was established when President Bill Clinton reauthorized the Safe Drinking Water Act (SDWA) on August 6, 1996. The program is managed by the U.S. Environmental Protection Agency (EPA) Office of Water. The fund guarantees each state 1 percent of approximately $9.5 billion available until 2003 for water quality projects. Specific grants to individual projects are made on the basis of need, as determined through the EPA's periodic Drinking Water Infrastructure Needs Survey.

4615. Law by the U.S. government to subject bottled water to the same standards as tap water was the Safe Drinking Water Reauthorization Act signed by President Bill Clinton on August 6, 1996. Under the Safe Drinking Water Act of 1974, the monitoring of contaminants in public water supplies is regulated by the Environmental Protection Agency (EPA). The reauthorization act characterized bottled water as a food product and assigned regulatory responsibility for its quality to the Food and Drug Administration.

WATER SUPPLY AND PURIFICATION—*continued*

4616. United Nations convention on transboundary water supplies was the Convention on the Law of the Non-Navigational Uses of International Watercourses, an attempt by signatory nations to preserve, conserve, manage, and share information about surface and groundwater sources shared by more than one country. The convention, which also provided a mechanism for arbitrating disputes over water supplies, was adopted by the UN General Assembly and opened for signatures on May 21, 1997.

4617. Clean Water Hardship Grant in the United States was awarded to the state of Connecticut by the Environmental Protection Agency (EPA) on May 22, 1997. The grant was created to assist rural or impoverished communities of 3,000 or fewer inhabitants with wastewater treatment problems. The EPA awarded $452,600 to Connecticut to build a collection system to eliminate leaking septic tanks that were contaminating well and lake water in Middlefield, CT.

WATER SUPPLY AND PURIFICATION—RESERVOIRS

4618. Reservoir to supply New York City was created in 1776 on Broadway between Pearl and White streets. Water from the reservoir was pumped to residents through hollowed logs.

4619. Reservoir to supply Hong Kong was the Pok Fu Lam Reservoir. It was built in 1863 by the British colonial government in an attempt to meet Hong Kong's rapidly growing water needs, but was soon an insufficient source for the expanding local population.

4620. Reservoir on the U.S.-Mexico border was the Amistad Reservoir, created in 1968 by construction of the Amistad Dam on the Rio Grande River northwest of the cities of Del Rio in Texas and Villa Acuña in Coahuila. It is a major recreational area whose U.S. side is administered by the National Park Service.

WHALING

4621. Commercial whaling in North America was done in the waters off Newfoundland and Labrador by Basque whalers in the 1500s. Basque whaling stations were established in Newfoundland but were abandoned when the industry declined.

4622. Whaling industry established by an American town was organized on March 7, 1644, by Southampton, NY, located on the coast of Long Island where whales often washed ashore. The town was divided into four wards of 11 persons each. Two persons from each ward were employed to cut up the stranded whales and divide them equally among the inhabitants.

4623. Sperm whale captured at sea by an American ship was caught by a vessel out of Nantucket, MA, in 1711. By 1846, the U.S. whaling fleet numbered more than 700 ships.

4624. Recorded whaling expedition from an American colony sailed from Nantucket, MA, circa 1715. Some 600 barrels of oil and 11,000 pounds of bone worth about £1,100 sterling were brought back by six sloops, of 30 to 40 tons burden each. Prior to this expedition, there had been whaling voyages by single vessels.

4625. Exploding grenade harpoon was developed in the 1860s by Norwegian whaler Sven Foyn. The invention, which was fired from a cannon, swiftly replaced hand-thrown harpoons in commercial whaling and vastly increased the number of whales killed.

4626. American steam whaler was the *Mary and Helen*. Although it was used as a whaling ship after being launched from New Bedford, MA, on July 30, 1879, the ship was purchased in 1881 by the U.S. Navy, which renamed it the USS *Rodgers* and used it as an Arctic rescue vessel.

4627. International agency formed specifically to regulate whaling was the International Whaling Commission (IWC). The IWC was created by signatory nations of the International Convention for the Regulation of Whaling, which was signed in Washington, DC, on December 2, 1946. The organization's stated goal was to set rules to safeguard whale stocks and protect the whaling industry through proper regulation. Policy disagreements and loopholes in IWC regulations, however, resulted in significant controversies over the IWC's effectiveness during its first decades of existence.

4628. American Indian tribal whale hunt in modern times during an International Whaling Commission (IWC) moratorium on whale hunting was conducted by the Makah tribe of Washington State. Under the terms of an 1855 treaty, the Makah are the only tribe guaranteed whale hunting rights under U.S. law. On October 23, 1997, the IWC permitted the hunt under a provision allowing subsistence whale hunting by aboriginal peoples. Despite protests by anti-whaling groups, Makah tribesmen killed a gray whale on May 17, 1999, before towing it ashore to be rendered into food and fuel products.

WILDERNESS

4629. European discovery of Iguaçu Falls occurred in 1541. Spaniards, led by Núñez Cabeza de Vaca, encountered the massive falls of South America's third largest river system while exploring the wilderness along what is now the border between Argentina and Brazil. Both modern nations have established national parks adjoining the area to protect the region's rich ecosystem.

4630. Descriptions of Mount Kilimanjaro and Mount Kenya to reach Europe were reports from German traveling missionaries Johannes Rebmann and Johann Ludwig Krapf. Rebmann encountered Kilimanjaro in 1848, while Krapf found Kenya the following year. Their descriptions of the massive snow-covered mountains in equatorial Africa were met with intense skepticism.

4631. Law establishing a protected wilderness in the Adirondack Mountains was signed by New York governor David B. Hill on May 15, 1885. The law permanently designated 715,000 acres as a protected wilderness area, the New York State Forest Preserve. This became the core of the Adirondack Forest Preserve, given state constitutional protection in 1894, after various interests attempted to weaken the law.

4632. Landscape architect employed by the U.S. Forest Service (USFS) was Arthur H. Carhart. In 1919 Carhart was hired by the USFS to survey federal land for recreational use. During and after his tenure with the service, he was a strong advocate of managing wilderness lands like Trappers Lake in Colorado and the area that later became Boundary Waters Canoe Area Wilderness in Minnesota, rather than opening them to commercial development.

4633. Wilderness area protected by the U.S. Forest Service (USFS) was the Trappers Lake area in Colorado's White River National Forest. In 1919, USFS landscape architect Arthur H. Carhart convinced the service that the area should remain wild instead of being developed for summer homes.

4634. Federally designated wilderness area in the United States was the Gila Wilderness Area in southwestern New Mexico. In 1924, at the urging of an employee, Aldo Leopold, the U.S. Forest Service decided to administer the watershed of the Gila River as a wilderness area that would remain immune from human development. The establishment of the Gila Wilderness became the model for federal designations of other wilderness environments, particularly those with rivers.

4635. Federal regulation in the United States for designation and management of wild areas was called an L-20 regulation. The L-20 rules were first issued in 1929 by the U.S. Forest Service to prevent commercial exploitation of private lands in federal wilderness or "primitive areas" until management plans could be completed.

4636. U.S. federal law to protect a wilderness area was the Shipstead-Newton-Nolan Act. Also known as the Shipstead-Nolan Act, the law was enacted on July 10, 1930. It protected over 1 million acres in the Superior Primitive Area in Minnesota, which later became the Boundary Waters Canoe Wilderness Area. Logging within 400 feet of natural shorelines and alteration of natural water levels, except by special act of Congress, were prohibited. The legislation was passed to prevent a timber developer who owned land within the Superior Primitive Area from flooding the entire region for business purposes.

4637. Snowmobile was invented in 1937 by Canadian engineer Joseph A. Bombardier, who also invented the lighter-weight Ski-Doo in 1959. Originally conceived as transportation aids for snowbound northern regions, the machines were used as recreational vehicles and later became a contentious issue between sports-related industries and conservationists concerned about their effects on wilderness, wildlife, and safety in North American national parks.

4638. Official use of the term *wilderness* to classify U.S. federal lands occurred in 1939. The U.S. Department of Agriculture replaced the U.S. Forest Service's L-20 designations of "primitive areas" with U-1 or U-2 designations, under which federal lands could be classified by size as "wilderness," "wild," or "roadless."

WILDERNESS—*continued*

4639. Ban of aircraft access to Superior National Forest in northern Minnesota was contained in an Executive Order signed by President Harry S. Truman on December 17, 1949. The regulation, which mainly affected hydroplanes filled with tourists, was intended to help maintain the wilderness character of the region by limiting access to environmentally unobtrusive means like canoes.

4640. U.S. National Wilderness Area in New England was Great Gulf, a cirque or steep-walled mountain basin within the Presidential Range of New Hampshire's White Mountain National Forest. The 5,552 acres of the Great Gulf Wilderness are bordered by Mount Washington and Mount Madison. The area was designated a National Wilderness by Congress on September 3, 1964 and is administered by the U.S. Forest Service.

4641. Rivers managed by the U.S. National Park Service were the Current and Jacks Fork rivers in Missouri. Congress directed the National Park Service to manage recreational uses, development, and conservation of the waterways and surrounding area by creating the Ozark National Scenic Riverways on August 27, 1964. National Park Service management programs had previously been limited to land use.

4642. U.S. National Wilderness Area in the Superior Upland region was Boundary Waters Canoe Wilderness Area. It was one of the first areas designated under the Wilderness Act on September 3, 1964. Entrance to and use of the northeastern Minnesota wilderness, which is bordered by Voyageurs National Park and Canada's Quetico Provincial Park, is regulated by the U.S. Forest Service due to the longstanding popularity of its trails and many lakes with boaters, hikers, and campers.

4643. U.S. National Wilderness Areas in the interior West were areas in Colorado, Idaho, Montana, and Wyoming designated for the National Wilderness Preservation System on September 3, 1964. They are still administered by the U.S. Forest Service.

4644. U.S. National Wilderness Areas in the Pacific Northwest were designated under the Wilderness Act on September 3, 1964. The U.S. Forest Service administers all nine of the areas originally designated in Oregon (Diamond Peak, Eagle Cap, Gearhart Mountain, Kalmiopsis, Mount Washington, Mountain Lakes) and Washington (Glacier Peaks, Goat Rocks, Mount Adams). All the wilderness areas are distinctive for their spectacular glacial or volcanic geological features.

4645. U.S. National Wilderness Areas in the Southeast were designated under the Wilderness Act on September 3, 1964. Linville River Gorge Wilderness and the mountain ridges of Shining Rock Wilderness are both in western North Carolina. Both areas are still administered by the U.S. Forest Service. Marjory Stoneman Douglas Wilderness covers more than a million and a quarter acres within Florida's Everglades National Park and is administered by the National Park Service.

4646. U.S. National Wilderness Areas in the Southwest were designated under the first Wilderness Act on September 3, 1964. Home to a variety of wildlife, all of the areas are rugged, mountainous, parts of two contiguous states: Arizona (Chiricahua, Galiuro, Kanab Creek, Matzatzal, Sierra Ancha) and New Mexico (Gila, Pecos, Wheeler Peak, White Mountain). All but one of the wilderness areas are administered by the U.S. Forest Service. The lone exception is Kanab Creek, which is divided into two areas administered by the U.S. Forest Service and U.S. Bureau of Land Management.

4647. Wilderness Act in the United States was approved on September 3, 1964. The law provided for the inclusion of federally owned land under the National Wilderness Preservation System and placed the power of designating such land as protected in the hands of Congress. Prior to passage of the bill, such designations were administrative decisions made by the Forestry Service.

4648. List of wilderness areas in the U.S. National Wilderness Preservation System (NWPS) was compiled by the U.S. Department of Agriculture (USDA). The system was established when the Wilderness Act was signed by President Lyndon B. Johnson on September 3, 1964, making all wilderness areas under the jurisdiction of the Forest Service part of the new NWPS. The act required the USDA to review all such areas and present their maps, acreages, and boundaries to the President, who would use the information to recommend wilderness areas to Congress for official designation under the act. The first affected areas were included in the initial assessment, the "Report of the Secretary of Agriculture on the Status of National Forest Units of the National Wilderness Preservation System," which was submitted to Congress by President Johnson on February 8, 1965.

4649. National Trail in Canada was the Bruce Trail, which covers 700 km along the Niagara Escarpment and was completed in 1967. It is the oldest hiking trail in Canada and is maintained by volunteer groups.

4650. U.S. National Wilderness Area to enclose a bison sanctuary was Wichita Mountains Wilderness in southwestern Oklahoma. Although the wilderness was designated by Congress on October 23, 1970, part of the area was established as the first game preserve in the United States in 1905 by President Theodore Roosevelt, as protection for the country's disappearing bison population. Part of the wilderness, which is administered by the U.S. Fish and Wildlife Service, is still closed to the public and reserved for bison and other wildlife.

4651. U.S. National Wilderness Areas in Alaska were designated by Congress on October 23, 1970. All are administered by the U.S. Fish and Wildlife Service. The first six designated areas—the Bering Sea, Bogoslof, Forrester Island, Hazy Islands, Saint Lazaria,Tuxedni—added 90,000 new acres to the National Wilderness Preservation System. They were a small addition, however, compared to the 56 million acres added a decade later under the Alaska National Interest Lands Conservation Act of 1980.

4652. Swampland designated as a U.S. National Wilderness Area was Okefenokee Wilderness, designated by Congress on October 1, 1974. Although the rivers and enormous peat bog of the swamp extend into Florida, all of the official wilderness area lies in Georgia. The swamp is a habitat for numerous birds, alligators, and other animals, who have been protected since 1936 on swampland designated as the Okefenokee National Wildlife Refuge. Both the refuge and the wilderness area are administered by the U.S. Fish and Wildlife Service.

4653. Federal law to create wilderness areas in the eastern United States was the Eastern Wilderness Act. Enacted January 3, 1975, the law added 16 new wildernesses east of the 100th meridian to the National Wilderness Preservation System. The additions were smaller than their generally large counterparts in the American West. Attempts to amend the law in subsequent years were also called Eastern Wilderness Acts.

4654. U.S. National Wilderness Area in the Great Lakes region was West Sister Island Wilderness in Ohio. The restricted 77-acre area is reserved for wildlife, including herons and egrets who feed in Lake Erie wetlands. It was designated by Congress on January 3, 1975, and is administered by the U.S. Fish and Wildlife Service.

4655. U.S. National Wilderness Area in the Mid-Atlantic states was the Brigantine Wilderness on the southern coast of New Jersey, designated by Congress on January 3, 1975. The tidal wetlands and barrier beaches of the area are an important habitat for rare and migratory birds. The protected wilderness area lies within the larger Edwin B. Forsythe National Wildlife Refuge.

4656. U.S. National Wilderness Area in the Ozark Plateau region was Upper Buffalo Wilderness in northern Arkansas, designated by Congress January 3, 1975. The area includes headwaters of Buffalo National River and abandoned farmland being reclaimed by woodlands, making it a significant habitat for forest mammals. The wilderness is administered by the U.S. Forest Service.

4657. U.S. National Wilderness Areas in the Great Plains were both North Dakota areas designated by Congress on January 3, 1975. Chase Lake Wilderness encompasses most of Chase Lake National Wildlife Refuge, the second oldest wildlife refuge and largest white pelican nesting area in the United States. Lostwood Wilderness is the northern portion of Lostwood National Wildlife Refuge, whose "prairie pothole" wetlands are important wildfowl habitats. Both wilderness areas are administered by the U.S. Fish and Wildlife Service.

4658. Australian Heritage Commission (AHC) Act became law on June 19, 1975. The AHC was established to advise the administrator of the Environmental Protection Act on the conservation and improvement of significant natural, aesthetic, historical, social, or cultural sites. Natural sites listed by the AHC in the Register of the National Estate include wilderness areas and habitats for Australia's unique wildlife.

4659. Coastal wetlands designated as a U.S. National Wilderness Area were the freshwater marshes of Lacassine Wilderness in southwest Louisiana. The wilderness is the southern portion of Lacassine National Wildlife Refuge and is an important winter habitat for ducks and geese, as well as regional species of mammals

WILDERNESS—*continued*

and birds. No recreational activities are allowed in the 3,345-acre wilderness, which was designated by Congress on October 19, 1976 and is administered by the U.S. Fish and Wildlife Service.

4660. U.S. National Wilderness Area in Hawaii was Haleakala Wilderness, designated by Congress on October 20, 1976. The area includes the geological features and rare flora of the dormant Haleakala volcano and its surrounding ecosystem on the island of Maui. It is administered by the National Park Service.

4661. Trans-Canada national foot trail entered the planning stages in 1971. Boosters of the Sentier National Trail, which was officially registered as a national organization on August 23, 1977, planned to link wilderness areas, national parks, and other routes to create a 10,000-km trail from Ontario to New Brunswick. It was planned to be used primarily for hiking and other nonmotorized recreation.

4662. Federal law to allow designation of formerly developed lands as wilderness areas was the Endangered American Wilderness Act (EAWA) of 1978. The Wilderness Act of 1964 had provided for the designation of pristine undeveloped lands as wilderness areas. The EAWA broadened the rule to allow lands that might once have been used by humans but were returning to their natural state to be included in the National Wilderness Preservation System. When the EAWA was enacted on February 24, 1978, it added 16 new wildernesses to the system.

4663. U.S. National Wilderness Area in the Gulf of Mexico was Gulf Islands Wilderness, designated by Congress on November 10, 1978. The chain of barrier islands lies several miles off the coast of Mississippi and is within Gulf Islands National Seashore, established in 1971. The islands are administered by the National Park Service and are open for recreational uses like camping and fishing.

4664. Tax exemptions in the United States for conservation easements were approved on December 17, 1980. The regulation allowed a taxpayer to take a deduction for a "qualified real property interest" contributed to a charitable organization exclusively for conservation purposes protected in perpetuity. The donor is not permitted to retain mineral interests in the land if the minerals can be extracted by surface mining.

4665. Colorado Wilderness Act was passed on December 22, 1980 and added 1.4 million acres to the National Wilderness Preservation System. The law also established regulations for livestock grazing on National Forest lands and forbade the government establishment of "buffer zones" that could be used to restrict commercial development around the borders of designated wilderness areas. Subsequent revisions of the law were also called Colorado Wilderness Acts.

4666. Permanent ban on oil and gas leasing on U.S. wilderness land was approved by Congress on December 19, 1982. Congress acted to forestall a plan by Interior Secretary James G. Watt that could have opened government-owned wilderness to logging and mining development in the year 2000, contrary to the provisions of the Wilderness Act of 1964. On December 30, President Ronald Reagan signed the ban, which allowed environmentally friendly surveying and reserved the right of Presidents to develop natural resources on wilderness land in case of national emergency.

4667. Law in the United States extending wilderness protection to land administered by the Bureau of Land Management (BLM) was the Arizona Wilderness Act. The legislation designating over one million acres in Arizona and southern Idaho as national wilderness was passed by Congress on August 10, 1984. Previously, wilderness designations had usually been given to areas situated on lands administered by other agencies like the Forestry or National Parks Services.

4668. National Wilderness Inventory of Australia began in 1986 and was conducted by the Australian Heritage Commission. The commission's surveys were undertaken to collect modern data on the condition of the continent's significant percentage of wild areas, to inform policy and management issues, and to make recommendations regarding protected wilderness designations.

4669. River protected by Iowa's Protected Water (PWA) program was the Boone River. The state's PWA program was begun in 1987 to prevent scenic and environmental damage to the natural ecosystems of selected rivers, maintain or improve water quality, protect fish and wildlife, and coordinate the work of management agencies.

4670. Federal laws in the United States to protect caves were included in the Federal Cave Resources Protection Act, enacted on November 18, 1988. The act established a permit system for removal of natural resources such as plant and animal life, fossils, or minerals from significant caves on public lands. It also provided criminal penalties for vandalism of designated caves and created a cave research program in the National Park Service.

4671. Trans-Canada multi-use trail was begun in 1992. Unlike the Sentier National Trail, whose route was plotted in the 1970s, the Trans-Canada Trail was designed to allow use by vehicles like snowmobiles where appropriate, as well as nonmotorized activities like walking, cycling, horseback riding, and cross-country skiing.

4672. Law establishing the deepest cave in the United States as a protected area concerned Luchugilla Cave in Carlsbad Caverns National Park, NM. When oil and mineral drilling near the park called the security of the cave's many miles of chambers into question, Congress placed it under federal protection. President Bill Clinton signed the bill into law on December 2, 1993.

4673. Exemption for U.S. military flyovers of National Wilderness deserts in California was Title VIII of the California Desert Protection Act (CDPA), signed on November 1, 1994. Because no suitable substitute airspace or training areas could be found for aircraft based in the California deserts, the act allowed continued military use of the areas. The act did, however, restrict low-level flights over national parks. This portion of the CDPA was also called the California Military Lands Withdrawal and Overflights Act of 1994.

4674. Wilderness protection act for California deserts was the California Desert Protection Act. The law was enacted November 1, 1994. It created the Imperial Refuge and Havasu Wilderness areas for management by the U.S. Fish and Wildlife Service and designated 69 more desert locales as wilderness areas to be administered by the U.S. Bureau of Land Management. The act also established the Death Valley and Joshua Tree National Parks and the Mojave National Preserve.

4675. Ban on motorized watercraft on Wyoming's Snake River became effective March 24, 1997. The Wyoming Game and Fish Commission closed eight miles of the river below Jackson to motorboats and Jet Skis because of safety concerns linked to the growing popularity of floating craft and nonmotorized recreation.

4676. Bioprospecting clause in U.S. Fish and Wildlife Service Special Use specimen-collecting permits was required after March 30, 1999. The rule allowed collection of specimens on wilderness lands administered by the U.S. Fish and Wildlife Service (FWS) for scientific or educational purposes, but forbade their use in commercial research. Sale or transfer of specimens to third parties was also prohibited. Commercial use of specimens found on wilderness land was allowed if the collectors were parties to a bioprospecting contract with the FWS known as a Cooperative Research and Development Agreement (CRADA). Violators were subject to a 20 percent royalty on any commercial sales in addition to possible damage lawsuits.

4677. Haze regulations concerning U.S. wilderness areas and national parks were signed by U.S. Environmental Protection Agency (EPA) Administrator Carol Browner on April 22, 1999. The EPA regulations called upon state governments to improve the visibility and scenic quality of wilderness areas and national parks by developing long-term strategies for reducing emissions of visible air pollutants.

4678. Comprehensive plan to redesign the water-flow system in and around the Everglades resulted from the bipartisan Water Resources Development Act, signed by President Bill Clinton on December 11, 2000. The Act contained $7.8 billion in funding for the project, which was intended to restore a natural flow of water through the increasingly parched wilderness.

WILDLIFE

4679. Description of chimpanzees known to Europeans appears in the account of the voyage of Hanno the Carthaginian, who sailed along the coast of West Africa around 525 B.C.. The navigator's short description of his travels includes an encounter with "forest people" with "shaggy pelts." Hanno's interpreters identified the creatures as gorillas, but later scholars identified them as chimpanzees.

4680. Reference work on Brazilian wildlife was Georg Marcgrav's *Historiae Rerum Naturalium Brasiliae* (Natural history of Brazil). Marcgrav, who recognized the unique character of the species he encountered in Brazil, described much South American fauna to Europeans for the first time. The German naturalist's accurately illustrated work was published in Amsterdam in 1648.

WILDLIFE—*continued*

4681. Sea otters encountered by Europeans were discovered in 1741 off the coast of Alaska by members of the Bering Expedition. Overhunting of sea otters for their valuable pelts nearly resulted in the extinction of the species by the 20th century. Since then the otters have made a moderately successful recovery.

4682. Naturalist to describe the wildlife of Australia was English botanist Sir Joseph Banks, a member of Captain James Cook's 1768 expedition to the South Pacific. Banks's descriptions of marsupial species and the collection of samples with which he returned to England were extremely influential in arousing English interest in Australia.

4683. American explorer to describe prairie dogs was William Clark of the Lewis and Clark expedition. While traveling along the Missouri River in what would later become South Dakota on September 7, 1804, the expedition encountered a small animal Clark likened to a "ground rat." A few days later, he described the plentiful animals as "barking squirrels" after the noise they made when disturbed.

4684. Description of the "black-tailed deer" by William Clark was on his expedition with Meriwether Lewis along the Missouri River above the mouth of the Niobrara River. Clark saw the deer on September 7, 1804, but noted it in his journal on September 17.

4685. Entomologist in the United States was German-born insect collector Frederick Melsheimer of Hanover, PA. He amassed one of the largest collections of entomological specimens in American history and published *Insects of Pennsylvania* in 1806.

4686. Significant study of the behavior of insects was French entomologist and writer Jean Henri Fabre's ten-volume *Souveniers entomologique*, which he began in 1879 and completed in 1907. In studies of insects he found on the rough land surrounding his home in Sérignan, Fabre was more concerned with the behavior of insects than earlier entomologists, who frequently confined their studies to identification and classification.

4687. U.S. government body to study the effects of insects on nature was the Entomological Division of the Department of Agriculture, created by a congressional act on March 3, 1885, to promote economic ornithology, the study of the interrelation of birds and agriculture. Insects and plant life were studied by the Entomological Division primarily in the context of their roles as food sources for birds.

4688. Game warden in the United States to be paid a salary was William Alden Smith of Grand Rapids, MI. A state law approved March 15, 1887, provided for the appointment of a game and fish warden to enforce state laws designed for the preservation of moose, wapiti, deer, birds, and fish. Smith was appointed to a four-year term at an annual salary of $1,200 plus expenses.

4689. Law in Wisconsin authorizing the pay of a game warden was approved on April 12, 1887. The law authorized the appointment of four game wardens for two-year terms at an annual salary of $600 with a maximum of $250 for expenses.

4690. Okapi known to Western science was discovered through specimens obtained by British explorer Sir Harry Hamilton Johnston in 1901, while he was consul general of Uganda Protectorate. The skin and bone specimens Johnston sent to the British Museum for analysis enabled zoologists to establish that the okapi was a rare relative of the giraffe, whose existence was previously unknown to science.

4691. Law in the United States to prohibit the transportation of wildlife across state lines if they were taken in violation of state law was the Lacey Act of 1900, signed on May 25.

4692. Effort for organized wildlife protection in Alabama began in 1907 with the passage of several wildlife protection laws.

4693. Cement gun for mounting wildlife specimens was invented by American Carl Ethan Akeley. A naturalist, photographer, and sculptor with a background in taxidermy, Akeley developed a device capable of applying adhesive under an animal's skin after it had been mounted on an anatomically accurate model. Akeley received an award from the Franklin Institute in 1911 for the invention, which was later used in repair work to prevent liquid seepage.

4694. Elk refuge in the United States was established in 1912 north of Jackson, WY. The refuge continues to aid the survival of elk herds in winter, when food supplies can be scarce, and remains the only elk management refuge in the United States.

4695. Pygmy hippos known to Western science were examples obtained in Liberia by German collector Hans Hermann Schomburgk in 1913. Previous evidence of the animal's existence suggested that it might have been a large type of hog that became extinct. The small West African hippopotamus approached endangered species status in the decades after its discovery due to its value as food and as an illegal source of ivory, and because of its occasional role as an agricultural pest.

4696. Zoological Survey of India (ZSI) was created on July 1, 1916. The organization expanded on earlier research and museum collections of Indian wildlife by increasing fieldwork, surveys, education, and taxonomic activities. Today the survey also promotes studies in zoo geography, bioethics, animal behavior and ecology, and marine fauna. ZSI's main headquarters are in Calcutta. Numerous regional field stations are located within India.

4697. Wildlife films of the fauna of Borneo were made by American photographers and naturalists Martin and Osa Johnson. Although the Johnsons became known for the popular films of African wildlife they made for the Museum of Natural History in the 1920s, their trips to the South Pacific resulted in a 1917 film about the wildlife of North Borneo.

4698. Wildlife films of mountain gorillas were made by American naturalist Carl Ethan Akeley near Lake Kivu in eastern Congo in 1921 and 1922. He used the Akeley camera, which he had invented for use by naturalists and patented in 1916. The experience inspired Akeley to help convince King Albert of Belgium to establish the first wildlife reserve in Africa, Albert National Park, later known as Virunga National Park.

4699. Duck-billed platypus successfully transported outside Australia was imported into the United States by an animal dealer, who arrived at the New York Zoological Park with one specimen on July 14, 1922. The platypus was a male and did not reproduce, but it was the first such animal seen alive outside of its native continent.

4700. Federal law enforcement agency for wildlife in the United States was the Division of Game Management, established on July 1, 1934 as part of the Department of Agriculture's Bureau of Biological Survey. Since 1972, federal wildlife laws have been enforced by the U.S. Fish and Wildlife Service's Division of Law Enforcement.

4701. Wildlife conference in North America was the North American Wildlife and Natural Resources Conference held in 1936.

4702. New Jersey program authorized by the Federal Aid in Wildlife Restoration Act (Pittman-Robertson Act) began in 1937 when its Division of Fish, Game and Wildlife developed state wildlife management areas.

4703. South Dakota program authorized by the Federal Aid in Wildlife Restoration Act (Pittman-Robertson Act) began in 1937. It was one of the first wildlife management programs authorized by the act.

4704. Legislation for cooperative wildlife management was the Federal Aid in Wildlife Restoration Act of 1937, called the Pittman-Robertson Act for its two sponsors, Senator Key Pittman and Congressman A. Willis Robertson. The act, signed on September 2, 1937, diverted receipts from excise taxes on guns and ammunition into a special fund to be distributed to the states for wildlife restoration and to better manage their conservation programs. The act was the result of cooperation among the federal government, the states, private conservation groups, and the sporting arms and ammunition industry.

4705. Tennessee program authorized by the Federal Aid in Wildlife Restoration Act (Pittman-Robertson Act) began in 1938. The act's funds were used as a first payment for 18,108 acres that became part of the 20,000-acre Cheatham Wildlife Management Area, owned entirely by the Tennessee Wildlife Resources Agency.

4706. Texas program authorized by the Federal Aid in Wildlife Restoration Act (Pittman-Robertson Act) began in 1938.

4707. Wildlife management project in West Virginia began in 1938 with the purchase of Nathaniel Mountain with funds provided from the Federal Aid in Wildlife Restoration Act.

4708. Project after the 1938 appropriation by the U.S. Congress to assist state wildlife restoration projects was the Utah Fish and Game Commission's plan to stabilize its water levels on some 2,000 acres of land bordering the Great Salt Lake. The plan was approved on July 23, 1938, by the U.S. Fish and Wildlife Service.

WILDLIFE—*continued*

4709. Vermont program authorized by the Federal Aid in Wildlife Restoration Act (Pittman-Robertson Act) began on September 27, 1938, when the Vermont Wildlife Survey was initiated. The survey, conducted by all eighteen of the Vermont Fish and Game Service wardens, was designed to determine the status and distribution of game populations.

4710. Indiana program authorized by the Federal Aid in Wildlife Restoration Act (Pittman-Robertson Act) began in 1939.

4711. Nebraska program authorized by the Federal Aid in Wildlife Restoration Act (Pittman-Robertson Act) began in 1939 when the state received its first apportionment of funds.

4712. Rhode Island project initiated by the Federal Aid in Wildlife Restoration Act (Pittman-Robertson Act) was a wildlife demonstration area developed in 1939.

4713. State in the United States to permanently employ wildlife pathologists was California. The state's division of Fish and Game hired its first staff of wildlife disease investigators in 1939.

4714. Montana program authorized by the Federal Aid in Wildlife Restoration Act (Pittman-Robertson Act) began in 1941. Assenting legislation also prohibited diversion of hunting license fees for purposes other than wildlife restoration.

4715. Federally funded wildlife management area in Georgia was the Cedar Creek and Coastal Flatwood Wildlife Management Area established in 1944.

4716. Law to prevent damage to fish and wildlife was Public Law 732, the Fish and Wildlife Coordination Act of 1946, signed on August 14. The law officially committed the U.S. government to the policy that all new federal water projects would, if possible, include provisions to prevent loss of or damage to fish and wildlife.

4717. Nevada program authorized by the Federal Aid in Wildlife Restoration Act (Pittman-Robertson Act) began in 1947 when its Fish and Game Commission was reorganized as a statewide wildlife agency.

4718. Wildlife research unit in the United States was the Sybille Wildlife Research Unit, west of Wheatland, WY, started in the 1950s. It has served as a model for wildlife agencies in the United States and in foreign countries in the development of similar facilities.

4719. Government survey of wildlife-related recreation activities by the U.S. was the Survey of Fishing, Hunting & Wildlife—Associated Recreation. The first survey was completed in 1955 and has been updated every five years by the U.S. Census Bureau for the Fish and Wildlife Service. The survey offers data on wildlife-related recreation ranging from hunting and fishing to photography, as well as on wildlife losses due to environmental contamination.

4720. Survey of wildlife-related recreation on a nationwide basis in the United States was conducted in 1955 as part of the National Survey of Fishing, Hunting, and Wildlife-Associated Recreation. The survey is conducted every five years by the Census Bureau for the Fish and Wildlife Service, which maintains collected data on activities like bird-watching, wildlife photography, nature hiking, expenditures on such pursuits, and donations to wildlife-related organizations.

4721. Program in Guam authorized by the Federal Aid in Wildlife Restoration Act (Pittman-Robertson Act) began in 1957, the first such program in a U.S. territory.

4722. Marine mammals studied by the U.S. military were dolphins. The U.S. Navy began observing dolphins in 1959 at Marineland of the Pacific, in Palos Verdes, CA. A white-sided dolphin was studied to devise improvements in the hydrodynamic design of torpedoes, ships, and submarines. During the 1960s, the Navy's Marine Mammal Program led to both the military use of dolphins and research in marine acoustics, diving physiology, anatomy, and medicine.

4723. Military uses of marine mammals were developed in secret programs during the 1960s by the navies of the United States and the Union of Soviet Socialist Republics. The U.S. Navy Marine Mammal Program trained dolphins to find and retrieve objects, mark undersea mines, protect military bases and vessels from enemy divers, and facilitate underwater photography. Sea lions and beluga whales were also trained for such tasks.

4724. Significant modern studies of elephant behavior in the wild were begun in the late 1960s by Iain Douglas-Hamilton. The Scottish-born field biologist studied the social life of African elephants in northeastern Tanzania's Lake Manyara National Park. Douglas-Hamilton later founded the conservation group Save the Elephants.

4725. Significant studies of lions in the wild were begun in 1966 by American field biologist and writer George B. Schaller. He conducted his research on the predatory habits of lions and other behavior patterns while based at the Serengeti Research Institute in Tanzania's Serengeti National Park. His book *The Serengeti Lion* was published in 1972.

4726. Law in the United States to protect other nations' wildlife was the Endangered Species Act of 1969, signed on December 5. This law widened the obligations of the Secretary of the Interior to ensure that the United States did not contribute to the depredation of other nations' wildlife.

4727. Law to protect future endangered fish and wildlife was the Endangered Species Act of 1973, signed on December 28. The law extended protection to fish and wildlife that were likely to become endangered, as well as those already officially listed as endangered by the U.S. Department of the Interior.

4728. U.S. National Wildlife Health Center was established in 1975 in Madison, WI. Its laboratory monitors and attempts to identify the causes of diseases that result in wildlife population losses both in the United States and internationally. The research and veterinary education facility is attached to the Biological Resources Division of the U.S. Geological Survey.

4729. Discovery of the monarch butterfly's winter home was made by Kenneth C. Brugger, an American textile engineer and amateur naturalist living in Mexico. On January 2, 1975, Brugger and his wife found millions of monarchs in the volcanic mountains of eastern Michoacán, southwest of Mexico City. Brugger relayed the information to Canadian scientist Fred A. Urquhart, who established that the area was indeed the southern terminus of the migration route taken by the butterflies.

4730. Ports in the United States to be monitored by U.S. Fish and Wildlife Service agents were Honolulu, Los Angeles, Seattle, San Francisco, Chicago, Miami, New Orleans, and the combined New York–Newark point of entry. Federal agents were first assigned in July 1975 to regulate imports and exports of fish and wildlife.

4731. Program in Colorado in which taxpayers could make a contribution to a wildlife program was instituted in 1977. Contributors could check it off on their 1978 state income tax forms. Nearly $350,000 was donated in the program's first year.

4732. Wildlife biologists on the Northern Mariana Islands were hired in 1983 by the Division of Fish and Wildlife on Saipan.

4733. Rewards offered by the U.S. Fish and Wildlife Service for information about wildlife crimes were authorized by a December 31, 1982, amendment to the Fish and Wildlife Improvement Act of 1978. The amendment allowed the service to use part of its appropriated budget to pay for evidence, information, rewards, and undercover businesses and operations.

4734. U.S. Fish and Wildlife Service Environmental Contaminants Program began as the Resource Contaminant Assessment Program in 1983. Prior to the RCA's habitat assessment, cleanup guidance, and biomonitoring work, investigations of contaminants affecting wildlife were handled individually by different agencies that eventually became the U.S. Fish and Wildlife Service. The Resource Contaminant Assessment Program became the Environmental Contaminants Program in 1987.

4735. Wild horse training program in a U.S. prison was established in 1986 at the Colorado State Correctional Facility at Canon City, CO. The program of wild horses being trained by inmates had been recommended by the Wild Horse and Burro Advisory Board, created that year by the Secretaries of Agriculture and the Interior.

4736. U.S. National Fish and Wildlife Foundation was created by Congress through the National Fish and Wildlife Foundation Establishment Act on November 14, 1988. The foundation is a private, nonprofit organization that administers federal grants benefiting programs in conservation education, fisheries conservation and management, neotropical migratory bird conservation, restoration and acquisition of wetlands and private lands, and wildlife and habitat management. While the foundation's programs are federally funded, its operating expenses come from private donations.

4737. International organization to monitor the disappearance of amphibians was the Declining Amphibian Populations Task Force (DAPTF). The program is a global network of scientists and conservationists who monitor frogs and other amphibians in an effort to discover and reverse the decline in their populations. DAPTF was established in 1991. Although its participants are active in more than 90 countries, the group has offices at the Open University Department of Biology in Milton Keynes, England.

WILDLIFE—*continued*

4738. North American organization to monitor the disappearance of amphibians was the North American Amphibian Monitoring Program (NAAMP), which covers Canada, Mexico, and the United States. Through species monitoring and population surveys, NAAMP groups study the decline in populations of frogs, salamanders, and other amphibians as part of the global Declining Amphibian Populations Task Force (DAPTF) program. The NAAMP is based at the U.S. Geological Survey's Patuxent Wildlife Research Center in Laurel, MD.

4739. Evidence of the existence of the Vu Quang ox were three pairs of horns discovered during a May 1992 survey of the Vu Quang Nature Reserve by the Vietnam Ministry of Forestry and the World Wide Fund for Nature. A live specimen of this rare nocturnal animal—also known as the saola—was captured in June 1994. Logging in the ox's habitat along the Vietnam-Laos border was cancelled to protect the species.

4740. Field station for the U.S. National Wildlife Health Center was established in Honolulu, HI, in 1992. The facility was created to aid in wildlife disease research work in the Hawaiian Islands, which are home to a large number of indigenous endangered species.

4741. Woman to direct the U.S. Fish and Wildlife Service was Mollie H. Beattie. Nominated by President Bill Clinton and confirmed to lead the agency by Congress on September 10, 1993, she advocated ecosystem management programs, endangered species protection, and expansion of federal wildlife refuge lands. Beattie resigned shortly before her death on June 27, 1996.

4742. U.S. Fish and Wildlife Service (FWS) guidelines for distributing eagle parts were issued on March 30, 1994. Legislation including the Eagle Act, Migratory Bird Treaty, and Endangered Species Act allowed the FWS to distribute eagle carcasses, feathers, and other parts to Native Americans for religious purposes. The guidelines directed FWS employees to salvage and send such material only to the National Eagle Repository for distribution to permit-holding Native Americans.

4743. Idaho Wildlife DNA Forensics Laboratory opened in Caldwell, ID, in 1995. The lab was designed to aid the state's Department of Fish and Game in wildlife issues like disease identification and prevention, as well as antipoaching enforcement in Idaho and other western states.

4744. Asian elephant tracked with a satellite transmitter was Mek Penawar, who was released in Malaysia's Taman Negara National Park on October 10, 1995. After raiding a fruit plantation, the female elephant was captured and relocated by the Malaysian Wildlife Department. A satellite transmitter attached by the Malaysian Elephant Satellite Tracking Project helped researchers monitor the elephant's movements in rough terrain and gather information about her behavior.

4745. Large-scale relocation of elephants began on October 4, 2001, when 40 elephants were moved from South Africa's Kruger Park over the border to a protected area in Mozambique, the first of an estimated 1,000 elephants to be relocated in the process of creating a huge tri-national preserve, the Gaza-Kruger-Gonarezhou, consolidating wild lands in South Africa, Mozambique, amd Zimbabwe.

WILDLIFE—BREEDING

4746. Pronghorn antelope bred and reared in captivity was born in the City Park Zoo, Denver, CO, in 1903. The zoo director, Alfred Hill, was congratulated by President Theodore Roosevelt.

4747. Musk ox born in captivity was born on September 5, 1925, at the Bronx Zoo in New York City.

4748. Snow goose born and bred in captivity was hatched in 1934 in the City Park Zoo in Denver, CO.

4749. Ptarmigan born and bred in captivity was hatched on July 24, 1934, in Ithaca, NY, from one of 10 eggs obtained from Churchill, Manitoba, Canada, by Arthur Augustus Allen, a professor of ornithology at Cornell University. The ptarmigan died of enterohepatitis when it was 110 days old.

4750. Duck-billed platypus bred in captivity was bred by prominent Australian naturalist David Fleay in 1944. Fleay's successful experiments with platypus breeding provided the first scientific descriptions of the reproductive behavior of the rare semi-aquatic mammals.

4751. Gorilla born in captivity in the United States was born on December 22, 1956, at the Columbus, OH, zoo. Named Colo, she weighed 3.25 pounds. She was the offspring of Baron (11 years old, 380 pounds) and Christiana (9 years old, 260 pounds).

4752. Giant panda born in captivity was Ming Ming, born on September 9, 1963, at the Beijing Zoo in China.

4753. Woolly monkeys successfully bred in captivity were at the Monkey Sanctuary in Looe, England. The cooperative wildlife preserve was founded in 1964 by Leonard Williams as a sanctuary for the monkeys, whose natural habitat in the Amazon rainforest was being destroyed.

4754. Whooping crane born in captivity was Dawn, a six-inch-tall crane hatched on May 28, 1975, at the Rare and Endangered Bird Research Center in Laurel, MD.

4755. Elephant and rhinoceros orphanage in Kenya was located at the David Sheldrick Wildlife Trust, established in 1977 in Nairobi. The program extended Daphne Sheldrick's unprecedented success in raising orphaned baby elephants in the 1970s. Elephants and rhinos raised at the trust are eventually released in Tsavo National Park.

4756. Successful birth of an elephant in captivity in the Western Hemisphere was at the Knoxville Zoological Gardens in Knoxville, TN, in 1978.

4757. Giant panda born outside of China was Xeng Li, born August 11, 1980, at the Chapultepec Zoo in Mexico City. Unfortunately, Xeng Li's mother Ying Ying accidentally crushed the panda cub to death days later. A second cub born to Ying Ying at Chapultepec Zoo in July 1981, Tohui, was the first zoo-born panda outside of China to survive.

4758. Elephant captive breeding program in Sri Lanka was begun in 1982 at Pinnawela Elephant Orphanage. The orphanage, started by the Sri Lankan Department of Wildlife in 1975 to rescue abandoned or orphaned elephants, saw its first birth in 1984. The orphanage is open to visitors.

4759. Aye-aye born in captivity was born at the Duke University Primate Center in April 1992. Named Blue Devil, the rare primate was born to a female aye-aye who had been captured in Madagascar while pregnant. The center breeds aye-ayes in an effort to preserve the species.

4760. Rhinoceros preserve in Tanzania was the Mkomazi Rhino Sanctuary, established as part of the Mkomazi Game Reserve in 1994 to safeguard a breeding population of black rhinos from poachers. It is funded and operated by the Tanzanian government and the Tony Fitzjohn/George Adamson African Wildlife Preservation Trust.

4761. Wild dog breeding program in East Africa began in 1995 at the Mkomazi Game Reserve in Tanzania. The effort is concentrated on ensuring the health of a breeding population of the endangered species and reintroducing the animals to the wild in national parks. The program is operated by the Tony Fitzjohn/George Adamson African Wildlife Preservation Trust.

WILDLIFE—CONSERVATION

4762. State wildlife conservation agency in the United States was the California Board of Fisheries, created in 1870 to regulate fishing in state waters. It predated the first federal conservation agency, the U.S. Commission of Fish and Fisheries, by a year.

4763. Conservation agency established by the U.S. government was the Commission of Fish and Fisheries. Also known as the Fish Commission, it was created by Congress and approved by President Ulysses S. Grant on February 9, 1871. Its main purpose was to study and suggest policies to reverse the declining trend of food fish populations in both fresh and ocean waters. This agency was the original version of the modern National Marine Fisheries Service. Its first commissioner was Spencer F. Baird.

4764. Law to protect the Asiatic lion was a hunting ban ordered by the Nawab of Junagarh in 1900. Protection for the endangered species was later formalized by the Indian government through the creation of the Gir Wildlife Sanctuary and National Park in the western state of Gujarat. The sanctuary's forest is the only place where Asiatic lions are now found in the wild.

4765. Wildlife management law in the United States was the Lacey Game and Wild Birds Preservation and Disposition Act. Passed by Congress on May 25, 1900, the Lacey Act was prompted by interstate traffic in birds, whose numbers were being decimated by market hunters. The act forbade the interstate transportation of any game killed in violation of local laws.

4766. Wildlife warden killed in the line of duty in the United States was Guy Bradley, a warden employed by the Audubon Society for the Protection of Wild Birds. Bradley was shot to death by egret plume poachers at Cape Sable, FL, in 1905. A trail at Everglades National Park and a National Fish and Wildlife Foundation Award for conservation law enforcement officers were later established in his memory.

WILDLIFE—CONSERVATION—*continued*

4767. Wildlife protection law in India was the Wild Birds and Animals Protection Act, which became effective on September 18, 1912. The act allowed India's state governments to set hunting seasons to conserve listed species and confiscate any illegally taken wildlife. Animals killed in defense of persons or property were excepted.

4768. Federal agency charged with comprehensively monitoring and protecting all wildlife in the United States was the U.S. Fish and Wildlife Service. The agency is part of the U.S. Department of the Interior and was created in 1940 by merging the Bureau of Fisheries and Bureau of Biological Survey.

4769. Interagency government property transfers for wildlife conservation in the United States were allowed by the Transfer of Certain Real Property for Wildlife Conservation Purposes Act. The law was enacted on May 19, 1948, and was subsequently amended on June 30, 1949, and September 26, 1972. It stated that if the Administrator of the General Services Administration determines that government-owned real property is no longer needed by a federal agency, the land can be transferred without reimbursement to the Secretary of the Interior if it has particular value for migratory birds or to a state agency for other wildlife conservation purposes.

4770. Comprehensive U.S. national fish and wildlife policy was the Fish and Wildlife Act of 1956, which aided in the development of wildlife refuges.

4771. Federal law in the United States to ban the use of motorized vehicles in capturing wild horses was PL 86-234, also known as the Wild Horse Annie Act. Congress approved the act in 1959 after a public campaign organized by Nevadan Velma Johnston, also known as "Wild Horse Annie." The law's provisions included a ban on motorized vehicles, including aircraft, being used to catch wild mustangs and burros, and a ban on the intentional polluting of water holes for the purpose of killing or capturing wild horses.

4772. Wildlife conservation law for U.S. military property was the Sikes Military Reservation Act. Approved by Congress on September 15, 1960, the Sikes Act required military bases to plan, develop, and maintain conservation programs for wildlife, fish, and game. The act was expanded on October 18, 1974 to include lands administered by the Forest Service, Atomic Energy Commission, Bureau of Land Management, and National Aeronautics and Space Administration. The 1974 amendment also required protection for endangered species.

4773. Wildlife protection conference of modern African governments was the Symposium on the Conservation of Nature and Natural Resources in Modern African States. Also known as the "Arusha Conference," the September 1961 meeting was notable for its declaration that African states would henceforth protect wildlife as a resource capable of attracting tourism revenues rather than solely as an exportable product.

4774. Wildlife resource management college in Africa was the College of African Wildlife Management in Mweka, Tanzania. Opened in 1963, the college trains future professional wildlife park staff in conservation, park and wildlife management, natural sciences, and other related disciplines.

4775. International Conference on Polar Bears was held at the University of Alaska at Fairbanks in September 1965. This meeting and subsequent efforts by the World Conservation Union (IUCN) led scientific representatives from all of the circumpolar nations—Canada, Denmark, Norway, the Soviet Union, and the United States—to establish a structure for sharing information in polar bear research.

4776. Federal law in the United States to protect fur seals was the Fur Seal Act, enacted on November 2, 1966. It prohibited the taking, transportation, importing, or possession of fur seals and sea otters. American Indians, Aleuts, and Eskimos of the North Pacific coastal regions were exempted from the law, as were scientists authorized by the Secretary of the Interior.

4777. United Nations convention to protect the vicuña was the UN Convention for the Conservation of the Vicuña. The document was signed in La Paz, Bolivia, by the host government and by Argentina, Chile, Ecuador, and Peru on August 16, 1969. With the exception of Argentina, the signatories agreed to a modified version of the treaty, the Convention for the Conservation and Management of the Vicuña, in Lima, Peru, on December 20, 1979. In an effort to slow the decline of vicuña populations and preserve the Andean animal as a resource, the convention established a moratorium on vicuña hunting and the sale of vicuña products until December 31, 1989.

4778. Law in the United States allowing teenagers to work on federal conservation programs was the Youth Conservation Corps (YCC) Act, approved by Congress on August 13, 1970. The law established permanent programs within the Departments of Interior and Agriculture for young adults between the ages of 15 and 19, enabling them to perform various tasks on national wildlife refuges, national fish hatcheries, research stations, and other facilities. The program is administered by the U.S. Fish and Wildlife Service.

4779. U.S. Environmental Protection Agency (EPA) role in wildlife conservation resulted from Reorganization Plan No. 3, which became effective on December 2, 1970. The EPA became responsible for studies of the effects of insecticides, herbicides, fungicides, and pesticides upon fish and wildlife resources. Such studies had previously been the responsibility of the U.S. Fish and Wildlife Service.

4780. Federal management policies for wild horses in the United States were contained in the Wild Horses and Burros Act, approved on December 15, 1971, to control, manage, and protect wild mustangs and burros, whowere being killed and sold to rendering plants and slaughterhouses. Horses and burros on Bureau of Land Management or U.S. Forest Service lands are now managed by those agencies.

4781. Comprehensive wildlife protection legislation in India was the Wild Life (Protection) Act, which became effective on September 9, 1972. The act allowed the government to appoint a wildlife protection directorate, wildlife wardens, and a Wild Life Advisory Board. It also set rules, record-keeping procedures, licensing regulations, and seasons for hunting and trapping. Killing of young or female animals was forbidden except in defense of persons or property. The use of aircraft or motorized vehicles in hunting was also prohibited.

4782. Marine mammal conservation and management law in the United States was the Marine Mammal Protection Act, passed by Congress on October 21, 1972. The act charged the U.S. Departments of Commerce and the Interior with overseeing a moratorium on the taking and importation of polar bears, sea otters, walruses, dugongs, and manatees. American Indians were exempted from the ban on hunting and trafficking in such species and their products. The act also created the Marine Mammal Commission.

4783. International organization to manage Antarctic wildlife was the Commission for the Conservation of Antarctic Marine Living Resources, established by the contracting parties of the Convention on the Conservation of Antarctic Marine Living Resources, which was signed in Canberra, Australia on May 20, 1980. This convention expanded the terms of the Antarctic Treaty (1959) and related wildlife agreements. The commission established a scientific committee to study and conserve species in Antarctica and the surrounding seas.

4784. Wildlife Diversity Program in Oklahoma was created in 1981, when the Oklahoma legislature allowed residents to donate any portion of their state income tax refund to wildlife programs. The program was known as the Nongame Wildlife Fund. Subsequent laws allowed donations by businesses and the sale of vehicle license plates to help fund wildlife management, species surveys, and other research, which collectively became known as the Wildlife Diversity Program in November 1996. The program is funded by donations and receives no assistance from state tax revenue.

4785. Interagency Grizzly Bear Committee (IGBC) was created in 1983 to aid recovery of the grizzly bear in the lower 48 U.S. states. It includes representatives of nine federal and state wildlife agencies from the western United States and Canada. Its advisory policies address grizzly management, responses to nuisance bears, relocation, habitat quality, and biological reports.

4786. Texas law to regulate game and fisheries statewide was the Wildlife Conservation Act, which became effective August 29, 1983. Prior to passage of the law, which is also known in Texas Parks and Wildlife Code as the Uniform Wildlife Regulatory Act, county commissioners controlled their own wildlife laws and in some cases could veto hunting or fishing regulations set forth by the state. The 1983 law gave all such regulatory powers to the state in an effort to consolidate conservation efforts.

4787. Wildlife crime lab in the United States was the National Fish and Wildlife Forensics Laboratory, established in Ashland, OR, in September 1988. The center provides forensic assistance to conservation and customs agencies in determining if crimes have been committed under federal wildlife protection and endangered species laws.

WILDLIFE—CONSERVATION—*continued*

4788. Mass burning of poached African ivory took place in Kenya on July 19, 1989. Kenyan President Daniel Arap Moi ignited a pile of $3 million worth of confiscated tusks to increase international awareness of the elephant's endangered status, which was worsening because of poaching. The publicity event was organized by naturalist Richard Leakey of Kenya's Wildlife Conservation Department.

4789. Junior Duck Stamp program in the United States was created through the Junior Duck Stamp Conservation and Design Program Act, passed on October 6, 1994. Junior Duck Stamps were modeled after the annual hunting permit stamps sold under the Migratory Bird Hunting and Conservation Stamp Act of 1934 or "Duck Stamp Act." The Junior Duck Stamp program sponsors an annual competition that solicits designs relating to migratory bird conservation from elementary and secondary school students. Sales of the stamps are used to provide scholarships for the winning designers and funding for conservation education programs. The Junior Duck Stamp Program began in 1995 and is administered by the Department of the Interior.

4790. Wildlife Habitat Incentive Program (WHIP) in the United States was authorized as a provision of the Federal Agriculture Improvement and Reform Act of 1996, also known as the 1996 Farm Bill. WHIP was designed to provide technical and financial assistance to landowners wishing to develop and implement management practices to improve wildlife habitat. The Farm Bill was signed by President Bill Clinton on April 4, 1996. WHIP was administered by the Department of Agriculture's Natural Resources Conservation Service.

4791. State in the United States to implement a multiple-species conservation plan statewide was Washington on January 30, 1997. Representatives of the U.S. Fish and Wildlife Service, the National Marine Fisheries Service, and Washington's Department of Natural Resources signed a Habitat Conservation Plan (HCP) for 1.6 million acres of forested Washington land. The renewable plan's original term would last 70 years. Unlike many HCPs, the Washington plan would protect more than one species.

4792. Modern wildlife resource crime laws in China are contained in Chapter Six of China's Criminal Code. Revisions of the code, effective October 1, 1997, made harvesting wild or aquatic animals by illegal methods or out of season punishable by prison terms of up to three years, fines, or both. Killing, buying, selling, or transporting state-protected wild animals or wildlife products are offenses punishable by up to five years imprisonment and fines.

4793. Elephant remote tracking program in Thailand began on March 10, 1998, with the release of five female Asian elephants in Dai Pha Muang Wildlife Sanctuary. The experimental release was an attempt to reintegrate elephants domesticated for use in the Thai timber industry into the wild. The first five animals were equipped with radio and satellite transmitters by researchers from the Thai Forest Industry Organization's Elephant Conservation Center and the Smithsonian Institution Conservation and Research Center.

4794. Major auction of confiscated wildlife products in the United States was held on June 4, 1999, in Denver, CO, by the National Wildlife Property Repository. Thousands of sale items made of coral, seashells, lizard, and snakeskin had been seized because of violations of import laws. Their sale did not contravene any endangered species or wildlife protection statutes. Proceeds benefited the Repository's conservation education program and the National Eagle Repository, which distributes feathers to Native Americans for religious uses. Proceeds were also earmarked for the care of live animals acquired by the Fish and Wildlife Service through enforcement of wildlife protection laws and for the service's wildlife crime information incentive Reward Fund.

WILDLIFE—CONTROL

4795. Wolf bounty enacted in what is now the United States was passed by the Virginia General Assembly on October 5, 1646, and signed by Governor Sir William Berkeley. It provided that "what person soever shall after publication hereof kill a wolfe, and bring in the head to any commissioner, upon certificate of the said communication to the county court, he or they shall receive one hundred pounds of tobacco for so doing to be raised out of the County where the wolf is killed."

4796. Colony in America to require the killing of wolves was Delaware. By law, in 1654, all Native Americans living within the boundaries of Delaware were required to kill at least two wolves annually.

4797. Failed U.S. Rocky Mountain mule deer management policy began in 1906 when President Theodore Roosevelt created a national game preserve on the Kaibab Plateau in northern Arizona. To preserve the deer for hunting, an unrestrained predator control policy virtually emptied the region of coyotes, mountain lions, and wolves. With no natural predators to balance the growth of mule deer herds, their population multiplied from 3,000 to 100,000 within 15 years, causing their mass starvation and severe stress on the local environment. The incident became a classic case study in wildlife management.

4798. Brucellosis in the Yellowstone National Park buffalo herd was discovered in 1917. Uncertainty over whether or not the bacterial disease might spread to domestic cattle led to the controversial killing of more than 1,000 buffalo.

4799. Law in the United States to allow federal control of wildlife on both public and private lands was the Animal Damage Control Act. Approved on March 2, 1931, it empowered the Department of Agriculture to destroy birds, rodents, and predators like wolves in the interests of managing wildlife and reducing losses to agriculture, ranching, and other human activities. In 1985, such authority was transferred to the Department of the Interior.

4800. Waterfowl Depredations Prevention Act in the United States was enacted on June 3, 1956. It allowed the Secretary of the Interior to distribute surplus grain owned by the federal Commodity Credit Corporation to lure waterfowl away from agricultural areas, thus reducing crop damage. The grain was made available to individual farmers as well as local and state government agencies.

4801. Wild horse range overseen by the U.S. government was the Nevada Wild Horse Range. Established in 1962 and administered by the Bureau of Land Management, the range is located within the boundaries of Nellis Air Force Base in southern Nevada. Thirty years after the range's founding, its habitat was already being strained by overpopulations of wild mustangs, despite periodic gathering and relocation efforts.

4802. Wild horse adoptions in the United States took place at the Pryor Mountain Wild Horse Range in Montana in 1973, three years before the official U.S. Bureau of Land Management adoption program began.

4803. Federal "Adopt a Horse or Burro" program in the United States was established by the Bureau of Land Management (BLM) in 1976. The national program was created to find private owners for wild horses and burros gathered on public lands during herd management roundups. The BLM retains title to adopted horses for one year to ensure that they are not being mistreated.

4804. Identification system for wild horses and burros by the U.S. Bureau of Land Management was instituted in 1978. The BLM started freeze-marking animals under their administration with individual identification marks to differentiate wild animals from domestic livestock.

4805. Vacuum-powered prairie dog extraction device was invented by Gay Balfour in Cortez, CO, in 1991. The machine, which Balfour called Dog-Gone, used wide plastic tubing to suck prairie dogs out of their burrows into a padded truck so that they could be relocated or sold rather than poisoned.

4806. Fertility control program for wild horses in the United States on public lands was launched in Nevada in December 1992 by the Bureau of Land Management. The pilot program used a vaccination procedure called immunocontraception as a humane alternative to previous wild horse and burro management options, which included relocation or destruction.

4807. Intentional poisoning of Lake Davis in California was undertaken by the Lake Davis Pike Eradication Project. In a controversial action to rid the lake of illegally introduced Northern pike that were preying on native trout and could threaten other species in neighboring waters, the California Department of Fish and Game poisoned Lake Davis with rotenone on October 15, 1997. Nine months later, after the poison was considered to have safely degraded, the fishery was restored by restocking the lake with approximately one million trout.

WILDLIFE—PRESERVES AND RESTORATION

4808. Duck sanctuary was established on the Farne Islands off the northeastern coast of England in the late 6th century by St. Cuthbert, who protected the eiders nesting around his hermitage there. Common eiders are known in the Farne Islands as St. Cuthbert's doves or Cuddy doves.

WILDLIFE—PRESERVES AND RESTORATION—*continued*

4809. National Wildlife Refuge in the United States was established on Pelican Island off the east coast of Florida in 1903 by President Theodore Roosevelt in response to the devastation wrought by the plume trade. Prior to the Civil War, Pelican Island had been covered by a thick mangrove forest that supported large colonies of heron, white ibis, and roseate spoonbill. A severe frost in 1886 killed the mangroves, and plume hunters then decimated the bird population. By 1903, the island was barren except for nesting brown pelicans.

4810. Refuge established west of the Mississippi was Three Arch Rocks National Wildlife Refuge in Oregon, a mere 15 acres that consists of 3 large rocks and smaller adjacent rocks. Established on October 14, 1907, Three Arch Rocks is home to several colonies of large marine birds and a birthing area for northern sea lions.

4811. Buffalo preserve in the United States was the National Bison Range in the Flathead Valley of western Montana. At the urging of the American Bison Society, the land purchase was authorized by Congress and President Theodore Roosevelt, who designated the area as a permanent range on May 23, 1908.

4812. Land bought by the U.S. government specifically for wildlife was the National Bison Range in the Flathead Valley of western Montana. To preserve the dwindling buffalo population, which had declined from millions to only a few hundred bison due to overhunting, Congress and private donors purchased the land. President Theodore Roosevelt designated the area as a permanent range on May 23, 1908.

4813. National wildlife refuge specially authorized by Congress was National Bison Range, which was created by an act of the U.S. Congress on May 23, 1908.

4814. National preserve created to protect the Javan rhinoceros was the Ujung Kulon Nature Reserve, established in western Java in 1921 to prevent the extinction of the endangered rhino. The reserve is also Indonesia's first national park and the protected home of many of the region's other endangered animals and birds.

4815. Wildlife preserve in Ukraine was established by the Soviet government in 1921 at Askania-Nova. The nature preserve's staff has played a major role in breeding the endangered Przewalski's horse.

4816. Refuge acquired by the U.S. National Wildlife Refuge System specifically for the management of waterfowl was the Upper Mississippi River Wildlife and Fish Refuge, located in Wisconsin and Illinois. It was established in 1924.

4817. Wildlife sanctuary in Tanganyika was Serengeti National Park, established on November 29, 1929, as a game reserve from part of the Serengeti Plain to save one of Africa's finest wildlife herds from extinction.

4818. Law in the United States to allow government fees to be levied to establish and maintain wildlife refuges was enacted by Congress on March 16, 1934. The Migratory Bird Hunting Stamp Act supported the Migratory Bird Conservation Act by using the funds generated to create sanctuaries and breeding grounds for migratory birds. Jay Norwood ("Ding") Darling, a cartoonist and conservationist, drew the first duck stamp.

4819. Wildlife research center in the United States was the Patuxent Research Refuge, established in 1936 in North Dakota. The center conducts research in almost every state in the union and has cooperative agreements with several foreign countries.

4820. Firearms tax in the United States to specifically benefit wildlife habitats resulted from the Federal Aid in Wildlife Restoration Act, signed into law on February 2, 1937, by President Franklin D. Roosevelt. Also known as the Pittman-Robertson Act, the law levied an excise tax on guns, ammunition, and other hunting supplies. Proceeds are disbursed to states for the acquisition and restoration of wildlife habitats.

4821. Game preserve appropriation by Congress to assist state wildlife restoration projects was the Federal Aid in Wildlife Restoration Act of 1937, called the Pittman-Robertson Act for its two sponsors, Senator Key Pittman and Congressman A. Willis Robertson. The act, signed on September 2, 1937, diverted receipts from excise taxes on guns and ammunition into a special fund to be distributed to the states for wildlife restoration and to better manage their conservation programs. One million dollars was appropriated on June 16, 1938; the federal government paid 75 percent of the costs while the state contributed the remaining 25 percent.

4822. U.S. preserve to feature indigenous wildlife in a natural setting was the Jackson Hole Wildlife Park, which opened on July 19, 1948. The 1,500-acre park near Moran, WY, allowed humans to view protected wildlife in the natural setting of the Rocky Mountains. The park later became part of Grand Tetons National Park.

4823. Modern nature reserve in China is Dinghu Mountain Reserve, established in 1956 near the southern city of Zhaoqin, Guangdong Province. The forest reserve provides a protected home for several thousand plant and animal species, including rare animals like civet cats, pangolin, and silver pheasants.

4824. National preserve created to protect the giant sable antelope was the Luando Natural Integral Reserve, established in northern Angola in 1957. Conservation efforts stabilized the declining numbers of the endangered species, which managed to survive overhunting, poaching, and decades of civil war across its natural range.

4825. Federal legislation in the United States authorizing recreational uses of wildlife conservation areas managed by the Department of the Interior was the Refuge Recreation Act of 1962.

4826. State nature preserve in North Carolina was dedicated in 1963 at Weymouth Woods, near Southern Pines. The Weymouth Woods Sandhills Nature Preserve protects the region's coastal pine barren flora and bird life, including the endangered red-cockaded woodpecker.

4827. Wilderness refuge system in the United States was created in 1964 when the Wilderness Act was signed into law, directing a review of national forest primitive areas and of the National Park System and National Wildlife Refuge System lands. By the end of 1998, Congress had enacted 88 laws designating new wilderness areas or adding to existing ones.

4828. Consolidation of various types of lands into a single U.S. National Wildlife Refuge System was in 1966, with passage of the National Wildlife Refuge Administration Act.

4829. Wild horse range established by the U.S. Bureau of Land Management was the Pryor Mountain Wild Horse Range, created in south central Montana in 1968.

4830. Tiger preserve in India was established at Corbett National Park. Founded as Hailey National Park in 1936 at the urging of hunter and naturalist Jim Corbett, it is also India's first national park. The World Wildlife Fund launched its Project Tiger protection campaign there on April 1, 1973.

4831. Federal legislation in the United States to tax hunting and fishing equipment in order to fund improvements in wildlife habitat was the Federal Aid in Wildlife Restoration Act of 1974.

4832. Wildlife preserve to protect golden lion tamarins was the Poco das Antas Biological Reserve in southeastern Brazil, established in 1974. Protection of wild tamarins and reintroduction of specimens bred in zoos are among the conservation activities carried out by zoologists at the forest preserve.

4833. U.S. federal funding to preserve the Tule elk was appropriated under the Tule Elk Preservation Act, enacted on August 14, 1976. The act limited the number of elk in the Owens River Watershed area, but Congress directed federal agencies to open public lands under their administration to the state of California to promote Tule elk restoration efforts.

4834. State nature preserve in Georgia was Ossabaw Island, which became the state's first Heritage Preserve program site in 1978. Wildlife species on the preserve include unique Ossabaw Island hogs, which are direct descendants of hogs imported from Spain by Christopher Columbus.

4835. Permanent Bureau of Land Management holding facility for wild horses and burros in the eastern United States was established in 1979 in Cross Plains, TN.

4836. Refuge in the United States to protect endangered plants and insects was established in 1980 as Antioch Dunes National Wildlife Resort in California. The refuge also harbors 78 species of birds and 8 different mammals.

4837. Wild horse range established in Colorado by the U.S. Bureau of Land Management was the Little Bookcliffs Herd Management Area. The range, which is located northeast of Grand Junction, was established in 1980.

4838. Jaguar preserve was created in November 1984 when the government of Belize banned hunting in the Cockscomb Basin and declared the area a forest preserve. The rainforest of the Cockscomb Basin Wildlife Sanctuary is also a habitat for other rare cat species and Central American mammals.

WILDLIFE—PRESERVES AND RESTORATION—*continued*

4839. Native American organization dedicated to preserving buffalo was the InterTribal Bison Cooperative (ITBC). Member tribes are "committed to reestablishing buffalo herds on Indian lands in a manner that promotes cultural enhancement, spiritual revitalization, ecological restoration, and economic development." The organization facilitates education and training programs, develops marketing strategies, coordinates the transfer of surplus buffalo from national parks to tribal lands, and provides technical assistance in developing management plans. The non-profit ITBC was formally organized in April 1992 at a meeting of tribal representatives in Albuquerque, NM. Its headquarters is located in Rapid City, SD.

4840. Law in the United States to allow combining federal and private funds for wildlife program grants was the Partnerships for Wildlife Act, enacted on November 4, 1992. It authorized state and federal government agencies, private foundations, and individuals to combine funds into a national Wildlife Conservation and Appreciation Fund, which would then disburse wildlife program grants to the states. The law was later altered by the North American Wetlands Conservation Act Amendments of 1994, which removed a requirement that matching funds from private sources be directed through the Fish and Wildlife Service.

4841. U.S. Navy report on the reintroduction of military dolphins to the wild was Technical Report 1543, presented in October 1993 and entitled "Reintroduction to the Wild as an Option for Managing Navy Marine Mammals." It addressed the question of whether or not dolphins trained for military tasks could safely be introduced to the open oceans. The report concluded that reintroduction was not cost effective and recommended that the navy continue to care for the marine mammals it had trained unless improved reintroduction techniques or sufficient funding became available.

4842. Transfrontier protected area for marine turtles was Turtle Island Heritage Park, created by the governments of the Philippines and Malaysia on May 31, 1996. The park conserves green and hawksbill turtles through fisheries management, egg protection, research, and local educational programs.

4843. Executive Order in the United States to protect the wildlife of the Midway Islands was Executive Order 13022, signed by President Bill Clinton on November 1, 1996. The order transferred jurisdiction over the islands from the U.S. Navy to the Department of the Interior, creating the Midway Atoll National Wildlife Refuge. Responsibility for administering the area was thus transferred to the Fish and Wildlife Service, which was charged with maintaining and restoring the biological diversity in the islands, conserving fish and wildlife, overseeing scientific research opportunities, and managing wildlife-related recreation programs. The goals were to be met in ways compatible with the islands' management as a historic World War II battle site.

4844. State nature preserve in Massachusetts was dedicated in September 1998 at Poutwater Pond. Located in Holden in central Massachusetts, the pond and the rare bog that surrounds it are protected by the state, as are the wildlife and plants that depend on the site's ecosystem.

4845. Transfrontier wildlife park in Africa was Kgalagadi Transfrontier Park, which straddles the border between Botswana and South Africa. The park was created on April 7, 1999, when the two governments agreed officially to manage the wildlife area as one natural ecosystem rather than as two national parks.

WIND POWER

4846. Windmills were used in Persia in the 7th century. Unlike their later European counterparts, early Middle Eastern windmills were horizontal.

4847. Windmills in Europe were built in the late 12th century. Unlike the earlier windmills of the Middle East, they were constructed vertically.

4848. Self-regulating windmill sails were invented by English engineer William Cubitt. After apprenticing as a woodworker and an agricultural machinist, Cubitt developed a system that used shuttered sails and a fan tail to ensure that a windmill would receive an evenly regulated supply of wind power. He patented his design in 1807 and was later knighted for his work as a civil engineer.

4849. Wind turbine ship to cross the Atlantic Ocean was the *City of Ragusa*. The small ship used a wind turbine to drive a propeller, enabling the vessel to travel from England to the United States in 96 days in the summer of 1870.

4850. Windmill to generate electricity was built by American electrical engineer and arc lighting pioneer Charles Francis Brush in 1888 in Cleveland, OH. The windmill was large and relatively inefficient, but it supplied enough power to light Brush's mansion.

4851. Aerodynamically efficient windmills to generate electricity were invented by engineer Poul la Cour in Askov, Denmark, in 1891.

4852. Wind turbine to generate energy for a central power system producing alternating current went into service at Grandpa's Knob, VT, on October 19, 1941, when it was phased into the Central Vermont Public Service Corporation's system. Synchronized operation continued for two hours, during which a maximum output of 800 kilowatts was delivered. The wind velocity producing this output was measured at 26 miles per hour. The turbine was invented by Palmer Cosslet Putnam.

4853. Defense laboratory operated by the U.S. government to study wind power was Sandia Laboratories near Albuquerque, NM. While the laboratory's previous experiments had been confined to nuclear weaponry and other defense-related research, it began studying renewable energy sources like wind, solar, and photovoltaic power in 1973 in response to concerns over the U.S. energy supply.

4854. National trade organization for wind power in the United States was the American Wind Energy Association (AWEA). The group represents wind power plant developers, wind turbine manufacturers, utilities, researchers, and others involved in the industry in the United States and abroad. The AWEA was established in 1974 to share technological developments and promote the use of wind power to the public and to government policy makers. Its headquarters is in Washington, DC.

4855. California state agency to regulate wind power and other renewable energy sources was the California Energy Commission. The commission was created in 1974 by passage of the Warren-Alquist State Energy Resources Conservation and Development Act and was officially established on January 1, 1975. California is one of the world's greatest producers of wind-generated electricity.

4856. Federal tax credits for wind power investments in the United States became national tax policy under the Public Utility Regulatory Policies Act (PURPA), enacted on November 9, 1978. The tax breaks included a 10 percent credit on the value of installed wind energy equipment and a 15 percent energy investment credit. The credits expired in 1985.

4857. National surveys of wind as an energy source in the United States were compiled in 1979 by the U.S. Department of Energy's Pacific Northwest Laboratory in Richland, WA. The surveys amounted to a national inventory of wind resources, measuring annual and seasonal wind averages on a regional and state level. They also included a wind certainty rating system, with conclusions based on the percentage of land area suitable for wind energy development. The surveys were precursors to later "wind atlases" of the United States.

4858. Law in the United States to promote wind power as a renewable resource was the Wind Energy Systems Act, enacted on September 8, 1980. The law stated that the U.S. government should assist in research, development, demonstration, and technological programs for converting wind energy into electricity and mechanical energy.

4859. Wind farm in Australia began operating in 1986 in Esperance, Western Australia. The facility took advantage of the steady winds along the continent's southern coast to supply the township of Esperance with electricity.

4860. Offshore wind power plant or commercial "wind farm" was located near Vindeby, a town on the island of Lolland in southeastern Denmark. The facility began operating in 1991.

4861. Wind farm in England began operating in December 1991 at Delabole in Cornwall. In addition to providing electrical power, it is a tourist attraction.

4862. Minnesota wind power tax law exempted wind energy conversion systems placed in service after January 1, 1991 from property taxes. After sporadic inclusion in the state's tax code by the Minnesota legislature, a sales tax exemption for wind-produced energy was made permanent by executive order in 1998.

4863. Federal income tax credit for wind energy use in the United States was created through the Energy Policy Act of 1992, which enacted a Production Tax Credit as Section 45 of the Internal Revenue Code. This provided a 1.5-cent per kilowatt-hour tax credit for electricity generated from wind plants put into service after December 31, 1993, and before July 1, 1999.

WIND POWER—*continued*

4864. Government agency of India to promote wind power was the Ministry of Non-Conventional Energy Sources (MNES), a former department that achieved ministerial status in July 1992. MNES works with both the commercial and scientific communities to develop energy from a variety of renewable sources. India is one of the leading producers of wind energy in the world.

4865. Feasibility studies of wind-generated energy use in U.S. government buildings resulted from Executive Order 12902, signed by President Bill Clinton on March 8, 1994. The directive was entitled "Energy Efficiency and Water Conservation at Federal Facilities." It ordered audits that would produce recommendations for the acquisition and installation of energy conservation measures, including the use of wind power. Consideration of other renewable energy sources, such as geothermal and solar power, was also ordered.

4866. North American conference on bird mortality and wind power was the National Avian–Wind Power Planning Meeting held July 20–21, 1994 in Denver, CO. The public meeting was cosponsored by the National Renewable Energy Laboratory, the Department of Energy, the American Wind Energy Association, the Electric Power Research Institute, the National Audubon Society, and the Union of Concerned Scientists. The meeting was called to address concerns about the effects of wind turbines on avian safety, particularly among endangered species. The attendees agreed to develop standard methods for conducting population and other field studies in order to search out possible solutions.

4867. Wind farm in Colorado was the Ponnequin Wind Farm, owned by Public Service Company of Colorado. The facility in rural Weld County, northeast of Denver, began producing electricity for residential customers in April 1998.

4868. Wind farm in Africa was a 50-megawatt facility in the Tetuan province on Morocco's northeastern Mediterranean coast, near the Strait of Gibraltar. A contract to construct the facility was signed on November 3, 1998, by its designer and builder, a French consortium, and Morocco's government-owned utility, the National Office of Electricity.

4869. Wind-powered brewery in the United States was the New Belgium Brewing Company of Fort Collins, CO. On January 29, 1999, the brewery signed a ten-year contract agreeing to buy wind-generated electricity from the city of Fort Collins. The deal allowed the city to purchase an additional wind turbine and substantially reduced the amount of coal-fire emissions released into the atmosphere by the municipal utility.

4870. Wind energy–generating station in southern Africa was proposed to be located in Lüderitz, Namibia. On March 4, 1999, the Namibian Ministry of Mines and Energy announced plans to construct a 10MW power station to be run by winds approaching the small coastal town from the Atlantic Ocean.

Z

ZOOS, AQUARIUMS, AND MUSEUMS

4871. Man-made salt-water environment to contain over 5.5 million gallons of water was located at the Living Seas exhibit in Walt Disney World's Epcot Center in Lake Buena Vista, FL.

4872. Chinese animal park was created by Wen-Wang, the ruler of Chou, in the 11th century B.C. Called Intelligence Park, it included animals, a tower, a pond, and musicians and was open to the people.

4873. Illustrations of animals in a zoo-like setting appeared on the walls of Mereruka's tomb, built circa 2200 B.C. in Egypt. Antelope, geese, and hyenas are shown being fed and cared for by attendants.

4874. Large carnivores to be held in captivity were in Sumer circa 2050 B.C. Records from the period of King Shulgi (of the third dynasty of Ur) show that lions were held there in cages and pits.

4875. Acclimatization garden was created by Egyptian Queen Hatshepsut of the 18th dynasty circa 1400 B.C. The Garden of Ammon included some trees from the animals' natural habitat and thus gave the animals a place to adjust to the conditions of the new land.

4876. Importation of exotic animals for a royal collection was in 1400 B.C. by Egyptian Queen Hatshepsut of the 18th dynasty. Her ships returned with monkeys, leopards, a giraffe, birds, and even whole trees for her private zoo.

4877. Written work of significance to use animals in captivity as subjects of a study was *Historica animalium (History of animals),* written by Aristotle in the 4th century B.C. This systematic zoological survey described about 350 vertebrates.

4878. Exotic animal spectacle in the Roman Empire occurred in 275 B.C. as a result of the victory over Pyrrhus. Four elephants were displayed.

4879. Royal menagerie known to exist in England was maintained by King William II in Woodstock, England, during the late 11th and early 12th centuries. Chroniclers record that there were bears, lions, leopards, lynxes, and camels. When William II died, the menagerie was continued by his brother, King Henry I.

4880. Royal European menagerie with a research component was built by Louis XIV at Versailles in 1664. The menagerie was open to both scholars and the public. Research at the menagerie contributed to the developing science of comparative anatomy. From 1669 to 1690, members of the Academy of Sciences performed dissections of animals that had died there.

4881. Public museum in Great Britain was the Ashmolean Museum in Oxford. It was named after Elias Ashmole, an alchemist, botanist, and antiquary who donated his natural history collections and those of naturalist John Tradescant to Oxford University, which opened the museum to the public on May 24, 1683.

4882. Surviving zoo is the Schönbrunn in Vienna, Austria, founded in 1752 by the Holy Roman Emperor Francis I, husband of the Austrian Empress Maria Theresa. There had been previous menageries at Schönbrunn, but Francis considered the old cages much too small and had the zoo completely redesigned. It was opened to the public in 1779.

4883. Natural history museum foundation in the United States was laid on January 12, 1773 by the Charleston Library Society in South Carolina. The work was interrupted by the Revolutionary War. Because of the interruption, Charles Willson Peale's Museum in Philadelphia, PA, was finished before the one in Charleston and has the honor of being the first natural history museum in the United States.

4884. Animal exhibits in a U.S. museum to be in natural settings and groups were displayed at Peale's Museum in Philadelphia, PA, in the 1790s.

4885. Museum exhibits with labels and descriptions were displayed in Peale's Museum in Philadelphia, PA, during the 1790s. Charles Willson Peale wanted his museum to be for everyone, not just scholars. The labels and descriptions reflect the educational mission of his museum.

4886. Natural history museum in North America was Peale's Museum in Philadelphia, Pennsylvania. American artist Charles Willson Peale's educational collection of plants, animals, minerals, and artwork depicting natural phenomena opened for public viewing in 1791.

4887. Rally to free animals in a menagerie in France occurred in 1792. A group of Jacobin sympathizers marched up to Versailles. Once there they demanded that the director of the menagerie release the animals. When released, some of the animals adapted to the environment.

4888. Salaried employee at the Western Museum of Natural History in Cincinnati, OH, was John James Audubon, the noted ornithologist and artist. The museum, the first of its kind west of the Appalachians, was based on private collections and opened in 1818. Audubon was hired to stuff birds and animals and paint the backgrounds for displays. The Western was the forerunner of today's Cincinnati Museum of Natural History and Science.

4889. Zoo explicitly set up as a scientific institution was the Zoological Society of London, founded in 1826 by Sir Stamford Raffles. The zoo's founders stated that the role of the zoo was to domesticate and acclimatize animals for stocking farmyards and pleasure parks. An additional role was to allow the study of the animals. Darwin studied monkeys' reactions to snakes at the Zoological Society of London.

4890. Modern public zoo was the Regent's Park Zoological Garden in London, England, opened by the Zoological Society of London in 1828. Public access to the study of animal life was a departure from earlier English and European menageries, which were usually private collections accumulated for the amusement of royalty.

4891. Natural history museum in Canada was established in 1842 at the Mechanics Institute, St. John, New Brunswick. The museum featured the collections of its founder, geologist Abraham Gesner, the inventor of kerosene.

4892. Zoo to have an eagle aviary was the Philadelphia Zoological Garden, in Philadelphia, PA, in 1847.

4893. Public aquarium for exhibition of fish opened in Regents Park in London in 1853.

ZOOS, AQUARIUMS, AND MUSEUMS—*continued*

4894. Popular use of the word *zoo* occurred in a song performed by "The Great Vance" in an English music hall one night in 1867. He sang about walking in the zoo on a Sunday, referring to the Zoological Society of London.

4895. Public aquarium in the United States was the National Aquarium in Washington, DC, established in 1873. Until 1980 the U.S. Fish and Wildlife Service administered it. Budget cuts threatened the aquarium and led to the formation of the National Aquarium Society, which now operates and supports the aquarium.

4896. Zoo in the United States was the Philadelphia Zoological Garden, under the management of the Zoological Society of Philadelphia, PA. The garden was opened to the public on July 1, 1874. Featured attractions were the bear pit and the lion house.

4897. Surviving zoo in Ohio was the Zoological Society of Cincinnati, better known as the Cincinnati Zoo, in Cincinnati, OH, in 1875.

4898. Zoo to exhibit a great Indian rhinoceros was the Philadelphia Zoological Garden, in Philadelphia, PA, in 1875.

4899. Zoo to exhibit rare black apes from the Indonesian island of Celebes (Sulawesi) was the Philadelphia Zoological Garden, in Philadelphia, PA, in 1879.

4900. Public aquarium in Japan was created at Tokyo's Ueno Zoological Gardens in 1882. In 1989 the aquarium was moved to Tokyo Bay and renamed Tokyo Sea Life Park.

4901. Zoo that started as a sugar plantation was the Audubon Zoological Garden in New Orleans, LA, which traced its origins to the 1884 World's Industrial and Cotton Centennial Exposition in Audubon Park. The zoo was once part of the former Foucher-Bore sugar plantation. The Horticulture Hall at the exposition contained the antecedents of the present zoo and continued to operate after the fair ended.

4902. Marine aquarium built primarily for education and research was the Fisheries Aquarium at the Northeast Fisheries Science Center founded by the U.S. Commission of Fish and Fisheries (now the National Marine Fisheries Service) in Woods Hole, MA, in 1885.

4903. Zoo to exhibit adult bull elephants was the Philadelphia Zoological Garden, in Philadelphia, PA, in 1888.

4904. Natural history museum in the United States named for a descendant of royalty was the Bernice P. Bishop Museum in Honolulu, HI. It was founded in 1889 by Charles Reed Bishop as a memorial to his wife, Princess Bernice Pauahi Bishop, great-granddaughter and last surviving descendant of Kamehameha I, unifier of the Hawaiian Islands.

4905. Surviving zoo established by the U.S. Congress was the National Zoological Park, a part of the Smithsonian Institution, in Washington, DC, in 1889.

4906. Zoo in Rhode Island to continuously operate was the Roger Williams Park Zoo, in Providence, Rhode Island, opened in 1891.

4907. Aquarium with an inland salt-water environment was installed at the 1893 Columbian Exposition in Chicago, IL, by Marshall McDonald, who was awarded medals by Belgium, Britain, France, Germany, and Russia for improving fish hatching and propagation.

4908. Zoo in Nebraska was the Henry Doorly Zoo, established in Omaha in 1894.

4909. Public aquarium in the United States funded as a municipal facility was the New York Aquarium, opened in 1896, in Manhattan's Battery Park. The New York Aquarium is now housed in Brooklyn.

4910. Zoo to have a research institution was the Philadelphia Zoological Garden, in Philadelphia, PA, which started the Penrose Research Laboratory in 1901.

4911. Natural history museum in western Canada was the Banff National Park Museum, located in Canada's first national park. Built in 1903, the museum exhibits featured the flora and fauna of the Rocky Mountains. The museum still operates as a tourist attraction.

4912. Aquarium in Hawaii was the Waikiki Aquarium, designed in 1904 as a commercial display of Pacific marine life. The aquarium's exhibits and research on Pacific marine ecosystems have operated as part of the University of Hawaii since 1919.

4913. Zoo to display animals in a naturalistic setting was built by Carl Hagenbeck in Hamburg, Germany, and opened in 1907. The Hagenbeck Tierpark used a system of hidden moats to separate predatory species from other animals, thus creating illusions of wild animals as they would appear in their natural ecosystems. Hagenbeck's designs became influential in the construction of zoos throughout the world.

4914. Zookeeper to recognize the benefits of fresh air for animals in captivity was Carl Hagenbeck in 1907. Hagenbeck housed his animals in open-air exhibits during the daytime in Tierpark at Stallingen, Germany.

4915. Continuously operated zoo in Texas was the Fort Worth Zoological Park in Fort Worth, founded in 1909.

4916. Oceanographic museum is considered to be the Musée Océanographique de Monaco in Monte Carlo. The museum was founded in 1910 by Monaco's Prince Albert I, a devoted marine biologist. Its laboratory was later directed by the renowned marine biologist Jacques-Yves Cousteau.

4917. Zoo habitat constructed of simulated rock formations without bars was built at the City Park Zoo, Denver, CO, between 1915 and 1918 at a cost of $60,000. Tinted concrete and steel were used to simulate a mountain habitat. The enclosure was designed by Victor Borcherdt, the director of the zoo.

4918. Zoo with a veterinary hospital was the Bronx Zoo in New York City. The zoo's original animal hospital opened in 1916. It was succeeded in 1985 by the Wildlife Health Center, which serves as the main health care and research facility for the five zoos located in New York.

4919. Aquarium to be part of a natural history museum was the Steinart Aquarium, located in the California Academy of Sciences in San Francisco and founded in 1923. San Francisco architect Louis Hobart designed the aquarium to resemble a European zoological garden.

4920. Outdoor museum with nature trails in the United States was established in 1925 in the Ramapo Mountains in Tuxedo Park, NY, at the Station for the Study of Insects. The nature trails were developed by Frank Eugene Lutz under the auspices of the American Museum of Natural History in New York City and the Palisades Interstate Park Commission. Two trails, each half a mile long, were posted with signs describing the trees, shrubs, flowering plants, and insects.

4921. Zoo to have an African forest elephant was the Philadelphia Zoological Garden, in Philadelphia, PA, in 1925.

4922. Zoo to breed a Bornean orangutan was the Philadelphia Zoological Garden, in Philadelphia, PA, in 1928.

4923. Zoo to breed a captive chimpanzee was the Philadelphia Zoological Garden, in Philadelphia, PA, in 1928.

4924. Zoo to use bar-less exhibits extensively was the Detroit Zoological Park, which opened on April 1, 1928. A system of dry or water-filled moats separated animals from human spectators.

4925. Zoo to exhibit a pair of African buffalo was the Philadelphia Zoological Garden, in Philadelphia, PA, in 1929.

4926. Zoo to have an outdoor gibbon cage was the Philadelphia Zoological Garden, in Philadelphia, PA, in 1929.

4927. Contemporary American science and technology museum was the New York Museum of Science and Industry, which opened in 1930. The museum closed in the mid-1940s from lack of support.

4928. Radio broadcast from a zoo was from the Bronx Zoo in New York City on April 21, 1930. The program was broadcast by National Broadcasting Company radio station WEAF. Master of ceremonies was Claude Willard Leister, curator of educational activities.

4929. Modern animal park was Whipsnade, created for the Zoological Society of London by Chalmer Mitchell in 1931. When Whipsnade opened, this Bedfordshire Downs park contained 500 acres to allow the animals to roam in large enclosures.

4930. Zoo to have a model dairy barn was the Philadelphia Zoological Garden, in Philadelphia, PA, in 1936.

4931. Zoo to have a regular radio program was the Philadelphia Zoological Garden, in Philadelphia, PA, in 1936.

4932. Underwater action picture studio and marine attraction was Marineland of Florida, located in Marineland, FL. The attraction opened in 1938.

4933. Zoo to have a "baby pet zoo" was the Philadelphia Zoological Garden, in Philadelphia, PA, in 1938.

4934. Aquarium in the United States for large marine animals was Marineland, 18 miles south of St. Augustine, FL, dedicated and formally opened on June 23, 1938. Built at an approximate cost of $500,000, the facility consisted of two adjacent open-air steel and concrete tanks (one rectangular, 100 feet by 40 feet, and 18 feet deep; the other circular, 75 feet in diameter and 11 feet deep) with 200 portholes for viewing.

ZOOS, AQUARIUMS, AND MUSEUMS—*continued*

4935. Public photography at the Bronx Zoo was allowed in 1940. From the zoo's opening in 1899 until the ban was lifted in 1940, the New York Zoological Society prohibited the taking of any photographs except by official zoo photographers. Sales of their photographs raised funds for the zoo.

4936. Energy museum that began in a cafeteria was the American Museum of Science and Energy in Oak Ridge, TN, opened in 1949 as the American Museum of Atomic Energy.

4937. Science and technology museum that began as a state agricultural exhibition hall was the California Museum of Science and Industry, opened in 1951. The hall was part of a Los Angeles exposition in 1880, and the original building became the museum in 1951.

4938. Zoo to breed a cheetah was the Philadelphia Zoological Garden, in Philadelphia, PA, in 1956.

4939. Zoo to breed a Chilean flamingo was the Philadelphia Zoological Garden, in Philadelphia, PA, in 1957.

4940. Safari park was Longleat Safari Park, founded by Jimmy Chippenfield and Lord Bath. It opened in Bath, England, in 1966.

4941. Modern aquarium is considered to be the New England Aquarium in Boston, MA, which opened in 1969. The aquarium's combination of research work, educational programs, and technological features with marketing as a tourist attraction made it a model of scientific and commercial success, emulated by other modern aquariums.

4942. Aquatic environment built to revitalize urban waterfronts was the New England Aquarium in Boston, MA, which opened in June 1969.

4943. Museum devoted to forestry education that started in a world's fair was the World Forestry Center, originally the Forestry Building at the Lewis and Clark Centennial and American Pacific Exposition and Oriental Fair, held in Portland, OR, in 1905. After it was destroyed by fire in 1964, forestry leaders created the Worldlife Forestry Center in 1971.

4944. Museum established by a university to support its own programs was the MIT Museum in the Massachusetts Institute of Technology in Cambridge, MA, founded in 1971 to collect, preserve, and exhibit materials associated with the development of science and technology as it interrelates with MIT research.

4945. Zoo with twilight conditions was dedicated on April 3, 1973, in Highland Park, Pittsburgh, PA, by the Pittsburgh Zoological Society. The zoo contained six ecological niches depicting nocturnal and diurnal scenes from various regions around the world. The same degree of lighting prevailed both day and night.

4946. International organization to manage breeding animals in captivity was the International Species Inventory System (ISIS), founded by Dr. Ulysses Seal in 1974 as a result of various inbreeding studies. The system was set up to alleviate the problem of inbreeding in zoos and animal parks. Managers can search the database for suitable matches for their animals. The organization is located in Apple Valley, MN.

4947. Discovery room in a natural history museum in the United States was in the National Museum of Natural History (NMNH) at the Smithsonian Institution in Washington, DC. It opened on March 5, 1974, as part of the museum's Office of Exhibits to study visitor response to tactile displays by encouraging close-up observation and self-discovery in an informal atmosphere. NMNH's Discovery Room became a model for other museums.

4948. Aquarium in North America to be accredited by the American Association of Zoological Parks and Aquariums was the Vancouver Public Aquarium, which received accreditation in September 1975. The association was founded in 1924 for the advancement of North American zoos and aquariums.

4949. Learning lab in a national zoo was established at the National Zoological Park (NZP) at the Smithsonian Institution in fall 1977 in Washington, DC. The lab was developed under the direction of Judith White of NZP's Education Office following her involvement with the Discovery Room at the Smithsonian's National Museum of Natural History.

4950. Learning lab in the United States in a national zoo specifically related to birds was Birdlab, established in the Bird House at the National Zoological Park (NZP) of the Smithsonian Institution in October 1978 in Washington, DC.

4951. Aquarium with a salmon ladder was the Seattle Aquarium, which opened in 1979. Pacific salmon raised at the aquarium are released into Washington's Puget Sound and return to spawn. The salmon run is one of the aquarium's many salmon-related features.

4952. National zoo licensing legislation was passed in Great Britain in 1981. The Zoo Licensing Act was passed as a result of lobbying by the National Zoological Association and the Federation of Zoos in Britain.

4953. Reproductive research foundation sponsored by a zoo, a university, and a wild animal park was started in 1981 by Betsy Dresser in Cincinnati, OH. The Cincinnati Zoo, the University of Cincinnati, and King's Island Wild Animal Park formed the Cincinnati Wildlife Research Foundation to support Dresser's research in exotic animal reproduction.

4954. Learning lab related to reptiles in a national zoo in the United States was HERPlab. It was integrated into the renovated Reptile House at the National Zoological Park (NZP) at the Smithsonian Institution in October 1982, in Washington, DC. HERPlab was both an expansion and refinement of NZP's two earlier labs, Zoolab and Birdlab. In its developmental stages, the John Bull Zoological Gardens in Grand Rapids, MI, and the Philadelphia Zoo served as field test sites for HERPlab.

4955. Successful embryo transfer between females of endangered species was implanted by Betsy Dresser in August 1983. The Cincinnati Wildlife Research Foundation sent Dresser to the Los Angeles Zoo to implant a bongo embryo into a surrogate bongo mother. A healthy baby bongo was born in June 1984.

4956. Mall to feature aquatic displays was the West Edmonton Mall in Edmonton, Alberta, Canada. In 1985, three displays were constructed that show a Deep Sea Adventure, Sea Life Caverns, and Dolphin Pools.

4957. Zoo with a museum devoted to elephants was the Oregon Zoo in Portland. The Lilah Callen Holden Elephant Museum opened in December 1985, opposite the zoo's live elephant exhibits and breeding facility. The museum features exhibits relating to the history of the species and its interactions with humans.

4958. Aquatic theme park mixing education, research, and entertainment to be owned by a nonprofit organization was Marine World, Africa, USA, which opened to the public on June 16, 1986, in Vallejo, CA. The park is owned and operated by the Marine Wildlife Foundation.

4959. Natural history museum in Egypt was opened in 1988 in a suburb of Cairo. The nonprofit museum was designed after consultation with Smithsonian Institution officials who visited the museum's organizers in Egypt in 1987.

4960. Aquarium sleepovers were offered at the Vancouver Public Aquarium in 1989. At first only members could stay the night and "Sleep with the Whales," but the program was so popular the acquarium began offering it to other groups as well.

4961. Aquarium devoted to freshwater life was the Tennessee Aquarium, which opened in Chattanooga in 1992. It is dedicated to exhibits featuring freshwater ecosystems related to the Tennessee River, which range from the Appalachian Mountains to the Gulf of Mexico.

4962. Biotechnology museum was established by the DNA Learning Center, the educational facility created by the Cold Spring Harbor Laboratory in Cold Spring Harbor on Long Island, NY, in 1992. The public museum is devoted to teaching students about the history of genetics, DNA research, and biotechnology.

4963. Oceanographic exhibit with more than 3,000 fish opened in 1992 at the Stephen Birch Aquarium and Museum in La Jolla, CA. The new aquarium building was four times the size of the previous aquarium.

4964. Biodome was the Biodome de Montreal, opened on June 19, 1992. A biodome is an environmental museum with complete ecosystems. This one was a whole new approach to public display of flora and fauna. It contains four complete ecosystems, from a tropical forest to a polar world. *Biodome* means "house of life."

4965. Zoo built for nocturnal viewing was Night Safari, part of the Singapore Zoological Gardens in Malaysia. Between 7:30 P.M. and midnight, people could take a train, tram, or walking trail in the large, heavily forested park to view more than 1,000 animals, most of which are nocturnal species. Night Safari opened on May 26, 1994.

4966. Self-sustaining greenhouse in a museum was Wintergarden, opened in 1997 at the Children's Museum in South Dartmouth, MA. To expand participation in its agricultural programs, the museum approached local students and teachers to help design the greenhouse in which the Children's Museum can grow vegetables and plants year-round without using fossil fuels.

4967. Zoo in the United States to exhibit a pair of aye-ayes was the San Francisco Zoo. On August 28, 1997, the zoo received a pair of the extremely rare Malagasy lemurs from the Duke University Primate Center, the first facility to breed aye-ayes in captivity.

ZOOS, AQUARIUMS, AND MUSEUMS—*continued*

4968. Cyber animal exhibit that allows people to view bears in their natural habitat was Wild Bear Cam at *National Geographic*'s Web site. Daniel Zatz, founder of See More Wildlife, and *National Geographic* set up the camera at the McNeil River Sanctuary near Homer, AL. People all over the world can watch bears prepare for hibernation at the river.

ZOOS, AQUARIUMS, AND MUSEUMS—PLANETARIUMS

4969. Planetarium constructed by a school system in reponse to a space launch was the Chesapeake Planetarium, built by the Chesapeake, VA, school system with funds provided under the National Defense Education Act after the launch of *Sputnik 1*, which stimulated interest in science programs and facilities throughout the United States. The planetarium was completed in 1963 and is open to the public as well as to students.

Subject Index

The Subject Index is an alphabetical listing of the subjects mentioned in the main text of the book. Each index entry includes key information about the "first" and a 4-digit number in italics. That number directs you to the full entry in the main text, where entries are numbered in order, starting with 1001. To find the full entry, look in the main text for the entry tagged with that 4-digit number.

Note that most names of people and places are not included in this index, but instead are listed separately in the Personal Name Index and the Geographical Index.

In addition, major environmental issues such as air pollution or nuclear power can be investigated under their appropriate headings in the main text, which are listed in full in the front matter and also cross referenced here.

For more information, see "How to Use This Book," on page ix.

A

Abu Dhabi Marine Areas company, *3657*
Abu Simbel temples, *2105*
Academy of Natural Sciences (Philadelphia), *2260*
Academy of Sciences (Paris), *1887, 4880*
Acadia National Park, *3791, 3792*
***Acid and Rain*,** *1437*
acid rain, *1324, 1327, 1413, 1414, 1415, 1435, 1437, 1441, 1450, 1453, 1456, 1460, 1463, 1464, 1467, 1469, 1474, 1477, 1481, 1818, 3038, 3042, 3400*
***Acres U.S.A.*,** *4042*
activist movements, *1319, 2617, 2644, 3520, 3705*
See also
"Activist Movements" heading in main text
"Adaptation of the Explorer to the Climate of Antarctica", *1941*
Adirondack Forest Preserve, *3715, 4631*
Adirondack Mountains, *1413, 3242, 4631*
Adirondack Park, *2711*
Adopt-a-Highway Day, *3290*
Advanced Genetic Services, *2761*
aerosol cans, *1321*
***Aeterni Regis* (papal bull),** *3007*
African Convention on the Conservation of Nature and Natural Resources, *3378*
African Elephant Conservation Act, *2340*
African Elephant Conservation Fund, *2340*
African Population Commission, *3899*
Agenda 21, *4310, 4313*
"An Agenda for the Future", *1081*
Agent Orange, *1289, 1295, 1304*
Agreed Measures for the Conservation of Antarctic Fauna and Flora, *3026, 3031*
Agreement . . . Concerning Cooperation in Measures to Deal with Pollution of the Sea by Oil, *3583*
Agreement for Cooperation Concerning Peaceful Uses of Nuclear Energy, *3523*
Agreement on Conservation of Polar Bears, *3034*
Agreement to Promote Compliance with International Conservation and Management Measures by Fishing Vessels on the High Seas, *2484*
Agricultural Adjustment Act, *1225*
Agricultural Conservation Program, *4038*
Agricultural Hall of Fame, *1135*
agriculture and horticulture, *3067, 4037, 4058, 4060, 4061*
crops, *3120*
legislation, *3419*
organizations, *3370*
pest and weed control, *4799, 4800*
See also
"Agriculture and Horticulture" heading in main text
agroforestry, *1043*
AIDS, *3957*
Air and Waste Management Association, *1336*
air pollution, *1845, 2064, 2147, 4677*
effect on trees, *2606*
electric car to be mass-produced, *1609*
See also
"Air Pollution" heading in main text

Air Pollution Control Association, *1336*
Air Quality Accord, *1477, 3042*
Air Quality Act, *1401*
Air Quality Improvement Program, *1585*
Airborne Antarctic Ozone Experiment, *1988*
Airborne Hunting Act, *2959*
aircraft, *3745, 4639*
crop dusting, *1245*
forest fire air patrol, *2615*
noise, *3430, 3438, 3445, 3450*
solar, *4118*
sonic boom, *3447*
supersonic, *1973*
airport noise, *3432*
Akeley camera, *4698*
Akeley Hall of African Mammals, *2281*
Alamogordo Bombing Range, *3453*
Alar, *1305*
Alaska Highway, *1564*
Alaska National Interest Lands Conservation Act, *3388, 3999, 4651*
Alaska Permanent Fund, *3384*
Alaska Permanent Fund Corporation, *3384*
Alaska Pipeline, *3624*
albatross, *1663, 3562*
Alcan Military Highway, *1564*
alfalfa, *1188, 1194*
alfalfa blotch leaf miners, *1194*
alfalfa weevils, *1188*
alien species, *1162, 1184, 1188, 1189, 1192, 1193, 1194, 1283, 1655, 2461, 2494, 2590, 2596, 2600, 2602, 2608, 3426, 3907, 3972*
See also
"Alien Species and Species Migration" heading in main text
alkali, *1434*
Alkali and Works Act (England), *1382, 1383*
Alkali Inspectorate, *1381*
Allagash River, *3986*
Allagash Wilderness Waterway, *3986*
Alliance for Acid Rain Control, *1327*
Alliance of Small Island States (AOSIS), *1090*
Alligator River National Wildlife Refuge, *2434*
alligators, *2413*
alligatorweed, *4500*
alpacas, *1148*
alternative fuels, *1611*
Alternative Motor Fuels Act, *4333*
Alvaro Obregon Dam, *2100*
***Alvin* (research vessel),** *3580*
***America* (steamship),** *1937*
America United Fruit Company, *1181*
American Academy of Otolaryngology–Head and Neck Surgery, Inc., *3969*
American Anti-Vivisection Society, *1010*
American Association for the Advancement of Science, *3257*
American Association of State Highway Officials, *1561*
American Association of Zoological Parks and Aquariums, *4948*
American Bird Banding Association, *1716*
American Bison Society, *4811*
American Civil War, *2529*
***The American Coast Pilot . . .*,** *3235*
American Farm Bureau, *2440*
American Fisheries Society, *2452*
American Forest Council, *2700*
American Forest Institute, *2704*
American Forest Products Industries Inc., *2704*
American Forestry Association, *1028, 2691, 2693, 2713*
American Forestry Conference, *1055*
American Forestry Congress, *2691*
American Forests, *2731*
American Forks Canyon, *3781*
American Geological Society, *2799*
***The American Herd Book*,** *3240*
American Indian Environmental Office (AIEO), *3411*
***American Journal of Science*,** *2798*
American Meteorological Society, *1834, 1928*
***American Mineralogist Journal*,** *2793*
American Museum of Atomic Energy, *3458, 4936*
American Museum of Natural History, *1697, 1733, 4920*
American Museum of Science and Energy, *3458, 4936*
American Philosophical Society of Philadelphia, *2796*
American Plastics Council, *1607*
American Red Cross, *2144*
American Revolution, *3304*
American Society of Heating, Refrigeration and Air-Conditioning Engineers, *1363*
American Tree Farm System, *2700*
American vine aphids, *1271, 1505*
American Water Works Association, *2844*
American Wind Energy Association, *4854, 4866*
American Zoo and Aquarium Association, *2335*
AmeriFlora '92, *2316*
Amistad Reservoir, *4620*
***Amoco Cadiz* (tanker),** *3639*

Amoenitatum Exoticarum, *1763*
amphibians
decline, *4417, 4737*
organization to monitor, *4738*
Pine Barrens tree frog, *2409*
Vegas Valley leopard frog, *2371*
Anadromous Fish Conservation Act, *2463*
analine-transfer RNA, *2754*
***Anatome plantarum idea*,** *1759*
Anderson School of Natural History, *3561*
Androscoggin River, *4406*
angled barometers, *1865*
Anglo-Persian Oil Company, *3695*
Animal Damage Control Act, *4799*
animal-drawn water wheel, *1232*
animal ecology, *1053, 1053*
***Animal Liberation*,** *1019*
Animal Liberation Front, *1021*
animal rights
See
"Activist Movements—Animal Rights" heading in main text
Animal Welfare Act, *1018, 2287*
Animal Welfare Act Amendments, *1020*
Animal Welfare Institute, *1017*
***Animals' Rights Considered in Relation to Social Progress*,** *1012*
animals used in research, *1002, 1009, 1010, 1017, 1018, 2287, 2762*
Annapurna Conservation Area Project, *2248*
***Antarctic* (ship),** *3839*
Antarctic Treaty, *3025, 3026, 3046, 4783*
Antarctic Weather Central, *3865*
antelopes, *2947*
depicted, *2922*
giant sable antelope, *4824*
anthracite coal, *2042*
"An Anti-Environmentalist Manifesto", *1025*
antibiotics, *1172, 1173, 1176, 3932*
anticyclone, *1907*
Antioch Dunes National Wildlife Resort, *4836*
Antiquities Act, *3746*
Apache National Forest, *2444*
***Apollo 17*,** *2825*
Apollo Space Program, *2229*
Appalachian Mountain Club, *1029*
Appalachian Mountains, *1029, 4214*
Appalachian National Scenic Trail, *3723, 3981*
***Appeal to Reason*,** *3249*
apples, *1276*
Appleton Layer, *1933*
***Applications Explorer Mission-B* (satellite),** *1981*
aquaculture, *2476*
aqualung, *3573*
aquariums
See
"Zoos, Aquariums, and Museums" heading in main text and names of specific aquariums
aqueducts, *4568, 4585*
aquifers, *4608, 4609*
protection, *4425, 4599*
Aransas National Wildlife Refuge, *1724*
Arbor Day, *4049*
***Archaeopteryx*,** *1681, 1732*
ARCO Petroleum Company, *1479, 3626*
Arctic Chemical Network, *1472*
Arctic Council, *4319*
Arctic haze, *1468*
Argonne National Laboratory, *1607*
Arizona Wilderness Act, *4667*
Arkansas River, *2563*
Armour Institute of Technology, *3860*
Arnold Arboretum, *1786*
arsenic, *1268, 2854, 4478*
art and artists, *1800, 3162*
See also
"Literature and Arts—Art and Artists" heading in main text
***Art of Falconry*,** *2898*
Arthur Kill, *4406*
artificial fertilizer, *4061*
Arusha Conference, *4773*
asbestos, *1315, 1317, 1328, 1333, 1335, 1348, 1365, 2982, 3962, 4147*
Asbestos Hazard Emergency Response Act, *1365, 3962, 4147*
Asbestos School Hazard Abatement Act, *2982*
asbestosis, *1335*
ash, *2894*
Ash Fork Dam, *2089*
Ash Meadows National Wildlife Refuge, *2339*
Ashmolean Museum, *4881*
Asian long-horned beetles, *2608*
Asiatic chestnut blight, *2602*
asphalt, *1553*
Assam Railway and Trading Company, *3690*
Association of Dam Safety Officials, *2074*
Association of Indigenous Minorities, *4319*
Association of International Horticultural Producers, *2316*
Astrakhan Preserve, *3796*
Aswan High Dam, *2105, 3033*
Atlantic Coastal Fisheries Cooperative Management Act, *2467*

Atlantic Ocean, *3552, 3569*
charts, *3554*
deadliest hurricane, *4238*
Atlantic salmon, *2466, 2492*
Atlantic striped bass, *2465*
Atlantic Striped Bass Conservation Act, *2465*
Atlantis, *4353*
***Atlantis* (research vessel),** *3571*
Atlas Mountains, *1628*
atmospheric echoes, *1938*
Atomic Energy Act of 1946, *2840, 3455*
Atomic Energy Act of 1954, *3465, 3468*
Atomic Energy Bureau (South Korea), *3486*
Atomic Energy Commission (Japan), *3476*
Atomic Energy Control Board, *2840*
"Atoms for Peace" program, *3464*
Attwater Prairie Chicken National Wildlife Reserve, *2329*
Auditor General Act (Canada), *4316, 4317*
Audubon Society, *1712*
Audubon Society for the Protection of Wild Birds, *4766*
Audubon Zoological Garden, *4901*
Australian Aborigines, *3525*
Australian Atomic Energy Commission, *3463*
Australian Bureau of Meteorology, *1924*
Australian Coral Reef Society, *3575*
Australian Heritage Commission, *4668*
Australian World Solar Challenge, *4121*
Autobahn, *1562*
automatic weapons, *2917*
automobile emissions, *1573, 1582, 1583, 1598, 1605, 1615, 2606*
automobile racing, *1532*
automobiles, *1651, 4142*
solar-powered, *4110*
steam-powered, *1587*
automotive industry, *3106, 3142, 3619*
hybrid electric vehicles, *4337*
legislation and regulation, *3443*
See also
"Automotive Industry" heading in main text
Automotive Recyclers Association, *1607*
avalanches, *4196, 4198, 4202*
***Avicultural Magazine*,** *1713*
Avicultural Society, *1713*
Awakening School, *1794*
aye-ayes, *4759, 4967*

B

B. F. Goodrich Company, *2971*
Bach Ho oil field, *2067*
Back Bay, *3145*
Background Air Pollution Monitoring Network, *1465*
bacteria, *3644*
***Bacterium tularense*,** *3926*
Baffin Bay, *3550*
Bahama swallowtail butterflies, *2411*
Bailor Manufacturing Company, *4327*
Bald Eagle Protection Act of 1940, *1694, 2384*
Baldwin Dam, *2119*
Bali Barat National Park (Indonesia), *2433*
balloon logging, *2663*
Baltimore & Ohio Railroad, *3766*
Bañados del Este, *1624*
bananas, *1177*
Banff National Park Museum, *4911*
barbed wire, *1163, 1163*
Barcelona Convention for the Protection of the Mediterranean Sea Against Pollution, *3040*
barium, *4478*
***Bark* (ship),** *2881*
bark beetles, *2601*
barley, *1120*
barnacles, *2504*
barometers, *1864, 1892*
angled barometer, *1865*
Fitzroy barometer, *1902*
mercury barometer, *1863*
Base Cartographic Data Program, *3405*
Basel Convention on the Control of Transboundary Movements of Hazardous Wastes and Their Disposal, *2885, 2887, 2891, 3045*
Baseline Water Quality Status and Trends Project, *3406*
"Basketmakers", *3788*
Basques, *4621*
bass
Atlantic striped bass, *2465*
black bass, *2458*
bats, *2386*
battery acids, *2888*
bayberry whiteflies, *1301*
beach closings, *4428*
***Beagle* (ship),** *1902, 2200, 3238*
Bear Valley National Wildlife Refuge, *2332*
bears, *2442*
grizzly bears, *2387, 4785*
polar bears, *2956, 3034, 4775*
Beaufort Wind Scale, *1882*
beer cans, *4127*

bees
honeybee, *1158*
killer bee, *1529, 3968*
beetles
Asian long-horned beetle, *2608*
bark beetle, *2601*
cereal leaf beetle, *1192*
Colorado potato beetle, *1182, 1185, 1274, 1519*
Japanese beetle, *1189*
Belize Audubon Society, *3594*
Belovezhskaya Pushcha, *3996*
benchmarks, *2226*
benzaldehyde, *1310*
benzene, *3952, 3952*
Bergen Geophysical Institute, *4248*
Bering Expedition, *4681*
Bering Land Bridge National Preserve, *3750*
Bernice P. Bishop Museum, *4904*
Beschreibung Allerfurnemisten mineralischen ertzt und bergwerksarten, *3297*
Betampona Nature Reserve, *2445*
Beyond the Limits, *3261*
Big Basin Park, *1707*
Big Bend National Park, *3993*
"Big Cut", *2651*
Big Cypress National Preserve, *4003*
Big Sur Coast Highway, *4013*
"Big Wheel", *2655, 2657*
billfish, *2475*
Bills of Mortality (England), *1331*
Biltmore, *2695*
binary nomenclature, *1764*
biodiesel fuel, *4345*
biodiversity, *3873*
See also
"Biodiversity" heading in main text
Biodiversity Convention Office, *1640*
Biodome de Montreal, *4964*
Biodynamic Farming Association, *1263*
biological controls, *1175, 1270, 1273, 1275, 1279, 1293, 1301, 1311, 1513, 2594, 2607, 4503*
biological diversity, *3426*
biological resources, U.S., assessed, *3426*
Biological Resources Division of the U.S. Geological Survey, *3408*
Biologics Control Act, *1281*
Biologie oder Philosophie der Lebenden Natur, *2259*
biology, *2259*
Biomass Feedstock Development Program, *4329*
Biomass Resource Assessment Task, *4339*
biomes, *3399*
African, *1628*
Arctic and subarctic, *3398*
Atlantic-Gulf coast, *3416*
Australian, *1629*
coniferous forest, *3404*
Eastern deciduous forest, *3400*
Pacific coast, *3401*
prairies and grasslands, *3410*
tropical-subtropical, *3417*
bionics, *2285*
bioprospecting, *4676*
bioregionalism, *1066, 1068*
biosphere, *1618*
Biosphere Conference, *1060*
biotechnology, *1208, 1255, 1256, 1302*
birch, roundleaf, *2336*
bird banding, *1716, 1728*
Bird Lore, *1733*
bird watching, *3253*
Birdlab, *4950*
birds, *1221, 4687, 4799, 4866*
canaries in mines, *2034*
deaths due to oil spills, *3630*
endangered species, *2408*
extinction of a native American animal caused by humans, *2368*
legislation and regulation, *2498*
See also
"Birds" heading in main text
The Birds of America, *3160, 3263*
Birdseye Seafoods, *2473*
birth control, *3255, 3878, 3879, 3880*
"The Birth of a Plant", *3266*
Bishbios sect, *1040*
bison, *2947, 4811, 4811, 4812, 4813, 4839, 4925*
black apes, *4899*
black bass, *2458*
Black Bass Act, *2458*
Black Canyon, *3741*
Black Environmental Scientific Trust, *2886*
Black Hills, *3787*
Black Sea, *2516*
"Black Smoke", *1384*
"Black Sunday", *2150*
"Black Wednesday", *1449, 1454*
blowdown, *2587*
Blue Ridge Mountains, *3754*
Blue Ridge Parkway, *4012*
bluefin tuna, *2486*
Board of Fish Commissioners (California), *2503, 2953*
Board of Trade (Great Britain), *1898, 1905*
Boke of St. Albans, *2505*

boll weevil, *1284, 3164*
boll weevils, *1190*
Bonn Convention, *3062*
Boone River, *4669*
bordeaux mixture, *1278*
Bosque del Apache Refuge, *1686*
Boston Common, *3816*
Boston Public Garden, *3819*
botanical gardens, *1754*
Botanical Society of South Africa, *1792*
botany, *2256, 2257, 3408, 3409, 3413*
See also
"Botany" heading in main text
Boulder Dam, *4560*
Boundary Waters Canoe Area Wilderness, *4636, 4642*
bowerbirds, *1708*
bows and arrows, *2926, 2928*
boycotts, *2720*
Brandywine Creek, *4471*
Brazilwood, *3344*
Brigantine Wilderness, *4655*
brightfields, *3005*
British Broadcasting Company (BBC), *1932, 1952*
British Ecological Society, *2273*
British Gas Corporation, *2052*
British Meteorological Office, *1958*
British Meteorological Society, *1899*
British Parliament, *3917*
Brittany Dam, *2104, 4562*
bromide oxide, *1997*
Bronx River Parkway, *1560*
Bronx Zoo, *4918, 4928, 4935*
Brown Creek Soil Conservation District, *4040*
brown trout, *2494*
brownfields, *3005*
Brownfields Economic Development Initiatives, *2992*
Bruce Trail, *4649*
Brundtland Commission, *1842, 4305*
"bubble concept", *1412*
bubonic plague, *3903, 3904, 3905*
Buck Island Reef National Monument, *3417*
"bucket-and-pole system", *1231*
buckwheat, *1202*
buffalo
See
bison
Buffalo Mining Company, *2556*
Bukom refinery, *3704*
***Bulletin Volcanologique*,** *4378*
bullfighting, *1001*
Bunsen burner, *2045*
buntings, *1709*
Bureau of International Expositions, *2312, 2316*
Burger King, *2720*
Burma Oil Company, *3690*
burros, wild, *4771, 4780, 4803*
bush shrike, *1687*
butadiene, *1449*
butterflies
Bahama swallowtail, *2411*
mission blue, *2337*
monarch, *4729*
San Francisco silverspot, *2337*
Schaus swallowtail, *2407*

C

Cabrillo National Monument, *3739*
cadmium, *2974*
cahow, *1657, 2319*
Cairo Guidelines and Principles for the Environmentally Sound Management of Hazardous Wastes, *2885*
calcium arsenate, *1284*
California Board of Fisheries, *4762*
California condor, *1684*
California Department of Fish and Game, *4807*
California Desert Protection Act, *3813, 4673, 4674*
California Energy Commission, *4855*
California Fish Commission, *2491*
California Geological Survey, *3769*
California Gold Rush, *3311*
California Heritage Trout Program, *2513*
California Institute of Technology, *1451, 3874*
California Military Lands Withdrawal and Overflights Act, *4673*
California Museum of Science and Industry, *4937*
California red scale, *1275*
California Waste Management Agency, *2869*
camels, *1160*
Bactrian camel, *1146*
dromedary camel, *1146*
Campaign for Lead-free Air (CLEAR), *1549*
Campbell-Stokes sunshine recorder, *1901*
Canadian Dam Association, *2076*
Canadian Dam Safety Association, *2076*
Canadian Environment Ministry, *1324*
Canadian Environmental Protection Act, *2882*

Canadian Institute of Forestry, *2697*
Canadian National Energy Board, *3521*
Canadian Parliament, *2151*
Canadian Society of Forest Engineers, *2697*
cancer, *1294, 1333, 1368, 1369, 1476, 2744, 2762, 2969, 2971*
Candlestick Park, *2233*
cannibalism, *2132, 2137*
Canyonlands National Park, *3764, 3770*
Cape Charles Sustainable Technology Park, *4307*
Cape Cod National Seashore, *3416, 3804*
Cape Hatteras National Seashore, *3800*
captive breeding, *1665*
butterflies, *2407*
cheetahs, *4938*
Chilean flamingos, *4939*
chimpanzees, *4923*
orangutans, *4922*
wolves, *2434*
See also
"Wildlife—Breeding" heading in main text
carbon dating, *2820*
carbon dioxide, *1438, 1459, 1828, 1836*
carbon tetrachloride, *3961*
Carey Irrigation Act, *1031, 1219*
caribou, *2948*
Carlsbad Caverns National Park, *4672*
cartography, *1885, 1906, 1908, 3600*
cast-iron plows, *1237*
catastrophism, *2801*
Category 5 hurricanes, *4251*
catfish, *2476*
cattle, *1152, 1156, 1173, 1343*
Caustic Poison Act, *1285*
cave painting, *3159*
"Cavendish Experiment", *2791*
caves, *3781, 3787, 3809, 4670, 4672*
Cedar Creek and Coastal Flatwood Wildlife Management Area, *4715*
cell theory, *1780*
cells, *1758*
celluloid, *4125, 4125*
cement guns, *4693*
cement industry, *1332*
Center for Auto Safety, *1542*
Centers for Disease Control (CDC), *3941*
Central Meteorological Bureau, *1921*
Central Pacific Railroad, *3802*
Central Pollution Control Board (India), *2976*
Central Vermont Public Service Corporation, *4852*
cereal leaf beetle, *1192*
Cernavoda-1, *3536*
chachalacas, *1660*
Chalk River Nuclear Laboratories, *2859*
***Challenger* (ship),** *3560*
Champa rice, *1122*
"The Changing Atmosphere: Implications for Global Security" (international conference), *2162*
Channel Islands National Park, *3401*
Charbonnages de France, *2066*
Charge Up To Recycle! (Canada), *4171*
***Charities and the Commons*,** *2270*
Charleston Library Society, *4883*
charlock weed, *1280*
Chase Lake Wilderness, *4657*
Chatham Islands black robin, *1680*
Chattahoochee National Forest, *3988*
Chattooga River, *3988*
Cheatham Wildlife Management Area, *4705*
cheetahs, *2353, 4938*
Chelyabinsk-40, *3482*
chemical pollutants, *4478*
Chemical Substance Law (Germany), *3958*
Chemical Waste Sites Act (Denmark), *3945*
Chemstrand Corporation, *3725*
Chequamegon National Forest, *2441*
Chesapeake & Ohio Canal National Historic Park, *3766*
Chesapeake Bay Program, *3593*
Chesapeake Planetarium, *4969*
chestnut blights, *2592, 2602*
Chevrolet Corvair, *1547*
Chevrolet Lumina Variable Fuel Vehicle, *4336*
Chevron Oil Company, *2053*
Chicago River, *4530*
Chicago Sanitary and Ship Canal, *4530, 4531*
chickens, *1142*
Children's Museum, *4966*
chimpanzees, *4679, 4923*
China National Nuclear Corporation, *3526*
Chinshan-1 nuclear plant, *3512*
Chipko Movement, *1041, 1094*
Chippewa National Forest, *3716*
Chippewa River, *2457*
chlorofluorocarbons (CFCs), *1339, 1340, 1409, 1410, 1418, 1426, 1466, 1470, 1846, 1970, 1971, 1974, 1977, 1978, 1982, 1983, 1984, 1987, 1988, 1995, 3043, 3951, 4611*
chlorophyll, *1775*
cholera, *3913, 3917, 4420*
Cholla I Station, *2295*
***Chou Li*,** *4074*

chronometer, *3545*
chrysanthemum powder, *1269*
Chrysler Clean Air Package, *1599*
Chrysler Corporation, *1548, 1550, 1595, 1599, 1603, 1606, 1610, 1613, 1614*
Chrysler Six, *1595*
chrysolite, *1333*
chukar, *1666*
chum salmon, *2500*
Ciba Seeds, *2768*
cigarettes, *1130, 1357, 1370*
Cincinnati Museum of Natural History and Science, *4888*
Cincinnati Wildlife Research Foundation, *4953, 4955*
Cincinnati Zoo, *1656, 4897, 4953*
cinemicrography film, *3266*
cities
- boards of health in, *3909*
- drinking water supply in the United States to be chemically treated with chlorine, *4583*
- Housing and Slum Clearance Act, *3085*
- planned, *3095*
- railway transit, *3072*
- street-cleaning machine in the United States, *4175*
- street-sweeping service in the United States, *4174*
- underground city sewer system in the United States, *4523*
- waterworks system in the United States to use cast-iron pipes, *4573*
- zoning ordinance in the United States, *3153*

citrus fruits, *1199*
City Beautiful Movement, *3098*
City of London Sewers Bill, *1378*
***City of Philadelphia* v. *New Jersey*,** *4145*
***City of Ragusa* (wind turbine ship),** *4849*
City Park Zoo, *4917*
civet cats, *4823*
Civilian Conservation Corps, *2683, 4035, 4053*
clams, *4418*
Clarke-McNary Act of 1924, *2536, 2616*
classification systems, biological, *1617, 1764, 1774, 3413*
Clean Air Act Amendments of 1966, *1576*
Clean Air Act Amendments of 1970, *1405, 1406*
Clean Air Act Amendments of 1990, *1418, 1421, 1422, 1426*
Clean Air Act of 1963, *1096, 1395, 1398*
Clean Coal Technology Demonstration Program, *2064*
Clean Fuel Vehicles (CFVs), *4338*
Clean Vessel Act, *3598*
Clean Water Act, *1096, 3595, 4482, 4600*
Clean Water Amendments, *4467*
Clean Water Hardship Grant, *4617*
clear-cutting, *2639, 2665*
Climate and Global Change Program, *1833*
climate and weather, *1048, 1318, 1482, 2271, 4239*
  See also
  "Climate and Weather" heading in main text
Climate Change Treaty, *3050*
Climate Prediction Center, *2154*
climatology, *1923*
Climatology Weather Bureau, *1923*
Climax solar water heater, *4097*
cloning, *2769*
***The Closing Circle*,** *3260*
clouds
- classification, *1880*
- satellite photographs, *1825*
- seeding, *2620*

Club of Rome, *3261*
coal and natural gas, *1375, 3295, 3708*
  See also
  "Coal and Natural Gas" heading in main text
Coal Mine Inspection Act, *3327*
Coal Mines Act (United Kingdom), *3326*
Coal Smoke Abatement Society, *1443*
Coastal Barrier Resources Act, *3389*
Coastal Barrier Resources System, *3389, 3394*
coastal waters, *4455*
Coastal Zone Management Act, *3088*
Cockscomb Basin Wildlife Sanctuary, *4838*
cocoa, *1179*
Coconino National Forest, *2737, 4006*
cod, *1804*
coffee, *1178, 1187*
Cold Spring Harbor Laboratory, *2305, 4962*
cold war, *2155*
College of African Wildlife Management, *4774*
***Collier's* magazine,** *3250*
Colonial Williamsburg, *3108*
colonists, English, *4229, 4229*
Colorado Department of Health, *2852*
Colorado potato beetles, *1182, 1185, 1274, 1519*
Colorado River, *2562, 2810*
Colorado Wilderness Act, *4665*
Columbia River, *2097, 2495, 4408*
Columbia River Highway, *3975*
Columbian Exposition, *4907*

Comité International du Bois, *2702*
Commission for the Conservation of Antarctic Marine Living Resources, *4783*
Commission for the Control of Desert Locusts, *3059*
Commission of Fish and Fisheries, *2453, 2454, 2493, 4763*
Commissioner of the Environment and Sustainable Development (Canada), *4317*
Committee on Groundwater Supplies, *2835*
Commodity Credit Corporation, *4800*
Commonwealth Endangered Species Act (Australia), *2424*
Complete Manual for the Cultivation of the Cranberry, *3241*
compounds, *4583*
Comprehensive Environmental Response, Compensation, and Liability Act (CERCLA), *2863, 2980, 2981, 2985, 3956*
Compressed Natural Gas, *4331*
Comstock Law, *3879*
Concerning Stones, *2775*
Concord Resources, *2879*
Concrest, *3074*
concrete buttress dam, *2091*
condensation, *1879*
Conference on the Conservation of Natural Resources, *1033, 3365*
Conference on the Human Environment, *2292*
Congo peacock, *1697*
conifers, *1495, 1784, 2577, 2588, 2595, 2596, 2597, 2599, 2605, 2610, 2627, 2638, 2643, 2645, 2645, 2647, 2650, 2698, 2709, 2740, 3404, 4826*
Connecticut River, *2528*
Connecticut River Valley, *2803*
Conowingo Dam, *2096*
conservation, *1714, 1723*
history, *3812*
See also
"Activist Movements—Conservationism" heading in main text
Conservation Foundation, *1061*
Conservation International, *3393*
Conservation Needs Inventory, *3372, 3385*
Consumer Power, *1045*
contaminants, *4734*
Continental Congress, *2938*
Continental Divide National Scenic Trail, *3989*
continental drift, *2217, 2819, 3867*
Convention Concerning the Protection of World Cultural and Natural Heritage, *3033, 3987*
Convention for the Conservation and Management of the Vicuña, *4777*
Convention for the Conservation of Antarctic Seals, *3031*
Convention for the Prevention of Marine Pollution by Dumping from Ships and Aircraft, *3587, 4494*
Convention for the Prevention of Marine Pollution from Land-Based Sources, *3588*
Convention for the Protection of the Marine Environment of the North-East Atlantic, *3588*
Convention on Civil Liability for Oil Pollution Damage Resulting from Exploration for and Exploitation of Sea Bed Mineral Resources, *3591*
Convention on International Liability for Damage Caused by Space Objects, *3032, 3035*
Convention on International Trade in Endangered Species of Wild Fauna and Flora (CITES), *2343, 2422, 3036*
Convention on Long-Range Transboundary Air Pollution, *1411, 3039*
Convention on Migratory Species (CMS), *3062*
Convention on Nature Protection and Wildlife Preservation in the Western Hemisphere, *3021, 3022*
Convention on Registration of Objects Launched into Outer Space, *3035*
Convention on the Conservation of Antarctic Marine Living Resources, *4783*
Convention on the Law of the Non-Navigational Uses of International Watercourses, *4616*
Convention on the Protection of the Alps, *4309*
Convention on the Regulation of Antarctic Mineral Resource Activities, *3044*
Convention on the Transboundary Effects of Industrial Accidents, *3047*
Convention on Wetlands of International Importance Especially as Waterfowl Habitat, *1744, 3016, 3016, 3028, 3150, 3418*
Convention Relating to Civil Liability in the Field of Maritime Carriage of Nuclear Material, *3585*
Convention Relative to the Preservation of Fauna and Flora in Their Natural State, *3019*
Cooperative Extension Service, *1266*
Cooperative Forest Management Act of 1950, *2637*
Cooperative Research and Development Agreement, *4676*

Cooperative Wildlife Research Unit Program, *2279*
Coordinating Committee for the Ozone Layer, *1977*
copper, *3309, 3811*
mines, *3292*
sulphate, *1280*
coral reefs, *3417, 3553, 3575, 3579, 3589, 3594, 3601, 3728, 3799, 3803, 4000*
Corbett National Park, *4830*
cormorants, *1646, 1730, 2357*
corn (maize), *1123, 3123*
Cornell University, *2678*
Coronado National Monument, *3748*
Cosiguina, *4368*
***Cosmos 954* (satellite),** *1326*
Costa Rica Expeditions, *2243*
cotton, *1515*
cotton gin, *1236*
cottony-cushion scale, *1184*
Council of Energy Resource Tribes (CERT), *3382*
Council on Environmental Quality, *1062, 3113*
coyotes, *1175, 4797*
cranes (birds), *1741*
hybrid "whoop-bill" crane, *1686*
sandhill crane, *1686, 1726*
wattled crane, *1741*
whooping crane, *1686, 1726*
cranes (machines), *2749*
Crater Lake National Park, *3751, 4021, 4352*
crawfish, *2476*
Creole Nature Trail, *4018*
Cretaceous extinction, *2354*
crimes, *1020, 1084, 2679, 2889, 2961, 3004, 3422, 3974, 4154, 4607, 4670, 4733, 4766, 4787, 4792*
Croajingolong National Park, *1629*
crocodilians, *2394*
American alligator, *2413*
Crop Insurance Act, *1228*
crop rotation, *4028, 4037*
Cross-Florida Barge Canal, *3071*
crosscut saws, *2656*
Cruelty to Animals Act (Great Britain), *1009*
Crystal Palace, *1900*
Crystal River National Wildlife Refuge, *2338*
cuddy doves (eider), *4808*
Cumberland Plains Woodland (Australia), *2424*
Cumberland Road, *1552*
cup anemometers, *1893*
curlew, bristle-thighed, *1699*
currasows, *1662*
Current River, *3978, 4641*
Cuyahoga River, *4406*
cyclones, *4243, 4248*
See also
"Storms—Hurricanes, Cyclones, and Typhoons" heading in main text
cyclonic depression, *1908*

D

Dai Pha Muang Wildlife Sanctuary, *4793*
Daimler-Benz, *1608*
Daimler Engine Company, *1535*
dairy barns, *4930*
dairy products, *1147*
Dam Safety Act, *2077*
"damping off" disease, *2589*
dams, *1210, 2466, 2539, 2550, 2556*
See also
"Dams" heading in main text
Daniel Johnson Dam, *2106*
Daniell dew-point hygrometer, *1815*
Danish Geological Survey, *2014*
Danish Underground Consortium, *3660*
Danube River, *3011*
darters
Maryland darter, *2388*
snail darter, *2333, 2388*
watercress darter, *2334*
DAT (car), *1538*
Datsun, *1538, 1546*
Davey Tree Expert Company, *2681*
David Sheldrick Wildlife Trust, *4755*
Day and Night (water heater), *4101*
Dayton Engineering Company (Delco), *1592*
DDT (dichlorodiphenyl-trichloroethane), *1287, 1288, 2393, 2425, 2603, 2605, 3931, 4515*
De historia plantarum, *1746*
De la Beche–Playfair report, *1377*
De lapidibus, *2775*
De Magnete, Magneticisque Corporibus, *2777*
***De natura animalum* (On the nature of animals),** *2448*
De Plantis, *1753*
De Re Metallica, *3296*
De re rustica, *1747*
De Venandi cum Avibus, *3227*
Dead Sea, *1997*
Death Valley Scenic Byway, *4022*
"debt for nature", *2719*
Declining Amphibian Populations Task Force, *4737, 4738*

deep ecology, *1082*
Deep Space 1 **(satellite),** *4123*
deer, *2947*
black-tailed deer, *4684*
Columbian white-tailed deer, *2330, 2389*
hunting, *2905*
key deer, *2327, 2389*
mule deer, *4797*
deforestation, *1048, 2642, 4047*
See also
"Forests and Trees—Deforestation" heading in main text
"Delaney Amendment", *3938*
Delaware River, *3992*
Delhi Sands
dunes, *2351*
fly, *2344, 2351*
Delta Project, *2546*
Denali National Park and Preserve, *3398*
dendroclimatology, *2013*
Densmore rotary oil car, *3611*
Department of Environment (Iran), *2353*
Department of Natural Resources (Washington State), *4791*
Derwent, River, *4456*
Description of Leading Ore Processing and Mining Methods, *3297*
Desert Land Act, *1217*
desertification, *2743, 2745*
See also
"Land Use and Development—Desertification" heading in main text
deserts
Chihuahuan, *3993*
Gobi, *1742*
Great American, *2135*
in California, *3089, 4673, 4674*
Kalahari, *4212*
Mojave, *3813*
Sonoran, *4006*
detergents, *4397, 4404, 4405, 4407, 4411, 4489, 4507*
Detroit River, *4416*
Detroit Zoological Park, *4924*
developing nations, *1090, 1846, 1989, 2291, 2294, 4613*
Devils Tower National Monument, *3743, 3746*
Devonian extinction, *2354*
diamond mining, *3313*
Diamond v. Chakrabarty, *2757*
diatoms, *3874*
Diazinon, *1309*
The Digest of Justice, *3006*
Dingell-Johnson Act, *2507*
Dinghu Mountain Reserve, *4823*
Dinosaur National Monument, *3740*
dinosaurs, *2363, 2376, 2803*
dioxin, *1300, 3946, 4427*
diphtheria, *3941*
diquat, *1291*
Directive 631, *3960*
discomfort index, *1955*
Discovery **expedition,** *3844, 3846*
disposable diapers, *4129, 4537*
Dittmar's Principle, *3563*
Diversa Corporation, *2770*
DNA, *2756*
testing, *1687*
DNA Learning Center, *2305, 4962*
Dodge **(satellite),** *2823*
dogs, *1138*
dogwood, *2326*
Dolomites, *4202*
Dolphin Protection Consumer Information Act, *2479*
dolphin-safe nets, *2481*
dolphins, *2479, 2481, 4722, 4723, 4841*
Chinese whitefin dolphin, *2414*
Indus River dolphin, *2414*
Dominion Forest Reserve and Parks Act (Canada), *3735*
Donner party, *4197*
Doppler radar, *4294, 4297*
Dounreay Fast Reactor, *3485*
Dow Chemical Company, *1045, 4128*
Dresden-1 Nuclear Power Station, *3489*
drift nets, *2480*
Driftless Area National Wildlife Refuge, *2341*
Drinking Water Infrastructure Needs Survey, *4612, 4614*
drought-resistant rice, *1122*
droughts
See
"Droughts and Water Shortages" heading in main text
dry farming, *1243*
Dry Tortugas National Park, *3417, 3799*
Dryden Flight Research Center, *4113, 4119*
Du Pont Chemical Company, *1340*
Du Pont de Nemours, *4429, 4433*
"Duck Stamp Act", *2933*
duck stamps, *1693, 2933, 3195, 3208, 3373*
artist to design, 2 times, *3180*
artist to design, 3 times, *3199*
artist to design, 4 times, *3206*
artist to design, 5 times, *3208*
first, *3165, 4818*
junior, *3224, 3225, 4789*
price for, *3166, 3185, 3198, 3210, 3213, 3215, 3216, 3219*

duck stamps—*Continued*
reproduction of, *3196*
sale of, to collectors, *3168*
showing American eider, *3194*
showing black-bellied whistling duck, *3218*
showing black duck, *3173*
showing blue goose, *3192*
showing blue-winged teal, *3190*
showing bufflehead, *3184*
showing Canada goose, *3169, 3195*
showing canvasback, *3167*
showing cinnamon teal, *3208*
showing decoy, *3212*
showing dog with mallard, *3197*
showing empress goose, *3209*
showing fulvous duck, *3214*
showing gadwall, *3188*
showing goldeneye, *3185, 3186*
showing greater scaup, *3170*
showing green-winged teal, *3172*
showing harlequin duck, *3189*
showing Hawaiian nene, *3202*
showing hooded merganser, *3205*
showing king eider, *3220*
showing lesser scaup, *3217*
showing mallard, *3166*
showing merganser, *3193*
showing old squaw duck, *3204*
showing Pacific brant, *3200*
showing pintail, *3171*
showing red-breasted merganser, *3223*
showing redhead, *3182*
showing ring-necked duck, *3191*
showing Ross's goose, *3207*
showing ruddy duck, *3174*
showing shoveler, *3181*
showing snow goose, *3183*
showing species twice, *3195*
showing spectacled eider, *3222*
showing Steller's eider, *3211*
showing surf scoter, *3226*
showing trumpeter swan, *3180, 3187*
showing whistling swan, *3203*
showing white-fronted goose, *3178, 3180*
showing white-winged scoter, *3206*
showing widgeon, *3176*
showing wood duck, *3177*
state, *1677*
woman artist for, *3221*
ducks, *1734*
black duck, *2943*
Campbell Island flightless brown teal, *1673*
common eider, *4808*
Laysan duck, *2323*
Long Island duckling, *1162*
mallard duck, *1701*
Mexican duck, *2408*
Pekin duck, *1162*
ring-necked duck, *3191*
wood duck, *2943*
Duke University, *3726*
Duryea Motor Wagon Company, *1533*
Dust Bowl, *2144, 2147, 2148, 2149, 2150, 2152, 2166, 3254, 4037, 4052, 4054, 4055*
Dutch elm disease, *1523, 2593, 2600, 2601*

E

Eagle Act, *4742*
eagle aviary, *4892*
eagles, *1694, 1720, 1721, 4742*
bald eagle, *1688, 2332, 2384, 2425, 3161, 4416*
golden eagle, *2384*
Earth Day, *1069*
Earth First! *1075, 1079, 1083, 1084*
Earth Science, *2789*
"Earth Summit", *3049, 4310*
Earthlaw, *2348*
earthquakes
See
"Earthquakes" heading in main text
Earthwatch, *1071*
East Carpathians Biosphere Reserve, *1642*
East India Company, *1206*
Eastern Wilderness Act, *4653*
Eastman Kodak Corporation, *2998*
EBDC fungicide, *1308*
Ebola virus, *3950*
EC-1 Regular, *1479, 3626*
Echo Park, *3740*
Ecodefense: A Field Guide to Monkeywrenching, *1083, 1084*
ecofeminism
See
"Activist Movements—Ecofeminism" heading in main text
ecological economics, *2309*
Ecological Society of America, *2274*
ecology, *1049, 1052, 1076, 1087, 3233*
Ecology Party (England), *1109*
Economic Summit of Industrialized Nations, *1848*
ecosystem, *1056, 1621*
"ecotage", *1058, 1075*
ecotheology, *1076, 1086, 1087, 3264*
ecotourism
See
"Ecotourism" heading in main text
Ecotourism Conservation Program, *2247*
Ecowash, *4411*
EcoWriter, *2716*

education and research, *2932*
educational programs
for anglers, *2511*
for hunters, *2925*
for timber owners, *2677*
forestry school at a U.S. college, *2678*
See also
"Education and Research" heading in main text
Edwards Aquifer, *4425, 4600*
Edwards Dam, *2080*
Edwin B. Forsythe National Wildlife Refuge, *4655*
egrets, *4654*
Ekati Diamond Mine, *3338*
Ekofisk Field, *3638, 3662*
El Chichon (volcano), *4373*
El Niño, *1833, 1859, 1968*
El Oviachic Dam, *2100*
elastic rebound theory, *2218*
electric cars, *1604, 1609, 1612, 1614, 1616*
Electric Consumers Protection Act of 1986 (ECPA), *4567*
Electric Power Research Institute, *1550, 4866*
Electric Vehicle Tax Credit, *4337*
electron capture detector, *1971*
electrostatic precipitation, *1447*
electrostatic precipitator, *2970*
***Elements of Forestry*,** *3247*
Elephant Conservation Center, *4793*
elephants, *4756, 4903, 4957*
African, *2340, 2406, 4724, 4745, 4755, 4788*
African forest, *4921*
Asian, *2406, 4744, 4793*
Eli Lilly Company, *2759*
elk, *2947*
Tule elk, *4833*
Elm Park, *3817*
elm trees, *1523, 2601*
embryology, *2280*
embryonic stem cells, *2774*
Emergency Conservation Work, *4035*
Emergency Gas Act, *2056*
Emergency Planning and Community Right-to-Know Act (EPCRA), *1353, 2984, 2986, 2987, 3964*
emissions standards, *1572, 1574*
***Empire of Dust*,** *3254*
encephalitis, *3972*
Endangered American Wilderness Act, *4662*
Endangered Species Act, *1696, 2333, 2337, 2348, 2349, 2386, 2387, 2388, 2389, 2390, 2391, 2392, 2393, 2394, 2399, 2400, 2401, 2402, 2403, 2404, 2405, 2407, 2417, 2423, 3021, 4726, 4727, 4742*
"Energy Efficiency and Water Conservation at Federal Facilities", *4865*
Energy Policy Act of 1992, *4337, 4338, 4863*
Energy Reorganization Act of 1974, *3949*
Energy Tax Act, *4330*
English Channel, *4399*
Enovid, *3880*
***Enterprise* (nuclear-powered aircraft carrier),** *3487*
entomology, *2265*
Environment Agency of Japan, *1578*
Environment Canada, *1993*
***Environmental Changes in Lake Erie*,** *4510*
Environmental Defense Fund, *4424, 4514*
***Environmental Diplomacy: The Environment and U.S. Foreign Policy*,** *2308*
environmental disputes, *3066*
environmental impact statements, *1064, 3086*
environmental law, *1061*
***The Environmental Law Forum*,** *1093*
environmental marketing, *1092*
environmental organizations, *1063*
U.S., *1055*
Environmental Protection Act (Australia), *4658*
Environmental Sabbath, *1086*
Environmental Support Center, *1102*
***Ephemerides*,** *1876*
Erawan field, *2062*
***Erebus* (ship),** *3552*
Erie Canal, *1817*
erosion, *2123, 2149, 2581, 3390*
See also
"Soil Resources—Erosion" heading in main text
Eskom, *3543*
Esrange Satellite Station, *1962*
***Essay on the Immigration of the Norwegian Flora*,** *2011*
***An Essay on the Principle of Population*,** *3876*
***Essays upon Field-Husbandry in New England*,** *3232*
Estes Park, *3768*
estuaries, *3595*
Chesapeake Bay Program, *3593*
Estuary Protection Act, *4470*
ethanol, *4326, 4334*
Ethyl Corporation, *1594*

Etudes sur les glaciers (Study of Glaciers), *2010*
Euphrates River, *1212*
European Community, *1986, 1991*
European Convention for the Protection of Animals for Slaughter, *3037*
European Economic Community, *3954, 3960*
European Environment Agency (EEA), *4312*
European Free Trade Association, *1991*
European Organization for the Exploitation of Meteorological Satellites, *1964, 1967*
European Ozone Research Coordinating Unit, *1991, 2162*
European Rivers Network, *2075*
European Space Agency, *1964, 1967*
European Space Research Organization, *1962*
European Union, *4312*
eutrophication, *4405, 4409, 4476*
EV-1, *1609, 1612*
Everglades, *1095, 1244, 2352, 4487, 4678*
Everglades National Park, *3146, 4645, 4766*
Exclusive Economic Zone, *3052*
Executive Order 11296 (evaluating flood danger), *2552*
Executive Order 11988 (flood plain protection), *2553*
Executive Order 12843 (to prevent ozone depletion), *1996*
Executive Order 12873 (Federal Acquisition, Recycling, and Waste Reduction), *4158*
Executive Order 12898 ("Federal Actions to Address Environmental Justice in Minority Populations and Low-Income Populations"), *2995*
Executive Order 12902 ("Energy Efficiency and Water Conservation at Federal Facilities"), *4122, 4865*
Executive Order 13089 (protecting U.S. coral reefs), *3601*
Executive Order 13101 ("Greening the Government Through Efficient Energy Management"), *4122*
Executive Order 13134 (developing biobased products and bioenergy), *4347*
Exelon, *3543*
Experimental Breeder Reactor I, *3461*
experimental farm, *1126*
***Exploration of the Colorado River*,** *2810*
***Explorer 1* (satellite),** *1951*
***Explorer 7* (satellite),** *1956*
Explosives and Combustibles Act, *2830*
Expo '74 (The International Exposition on the Environment), *2310*
Expo '93 (Taejon International Exposition), *2317*
expositions
See
"Education and Research—Expositions" heading in main text and names of specific expositions
extinct and endangered species, *1048, 1650, 1656, 1658, 1673, 1680, 1694, 1726, 3603, 3994, 4727*
See also
"Extinct and Endangered Species" heading in main text
***Exxon Valdez* (tanker),** *3627, 3643*

F

Faber-Castell, *2716*
Fahrenheit temperature scale, *1869*
fairs
See
"Education and Research—Expositions" heading in main text
falconry, *2898*
falcons
Aplomado falcons, *2446*
peregrine falcon, *2393*
fallout
See
radioactive air pollution
***Familles des Plantes*,** *1767*
family planning, *3255, 3881, 3897*
famine, *1183, 1819, 2120, 2124, 2129, 2131, 2132, 2137, 2158, 3903, 4046, 4221, 4365, 4366*
***The Farmer's Almanack*,** *3234*
farmland preservation, *1136, 1229, 1313, 3110, 3113*
fast-breeder reactor, *3520*
fault theory, *2208*
faunal succession, *2788, 2795*
Federal Agriculture Improvement and Reform Act of 1996, *3419, 4790*
Federal-Aid Highway Act, *1565, 1566*
Federal Aid in Sport Fish Restoration Act of 1950, *2507, 2510*
Federal Aid in Wildlife Restoration Act of 1937, *1664, 1671, 1677, 1695, 1722, 1725, 2925, 2926, 4702, 4703, 4704, 4705, 4706, 4707, 4709, 4710, 4711, 4712, 4714, 4717, 4721, 4820, 4821*
Federal Aid in Wildlife Restoration Act of 1974, *4831*
Federal Aid in Wildlife Restoration Project 2-R, *1739*

Federal Aid Road Act, *1557, 1570*
Federal Aviation Act, *3438*
Federal Cave Resources Protection Act, *3809, 4670*
Federal Coal Mine Health and Safety Act, *3334*
Federal Facility Compliance Act, *4134*
Federal Food, Drug, and Cosmetic Act, *1227, 3938*
Federal Guidelines for Dam Safety, *2073*
Federal Highway Act, *1558*
Federal Insecticide, Fungicide, and Rodenticide Act, *1290*
Federal Land Policy and Management Act, *3089*
Federal Metal and Nonmetallic Mine Safety Act, *3333*
Federal Office for Occupational Health (Germany), *3958*
Federal Power Act, *4557, 4567*
Federal Water Power Act, *2095*
Federal Water Project Recreation Act, *4561*
feed additives, *1172*
Ferrel Maxima and Minima Tide Predictor, *3564*
ferrets, *2435*
fertilizers, *2672*
See also
"Soil Resources—Fertility and Fertilizers" heading in main text
FIAT Topolino, *1543*
Field Act (California), *2221*
***Field and Stream*,** *1030*
***Field Guide to the Birds*,** *3253*
field rotation, *1234*
figs, *1197*
Finance Act of 1996 (England), *4153*
First Arctic Ministerial Conference, *3873*
First International Air Pollution Conference, *1461*
First International Meeting of Experts, *1968*
First International Meeting of People Affected by Dams, *2078, 2079, 2081*
First San Diego Aqueduct, *4592*
fish and fishing, *1297, 1804, 2953, 3015, 3565, 3603, 4413, 4415*
See also
"Fish and Fishing" heading in main text and names of specific fish
Fish and Oyster Commission, *2455*
Fish and Wildlife Act, *4770*
Fish and Wildlife Improvement Act, *4733*
***Fish Hawk* (hatchery ship),** *2493*
Fish-Rice Rotation Farming Program Act, *2496*
The Fisheries Aquarium at the Northeast Fisheries Science Center, *4902*
Fisheries Management Plan for Sharks of the Atlantic Ocean, *2482*
Fishing Capacity Reduction Initiative, *2483*
Fitzroy barometer, *1902*
flamingos, *4939*
puna flamingo (James' flamingo), *1700*
Flavr Savr tomato, *2767*
flint mines, *3293*
Flood Control Act of 1928, *2537*
Flood Control Act of 1936, *2539*
Flood Control Act of 1946, *2542*
Flood Damage Reduction Program (Canada), *2557*
floods and flooding, *2118*
See also
"Floods and Flood Control" heading in main text
***A Flora of Leicestershire*,** *1783*
Florida Federation of Women's Clubs, *1095*
"Florida holly", *1496*
Florida land boom, *3141*
Florida National Scenic Trail, *4003*
Florida panther, *2347*
fluoridation, *4591*
fluoride, *4478*
Fog and Smoke Committee, *1430, 1431*
Food and Agriculture Organization, *1267*
Food Security Act, *1229, 3420, 4057*
Ford Motor Company, *1541, 1543, 1550, 1591, 1603, 1605, 1606*
forecast, *1898*
Forest and Rangeland Ecosystem Science Center, *3404*
Forest and Rangeland Renewable Resources Planning Act, *2640*
forest cover, *2735*
forest fires
See
"Forests and Trees—Fires and Fire Prevention" heading in main text
Forest Pest Control Act, *2604*
Forest Plan, *3414*
Forest Reserve Act, *2631, 3734*
***Forest Trees of California*,** *1784*
***Forestry Abstracts*,** *2741*
***Forestry Quarterly*,** *2736*
forests and trees, *1320, 1475, 1786*
diseases and pests, *1511, 1523*
logging, *1040, 1041*
reforestation and planting, *1098*
See also
"Forests and Trees" heading in main text

The Formation of Vegetable Mould Through the Action of Worms with Observations on Their Habits, *4070*
Fort Gratiot Sanitary Landfill v. *Michigan Department of Natural Resources*, *4150*
Fort Jefferson National Monument, *3799*
Fort Laramie National Historic Site, *3762, 3765*
Fort Worth Zoological Park, *4915*
42nd Amendment (India), *2978, 3386*
fossils, *1681, 1732, 2359, 2788, 2803, 3740, 3838, 3874*
Foundation House, *2826*
Fountain Grove Wildlife Management Area, *1725*
fox hunting, *2910, 2963, 2964, 2965*
foxes, *2390*
Frankfurt International Motor Show, *4343*
Fraser River, *2459*
***Freedom of the Seas*,** *4488*
Freeport Sulphur Company, *3331*
Freon, *1340, 1975*
"Frost Fair", *1802*
Frostban, *2761*
fruit, *1277*
fruit
See
"Agriculture and Horticulture—Food Crops" heading in main text
***Fruit Walls Improved*,** *4087*
fuel cells, *4324*
Fujita-Pearson Tornado Scale, *4293*
***Fumitugium: or the Inconvenience of the Aer and Smoake of London*,** *1429*
Fund for Animals, *1016*
fungicides, *1278, 1292, 1296, 1298, 1308*
Fur Seal Act, *4776*

G

Gaia, *1050*
Gaia hypothesis, *1828*
Galápagos Islands, *2252, 3995*
Galápagos tortoise, *2400*
Game Conservation Training School, *2278*
***Game Management*,** *3252*
game wardens, *2914, 4688, 4689*
"garden cities", *3099*
Garden of Ammon, *4875*
gas chromatograph, *2843*
Gas Light and Coke Company, *2037*
gas pipelines, *2052*
Gaz de France, *2066*
Gaza-Kruger-Gonarezhou trans-frontier park, *4745*
gazelles, *1137*
geese, *1145, 1698*
nene, *2426, 2426*
Ross's goose, *1738*
snow goose, *4748*
gene, *2752*
genecology, *1793*
Genentech, Inc., *2756*
***Genera of American Plants*,** *1776*
General Directorate for Environmental Quality (Portugal), *3959*
General Electric Company, *1947, 2856, 4205*
General Exchange Act, *2633*
General Long-term U.S.S.R. State Program of Environmental Protection and Rational use, *1843*
General Motors Corporation, *1022, 1340, 1534, 1537, 1541, 1544, 1545, 1547, 1550, 1590, 1594, 1597, 1603, 1606, 1609, 1612, 1970, 3940, 4336*
solar-powered car, *4110*
General Motors Research Laboratories, *1462*
genes, artificial, *2754*
***The Genesis Flood*,** *2547*
genetic code, *2751*
genetic engineering and biotechnology, *2305, 2500*
See also
"Genetic Engineering and Biotechnology" heading in main text
genetic pollution, *1257*
genetic recovery program, *2347*
Geochronological Institute, *2004*
***Geography*,** *2175*
geological maps, *2780, 2782, 2795*
geology and geophysics
See
"Geology and Geophysics" heading in main text
***Geology of Canada*,** *2806*
Geophysical Fluid Dynamics Laboratory, *1958*
George Washington **(submarine),** *3490*
Geosat, *3600*
geothermal heat exchange theory, *2800*
Geothermal Steam Act of 1970, *3379*
geothermics, *2800, 2812, 2815, 2826, 3379, 4376, 4385, 4390*
Gesandte **(ship),** *3824*
giant squid, *2451*
gibbons, *4926*
Gifford Pinchot National Forest, *2749*

Gila National Forest, *2682, 3797*
Gila National Wilderness, *4634*
Gilgamesh epic, *2518*
Ginkgo, *1763*
Gir Wildlife Sanctuary and National Park, *4764*
giraffes, *4876*
glaciation, *1999*
Glacier Bay National Park and Preserve, *3776*
Glacier National Park, *3798*
***glasnost*,** *2232*
Glen Canyon, *3245*
Glen Canyon Dam, *1079, 2562*
Glen Canyon–Lake Powell National Park, *2238, 3755*
Global Change Research Act, *1847*
Global Environment Monitoring System Freshwater Quality Program, *4601*
global warming, *1090, 1340, 1425, 1426, 1448, 1452, 1830, 2162, 2731, 3012*
See also
"Climate and Weather—Global Warming" heading in main text
Globus II, *3289*
***Glomar Challenger* (research vessel),** *2824*
Gloucester Fox Hunting Club, *2964*
goats, *1139, 1311, 1499*
mountain, *2946*
Gobi Desert, *1742*
Goddard Space Flight Center, *1992*
gold, *3340, 3341, 3349*
Golden Gate National Recreation Area, *3821*
Golden Spike National Historic Site, *3802*
gorillas, *2395, 4698, 4751*
***Gossamer Penguin* (solar aircraft),** *4119*
Governors Conference, *3365*
graffiti, *3279, 3284*
Grand Canyon, *1420, 2810, 3753*
Grand Canyon National Park, *3445, 3815, 4025*
Grand Coulee Dam, *2097*
Grand Rapids Electric Light and Power Company, *4548*
Grand Rounds Scenic Byway, *4024*
Grand Tetons National Park, *4822*
Grande Dixence Dam, *2103*
Grant-Kohrs Ranch National Historic Site, *3807*
grasshoppers, *1186*
Great American Desert, *2135*
"The Great American Fraud", *3250*
great auk, *1650, 2361, 2364, 2375*
Great Barrier Reef, *3579, 4000*
Great Barrier Reef Committee, *3575*
Great Chicago Fire, *2136*
Great China Flood, *2532*
Great Depression, *2152*
Great Exhibition of 1851, *1900*
Great Gulf Wilderness, *4640*
Great Lakes, *2461, 3029, 3070, 3394, 4016, 4184, 4441, 4476, 4532*
Lake Erie, *1501, 2374, 4414, 4416, 4428, 4465, 4474, 4502, 4510, 4511, 4539, 4654*
Lake Huron, *4412*
Lake Michigan, *3712, 4428, 4499, 4531*
Lake Ontario, *1501, 4511*
Lake Superior, *1526, 4401, 4442, 4512, 4636*
Great Lakes Coastal Barrier Act, *3394*
Great Plains Prairie Cluster, *3410*
Great Smokey Mountains, *3400*
"The Great Stink", *4525*
Great Swamp National Wilderness Area, *3730*
"Great Tri-State Tornado", *4286*
Great White Heron National Wildlife Refuge, *2327*
Green Bay, *4499*
"green belt", *3083, 3099, 3100, 3101*
Green Belt Movement, *1043, 1098*
Green Lights energy efficiency program, *1423*
Green Mountain Know Your Power Festival, *4345*
Green parties, *1106*
Australia, *1105, 1108*
Belgium, *1107, 1111*
Finland, *1117*
France, *1107*
Germany, *1105, 1110, 1112, 1119*
Great Britain, *1105, 1109*
Netherlands, *1105*
New Zealand, *1105, 1107*
Norway, *1114*
Sweden, *1105*
Switzerland, *1105, 1110, 1111*
Taiwan, *3512*
United Kingdom, *1107*
United States, *1113, 1115, 1118*
Green Revolution, *1132*
Greenfreeze, *1995*
greenhouse gases, *3050*
See also
"Climate and Weather—Global Warming" heading in main text
greenhouses, *4086*
Greenpeace, *1063, 1106, 1995, 2878*
Griffith University, *2254*
groundwater, *2776, 2835, 2839, 2852*

Group Against Smog and Pollution (GASP), *1404*
Group of Experts on the Scientific Aspects of Marine Pollution (GESAMP), *4495*
grouse, *2935*
Attwater prairie chicken, *2329*
heath hen, *2935*
Growers' Wool Cooperative, *1175*
guano, *1128*
guans, *1661*
A Guide to Green Government **(Canada),** *4316*
Gulf Islands National Seashore, *4003*
Gulf Islands Wilderness, *4663*
Gulf Oil Company, *3652, 3693*
Gulf Refining Company, *3619*
Gulf War, *3645, 3646*
Gunnison National Monument, *3741*
gunpowder, *3092*
gypsum dune field, *3722*
gypsy moths, *1511, 2590, 2591, 2594, 2603, 2607*

H

Habitat Conservation Plan (HCP)
San Francisco, CA, *2337*
Washington State, *4791*
Habitat II, *3901*
Hackensack Meadowlands Development Commission, *3377*
Hackensack Meadowlands Reclamation and Development Act, *3377*
Hagenbeck Tierpark, *4913*
Haleakala Wilderness, *4660*
haplochromines, *1524*
Harmony Resort, *4301*
harpoons, *4625*
Harvard College, *1807*
Harvard University, *1786*
Harwell laboratory, *3456*
Hatch Act, *1224*
Haughley Farm, *1262*
Hawaii Volcanoes National Park, *4377*
Hawaiian Islands Biosphere Reserve, *1634*
Hawk Mountain Sanctuary, *1720, 1721*
hawks and hawking, *1720, 1721, 2898, 3227*
Hazard Ranking System, *2868, 2981*
Hazardous and Dangerous Waste Law (Spain), *3965*
Hazardous Materials Transportation Act, *2855*
hazardous waste, *2977, 2981, 2984, 2998, 3465, 3522, 3530, 3541, 4143, 4144, 4427*
See also
"Hazardous Waste" heading in main text
Headwater Forest, *1088*
"Heartbreak of America", *1022*
heat waves, *1823, 1824*
Helicopter Noise Control and Safety Bill, *3450*
Henry Doorly Zoo, *4908*
Heppner Disaster, *2534*
herbicides, *1289, 1291, 1295, 1304, 3256*
effects on fish and wildlife, *4779*
Heron Island Research Station, *3575*
herons, *1727, 4654*
HERPlab, *4954*
Hetch Hetchy Dam, *2090*
hexacholorobenzene, *2841*
High Seas Compliance Act of 1995, *2484*
Highway 40, *1566*
Highway Beautification Act, *3980*
Highway Performance Monitoring System, *1569*
Highway Revenue Act, *1565*
Hindenburg**,** *4275*
Historia naturalis**,** *1748*
Historia Plantarum**,** *1760*
Historiae Rerum Naturalium Brasiliae **(Natural History of Brazil),** *2450, 4680*
historic preservation, *1312, 3082, 3112, 3116*
See also
"Land Use and Development—Preservation" heading in main text
Historica animalium, **(History of animals),** *4877*
"The Historical Roots of Our Ecological Crisis", *3257*
The History of Japan and Siam**,** *1763*
History of Ocean Basins**,** *2818*
Hol Chan Marine Reserve, *3594*
holidays, *2724*
Homestead Act, *3077, 3080, 3132, 3749*
Homestead National Monument, *3749*
Honda, *1615*
honey, *1144*
honeybees, *1158*
Hooker Chemicals and Plastics Corporation, *1099, 4426*
Hoover Dam, *2099, 2222, 4560*
Horsecollar Ruins, *3742*
horses, *1143, 1149, 1151, 1152*
Przewalski, *2404*
reintroduced in the Americas, *1485, 1488*

wild, *1153, 4771, 4780, 4801, 4802, 4803, 4829*
Horticultural Society of London, *1504*
Hot Springs National Park, *3772*
housing, *3918, 3924*
Housing and Slum Clearance Act, *3085*
Houston Ship Canal, *4406*
Hubbard Brook Experimental Forest, *1463*
"Huckleberries" (essay), *2710*
Hudson-Mohawk-Erie Canal system, *1501*
Hudson River, *1808, 1816, 2856, 3104, 3737, 4538*
Hughes Tool Company, *3617*
Huguenots, *3301*
***Hull* (warship),** *4254*
human cloning, *2771, 2773*
humane slaughter, *1015, 3037*
Humane Slaughter Law, *1015*
Humane Society of the United States, *1014, 1016, 1020, 1023*
Humphrey-Rarick Act, *2640*
Humulin, *2759*
Hunter Safety Education Coordinators Workshop, *2924*
***Hunters in the Snow*,** *1800*
hunting and trapping, *1666*
legislation and regulation, *3388*
See also
"Hunting and Trapping" heading in main text
"Hurricane Hunters", *4255*
"Hurricane of Independence", *4237*
hurricanes, *1954, 2541, 4228, 4237, 4244, 4259, 4263*
Alicia, *4267*
Andrew, *4269*
Audrey, *4261*
Carol, *4260*
David, *4265*
Frederic, *4266*
Ginger, *4262*
term, *4223*
tracking, *1827, 4236*
See also
"Storms—Hurricanes, Cyclones, and Typhoons" heading in main text
***The Husbandman's Guide*,** *3231*
"Hutchison Rider", *2348*
Hwai River, *2545*
hydrocarbon emissions, *1573, 1577*
hydrochloric acid, *1383, 1434*
hydrochlorofluorocarbons (HCFCs), *1995*
hydroelectric power, *1108, 2069, 2070, 2071, 2087, 2090, 2095, 2096, 2097, 2110, 3366, 4547, 4550, 4552, 4553, 4556*
hydrofluorocarbons (HFCs), *1995*
hydrogen bomb, *1322*
hydrography, *1895*
hydroponics, *1247, 1248, 1249, 1250, 1251*
hygrometers
Daniell dew-point hygrometer, *1815*
using human hair, *1877*
Hyundai Oil Refining Company, *3706*

I

***Ibbetsonia genistoides*,** *1770*
ibex, *3990*
Ice Age, *2001, 2012, 2805*
Ice Age National Scenic Trail, *3997*
ichnology, *2803*
Idaho potato, *1207*
Idikki dam, *2107*
Iguaçu Falls, *4629*
Iguaçu River, *4629*
Illinois and Michigan Canal, *3712*
Illinois River, *3712*
Illinois System, *4071*
imidaclopride (pesticide), *2608*
immunocontraception, *4806*
***Index to the Geology of the Northern States*,** *2797*
Indian Environmental General Assistance Program Act, *4151*
Indian Forest Act (India), *3369*
Indian Forest Service, *2673*
Indian Nonintercourse Act, *3078*
Indian Ocean, *1482*
Indian Reorganization Act, *1226*
Indian River, *1719*
indigenous peoples, *3873*
See also
Native Americans
indigo, *3347*
Indoor Radon Abatement Act, *1367*
Indus River, *2414*
***Industrial Poisons in the United States*,** *2276, 2832*
industrial pollution, *1099, 1314, 1320, 1351, 1353, 1354, 1416, 1452, 1818, 2836, 2838, 2851, 2867, 2868, 3522, 4396, 4423, 4426, 4472, 4475, 4512*
See also
"Industrial Pollution" heading in main text
industrialization, *1818, 2145*
influenza epidemic of 1918, *3927*
***Inquiry into the Causes and Effects of the Variolae Vaccinae*,** *3912*

insects, *4686, 4687*
endangered species, *2411*
entomologist hired by the U.S. government, *1272*
monuments to, *3164*
quarantine, *1282*
See also
names of specific insects
Insects Injurious and Beneficial to Vegetation, *1272*
Insects of Pennsylvania, *4685*
Institute for Advanced Study (Princeton), *1950, 1953*
Institute for Environmental Conflict Resolution (IECR), *3066*
Institute for Scrap Recycling Industries, *1607*
Integrated Flood Observing and Warning Systems (IFLOWS), *2558*
Integrated Solid Waste Collection System (Philippines), *4133*
Integrated Waste Management Act (California), *4160*
Intelligence Park, *4872*
Inter-American Convention for the Protection and Conservation of Sea Turtles, *3054*
Inter-American Institute for Global Change Research, *3064*
Inter Caetera **(Papal bull),** *3008*
intercropping, *1043*
Intergovernmental Panel on Climate Change, *1831, 1844*
Intermodal Surface Transportation Efficiency Act, *1581, 4010*
internal-combustion engine, *1588, 1589*
International Association of Fish & Wildlife Agencies, *2512*
International Association of Volcanology, *4378*
International Atomic Energy Agency, *3478*
International Boundary Treaties Act, *3366, 4555*
International Centre for Ecotourism Research, *2254*
International Centre for Integrated Mountain Development (ICIMOD), *4304*
International Civil Aviation Organization (ICAO), *3933*
International Commission for the Conservation of Atlantic Tunas (ICCAT), *2475, 2486, 3060*
International Committee of Large Dams, *2068*
International Conference on Population and Development, *3900*
International Convention for the Prevention of Pollution from Ships (MARPOL), *3024, 4148*
International Convention for the Prevention of Pollution of the Sea by Oil (OILPOL), *3024*
International Convention on Civil Liability for Oil Pollution Damage, *3582, 3586*
International Convention on the Establishment of an International Fund for Compensation (for oil pollution damage), *3586*
International Council for Bird Protection (ICBP), *3374*
International Council for the Exploration of the Sea (ICES), *3565, 3566, 3596*
International Court of Justice, *3011*
International Crane Foundation, *1741*
International Day of Action Against Dams: For Rivers, Water and Life, *2081*
International Directory of Environmental Education Institutions, *2282*
International Dolphin Conservation Act, *2481*
International Drought Information Center, *2160*
International Garden and Greenery Exposition, *2315*
International Great Lakes Water Quality Agreement, *3029, 4476*
International Hydrological Decade, *4596*
International Hydrological Program, *4596*
International Institute of Agriculture, *2699, 2701*
International Joint Commission, *3029, 3366*
International Log Rule, *2658*
international marine conference, *1895*
International Maritime Organization (IMO), *3024, 3574*
International Meteorological Organization, *1882, 1914*
International Network on Water, Environment and Health, *4613*
International Ocean Exposition, *2311*
International Oil Pollution Compensation Fund, *3586*
International Ozone Conference, *1990*
International Pacific Salmon Fisheries Commission, *2459*
International Planned Parenthood Federation (IPPF), *3884, 3897*
International Plant Protection Convention, *3023*
International Program for Climate Change (IPCC), *1832*
International Register of Potentially Toxic Chemicals (IRPTC), *2292, 2975*
International Rice Research Institute, *1133*
International Rivers Network, *2075*
International Sanitary Congress, *3925*

International Seabed Authority (ISA), *3065*
International Species Inventory System, *4946*
International Standards Organization (ISO), *4318*
International Statistical Institute, *3882*
International Tourism Exchange Berlin, *2255*
International Tropical Timber Association, *2687*
International Union for the Protection of Nature, *3058*
International Union for the Scientific Study of the Population, *3882*
International Union of Forestry Research Organizations, *2694*
International Union of Geodesy and Geophysics (IUGG), *4378*
International Union of Societies of Foresters, *2706*
International Whaling Commission, *4627, 4628*
International Wildfowl Research Bureau (IWRB), *3374*
Interstate 70, *1566*
Interstate Air Pollution Abatement Conference, *1399*
Interstate Quarantine Act of 1893, *4581*
InterTribal Bison Cooperative (ITBC), *4839*
An Introduction to Botany in a Series of Familiar Letters, *1771*
Introduction to the Principles of Morals and Legislation, *1003, 1004*
Intrusion Act, *3136*
Inuit Circumpolar Conference, *4319*
inventions
See
names of specific inventions
Freon, *1339*
Inventory and Monitoring Program, *3402*
Io (moon of Jupiter), *4387*
ion exchange, *4505*
ionosphere, *1920*
Iowa's Protected Water, *4669*
iron industry, *2028*
irradiation, *1209*
irrigation, *1031, 1240, 2090, 2102, 2151, 3073, 3366, 4098, 4408, 4553, 4556, 4569*
See also
"Agriculture and Horticulture—Irrigation" heading in main text
irrigation wheels, *1213*
Iset River, *4437*
island nations, *1090*
ISO 14000, *4318*
isothermal map, *1885*
IT Corporation, *2870*
Itai Itai disease, *2974*
Italian State Forest Service, *2436*
IUCN Red List of Threatened Plants, *2421*
ivory, *4788*

J

Jacks Fork River, *3978, 4641*
Jackson Hole Wildlife Park, *4822*
Japanese beetles, *1189*
Jefferson Parish Christmas Tree–Marsh Restoration Program, *4308*
Jenkins Laboratory, *1934*
jet stream, *1929*
Jinzu River, *2974*
John Bull Zoological Gardens, *4954*
John Day Fossil Beds National Monument, *3744*
John Pennekamp Coral Reef State Park, *3728*
Johns Hopkins University School of Medicine, *2774*
Johnstown Flood, *2112*
Joint Arctic Weather Stations, *1948*
Joint Commission on Boundary Waters, *4511*
Joint Declaration on the Protection of the Wadden Sea, *3010*
Joint Oceanographic Institutions for Deep Earth Sampling, *2824*
Joint Report of the Special Envoys on Acid Rain, *3042*
Journal of Forestry, *1054*
Journal of Industrial Hygiene and Toxicology, *2837*
Journal of Researches into the Geology and Natural History of the Various Countries Visited by the H.M.S. Beagle, *3238*
JPDR experimental reactor, *3505*
Juan Bautista de Anza National Historic Trail, *3711*
Julia Butler Hansen National Wildlife Reserve, *2330*
The Jungle, *1168, 3249*
Junior Duck Stamp Conservation and Design Program Act, *4789*
Junked Auto Disposal Bill, *4142*
Jupiter, *4387*
Justinian Code, *4083*

K

Kaibab Plateau, *4797*
Kaibab Plateau–North Rim Parkway, *4025*
Kaiko **(submersible),** *3599*
Kakadu National Park, *4000*
Kalahari Desert, *4212*
Kampinos Forest, *2428*
Kanawha River, *4433*
kangaroos, *2420*
Kariba Dam, *2071*
Keep America Beautiful, *3270*
Keep Britain Tidy Group, *3271*
Keep Green movement, *2617*
Kenai National Wildlife Refuge, *4023*
Kennebec River, *1801, 2080*
Kennelly-Heaviside layer, *1920*
Keokuk Dam, *2093*
kerosene, *3679, 4891*
Kerr-McGee Co., *3513, 3653*
Kesterson Refuge, *1221*
Kew Gardens (England), *1782*
Keweenaw National Historic Park, *3811*
Key Largo Coral Reef Preserve, *3803*
Kgalagadi Transfrontier Park, *4845*
Khao Yai National Park, *3727*
Khian Sea **(ship),** *2876*
killer bees, *1529, 3968*
king salmon, depicted, *2922*
Kings River, *3760*
KINS Act (South Korea), *3966*
Kirstenbosch Gardens, *1792*
Kirtland Air Force Base, *3510*
Kisatchie National Forest, *4007*
kiwis (birds), *1665*
Klondike gold rush, *3320*
Knoxville International Energy Exposition, *2312, 2313*
Knoxville Zoological Gardens, *4756*
Korea Institute of Nuclear Safety, *3966*
Krakatoa (volcano), *4371*
Kruger Park, *4745*
Kubutu oil well, *3709*
kudzu, *4050*
Kukdong Oil Company, *3706*
Kyoto Protocol, *3012, 3056*

L

L-20 regulation, *4635, 4638*
Laboratory Animal Welfare Act, *1017*
labyrinthodont, *3867*
Lacassine National Wildlife Refuge, *4659*
Lacassine Wilderness, *4659*
Lacey Act Amendments of 1981, *2458*
Lacey Game and Wild Birds Preservation and Disposition Act of 1900, *1013, 2458, 4691, 4765*
Lake Davis Pike Eradication Project, *4807*
Lake Manyara National Park, *4724*
lakes, *4405*
Baikal, *1065*
Constance (Bodensee), *1826*
Crater, *3767, 4370*
Davis, *4807*
Erie, *1501, 4414, 4416, 4428, 4465, 4474, 4502, 4511, 4539, 4654*
Great Salt, *4708*
Huron, *4412*
Kenyir, *2250*
Kivu, *4698*
Lukajno, *1625*
Mead, *2099, 2222, 3806*
Michigan, *4428, 4499, 4531*
Monoun, *4392*
Nyos, *4392*
Ohrid, *3996*
Okeechobee, *2538*
Ontario, *1501, 4511*
Pedder, *1108*
Plitvice, *3996*
Pyramid, *4017*
Saint Clair, *4474*
Superior, *1526, 3729, 4401, 4442, 4512*
Tahoe, *3421*
Tungting, *2545*
Victoria, *1524*
Willandra, *4000*
See also
Great Lakes
Laki fissure, *4365*
Lalonde Report, *3948*
lampreys, *1501, 1526, 2461, 4502*
Land and Water Conservation Fund, *3658*
Land Conservation and Development Commission (Oregon), *3107*
land-grant colleges, *1224*
Land Ordinance of 1785, *3077, 3132*
The Land Problem, *1050*
land reclamation, *3337*
land use and development, *1230, 2579, 2582, 3389, 3394, 3483*
See also
"Land Use and Development" heading in main text
Landsat **(satellite),** *2293, 3381*
larch sawfly, *2598*
Lassen Peak (volcano), *3793*

Lassen Volcanic National Park, *3793*
law of relative proportion, *3563*
law of superposition, *2781*
Law on Population, Development, and Prosperous Family of 1992 (Indonesia), *3889*
Law on the Disposal of Oil and Chemical Wastes (Denmark), *3945*
***Le feminisme ou la mort* (Feminism or Death),** *1042*
lead, *1358, 1580, 2514, 2827, 2888, 2962, 3944, 3955, 4478, 4588, 4606*
Lead Contamination Control Act, *4607*
leaded gasoline, *1339, 1970*
leeches, *4515*
lemurs, *2445*
***Lenin* (nuclear-powered ship),** *3504*
leopards, *4876*
levees, *2539, 2550, 3268*
See also
"Floods and Flood Control—Levees" heading in main text
Lewis and Clark Expedition, *1812, 4683, 4684*
lichen, *1785*
Lick Observatory, *2207*
life-history investigations, *1736*
***Life Zones and Crop Zones of the United States*,** *1619*
lighthouses, *3500*
lighting
alcohol and turpentine, *4323*
electric, *2048, 3359, 3360, 4548, 4550, 4850*
energy efficient, *1423, 4322*
gas, *2019, 2032, 2038, 2048*
in zoos, *4945*
oil, *3673*
Lilah Callen Holden Elephant Museum, *4957*
***The Limits of Growth*,** *3261*
Limnological Institute, *1065*
Linnaean system, *1771*
Linnean Society of London, *1769, 1789*
Linville River Wilderness, *4645*
lions, *4725*
Asiatic lion, *2396, 4764*
literature and arts
See
"Literature and Arts—Books and Writers" heading in main text and titles of specific works
Litter and Recycling General Laws (Rhode Island), *3276*
Litter Control and Beautification Act (New Mexico), *3278*
Litter Control Participation Permit (Rhode Island), *3280*
Litter Reduction and Recycling Fun (Nebraska), *3281*
littering
See
"Littering" heading in main text
Little Bookcliffs Herd Management Area, *4837*
Little Ice Age, *1799, 1804, 1811, 1826*
Little Miami River, *3985*
Little Tennessee River, *2333*
Live Aid, *2158, 2159*
living farm network, *1134*
Living Planet Campaign, *1039*
Living Seas Exhibit, *4871*
***The Living Soil*,** *1262, 4065*
***Living the Good Life*,** *1036*
llamas, *1148, 1175*
logging, *2586, 2627, 2630, 2632, 2635, 2639, 2677, 2687, 2688, 2689, 2711*
See also
"Forests and Trees—Logging" heading in main text
London Amendment, *1994*
London Dumping Compact of 1990, *4485*
London Dumping Convention of 1972, *3061*
London Electric Telegraph Company, *1900*
***London Fogs*,** *3246*
London Green Belt Act, *3083*
London University, *1933*
***Long Beach* (nuclear-powered warship),** *3487*
Longleat Safari Park, *4940*
Los Angeles Aqueduct, *4585*
Los Angeles Zoo, *4955*
Lost Cabin Gold Mine, *3767*
Lostwood Wilderness, *4657*
Louisiana Purchase, *3135*
Louisiana World Exposition, *2314*
Love Canal Homeowners Association, *4426*
low-income communities, *3412*
low-level radioactive waste, *2845*
Low-Level Radioactive Waste Policy Act, *2865, 3516*
Luando Natural Integral Reserve, *4824*
Luchugilla Cave, *4672*
Lumsden Lake, *1453*
lung cancer, *1348*
Lyceum, *1746*
lynx, *2437, 2442, 2447*

M

Maas River, *2546*

Mackenzie River, *3625*
Magenta Petrel, *1670*
Magnox, *3460*
Mahazat as-Sayd Protected Area, *2438*
maize, *1123, 3123*
Malabar Farm, *1131*
malaria, *1288, 3939, 3941, 4374*
malathion, *1303*
Malaysian Elephant Satellite Tracking Project, *4744*
Man and Nature; or Physical Geography as Modified by Human Action, *1048, 3073, 3244*
Man and the Biosphere Programme, *1627, 1635*
Manara Nord Biosphere Reserve, *1638*
manatees, *2338*
Manhattan Project, *3452, 3453*
Manicouagan River, *2106*
manioc, *1204*
Mapimi Reserve, *1630*
maps and mapping, *1895*
  geological, *2792*
  in Antarctica, *3861*
MAR Conference, *3374*
Mare Liberum **(Freedom of the Seas),** *3229, 3546, 4488*
Mariana Trench, *3578, 3599*
Marine Mammal Commission, *4782*
Marine Mammal Protection Act, *4782*
marine mammals, *4782*
marine pollution, *3024, 3040, 3061, 3574, 3577, 3582, 3583, 3586, 3587, 3588, 3594, 3936, 4410, 4434, 4479, 4480, 4485*
  See also
  "Water Pollution—Marine Pollution" heading in main text
Marine Resources and Engineering Development Act, *3581*
Marine Wildlife Foundation, *4958*
Marine World, Africa, USA, *4958*
Marineland (Florida), *4932, 4934*
Mariner 9**,** *4383*
Mariposa Grove, *3713*
Marjory Stoneman Douglas Wilderness, *4645*
market hunting, *2951, 2952, 2954*
marlin, *2475*
Mars, *2230, 4383, 4395*
Mars Pathfinder**,** *4395*
Marsh-Billings-Rockefeller National Historic Park, *3812*
Mary and Helen **(whaler),** *4626*
Maryland darter, *2388*
Mashal la Kadmoni**,** *2901*
Massachusetts Board of Health, *4419, 4421*
Massachusetts Environmental Trust, *2307*
Massachusetts Hazardous Waste Facility Siting Act (1980), *2864, 2870*
Massachusetts Institute of Technology, *4120, 4419, 4944*
Massachusetts State Board of Health, *3919*
Masters Ranch, *2245*
Mauna Kea Observatories, *4382*
Mauna Loa, *3759, 4377*
Mauna Loa Observatory, *1839*
Mauna Loa Research Station, *1459*
Maxine's Tree**,** *2665*
Maxwell Motor Corporation, *1595*
mayflies, *2374, 2374*
mayflowers, *2322, 2326*
Mazatzal Wilderness Area, *4006*
Mazda Motor Corporation, *1545*
McKillop Stewart Company, *3689*
McMurdo Station, *1988*
McSweeney-McNary Research Act, *2634*
measurements, *3545*
  Beaufort Wind Scale, *1882*
  elevation benchmarks, *2226*
  Fitzroy barometer, *1902*
  of age of trees, *1759*
  of atmospheric carbon dioxide, *1452, 1839*
  of atmospheric pressure, *1863*
  of distance to earth's core, *2219*
  of drought severity, *2154*
  of earthquake strength, *2212, 2223*
  of Earth's mean density, *2791*
  of humidity, *1877*
  of hurricane damage potential, *4263*
  of mortality, *3908*
  of rain, *1866*
  of seismic waves, *2177*
  of smoke density, *1457*
  of sunshine, *1901*
  of temperature, *1869*
  of tides, *1868*
  of timber, *2652, 2658*
  of tornado intensity, *4293*
  of wind chill, *3859*
  of wind speed, *1893*
  pH scale, *1325, 1444*
  rectangular system of land surveys, *3132*
  UV index, *1849*
Meat Inspection Act of 1906, *3249*
Mechanics Institute, *4891*
medical waste, *2884, 4149, 4428*
Medical Waste Tracking Act, *2883*
Mediterranean Action Plan, *3040*
Mediterranean fruit fly, *1191, 1283, 1299, 1303, 1522*
Megalopolis**,** *3143*
menageries, *4879, 4880*

"Mendocino Tree", *2580*
Mercalli Scale, *2212*
Mercedes-Benz, *1535*
Merck and Company, *1639*
mercury, *1268, 1296, 1298, 2972, 3352, 3936, 4152, 4434, 4440, 4468, 4474, 4478, 4513*
mercury barometers, *1863*
mercury thermometers, *1869*
Merz process, *4140*
Mesa Verde, *2125*
Mesa Verde National Park, *3788, 3973*
Mesabi Iron Range, *3318*
mesothelioma, *3963*
metallurgy, *1383*
***Meteor* (ship)**, *3570*
meteorites, *3853*
Meteorographica, *1907*
Meteorological Office (England), *1926, 1930, 1940, 1946, 1952*
Meteorological Research Flight (England), *1946*
meteorologists, *1891, 1902, 1908*
meteorology
See
"Climate and Weather—Meteorology" heading in main text
Meteosat, *1964, 1967*
Meteosat Second Generation, *1964*
methane gas, *2054*
methyl bromide, *1310*
methyl chloroform, *1994*
methylcyclopentadienyl manganese tricarbonyl (MMT), *3971*
Mexican environmental protection agency, *2994*
"Mexico City Policy", *3884*
Miami Conservancy District, *2535*
Michilia Reserve, *1630*
Micrographia, *1758, 2750*
Midway Atoll National Wildlife Refuge, *4843*
migrant workers, *1252*
Migratory Bird Act, *1689, 2955*
Migratory Bird Conservation Act, *1691, 4818*
Migratory Bird Conservation Commission, *1691*
Migratory Bird Hunting and Conservation Stamp Act (1976), *4789*
Migratory Bird Hunting Stamp Act (1934), *1693, 2933, 3165, 4818*
Migratory Bird Treaty Act of 1918, *1690, 3021*
Migratory Bird Treaty of 1916, *1743, 3017, 3018, 4742*
Mile-High Post, *3862*
Mill Creek Dam, *2111*
Mill Creek Flood, *2530*
millet, *1196*
Minamata disease, *2972, 3936, 4434*
Mineral Springs Resort, *3780*
mineralogy, *2258*
"Miner's friend", *2027, 3300*
mines and mining, *3811, 4198*
See also
"Mines and Mining" heading in main text
Minidoka Dam, *4554*
Ministry of Non-Conventional Energy Resources (India), *4864*
minks, *1161*
Minnesota Forestry Association, *2692*
Minnesota* v. *Clover Leaf Creamery Co., *4146*
Minoan civilization, *4353*
minority groups, *3412*
Minot Light, *4181*
"misery whip", *2656*
mission blue butterflies, *2337*
Mississippi River, *1810, 2161, 2197, 2457, 2524, 2531, 2537, 2551, 2567, 2643, 3003*
Mississippi Sandhill Crane National Wildlife Refuge, *2331*
Mississippi sandhill cranes, *2331*
Missouri River, *4683*
MIT Museum, *4944*
Mitchell Act, *2495*
Mitre Corporation, *1841*
Mitsui Mining and Smelting, *2974*
Mkomazi Game Reserve, *4761*
Mkomazi Rhino Sanctuary, *4760*
Model T Ford, *1591, 4326*
Mojave Desert, *3813*
mollusks, *4467*
***Monahan* (warship)**, *4254*
monarch butterflies, *4729*
***Monitor* (warship)**, *4095*
Monitor United States National Marine Sanctuary, *3590*
The Monkey Wrench Gang, *1075*
monkeys, *4753, 4876*
monkeywrenching, *1083, 1084*
Mononychus, *1742*
Monsanto, *2849*
monsoons, *2564*
Mont-St.-Hilaire Biosphere Reserve, *1631*
Montana Wildlife, *4508*
Montreal 2000 Electric Vehicles Project, *1616*
Montreal Botanical Gardens, *1794*
Montreal Hunt Club, *2965*

Montreal Protocol on Substances That Deplete the Ozone Layer, *1977, 1987, 1989, 1990, 3043*
London Amendment, *1994*
Moon, *2229, 2825*
moose hunting, *2941*
Morrill Act, *1223*
Moscow Agreement, *1397*
Mother Earth, Journal of the Soil Association, *4067*
motorwagen, *1589*
Mount Baker, *3757*
Mount Cotopaxi (volcano), *4362*
Mount Desert Island, *3791*
Mount Elias, *3784*
Mount Erebus (volcano), *3870, 4369*
Mount Etna (volcano), *4354*
Mount Everest, *3994*
Mount Fuji (volcano), *4349*
Mount Garibaldi (volcano), *4350*
Mount Ilopango (volcano), *4358*
Mount Kelut (volcano), *4360*
Mount Kenya, *4630*
Mount Kilauea (volcano), *4377, 4385, 4390*
Mount Kilimanjaro, *4630*
Mount Llaima (volcano), *4351*
Mount Mazama (volcano), *4352*
Mount McKinley National Park, *3795*
Mount Pelee (volcano), *4375*
Mount Rainier, *3771, 3783*
Mount Rainier National Park, *3721, 3763, 3771, 3785*
Mount Rushmore National Park, *3175, 3775*
Mount Saint Helens (volcano), *4388, 4389*
Mount Tambora (volcano), *1318, 1814, 4366*
Mount Taupo (volcano), *4357*
Mount Terror (volcano), *4369*
Mount Vesuvius (volcano), *4356, 4379*
Mount Washington, *1913*
Mount Whitney, *3769*
mountain goats, *2947*
Mountain Iron Company, *3318*
mountain laurel, *2326*
mountain lions, *2347, 4797*
Mountain States Legal Foundation, *2440*
The Mountains of California, *3248*
mousetraps, *2901*
Muir Glacier, *1026, 2209, 3776*
Muir Woods National Monument, *1026, 3747*
Multilateral Fund, *1989*
Multiple-Use Strategy Conference, *1024*
Multiple-Use–Sustained Yield Act of 1960, *2685*
municipal boards of health, *3909*
Murchison Mountains, *1667*
***Murex* (tanker),** *3613*
"Murk", *1457*
Musée Océanographique de Monaco, *4916*
museums
See
"Zoos, Aquariums, and Museums" heading in main text and names of specific museums
mussels, *2504, 4418, 4493, 4515*
Sampson's pearly mussel, *2380*
zebra mussel, *1528*
***Mutsu* (merchant ship),** *3506*
mycology, *1789*
mynah birds, *1270*
Rothschild's mynah, *2433*
myxomatosis, *1513*

N

Nahanni National Park, *3991*
Nanhaizi Zoo, *2432*
Nanisivik lead-zinc mine, *3335*
Nantahala National Forest, *3988*
Natchez Trace National Scenic Trail, *4002*
Natchez Trace Parkway, *4015*
Nathaniel Mountain, *4707*
National Academy of Sciences, *1975*
National Aeronautics and Space Administration (NASA), *1833, 1845, 1958, 1961, 2381, 3496, 4113, 4116, 4118, 4119, 4123*
Mission to Planet Earth, *1992*
National Air Pollution Symposium, *1451*
National Aquaculture Board, *2497*
National Aquarium, *4895*
National Aquarium Society, *4895*
National Association of Physicians for the Environment, *3969*
National Atmospheric Deposition Program, *1473*
National Atomic Museum, *3510*
National Audubon Society, *1081, 1714, 4866*
National Avian–Wind Power Planning Meeting, *4866*
National Biodiversity Institute (INBio), *1639*
National Biological Survey, *1797, 3413*
National Bison Range, *4811, 4812, 4813*
National Cancer Institute, *2744*
National Center for Atmospheric Research, *1834*
National City Planning Association, *3103*
National Climate Data Center, *2154*
National Coastal Ecosystems Team, *3383*

National Commission for the Knowledge and Use of Biodiversity (CONABIO), *1641*
National Council on Marine Resources and Engineering Development, *3581*
National Dam Inspection Act, *2072*
National Defense Education Act, *4969*
National Drought Mitigation Center, *2165*
National Drought Policy, *2164*
National Eagle Repository, *4742, 4794*
National Earthquake Information Center, *2228*
National Emission Standards for Hazardous Air Pollutants, *2997*
National Environment Tribunal Act (India), *2999*
National Environmental Action Plan (Madagascar), *3395*
National Environmental Management Act (South Africa), *3425*
National Environmental Policy Act, *1062, 3086*
National Environmental Protection Act (NEPA), *3414, 3508*
National Environmental Supercomputing Center (NESC), *2306*
National Estuary Program (NEP), *3593, 3595*
National Farmers Memorial, *1135*
National Fish and Wildlife Forensics Laboratory, *4787*
National Fish and Wildlife Foundation Establishment Act, *4736*
National Fisheries Center and Aquarium, *2462*
National Fisheries Center and Aquarium Act, *2462*
National Fishing Week, *2511*
National Flash Flood Program Development Plan, *2558*
National Forest System, *3734*
National Forum on BioDiversity, *1636*
National Gas Policy Act, *2063*
National Gas Wellhead Decontrol Act, *2063*
***National Geographic*,** *4968*
National Geographic Society, *3602*
National Health Service, *3935*
National Highway Traffic Safety Administration, *1584*
National Historic Preservation Act, *3111*
National Human Genome Research Institute, *2763*
National Hunting and Fishing Day, *2927*
National Hurricane Center, *1954*
National Ideals Foundation, *1135*
National Institute for Medical Research (England), *2843*
National Institute for Occupational Safety and Health, *1359*
National Institute of Environmental Health Sciences (NIEHS), *3942, 3969*
National Institutes of Health, *1255, 2763, 3969, 4423*
  Division of Environmental Health Sciences, *3942*
National Inventory of Dams, *2077*
National Irrigation Association, *1031*
National Key Deer Refuge, *2327*
National Marine Fisheries Service (NMFS), *2402, 2464, 2467, 2482, 2483, 2485, 4763, 4791, 4902*
National Marine Sanctuaries, *3589, 3589, 3590, 3602*
National Museum of Natural History, *4947*
National Observatory, *3551*
National Oceanic and Atmospheric Administration (NOAA), *1530, 1833, 1988, 2015, 2154, 2166, 2289, 3549, 3597, 3602*
National Park Inventory and Monitoring Program, *3407*
National Park Service, *2126, 2456, 3248, 3405, 3409, 3413, 3430, 3711, 3733, 3794, 3976, 3981, 3982, 3998, 4620, 4660, 4670*
National Park Service Organic Act, *2456, 3790, 3976*
National Park Service Water Resources Division, *3406*
National Park Service's Geologic Resources Division, *3424*
National Park System, *3379, 3732, 3795, 4827*
national parks
  biogeographic inventory sites in, *3399*
  international treaty on, *3022*
  inventory of geologic resources in, *3424*
  inventory of resources in, *3402*
  inventory of vegetation in, *3408*
  inventory of water resources in, *3406*
  maps of resources in, *3405*
  strategy for resource management in, *3397*
    See also
    "Parks—National Parks" heading in main text and names of specific national parks
National Parks Act (Canada), *3735, 3782*
National Parks Association, *1035, 3718*
National Parks Conservation Association, *1035, 2347, 3718*
National Physical Laboratory, *1938*
National Population Policy for Development, Progress, and Welfare (Sierra Leone), *3894*
National Priorities List (NPL), *2868*

National Program of Inspection of Dams, *2072*
National Recovery Act, *2635*
National Renewable Energy Laboratory, *4866*
National Resource Bibliography, *3407*
National Resources Defense Council, *1305*
National Resources Inventory, *3385*
National Rifle Association, *2924*
National Road, *1552*
National Science and Technology Council, *1847*
National Science Foundation, *1988, 2283, 2302, 2824, 4347*
National Science Foundation Act, *2283*
National Severe Storms Forecast Center, *4186*
National Severe Storms Laboratory, *4294, 4297*
National Survey of Fishing, Hunting, and Wildlife-Associated Recreation, *2508, 2921, 4720*
National System of Interstate and Defense Highways, *1565*
National Threatened Species Day (Tasmania), *2350*
National Traffic and Motor Vehicle Safety Act of 1966, *1542, 1575, 3940*
National Trails System Act, *3982*
National Trust for Historic Preservation, *3109*
National Water-Use Information Program, *4602, 4608*
National Weather Service, *1915, 1916, 2558, 4293*
National Wetlands Inventory (NWI), *3387*
National Wetlands Policy, *4486*
National Wetlands Research Center, *3383*
National Wild and Scenic Rivers Act, *3978, 4469*
National Wilderness Preservation System, *4647, 4653, 4662, 4665*
National Wildlife Federation, *4416*
National Wildlife Property Repository, *4794*
National Wildlife Refuge and Administration Act, *4828*
National Wildlife Refuge System, *1692, 3165*
National Zoological Park (U.S.), *4905, 4949, 4950, 4954*
Native Americans, *3133, 3368, 3382, 3391, 3411, 3412, 4151*
- Anasazi, *2125*
- and buffalo, *4839*
- and eagles, *4742*
- and settlers, *3078, 3129*
- environmental activism by, *4408*
- fishing rights, *2478*
- Hohokam, *1214, 4569*
- Hopi, *2020*
- hunting grounds, *2939*
- hunting rights, *2958, 4776, 4782*
- Inuit, *4319*
- Inupiat Eskimo, *3483*
- Makah, *4628*
- Norridgewock, *2903*
- Paiute, *4017*
- Plains tribes, *3162*
- Tulalip, *2500*

Natural and Political Observations on the Bills of Mortality, *3908*
Natural Bridges National Monument, *3742, 4120*
Natural Energy Laboratory of Hawaii (NELH), *2297, 4564*
Natural Energy Laboratory of Hawaii Authority (NELHA), *2297, 4306*
See also
"Coal and Natural Gas" heading in main text
Natural Gas Act of 1938, *2051*
Natural Gas Policy Act of 1978, *2058*
Natural History and Antiquities of Selborne, *3233*
Natural History Museum of Iceland, *2375*
natural resources, *1033, 3010*
See also
"Natural Resources" heading in main text
Natural Resources Conservation Service, *3372, 3415*
Natural Resources Defense Council, *2290, 3540*
Natural Resources Inventory & Analysis Institute (NRIAI), *3415*
Nature, *1257*
Nature Conservancy (Great Britain), *1053*
Nature Conservancy (U.S.), *1038, 3413*
nature reserves, *3796*
nature trails, *4920*
Nautilus **(nuclear-powered submarine),** *3466, 3480*
Navajo Generating Station, *1419*
NECAR 3, *4343*
Neem (natural pesticide), *1306*
nematodes, *2607*
Neptune, *4394*
neptunism, *2801*
Nereis Boreali-Americana, *3555*
Nevada del Ruiz (volcano), *2576, 4391*
Nevada Test Site (NTS), *3462*
Nevada Wild Horse Range, *4801*
New Belgium Brewing Company, *4869*
New Croton Dam, *2092*

New Deal, *1225, 2635, 2683, 3929, 4035, 4053*
New England Aquarium, *4941, 4942*
New England Botanical Club, *1788*
New Jersey Department of Environmental Protection, *3440*
New Jersey Major Hazardous Waste Facilities Siting Act, *2866*
New Jersey Office of Sustainability, *4321*
New Jersey Pinelands, *3732*
New River, *4443*
A New System of Minerology**,** *2794*
New Theoretical and Practical Treatise on Navigation**,** *3237*
New York Aquarium, *4909*
New York Association for the Protection of Game, *1027, 2967*
New York Botanical Gardens, *1790*
New York City Bureau of Noise Abatement, *3428*
New York Department of Environmental Conservation, *2861*
New York Harbor, *1808, 3104*
New York Museum of Science and Industry, *4927*
New York Sportsmen's Club, *2967*
New York State College of Forestry, *2678*
New York State Forest Preserve, *2711, 3715, 4631*
New York Zoological Park, *4699*
New Zealand Values Party, *1107*
Newlands Reclamation Act, *1031*
Niagara Falls, *2807, 4555*
Niagara Mohawk Power Corporation, *3472*
Niagara River, *4555*
nicotine, *1370, 3970*
Niger, River, *2525*
Night Safari, *4965*
Nile River, *2007*
Nile River Valley, *1231, 1240*
NIMBY (not in my backyard), *1100*
Nimrod **expedition,** *3848*
"1934 National Erosion Reconnaissance Survey", *3371*
Nissan Motor Company, *1538, 1546*
nitrophenols, *1286*
"no net loss" theory, *4486*
Nobel Prize, *1972, 2268, 2753, 2816*
Noise Abatement Commission (New York, N.Y.), *3427*
Noise Act (England), *3448*
Noise Control Act of 1971 (New Jersey), *3440*
Noise Control Act of 1972, *3429, 3441, 3442, 3443, 3445, 3451*
noise pollution
See
"Noise Pollution" heading in main text
noise suppression standards, *3437*
Nongame Wildlife Fund, *4784*
Nonindigenous Aquatic Nuisance Prevention and Control Act, *1530*
nonpoint sources of pollution, *4482*
nonviolent protests, *1040, 1041, 1094, 1103*
Nordstrand Island, *4230*
Norek Dam, *2108*
Norman Wells Pipeline, *3625*
Norris-Doxey Farm Forestry Act, *2636, 2728*
North American Amphibian Monitoring Program, *4738*
North American Conservation Conference, *1034, 3367*
North American Free Trade Agreement (NAFTA), *2348, 3971*
North American Wetlands Conservation Act, *1691, 3090, 4840*
North American Wildlife and Natural Resources Conference, *4701*
North Carolina Biotechnology Center in Research Triangle Park, *2758*
North Carolina State University, *3726*
North Country National Scenic Trail, *3998*
North Pacific Marine Science Organization (PICES), *3063, 3596*
North Pole, *1709, 3823, 3849, 3864, 3871*
North Sea Flood, *2544*
Northeast Passage, *3836*
Northeastern Interstate Forest Fire Protection Compact, *2621*
Northwest Passage, *3550*
Noumea Convention, *4497*
NRX reactor, *3459*
nuclear accidents, *1326, 3459, 3479, 3480, 3482, 3494, 3495, 3499, 3503, 3504, 3507, 3515, 3518, 3524, 3528, 3531, 3538, 3580, 3585*
nuclear power
See
"Nuclear Power" heading in main text
nuclear propulsion, *3487, 3506*
Nuclear Regulatory Commission, *3455, 3949*
Nuclear Test Ban Treaty, *1397*
nuclear testing, *1397*
nuclear transfer cloning, *2280*
nuclear waste, *3533*
Nuclear Waste Fund, *3522*
Nuclear Waste Policy Act (NWPA), *3522*
Nuclear Waste Policy Amendments Act, *3530*

nuclear weapons, *1342, 3462, 3490, 3492, 3502, 3509, 3517, 3525, 3542*
testing, *1323, 3484*
numerical modeling, *1931*
Nyiragongo (volcano), *4386*

O

oak trees, *2609*
observatories, *1813, 1839, 1890, 1906, 1910, 1913, 2143, 2207, 3551, 4382*
satellite, *1825, 2822*
Occidental Petroleum Company, *3642*
Occupational Safety and Health Act, *3943*
Ocean Dumping Act, *4479*
Ocean Ranger (oil rig), *3641*
Ocean Thermal Energy Conversion (OTEC), *4306, 4549, 4564, 4565*
Ocean Thermal Energy Conversion Act, *4565*
oceans and oceanography, *2269, 3010, 3237, 4488, 4916*
See also
"Oceans and Oceanography" heading in main text
Office of Aquaculture Coordination & Development, *2497*
Office of Atomic Energy (South Korea), *3486*
Office of Civilian Radioactive Waste Management, *3522*
Office of Coast Survey (OCS), *3549*
Office of Noise Abatement and Control, *3429, 3451*
Office of Public Roads, *1556*
Office of the Nuclear Waste Negotiator, *3530*
Office of Underground Storage Tanks (OUST), *2872*
offshore waters, national claim to, *3547*
Ogden Bay Waterfowl Management Area, *1722*
***OGO 1* (satellite),** *2822*
Ohio and Erie Canal, *4019*
Ohio Division of Recycling and Litter Prevention, *3288*
Ohio River, *1820, 2531, 2540, 4406*
Ohio River Valley, *4457*
oil and petroleum industry, *3577*
legislation and regulation, *2872*
See also
"Oil and Petroleum Industry" heading in main text
oil cake, *1159*
Oil Pollution Act of 1990, *3627, 3643*
Oil Pollution Control Act, *4455*
oilbird, *1647*
Okefenokee Wilderness, *4652*
***Old Farmer's Almanac*,** *1878, 3234*
"The Old Swimmin' Hole", *4471*
Olds Motor Vehicle Company, *1534, 1590*
Olin Corporation, *4426*
olives, *1197*
Olympic National Park, *3404, 3758*
Omnibus Reconciliation Act, *3389*
***On the Burning Mirror*,** *4078*
***On the Germination of the Spore of Agaricineae*,** *1789*
***On the Law of War and Peace*,** *3009*
***On the Modifications of Clouds*,** *1880*
***On the Ocean*,** *3294*
***On the Structure of Moving Cyclones*,** *4248*
oncomouse, *2762*
***100,000,000 Guinea Pigs*,** *3251*
open-air thermometers, *1860*
Operation Amazonia, *4299*
Operation Oryx, *2431*
Oraefajokull, *4359*
orangutans, *4922*
Ordinance of Waters and Forests, *2626*
Ordovician extinction, *2354*
Oregon Trail, *3762*
Oregon Zoo, *4957*
Organic Act of 1897, *2632, 2676*
organic certification, *1264*
organic farming, *4067*
See also
"Agriculture and Horticulture—Organic Farming" heading in main text
***Organic Gardening and Farming*,** *1261*
Organization for Economic Cooperation and Development, *2880*
Organization of African Unity (OAU), *3899*
Organization of Petroleum Exporting Countries (OPEC), *3622*
organohalogen, *4490*
***Orientation of Buildings or Planning for Sunlight*,** *4102*
***The Origin of Continents and Oceans*,** *2817*
ornithological illustration, *3160*
ornithology, *1711*
***The Ornithology of Francis Willughby*,** *3230*
Osirak Nuclear Reactor, *3519*
Oslo Convention, *4494*
osmosis, *1777*
Ossabaw Island, *4834*
ostrich farming, *1512*
ostriches, *1648*
otters, *4681, 4776*

Our Common Future*, *3262
Our Wandering Continents*, *2819
Out of the Earth*, *1131
outdoor museum, *4920*
Outer Space Treaty, *3027*
owls, *1685*
Mexican spotted owl, *2418*
Oxford Physic Garden, *1756*
Oxford University, *1813*
oysters, *2488, 2489, 2501, 2504, 4396, 4418*
Ozark National Scenic Riverways, *3805, 3978*
ozone layer, *1339, 1340, 1480, 1830, 3041, 3043*
See also
"Climate and Weather—Ozone Layer" heading in main text
Ozone Secretariat, *1991*
Ozone Watch*, *1993

P

Pacific Crest Scenic Trail, *3981*
Pacific Lumber Company, *2666*
Pacific salmon, *4951*
Pacific Science Center, *2286*
pack ice, *1805*
paddy fields, *1198*
paint
lead-based, *1358, 3944, 3955*
tin-based, *2504*
Paleoclimatology Program, *2015*
Paleozoic Era, *2000*
Palisades Interstate Park Commission, *4920*
Palmer Drought Severity Index, *2154*
Pampers, *4537*
Pan American Sanitary Bureau, *3925*
pandas, *2957, 4752, 4757*
pangolins, *4823*
paper
recycling, *1851, 4156, 4158, 4162, 4166, 4170*
use reduction, *4155*
Papnadayan (volcano), *4364*
parabolic trough solar reflectors, *4095*
paradise birds, *2369*
Paradise in a Sea of Sorrows*, *2972
paraffin oil, *3678*
paralytic shellfish poisoning (PSP), *4418*
Paris green, *1274, 1276*
Paris Peace Treaty of 1783, *3014*
parks, *3589, 3590, 4745*
See also
"Parks" heading in main text and names of specific parks
***Parks* magazine,** *3810*
parrots, *1717*
Carolina parakeet, *2367*
Puerto Rican parrot, *2391*
Partnership for a New Generation of Vehicles, *1606*
Partnerships for Wildlife Act, *4840*
passenger pigeon, *2368*
Passive Seismic Experiment, *2229*
patents, *2757*
Patrick J. Buchanan . . . From the Right*, *1025
Patuxent Research Refuge, *4819*
PCBs (polychlorinated biphenyls), *2829, 2843, 2847, 2849, 2851, 2856, 2857, 2977, 4430, 4512*
PCV valve, *1602*
Peace with God the Creator, Peace with All of Creation*, *3264
peacocks, *1697*
Peale's Museum, *4883, 4884, 4885, 4886*
Pearl River, *2559*
pebble-bed reactors, *3543*
pelagic birds, *1705*
***Pelicano* (ship),** *2876*
pelicans, *2412*
Penal Code of 1860 (India), *4578*
penguins, *1735*
penicillin, *3932*
Pennsylvania Hospital, *3910*
Pennsylvania Rock Oil Company, *3609*
Pennsylvania Scenic Rivers Act, *4001*
Pennsylvania School of Conservation, *2303*
Pennsylvania Turnpike, *1563*
Penrose Research Laboratory, *4910*
People for the Ethical Treatment of Animals (PETA), *1016, 1022*
perch, *4474*
Peregrine Fund, *2446*
Permian extinction, *2354*
Permian Period, *2000*
Perry Power Plant, *3527*
pesticides, *1290, 4779*
"Petaluma Plan", *3157*
Peter Norbeck Scenic Byway, *4011*
petrels, *1682*
Bermuda petrel, *1657, 2319*
Petroleos Mexicanos, *3699*
pH scale, *1325, 1444*
phalaropes, *1704*
pheasants, *1674*
imperial, *1659*
ring-neck, *1514*
silver, *4823*

Philadelphia Zoological Garden (Philadelphia Zoo), *4892, 4896, 4898, 4899, 4903, 4910, 4921, 4922, 4923, 4925, 4926, 4930, 4931, 4933, 4938, 4939, 4954*
Phillips Petroleum Company, *3661, 3662, 3707*
Philosophical Transactions, *1872*
Philosophico-Choreographical Chart of East Kent, *2782*
Philosophy of Storms, *4242*
phosphates, *4404, 4405, 4407*
photography, *1825, 3265, 3266, 3568, 4698*
Phylloxera vastatrix, *1271*
The Physical Geography of the Sea, *3556, 3557*
"Physicians and the Environment" (conference), *3969*
Phytologia Britannica, *1757*
Phytophthora, *2609*
Pictured Rocks National Lakeshore, *3729*
pigeons, *1643*
passenger pigeon, *1656*
pigs, *1140, 4443*
Piper Alpha (oil rig), *3642*
Piscataway Park, *3820*
Pitch-In Canada, *3274*
Pittman-Robertson Act, *1664, 1671, 1677, 1695, 1722, 1725, 4702, 4703, 4704, 4706, 4709, 4710, 4711, 4712, 4714, 4717, 4721, 4820, 4821*
Pittsburgh Zoological Society, *4945*
Planet Drum, *1068*
planetariums, *4969*
plants
classification, *1767, 3413*
endangered, *2320, 2321, 2322, 2326, 2336, 2403, 2415, 2424*
genetically altered, *2760*
international horticultural fair, *2316*
National Herbarium, *1787*
shipments across borders, *3023*
plastics, *3291, 4125, 4125, 4146, 4148*
plate tectonics theory, *4380*
platypuses, *4699, 4750*
Pleasant Valley, *1131*
Pleistocene Epoch, *4348*
Plitvice Lakes National Park (Croatia), *3996*
plows
cast-iron, *1237*
sulky plow, *1242*
Plualuang National Forest Reserve (Thailand), *1623*
plume trade, *1648, 4766, 4809*
pluviculture, *2620*
See also
"Droughts and Water Shortages—Pluviculture" heading in main text
Plymouth Plantation, *2167, 2625*
poaching, *2395, 2398, 2422, 2433, 2911, 3390, 4788*
Poco das Antas Biological Reserve, *4832*
Poij Bara (Skull Famine), *2132*
Pok Fu Lam Reservoir, *4619*
Polar Dreams, *3872*
polio, *3937, 3941*
Polish Ecological Club, *2979*
pollen, *1749, 2011*
pollination, *1766*
Pollution of the Sea by Oil Convention, *3577*
Pollution Prevention Act, *1417, 2989, 2990*
Pollution-Related Health Injuries Compensation Law, *3947*
polybrominated biphenyl (PBB), *2850*
Ponnequin Wind Farm, *4867*
Poor Richard's Almanac, *1871*
population growth
See
"Population Growth" heading in main text
porcupines, *2595*
Port Authority of New York and New Jersey, *3104*
Port of London Authority, *3104*
Port of New York Authority, *3104*
Portage Glacier Recreation Area, *4023*
postage stamps, *2922, 3161*
potatoes, *1121, 1123, 1203, 1205*
blight, *1183*
introduced in Europe, *1205*
Potomac Heritage National Scenic Trail, *4004*
Potomac River, *3766, 3820, 4400, 4533, 4536*
Power Demonstration Reactor Program, *3470*
Power Reactor and Nuclear Fuel Development Corporation, *3538*
Praecipuarum Quarum apud Plinim et Aristotelem Mentio est Brevis & Succincta, *3228*
prairie dogs, *4683, 4805*
Mexican prairie dog, *2397*
Utah prairie dog, *2397*
Prairie Farm Rehabilitation Administration (Canada), *2102, 2151*
prairies, *2151, 3410*
alien introductions, *1497*
restoration, *3076, 3749*
Pre-emption Act of 1830, *3138*
Precambrian rocks, *2808*

Prescott National Forest, *4006*
President's Council on Environmental Quality, *1841, 4314*
President's Council on Sustainable Development, *4314*
Prevention and Control Exchange (PACE), *3431*
primates
See
names of specific types of primates
"primitive area", *2714*
"primitive forest", *2710*
Princeton University, *1949*
Principle of Original Horizontality, *2779*
Principles of Geology**,** *2801*
Private Ownership of Special Nuclear Materials Act, *3501*
Proctor & Gamble Company, *4129, 4537*
Prodromus to a Dissertation Concerning a Solid Naturally Enclosed within a Solid**,** *2779*
Project Chariot, *3483*
Project Cirrus, *2620*
Project Tiger, *2241*
Protection of Wild Mammals (Scotland) Bill, *2963*
Protoavis**,** *1681*
Protocol on Environmental Protection to the Antarctic Treaty, *3044, 3046*
Protocol on Substances That Deplete the Ozone Layer, *1986*
protoplasm, *1773*
Prototype Fast Reactor, *3485*
Pryor Mountain Wild Horse Range, *4802, 4829*
PsittaScene**,** *1717*
Public Health Act of 1848 (Great Britain), *3915, 3917*
Public Health Act of 1875 (England), *1384*
Public Health Act of 1936 (Great Britain), *2834*
public health and safety, *1173, 1227, 1357, 1359, 1364, 1366, 1368, 1369, 1403, 1416, 1427, 1449, 2049, 2974, 2982, 3319, 3324, 3444, 3542, 4419, 4428*
See also
"Public Health and Safety" heading in main text
Public Health Congress (England), *1439*
Public Health Service Act, *4452*
Public Land Review Commission, *3081*
public lands, U.S., *3139*
Public Lands Division, *3368*
Public Liability Insurance Act (India), *2991*
Public Utility Regulatory Policies Act, *4115, 4856*
Purdue Industrial Wastes Conference, *2838*
Pure Food and Drug Act of 1906, *1281, 3249, 3250*
Pure Food Campaign, *2764*
purse-seine nets, *2479*
Pyramid Lake Scenic Byway, *4017*

Q

Quabbin Reservoir, *2101*
quagga, *2365*
quail, *2935*
quails
bobwhite quail, *1736*
common quail, *1520*
quarantine programs, *1277, 1282, 1283, 3913, 4581*
Queensland Weather Bureau, *4245*
Quiet Communities Bill, *3451*
quinine, *1507*

R

rabbits, *1150*
Radcliffe Observatory, *1813*
radiation "hot spot", *1346*
radio waves, *1963*
radioactive air pollution, *1322, 1323, 1326, 1342, 3528*
Radioactive Substances Act (United Kingdom), *3491*
radioactivity, *2268*
Radiological Emergency Protection Plan (Ireland), *3531*
radionuclide emissions, *2997*
"radiosonde" sensors, *1940*
radon, *1352, 1367, 4610*
Radon Gas and Indoor Air Quality Research Act, *1352*
railroads, *2654, 2830, 3139, 3766*
railways (urban), *3072*
rain, *1866, 1959*
Rainforest, Inc., *2246*
Rainforest Action Network, *2720*
Raise the Stakes**,** *1068*
Ramsar Convention, *1744, 3016, 3028, 3150, 3418*
Rance River, *2104, 4562*
Rancho Seco Nuclear Power Plant, *3532*
Rara Avis, *2246*

Rayleigh Wave, *2211*
reapers, *1238, 1239*
Reclamation Act, *1220, 4553*
Reclamation Service, *4553*
Recreational Hunting Safety and Preservation Act, *2961*
recycling, *1851, 2284, 2295*
See also
"Solid Waste—Recycling" heading in main text
Recycling Economic Development Advocate project, *4159, 4160, 4161, 4162, 4163, 4164, 4165, 4166, 4167, 4168, 4169, 4170*
Red Cross, *2531*
"Red List" of threatened and endangered species, *3058*
Red List of Threatened Plants, *2421*
Red River, *2563*
Redoubt (volcano), *4393*
Redwood Summer of 1990, *1088*
Redwoods National Forest, *2698*
Redwoods National Park Act, *2638*
Refiners Oil Company, *1540*
reforestation, *2536, 2634*
See also
"Forests and Trees—Reforestation and Planting" heading in main text
Reforestation and Relief Bill, *2683*
Reform Club, *2043*
Refuge Recreation Act, *4825*
Refuse Act of 1886, *4448*
Refuse Act of 1899, *4472, 4473*
regenerative farming, *1254, 4027*
Regent's Park Zoological Garden, *4890*
Regional Biomass Energy Program, *4332*
Regional Directorate of Industry and Research (France), *3953*
"Reintroduction to the Wild as an Option for Managing Navy Marine Mammals", *4841*
ReLeaf, *2731*
Relief Coordination Branch, *3002*
Rensselaer Polytechnic Institute, *2261, 2262, 2309*
Reo Motor Car Company, *1590*
Reorganization Plan No. 3, *4779*
Reorganization Plan No. 4, *2477*
Report on an Inquiry into the Sanitary Conditions of the Labouring Population of Great Britain, *3915, 3916*
Report on the Arid Regions of the United States, *2138*
Report on the Chemical Analysis of Grapes, *3243*
Report on the Exploration of the Colorado River and Its Tributaries, *3245*
A Report on the Sandstone of the Connecticut Valley Especially Its Fossil Foot Marks, *2803*
reptiles, *4954*
Research Grants Act, *2288*
Research Triangle Institute, *3726*
research vessels
Alvin, 3580
Atlantis, 3571
Glomar Challenger, 2824
Resolution of 1993 on Policy Concerning the Population and Family Planning Work (Vietnam), *3892*
Resource Assessment Program, *4339*
Resource Conservation and Recovery Act (RCRA), *2858, 2889, 2983, 2998, 4131, 4132, 4143, 4144*
Resource Contaminant Assessment Program, *4734*
Restatement of Torts laws, *2836*
Reston Chlorofluorocarbon Laboratory, *4611*
Revenue Act of 1932, *1571*
Rhine River, *1297, 2546*
Rhinoceros and Tiger Conservation Act, *2346*
Rhinoceros and Tiger Product Labeling Act, *2346*
rhinoceroses, *2346, 2398, 4755*
great Indian, *4898*
Javan rhinoceros, *4814*
Rhodora, *1788*
rice, *1253*
Richfield Oil Company, *3703*
Richter-Gutenberg scale, *2219*
Richter scale, *2219, 2223*
***Riesen durch Sibirien* (Journey through Siberia),** *1806*
Ringer Corporation, *1306*
Rio Grande, *3983, 3993*
Rio Platano Biosphere Reserve, *1633*
river rafting, *2238*
rivers, *2314, 2539, 2542, 2550, 3268, 3273, 4403, 4405, 4436, 4449, 4470, 4609*
American Heritage, *3423*
of New England, *2466*
stocked with trout, *2513*
wild and scenic, *3731, 3978, 3983, 3986, 3988, 3992, 3993, 3999, 4001, 4006, 4009, 4454, 4469*
Rivers, Trails and Conservation Assistance Program, *3733*
Rivers and Harbors Appropriations Act of 1899, *4141, 4449, 4450*
Road to Survival, *3255*

Rockefeller farm, *1136*
"rockoon", *1951*
Rocky Mountain grasshoppers, *1186*
Rocky Mountain National Arsenal, *2842, 2852*
Rocky Mountain National Park, *3768, 3774, 4014*
Rocky Mountains, *3239*
Rocky Mountains Park (Canada), *3782*
Rocky River hydroelectric station, *4558*
Rodale Institute, *4027*
Rodgers **(rescue vessel),** *4626*
Roger Williams Park Zoo, *4906*
roller-compacted concrete, *2109*
Rolls-Royce Motor Cars, *1536*
Rose Elementary School, *4106*
Ross Leffler School of Conservation, *2278*
Royal Geographical Society, *1851*
Royal Meteorological Society, *1448, 1899*
Royal Observatory, *1890*
Royal Society of London, *1759, 1796*
rubber, *1180*
Ruhr River, *4453*
Russell Cave National Monument, *3738*
Rydberg milk vetch, *2415*

S

Saami Council, *4319*
sabal palm, *2748*
sabotage, *1058, 1075, 1083, 1084*
Safaniya oil deposits, *3654*
Safe Drinking Water Act Amendment of 1986, *4606*
Safe Drinking Water Act of 1974, *4424, 4425, 4477, 4478, 4514, 4598, 4599, 4615*
Safe Drinking Water Reauthorization Act of 1996, *4610, 4614, 4615*
Saffir-Simpson Hurricane Damage Potential Scale, *4263*
Sagarmatha National Park (Nepal), *3994*
Sage Fellowship in Entomology and Botany, *2266*
Saginaw River, *4412*
Sahel region, *4044*
Saint Cuthbert's doves, *4808*
Saint Elmo's fire, *4195, 4275*
Saint Francis Dam, *2116*
Saint Johns National Wildlife Refuge, *2328*
Saint Lawrence River, *3712, 4016, 4511*
Saint Lawrence River Basin, *4515*
Saint Mary Dam, *2102*
Saint Mary Hydroelectric Plant, *2102*
Saint Mary River, *2102*
Sakaerat Biosphere Reserve (Thailand), *1623*
Salem Maritime National Historic Site, *3801*
Saline Bayou, *4007*
Saline Water Act, *4459*
saline water conversion, *4459, 4462*
salinization, *1212*
salmon, *2080, 2476, 2478, 2490, 2495, 2504, 2509, 4951*
  Atlantic salmon, *2466, 2492*
  chum salmon, *2500*
  king salmon, depicted, *2922*
  sockeye salmon, *2417, 2459*
salmon ladder, *4951*
Salt River, *4556*
San Bernardino National Forest, *2606*
San Diego Marine Biological Station, *2269*
San Francisco silverspot butterflies, *2337*
San Francisco water project, *2090*
San Francisquito Canyon, *2116*
San Juan **(ship),** *3628*
San Juan National Forest, *4014*
San Juan Skyway, *4014*
A Sand County Almanac**,** *1054*
sand filtration, *4504*
Sandia Laboratories, *4853*
The Sanitation Manual for Public Groundwater Supplies**,** *2839*
Santa Ana wind, *4183*
Santa Maria (volcano), *4374*
Santa Ynez Mountains, *2115*
Saratoga Spring Water Company, *4580*
satellites, *1326, 1825, 1956, 1960, 1961, 1964, 1965, 1981, 2293, 2822, 4969*
  orbiting geophysical observatory, *2822*
Savannah **(nuclear-powered ship),** *3488*
Save the Elephant, *4724*
Save-the-Redwoods League, *2698*
scabland, *2515*
scale insects
  California red scale, *1275*
  cottony-cushion scale, *1184*
scenery, *4629, 4677*
  See also
  "Scenery" heading in main text
Scenes in the Rocky Mountains**,** *3239*
Scenic Byway program (Ohio), *4019*
"Scenic Hudson Case", *3737*
Scenic Hudson Preservation Conference v. Federal Power Commission (FPC)**,** *3979*
Schaus swallowtail butterflies, *2407*
Schelde River, *2546*
Schemnitz Mining Academy, *3303*

Schönbrunn, *4882*
Schuylkill Fishing Company, *2506*
scientific procedures
for classifying clouds, *1880*
for describing species, *1687*
screw propellers, *4095*
screwworms, *1293*
Scribner Log Rule, *2652*
Scripps Institution for Biological Research, *2269*
Scripps Institution of Oceanography, *1458, 1838, 1839, 1841, 2269, 3567*
Sea-Bed Arms Control Treaty, *3584*
***Sea Empress* (tanker),** *3649*
sea lamprey, *1501, 1526, 2461, 4502*
***Sea Star* (tanker),** *3636*
***Sea Venture* (ship),** *4229*
Sea World, *1678*
seal hunting, *2958*
sealed thermometers, *1862*
seals, *3827, 3829*
Antarctic seal, *3031*
fur seals, *1016, 4776*
harp seal, *2958*
Hawaiian monk seal, *2405*
northern sea lion, *4810*
sea lions, *4723*
Sears, Roebuck and Co., *4405*
seat belts, *1600*
Seattle Aquarium, *4951*
***Seattle Audubon Society v. Lyons*,** *3414*
Seattle World's Fair, *2286*
Seaway Trail Scenic Byway, *4016*
seaweed, *4532*
Second United Nations Conference for Human Settlements, *3901*
Secretariat of the Environment, Natural Resources, and Fisheries (Mexico), *2994*
sedentary agriculture, *1139*
sedentary ranching, *1157*
Seine River, *4522*
seismic map, *2203*
seismic waves, *2177*
Seismological Committees of the British Association for the Advancement of Science, *2213*
Seismological Society of America, *2216*
Seismological Society of Japan, *2205*
seismology, *2204, 2219, 2229, 2230*
selenium, *4026*
Sellafield nuclear power plant, *3475, 3479, 3524*
Seney National Wildlife Reserve, *2325*
Sentier National Trail, *4661*
Sequoia and Kings Canyon National Park, *3760, 3769*
***The Serengeti Lion*,** *4725*
Serengeti National Park, *4725, 4817*
Serengeti Plain, *4817*
Serengeti Research Institute, *4725*
Servicemen's Readjustment Act, *3084*
Sespe Condor Sanctuary, *1723*
Seven Cities of Cibola, *3748*
Severe Local Storms Forecasting Unit, *4186*
sewage, *4479*
Seward Highway, *4023*
sewers and human waste, *4410, 4446*
See also
"Water Pollution—Sewers and Human Waste" heading in main text
shad, *2491*
shaduf, *1231*
shale oil industry, *3678*
sharks, *2482*
shearwaters, *1702*
sheep, *1139, 1175*
bighorn, *2947*
cloned, *2769*
mountain, *2946*
Sheffield Dam, *2115*
Shell-BP, *3702*
Shell Oil Company, *3613*
Shell Singapore Company, *3704*
Shenandoah National Park, *3400, 3754*
Shide Circulars, *2213*
Shining Rock Wilderness, *4645*
shipbuilding, *2583, 2643, 2669*
shipping bulletins, *4185*
Shippingport Atomic Power Station, *3481, 3529*
Shipstead-Newton-Nolan Act, *4636*
shorebirds, *1683*
Shoshone National Forest, *2712, 3714*
shrikes, *1687*
Si a Paz Park, *1637*
"sick building syndrome", *1361*
Sierra Club, *1026, 1051, 1057, 1081, 1101, 3737*
Sikes Military Reservation Act, *4772*
***Silent Spring*,** *1288, 3256*
silver, *3343, 3348, 3352*
Simien National Park, *3990*
Sinclair Oil Corporation, *3701*
Singapore Zoological Gardens, *4965*
***The Sinking Ark*,** *1632*
Sisquac Condor Sanctuary, *1723*
Sizewell B reactor, *3534*
skipjack tuna, *2460*

SLAPP (Strategic Lawsuit Against Public Participation), *1077*
slavery, *2124*
"Sleep with the Whales", *4960*
smallpox, *3906, 3912*
Smith-Lever Act, *1266*
Smithsonian Institution, *1787, 1894, 1896, 1897, 1903, 1904, 2403, 3555, 4282, 4905, 4947, 4949, 4950, 4954, 4959*
Conservation and Research Center, *4793*
smog, *1345, 1439*
deaths related to, *1341*
Smoke Abatement Exhibition, *1431*
Smoke Nuisance Abatement Act (England), *1379*
smoke pollution, *1338, 1380, 1388*
Smoke Prevention Association, *1336*
Smokey Bear, *2618, 2619, 3179*
smoking, *1366, 1368, 1369, 1370, 1476*
restrictions, *1362*
snail darter, *2333, 2388*
snails
Iowa pleistocene snail, *2341*
Manus Island tree snail, *2399*
Snake River, *2417, 4675*
snakes, *2392*
snipe, *2943*
***Snow Crystals*,** *4203*
Soap & Detergent Association, *4507*
Social Security Act of 1935, *3929*
Societas Meteorologica Palatina, *1876*
Society for the Prevention of Cruelty to Animals, *1007*
Society for the Protection of Animals, *1006*
sockeye salmon, *2417, 2459*
Sockeye Salmon Fishery Act of 1947, *2459*
Sodium Reactor Experiment, *3477*
***The Soil and Health*,** *4066*
Soil Association (England), *1262*
soil conservation, *1262, 1263, 2536, 2628*
See also
"Soil Resources" heading in main text
Soil Conservation Service, *4055*
solar collectors, *4089*
solar energy, *1604, 4345*
See also
"Solar Energy" heading in main text
Solar Energy Coordination and Management Project, *2298*
Solar Energy Research, Development, and Demonstration Act, *2298*
Solar Energy Research Institute, *2298*
solar hot-air engines, *4095*
Solar Motor Company, *4098*
Solar Park, *4105*
solar-powered automobile, *4110*
solar siphon, *4081*
solar still, *4091*
solid waste, *2894, 3286*
See also
"Solid Waste" heading in main text
sonar, *3570*
Sotheby's Auction House, *3263*
South African National Center for Occupational Health, *3963*
South Central Interstate Forest Fire Protection Compact, *2622*
South Coast Air Quality Management District, *1427, 2065*
South Fork Dam, *2112, 2113*
South Pole, *3852*
Southern California Gas Company, *2065*
***Souvenirs entomologique*,** *4686*
space exploration, *3027*
space reactor, *3469*
Spanish-American War, *3923*
Spanish Armada, *4176*
sparrows
Cape Sable sparrow, *1658*
dusky seaside sparrow, *2328, 2377, 2381*
English sparrow, *1506, 1651*
Eurasian tree sparrow, *1653*
Lincoln's sparrow, *1654*
Santa Barbara song sparrow, *2379*
***Species Plantarum*,** *1764*
speciesism, *1019*
***Spence* (warship),** *4254*
Spindletop oil well, *3693*
spinning machines, water-powered, *4544*
sponge fishing, *2472*
spruce budworm, *2599, 2605*
***Sputnik 1* (satellite),** *4969*
SRI International, *1451*
Sri Lankan Department of Wildlife, *4758*
Standard State Soil Conservation District Laws, *4039*
Stanford Research Institute, *1451*
Stanford University, *3724*
Stapleton Development Corporation, *4315*
Stapleton International Airport, *4315*
starfish, *3579*
starlings, *1516, 2433*
Station for the Study of Insects, *4920*
"Status and Trends of the Nation's Biological Resources", *3426*
Stazione Zoologica, *3559*
steam engines, coal-burning, *2030*
steam power, *1433*
steam-powered automobiles, *1587*
steelhead trout, *2423*

Steinart Aquarium, *4919*
Steller's sea cow, *2358*
Stephen Birch Aquarium and Museum, *4963*
Stephen Mather Memorial Parkway, *4020*
Stiger gun, *4215*
"Stillman's Journal", *2798*
"stinkerds", *4444*
"The Storm King", *4242*
Storm King State Park, *3737*
storms, *1888*
See also
"Storms" heading in main text
Strait of Juan de Fuca, *3757*
Stratospheric Aerosol and Gas Experiment (SAGE), *1981*
street lighting, *2035*
***The Structure and Distribution of Coral Reefs*,** *3553*
***Studies in Glaciers*,** *2805*
sturgeon, *2422*
***Sturgis* (nuclear power barge),** *3498*
Styrofoam, *4128*
submarines, nuclear-powered, *3466, 3499*
suburban development, *3084*
suction pumps, *2027*
"sudden oak death", *2609*
sugar cane, *1200*
Sulawesi apes, *4899*
sulfur, *1268*
sulfur dioxide emissions, *1414*
sulky plows, *1242*
sulphur-crested cockatoo, *1679*
Sumter National Forest, *3988*
Sun Day, *4114*
SunRaycer, *4121*
***Sunrise II* (solar aircraft),** *4113*
Super-Phénix nuclear reactor, *3520*
Superfund, *2863, 2868, 2996, 3956*
Superfund Amendments and Reauthorization Act (SARA), *2985, 3964*
Superior National Forest, *4639*
Superior Primitive Area, *4636*
superphosphate, *4059*
Surface Mining Control and Reclamation Act, *3336, 3337*
Surface Water Improvement and Management Act (SWIM), *4483*
Survey of Fishing, Hunting & Wildlife—Associated Recreation, *4719*
surveying, *3132*
Susquehanna River, *2096*
sustainable development, *1622, 2640, 2668, 2685, 2715*
See also
"Sustainable Development" heading in main text
Sustainable Seas Expedition, *3602*
Sutter's Mill, *3311*
swallows, *1644*
Swan Hills Special Waste Treatment Center, *2871, 2877*
Swann Committee, *1173*
swans
black-necked swan, *1710*
trumpeter swan, *2324*
Swanson River, *3703*
Swedish Ministry of Agriculture, *1982*
Swedish Space Corporation, *1962*
Swift River Act of 1927 (Massachusetts), *2101*
swordfish, *2475, 2485, 4513*
Sybille Wildlife Research Unit, *4718*
Symposium on the Conservation of Nature and Natural Resources in Modern African States, *4773*
Syncor International Corporation, *2997*
synthetic fuel plant, *2060*
synthetic fuels
See
"Synthetic Fuels" heading in main text
***Systema Naturae*,** *1617*

T

Taejon International Exposition (Expo '93), *2317*
Tahoe Federal Interagency Partnership, *3421*
taiko, *1670*
Taiwan Environmental Protection Union, *3512*
takahe, *1667*
Taman Negara National Park, *4744*
tamarins, *4832*
tamarisk, *1503*
Tanner Act (California), *2875*
Tara River Basin, *1625*
Tasmanian tiger (thylacine), *2350*
Tata Conference, *1846*
Tax Reform Act of 1976, *3112*
taxidermy, *4693*
taxonomy, *1617, 1774*
Taylor Grazing Act, *1169, 1170, 4036*
Tbilisi Declaration, *2300*
tea cultivation, *1206*

teak, *2584, 2673*
TEAS (Turkish power authority), *3544*
Technical Report 1543, *4841*
Tecopa pupfish, *2378*
telegraph, *1897*
television, *2233*
Tellico Dam, *2333*
Tennessee Aquarium, *4961*
Tennessee Valley Authority, *1422, 2070, 2094*
Tennessee Wildlife Resources Agency, *4705*
Tera River Dam, *2117*
terrariums, *1779*
Teton Dam, *2073*
tetraethyl lead, *1540*
Texaco Oil Company, *3693*
Texas Game Department, *2916*
Texas Parks and Wildlife Department, *2455, 2916*
Texasgulf, Inc., *3335*
TFM, *2461*
Thai Forest Industry Organization, *4793*
Thames, River, *1802, 2509, 2561, 4520, 4525*
Thames Barrier, *2561*
Theodore Roosevelt Dam and Powerplant, *4556*
***Théorie élémentaire de la Botanique*,** *1774*
***Theory of the Earth*,** *2789*
thermokarts, *1564*
thermometers, *1860*
mercury thermometer, *1869*
sealed thermometer, *1862*
"Think globally, act locally", *1073*
"30 Percent Club", *1414*
Three Arch Rocks National Wildlife Refuge, *4810*
Three Mile Island, *3514, 3515*
***Thresher* (nuclear submarine),** *3499*
thrushes, *2360*
thunderstorms, *1478*
See also
"Storms—Thunderstorms" heading in main text
Tianezhou Natural Reserve, *2432*
Tiber River, *2519, 4568*
***Ticonderoga* (aircraft carrier),** *3503*
tidal power, *2104, 4562, 4563, 4566*
tide-predicting machines, *3564*
Tidy Britain Group, *3271*
Tierpark, *4914*
tigers, *2241, 2346, 4830*
Tigris River, *1212*
tilapia, *2476*
Timber Culture Act of 1873, *2725*
Timber Cutting Act of 1878, *2630*
Timpanogos National Monument, *3781*
tin cans, *4124*
tinamou, *1655*
***Tiros 1* (weather satellite),** *1825, 1956, 1960*
***Tiros 10* (weather satellite),** *1961*
Title III of the Superfund Amendments and Reauthorization Act of 1986 (SARA), *2984*
tobacco, *1124, 1125, 1130, 2760*
Toka No. 1, *3505*
Tokai Reprocessing Plant, *3538*
Tokyo Sea Life Park, *4900*
tomatoes, *1201*
Tonto National Forest, *4006*
tornadoes
See
"Storms—Tornadoes" heading in main text
***Torrey Canyon* (tanker),** *3630, 4489*
"Torricellian tube" (or "vacuum"), *1863*
torsion balances, *2791*
Total Ozone Mapping Spectrometer, *1979*
towhees, *1731*
town relocation, *3432*
Toxic Substances Control Act, *2857, 2977, 3951*
Toxics Release Inventory (TRI), *2984*
Toy Safety and Child Protection Act, *3439*
tractors, *1246, 2659*
trade winds, *1895*
Trail Ridge Road/Beaver Meadow Road, *4014*
Trans-Canada Trail, *4671*
***Transactions of the American Fisheries Society*,** *4510*
transatlantic crossing, *4334*
transboundary pollution, *1468, 1471, 1472*
transcendentalism, *1047*
Transfer of Certain Real Property for Wildlife Conservation Purposes Act, *4769*
transgenic corn, *2768*
***Transit 4A* (satellite),** *3496*
Transportation Equity Act for the 21st Century, *1585*
transuranic wastes, *3533*
Trappers Lake, *4633*
Treaty of Tordesillas, *3013*
"Treatyse of Fysshynge wyth an Angle", *2505*
tree farming, *2725*
tree rings, *1759*
Trent, River, *4456*
Triassic extinction, *2354*
tributylin (TBT), *4492*
tricalcium phosphate, *4059*

***Trieste* (bathyscaph),** *3576, 3578*
Trinity River, *2563*
Trinity Test, *3453*
Triton (moon of Jupiter), *4394*
tropopause, *1929, 1929*
troposphere, *1921*
trout, *2476, 2513*
brown trout, *2494*
steelhead trout, *2423*
Truckee River, *4017*
Tsavo National Park, *4755*
tsunamis, *2184, 2186, 2188, 2193, 2224, 2523, 2543, 2549, 4177, 4247, 4372*
Tulane Environmental Law Clinic, *3003*
tularemia, *3926*
Tule Elk Preservation Act, *4833*
tuna, *2475, 3060, 4513*
Atlantic tuna, *3060*
bluefin tuna, *2486*
skipjack tuna, *2460*
yellowfin tuna, *2460*
tuna industry, *2460, 2475, 2481, 2486*
Tuolomne Meadows, *3786*
Tuolumne River, *2090*
turkeys, *1671, 1739, 2935, 3160*
turkeys, wild
depicted, *2922*
Turkish Atomic Energy Commission, *3474*
Turtle Island Heritage Park, *4842*
turtles
excluder device, *2464*
green, *4842*
hawksbill, *2401, 4842*
Indian flap-shelled turtle, *2410*
leatherback, *2401*
sea turtles, *2402, 2464, 3054*
western swamp turtle, *2401*
See also
Galápagos tortoise
2,4-D (2,4-dichlorophenoxyacetic acid), *1289*
Twyford Down Alert! *1103*
Tyler Dam, *2088*
typhoid, *3941*
typhoons
See
"Storms—Hurricanes, Cyclones, and Typhoons" heading in main text

U

U.S. 40, *1552*
U.S. Agency for International Development, *3882*
U.S. Agricultural Research Service, *1310*
U.S. Air Force, *1295, 1944, 4256, 4258*
U.S. Antarctic Service Expedition, *3859, 3862*
U.S. Army Corps of Engineers, *1530, 2072, 3069, 3149, 4472*
U.S. Army Nuclear Power Program, *3498*
U.S. Army Signal Corps, *1910, 1911, 1916, 4244, 4289*
U.S. Army Signal Service, *1915*
U.S. Atomic Energy Commission, *3455, 3457, 3470, 3472, 3483, 3488, 3493, 3508, 4772*
U.S. Biological Survey, *1619*
U.S. Bureau of Biological Survey, *1736, 3166*
Division of Biological Survey, *3361*
Division of Game Management, *4700*
U.S. Bureau of Commercial Fisheries, *2477*
U.S. Bureau of Land Management, *1174, 3663, 3982, 3999, 4646, 4667, 4674, 4772, 4780, 4801, 4802, 4803, 4804, 4806, 4835, 4837*
U.S. Bureau of Mines, *1442, 3324, 3325, 3327, 3333*
U.S. Bureau of Outdoor Recreation, *2686*
U.S. Bureau of Reclamation, *1221, 4553, 4556*
U.S. Bureau of Soils, *4033*
U.S. Bureau of Sport Fisheries and Wildlife, *2477*
U.S. Census Bureau, *4719*
census to include natural resources, *3357*
U.S. Chemical Manufacturers Association, *1988*
U.S. Clean Air Act, *1416, 1582*
U.S. Coast and Geodetic Survey, *3549*
U.S. Coast Guard, *1530*
U.S. Commerce Department, *2485*
U.S. Commission of Fish and Fisheries, *3562, 4762*
U.S. Congress
African elephant law is passed, *2340*
air pollution legislation, *1398*
approves express nationwide highway system, *1565*
approves permanent ban on oil and gas leasing on U.S. wilderness land, *4666*
creates government body for mediating environmental disputes, *3066*
establishes Bureau of Mines, *3325*
Flood Control Act, *2537*
hunting and trapping legislation, *4765*
legislation designating new wilderness areas, *4827*
legislation for significant mass-transit funding, *1581*
legislation outlining federal guidelines for disposing of hazardous waste, *4143*

legislation to create conservation agency, *4763*
legislation to prohibit the dumping of solid waste in U.S. waters, *4141*
legislation to protect the bald eagle, *2384*
legislation to rent or purchase lands to be set aside for migratory birds, *1691*
legislation to restrict the dumping of hazardous waste in the U.S., *2977*
marine mammal conservation and management legislation, *4782*
nuclear waste clean-up legislation, *3516*
nuclear waste removal legislation, *3533*
oil spills clean-up legislation, *3643*
orders first survey of the coast of the United States, *3549*
passage of automobile emissions legislation, *1574*
passage of ban on the sale of leaded gasoline, *1551*
passage of coal mining safety law, *3319*
passage of comprehensive national policies for protecting the environment, *1062*
passage of dam inspection legislation, *2072*
passage of dam safety code legislation, *2077*
passage of first asbestos legislation, *3962*
passage of first coal mine inspection law, *3327*
passage of first hazardous waste comprehensive cleanup program, *3956*
passage of first mining law releasing federal land for public exploitation, *3316*
passage of first quarantine law for plants, *1283*
passage of first significant automobile legislation, *1575, 3940*
passage of flood control legislation, *2542*
passage of flood legislation emphasizing the environment, *2550*
passage of forest conservation legislation, *2631*
passage of gasoline tax, *1571*
passage of law against shipping wild animals taken in violation of state laws, *1013*
passage of law calling for reforestation and soil conservation as flood control measures, *2536*
passage of law for federal-state funding of highways, *1570*
passage of law permitting timber cutting in forest preserves, *2676*
passage of law protecting and preserving rivers, *4469*
passage of law regulating explosives and flammable materials, *2828*
passage of law requiring laboratories to furnish adequate food and shelter, *1018*
passage of legislation allowing private ownership of Uranium fuel, *3501*
passage of legislation banning the discharge of sewage sludge in U.S. waters, *4479*
passage of legislation banning the fencing of the public domain, *1166*
passage of legislation creating a national wildlife refuge, *4813*
passage of legislation deregulating the price of natural gas, *2063*
passage of legislation establishing a comprehensive U.S. national fish and wildlife policy, *4770*
passage of legislation for hydroelectric power-plant licenses, *2095*
passage of legislation protecting the bald eagle, *1694*
passage of legislation protecting the public from hazardous substances, *1281*
passage of legislation providing funds for protection and restoration of wetlands, *3090*
passage of legislation regulating pesticide use, *1290*
passage of legislation regulating the use of animals in medical and commercial research, *1017*
passage of legislation requiring humane slaughter of animals, *1015*
passage of legislation requiring public disclosure of pollutants through the Toxics Release Inventory, *2984*
passage of legislation to control acid rain, *1413*
passage of legislation to regulate the discharge of petroleum products into coastal waters, *4455*
passage of legislation to regulate the disposal of hazardous medical materials, *2883*
passage of legislation to remove ecologically fragile farmland from production, *1229*
passage of legislation withdrawing land from cultivation, *1225*
passage of mine safety law applying to noncoal mines, *3333*
passage of national standards for air polluting motor vehicles emissions, *1406*
passage of natural gas emergency legislation, *2056*
passage of rhinoceros and tiger protection law, *2346*
passage of "Superfund" environmental cleanup legislation, *2980*
passage of taxes on boats and motors, *2510*
passage of the Nonindigenous Aquatic Nuisance Prevention and Control Act, *1530*
passage of toxic substances law, *3951*
passage of water pollution legislation, *4448*
passage of wilderness protection legislation, *4667*
passes law establishing federal highway system, *1558*

U.S. Congress—*Continued*
passes law requiring catalytic converters for automobiles, *1576*
passes legislation to create United States Geological Survey, *2811*
provides funds for Antarctic expedition, *3835*
public health legislation, *3913, 3920, 3929*
ratifies Convention on Nature Protection and Wildlife Preservation in the Western Hemisphere, *3021*
rivers and harbors anti-pollution law, *4450*
water pollution legislation, *4458*
water quality standards legislation, *4466*
wildlife legislation, *2959, 4772, 4811*

U.S. Consumer Product Safety Commission, *1978, 3955*

U.S. Coral Reef Task Force, *3601*

U.S. Department of Agriculture, *1129, 1222, 1224, 1321, 1689, 1736, 1787, 1795, 1911, 1917, 2165, 2496, 2615, 2677, 2772, 2955, 3113, 3372, 3380, 3414, 3415, 3419, 3984, 4347, 4638, 4648, 4700, 4778, 4799*
Animal and Plant Health Inspection Service, *2608*
Division of Economic Ornithology and Mammalogy, *3361*
Entomological Division, *4687*
Natural Resources Conservation Service, *3370*
Soil Cartographic Unit, *3381*
Soil Erosion Service, *3370*

U.S. Department of Commerce, *1911, 2165, 2228, 2465, 2467, 2477, 2484, 2497, 3154, 4347, 4782*

U.S. Department of Defense, *4134*

U.S. Department of Energy, *1550, 1833, 2060, 2763, 3005, 3396, 3515, 3523, 3529, 3530, 3540, 3541, 4134, 4332, 4339, 4347, 4857, 4866*
Pacific Northwest Laboratory, *4857*

U.S. Department of Health, Education and Welfare, *1574*

U.S. Department of Health and Human Services, *3914, 3920*

U.S. Department of Interior, *4347*

U.S. Department of Justice, *3368*

U.S. Department of Labor, *3943*

U.S. Department of State, *2308*

U.S. Department of the Interior, *1669, 2288, 2352, 2385, 2456, 2462, 2495, 2496, 2512, 2712, 2811, 2933, 3079, 3375, 3389, 3394, 3414, 3445, 3721, 3794, 3976, 3984, 4036, 4727, 4768, 4776, 4778, 4782, 4789, 4799, 4800, 4825, 4843*

U.S. Division of Forestry, *2736*

U.S. Environmental Protection Agency, *1067, 1074, 1080, 1116, 1287, 1290, 1298, 1308, 1328, 1368, 1412, 1419, 1421, 1423, 1428, 1530, 1576, 1978, 2304, 2306, 2463, 2768, 2854, 2868, 2893, 2977, 2981, 2982, 2986, 2987, 2988, 2989, 2992, 2993, 2996, 2997, 3003, 3113, 3391, 3403, 3406, 3413, 3429, 3441, 3445, 3451, 3951, 3956, 3961, 3962, 4131, 4134, 4135, 4149, 4151, 4159, 4167, 4347, 4427, 4439, 4477, 4478, 4512, 4514, 4598, 4599, 4600, 4607, 4610, 4612, 4614, 4617, 4677*
Chesapeake Bay Program, *3593*
Office of Underground Storage Tanks (OUST), *2872*

U.S. Federal Aviation Administration, *3430, 3438, 3447, 3450*

U.S. Federal Energy Regulatory Commission, *2063, 4557, 4567*

U.S. Federal Highway Administration, *1569*

U.S. Federal Power Commission (FPC), *2051, 2095, 4557*

U.S. Federal Trade Commission, *1092*

U.S. Federal Water Pollution Control Administration, *4460*

U.S. Federal Water Quality Administration, *4509, 4542*

U.S. Fish and Wildlife Service, *1530, 1692, 2336, 2347, 2348, 2393, 2402, 2403, 2459, 2463, 2466, 2468, 2498, 2508, 2512, 2921, 2933, 3076, 3147, 3358, 3383, 3387, 3403, 3413, 4567, 4650, 4651, 4652, 4654, 4657, 4659, 4674, 4676, 4708, 4719, 4720, 4733, 4734, 4741, 4768, 4779, 4791, 4794, 4840, 4843, 4895*
Division of Law Enforcement, *4700*

U.S. Food and Drug Administration, *1176, 1227, 1370, 1978, 2764, 2765, 3439, 3880, 3970, 4422, 4615*

U.S. Forest Service, *2618, 2658, 2674, 2679, 2680, 2688, 2696, 2708, 2714, 2738, 2747, 3179, 3364, 3413, 3717, 3719, 3720, 3797, 3981, 3982, 3989, 3999, 4007, 4009, 4632, 4633, 4634, 4635, 4638, 4640, 4642, 4643, 4644, 4645, 4646, 4656, 4772, 4780*

U.S. General Land Office, *3137*

U.S. Geological Survey, *1218, 1481, 2049, 2228, 2809, 2811, 2813, 3358, 3404, 3405, 3409, 3424, 3426, 3533, 4026, 4602, 4605, 4609, 4738*
Toxic Substances Hydrology Program, *4603*
Water-Use Information Program, *4604, 4608*

U.S. Global Change Research Program, *1847*

U.S. Green Building Council, *4311*

U.S. Greens, *1113, 1118*

U.S. Home Department, *3358*
U.S. House of Representatives, *3450*
U.S. Human Genome Project's Ethical, Legal, and Social Issues Group, *2766*
U.S. Maritime Administration, *3488*
U.S. Mussel Watch Program, *4493*
U.S. National Park Service, *1797, 3358, 3789, 3999, 4463, 4641*
U.S. National Trails System, *3983*
U.S. National Wild and Scenic Rivers System, *4007, 4008*
U.S. National Wilderness Preservation System, *4648*
U.S. National Wildlife Health Center, *4740*
U.S. National Wildlife Refuge System, *4816, 4828*
U.S. Natural Resources Conservation Service, *3385*
U.S. Navy, *4722, 4843*
Marine Mammal Program, *4722, 4723*
U.S. Occupational Safety and Health Administration, *3444, 3943, 3952, 3964*
U.S. Office of Management and Budget, *4155*
U.S. Office of Naval Research, *3576*
U.S. Office of Solid Waste (OSW), *4131*
U.S. Office of Surface Mining Reclamation and Enforcement, *3336*
U.S. Park Police, *3814*
U.S. Park Service, *4276*
U.S. Peace Corps, *2707*
U.S. Postal Service, *4331*
U.S. Public Health Service, *1445, 1455, 2839, 3324, 3914, 3920, 4415, 4435, 4451, 4581, 4584, 4586, 4588, 4590*
Division of Water Pollution Control, *4535*
U.S. Public Lands Commission, *1165*
U.S. Signal Corps, *4184*
U.S. Soil Erosion Service, *4054*
U.S. State Department, *4511*
U.S. Steel, *1058, 4475*
U.S. Supreme Court, *1093, 3116, 4145, 4146, 4150*
international wildlife treaty, *3018*
ruling against NRA, *2635*
ruling on incinerator ash, *2894*
ruling on patenting lifeforms, *2757*
ruling on radioactive waste, *2890*
ruling on sponge harvesting, *2472*
Tennessee Valley Authority (TVA) v. Hill, 2333
U.S. Threatened and Endangered Species List, *2425*
U.S. Treasury Department, *3137, 3373*
U.S. v. Carborundum Company et al, *2996*
U.S. Water Resources Council, *2550*
U.S. Weather Bureau, *1911, 1918, 1919, 1934, 1942, 1943, 1953, 1954, 1955, 4184, 4186, 4216, 4256, 4257, 4258, 4287, 4289, 4292*
U.S. Weather Service, *4261*
U.S. Wild and Scenic Rivers Act, *3983*
U.S. Wild and Scenic Rivers System, *3985, 3999*
ultraviolet radiation, *1849, 1975*
Uluru National Park, *1629*
underwater film, *3568*
underwater photograph, *3558*
UNEP Chemicals, *2975*
Unified National Program for Flood Plain Management, *2550*
uniformitarianism, *2790, 2801*
Union Carbide, *1353*
Union Electric Company, *4130*
Union of Concerned Scientists, *4866*
Union Pacific Railroad, *3802*
United Auto Workers, *1544*
United Bowmen of Philadelphia, *2966*
United Mine Workers of America, *3329*
United Nations, *1985, 3464, 3478, 3582, 3586, 3933*
United Nations Commission on Sustainable Development (CSD), *4313*
United Nations Conference on Desertification, *4045*
United Nations Conference on Environment and Development (UNCED), *1640, 1850, 2999, 3049, 4310, 4318*
United Nations Conference on the Human Environment, *1070, 1073, 1840, 2292, 3030, 3509*
United Nations Conference on the Law of the Sea, *3052, 4488*
United Nations Convention for the Conservation of the Vicuna, *4777*
United Nations Convention on Biological Diversity, *3048*
United Nations Convention on the Law of the Sea, *3051, 3053, 3065, 3592*
United Nations Declaration on Population, *3881*
United Nations Department of Humanitarian Affairs, *3002*
United Nations Development Fund, *2705*
United Nations Earth Summit, *2715*
United Nations Educational, Scientific, and Cultural Organization (UNESCO), *2282, 2300, 3033, 3374, 3977, 3987, 4304, 4596*
Conference on the Conservation and Rational Use of the Biosphere, *1622*
Man and the Biosphere program, *1622*

United Nations Environment Programme (UNEP), *1070, 1072, 1829, 1831, 1832, 1844, 1846, 1848, 1977, 2160, 2300, 2975, 3002, 3030, 3040, 3045*
United Nations Environment Programme Governing Council, *1976, 2885*
United Nations Environmental Conference, *2294*
United Nations Food and Agriculture Organization, *2474, 2484, 2746*
United Nations Framework Convention on Climate Change, *1850, 3050*
United Nations Fund for Population Activities, *3882, 3884*
United Nations glossary of forest terms, *2742*
United Nations International Conference on the Peaceful Uses of Atomic Energy, *3473*
United Nations Sofia Agreement, *1415*
United Nations Task Force on El Niño, *1968*
United Nations University Institute for Natural Resources in Africa, *3392*
United Nations University system, *4613*
United Nations World Charter for Nature, *1059*
United Nations World Population Conference, *3898*
United States Advanced Battery Consortium, *1550*
United States Council for Automotive Research (USCAR), *1603, 1606, 1607*
United Tasmanian Group, *1108*
Universal Declaration on the Human Genome and Human Rights, *3055*
Universal Waste Rule, *4135*
University of California, *1451*
at Berkeley, *3744*
at Davis, *2609*
School of Veterinary Medicine, *2468*
University of Cambridge, *1991*
University of Iowa, *2920*
University of Kansas, *3869*
University of Michigan, *2277*
University of Nebraska, *2165*
University of North Carolina, *1085, 3726*
University of Oxford Botanic Garden, *1756*
University of Padua, *2256*
University of Wisconsin, *2774, 2932*
***Unsafe at Any Speed*,** *1542, 1547, 1575, 3940*
Upper Atmosphere Research Satellite, *1992*
Upper Buffalo Wilderness, *4656*
Upper Mississippi River Wildlife and Fish Refuge, *2457, 2919, 4816*
Uppsala University, *3874*
Ural Mountains, *3482*
uranium, *3330, 3501, 3535*
Urban Earthquake Hazard Reduction Workshop, *2235*
Urban Park and Recreation Recovery Act, *3822*
urban parks, *3130*
urban transportation, *3072*
urea-formaldehyde foam insulation, *1356, 1364*
***Urquiola* (oil tanker),** *3637*
UV Index Program, *1849*

V

vaccination, *3912*
Vail Agenda, *3397*
Vaiont Dam, *2118*
***Valbanera* (ocean liner),** *4249*
Valhalla Provincial Park, *1097*
Vancouver Public Aquarium, *4948, 4960*
***Vanguard 1* (satellite),** *4112*
***Vanguard 2* (satellite),** *1957*
"variable plot cruising", *2662*
Vedalia beetle, *1279*
"veggie van", *1611*
Vehicle Recycling Development Center (VRDC), *1607*
Verde River, *4006*
Verification of the Origin of Rotation in Tornadoes (VORTEX), *4297*
Vermilion River, *4008*
Vermont Wildlife Survey, *4709*
***The Vessel "Abby Dodge" v. United States*,** *2472*
VF Corporation, *1265*
vicuña, *2318, 4777, 4777*
Vienna Convention for the Protection of the Ozone Layer, *1985, 3041*
Vietnam War, *1304*
***Viking 2*,** *2230*
vines, *1197*
vineyards, *1500, 1505*
vinyl chloride gas, *2971*
Violent Crime Control and Law Enforcement Act, *2961*
Virgin Islands National Park, *3417*
Virgin Islands–Southern Florida Cluster, *3417*
Virginia Cooperative Wildlife Research Unit, *1739*
Virunga National Park, *4698*
vivisection, *1008*
Volcanic Legacy Scenic Byway, *4021*

volcanoes, *1318, 1809, 2373, 2576, 3870*
active, in U.S. national park, *3793*
forming lakes, *4352*
See also
"Volcanoes" heading in main text and names of specific volcanoes
Volcanoes National Park, *3759, 4367*
Volcanology Section (IUGG), *4378*
Volga River, *2069, 3796*
Volkswagen, *1596*
***Voyager 2*,** *4387, 4394*
***Voyages dans les Alpes*,** *2785, 2787*
Vu Quang ox, *4739*
vultures
condor, *1723*
griffon, *2439*
Rueppell's griffon, *1706*

W

Wadden Sea, *3010*
Waikiki Aquarium, *4912*
***Wakefield's Botany*,** *1771*
walleye, *4474*
Wallop-Breaux amendments, *2510*
Walt Disney World, *4871*
Wankel engine, *1545*
warblers
Kirtland's warbler, *2325, 3201*
monument to, *3201*
Virginia warbler, *1652*
"Wardian cases", *1779*
Ware-Colebrook Tunnel, *4589*
Warren-Alquist State Energy Resources Conservation and Development Act (California), *4855*
Wasatch National Forest, *3707*
***Washed Soils: How to Prevent and Reclaim Them*,** *4032*
Waste Import Restrictions in Michigan's Solid Waste Management Act, *4150*
Waste Isolation Pilot Plan facility, *3541*
Waste Isolation Pilot Plant Land Withdrawal Act, *3533*
Waste Reduction and Recycling Incentive Fund (Nebraska), *3281*
Water Act of 1974 (India), *2976, 4597*
Water Bank Act, *3148, 3380*
water frames, *4544*
water pollution, *2872, 2974, 3630, 4479, 4568, 4603*
See also
"Water Pollution" heading in main text
Water Pollution Control Act, *4458, 4460, 4535*
water power
See
"Water Power" heading in main text and individual types of water power
Water Quality Act, *4466*
Water Resources Planning Act of 1965, *2550*
Water Resources Research Act, *4402*
water supply and purification, *3091*
See also
"Water Supply and Purification" heading in main text
Watercress Darter National Wildlife Refuge, *2334*
watercress darters, *2334*
waterfowl, *1690, 1744*
See also
specific kinds of water birds
***Waterfowl Tomorrow*,** *3375*
Watershed Protection and Flood Prevention Act, *4593*
watersheds, *2713*
Waterton-Glacier International Peace Park, *3756, 3798*
Waterton Lakes National Park (Canada), *3798*
weather balloons, *1881, 1951*
Weather Bureau, *1917*
Weather Channel, *1966*
weather forecasting, *1849, 1902, 1927, 1931, 1939, 4186, 4257, 4288*
weather instruments, *1815, 1853, 1856, 1860, 1863, 1865, 1877, 1892, 1893, 1901*
weather maps, *1896, 1909*
***Weather Prediction by Numerical Processes*,** *1931*
***Weather Proverbs*,** *1916*
weather records, *1867, 1875, 4200*
weather reports, *1873*
weather satellites, *4388*
weather vanes, *1856, 1870*
weeds, *1497*
Weeks Act, *2614, 2713*
Weeks-McLean Act of 1913, *1689, 2955*
weevils
alfalfa weevil, *1188*
boll weevil, *1190, 1284, 3164*
Welland Canal, *1501, 3070, 4502*
wells, *2839*
Weminuche Wilderness Area, *2447*
West Antarctic Ice Sheet, *3874*
West Edmonton Mall, *4956*
West India Company, *3267*
West Jersey Game Protective Association, *2950*

West Nile virus, *3972*
West Sister Island Wilderness, *4654*
Western Museum of Natural History, *4888*
wetlands, *1048, 1498, 1692, 2919, 3016, 3090, 3373, 3374, 3375, 3377, 3380, 3383, 3387, 3418, 3420, 4484, 4657, 4659*
See also
"Land Use and Development—Wetlands" heading in main text
Wetlands Loan Act, *3373*
"Wetlands of the United States", *3147*
Wetlands Reserve Program, *3420*
Weymouth Woods Sandhills Nature Preserve, *4826*
whale-watching, *2239, 2244, 2253*
whales and whaling, *3020, 3057, 3847, 4723*
gray whale, *2419*
See also
"Whaling" heading in main text
Wheeler Dam, *2094, 2098*
whippoorwills, *1740*
Whipsnade, *4929*
whirling disease, *2468*
white pine, *2588, 2645, 2647, 2650*
white pine blister rust, *1283, 2596, 2597*
White River National Forest, *3719, 4633*
White Sands National Monument, *3722*
whiteflies
bayberry whitefly, *1301*
wooly whitefly, *1193*
***The Whole Earth Catalog*,** *3258, 3259*
***Whole Earth News*,** *3259*
Wichita Mountains Wilderness, *4650*
Wild and Scenic Rivers Act of 1915 (Wisconsin), *4454*
Wild and Scenic Rivers Act of 1968, *3731, 3733, 3805, 3984*
Wild Animal and Plant Protection and Regulation of International and Interprovincial Trade Act (WAPPRIITA), *2343*
Wild Bear Cam, *4968*
Wild Birds and Animals Protection Act (India), *4767*
Wild Flower Preservation Society of America, *1790*
Wild Horse Annie Act, *4771*
Wild Horses and Burros Act, *4780*
Wild Life (Protection) Act (India), *4781*
***Wild Northern Scenes*,** *3242*
***The Wild Turkey in Virginia, Its Status, Life History and Management*,** *1739*
Wildcat Brook, *4009*
wildcat drilling, *3692*
"wildcatter", *3697*
wilderness, *3242*
defined, *2714*
See also
"Wilderness" heading in main text
Wilderness Act, *3736, 4642, 4644, 4645, 4646, 4648, 4662, 4666, 4827*
wilderness areas, *2682, 3730, 3797, 4652*
Wilderness Society, *1055*
wildflowers, *2341*
wildlife, *2353, 3354, 3361*
conservation, *1717, 3390, 3590*
management, *3252*
See also
"Wildlife" heading in main text
Wildlife Conservation Act (Texas), *4786*
Wildlife Conservation and Appreciation Fund, *4840*
Wildlife Conservation Society, *2353, 3594*
Wildlife Coordination Act of 1946, *4716*
Willandra Lakes Region, *4000*
Willow Creek Dam, *2109*
Willsie Sun Power Company in Needles, *4100*
Wilson Dam, *2094*
Wind Cave National Park, *3778, 3779, 3787*
wind chill, *1941*
index, *3859*
Wind Energy Systems Act, *4858*
wind power
See
"Wind Power" heading in main text
Wind River Experimental Forest, *2749*
wind turbine, *4852*
windmill sails, self-regulating, *4848*
windmills, *4846, 4847, 4848, 4851*
Windscale Nuclear Reactor, *3475, 3479*
Windsor Dam, *2101*
Winter of the Deep Snow, *4192*
Wintergarden, *4966*
"wise use", *3364*
movement, *1024*
***The Wise Use Agenda: The Citizen's Policy Guide to Environmental Resource Issues*,** *1024*
Wizard Island, *4352*
wolves, *2429, 2434, 2440, 2442, 2444, 4796, 4797, 4799*
women and girls
awarded Sage Fellowship, *2266*
barred from mine work, *2044*
botanist of significance in U.S., *1761*
college professor with full rights, *2264*
director of Antarctic research station, *3868*
director of Environmental Protection Agency, *1116*
director of Fish and Wildlife Service, *4741*
environmental activist in women's hall of fame, *1045*

environmental health organization for, *3967*
experimentalist in botany, *1770*
falconer of note, *2899*
graduate in geology, *2814*
graduate of Pennsylvania School of Conservation, *2303*
in Antarctica, *3861, 3863*
international conference on environment, *1046*
major, in U.S. Park Police, *3814*
oceanographer of note, *3597*
on Oregon Trail, *3765*
on science staff at university, *1791*
pioneer in mycology, *1789*
pioneer in recycling rubber, *2284*
professor at Harvard, *2275*
researcher in occupational health, *2270*
rights linked to population growth, *3900*
salaried head of major U.S. conservation group, *1039*
scientist employed by U.S. Geological Survey, *2813*
to advocate organic gardening, *4065*
to climb Long's Peak, *3774*
to climb Mount Rainier, *3783*
to fly over North Pole, *3864*
to join Royal Society, *1796*
to travel solo to Pole, *3872*
to win duck stamp contest, *3221*
to win junior duck stamp contest, *3225*
to write an introduction to botany, *1771*
WorldWIDE Network Inc., *1044*
Women in Europe for a Common Future, *3967*
Women's Environment and Development Organization, *1046*
Women's Institute, *3271*
woodcocks, *2943*
woodpeckers, *4826*
Woods Hole Biological Laboratory, *1841*
Woods Hole Oceanographic Institution, *3571, 3580*
wooly whiteflies, *1193*
Works Progress Administration (WPA), *4431*
World Bank, *2721*
World Center for Birds of Prey, *2446*
World Climate Program, *1829*
World Commission on Dams, *2079*
World Commission on Environment and Development (WCED), *1842, 3262, 4305*
World Commission on Protected Areas (WCPA), *3810*
World Conservation Union (IUCN), *1037, 2079, 2247, 2353, 2395, 2421, 3028, 3058, 3374, 4775*
World Day of Peace, *3264*
World Fertility Survey, *3882*
World Forestry Center, *4943*
World Forestry Congress, *2739*
***World Glory*,** *3631*
World Health Organization (WHO), *3431, 3934, 4594*
World Heritage program, *3987*
sites, *3808, 3973, 3990, 3991, 3994, 3995, 3996*
World Meteorological Organization, *1465, 1822, 1829, 1831, 1832, 1844, 1848, 1914, 1948*
World Meteorological Organization Commission for Climatology, *1923*
World Parrot Trust, *1717*
World Population Conference, *3896*
World Resources Institute of the U.S., *1846*
***World Rivers Review*,** *2075*
World War I, *2659, 4202*
World War II, *1288, 1321, 1940, 2683, 3084, 3454, 3667, 3931, 4379*
World Wetlands Day, *3418*
World Wildlife Fund, *1039, 2442, 2719, 4830*
Worldlife Forestry Center, *4943*
World's Columbian Exposition, *3098*
World's Industrial and Cotton Centennial Exposition, *4901*
WorldWIDE Network Inc., *1044*
Wrangell–Mt. Elias National Park and Preserve, *3784*
wrens
San Benedicto rock wren, *2373*
Stephen Island wren, *2366*
Zapata wren, *1737*
Wright Brothers National Monument, *3745*

Y

Yacare Caiman, *2394*
Yakuta Glacier, *2209*
Yangtze River, *2110, 2414, 2527, 2545, 2564*
Yankee Atomic Electric Company, *3494*
Yaqui River, *2100*
"year without summer", *1318*
yellow fever, *3907, 3913, 3923*
Yellow River, *2156, 2517, 2527, 2532, 2566*
yellowfin tuna, *2460*
Yellowstone National Park, *2440, 2623, 2712, 2770, 3265, 3761, 3772, 3773, 3777, 3808, 3973, 4348, 4798*
Timberland Reserve, *2712, 3714*
Yosemite National Park, *1026, 1051, 2090, 3818*
Yosemite Valley, *3248, 3713*
***Young Gardener's Assistant*,** *3236*

Young Naturalists Club, *1794*
Youth Conservation Corps Act, *4778*

Z

Zambesi River, *2071*
Zero-Level Emission Vehicle (ZLEV), *1615*
Zion National Park, *3752*
zoning, *3115, 3117*
See also
"Land Use and Development—Zoning" heading in main text
Zoo Licensing Act (England), *4952*
Zoological Society of Cincinnati, *4897*
Zoological Society of London, *4889, 4890, 4894, 4929*
zoos
See
"Zoos, Aquariums, and Museums" heading in main text and names of specific zoos

Index by Year

The Index by Year is a chronological listing of key information from the main text of the book, organized by year starting with the earliest. Each index entry includes key information about the "first" and a 4-digit number in italics. That number directs you to the full entry in the main text, where entries are numbered in order, starting with 1001. To find the full entry, look in the main text for the entry tagged with that 4-digit nun

Note that entries containing specific dates (such as March 14) are also indexed separately in the Index by Month and Day.

For more information, see "How to Use This Book," on page ix.

2,700,000,000 BC

Glacial epoch that is verifiable, *1999*

440,000,000 BC

Mass extinction of species, *2354*

250,000,000 BC

Ice age to cover tropical areas, *2000*

230,000,000 BC

Conifer forests, *2577*

50,000,000 BC

Broadleaf deciduous forests, *2578*

2,000,000 BC

Volcano eruption known in North America, *4348*

100,000 BC

Climate change to lead to the peopling of a continent, *2001*

50,000 BC

Bow and arrow, *2895*

30,000 BC

Evidence of widespread forest clearing by felling and the use of fire, *2667*

18,000 BC

Human herding of semi-domesticated gazelles, *1137*

15,000 BC

Artistic representation of a bird or bird parts, *3159*

12,000 BC

Animal to be domesticated, *1138*
Sanitary sewers, *4516*

11,000 BC

Humans to cross the Bering Land Bridge from Asia to North America, *3750*

10,000 BC

Chinese animal park, *4872*
Reversal of the post-glacial warming trend in China, *2002*

9000 BC

Domestication of sheep and goats, *1139*
Extinctions of North American animals caused by humans, *2355*
Human practice of sedentary agriculture, *1230*

8540 BC

Volcano eruption of significance in Japan, *4349*

8500 BC

Semi-permanent human settlements, *3118*

8055 BC

Volcano eruption of significance in Canada, *4350*

8000 BC

Domesticated plants genuinely cultivated, *1195*
Glacial floods of major significance known in North America, *2515*
Large-scale human settlement, *3119*
Stabilization of climates of the present-day United States to their current levels, *2003*

7000 BC

Domestication of the grain barley, *1120*
Flood known in Asia Minor, *2516*
Human resident in what is now Russell Cave National Monument in Alabama, *3738*

6880 BC

Volcano eruption of significance in Chile, *4351*

6750 BC

Domestication of pigs, *1140*

6740 BC

Ice Age calculations done scientifically, *2004*

6000 BC

Abandonment of villages because of deforestation and resulting soil erosion, *4046*
Domestication of cattle, *1141*
Domestication of millet, a type of cereal grass, *1196*
Domestication of the chicken, *1142*
Farming communities in China, *3120*
Smelting of copper, *1329*

5700 BC

Human beings to witness volcanic eruptions of Mount Mazama in what is now Crater Lake National Park, *3751*

5500 BC

Agricultural settlements in the fertile Nile Valley of Egypt, *3121*
Irrigation, as distinguished from dry farming, *1210*

5000 BC

Clearing of the natural forests of northern Europe, *2579*
Metal mines, *3292*
Volcano eruption to create a lake in North America, *4352*

4004 BC

Dating of the origin of Earth through biblical analysis, *2778*

4000 BC

Cultivation of vines, olives, and figs, *1197*
Evidence of long-term cooling and drying of Australia, *2005*

3500 BC

Domestication of the horse, *1143*
Drought recorded known to have caused a famine, *2120*
Known use of honey from honeybees as a food, *1144*
Post-glacial cold spell of significance in the Northern Hemisphere, *2006*
Settlement of the Indus Valley of the Indian subcontinent, *3122*

3300 BC

Mines known in Great Britain, *3293*

3100 BC

Surviving year-by-year record of flood levels in the Nile Valley in Egypt, *2007*

3000 BC

Dams built whose remains have been discovered, *2082*
Diversion of the Yellow River, *2566*
Domestication of the common goose, *1145*
Evidence that dromedary (one-humped) camels were domesticated in Arabia, *1146*
Retreat of human settlement from the North African and Arabian deserts, *2008*

2950 BC

Dam constructed in Ancient Egypt, *2083*

2500 BC

Domestication of the root vegetable potato, *1121*
Drainage systems that ran into brick-lined sewers in Indus Valley cities, *4517*
Environmental damage of significance resulting from irrigation, *1211*
Evidence that dromedary (one-humped) camels were domesticated in Arabia, *1146*
Indications of private landowning, *3114*
Use of cattle for dairy products, *1147*

2297 BC

Recorded flood on China's Yellow River, *2517*

2200 BC

Illustrations of animals in a zoo-like setting, *4873*

2050 BC

Large carnivores to be held in captivity, *4874*

2000 BC

Evidence of bullfighting as a sport, *1001*
Large animals domesticated in the Western Hemisphere, *1148*
Settled communities in Mesoamerica, *3123*
Surviving account of a great flood covering the Earth, *2518*

Use of sulfur in an industrial process, *1314*

1900 BC

Evidence of extended drought causing the decline and collapse of a great civilization, *2121*

1645 BC

Volcano eruption known on a Mediterranean island, *4353*

1500 BC

Volcano eruption that was documented, *4354*

1400 BC

Acclimatization garden, *4875*
Importation of exotic animals for a royal collection, *4876*

1340 BC

Use of the shaduf or bucket-and-pole system of watering fields in the Nile Valley of Egypt, *1231*

1200 BC

Recorded use of pigeons as messengers, *1643*

1177 BC

People to keep regular records and collect data on earthquakes, *2171*

1000 BC

Commercial use of coal, *2016*
Fish farms, *2487*
Use of natural gas as a fuel, *2017*

900 BC

Agricultural collapse caused by irrigation, *1212*

750 BC

Biblical prophet to use a famous earthquake to date his prognostications, *2172*

700s BC

Evidence of humans using birds to hunt, *2896*

650 BC

Signs of large-scale erosion because of deforestation in Greece, *4047*

600 BC

Sewers of Rome, *4518*

525 BC

Description of chimpanzees, *4679*

500s BC

Written description of the medical uses of plants, *1745*

500 BC

Appearance of wet-rice cultivation in paddy fields, *1198*
Recorded scientific theories explaining earthquakes, *2174*

493 BC

Aedile office in Rome, *4519*

464 BC

Earthquake known to cause massive deaths throughout Greece, *2173*

450 BC

Description of an oil production industry, *3670*

413 BC

Recorded flood in Europe, *2519*

400s BC

Recorded scientific theories explaining earthquakes, *2174*
Use of solar architecture by the Greeks, *4075*
Written work of significance to use animals in captivity as subjects of a study, *4877*

400 BC

Municipal dump in the Western world, *4136*
Recorded mention of a rain gauge, *1853*
Volcano eruption known to occur in Mexico, *4355*

373 BC

Precise description of an earthquake, *2175*

330 BC

First recorded polar expedition, *3823*

320 BC

Botanical study of significance, *1746*
Tin mines in ancient Britain, *3294*

312 BC

Aqueduct built to carry water into Rome, *4568*

300s BC

Greek city to include solar designs in the original plans, *4076*
Western man to build a parabolic or "burning" mirror, *4077*

300 BC

275 BC

250 BC

200s BC

100 BC

95 BC

55 BC

48 BC

1st century

1

65

79

100

103

132

150

165

200

245

265

285

347

400

500s

Duck sanctuary, *4808*
Recorded statement of the law of the sea, *3006*

526

Earthquake in which a massive loss of life was recorded, *2178*

527

Dam known to have a curved shape, *2084*

534

"Sun rights" laws to guarantee buildings access to solar warmth and light, *4083*

600s

Windmills, *4846*

615

Discovery of "burning water" (petroleum) in Japan, *3604*

618

Commerce in cave swiftlet nests, *1645*

700s

English falconer of record, *2897*

800s

Weathercock wind vanes, *1856*

800

Adoption of the three-field rotation system, *1234*
Drought of major significance in Central America, *2123*

874

Colonization of Iceland from Norway, *3124*

900s

Introduction of orange, lemon, and lime trees into Spain and the western Mediterranean, *1199*
Introduction of sugar cane from Mesopotamia into the Levant, Egypt, and Cyprus, *1200*
Natural resources exploited in Indonesia, *3339*

900

Evidence of the use of shoes for horses, *1149*
Major land reclamations in low-country Flanders and Zeeland, *3067*
Use of natural gas for illumination, *2019*

1000s

Evidence of solar architecture, *4084*
Royal menagerie known to exist in England, *4879*

1000

Coal use known in North America, *2020*
Introduction of drought-resistant rice, *1122*

1012

Commercial logger on the East Coast of North America, *2641*

1069

Drought in England known to have created slavery, *2124*

1091

Written record of a devastating tornado in London, *4278*

1100s

Introduction of rabbits into England, *1150*
Windmills in Europe, *4847*

1100

Widespread use of the horse as a plough animal in western Europe, *1151*

1195

Recorded reference to a tornado as a "black horse", *4279*

1200s

Falconry book, *2898*
Monastic and manorial records of the coal mining industry in England, *3295*
Natural resource exploited in Tanzania, *3340*
Regulations designed to curb the ill effects of coal smoke, *1372*

1200

Acute timber shortages in north China, *2582*
Coal use documented in Europe, *2021*
Laws protecting the vicuña, *2318*

1201

Earthquake to kill more than 1 million people, *2179*

1228

Flood known to cause massive loss of life in Europe, *2520*

1230

Recorded imports of timber into England, *2642*

1236

Recorded evidence of serious pollution of the River Thames in London, *4520*

1240

True book devoted to birds, *3227*

1264

Recorded mention of oil in Persia, *3605*

1273

Statute in England against air pollution, *1373*

1275

Report to Europe of vast coal deposits in China, *2022*

1276

North American civilization known to end due to a drought, *2125*

1281

Typhoon known to have prevented an invasion of Japan, *4222*

1287

Commission of inquiry to investigate complaints about smoke levels in London, *1374*

1293

Earthquake in Japan causing massive deaths, *2180*

1300s

Female falconer of note, *2899*
New technology to modify city design in late medieval Europe, *3092*
Substantial waterborne export trade in British coal, *2023*

1300

Widespread cultivation of legumes that fixed nitrogen and thus improved soil fertility, *4058*

1307

Royal proclamation in England to ban the burning of coal, *1375*

1337

Known journal about the weather, *1857*

1346

Black Death (bubonic plague) outbreak, *3903*

1347

Black Death (bubonic plague) outbreak in Europe, *3904*

1348

Black Death epidemic in England, *3905*

1362

Flood to cause massive deaths in what is now Germany, *2521*
Volcano eruption of significance in Iceland, *4359*

1388

Waste disposal law in England, *4138*

1390

Game laws in England restricting hunting to the nobility, *2900*

1400s

European use of the term *hurricane*, *4223*
Person to imagine the possibilities of power-driven vehicles, *1531*
Signs of a timber shortage caused by the deforestation of western Europe, *2583*

1420s

Rabbit infestation due to introduction, *1483*

1421

Flood to cause massive deaths in the Dordrecht region of Holland, *2522*

1450s

Sugar cane cultivation in the Madeiras Islands of the Atlantic, *1235*

1450

Written reference to a mousetrap, *2901*

1481

Papal declaration regarding world exploration, *3007*

1482

Natural resource exploited in Ghana, *3341*

1489

Tsunami (giant wave) with great loss of life recorded in Japan, *2523*

1492

Hurricane to destroy a European colony in the Western Hemisphere, *4224*

1493

Cattle in the Western Hemisphere, *1484*

East-west division of the world for exploration, *3008*
Horses reintroduced in the Americas, *1485*
Introduction of cattle and reintroduction of horses into the Americas, *1152*
Pigs in the Western Hemisphere, *1486*
Sheep in the Western Hemisphere, *1487*

1494

Treaty between nations dividing the world for exploration, *3013*

1495

Hurricane reported in the Americas by a European explorer, *4225*

1496

Sport fishing guide written in English, *2505*

1500s

Almanacs with yearlong weather forecasts, *1858*
Appearance in Europe of American corn and potatoes, *1123*
Commercial whaling in North America, *4621*
Controlled forest management, *2668*
Development of the science of forestry, *2669*
Forest fire ordinances, *2611*
Large-scale air pollution problems in London, *1330*
Natural resource exploited in Argentina, *3342*
Natural resource exploited in Bolivia, *3343*
Natural resource exploited in Brazil, *3344*
Natural resource exploited in Cuba, *3345*
Person known to have studied snow crystals, *4190*
Use of the word *pollen*, *1749*
Wild horses appeared in America, *1153*

1500

Masonry dam built that is still in use, *2085*
Street lamps lighted by using oil, *3606*

1508

Statement of the theory of groundwater, *2776*
Trash collection tax in Paris, *4173*

1511

Natural resource exploited in Panama, *3346*

1516

Introduction of bananas into the Americas, *1177*

1518

Cases of smallpox to reach the Caribbean island of Hispaniola, *3906*
Dodos encountered by Europeans, *2356*

1519

Horses reintroduced to the American mainland, *1488*

1520

Wheat plants grown in the Americas, *1489*

1522

Silkworms in the Western Hemisphere, *1490*

1528

Known military expedition in North America sunk by a hurricane, *4226*
Natural resource exploited in El Salvador, *3347*

1531

Earthquake causing thousands of deaths in Portugal, *2181*

1532

Recorded mention of the effects of El Niño, *1859*

1535

European to systematically record a North American winter, *1798*
Natural resource exploited in Peru, *3348*

1536

Natural resource exploited in Honduras, *3349*

1539

European mission of significance to the interior of North America, *3710*
Natural resources exploited in Colombia, *3350*
Trash collection tax in Paris, *4173*

1540

European to see the Grand Canyon, *3753*
Natural resource exploited in Chile, *3351*

1541

European discovery of Iguaçu Falls, *4629*
Mudslide catastrophe in Guatemala caused by a volcano, *2570*

1542

English law to support North American fisheries, *2469*
European to set foot on what would later become the west coast of the United States., *3739*

1543

Botanical garden in Europe, *1750*

1543—*continued*

Flood recorded by Europeans in North America, *2524*
Spanish ships built from pine trees felled along the southern reaches of the Mississippi, *2643*

1544

Book devoted entirely to birds, *3228*

1546

Natural resource exploited in Mexico, *3352*

1550

Appearance of the "Little Ice Age,", *1799*
Introduction of tomatoes to Europe, *1201*
Widespread cultivation of buckwheat in northern Europe, *1202*

1553

Hurricane to sink a fleet in the Gulf of Mexico, *4227*

1556

List of earthquake deaths, *2182*
Treatise on mining that was scholarly, detailed, and systematic, *3296*

1559

Hurricane to wreck a colonists' ship in North America, *4228*
Introduction of tobacco plants into Europe, *1124*

1561

Professorship of botany in Europe, *2256*
Record of using solar energy to make perfume, *4085*

1563

Herbarium, *1751*
Official "fish days" in England, *2449*

1564

Natural resource exploited in Costa Rica, *3353*

1565

Oil spill of significance recorded, *3628*
Permanent Spanish settlement in what is now the United States, *3125*
Severe winter said to have introduced a new artistic approach, *1800*

1570

Introduction of the potato into Europe, *1203*

1574

Mining book detailing methods used by miners, *3297*

1580

Death in England known to have been caused by an earthquake, *2183*
Western botanist to note the sexual identity of plants, *1752*

1582

Order to Dutch linen bleachers barring the dumping of wastes in canals, *4444*

1583

Plant taxonomist, *1753*

1586

Volcano eruption known to cause massive deaths in Indonesia, *4360*

1587

Drought of significance in colonial America, *2126*

1588

Abandonment of a city in India because of a failure of the water supply, *2127*
Invasion of England to be halted by gale-force winds, *4176*

1590

Scientific botanical garden in Western Europe, *1754*

1592

Great flood at Tombouctou on the River Niger in present-day Mali, *2525*

1593

Tulips in Holland, *1755*

1594

Oil wells in Persia, *3672*

1596

Europeans known to have wintered in the extreme north, *3824*
True book devoted to birds, *3227*

1599

Practical modern greenhouse, *4086*

1600s

Forest replanting schemes promoted by the British Admiralty, *2723*
Fruit walls, *4087*
Introduction of manioc into Africa from Brazil, *1204*

Mass death of "tree huggers", *1094*
Pacifist ecological protests in India, *2644*
Tree species in North America to be widely harvested commercially, *2645*

1600

Natural resource exploited in North America, *3354*
Terrestrial magnetism theory, *2777*

1606

Law binding Scottish coal miners to their employers, *2024*

1607

Case of winter cold causing an English colony in New England to fail, *1801*
"Frost Fair" on the frozen Thames river in London, *1802*
Open-air thermometer, *1860*
Permanent English settlement in America, *3126*
Person to reach latitude 80 degrees north in the arctic region, *3825*
Proclamations against pollution caused by London starchmakers, *4445*
Sawmill established in America, *2646*

1608

Permanent French settlement in North America, *3127*
Use of the term *mare liberum* or "freedom of the seas", *4488*

1609

Freedom-of-the-seas doctrine, *3546*
Hurricane that caused islands to be settled, *4229*
Sheep imported into an American colony, *1154*

1611

Scientist to describe the characteristic six-sided symmetry of snow crystals, *4191*

1612

Commercial cultivation of tobacco, *1125*

1613

Introduction of the potato into North American colonies, *1205*

1616

Endangered species laws in the Western Hemisphere, *2319*
Land grants of 100 acres to each Virginia colony settler, *3128*

1618

Landslide known to kill more than 2,400 people, *2571*

1621

Botanic garden in England, *1756*
Drought affecting English colonists in Massachusetts, *2128*
Endangered species laws in the Western Hemisphere, *2319*
Furs exported from an American colony, *2902*
Introduction of the potato into North American colonies, *1205*

1622

Area in Europe to become industrialized, *2968*
Honey bees carried into the Western Hemisphere, *1155*

1623

Christian rainmaking ceremony in New England, *2167*

1624

Cattle imported into a North American colony, *1156*

1625

Statement of the idea that nations should be held responsible for their actions just as individuals were, *3009*

1626

Forestry law enacted by a British colony, *2625*
Natural gas sighting by Europeans in North America, *2025*

1627

Written mention of petroleum in what would become the United States, *3607*

1628

Fur trading post in the American colonies, *2903*

1629

Commercial fishery in an American colony, *2470*
Hunting law enacted by an American colony, *2904*

1631

Exports of New England white pine to England, *2647*

1633

Court ruling in colonial New England barring settlers from direct purchase of American Indian lands, *3129*
Scholarly text on the international law of the sea, *3229*

1634

Storm on record to demolish an entire Dutch island, *4230*
Urban park in the United States, *3816*

1635

Hurricane recorded in an American colony, *4231*

1639

Closed season on deer hunting in the United States, *2905*

1640s

Iron mines in the American colonies, *3298*

1640

Observer to realize that the atmosphere becomes colder at the upper levels nearer the heat-giving sun, *1861*

1641

Sealed thermometer, *1862*

1642

Flood deaths of major significance deliberately caused by humans, *2526*

1643

Instrument to measure atmospheric pressure, *1863*

1644

Weather records kept continuously in the United States, *1803*
Whaling industry established by an American town, *4622*

1646

Wolf bounty enacted in what is now the United States, *4795*

1647

Hunting privileges in a North American colony, *2906*

1648

Littering law in New York City, *3267*
Reference work on Brazilian wildlife, *4680*
Reference work to identify Brazilian fish, *2450*
Yellow fever epidemics in the New World, *3907*

1650

Flora of the British Isles, *1757*
Hunting hounds in North America, *2907*
Hurricane to kill thousands in the Lesser Antilles, *4232*
Map showing a "fountain of bitumen" (petroleum) in what would become the U.S., *3608*

1652

Natural resource exploited in South Africa, *3355*
New England pines for Royal Navy masts, *2648*
Water supply system built for a U.S. city, *4570*

1654

Colony in America to require the killing of wolves, *4796*
Dating of the origin of Earth through biblical analysis, *2778*

1657

Municipal garbage dumps in New York City, *4139*

1660

Weather forecast using a barometer, *1864*

1661

Major anti–air pollution tract, *1429*

1662

Systematic quantitative study of death in populations, *3908*
Use of London's weekly Bills of Mortality to establish a link between death rates and the burning of coal, *1331*

1664

Royal European menagerie with a research component, *4880*

1665

Cell description and use of the term *cell*, *2750*
Use of the term *cell* to describe microscopic units of plants, *1758*

1667

Hurricane report in colonial Virginia, *4233*

1669

Drought in India to cause massive deaths by starvation, *2129*
Formulation of the Principle of Original Horizontality, *2779*
Significant legislation to manage French forests to assure a continuous yield of timber, *2626*
Written account of a journey through the area of the Blue Ridge Mountains that later became Shenandoah National Park, *3754*

1670

Angled barometer, *1865*

1672

Coal mined in Canada, *3299*
Law to assert control of offshore waters, *3547*

1673

Coal discovery recorded in North America, *2026*

1675

Long-term failure of the cod fishery in and around the Norwegian Sea, *1804*
Recognition of the correlation between cellular rings in trees and age, *1759*

1676

Rain gauge, *1866*

1677

Person to keep a continuous record of rainfall in England, *1867*
Public water supply in New York City, *4571*

1678

Book devoted to birds and written in English, *3230*

1682

Identification of some disabling industrial diseases, *1354*
Tide gauge, *1868*

1683

Proposal to create a geological map, *2780*
Public museum in Great Britain, *4881*

1685

Coal mined in Canada, *3299*
Introduction of the potato into North American colonies, *1205*

1686

Botanical account, in detail, of the plants of North America, *1760*

1688

Magnolias to arrive in England, *1491*

1690

Recycled paper in the United States, *4156*

1691

Edict setting aside for Royal Navy use all Massachusetts trees more than 24 inches in diameter, *2670*
Environmental law in the American colonies, *2627*

1692

Archaeology site preserved by an earthquake, *2185*
City plan in America, *3093*
Earthquake to cause massive loss of life in Jamaica, *2184*

1694

Closed hunting season in North America, *2934*
Volcano eruption in North America for which a date can be estimated, *4361*

1695

Recorded long-term blockade of Iceland by sea ice, *1805*

1698

Steam engine of high pressure used in a mine, *3300*

1699

Stratigraphy theory, *2781*
Suction pump to draw water from coal mine shafts, *2027*

1700s

Boards of health in American cities, *3909*
Cattle and sheep in Australia and New Zealand, *1492*
Ecological protests by women, *1040*
Female botanist of significance in the United States, *1761*
Large-scale urban middle-class movement from townhouses to "suburbs of privilege", *3130*
Mangos in the Western Hemisphere, *1493*
Natural resource exploited in Nigeria, *3356*
Widespread use of crop rotation to preserve soil health, *4028*

1701

Coal deposits recorded in the American colonies, *3301*

1703

Earthquake and tsunami (giant wave) to destroy Japan's capital, *2186*
Modern "artificial" capital built in Europe, *3094*
Person to keep a continuous record of rainfall in England, *1867*
Tsunami to cause more than 100,000 deaths in Japan, *4177*

1708

Closed season on bird hunting in North America, *2935*

1709

Legislation in an American colony regulating timber cutting to protect the soil, *2628*
Use of the process of smelting by coke, *2028*

1710

Agricultural book published in the United States, *3231*
Bird band recovery, *1727*
Community forest, *2709*
Prohibition against using camouflaged hunting boats in North America, *2936*

1711

Sperm whale captured at sea by an American ship, *4623*
Storm to defeat a naval invasion of Canada, *4178*

1712

Steam engine of high pressure used in a mine, *3300*

1714

Mercury thermometer, *1869*

1715

Hurricane to sink a fleet off North America, *4234*
Recorded whaling expedition from an American colony, *4624*

1716

Weather vane maker in Colonial America, *1870*

1718

Closed hunting term in North America, *2937*

1719

American colony to restrict oyster fishing, *2501*

1720

Coal mine used on a commercial basis in Canada, *3302*
Timber trespass laws in French Canada, *2629*

1721

Game law in Pennsylvania, *2908*

1724

Levees along the Mississippi River, *2567*

1727

Drought of significance recorded in the American colonies, *2130*

1728

Botanical garden in the United States, *1762*
Japanese flora description printed in English, *1763*

1730s

Successful introduction of the ring-neck pheasant in the United States, *1514*

1732

Edition of Benjamin Franklin's *Poor Richard's Almanac*, *1871*
Fishing club in the United States of any duration, *2506*

1733

Mining school, *3303*

1734

Fish protection law enacted by an American city, *2502*

1735

Agricultural experimental farm in an American colony, *1126*
Permafrost, *1806*
Trade-wind explanation, *1872*
Use of taxonomy to classify living organisms, *1617*

1736

Hydrangea in England, *1494*

1737

Continuous daily weather observations in the United States, *1873*
Storm surge on record known to kill many thousands in India, *4179*
Typhoon to destroy 20,000 ships in India, *4235*

1739

Colonial American game warden system, *2909*

1741

Colonial American game warden system, *2909*
Discovery of the spectacled cormorant, *1646*
Sea otters encountered by Europeans, *4681*
Spectacled cormorant description, *2357*

1742

Fox hunting pack in North America, *2910*
Weather reports kept continuously by a U.S. college, *1807*

1743

Geological map, *2782*

Hurricane known to be accurately tracked in colonial America, *4236*

1744

Volcano eruption of significance recorded in today's Ecuador, *4362*

1745

Bituminous coal commercially mined in the American colonies, *3304*

1746

Earthquake to cause massive loss of life in Peru, *2187*

1748

So-called "Winter of the Deep Snow" in New England, *4192*

1749

Upper-air temperature measure of significance, *1874*

1750

Appearance of modern sedentary ranching in the Western Hemisphere, *1157*

1752

Hospital in the United States for the sick, injured, and insane, *3910*
Lightning conductor in the United States, *4274*
Paper on the extinct volcanoes of Europe, *4363*
Scientist to fly a kite in a thunderstorm in hopes of attracting a lightning bolt, *4180*
Statement that rocks and minerals were not randomly scattered but deposited in bands, *2783*
Surviving zoo, *4882*

1753

Systematic use of binary nomenclature to describe plants, *1764*

1755

Description of an earthquake or aftershock as a wave, *2190*
Earthquake causing massive loss of life in Portugal, *2188*
Scientific investigation of an earthquake, *2189*
Water pumping plant in the United States, *4572*

1756

Description of two types of earthquake motion, *2191*
Theory on how to locate the epicenter of an earthquake, *2192*

1757

Earthquake causing thousands of deaths in Chile, *2193*
Street-sweeping service in the United States, *4174*

1760

Distinctively American agricultural book, *3232*

1761

Artificial plant hybrid, *1765*
Scientific discovery of the role of bees in pollination, *1766*

1762

Known case of the importation of a natural enemy to control an alien pest, *1270*

1763

Botanical classification based on the natural relationships between plants, *1767*
British government attempt to restrict American colonial settlement to the area east of the Appalachian Mountains, *3131*

1765

Organized strike against a colliery in England, *2029*

1766

Fox hunting club in the United States, *2964*

1767

Experiments to measure the amount of heat trapped by glass, *4088*
Solar collector, *4089*

1768

Animal hunted to extinction in North America, *2358*
Botany professor at a college, *2257*
Geological survey of Russia, *2784*
Naturalist to describe the wildlife of Australia, *4682*
School of forest practice, *2732*

1769

Automobile, *1586*
Charts of the Gulf Stream, *3548*
Coal-burning steam engine, *2030*
Great famine due to a drought recorded in detail, *2131*
Water frame, *4544*

1770

English law forbidding night poaching, *2911*

1771

"Moss flood" of significance in England, *2572*

1772

Volcano eruption of significance in Java, *4364*

1773

Earthquake to devastate Guatemala, *2194*
Natural history museum foundation in the United States, *4883*
Person to cross the Antarctic Circle, *3826*

1774

Coffee plantations in Brazil, *1178*
Geological survey of Russia, *2784*

1775

Historic trail designated in the Western Region of the U.S. National Park Service, *3711*
Hurricane of significance recorded in North America, *4237*
Recognition of cancer as an occupational hazard, *2969*

1776

Documented journey through the Glen Canyon–Lake Powell area, *3755*
Federal game law in the United States, *2938*
Reservoir to supply New York City, *4618*

1777

Simultaneous weather observations in the United States, *1875*

1778

Weather reports kept continuously by a U.S. college, *1807*

1779

Oyster farming under state auspices, *2489*
Use of the term *geology*, *2785*
Winter in which all the waters surrounding New York City froze solid, *1808*

1780s

Person to write about what later became Waterton-Glacier International Peace Park, *3756*

1780

Commercial sealing in the Antarctic region, *3827*
Hurricane in the Atlantic causing massive deaths, *4238*
Meteorological society, *1876*
Published presentation of the philosophy that humans should not be cruel to animals because animals can feel pain, *1003*
Use of the phrase "The question is not, Can they reason?, not, Can they talk?, but, Can they suffer?", *1004*

1782

Bird officially designated as the national bird of the United States, *1688*

1783

Hygrometer utilizing human hair, *1877*
Post–American Revolution "broad arrow" policy marking trees for naval use, *2671*
Treaty in which England recognized the fishing rights of the United States, *3014*
Volcano known to have caused significant deaths by starvation, *4365*

1784

Botanical expedition in the United States, *1768*
Observer to propose a link between volcanic eruptions and the weather, *1809*
Recorded instance of ice floes blocking the Mississippi River at New Orleans, LA, *1810*

1785

Introduction of the rectangular system of land surveys in the United States, *3132*
National law regulating the surveying and sale of U.S. public lands, *3077*
Patent on a device to enable a steam engine to consume its own smoke, *1433*

1786

College instruction course in mineralogy, *2258*
Earthquake intensity scale, *2195*

1787

Systematic work on American geology, *2786*

1788

Botanical society formed that is still in existence today, *1769*
Mining by European settlers in Australia, *3305*

1789

Textbook on ecology, *3233*
U.S. congressional "treaty" acquiring American Indian lands for the government, *3133*

1790s

Animal exhibits in a U.S. museum to be in natural settings and groups, *4884*
Extensive geological study of the Alps, *2787*
Female botany experimentalist of significance in England, *1770*
Museum exhibits with labels and descriptions, *4885*

1790

Discovery by Europeans of Mount Baker in Washington, *3757*

Drought known to have caused cannibalism in India, *2132*
Legislation by the United States barring private acquisition of American Indian lands, *3078*
Partridge propagation in the United States, *2930*
U.S. national census, *3875*

1791

City planned and built as a seat of national government, *3095*
Coal discovery in Australia, *2031*
Drought recorded in Australia, *2133*
Natural history museum in North America, *4886*

1792

Almanac with a continuous existence in the United States, *3234*
European settlement in the area that later became Olympic National Park, *3758*
Farmer's Almanac, *1878*
Gas used to light a house, *2032*
Rally to free animals in a menagerie in France, *4887*

1793

Paralytic shellfish poisoning (PSP) cases, *4418*
Water-powered cotton mill in an American colony, *4545*
Widespread use of the cotton gin, *1236*
Yellow fever outbreak to force as much as half the population of Philadelphia, PA, to flee to the countryside, *3911*

1794

Faunal succession theory, *2788*
Non-Hawaiian to climb Mauna Loa in the area that later became Volcanoes National Park, *3759*

1795

Observation of the disease known as damping off, *2589*
Publication of the geological principle that the present is the key to the past, *2789*
Scientist to establish "uniformitarianism" as a principle of geology, *2790*

1796

Coast survey book written and published in the United States, *3235*
Introduction to the botanical sciences written by a woman, *1771*
Portland cement patent, *1332*
Protection of American Indian hunting grounds by the U.S. Congress, *2939*
Scientist to establish the fact of extinction, *2359*
Successful vaccination against smallpox, *3912*

1797

Cast-iron plow, *1237*
Earthquake causing massive loss of life in Ecuador, *2196*
Soo Lock, to carry shipping over the rapids at Sault Ste. Marie from Lake Huron to Lake Superior, *3068*

1798

Calculation of the Earth's mean density, *2791*
Federal health program in the United States, *3913*
National public health agency in the United States, *3914*
Proposed theory that human population growth might create a global catastrophe, *3876*

1799

Coal exported from Australia, *2033*
Gold discovery of significance in the United States, *3306*
Oilbird, *1647*

1800s

Appearance of the "bosom friend," an article of warm clothing for women, *1811*
Australian pine in the United States, *1495*
Brazilian pepper in the United States, *1496*
Colorado potato beetles, *1182*
Demonstration of the alternation of generations, *1772*
Female botany experimentalist of significance in England, *1770*
Leafy spurge in the United States, *1497*
Purple loosestrife in North America, *1498*
Soil conservation advocate of prominence in the United States, *4029*
Solar water heaters on record, *4090*
Terraced or horizontal plowing in the United States, *4048*
Use of fertilizers in forests, *2672*
Use of the term *protoplasm* to describe the contents of plant cells, *1773*
Water purity tables, *4419*
Widespread use of canaries to detect deadly carbon monoxide gas in coal mines, *2034*

1800

City to reach one million in population, *3877*
Explanation of atmospheric condensation, *1879*
Local land-sales offices in the United States, *3134*
Naturalization of the honeybee in North America, *1158*
Snowstorm known to have dropped 18 inches of snow on Savannah, GA, *4193*

1801

Automobile using steam power in England, *1587*

1802

Use of the term *biology*, *2259*

1803

Acquisition of land of significance by the United States from a foreign government, *3135*
Bird banding in the United States, *1728*
Cloud classification of scientific significance, *1880*
Steam-powered sawmill, *2649*

1804

American explorer to describe prairie dogs, *4683*
Description of the "black-tailed deer" by William Clark, *4684*
Meteorological research done by balloon, *1881*
White Americans to systematically describe the weather of the western United States, *1812*

1806

Breeding bird census, *1729*
Drought to cause severe agricultural damage in Canada, *2134*
Entomologist in the United States, *4685*
European to lead an expedition into the area that later became Sequoia and Kings Canyon National Park, *3760*
Scale used internationally for wind forces, *1882*

1807

Gas lamps used to fully light a public thoroughfare, *2035*
Legislation permitting settlers to live on U.S. public land until they could pay for it or until someone else bought it, *3136*
Non–Native American to see the geysers of Yellowstone National Park, *3761*
Self regulating windmill sails, *4848*
Survey of the coast of the United States, *3549*

1808

Anthracite coal burned experimentally, *2036*

1809

Geological maps of the United States, *2792*

1810

Explorer to describe the High Plains of the western United States as the Great American Desert, *2135*
Geological publication in the United States, *2793*
Introduction of goats on St. Helena Island in the South Atlantic, *1499*
Reported use of a refined oil for lighting, *3673*
"Tin" can, *4124*

1811

Earthquake known to have changed the course of the Mississippi River, *2197*
Highway built in the United States entirely with federal funds, *1552*
Introduction of the famous "grid pattern" of New York City, *3096*
Tornado to cause a major death toll in a U.S. city, *4280*

1812

Continuously operating natural sciences institution in North America, *2260*
Government agency in the United States to keep track of the sale of federal land, *3137*
Recorded description of the area that would become the Fort Laramie National Historic Site at the mouth of the Laramie River, *3762*
Weather observations by the U.S. military, *1883*

1813

Gas supplied by a private company, *2037*
Scientist to explain the reasons for the value of adding manure and ashes to the soil, *4068*
Use of the term *taxonomy* to describe the science of classification, *1774*

1814

Classification of minerals by their chemical composition, *2794*
Geological publication in the United States, *2793*

1815

Documented worldwide climate change from volcanic eruptions, *1318*
Modern geological map, *2795*
National Disaster Relief Act in the United States, *2198*
Use of the name Green Bay for an arm of Lake Michigan along the Wisconsin shore, *4499*
Volcano to cause deaths by famine in Indonesia, *4366*
Weather observations in Britain that are still being taken, *1813*

1816

Recorded "year without summer" in New England, *1814*
Summer blizzard recorded in the United States, *4194*
Use of natural gas for commercial lighting in the United States, *2038*

Use of natural gas for street lamps in a U.S. city, *2039*
Weather map, *1884*

1817

Geological map of the eastern United States, *2796*
Geological maps of the United States, *2792*
"Luminous" snowstorm in Vermont and Massachusetts, *4195*
Map using isothermal lines to map temperatures, *1885*
Use of *chlorophyll* as a term to describe the green substance that colors plants, *1775*
Waterworks system in the United States to use cast-iron pipes, *4573*

1818

Botanist to explore the American West, *1776*
Classification of the layers of the Earth's crust in North America, *2797*
Evidence of life on the sea floor, *3550*
Geological periodical still published today, *2798*
Hunting law to protect non-game birds in the United States, *2940*
Oil well in the United States, *3674*
Planter to recognize that some plants could put acid in the soil, *4030*
Salaried employee at the Western Museum of Natural History, *4888*
Treaty to recognize the three-mile limit for fishing off the North American coast, *3015*

1819

Campaign of significance against air pollution in London, *1319*
Geological organization in the United States, *2799*
Parliamentary committee to investigate air pollution in England, *1376*

1820s

Observer to describe tropical storms as large vortices of air rotating around an "eye" of low pressure, *4239*
Widespread use of oil cake as an animal feed in England, *1159*

1820

Dew-point hygrometer, *1815*
Ship to bring home Antarctic sealskins, *3829*
Sighting of Antarctica, *3828*

1821

Explorer to land on Antarctica, *3832*
Men to winter in the Antarctic, *3830*
Natural gas well in the United States, *2040*
Person to see the Antarctic territories of Peter I Island and Alexander I Island, *3831*

1822

Animal protection legislation, *1005*
Highway built in the United States entirely with federal funds, *1552*

1823

Metal trap mass-produced in the U.S., *2912*
Meteorology society in Great Britain, *1886*
Non-Hawaiian to observe Kilauea Volcano erupting in the area that later became Volcanoes National Park, *4367*

1824

Animal welfare organization started on a national level, *1006*
Civil works project undertaken by the U.S. Army Corps of Engineers, *3069*
Discovery and naming of the process of osmosis, *1777*
Geothermal heat exchange theory, *2800*
Hurricane to kill many slaves in the United States, *4240*

1825

Degree program in natural science in the United States, *2261*
Forestry school in France, *2733*
Natural catastrophe theory about animal extinctions, *2009*
Natural gas used to light a U.S. house, *2041*
Non–Native Americans to enter what is now Dinosaur National Monument, *3740*
"Refreshment taverns" in the middle of the frozen Hudson River, *1816*

1826

Fox hunting club in Canada, *2965*
Internal combustion engine in the United States, *4323*
Summer research term at a U.S. college, *2262*
Widespread teak harvesting by British companies in Burma, *2584*
Zoo explicitly set up as a scientific institution, *4889*

1827

Bird pictured in John James Audubon's *The Birds of America*, *3160*
Greenhouse effect description, *1835*
Meteorological theory saying human activities affect the climate, *1887*
Water-powered turbine, *4546*

1828

Archery club in the United States, *2966*

1828—*continued*

Gold discovery in Georgia, *3307*
Modern public zoo, *4890*
Report of the extinction of the Bonin Islands thrush, *2360*
Water filtration system, *4574*

1829

American scientist to go to Antarctica, *3833*
Canal connecting Lake Ontario with the other four of the Great Lakes, *3070*
Professor of botany at the University of London, *1778*
Proposal to dig a canal across the Florida peninsula, *3071*
Terrarium, *1779*

1830s

Great "lumber town" in the United States, *2650*
Person to establish a vineyard in San Francisco, CA, *1500*

1830

Closed season on moose hunting in the United States, *2941*
Coal mines in England to use steam engines, *3308*
Game laws in England restricting hunting to the nobility, *2900*
Pre-emption act in the United States involving lands settled before the government had purchased and surveyed them, *3138*
Pure copper deposits discovery in Michigan, *3309*
Scientist to popularize uniformitarianism, *2801*

1831

Complete state geological survey in the United States, *2802*
Known earthquake on the historical record, *2199*
Occasion of ice closing the Erie Canal for an entire month, *1817*
Successful harvesting machine, *1238*

1832

Industrial distillation of oil from coal and shale, *3675*
Public health advocate of significance in Great Britain, *3915*

1833

Complete state geological survey in the United States, *2802*
Cultivation of tea in India for commercial use, *1206*
Man to reach 74 degrees 15 minutes south latitude, *3834*
Non–American Indian to visit the area that later became Mount Rainier National Park, *3763*
Patent on a practical reaping machine, *1239*

1834

Report of the death of the last great auk in the British Isles, *2361*

1835

Earthquake to be described in detail by Charles Darwin, *2200*
Gardener's manual written and published in the United States, *3236*
Logging crews to form the advance guard of what would become known as the Big Cut in the U.S. Great Lakes region, *2651*
Sea lamprey positively identified in Lake Ontario, *1501*
Volcano eruption mistaken for a war, *4368*

1836

Disastrous snow avalanche in England, *4196*
Effective treatment for stack emissions of hydrochloric acid in the alkali industry, *1434*
Environmental movement of significance in the United States, *1047*
European to discover what is now Canyonlands National Park, *3764*
Paper describing dinosaur footprints, *2803*
Scientist in the United States to conduct a detailed study of a single winter storm, *1888*
Text of modern oceanography detailing the trade winds and ocean currents, *3237*
Women to pass along the Oregon Trail, *3765*

1837

Brandt's Cormorant description, *1730*
Meteorologist to claim that lighting large forest fires would produce widespread rain, *1889*
Systematic catalog of earthquakes, *2201*

1838

Commercial ostrich farm, *1648*
Conservationist of note in the United States, *1026*
Presentation of the cell theory, *1780*
Prohibition against the use of gun batteries in hunting in the United States, *2942*

1839

Book on natural history written by Charles Darwin, *3238*
Botanical chart to be published, *1781*
Emperor penguin egg seen and collected by humans, *1649*
Fuel cell, *4324*
Geologist hired by the U.S. government, *2804*
Prickly pear in Australia, *1502*

Scientific description of a moa, *2362*

1840s

American naturalist to suggest that every town preserve a "primitive forest", *2710*
Appearance of the Scribner Log Rule for determining board-foot measurements of timber, *2652*
Camels imported into the United States, *1160*
Company forests in the United States, *2653*
Large-scale irrigation of land in the modern era, *1215*
Mass-produced condoms, *3878*
Saltcedar or tamarisk in the United States, *1503*
Scientist to be dubbed the "father of chemical agriculture", *4060*
Use of artificial irrigation in the Nile Valley of Egypt, *1240*
Wind and current analysis of significance, *3551*

1840

Antarctic expedition funded by the U.S. Congress, *3835*
Botany professorship in the United States, *2263*
Deep-sea soundings, *3552*
Discovery that tricalcium phosphate could be converted into a soluble plant fertilizer by treatment with sulfuric acid, *4059*
Hot-blast iron furnace fueled by anthracite coal in the United States, *2042*
Meteorological records officially kept in Great Britain, *1890*
Scientific evidence to support the theory of an ice age, *2010*
Statement of the theory of a world wide ice age, *2805*
Tornado in the United States to leave a city in ruins, *4281*
U.S. census to collect information on a natural resource, *3357*

1841

Director of Kew Gardens, *1782*
Lead mine in Australia, *3310*
"October Gale" to sink 40 Cape Cod fishing vessels off Nantucket Island, MA, *4241*
Use of natural gas for cooking, *2043*
Use of the word *dinosaur*, *2363*
Volcanoes discovered in Antarctica, *4369*

1842

Aqueduct to supply New York City, *4575*
Fertilizer factory, *4061*
Geological Survey of Canada, *2806*
Geological survey of Niagara Falls, *2807*
Geologist to classify rocks from Precambrian times, *2808*
Law in England banning women from working underground, *2044*
Meteorologist to discover the processes of cyclones and tornadoes, *4242*
Natural history museum in Canada, *4891*
Official meteorologist to work for the U.S. government, *1891*
Railroad traffic in the United States, *3766*
Study in England of significance into the effects of industrialization on public health, *3916*
Theory that coral reefs are founded on sinking underwater land masses like subsiding volcanoes and continental shelves, *3553*

1843

Account of the wonders of the Rocky Mountains to reach the East Coast of the United States, *3239*
Aneroid barometer, *1892*

1844

Bird species native to North America to be extinguished, *1650*
Botanical chart to be published, *1781*
Forsythia in England, *1504*
Game protection society in the United States, *2967*
Report of extinction of the great auk, *2364*
Wildlife conservation group in the U.S., *1027*

1845

British government inquiry into the practicability of laws to control smoke, *1377*
Evidence that human pollution had reduced fish spawns in the Saginaw River and Saginaw Bay of Lake Huron, *4412*

1846

American migrants to be forced into cannibalism after being trapped in the High Sierra by heavy snow, *4197*
Cup anemometer, *1893*
Famine as a result of potato blight, *1183*
Herd book for livestock compiled and published in the United States, *3240*
Law in the United States to outlaw spring shooting, *2943*

1847

Charts of the Atlantic Ocean winds and currents, *3554*
Exploitable petroleum "seepage" in England, *3676*
Law making water closet connections to surface streams compulsory in London, *4446*
Plow for pulverizing the soil, *1241*
Synthetic fertilizer, *1127*
Zoo to have an eagle aviary, *4892*

1848

Appeal for U.S. weather observers, *1894*

1848—*continued*

Appearance of the word *cyclone*, *4243*
Descriptions of Mount Kilimanjaro and Mount Kenya to reach Europe, *4630*
Drinking water conduit in the United States to be built underwater, *4576*
Gold deposit discovery of significance in California, *3311*
Jointure of the Gulf of Saint Lawrence and the Gulf of Mexico, *3712*
Maps of the trade winds, *1895*
Oil well with a modern design, *3677*
Public health legislation of significance, *3917*
Reference to the problem of acid rain, *1435*

1849

Cabinet-level agency concerned primarily with natural resource development and conservation in the United States, *3079*
Coordinated weather observations in the United States, *1896*
Department of the Interior in the United States, *3358*
Descriptions of Mount Kilimanjaro and Mount Kenya to reach Europe, *4630*
Use of the telegraph for reporting weather conditions, *1897*

1850s

American vine aphids or *Phylloxera vastatrix* in Europe, *1271*
Appearance of the American vine aphid in Europe, *1505*
Law in the United States requiring ingredient labels on fertilizers, *4062*
Researcher to prove the transmission of cholera via a contaminated water supply, *4420*
Serious warnings about a timber shortage in the United States, *2585*
Use of carbolic acids to treat foul-smelling sewers, *4521*
Use of the word *forecast* relating to meteorology, *1898*

1850

Complete flora of Leicestershire, England, *1783*
English sparrows imported into the United States, *1506*
Federal government action granting the U.S. states sovereignty over swamplands, *3144*
Government land grant to a railroad by the United States, *3139*
Meteorology society in Great Britain continuing through the 20th century, *1899*
Modern oil refining processes, *3678*
Widespread use of South American guano in European agriculture, *1128*

1851

Air pollution prosecutions under the smoke clause of the City of London Sewers Bill, *1378*
Floods known to have lasted 15 years in China, *2527*
Postage stamps depicting the American bald eagle, *3161*
Storm that destroyed a lighthouse in the United States, *4181*
Weather maps published for general sale, *1900*

1852

Aqueduct to supply Washington, DC, *4577*
Closed season on antelope and elk hunting in the United States, *2944*
Decision of the Paris authorities to ban the taking of water from the polluted Seine, *4522*
House (English) sparrows to thrive in the United States, *1651*
Hunting law in California, *2945*
Patent for distilling kerosene from crude oil, *3679*
Scientist to systematically study dispersal of industrial age air pollutants, *1436*
Significant study of North American marine algae, *3555*
Street railway in Boston, *3072*
Systematic study of *Sequoia gigantea*, the California sequoia, *1784*
Towhee named after an individual, *1731*
Use of a locomotive in logging operations, *2654*

1853

Air pollution law of significance in England aimed at actually ridding London of air pollution rather than making it easier to prosecute polluters, *1379*
Court ruling upholding Boston's right to regulate the boundary line for wharves, *3115*
English police force given power to enforce smoke abatement laws, *1380*
European to "discover" the volcanic Crater Lake in Oregon, *4370*
Female college professor with the same rights and privileges as her male colleagues, *2264*
Instrument to accurately measure the length of time the Earth's surface receives sunshine, *1901*
Non–American Indian to see Crater Lake, *3767*
Public aquarium for exhibition of fish, *4893*

1854

Bathymetric map of the North Atlantic Ocean, *3556*
Connecticut River flood crest at Hartford to reach 28 feet, 10.5 inches, *2528*
Entomologist hired by the U.S. government, *1272*

First systematic weather forecasts, *1902*
National storm warning service, *4182*
Oil company in the United States, *3609*
Oil wells in Europe, *3680*
Parkland purchased by a city in the United States, *3817*
Street-cleaning machine of importance in the United States, *4175*

1855

Forest policy for British India, *2673*
Natural gas invention for heating, *2045*
Oceanography textbook, *3557*
Oil refinery in the world, *3681*
Published research suggesting a wide range of uses and products for distilled petroleum, *3682*
Tourists in Yosemite National Park, *2237*

1856

Cranberry treatise, *3241*
National forecasting map in the United States, *1903*
National storm warning service, *4182*
Underground city sewer system in the United States, *4523*
Underwater photograph, *3558*
Use of biological control against an unwanted plant, *1273*

1857

Crude oil wells in Germany, *3683*
Description of the giant squid, *2451*
Dual sewer system in the United States for sewage and storm water, *4524*
Newspaper weather forecast in the U.S., *1904*
Significant statement about the value of wilderness that appeared in the United States, *3242*
Systematic catalog of earthquakes, *2201*
Theory stating that earthquakes come from the Earth's crust and not from outside forces working on the crust, *2202*
World seismic map of significance, *2203*

1858

Cancellation of the British House of Commons sittings because of the stench of human waste in the River Thames, *4525*
Description of the Virginia warbler, *1652*
Draining and filling of the marshy area of the Charles River near Boston, MA, *3145*
Oil well drilled intentionally and successfully in North America, *3684*

1859

Cinchona trees in India, *1507*
Comstock Lode exploration in Nevada, *3312*
Oil refinery, *3685*
Oil well that was commercially productive, *3686*
Rabbits in Australia, *1508*
Report of a Santa Ana wind to force temperatures to over 130 degrees, *4183*
Synthetic fertilizer, *1127*

1860s

Exploding grenade harpoon, *4625*
Formal government antipollution agency in England, *1381*

1860

Automobile with a practical internal-combustion engine, *1588*
Commercial oil refinery, *3687*
Game preserve in the United States, *2931*
National water quality laws in India, *4578*
Non–American Indian settler in what is now Rocky Mountain National Park, *3768*
Oil production on a large scale in the United States, *3610*

1861

Appearance of the two-wheeled sulky plow, *1242*
Brandy distilled by the heat of the sun, *4091*
Closed season on hunting mountain goats and mountain sheep in the United States, *2946*
Combination of a glass heat trap and a burning mirror, *4092*
Discovery of *Archaeopteryx*, a fossil proto-bird, *1732*
Fire in a U.S. oil well, *3629*
Weather forecasts available to the public in Great Britain, *1905*

1862

Agricultural bureau established by the U.S. government, *1222*
Agriculture Bureau scientific publication, *3243*
California quail in New Zealand, *1509*
Federal land grants for colleges that would further the "agricultural and mechanical arts", *1223*
Law to open vast areas of the Great Plains in the United States to systematic settlement, *3080*
Tornado observation instructions in the United States, *4282*

1863

Air pollution law in England granting British inspectors the right to enter factories during their investigations, *1382*
Forestry experiment stations, *2734*
Heavy rainstorm and local flooding to cause the complete failure of an offensive operation in the American Civil War, *2529*

1863—*continued*
Law to establish limits on emissions of hydrochloric acid, *1383*
Publication of modern weather maps, *1906*
Reservoir to supply Hong Kong, *4619*
Weather maps in the modern version, *1907*

1864

Comprehensive study of human impact on the environment, *3244*
Fish hatchery to breed salmon in the United States, *2490*
Hunting license required by a state, *2913*
Public preserve in the United States, *3713*
Published presentation of a more holistic view of U.S. public land management, *3073*
Research showing that human activity could cause irretrievable damage to the earth, *1048*
Scientific exploration of the area that later became Sequoia and Kings Canyon National, *3769*
State in the United States to protect bison, *2947*
State park in the United States, *3818*

1865

Forest experiment station to systematically study the effects of the removal of litter from the forest floor, *2735*
Natural scenery protection state law connected with advertising, *3974*
Oil-tank railroad car purposely built in the United States, *3611*
Solar motor, *4093*

1866

Diamond in South Africa, *3313*
Fur-bearing animals raised commercially, *1161*
Irrigation law passed by U.S. Congress, *1216*
Law in the United States regulating explosives and flammable materials, *2828*
Oil deposits in Trinidad, *3688*
Society for animal welfare in the U.S., *1007*

1867

Begonia sutherlandii to flower in England, *1510*
Building code to be adopted in the United States, *1355*
Director of the U.S. Geological Survey (USGS), *2809*
Formaldehyde synthesization, *1356*
Insecticide that was effective in the United States, *1274*
Popular use of the word *zoo*, *4894*
Public health housing law in a large U.S. city, *3918*
Research showing that lichen are two organisms, *1785*
Water tunnel to supply Chicago, *4579*

1868

Commercially available plastic, *4125*
Introduction of the citrus-damaging cottony-cushion scale insect in California, *1184*
Map showing movement of cyclonic depression, *1908*
Plan of significance for a suburban community in the United States, *3097*

1869

Anthracite coal mine disaster of major significance in the United States, *2046*
Board of health established in the United States by a state, *3919*
Detailed geologic and topographic information about the area that later became Canyonlands National Park, *3770*
Geological observations of the Grand Canyon, *2810*
Introduction of the gypsy moth in the United States, *1511*
National park dedicated to the first transcontinental railroad, *3802*
Oil drill offshore rig, *3650*
Ostrich in southeastern Australia, *1512*
State in the United States to protect an endangered plant species by law, *2320*
Use of the term *ecology*, *1049*
Weather bulletin in the United States, *1910*
Weather charts in the United States, *1909*

1870s

Development of the Burbank or Idaho potato, *1207*
Establishment of the concept that soils are individual and natural and have their own form and structure, *4069*
Introduction of cocoa into West Africa, *1179*
Ledger-art drawings of the Plains Indians, *3162*
Recognition of the citrus-attacking California red scale insect as a serious pest in the United States, *1275*
Use of asbestos as an insulator on a large scale, *1333*
Use of the "Big Wheel" in logging, *2655*

1870

"Blizzard" designation for a snowstorm, *1912*
Closed season on caribou hunting in the United States, *2948*
Coal-mine safety laws in a U.S. state, *3314*
Documented ascent of Mount Rainier, *3771*
Entomology professor at a U.S. college, *2265*
Eurasian tree sparrow in the United States, *1653*
Government weather service in the United States, *1911*

Mining law in Pennsylvania, *3315*
Professional association of fisheries scientists in the United States, *2452*
Public health department created by the U.S. government, *3920*
Solar-powered steam engine in the United States, *4094*
State fishing commission in the United States, *2503*
State wildlife conservation agency in the United States, *4762*
U.S. Signal Corps storm warning, *4184*
Use of asphalt for street paving in the United States, *1553*
Weather observations atop Mount Washington in New Hampshire, *1913*
Wind turbine ship to cross the Atlantic Ocean, *4849*

1871

Conservation agency established by the U.S. government, *4763*
Detailed geologic and topographic information about the area that later became Canyonlands National Park, *3770*
Drought to create the conditions for a devastating fire in a major U.S. city, *2136*
Forest fire to cause a large number of deaths in the United States, *2612*
Geological observations of the Grand Canyon, *2810*
Known successful introduction of a fish species within the United States, *2491*
List of wetlands of international importance, *3016*
Photographer to capture successfully the beauties of Yellowstone on film, *3265*

1872

Arbor Day celebration, *2724*
Bird refuge established by a U.S. state, *1718*
Commercially bottled water in the United States, *4580*
Effective promoter of tree planting to prevent soil erosion on the U.S. Great Plains, *4049*
Fish hatchery run by the U.S. government, *2492*
Hunting law establishing "rest days" for waterfowl in the United States, *2949*
Marine biology station of significance, *3559*
Mining law that freely released federal land for public exploitation, *3316*
National park in the United States, *3772*
Oceanographic expedition, *3560*
Public arboretum affiliated with a university, *1786*
Superintendent of Yellowstone National Park, *3773*
Treatise on the phenomenon of acid rain, *1437*
Use of the term *acid rain*, *1818*

1873

Dam failure of major significance in the United States, *2111*
Hurricane warning, *4244*
Legislation of significance by the U.S. government promoting tree farming, *2725*
Marine biology station in the United States, *3561*
Meteorological organization on an international scale, *1914*
Nonresident hunting license in the United States, *2950*
Pekin duck brought to the United States, *1162*
Public aquarium in the United States, *4895*
Smog-related deaths reported in London, *1334*
Water filtration system for bacterial purification of a U.S. city water supply, *4582*
Weather bulletins for U.S. farmers, *1915*
White men to see the great chasm at what later became Black Canyon of the Gunnison National Monument in Colorado, *3741*
Woman to climb Long's Peak in what is now Rocky Mountain National Park, *3774*

1874

American song written about a flood warning, *2530*
Appearance of the potato-destroying Colorado beetle on the east coast of the U.S., *1185*
Barbed wire product on the market, *1163*
DDT (Dichlorodiphenyl-Trichloroethane) synthesized for use as an insecticide, *1287*
Massively destructive eastward migration of the Rocky Mountain grasshopper, *1186*
Nucleic acids, *2751*
Official exploration of the area that later became Mount Rushmore National Park, *3775*
Rabbits in Antarctica, *1513*
Snow avalanche to destroy a mining camp near Alta, UT, *4198*
Widespread damage from coffee leaf disease in Ceylon (Sri Lanka), *1187*
Zoo in the United States, *4896*

1875

Antivivisection legislation, *1008*
Geological observations of the Grand Canyon, *2810*
Legislation explicitly barring industries in England from emitting "black smoke", *1384*
National forestry association, *2691*
Society emphasizing conservation in the United States, *1028*
State in the United States to restrict market hunting, *2951*
Surviving zoo in Ohio, *4897*
Written record of the discovery of Glen Canyon, *3245*
Zoo to exhibit a great Indian rhinoceros, *4898*

1876

British dam of significance built entirely of concrete, *2086*
Criminal statutes against water pollution in Great Britain, *4447*
Division of Forestry in the United States, *2674*
Drought known to have caused cannibalism in China, *2137*
Instrument to accurately measure the length of time the Earth's surface receives sunshine, *1901*
Introduction of kudzu into the United States for erosion control, *4050*
Local ordinance in the United States to regulate chimney height as a means of curbing air pollution, *1385*
National act regulating painful experimentation on animals, *1009*
Private and nonprofit land conservation organization in the United States, *1029*
State forestry association, *2692*
Use of pollen analysis and vegetation history to determine successive climate patterns, *2011*

1877

Copper mine in Australia, *3317*
El Niño known to have caused massive deaths, *1819*
Experimental rubber plantations in Malaya, *1180*
Legislation in the United States granting a homesteader land at bargain rates in return for a promise to irrigate, *1217*

1878

Bag limit on game bird hunting in the United States, *2952*
Botanical garden in a city opened to the public in the United States, *3819*
Cannery in Alaska, *2471*
Fruit spraying in the United States, *1276*
Game commission in the United States, *2953*
General hunting license in North America, *2954*
Geologist/explorer to state flatly that much of the western United States was too dry to farm, *2138*
Legislation allowing settlers and miners to cut timber on U.S. public lands for their own use, *2630*

1879

Agency created by U.S. government specifically for geological issues, *2811*
American steam whaler, *4626*
Electricity generated by Niagara Falls waterpower, *4547*
Explorer to sail through the Northeast Passage from Norway to the Bering Sea, *3836*
Federal government program to finance construction of flood control levees in the United States, *2568*
Non–American Indian to discover Glacier Bay and the Muir Glacier, *3776*
Public Land Review Commission in the United States, *3081*
Significant study of the behavior of insects, *4686*
Zoo to exhibit rare black apes from the Indonesian island of Celebes (Sulawesi), *4899*

1880s

Appearance of the two-man crosscut saw for felling timber, *2656*
Discovery by non–American Indians of the Horsecollar Ruins in what would become Natural Bridges National Monument in Utah, *3742*
Gypsy moth infestation in a U.S town, *2590*
Introduction of larger, meatier British cattle on U.S. ranches, *1164*
Large-scale practice of dry farming methods in the United States, *1243*
Mass production of cigarettes, *1357*
Recommendation to sell U.S. public lands in large parcels for grazing, *1165*
Warnings of structural problems at the South Fork Dam near Johnstown, PA, *2112*

1880

Air pollution book to dramatize London's exploding 19th-century air pollution problem, *3246*
Anti–air pollution activities by the English public interest group known as the Fog and Smoke Committee, *1430*
Blizzards to completely halt railroad traffic in the Great Plains for many weeks, *4199*
Electric generating station, *3359*
Floating hatchery run by the U.S. government in ocean waters, *2493*
Hydroelectric power plant to furnish arc lighting service in the United States, *4548*
Ranger hired in Yellowstone National Park, *3777*
Seismograph in its modern form, *2204*
Seismology association, *2205*
Sewage-disposal system in the United States separate from the city water system, *4526*

1881

Arachlor to be successfully synthesized, *2829*
Local ordinances curbing air pollution in the United States, *1386*
Person to enter into the cave at what would become Wind Cave National Park, *3778*
Quarantine law for plants by a state, *1277*

Recorded discovery of what would become Wind Cave National Park in South Dakota, *3779*
Renowned scientist to ennoble the earthworm as one of the most important animals in the history of the world, *4070*
Smoke reduction exhibition, *1431*
Theory of Ocean Thermal Energy Conversion (OTEC), *4549*

1882

American Forest Congress, *2693*
Electric generating plant in the United States, *3360*
Forestry book written and published in the United States, *3247*
Fungicide considered effective, *1278*
Hydroelectric power plant in the United States to furnish incandescent lighting, *4550*
International Polar Year, *3837*
Ostrich in southeastern Australia, *1512*
Public aquarium in Japan, *4900*
Recorded oil seepages in Alaska, *3612*
Successful introduction of the ring-neck pheasant in the United States, *1514*
Systematic study of *Sequoia gigantea*, the California sequoia, *1784*
U.S. oceanographic research ship, *3562*

1883

Brown trout in the United States, *2494*
Collection of weather adages gathered under the aegis of the U.S. Army Signal Corps, *1916*
Hotel in Mt. Rainier National Park, *3780*
Lincoln's sparrow, *1654*
National animal rights organization in the United States, *1010*
Report of the extinction of the quagga, *2365*
Vessel constructed solely for fisheries research, *2453*
Volcano eruption known to cause catastrophic tsunamis (giant sea waves), *4372*
Volcano eruption to be audible from 3,000 miles away, *4371*

1884

Burial of electric and other wires to be required by a state, *3921*
Female awarded the Sage Fellowship in Entomology and Botany, *2266*
Flood victims to receive aid from the hand of Clara Barton, *2531*
International Ornithological Congress, *1711*
Parabolic trough solar reflector, *4095*
Scientist to state the law of relative proportion in seawater, *3563*
Tide-predicting machine, *3564*
Tornadoes in the United States reported to cause a great loss of life, *4283*
Turbogenerator to produce hydroelectric power, *4551*
Zoo that started as a sugar plantation, *4901*

1885

Automobile that proved practical with an internal-combustion engine, *1589*
Ban on fencing the public domain in the United States, *1166*
Biology course offered at a U.S. college, *2267*
Broadbase terrace for farming, *4031*
Fish research laboratory operated by the U.S. government on a permanent basis, *2454*
Forest reserve set aside by a state, *2711*
Law establishing a protected wilderness in the Adirondack Mountains, *4631*
Marine aquarium built primarily for education and research, *4902*
Nonreflecting solar motor, *4096*
Successful introduction of a tinamou, *1655*
U.S. government body to study the effects of insects on nature, *4687*

1886

Bird preservation organization in the United States, *1712*
Oil well in India, *3689*
Severe earthquake to strike the U.S. east coast in modern times, *2206*
Skyline lead for logging, *2657*
U.S. government study into the effects of wildlife on natural resources, *3361*
Water pollution legislation passed by Congress, *4448*

1887

Agricultural experiment stations established by the U.S. government in individual states, *1224*
Cave discovered in American Forks Canyon, *3781*
Flood death toll to exceed 1.5 million, *2532*
Fur farm, *1011*
Game warden in the United States to be paid a salary, *4688*
Law in Wisconsin authorizing the pay of a game warden, *4689*
Meteorologist to name hurricanes, *4245*
National park in Canada, *3782*
Snowfall recorded in excess of 3.5 inches in San Francisco, *4200*
Year in which salaried game wardens were appointed in the United States, *2914*

1888

Blizzard to cause numerous deaths on the U.S. East Coast, *4201*
Boone and Crockett Club meeting, *1030*
Drought patterns discovered in the Southern hemisphere, *2139*

1888—*continued*

Drought to depopulate the Great Plains state of Nebraska by half, *2140*
Introduction of the scale insect–destroying Vedalia beetle in California, *1279*
Seismograph in the United States, *2207*
Water resources inventory of the arid western U.S., *1218*
Windmill to generate electricity, *4850*
Zoo to exhibit adult bull elephants, *4903*

1889

Banana plantations in Central America started by a U.S. company, *1181*
Dam built in the United States to run a hydroelectric plant, *2087*
Dam disaster in the United States with a high death toll, *2113*
Head of the Department of Agriculture to be a member of the President's Cabinet, *1129*
Hydroelectric power plant in the United States transmitting alternating current over a long distance, *4552*
Irrigation advocacy group in the United States, *1031*
Natural history museum in the United States named for a descendant of royalty, *4904*
Oil well in India that was commercially successful, *3690*
Sewer district established by a city in the United States, *4527*
Surviving zoo established by the U.S. Congress, *4905*
Warnings of structural problems at the South Fork Dam near Johnstown, PA, *2112*

1890s

Appearance of alligatorweed in the United States, *4500*
Boll weevils in the United States, *1515*
Important scientific discussion of the negative impact of chemical fertilizers on food crops, *1258*
Introduction of the water hyacinth in the United States, *4501*
Natural resource exploited in Honduras, *3349*
Scientist to use the term *biosphere*, *1618*
Use of copper sulphate to destroy the charlock weed, *1280*
Use of rapid sand filtration for treating industrial wastewater in the United States, *4504*

1890

Mining of significance in the iron-rich Mesabi Iron Range in northeast Minnesota, *3318*
National and state efforts to eradicate the foliage-destroying gypsy moth in the United States, *2591*
Petroleum discovery in Sumatra, *3691*
Sewage disposal system in the United States by chemical precipitation, *4528*
Weather service of the U.S. government not run by the military, *1917*
Woman to climb Mount Rainier, *3783*

1891

Aerodynamically efficient windmills to generate electricity, *4851*
Coal-mining safety law in the United States, *3319*
Commercially marketed solar water heater, *4097*
European starlings introduced to the United States, *1516*
Forest conservation legislation in the United States, *2631*
Forest management on a professional scale, *2675*
Forest planted by the U.S. government, *2726*
"Forest preserves" legislation in the United States, *3734*
Gas conversion law by a state, *2047*
National forest in the United States, *3714*
National forest reserve in the United States, *2712*
Scientific formulation of the "fault theory" as we know it, *2208*
State highway department in the United States, *1554*
Zoo in Rhode Island to continuously operate, *4906*

1892

Diesel engine fuels, *4325*
Eruption in modern times of the Mexican volcano El Chichon in the southeastern state of Chiapas, *4373*
Expression of the concept that the earth is a living organism, *1050*
Flood covered with burning oil in the United States, *2533*
Geothermal district heating system, *2812*
International forestry research agency, *2694*
National environmental organization in the United States, *1051*
Oil tanker built for long-distance conveyance, *3613*
Use of the term *animal rights*, *1012*
Use of the term *oekology* (ecology) in the United States, *1052*
Wood fossils found on the Antarctic, *3838*

1893

Aquarium with an inland salt-water environment, *4907*
Declaration that the American frontier had closed, *3140*
Formal ascent of Devils Tower, *3743*

Legal basis for U.S. federal drinking water standards, *4581*
National urban beautification movement in the United States, *3098*
Water filtration system for bacterial purification of a U.S. city water supply, *4582*

1894

Avicultural organization founded, *1713*
Dam built using the hydraulic fill process, *2088*
Federal land-grant program to promote irrigation in western U.S. states, *1219*
Government publication of importance issued by the United States telling farmers how to protect their soil, *4032*
Irrigation advocacy group in the United States, *1031*
Lands in the United States to have constitutional protection, *3715*
Municipal public health laboratory in the United States, *3922*
National Herbarium in the United States, *1787*
Person to write a book promoting the idea of a U.S. national parks system, *3248*
Species of bird to be eliminated by a single cat, *2366*
Widespread and catastrophic drought to afflict farmers in the U.S. Great Plains, *2141*
Zoo in Nebraska, *4908*

1895

Automobile race in the United States, *1532*
Forest fire lookout tower, *2613*
Game laws in Oklahoma, *2915*
Landing on Antarctica that initiated the "Heroic Age" of Antarctic exploration, *3839*
Meeting to create the New England Botanical Club, *1788*
Postcard weather-forecasting service in the United States, *1918*
Texas state agency to regulate fishing, *2455*

1896

Electricity generated by Niagara Falls waterpower, *4547*
Female scientist employed by the U.S. Geological Survey (USGS), *2813*
Gold diggings discovery on Klondike Creek in the Yukon Territory of Canada, *3320*
Land boom of significance in Florida, *3141*
Mass-produced automobile in the United States, *1533*
Mathematical equation to calculate atmospheric warming from carbon dioxide concentrations, *1438*
Natural resources exploited in Madagascar, *3362*
Oil wells in the ocean floor to be successfully drilled, *3651*
Public aquarium in the United States funded as a municipal facility, *4909*
Tourists to visit the Glen Canyon–Lake Powell area, *2238*
U.S. government study into the effects of wildlife on natural resources, *3361*
Use of the Stiger gun as a defense against hail, *4215*
Warning about global warming, *1836*

1897

Ascent of Mount Elias, *3784*
Female pioneer in the field of mycology, the study of mushrooms, *1789*
Garbage incinerator, *4140*
Law of significance to permit timber cutting in U.S. forest preserves, *2676*
National legislation mandating protection of timber and water resources in U.S. national forests, *2632*
Oil pipeline in Russia, *3614*
Successful U.S. passenger car manufacturer, *1534*

1898

Expedition to survive a winter in the Antarctic, *3840*
Female college graduate in geology, *2814*
Forestry school at a U.S. college, *2678*
Introduction of the "green belt" concept, *3099*
"Life zones" concept, *1619*
Mass-produced automobile in the United States, *1533*
"Master school" for forestry training in the United States, *2695*
Rubbish-sorting facility in New York City, *4126*
Sewer in Moscow, *4529*
Significant theory that Antarctica was a continent rather than a series of islands, *3841*
Steel dam in the United States, *2089*
Substantial federal effort to promote industrial forestry on private land in the United States, *2677*

1899

Automaker to use an assembly line, *1590*
Automobile in the area of Mt. Rainier National Park, *3785*
Geologist to discovery that several ice ages had existed, *2012*
Government Soil Survey in the United States, *4033*
Law making it illegal to dump waste in U.S. navigable waters except by special permission, *4449*
Law to prohibit the dumping of solid waste in U.S. waters, *4141*
Person to suggest starting the New York Botanical Garden, *1790*

1899—*continued*

Recorded incidence of an earthquake reversing the paths of two glaciers, *2209*
Recorded instance of ice floes blocking the Mississippi River at New Orleans, LA, *1810*
Rivers and harbors antipollution law in the United States, *4450*
Scientific expedition to what would later be designated the John Day Fossil Beds National Monument, *3744*
Scientific record of an Antarctic winter, *3842*
Two-day U.S. Weather Bureau forecasts, *1919*
Vertical surface displacement of more than 45 feet caused by an earthquake, *2210*

1900s

Australian melaleuca in the United States, *1517*
Major pollution damage to West Coast oyster beds, *4396*

1900

Agreement to finance development of a new high-performance car on the condition that the model be named for the financier's daughter, *1535*
Appearance of the alfalfa weevil in the United States, *1188*
Coal mining accident of significance in the United States, *3321*
Evidence conclusively showing that yellow fever was transmitted by a variety of mosquito, *3923*
Hurricane causing massive loss of life in the United States, *4246*
Introduction of blended tobacco, *1130*
Law in the United States to make shipment of wild animals a federal offense if the animals were taken in violation of state laws, *1013*
Law in the United States to prohibit the transportation of wildlife across state lines if they were taken in violation of state law, *4691*
Law to protect the Asiatic lion, *4764*
Naming of the disease asbestosis, *1335*
National Christmas Bird Count in the United States, *1733*
Natural resource exploited in Kenya, *3363*
Okapi known to Western science, *4690*
Reversal of course of the Chicago River to keep raw human waste from washing up on beaches in front of Chicago Gold Coast apartment buildings, *4530*
Scientist to describe the type of earthquake surface wave that acts like waves caused by a pebble tossed into a pond, *2211*
Soil survey of a state in the United States, *4071*
Standards for water quality in the United States, *4421*
Test to identify critical elements missing in the soil, *4034*
Time that natural gas was overtaken by electricity as the primary power source for lighting in the United States, *2048*
Use of the expression *wise use* in discussing U.S. resource issues, *3364*
Use of the term *green belt*, *3100*
Warden hired by a bird conservation society in the United States, *1714*
Wildlife management law in the United States, *4765*

1901

Attempts to cause rain with gun blasts into vapor-laden clouds, *2168*
City planned and built as a seat of national government, *3095*
Comprehensive public health housing law in New York City, *3924*
Crude oil production in Mexico, *3615*
Dam project to provoke major opposition from environmentalists, *2090*
National Audubon Society meeting, *1715*
Oil well in the United States with a massive petroleum deposit, *3693*
Prediction of the existence of the ionosphere, *1920*
Recorded wildcat oil drilling team in Alaska, *3692*
Sierra Club outing to a national park, *3786*
Zoo to have a research institution, *4910*

1902

Aerial photograph in Antarctica, *3845*
Hydropower agency of the U.S. government, *4553*
Intergovernmental agency devoted to marine and fisheries science, *3565*
International Council for the Exploration of the Sea (ICES), *3566*
International public health agency, *3925*
International seismographic recording network, *2213*
Irrigation advocacy group in the United States, *1031*
Issue of *Forestry Quarterly*, *2736*
Law in the United States that protected the public from hazardous substances, *1281*
Law to provide for federal development of irrigated agriculture in the United States, *1220*
Long-distance sledge journey on the Antarctic continent, *3843*
Person to suggest starting the New York Botanical Garden, *1790*
Polar explorer to ascend in a balloon in the Antarctic regions, *3846*
Scale widely used to measure earthquake intensity, *2212*
Scientific exploration of significance of Antarctica, *3844*
Stratosphere and troposphere discoveries, *1921*

Volcano causing massive deaths in the West Indies, *4375*
Volcano eruption recorded as a major disaster in the Western Hemisphere, *4374*

1903

Bird reservation established by the U.S. government, *1719*
Chestnut blight in the United States, *2592*
Flash flood to cause $100 million in damages, *2534*
Monument erected to commemorate the first flight by an airplane, *3745*
National park established to protect a cave, *3787*
National Wildlife Refuge in the United States, *4809*
Natural history museum in western Canada, *4911*
Planned green belt town, *3101*
Pronghorn antelope bred and reared in captivity, *4746*
Reinforced concrete buttress dam, *2091*
Solar power company, *4098*
Stratosphere and troposphere discoveries, *1921*
U.S. institute for study and research on oceans, *2269*
Use of fuel oil to power warships, *3666*
Woman to win a Nobel Prize, *2268*

1904

Aquarium in Hawaii, *4912*
Bark beetle in the United States, *2593*
Census of public roads in the United States, *1555*
"Cloud compeller" to claim success in bringing rain to California with ill-smelling chemical cocktails set in pans atop high wooden towers, *2169*
Geothermal power plant, *2815*
Road inventory in the United States, *1556*
Significant use of the natural heat of volcanic systems for electrical power in Italy, *4376*
Solar power plant using a low-boiling-point liquid to power an engine, *4099*
U.S. manufacturer to make cars in great quantities to reduce costs, *1591*
Weather predictions using scientific methods, *1922*
Whaling station in the Antarctic, *3847*
Widespread commercial logging in the Philippines, *2586*
Woman on the science staff of Manchester University, *1791*

1905

Appearance of "The Great American Fraud," Samuel Hopkins Adams's article on patent medicines, *3250*
Authorization of the U.S. Forest Service to make arrests for violations of its regulations, *2679*
Commercial use of the ion exchange method for treating industrial wastewater in the United States, *4505*
Discovery of a laboratory process for extracting liquid ammonia from free nitrogen in the air, *4063*
Introduction of gypsy moth-attacking parasites into the United States from Europe and Asia, *2594*
National Audubon Society meeting, *1715*
National insect quarantine program in the United States, *1282*
Office of chief forester in the United States, *2696*
Serialization of Upton Sinclair's *The Jungle*, *3249*
Use of the word *smog*, *1439*
Wildlife warden killed in the line of duty in the United States, *4766*

1906

Alkali Act (1863) revisions of significance, *1387*
Climatologist of repute, *1923*
Coal mining accident in Europe to cause catastrophic deaths, *3322*
Cultural park set aside in the U.S. National Park System, *3788*
Earthquake in the United States believed to have surpassed 8.0 on the Richter scale, *2215*
Earthquake organization formed because of a U.S. earthquake, *2216*
Earthquake to register an 8.9 on the Richter scale, *2214*
Electrostatic device to remove particles from industrial smoke, *2970*
Failed U.S. Rocky Mountain mule deer management policy, *4797*
Grazing fees assessment for use of U.S. federal lands, *1167*
International Log Rule for measuring timber yield, *2658*
Large-scale dredging effort to reclaim land in Florida's Everglades for farming, *1244*
Long-lasting partnership to build luxury cars in Great Britain, *1536*
Major exposé of unsanitary conditions in the U.S. meatpacking industry, *1168*
Natural national monument in the United States, *3746*
Oil pipeline of significance in Russia, *3616*
Smokeless coalite patent, *1440*
State in the United States to control littering on levees, *3268*
Typhoon to cause massive loss of life in Hong Kong, *4247*

1907

Building code in the United States to require new factories to reduce smoke production, *1388*
City planning commission in the United States, *3102*
Coal-mine explosion with massive loss of life in the United States, *3323*
Dam to have a maximum height of 295 feet, *2092*
Effort for organized wildlife protection in Alabama, *4692*
Electrostatic device to remove particles from industrial smoke, *2970*
Government study on U.S. coal industry conditions, *2049*
Major private organization to campaign against air pollution in the United States, *1336*
Refuge established west of the Mississippi, *4810*
State agency in Texas to regulate game, *2916*
State in the United States to ban use of automatic weapons in hunting, *2917*
Zoo to display animals in a naturalistic setting, *4913*
Zookeeper to recognize the benefits of fresh air for animals in captivity, *4914*

1908

Bounty for tree-damaging porcupines, *2595*
Buffalo preserve in the United States, *4811*
Carmaking company to produce a range of models, *1537*
Drinking water supply in the United States to be chemically treated with chlorine compounds on a large scale, *4583*
Establishment of U.S. Forest Service Field Districts, *2680*
Explorers credited with reaching the North Pole, *3849*
Federal effort to regulate hazardous materials on railroads in the United States, *2830*
Forest experiment station in the United States, *2737*
Governors' conference on U.S. conservation issues, *1033*
Land bought by the U.S. government specifically for wildlife, *4812*
Mass-produced car to run on ethanol, *4326*
Motorized transportation in a polar region, *3848*
National Conservation Commission in the United States, *1032*
National forest established east of the Mississippi, *3716*
National forest in the United States, *3714*
National monument in the United States created from land donated by a private individual, *3747*
National professional organization for foresters in Canada, *2697*
National wildlife refuge specially authorized by Congress, *4813*
Oil discovery in the Middle East, *3694*
Oil discovery of significance in Persia, *3695*
Oil-drilling equipment to penetrate hard rock, *3617*
Research conducted in the United States by a woman in the field of occupational health, *2270*
Solar-powered energy plant capable of operating at night, *4100*
U.S. government conference devoted to conservation issues, *3365*
Warning of air pollution dangers from coal-fired electrical generation, *1337*

1909

Canadian–U.S. International Boundary Treaties Act, *3366*
Canadian–U.S. treaty to regulate use of the Niagara River for water power, *4555*
Continuously operated zoo in Texas, *4915*
District zoning in the United States, *3152*
Environment and Natural Resources Division of the U.S. Department of Justice, *3368*
Explorers credited with reaching the North Pole, *3849*
Forestry school privately operated in the United States, *2681*
Hydroelectric power plant built by the U.S. government, *4554*
Intergovernmental conservation conference in North America, *3367*
Miners' health protection agency established by the U.S. government, *3324*
National U.S. bird banding association, *1716*
Oil discovery of significance in Persia, *3695*
Pennsylvania state law prohibiting aliens from owning firearms, *2918*
Reported period of more than 900 days without measurable precipitation in a southern California town, *2142*
Solar water heater and hot water storage system to provide high-temperature water around the clock, *4101*
Tree-killing white pine blister rust in the United States, *2596*
Tri-national conference on conservation in North America, *1034*
Use of the word *gene* to describe units of heredity, *2752*

1910

Animal disease of American origin, *3926*
Appearance of white pine blister rust on the Pacific coast of North America, *2597*
Climatology professor at a U.S. college, *2271*

Coal-mining agency in the U.S. government, *3325*
Forest products laboratory of the U.S. Forest Service, *2738*
Large-scale widely reported attack of the larch sawfly, *2598*
Large-scale widely reported outbreak of spruce budworm, *2599*
Magellan Goose on South Georgia Island, *1518*
National association for city planning in the United States, *3103*
Oceanographic museum, *4916*
Oil well in Malaysia, *3696*
Rainfall predictions in Australia using atmospheric pressure, *1924*
Research results in the United States on the use of sunlight through windows to generate heat, *4102*
Statewide survey of industrial poisons, *2272*
Theory of continental drift, *2217*

1911

Automatic starter for automobiles, *1592*
Cement gun for mounting wildlife specimens, *4693*
Coal-mine emergency legislation of significance passed by the British Parliament, *3326*
Deaths caused by air pollution in Scotland, *1338*
Explorers to reach the South Pole, *3852*
Federal program of significance to purchase forestland for watershed protection in the United States, *2713*
Hydropower facilities operated by the U.S. Bureau of Reclamation, *4556*
Japanese Antarctic expedition, *3851*
Legislation in the United States to provide federal funds to states for cooperation in forest fire control, *2614*
Monument to commemorate the first major exploration of the American Southwest, *3748*
Offshore oil well, *3652*
Radio gale warnings from Great Britain to ships in the eastern Atlantic, *4185*
Radio station in Antarctica, *3850*
Research showing that the acidity of rain decreased the farther one traveled from a city center, *1441*
Seismologist to advance the elastic rebound theory to explain earthquakes in tectonic, *2218*

1912

Cohesive theory of the concept of "continental drift", *2817*
Cosmic radiation discovery, *1925*
Elk refuge in the United States, *4694*
Federal drinking water safety regulations in the United States, *4584*
Federal research on ways to limit the release of smoke into the atmosphere in the United States, *1442*
Formal complaint about Missouri River pollution from stockyards and municipal sewers in St. Joseph, MO, *4451*
Legislation on health issues involving water pollution passed by the U.S. government, *4452*
Meteorite discovered in Antarctica, *3853*
Meteorological reports for British pilots, *1926*
Modern experiments on the effects of sunlight on buildings, *4103*
Oceanography institution, *3567*
Plantings of ornamental Japanese cherry trees along the Potomac River in Washington, DC, *2727*
Quarantine law for plants enacted by the U.S. Congress, *1283*
Treatise on mining that was scholarly, detailed, and systematic, *3296*
U.S. institute for study and research on oceans, *2269*
U.S. Supreme Court decision regarding a sponge harvesting law, *2472*
Wildlife protection law in India, *4767*
X-ray analysis of minerals, *2816*

1913

Aqueduct to supply Los Angeles, CA, *4585*
Bird hunting regulation enacted by the U.S. Congress, *2955*
Botanical garden dedicated to the native flora of a single country, *1792*
Discovery of the ozone layer, *1969*
Federal law in Germany to clean up the Ruhr river, *4453*
Gas station in the United States, *3619*
International seismographic recording network, *2213*
Legislation to protect migratory game and insectivorous birds, *1689*
Major dam on the Mississippi River below St. Paul, MN, *2093*
Monument to a bird in the United States, *3163*
Oil and gas conservation law by a state, *3618*
Pygmy hippos known to Western science, *4695*
Scenic highway in the United States, *3975*
Society of professional ecologists, *2273*
Tract housing development in the United States, *3074*

1914

Air pollution measurements taken in London, *1443*
Bird banding by a U.S. government agency, *1734*
Car produced by the firm that would become the Nissan Motor Company of Japan, *1538*
Carolina parakeet extinction report, *2367*

1914—*continued*

Extinction caused by humans of an animal endemic to North America in modern times, *2368*
Extinction of a native American bird in modern times, *1656*
Federal drinking water standards in the United States, *4586*
Frontal air mass theory of forecasting, *1927*
Law setting up the Cooperative Extension Service in the United States, *1266*
Person to compare tree ring width to climate and droughts, *2143*
Scientist to use records of seismic waves to demonstrate the existence of the earth's core and measure its distance from the surface, *2219*
Synthetic detergents for clothing, *4397*
Underwater movie seen by the public, *3568*
Use of tractors in timber harvesting, *2659*

1915

Government declaration of large stretches of the Hudson River as dangerous for recreation, *4398*
Society of professional ecologists in the United States, *2274*
State law in the United States to protect wild and scenic rivers from development, *4454*
Zoo habitat constructed of simulated rock formations without bars, *4917*

1916

Bird protection treaty, *3017*
Birth control clinic in the United States, *3879*
Dams on the Tennessee River near Muscle Shoals, AL, *2094*
Director of the U.S. National Park Service, *3789*
Federal legislation to establish the national parks system, *3790*
Fisheries management in U.S. national parks, *2456*
Gasoline additive tried, *1593*
Highway law in the United States for federal-state funding, *1570*
Highway program of significance in the United States granting federal funds for state roads, *1557*
International treaty to protect birds, *1743*
Japanese beetles in the United States, *1189*
Law in the United States requiring preservation of scenery within national parks, *3976*
Major opposition to chemical fertilizers, *4064*
National car industry to produce one million units a year, *1539*
National park east of the Mississippi River and the first located on an island, *3792*
National park in the United States made up entirely of lands donated by private citizens, *3791*
National park in the United States to contain an active volcano, *3793*
National parks management agency in the United States, *3794*
Report of the reappearance of the cahow, or Bermuda petrel, *1657*
Snow avalanche to kill Austrian and Italian troops during a war, *4202*
U.S. national park in a volcanic range, *4377*
Zoning ordinance in the U.S., *3153*
Zoo with a veterinary hospital, *4918*
Zoological Survey of India (ZSI), *4696*

1917

Brucellosis in the Yellowstone National Park buffalo herd, *4798*
"Ecological niche" concept, *1620*
Ice jam to close the Ohio River for nearly two months, *1820*
National park created after the establishment of the National Park Service, *3795*
U.S. Forest Service study of recreational opportunities and values in the national forests, *2682*
Water tunnel to supply New York City, *4587*
Wildlife films of the fauna of Borneo, *4697*

1918

Bird of a new type discovered in the 20th century in the United States, *1658*
Crop dusting by an airplane, *1245*
Nene (Hawaiian goose) breeding program, *2426*
Private organization for protecting original redwood forests, *2698*
Successful challenge to state-ownership doctrine, *1690*
Widespread deadly global influenza epidemic, *3927*

1919

Conservation group to focus on the eastern United States in addition to western states, *1035*
Cyclone theory, *4248*
Female professor at Harvard, *2275*
Forest fire air patrol, *2615*
Full-time landscape architect hired by the U.S. Forest Service, *3717*
Hurricane to sink an ocean liner off the United States, *4249*
Landscape architect employed by the U.S. Forest Service (USFS), *4632*
Monument to an insect in the United States, *3164*
National meteorological organization in the United States, *1928*

Nature reserve in the Soviet Union, *3796*
Private organization to support national parks in the United States, *3718*
Trophic-level system application to classify lakes, *4506*
Wilderness area protected by the U.S. Forest Service (USFS), *4633*

1920s

Fish-stocking program failures in acidic lakes in upstate New York, *4413*
Indications of a jet stream, *1929*
Large-scale peripheral suburban growth in the United States, *3142*
Long-standing survey of water quality conditions in the Western Approaches to the English Channel, *4399*
Successful control of the cotton-destroying boll weevil, *1284*
Use of the term *wildcatter* in oil exploration, *3697*

1920

Appearance of the Colorado potato beetle in Europe, *1519*
Application of the wilderness concept to U.S. Forest Service areas, *3719*
Hydroelectric power-plant licenses in the United States, *2095*
International wildlife treaty upheld by the U.S. Supreme Court, *3018*
Killer tornado to cause $3 million in damages in Chicago, *4284*
Law in the United States to license construction of commercial hydropower plants, *4557*
Long-range weather forecasting attempt using equations, *1930*

1921

Chemist to add tetraethyl lead to gasoline, *1339*
Common quail introduced to the Hawaiian Islands, *1520*
Dam burst to cause a disaster by chemical contamination, *2114*
Devastating ice storm in New England in the 20th century, *4270*
Devastating storm of freezing rain to strike Illinois in the 20th century, *4271*
Dutch elm disease description, *2600*
Federal highway system in the United States, *1558*
Interstate planning and development agency in the United States, *3104*
Law in Vermont to protect endangered plants, *2321*
National preserve created to protect the Javan rhinoceros, *4814*
Organization to gather international forestry statistics, *2699*
Sea lamprey in Lake Erie, *4502*
Widespread, widely reported blowdown in the United States, *2587*
Wildlife films of mountain gorillas, *4698*
Wildlife preserve in Ukraine, *4815*

1922

Appearance of the cotton-destroying boll weevil in Virginia, *1190*
Commercially marketed frozen fish, *2473*
Duck-billed platypus successfully transported outside Australia, *4699*
Emperor penguin egg collected for science, *1735*
Large-scale flood control project by a single U.S. state, *2535*
Large shopping center built on the outskirts of an urban area specifically to serve people with cars, *3105*
Legislation allowing exchanges of U.S. public domain forest land for other lands of equal value, *2633*
National guidelines for local zoning ordinances in the United States, *3154*
Outline of the mathematical concepts that would lead to numerical modeling for accurate weather forecasting, *1931*
Volcano information network of significance set up internationally, *4378*
Weather reports on radio in Britain, *1932*
Wildlife films of mountain gorillas, *4698*

1923

Aquarium to be part of a natural history museum, *4919*
Ethyl gasoline with tetraethyl lead, *1540*
Introduction of the pH scale to measure acidity in rain, *1444*
Use of tetraethyl lead for automobiles, *1594*
Use of the term *genecology* to describe the study of genetic variations within individual plant species as they relate to the environment, *1793*

1924

Aquarium in North America to be accredited by the American Association of Zoological Parks and Aquariums, *4948*
Car model with a high-compression six-cylinder engine, *1595*
Detailed life-history investigation of a major wildlife species, *1736*
Federally designated wilderness area in the United States, *4634*
Fishing legally allowed on a U.S. government wildlife refuge, *2457*
Hunting legally allowed on a U.S. government wildlife refuge, *2919*
Imperial pheasants collected from the wild, *1659*

1924—*continued*

Law to regulate the discharge of petroleum products into coastal waters, *4455*
Legislation of significance in the United States to promote reforestation and soil conservation as flood control measures, *2536*
Modern tollways, *1559*
National wilderness area in the world, *3797*
Provision for federal cooperation in fire control on private timber or forest-producing lands, *2616*
Refuge acquired by the U.S. National Wildlife Refuge System specifically for the management of waterfowl, *4816*
Small all-purpose tractor in the United States, *1246*
Wilderness area in New Mexico, *3720*
Year all the states in the United States had at least one law regulating hazardous wastes, *2831*

1925

Car model with a high-compression six-cylinder engine, *1595*
Dam in the United States to fail during an earthquake, *2115*
Federal chemical standards for drinking water in the United States, *4588*
Federal studies of the effects on the atmosphere of carbon monoxide from automobile exhaust in the United States, *1445*
Ionosphere direct measurements, *1933*
Law in Massachusetts to protect an endangered plant species, *2322*
Musk ox born in captivity, *4747*
Oceanographic study of an entire ocean, *3569*
Outdoor museum with nature trails in the United States, *4920*
Parkway in the United States, *1560*
Statement of the theory that constant exposure over time to small doses of toxins causes poisoning, *2832*
Textbook on industrial poisons in the United States, *2276*
Tornado in the United States with a death toll in excess of 650, *4285*
Tornado to last 3 hours in the United States, *4286*
U.S. institute for study and research on oceans, *2269*
Use of sonar (sound navigation and ranging) to measure ocean depths, *3570*
Zoo to have an African forest elephant, *4921*

1926

Emus on Kangaroo Island, *1521*
Highway-marking system in the United States, *1561*
Law passed to protect black bass in the United States, *2458*
National trade group to promote the U.S. timber industry's positions on environmental issues, *2700*
Polyvinyl chloride (PVC), *2971*
Species to be given a scientific name incorporating the first and last names of a naturalist, *1737*
Truly international forestry literature, *2739*
Tunnel in the Massachusetts water system, *4589*
U.S. city to decide to widen roads rather than build a subway system, *3106*
U.S. Supreme Court ruling affirming the legality of zoning, *3155*
Weather map to be telecast from a land sending station to a land receiving station, *1934*
World Forestry Congress, *2701*

1927

International conference to address world population problems, *3896*
Introduction of the discipline of animal ecology, *1053*
Land purchased for what would become Everglades National Park in Florida, *1095*
Law in the United States to require warning labels on hazardous substances, *1285*
Lawsuit against the state of Illinois and the city of Chicago claiming the diversion of Lake Michigan water to flush sewage down the Chicago Sanitary and Ship Canal was lowering water levels in the Great Lakes system, *4531*
Microclimatology research, *1935*
National natural-resource protection law in India, *3369*
Oil well that was productive in Iraq, *3698*
Report of the extinction of the paradise bird of Australia, *2369*
Restoration of a colonial-era town to museum status, *3108*
School of conservation, *2277*
Year in which General Motors vehicles outsold those of the Ford Motor Company, *1541*
Yellow fever vaccine, *3928*

1928

Aerial mapping photographs over Antarctica, *3854*
Catastrophic flood caused by a hurricane in the United States, *2538*
Chestnut-winged chachalaca introduced in the United States, *1660*
Chlorofluorocarbons (CFCs), *1970*
Crested guan introduced in the United States, *1661*
Dam of significance in the United States to collapse when completely filled, *2116*
Dam organization of international significance, *2068*

Flight across the Arctic Ocean, *3855*
Great currasows introduced into the United States, *1662*
Hydroelectric dam of significance in the United States financed by private capital, *2096*
Identification of the Congo peacock, *1697*
Mississippi River flood control project of significance, *2537*
Substantial federal legislation to provide funds for research into reforestation in the United States, *2634*
Theory that seafloor spreading was caused by convective heat in the Earth's mantle, *2818*
Zoo to breed a Bornean orangutan, *4922*
Zoo to breed a captive chimpanzee, *4923*
Zoo to use bar-less exhibits extensively, *4924*

1929

Airplane flight over the South Pole, *3856*
Blue goose nest discovered, *1698*
Citation for heroism ever bestowed by the U.S. Department of the Interior, *3721*
Description of hydroponic agriculture, *1247*
Electrostatic precipitation application, *1446*
Federal regulation in the United States for designation and management of wild areas, *4635*
Government commission to rent or purchase lands to be set aside for migratory birds in the United States, *1691*
Infestation in the United States of the Mediterranean fruit fly, *1191*
Mediterranean fruit fly infestation in North America, *1522*
Pumped storage hydroelectric plant of significance in the United States, *4558*
Reports of sharp decreases in dissolved oxygen levels in Lake Erie, *4414*
Scientist to use the term *biosphere*, *1618*
Substantial U.S. government funding for a study of causes of soil erosion and methods of erosion control, *4051*
Use of dendroclimatology as a means of determining historic yearly rainfall and temperatures by tree rings, *2013*
Wildlife sanctuary in Tanganyika, *4817*
Zoo to exhibit a pair of African buffalo, *4925*
Zoo to have an outdoor gibbon cage, *4926*

1930s

Aerial surveys of large parts of the United States to support soil conservation programs, *4072*
Corporate effort at pollution control at Du Pont de Nemours chemical plants in the United States, *4429*
Diebacks of significance to western white pine forests in the United States, *2588*
Dumping of highly toxic PCBs (polychlorinated biphenyls) in waterways, *4430*
Federal program of significance to seal up abandoned mines in the United States to prevent water entry and polluting drainage, *4431*
Hydroelectric dams on Russia's Volga river, *2069*
Hydroelectric dams on the Columbia River in the U.S. Pacific Northwest, *2097*
Instances of modern organic farming as a conscious rejection of traditional chemical methods, *1259*
Interstate gas pipelines in the United States, *2050*
Modern planned solar community in Switzerland, *4104*
Network of high-speed freeways in Germany, *1562*
Person to lobby for a botanical garden in Montreal, *1794*
Reports of widespread fouling of Great Lakes beaches by ill-smelling seaweed, *4532*
State government concerns about public landfills, *2833*
Symptoms of nutrient deficiencies in exotic pine plantations in Australia, *2740*
Use of the term *Dust Bowl* to describe drought-seared and wind-eroded U.S. southern Great Plains, *4052*
Use of the term *primitive area* by the U.S. Forest Service to describe wilderness, *2714*
Widespread deaths of trees in Moscow attributed to air pollution, *1320*
Widespread replacement of railroad logging by trucks in the United States, *2660*

1930

Authenticated death by hail recorded by the U.S. Weather Bureau, *4216*
Contemporary American science and technology museum, *4927*
Dutch elm disease in the United States, *1523*
Effective refrigerant for mass consumer use, *1340*
Extensive peacetime drought relief project in the United States, *2144*
Hurricane to cause massive deaths in the Dominican Republic, *4250*
Legislation in Canada that preserved parks for future generations, *3735*
Noise control commission in the United States, *3427*
Ocean Thermal Energy Conversion (OTEC) power plant, *4559*
Radio broadcast from a zoo, *4928*
Species of bird to be resuscitated from a single live specimen, *2323*
Teletype communications of regularity between U.S. weather stations, *1936*
U.S. federal law to protect a wilderness area, *4636*

1930—*continued*

Weather map to be telecast to a transatlantic steamer, *1937*

1931

Earthquake of significance in New Zealand, *2220*
Law in the United States to allow federal control of wildlife on both public and private lands, *4799*
Meteorologist to photograph ice crystals with a camera-equipped microscope, *4203*
Modern animal park, *4929*
Multi-purpose oceanographic research vessel in the United States, *3571*
Private hydroponic garden, *1248*
Protected historic district in the United States, *3082*

1932

Drought used for political purposes, *2145*
Gasoline tax levied by the U.S. Congress, *1571*
Hailstorm of significance known to have hit the western Hunan province of China, *4217*
Heath hen extinction report, *2370*
International organization to promote the timber trade, *2702*
League of Nations conference of world timber experts, *2703*
Noted "back to the land" conservationist in the United States, *1036*
Officially reported measurable snowfall in downtown Los Angeles, *4204*
Park to be designated an international peace park, *3798*
Position of director of game research, *2920*
School devoted exclusively to training wildlife conservation officers, *2278*

1933

Comprehensive scientific effort to conserve soil and water resources on a national scale, *4035*
Dust storm known to carry dust from Montana all the way to the U.S. eastern seaboard, *2146*
Employment program of significance for forestry and other natural resource projects, *2683*
Explosion seismology experiments in Antarctica, *3857*
Game management chair at a U.S. college, *2932*
International treaty to propose African wildlife preserves, *3019*
"Land Ethic" concept, *1054*
Legislation in the United States to withdraw substantial amounts of land from cultivation, *1225*
Major attack on the U.S. consumer products industry, *3251*
Multi-purpose dam development of significance in the United States, *2070*
National monument to commemorate the world's largest gypsum dune field, *3722*
Natural Resources Conservation Service of the U.S. Department of Agriculture, *3370*
Short-tailed albatross protection law, *2383*
Statewide construction and design standards for earthquakes in the United States, *2221*
Successful application of electrostatic precipitation to scrub pollutants from cement kiln stacks in Britain, *1447*
Textbook on wildlife management, *3252*
Watershed erosion control demonstration project, *4053*

1934

Annual license required of all U.S. waterfowl hunters 16 years of age or older, *2933*
Automobile safety advocate of note in the United States, *1542*
Bathysphere, *3572*
Bird sanctuary to protect migrating hawks and eagles, *1720*
Commercial production of plants in water instead of soil, *1249*
Confirmation that Dutch elm disease-carrying bark beetles deposit fungi spores on feeding wounds of trees, *2601*
Duck stamps for waterfowl hunters in the United States, *1692*
Dust storm known to drop massive quantities of Great Plains topsoil on Chicago, *2147*
Federal duck stamp in the United States, *3165*
Federal duck stamp in the United States to feature the mallard, *3166*
Federal law enforcement agency for wildlife in the United States, *4700*
Federal Lumber Code Authority in the United States, *2635*
Federal management policy for public grazing lands in the United States, *1169*
Field guide to birds, *3253*
Law in the United States designed to stop overgrazing and its adverse effects on the soil, *4036*
Law in the United States to allow government fees to be levied to establish and maintain wildlife refuges, *4818*
Legislation that provided regular federal funding for waterfowl management in the United States, *1693*
Legislation to regulate Indian grazing ranges in the United States, *1226*
Major windstorm of the Dust Bowl drought in the U.S. Midwest, *2149*
National natural resource inventory document in the United States, *3371*

National organization dedicated specifically to preserving wilderness (as opposed to wildlife) in the United States, *1055*
National soil erosion survey in the United States, *4054*
Oil pipeline of significance in Iraq, *3620*
Ptarmigan born and bred in captivity, *4749*
Raptor sanctuary, *1721*
Severe drought year affecting nearly one-half of the United States, *2148*
Snow goose born and bred in captivity, *4748*

1935

Arbitration ruling to assign national responsibility for air pollution damage, *1389*
Commercial hydroponicum on a large scale, *1250*
Commercial production of plants in water instead of soil, *1249*
Commercially marketed beer cans in the United States, *4127*
Drought rehabilitation program by the Canadian government, *2151*
Effort of significance by the United States government to develop soil conservation and rehabilitation programs for the southern Great Plains during the Dust Bowl, *4037*
Evidence of water reservoir-induced earthquakes, *2222*
Federal agency charged with preventing soil erosion, *4055*
Federal duck stamp in the United States to feature the canvasback, *3167*
Funding for multisector wildlife research, *2279*
Grazing district on public lands in the United States, *1170*
Hurricane in the United States rated Category 5 (most intense), *4251*
National organization dedicated specifically to preserving wilderness (as opposed to wildlife) in the United States, *1055*
National Park Service unit including coral reef resources, *3799*
Program to stop extinction of trumpeter swans, *2324*
Public health grants from the U.S. government to states, *3929*
Radar detection of air echoes from the lower atmosphere, *1938*
Sales of federal duck stamps to collectors in the United States, *3168*
Scale to accurately measure earthquake strength, *2223*
Scientist to use the term *ecosystem*, *1621*
Subcompact car designed to be fuel-efficient, *1596*
U.S. National Wildlife Reserve created specifically for Kirtland's warbler, *2325*
U.S. Soil Conservation Service chief, *4056*
Use of organic chemicals for weed control, *1286*
Use of the term *Dust Bowl* to describe the drought-devastated Southern Plains, *2150*
Use of the term *ecosystem*, *1056*
Woman to visit Antarctica, *3858*
Zoologist to win a Nobel Prize, *2280*

1936

Dam built as part of the Tennessee Valley Authority (TVA), *2098*
Dam to be constructed with a height greater than 152 meters (500 feet), *2099*
Federal duck stamp in the United States to feature the Canada goose, *3169*
Flood legislation by the U.S. government that called floods a national problem, *2539*
Fuel alcohol plant in the United States, *4327*
Law in Great Britain in the modern era addressing hazardous waste, *2834*
Legislation linking soil conservation and commodity policy in the United States, *4038*
Mass-produced FIAT Topolino, *1543*
Prairie restoration in the U.S. National Parks System, *3749*
Sit-down strikes to cripple General Motors car production, *1544*
Television weather forecast in Great Britain, *1939*
Use of the term *hydroponics*, *1251*
Wildlife conference in North America, *4701*
Wildlife dioramas exhibited in a museum that were accurate, *2281*
Wildlife research center in the United States, *4819*
Zoo to have a model dairy barn, *4930*
Zoo to have a regular radio program, *4931*

1937

Biological evidence that continents were once joined, *2819*
Brush charge of atmospheric electricity to destroy an airship, *4275*
California condor sanctuary, *1723*
Description of hydroponic agriculture, *1247*
Federal duck stamp in the United States to feature the greater scaup, *3170*
Firearms tax in the United States to specifically benefit wildlife habitats, *4820*
Game preserve appropriation by Congress to assist state wildlife restoration projects, *4821*
Great flood to maroon the entire city of Louisville, KY, *2540*
International agreement on whaling regulations, *3020*
Legislation authorizing federal-state cooperation in U.S. farm forestry, *2636*
Legislation for cooperative wildlife management, *4704*
National seashore in the United States, *3800*

1937—*continued*

National Trails system unit opened, *3723*
New Jersey program authorized by the Federal Aid in Wildlife Restoration Act (Pittman-Robertson Act), *4702*
Project in Utah approved under the Federal Aid in Wildlife Restoration Act (Pittman-Robertson Act), *1722*
Screwworm infestations in Florida and Georgia, *1171*
Sit-down strikes to cripple General Motors car production, *1544*
Snow avalanche to destroy a mining camp near Alta, UT, *4198*
Snowmobile, *4637*
Soil conservation district in the United States, *4040*
South Dakota program authorized by the Federal Aid in Wildlife Restoration Act (Pittman-Robertson Act), *4703*
Storm sewer building project of significance in the District of Columbia, *4533*
Two states to adopt President Franklin Roosevelt's Standard State Soil Conservation District Laws, *4039*
U.S. legislation providing federal aid for up to 50 percent of the cost of reforestation of farmlands on the Great Plains, *2728*
Warnings against tapping groundwater near industrial developments, *2835*

1938

Albatross colony located on a mainland, *1663*
Appearance of the Asiatic chestnut blight in Europe, *2602*
Aquarium in the United States for large marine animals, *4934*
Atmospheric carbon dioxide link to human activities, *1837*
Emergence of Oregon as the leading U.S. timber-producing state, *2661*
Extensive U.S. government standards for quality and labeling of foods, *1227*
Federal duck stamp in the United States to feature the pintail, *3171*
Federal law in the United States to develop salmon fisheries in the Columbia River Basin, *2495*
Flood in the United States to claim more than 600 lives, *2541*
Government-backed crop insurance in the United States, *1228*
Green belt legislation for London, *3083*
Levees destroyed in China to halt a foreign invasion, *2569*
Limited-access highway in the United States built like an interstate, *1563*
Model conservationist farm in the United States, *1131*
Nation to adopt building codes with noise standards, *3433*
National historic site in the United States National Park System, *3801*
Natural energy directly regulated by the U.S. government, *2051*
Oil wells seized from foreign companies, *3699*
Project after the 1938 appropriation by the U.S. Congress to assist state wildlife restoration projects, *4708*
Refuge for whooping cranes, *1724*
Strong claim that industrial burning of fossil fuels increased carbon dioxide levels in the atmosphere, *1448*
Tennessee program authorized by the Federal Aid in Wildlife Restoration Act (Pittman-Robertson Act), *4705*
Texas program authorized by the Federal Aid in Wildlife Restoration Act (Pittman-Robertson Act), *4706*
Underwater action picture studio and marine attraction, *4932*
Vermont program authorized by the Federal Aid in Wildlife Restoration Act (Pittman-Robertson Act), *4709*
Wildlife management project in West Virginia, *4707*
Zoo to have a "baby pet zoo", *4933*

1939

DDT (Dichlorodiphenyl-Trichloroethane) synthesized for use as an insecticide, *1287*
Edition of *Empire of Dust* by Lawrence Svobida, *3254*
Federal duck stamp in the United States to feature the green-winged teal, *3172*
Indiana program authorized by the Federal Aid in Wildlife Restoration Act (Pittman-Robertson Act), *4710*
Issue of *Forestry Abstracts*, *2741*
Law in Rhode Island to protect endangered plants, *2326*
Laws in the United States to include a cost-benefit analysis of industrial wastes, *2836*
Meteorological sensors on balloons in Great Britain, *1940*
Nebraska program authorized by the Federal Aid in Wildlife Restoration Act (Pittman-Robertson Act), *4711*
Official use of the term *wilderness* to classify U.S. federal lands, *4638*
Plant Materials Center (PMC) in the United States, *1795*
Research on the interaction of wind and cold on the human body, *3859*
Rhode Island project initiated by the Federal Aid in Wildlife Restoration Act (Pittman-Robertson Act), *4712*
Snow cruiser for the Antarctic research, *3860*

State in the United States to permanently employ wildlife pathologists, *4713*
Wind chill calculations, *1941*

1940s

Comparative study of organic and conventional farming, *1260*
Completely sun-oriented residential community in the United States, *4105*
Use of high-speed electronic computers for U.S. weather forecasts, *1943*
Widespread spraying of the pesticide DDT to eradicate gyspy moths in the United States, *2603*

1940

Detailed five-day U.S. Weather Bureau forecasts, *1942*
Edition of *Empire of Dust* by Lawrence Svobida, *3254*
Effective control measures for solid-fuel stationary sources of air pollution in the United States, *1390*
Federal agency charged with comprehensively monitoring and protecting all wildlife in the United States, *4768*
Federal duck stamp in the United States to feature the black duck, *3173*
Federal law to protect the bald eagle, *2384*
Federal protection for bald eagles in the lower 48 states of the United States, *1694*
Federation of Sewage Works Association meetings, *4534*
High altitude meteorological station in Antarctica, *3862*
Large-scale fire prevention publicity campaign in the United States, *2617*
Limited-access highway in the United States built like an interstate, *1563*
Maximum Allowable Concentrations (MACs) for hazardous substances in the United States, *2837*
Meteorologist to reproduce natural snow crystals in plastic, *4205*
Permanent American base in Antarctica, *3861*
Public photography at the Bronx Zoo, *4935*
Reservoir to cover four towns and affect seven others in the United States, *2101*
Return of normal rainfall to the area of the central United States hit by the Dust Bowl, *2152*
Ross's goose breeding grounds, *1738*
Treaty obliging the United States to maintain a list of endangered species, *3021*
Treaty to establish national parks and preserves and to identify protected flora and fauna in the Western Hemisphere, *3022*
Tree farm movement in the United States, *2729*

1941

Aerosol spray containers for insecticide that were portable, *1321*
Coal-mine inspection law by the U.S. government, *3327*
Federal duck stamp in the United States to feature the ruddy duck, *3174*
Federal recommendation that areas of the San Joaquin Valley in California be taken out of agricultural use because of naturally high selenium concentrations, *4026*
Montana program authorized by the Federal Aid in Wildlife Restoration Act (Pittman-Robertson Act), *4714*
Mountain range to become a work of art, *3175*
National timber industry trade group in the United States, *2704*
Proof that radar could be used to locate precipitation, *1944*
Televised weather forecast in the United States, *1945*
Tree farm in the United States, *2730*
Wind turbine, *4852*

1942

Amphibian to become extinct in U.S. history, *2371*
"Atomic city" in the United States, *3452*
Coal mining accident in the Far East causing catastrophic deaths, *3328*
Federal duck stamp in the United States to feature the widgeon, *3176*
Federal requirement in the United States for bacteriological analysis of drinking water systems, *4590*
Hazardous substances transportation law in France, *3930*
Meteorological data reported from airplanes in England, *1946*
Person known to survive being struck by lightning seven times, *4276*
Road link between Alaska and the lower 48 states, *1564*
Volunteer tornado spotters network established by the U.S. government, *4287*

1943

Aqualung or diving tank, *3573*
Attack on the German-controlled oilfields at Ploesti, Romania, by the U.S. Air Force, *3667*
Comprehensive study on wild turkeys, *1739*
Conference to discuss the scope of the newly established international Food and Agriculture Organization, *1267*
Federal duck stamp in the United States to feature the wood duck, *3177*
Hurricane report that was censored in the United States, *4253*

1943—*continued*
Hydroelectric power plant in the United States to produce a million kilowatts, *4560*
Indication of a new type of air pollution, *1449*
Pilot to fly into a hurricane intentionally, *4252*
Public health use of DDT, *3931*
Report of wild temperature variations on simultaneous readings from nearby stations in South Dakota, *1821*
Scientific paper demonstrating that hydrocarbons and nitric oxide react in sunlight to form ozone pollutants, *1454*
Widespread attempts to grow yeast on pulp wastes from paper manufacturing in the United States, *4432*
Woman to advocate organic gardening, *4065*

1944

Conference on industrial wastes, *2838*
Detailed model for determining prevailing temperatures in prehistoric northern Europe, *2014*
Duck-billed platypus bred in captivity, *4750*
Environmental public service symbol of the U.S. government, *2618*
Express nationwide highway system in the United States, *1565*
Federal duck stamp in the United States to feature the white-fronted goose, *3178*
Federally funded wildlife management area in Georgia, *4715*
High-yield crops were developed, *1132*
Housing program for U.S. military veterans subsidized by the federal government, *3084*
Oil discovery at Prudhoe Bay on Alaska's North Slope, *3700*
Programs in Georgia authorized by the Federal Aid in Wildlife Restoration Act (Pittman-Robertson Act), *1695*
Public service symbol created for the U.S. government, *3179*
Styrofoam, *4128*
Two-time winner of the U.S. federal duck stamp contest, *3180*
Typhoon to sink U.S. warships, *4254*
U.S. government manual with guidelines for placement of public wells and instructions about possible contamination from industrial sites, *2839*
Use of penicillin in general clinical practice, *3932*
Volcano eruption to end a World War II battle, *4379*

1945

Air pollution control director position for a U.S. city, *1391*
Commercial use of the pesticide DDT in the U.S., *1288*
Federal duck stamp in the United States to feature a shoveler, *3181*
Hurricane entered intentionally with a large aircraft, *4255*
Kiwi bred in captivity, *1665*
Modern herbicide in the United States, *1289*
Serial publication in the United States devoted to organic agriculture, *1261*
Significant pollution abatement at the giant Du Pont de Nemours chemical plant on the Kanawha River in Belle, WV, *4433*
Soil conservation district in the United States, *4040*
Test of an atomic weapon, *3453*
United Nations agency to develop environmental education and research programs, *2282*
Use of an atomic weapon in war, *3454*
Utilization of the Federal Aid in Wildlife Restoration Act (Pittman-Robertson Act) in Hawaii, *1664*
Water supply in the United States to be fluoridated in order to reduce tooth decay, *4591*

1946

Artificial snowstorm, *1947*
Dam with a height greater than 214 meters (800 feet), *2100*
Federal duck stamp in the United States to feature the redhead, *3182*
Flood control legislation of significance in the United States, *2542*
Hibernating bird discovered, *1740*
International agency formed specifically to regulate whaling, *4627*
International whaling commission, *3057*
Law to prevent damage to fish and wildlife, *4716*
Legislation in Canada to establish procedures for handling radioactive waste, *2840*
Municipal smokeless zones in England, *1392*
Nuclear energy development agency of the U.S. government, *3455*
Organic farming association in the United Kingdom, *1262*
Reservoir to cover four towns and affect seven others in the United States, *2101*
Tsunami of significance to hit a U.S. territory, *2543*
Use of nuclear power to produce electricity in Great Britain, *3456*
World Bank meeting, *4298*

1947

Acquisition of fragile wetlands on a large scale by the United States, *3146*
Aqueduct to supply San Diego, CA, *4592*
California condor sanctuary, *1723*
Federal duck stamp in the United States to feature a snow goose, *3183*
Forest fire battled using artificial rain, *2620*

Hurricane experiment to reduce cloud intensity, *4256*
Hurricane warning service to operate around the clock, *4257*
Ionosphere direct measurements, *1933*
Joint arctic weather stations, *1948*
Law of importance in the United States regulating pesticide use, *1290*
Modern-day large-scale planned U.S. suburban development to use assembly-line methods of construction, *3075*
Nevada program authorized by the Federal Aid in Wildlife Restoration Act (Pittman-Robertson Act), *4717*
Offshore oil to be accessed from a modern platform out of sight of land, *3653*
Open hunting season on the chukar in the United States, *1666*
Published warnings against using chemical fertilizers, *4066*
Radiocarbon or C-14 dating, *2820*
Research of significance by the U.S. government on the peaceful uses of nuclear energy, *3457*
Significant federal legislation attacking forest pests in the United States, *2604*
Smog-related deaths recorded in the United States, *1341*
U.S. Fish and Wildlife Service jurisdiction over sockeye salmon conservation, *2459*
Use by Smokey Bear of the fire prevention slogan "Remember—only you can prevent forest fires", *2619*
Waterfowl management area developed by the Missouri Conservation Commission, *1725*
Winter in Great Britain in the 20th century during which snow fell somewhere in the country every day from January 22 to March 17, *4206*
Women to spend the winter in the Antarctic, *3863*
Worldwide organization concerned with civil aviation safety, *3933*

1948

Acid rain measurement monitoring systems, *1450*
Agency to administer national water pollution legislation in the United States, *4535*
Air pollution control districts in a U.S. state, *1393*
Anglers' Co-operative Association request for an injunction against a water polluter under English Common Law, *4456*
Discovery of a bristle-thighed curlew's nest, *1699*
Federal duck stamp in the United States to feature a bufflehead, *3184*
Female botanist to become a member of the Royal Society in England, *1796*
House completely heated by solar energy in the United States, *4107*
Interagency government property transfers for wildlife conservation in the United States, *4769*
International agency to address pollution of the oceans by ships, *3574*
International organization of significance dedicated to protecting and conserving global resources, *1037*
International whaling commission, *3057*
Introduction of the discipline of animal ecology, *1053*
Multistate clean streams campaign for the Ohio Valley, *4457*
National Health Service in Great Britain, *3935*
Population control book published after World War II, *3255*
Public health agency of the United Nations, *3934*
Rediscovery of the takahe, *1667*
Seismic Tidal Wave Warning System, *2224*
Solar-heated school, *4106*
Tornado forecast for a specific U.S. location, *4288*
U.S. preserve to feature indigenous wildlife in a natural setting, *4822*
Version of the World Conservation Union (IUCN), *3058*
Water pollution law enacted by the U.S. Congress, *4458*

1949

Ban of aircraft access to Superior National Forest, *4639*
Charter for the National Trust for Historic Preservation in the United States, *3109*
Energy museum that began in a cafeteria, *4936*
Federal duck stamp in the United States to be priced at two dollars, *3185*
Federal duck stamp in the United States to feature a goldeneye, *3186*
Inter-American Tropical Tuna Commission, *2460*
Interagency government property transfers for wildlife conservation in the United States, *4769*
Interstate forest-fire protection agreement in the United States, *2621*
Legislation of significance in the United States promoting and funding mass public housing, *3085*
Multilingual forest terminology, *2742*
Museum dedicated to explaining the peaceful uses of atomic energy to the public, *3458*
Nuclear-weapons tests in the atmosphere in the Pacific, *1342*
Secondary sewage treatment program in the District of Columbia, *4536*

1949—*continued*

State in the United States to forbid the posting of advertising flyers on telephone poles, *3269*
Television weather forecast in Great Britain, *1939*
United States National Air Pollution Symposium, *1451*
Use of antibiotics as an animal feed additive, *1172*
Use of the term *desertification*, *4043*
Use of the term *desertification* to describe deforestation of tropical and subtropical Africa, *2743*
Weather prediction system using modern numerical systems, *1949*

1950s

Accidental distribution in Turkey of feed grain treated with toxic hexacholorobenzene, *2841*
Careful scientific monitoring of Lumsden Lake in Ontario, Canada, for acid rain, *1453*
Control method used on sea lampreys in the Great Lakes, *2461*
Coral reef field research station in Australia, *3575*
Instrument to measure carbon dioxide in the atmosphere in parts per million, *1452*
Nile perch in Lake Victoria, *1524*
Oral contraceptive, *3880*
Organic farming organization in Australia, *1263*
Sierra Club chapter formed outside of California, *1057*
Sri Lankan hydrilla in the United States, *1525*
Widespread incidence of poisoning on account of mercury pollution, *4434*
Wildlife research unit in the United States, *4718*

1950

Canadian–U.S. treaty to regulate use of the Niagara River for water power, *4555*
Coal mines to be fully automated in the United States, *3329*
Environmental public service symbol of the U.S. government, *2618*
Federal duck stamp in the United States to feature a trumpeter swan, *3187*
Important federal legislation to provide technical forestry services to private nonindustrial landowners, *2637*
Municipal composting plant in the United States, *4041*
Nuclear generator accident of significance, *3459*
Numerical weather forecast on an electronic computer, *1950*
Plutonium-producing nuclear reactor in England, *3460*
State to require the registration of foresters, *2684*
Tax on fishing equipment in the United States to specifically benefit wildlife habitats, *2507*
Two-time winner of the U.S. federal duck stamp contest, *3180*
U.S. National Science Foundation (NSF), *2283*
Year oil became the major energy source in the United States, *3621*

1951

Controls on the import and export of plants and produce, *3023*
Dam built by Canada's Prairie Farm Rehabilitation Administration, *2102*
Detergent industry research for an answer to the problem of foam in tap water, *4507*
Electric power from nuclear energy, *3461*
Evidence of toxic chemicals in groundwater near the U.S. government's Rocky Mountain arsenal in Colorado, *2842*
Federal duck stamp in the United States to feature a gadwall, *3188*
Hurricanes named by the U.S. Weather Bureau, *4258*
Ice storm to cause $100 million in damages in the United States, *4272*
Major oil producer to fully nationalize the oil industry, *3668*
National Health Service in Great Britain, *3935*
Oil discovery in the Persian Gulf, *3654*
Plutonium-producing nuclear reactor in England, *3460*
Private environmental organization of significance in the United States to purchase and preserve substantial natural habitats, *1038*
Research park in the United States, *3724*
Science and technology museum that began as a state agricultural exhibition hall, *4937*
Scientific paper demonstrating that hydrocarbons and nitric oxide react in sunlight to form ozone pollutants, *1454*
U.S. Soil Conservation Service chief, *4056*
Underground nuclear weapons test explosion at the Nevada Test Site (NTS) northwest of Las Vegas, *3462*
World Meteorological Organization (WMO) Congress, *1822*

1952

Atmospheric tests of a new, more powerful atomic weapon, the hydrogen or thermonuclear bomb, *1322*
Caribbean monk seal extinction report, *2372*
Casualties of the great London killer fog of December 1952, *1343*
European bison reintroduction, *2427*
Federal duck stamp in the United States to feature a harlequin duck, *3189*

Instigation in the United States of the use of easements to preserve parklands from urban expansion, *3820*
Instrument that could detect hazardous chemicals such as PCBs (polychlorinated biphenyls), *2843*
Introduction into the United States of the method of timber cruising called "variable plot cruising", *2662*
Large-scale federal legislation aimed at purifying salty waters in the arid western United States, *4459*
Model conservationist farm in the United States, *1131*
Severe storm forecasting service of significance in the United States, *4186*
Significant air pollution control program to be adopted in the United States, *1394*
Smog episode to cause widespread loss of life, *1344*
Species of bird to be eliminated by volcanic eruption, *2373*
Tornado designation publicly used by the U.S. Weather Bureau, *4289*
Tropical storm to hit the United States in February, *4187*
Upper-atmosphere study with a rocket launched from a balloon, *1951*
Widespread use of DDT to spray for the tree-killing spruce budworm in Canada, *2605*
Worldwide organization to advocate family planning, *3897*

1953

Accurate DNA structure description, *2753*
Bathyscaph, *3576*
Federal duck stamp in the United State to feature a blue-winged teal, *3190*
Flood leaving 1 million people homeless in the Netherlands, *2544*
Hurricanes to be given feminine names, *4259*
International convention on ocean pollution, *3577*
Mass fatalities from eating poisoned fish and shellfish, *3936*
Mass-produced car with a fiberglass body, *1597*
Minamata Disease or "cats dancing disease" cases, *2972*
National antilitter organization in the United States, *3270*
Nuclear energy proposal at an international level, *3464*
Nuclear power organization of the Australian government, *3463*
Oil refinery to achieve complete conversion of waste gasses into useful power with a carbon monoxide boiler, *3701*
Recorded blanket of killer smog to cover New York City, *1345*
Reports of the virtual disappearance of the mayfly in Lake Erie, *2374*
Storm tide warning service in Great Britain, *4188*
Substantial radiation "hot spot" to be observed far from the site of an atmospheric nuclear test, *1346*
Tornado in the eastern United States to claim 90 lives, *4290*

1954

Automobile rotary engine, *1545*
Federal duck stamp in the United States to feature a ring-necked duck, *3191*
Hurricane of significance with a feminine name to hit the United States, *4260*
International convention to address oil pollution of the oceans, *3024*
Interstate forest-fire protection agreement in the southern United States, *2622*
Law in the United States for watershed protection, *4593*
Legislation in the United States to specifically control the disposal of nuclear waste, *3465*
Live television reports of weather in England, *1952*
Long-range numerical weather forecasts by computer, *1953*
Major public outcry on atmospheric nuclear-weapons testing, *1323*
Mass-membership animal rights organization in the United States, *1014*
Nuclear power station, *3467*
Nuclear-powered submarine, *3466*
Nuclear technology access of significance by U.S. corporations, *3468*
Offshore oil wells in Trinidad, *3655*
Phase-contrast cinemicrography film to be shown on television, *3266*
Photovoltaic cell or solar battery to work effectively, *4108*
Scientist to study snow with modern X-ray equipment, *4207*
Time the Yangtze and Hwai rivers flooded to over 96 feet, *2545*

1955

Air pollution study sponsored by the U.S. government, *1455*
Appearance of the Colorado potato beetle in Europe, *1519*
Coffinite mineral, *3330*
Commercial whale-watching ventures, *2239*
Development by the United States of small nuclear reactors for use in space, *3469*
Eight-day stretch of 100-plus degree heat in Los Angeles, CA, *1824*
Electric power from nuclear energy used to illuminate a town, *3471*

1955—*continued*

Electric power generated by nuclear energy to be sold commercially, *3472*
Federal duck stamp in the United States to feature a blue goose, *3192*
Government survey of wildlife-related recreation activities by the U.S., *4719*
Heat wave in the United States known to have lasted for two months, *1823*
House with solar heating and radiation cooling in the United States, *4109*
Major hydroelectric power-producing dam on the Zambesi River in Africa, *2071*
Mass polio vaccination campaign, *3937*
National Air Pollution Control Act in the United States, *1395*
National antilitter organization in Great Britain, *3271*
National Hurricane Center in the United States, *1954*
Nuclear energy nonmilitary joint projects of the U.S. government and industry, *3470*
Pollution-control data swapped by U.S. car companies, *1598*
Rabbits in Antarctica, *1513*
Research showing acid-forming pollution could travel great distances, *1456*
Seagoing oil drilling rig for drilling in water more than 100 feet deep, *3656*
Smog emergency alert plan in Los Angeles County, *1347*
Solar-powered car, *4110*
State law to revoke camping permits for littering convictions, *3272*
State law to revoke fishing, hunting, and trapping licenses for littering convictions, *3273*
Survey of recreational fishing on a nationwide basis in the United States, *2508*
Survey of recreational hunting on a nationwide basis in the United States, *2921*
Survey of wildlife-related recreation on a nationwide basis in the United States, *4720*
United Nations meeting directly concerned with peaceful uses of nuclear energy, *3473*
Widely used chemical herbicide regarded as benign in the environment, *1291*
Woman to fly over the North Pole, *3864*

1956

Agency in the United States created to deal exclusively with the problem of water pollution, *4460*
Balloon logging experiments, *2663*
Bill to require humane slaughter of animals in the United States, *1015*
Comprehensive U.S. national fish and wildlife policy, *4770*
Express nationwide highway system in the United States, *1565*
Federal duck stamp in the United States to feature a merganser, *3193*
Federal grants-in-aid to U.S. municipalities for pollution cleanup and control, *4461*
Game animals on U.S. postage stamps in 1956, *2922*
Gorilla born in captivity in the United States, *4751*
Interstate section in the United States, *1566*
Large-scale epidemic of fungicide poisoning from treated grain seeds, *1292*
Modern nature reserve in China, *4823*
National survey of wetlands in the United States, *3147*
Nuclear power agency established by the Turkish government, *3474*
Nuclear power agency of the Japanese government, *3476*
Nuclear power plant in England, *3475*
Oil discovered in Nigeria, *3702*
Smoke density standard measurement in England, *1457*
Solar-heated office building, *4111*
Soviet law to ban polar bear hunting, *2956*
Waterfowl Depredations Prevention Act in the United States, *4800*
Zoo to breed a cheetah, *4938*

1957

Accident aboard the USS *Nautilus*, the world's first nuclear-powered submarine, *3480*
Announcement that Brazil would transfer its capital from Rio de Janeiro to a new city to be built in the undeveloped interior, *2717*
Atomic energy organization of the United Nations for peaceful nuclear power, *3478*
Classification of rivers, *4508*
Colony of puna flamingos to be discovered, *1700*
Federal duck stamp in the United States to feature an American eider, *3194*
Global warming report to gain international scientific attention, *1838*
Groundwater contamination inventory in the United States, *2844*
Hurricane reported to occur during early summer in the United States, *4261*
International Antarctic meteorological station, *3865*
National park dedicated to the first transcontinental railroad, *3802*
National preserve created to protect the giant sable antelope, *4824*
Nuclear accident of significance in the Soviet Union, *3482*
Nuclear power from a civilian reactor in the United States, *3477*
Nuclear power-plant accident of significance, *3479*

Nuclear power plant exclusively used for peaceful purposes, *3481*
Oil discovery of major significance in Alaska, *3703*
Potomac River pollution abatement conference, *4400*
Program in Guam authorized by the Federal Aid in Wildlife Restoration Act (Pittman-Robertson Act), *4721*
Statewide statutes authorizing preferential tax assessment of farmland in the United States, *3110*
U.S. National Wildlife Reserve created specifically for Key deer, *2327*
Wilderness bill introduced in the U.S. Congress, *3736*
Zoo to breed a Chilean flamingo, *4939*

1958

Air pollution ordinance enacted in Chicago, *1396*
Bathyscaph, *3576*
Blizzard to cause more than $500 million damage in the United States, *4208*
Carbon dioxide monitoring station, *1458*
Cloud-seeding demonstration, *2170*
Daily measurements of atmospheric carbon dioxide, *1839*
Elk reintroduced in Europe, *2428*
Eradication of the flesh-eating screwworm in the southeastern U.S., *1293*
Exports of Datsun cars from Japan to the U.S., *1546*
Federal fish-agriculture rotation program in the United States, *2496*
Female pioneer in the field of rubber recycling, *2284*
Food additive regulation passed by the U.S. Congress, *3938*
Government-built saline water conversion demonstration plants in the United States, *4462*
Harbor created by an atomic bomb, *3483*
International campaign to eradicate malaria, *3939*
International moratorium on above-ground nuclear weapons testing, *3484*
International organization to promote water quality monitoring, *4594*
Large-scale United Nations program to fund forestry projects, *2705*
Measurement of carbon dioxide levels in the atmosphere, *1459*
Natural resources inventory to employ statistical sampling in the United States, *3372*
Nuclear accident of significance in the Soviet Union, *3482*
Offshore oil drilling of significance in the Trucial States (now United Arab Emirates), *3657*
Public hearing on meatpackers' discharge of offal into the Missouri River sponsored by the U.S. Public Health Service, *4435*
Renovation of significance to the dike system in the Netherlands, *2546*
Reproduction of U.S. federal duck stamps, *3196*
Satellite instruments powered by solar energy, *4112*
TFM chemical releases into the lamprey-infested waters of Lake Superior, *1526*
Tourists to visit the continent of Antarctica, *2240*
Waterfowl species to appear twice on U.S. federal duck stamps, *3195*

1959

Artist to win the U.S. federal duck stamp contest three times, *3199*
Automobile emissions standards set by a U.S. state, *1572*
British four-lane limited-access highway, *1567*
Carefully documented reports of a link between acid rain and damage to fish populations, *1460*
Compact car built by General Motors, *1547*
Corporation to join the newly created Research Triangle Institute in North Carolina, *3725*
Dam disaster of significance in Spain in the 20th century, *2117*
Federal duck stamp in the United States to feature an image other than waterfowl alone, *3197*
Federal duck stamp in the United States to sell for $3.00, *3198*
Federal law in the United States to ban the use of motorized vehicles in capturing wild horses, *4771*
Index for temperature-humidity, *1955*
International conference on air pollution, *1461*
International treaty regarding the use of Antarctica, *3025*
Landslide in Mexico on record to kill more than 2,000 people, *2573*
Marine mammals studied by the U.S. military, *4722*
Nuclear energy produced by a fast-breeder reactor in Great Britain, *3485*
Nuclear power agency of the South Korean government, *3486*
Nuclear power plant in the United States built without government funding, *3489*
Nuclear-powered merchant ship, *3488*
Probable appearance of the cereal leaf beetle in the United States, *1192*
Public outcry of significance in the United States about cancer-causing chemicals in food, *1294*
Research park inside an "education triangle", *3726*

1959—*continued*
Satellite to offer useful weather data, *1956*
Satellite to transmit weather information to the earth, *1957*
Snowmobile, *4637*
Tests by the military of the defoliant Agent Orange, *1295*
U.S. nuclear-powered surface warship, *3487*
U.S. submarine to carry nuclear-armed ballistic missiles, *3490*

1960s

Attempts at climate modeling, *1958*
Commercial land disposal of low level nuclear waste in the United States, *2845*
Conclusive link between asbestos and lung cancer, *1348*
Exploits of "The Fox," the anonymous Chicago-area environmentalist, *1058*
Government programs in the United States to assist migrant workers and their families, *1252*
Helicopter logging experiments in the United States, *2664*
Military uses of marine mammals, *4723*
"Miracle rices", *1253*
Modern analysis of significance into the chemistry of rain, *1959*
Rankings by the U.S. government of the nation's most polluted cities, *1432*
Reports of decreases in commercial fishermen's catches in Lake Superior, *4401*
Scientific explanation for the distribution of volcanoes around the world, *4380*
Significant modern studies of elephant behavior in the wild, *4724*
Signs of damage to trees far from automobile pollution sources, *2606*

1960

All-weather satellite, *1960*
Broad-based animal welfare organization to oppose the slaughter of fur seals, *1016*
Census of fish kills by pollution in the United States, *4415*
DES hormone ban by the U.S. government, *4422*
Earthquake to register a magnitude of 9 or more on the Richter scale, *2225*
Government program in the United States to screen plant and animal species for natural products with anticancer properties, *2744*
International oil cartel, *3622*
Large-scale epidemic of fungicide poisoning from treated grain seeds, *1292*
Law establishing comprehensive control of radioactive material by the British government, *3491*
Law stating that U.S. forests would be exploited for a variety of recreational, commercial, and conservation purposes, *2685*
Manned descent to the greatest depth on earth, *3578*
Nuclear power station to operate on a commercial basis, *3494*
Nuclear reactor to power space vehicles, *3493*
Nuclear weapons test conducted by the French, *3492*
Research and development program for water pollution control sponsored by the U.S. government, *4509*
Satellite to provide cloud-cover photographs, *1825*
Scarlet ibis introduced into the United States, *1668*
Strict standards for maintaining stream temperatures in a U.S. state, *4436*
Sulfur mine offshore, *3331*
Undersea park established by the U.S. government, *3803*
Use of the term *bionics*, *2285*
Wildlife conservation law for U.S. military property, *4772*

1961

Automobile crankcase emissions law in the United States, *1573*
Deaths caused by an American nuclear reactor, *3495*
Disposable diapers, *4129*
Disposable diapers mass-marketed in the United States, *4537*
Formal identification of the line of continuous development along the U.S. eastern corridor from Washington to Boston, *3143*
International convention on ocean pollution, *3577*
International moratorium on above-ground nuclear weapons testing, *3484*
International treaty regarding the use of Antarctica, *3025*
National parkland acquired with congressionally authorized funds, *3804*
Oil refinery in Singapore, *3704*
Satellite to use nuclear power for electricity, *3496*
Scholarly article to show that Lake Erie was dying a slow death from pollution, *4510*
Seawater conversion plant in the United States on a practical scale, *4595*
Tests by the military of the defoliant Agent Orange, *1295*
Wetlands Loan Act in the United States, *3373*
Widely publicized version of the Christian fundamentalist theory of flood geology, *2547*
Wildlife protection conference of modern African governments, *4773*

1962

Advanced pollution-control system introduced by a U.S. car manufacturer, *1599*
Coal-mine fire to depopulate a U.S. city, *3332*
Dam with a height greater than 273 meters (900 feet), *2103*
Drought in modern United States to affect 28 percent of the population, *2153*
Establishment of a federal Bureau of Outdoor Recreation in the United States, *2686*
Federal legislation in the United States authorizing recreational uses of wildlife conservation areas managed by the Department of the Interior, *4825*
Flood of significance in Spain during the 20th century, *2548*
Industrial waste treatment program at an airport, *2973*
International conference on wetlands, *3374*
Law in China to outlaw hunting giant panda, *2957*
National fisheries center approved by Congress, *2462*
National park in Thailand, *3727*
Nuclear power plant in Antarctica, *3497*
Popular book to warn of pesticide threat, *3256*
Recommendations for protecting landscapes issued by the United Nations, *3977*
Smog chamber for air pollution research built by an industrial organization, *1462*
Theory that seafloor spreading was caused by convective heat in the Earth's mantle, *2818*
Wild horse range overseen by the U.S. government, *4801*
World Charter for Nature, *1059*

1963

Air pollution law of importance enacted by the U.S. Congress, *1398*
Bird known to fly at an altitude of 21,000 feet, *1701*
Complete freeze of Lake Constance (the Bodensee) in West Germany in more than two centuries, *1826*
Federal duck stamp in the United States to feature a Pacific brant, *3200*
General Fisheries Council for the Mediterranean, *2474*
Giant panda born in captivity, *4752*
Jetliner to be destroyed in flight by lightning, *4277*
Major destruction by starfish of coral in the Great Barrier Reef, *3579*
Modern disaster blamed on faulty geologic analysis, *2118*
Monument dedicated to a songbird, *3201*
Multilateral treaty to address an air pollution problem, *1397*
National Air Pollution Control Act in the United States, *1395*
Nuclear power barge, *3498*
Nuclear submarine from the U.S. Navy to be lost at sea, *3499*
Organization for the control of desert locusts, *3059*
Planetarium constructed by a school system in reponse to a space launch, *4969*
Plate tectonics theory verification, *2821*
Rocket trials for hailstone protection, *4218*
Science center originally designed for a world's fair in the United States, *2286*
Serious evidence of acid rain in the United States, *1463*
State nature preserve in North Carolina, *4826*
Televised dam collapse, *2119*
U.S. National Wildlife Refuge created specifically for the dusky seaside sparrow, *2328*
Underwater state park in the United States, *3728*
United Nations Fund for Population Activities (UNFPA), *3881*
Volcanic island scientifically studied during its creation, *4381*
Wildlife resource management college in Africa, *4774*

1964

Agreement governing treatment of Antarctic wildlife and plants, *3026*
Air pollution index broadcast on television, *1349*
Federal duck stamp in the United States to feature a Hawaiian nene goose, *3202*
Federally sponsored water resource research in the United States, *4402*
International Council for the Exploration of the Sea (ICES), *3566*
International Joint Commission on Boundary Waters study of Lake Erie, Lake Ontario, and the upper St. Lawrence River, *4511*
List of protected rivers in the United States, *4463*
National recreation area designated in the United States, *3806*
National Scenic Riverway law in the United States, *3978*
Nuclear-powered lighthouse, *3500*
Oil refinery construction halted by environmental protests in Japan, *3705*
Oil refinery in South Korea built with private funds, *3706*
Orbiting geophysical observatory, *2822*
Proposals for a federal tax on water polluters in the United States, *4464*
River to be declared a national Scenic Riverway in the United States, *3805*
Rivers managed by the U.S. National Park Service, *4641*
Satellite to provide high resolution night photographs of cloud-cover, *1827*

1964—*continued*

Tsunami to devastate the U.S. West Coast, *2549*
U.S. National Wilderness Area in New England, *4640*
U.S. National Wilderness Area in the Superior Upland region, *4642*
U.S. National Wilderness Areas in the interior West, *4643*
U.S. National Wilderness Areas in the Pacific Northwest, *4644*
U.S. National Wilderness Areas in the Southeast, *4645*
U.S. National Wilderness Areas in the Southwest, *4646*
Uranium fuel ownership by private U.S. companies, *3501*
Use of elevation benchmarks to measure settling or uplifting of the earth, *2226*
Variety of high-yield rice, *1133*
Wetlands conservation book published by the U.S. government for a general audience, *3375*
Wilderness Act in the United States, *4647*
Wilderness bill introduced in the U.S. Congress, *3736*
Wilderness refuge system in the United States, *4827*
Woolly monkeys successfully bred in captivity, *4753*

1965

Anadromous fish conservation law in the United States, *2463*
Automobile emissions bill approved by the U.S. Congress, *1574*
Automobile safety warning having national impact, *3940*
Compact car built by General Motors, *1547*
Discovery of the Hutton's shearwater breeding grounds, *1702*
Effort in the United States to address severe pollution in Lake Erie, *4539*
Environmental law case in which citizens were granted standing for a noneconomic interest, *3737*
Federal law in the United States to restrict use of billboards in scenic areas, *3980*
Federally sponsored interstate air pollution abatement conference, *1399*
Flood crest on the Mississippi River at St. Paul, MN, to exceed the previous record by 4 feet, *2551*
Flood legislation by the U.S. government to emphasize the environment, *2550*
Formal earthquake prediction research program in quake-prone Japan, *2227*
Generally known instance of a river catching fire because of pollution, *4437*
Index to scientifically measure droughts, *2154*
International Conference on Polar Bears, *4775*
Introduction of alligatorweed-eating adult flea beetles in Florida, *4503*
Iowa State Preserves System, *3376*
Lawsuit by a conservation group to overrule a decision by a U.S. federal regulatory agency, *3979*
Legislation in the United States to provide for use of fees from offshore oil drilling for land purchases and historic preservation, *3658*
List of wilderness areas in the U.S. National Wilderness Preservation System (NWPS), *4648*
Meteorological satellite funded by the U.S. Weather Bureau, *1961*
New York highway-noise emissions limits, *3434*
Oil exploration and production in a U.S. National Forest, *3707*
Organization for the control of desert locusts, *3059*
Proposal for a national network of living historical farms in the United States, *1134*
Public statement by a U.S. president that all of the country's major river systems were polluted, *4403*
Recorded nuclear weapons accident involving a U.S. warship at sea, *3503*
Reports that latrines at the U.S. Military Academy at West Point were seriously polluting the Hudson River, *4538*
"Save Lake Erie" petition, *4465*
Set of tornadoes to cause $500 million in damages, *4291*
State to own one million acres of game lands, *2923*
Titan II silo explosion, *3502*
U.S. law allowing consideration of recreation, fish, and wildlife enhancement as purposes of federal water development projects, *4561*
United Nations hydrological agency, *4596*
Water pollution national conference of significance sponsored by the U.S. Chamber of Commerce to address the issue of industrial waste, *4438*
Water quality standards established at the U.S. federal level, *4466*

1966

Appearance of the wooly whitefly in California, *1193*
Automobile safety advocate of note in the United States, *1542*
Automobile safety legislation of significance passed by the U.S. Congress, *1575*
Ban on mercury fungicides in agriculture, *1296*
Claim by the Centers for Disease Control that malaria had been eradicated in the United States, *3941*

Consolidation of various types of lands into a single U.S. National Wildlife Refuge System, *4828*
Earthquake data center established by the U.S. government, *2228*
Endangered species list by the U.S. government, *2385*
Endangered species list to include birds, *1669*
Environmental health organization of significance started by the U.S. government, *3942*
Exploratory offshore oil wells in Canada, *3659*
Federal duck stamp in the United States to feature a whistling swan, *3203*
Federal law in the United States closely regulating the use of animals in medical and commercial research, *1017*
Federal law in the United States to protect fur seals, *4776*
Flood-danger evaluations required by a U.S. presidential executive order, *2552*
International body to regulate conservation of Atlantic tunas, *3060*
International organization to regulate tuna fisheries, *2475*
Law in the United States to require laboratories to furnish adequate food and shelter for test animals, *1018*
Law passed by U.S. Congress specifically aimed at preserving endangered birds, *1696*
Law requiring the U.S. president to develop and conduct a comprehensive program of marine science activities, *3581*
Laws requiring catalytic converters for automobiles in the United States, *1576*
Major landslide of a slag heap in Wales, *2574*
Mine safety law in the United States applying to non-coal mines, *3333*
National Historic Preservation Act, *3111*
National lakeshore in the U.S., *3729*
Nuclear power commercial reactor in Japan, *3505*
Oceanographic research vessel to recover a lost hydrogen bomb, *3580*
Offshore oil discovered off Denmark, *3660*
PCB (polychlorinated biphenyl) testing to show that the chemical was highly toxic and dangerous, *2846*
Presentation of the essay "The Historical Roots of Our Ecological Crisis" by historian Lynn Townsend White Jr., *3257*
Program opening the Amazon region to economic development, *4299*
Proposed U.S. legislation regarding abandoned cars, *4142*
Reactor meltdown aboard a nuclear-powered ship, *3504*
Regulations on animal research in the United States, *2287*
Report on the effects of acid rain on fish populations, *1464*
Research grant program administered by the U.S. Department of the Interior, *2288*
Research showing that PCBs (polychlorinated biphenyls) were present in human and wildlife tissues, *2847*
Safari park, *4940*
"Shellfish clause" in the U.S. Clean Water Amendments, *4467*
Significant conference on the risks of chemical contaminants to future generations to be sponsored by the U.S. government, *4423*
Significant studies of lions in the wild, *4725*
U.S. National Sea Grant Program, *2289*
Use of the phrase *tornado watch* by the U.S. Weather Bureau, *4292*
Weather satellite monitoring station north of the Arctic Circle, *1962*
Workshops on hunter safety in the United States, *2924*

1967

Antilitter campaign in Canada, *3274*
Bat listing under the U.S. Endangered Species Act, *2386*
Bear listing under the U.S. Endangered Species Act, *2387*
Dam to harness tidal power, *2104*
Darter listed under the U.S. Endangered Species Act, *2388*
Deer listings under the U.S. Endangered Species Act, *2389*
Electric power project to harness tidal motion, *4562*
Evidence of prehistoric vertebrate life in Antarctica, *3866*
Federal duck stamp in the United States to feature an old squaw duck, *3204*
Flood-plain protection advocated by a U.S. presidential executive order, *2553*
Fossil found in Antarctica, *3867*
Fox listing under the U.S. Endangered Species Act, *2390*
Large U.S. industrial state to ban the burning of high-sulfur fuels, *1400*
Law in the United States calling for ways of controlling jet aircraft emissions, *1401*
Living historical farm in the United States, *1312*
Municipal office for noise control in the United States, *3428*
National Trail in Canada, *4649*
Natural gas piped ashore from the North Sea, *2052*
Parrot listed under the U.S. Endangered Species Act, *2391*
Pennsylvania state law prohibiting aliens from owning firearms, *2918*
Reported new sighting of the taiko, *1670*
Satellite to transmit color photographs of the full Earth face, *2823*

1967—*continued*

Snake listed under the U.S. Endangered Species Act, *2392*
Standards for organic farming in Great Britain, *4067*
State governor in the United States to implement a doctrine of conservation in balance with economic concerns, *1104*
Supertanker to founder and spill its cargo of oil, *3630*
Treaty to ban the introduction of nuclear weapons into orbit around the Earth, *3027*
Use of detergents and first-generation dispersants to clean up a beach after an oil spill, *4489*
Wandering albatross recorded ashore in the United States, *1703*

1968

Ban on phenylmercury as a slimicide in the pulp and paper industry in Sweden, *4468*
City in the U.S. to adopt a building code with internal noise suppression standards, *3437*
Classification of rivers and federally-owned adjoining lands to determine their use by the public, *3731*
Dam built to totally control the annual flooding of Egypt's Nile River, *2105*
Dam with a reservoir that could hold more than 100,000 cubic meters of water, *2106*
Emergence of desertification as an international issue, *4044*
Estuary protection funding in the United States, *4470*
Federal duck stamp in the United States to feature a hooded merganser, *3205*
Federal law in the United States protecting and preserving rivers, *4469*
International conference to discuss global environmental issues, *1060*
International Council for the Exploration of the Sea (ICES), *3566*
International monitoring system for air pollution, *1465*
Large-scale astronomical observatory on a volcano, *4382*
Law in California to limit motor-vehicle noise emissions on highways, *3435*
Law in California to limit motor-vehicle noise emissions through sales restrictions, *3436*
Legislation preserving a significant extent of coastal redwood forest in the United States, *2638*
Marine geology vessel, *2824*
National Scenic Trails in the United States, *3981*
National Trails System in the United States, *3982*
National Wilderness Area in New Jersey, *3730*
Oil spill of significance off South Africa, *3631*
Publication that offered "access to tools" information for the energy-efficient and promoted ecologically aware lifestyles, *3258*
Regional wetlands development commission in New Jersey, *3377*
Report of a shorebird at latitude 71 degrees south, *1704*
Reservoir on the U.S.-Mexico border, *4620*
Tidal power plant north of the Arctic Circle, *4563*
United Nations conference linking biodiversity and sustainable development, *1622*
United Nations convention on African natural resources, *3378*
Waterways in the U.S. National Wild and Scenic Rivers System, *3983*
Wild horse range established by the U.S. Bureau of Land Management, *4829*

1969

Air Pollution Control Alert Warning System for New York City, *1350*
Aircraft noise certification standards in the U.S., *3438*
Appearance of the alfalfa blotch leaf miner in the United States, *1194*
Aquatic environment built to revitalize urban waterfronts, *4942*
Assembly of significance of environmental lawyers, law school faculty and environmental leaders in the United States, *1061*
Ban on swimming in Indiana's Brandywine Creek because of bacterial contamination, *4471*
Canadian ban on hunting baby seals, *2958*
Coal-mining law with penalty fines imposed by the U.S. government, *3334*
Comprehensive national policies for protecting the environment in the United States, *1062*
Comprehensive reform of land-use planning laws in the United States, *3107*
Comprehensive report to conclude that subtherapeutic use of antibiotics in animal feed posed a significant health risk to humans, *1173*
Declaration of phosphates as a major source of water pollution issued jointly by two countries, *4404*
Environmental movement of significance in the Soviet Union, *1065*
Environmental organization to pursue direct action as a primary strategy, *1063*
Federal duck stamp in the United States to feature a white-winged scoter, *3206*
Hazardous waste exchange program in Europe, *2848*
Incineration of liquid organohalogen compounds at sea, *4490*
International organization to regulate tuna fisheries, *2475*

International professional forestry society, *2706*
Large U.S. states to file civil suits charging airlines with violating air pollution control, *1402*
Law in the United States to protect other nations' wildlife, *4726*
Law in the United States to require an environmental impact statement for all major federal projects that could significantly affect the environment, *3086*
Local private grassroots organization to successfully pressure governmental agencies to take steps to curb air pollution, *1404*
Mass fish kill attributed to an insecticide spill, *1297*
Modern aquarium, *4941*
Mudslides to cause property damage of over $135 million, *2575*
National Wild and Scenic Rivers System in the United States, *3984*
Nuclear power accident in Switzerland, *3507*
Nuclear-powered ship launched by Japan, *3506*
Offshore oil spill to cause widespread damage, *3632*
Ohio State Scenic River, *3985*
Public disclosure of a U.S. Army proposal to dump nerve gas and live explosives 200 miles off the New Jersey coast, *4491*
Recorded evidence of quakes on the Moon, *2229*
Regional wetlands development commission in New Jersey, *3377*
Regulation of toy guns as a noise hazard in the United States, *3439*
Requirement for an environmental impact statement for development projects in the United States, *1064*
State in the United States to enact legislation to curtail aircraft emissions, *1403*
State in the United States to mandate citizen participation in refuse collection systems, *3275*
United Nations compensation policy on marine oil pollution, *3582*
United Nations convention on African natural resources, *3378*
United Nations convention to protect the vicuña, *4777*
Use of wrecked cars from a junkyard to make a barrier against a flood, *2554*

1970s

Appearance of the environmental movement known as bioregionalism, *1066*
Artificial wetlands treatment of wastewater in the United States, *4540*
Citizen lawsuit provisions allowing people to file legal actions in U.S. courts to protect the environment, *1096*
Documentation of industrial pollutants in a lake on the Lake Superior wilderness preserve Isle Royale, *4512*
Eco-industrial park, *4300*
Evidence of significant acid rain damage in the Great Britain, *1467*
Introduction of the concept of "regenerative farming", *1254*
Measurements of chlorofluorocarbons (CFCs) in the atmosphere, *1971*
Municipality to adapt the Federal Water Quality Administration's tertiary treatment system to raw sewage rather than effluent that had undergone secondary treatment, *4542*
Notice of the effects of tributylin (TBT) on marine life, *4492*
Notice of the wintertime phenomenon known as "Arctic haze", *1468*
Organizational expression of the bioregional movement, *1068*
Reports of high concentrations of toxic metals in bivalves from the U.S. Mussel Watch Program, *4493*
Resort powered by renewable energy, *4301*
State in the United States to suspend vehicle registrations of litterers, *3276*

1970

Announcement that the U.S. Army Corps of Engineers would require mandatory disclosure of industrial effluent, *4472*
Artificial gene, *2754*
Automobile emissions control law of significance in the United States, *1577*
Cabinet-level regulatory environmental agency in the United States, *1067*
Catfish farms in the United States, *2476*
Company to voluntarily limit production of PCBs (polychlorinated biphenyls), *2849*
Conservation legal organization, *2290*
Crocodilians listed under the U.S. Endangered Species Act, *2394*
Decision to build a highway into Brazil's Amazon wilderness, *2718*
Direct measurements of chlorofluorocarbon (CFC) levels at the earth's surface, *1466*
Disabling smogs in Tokyo, *1351*
Earth Day to be celebrated nationwide in the United States, *1069*
Federal aid to hunter education programs in the United States, *2925*
Federal duck stamp in the United States to feature a Ross's goose, *3207*
Geothermal Steam Act in the United States, *3379*
Gorilla listing under the U.S. Endangered Species Act, *2395*
Involvement of the U.S. Department of Commerce with commercial fisheries, *2477*

1970—*continued*
Large-scale natural waste disposal system in the United States, *4541*
Law in the United States allowing teenagers to work on federal conservation programs, *4778*
Legal proceedings against Alabama polluters by the Bass Anglers Sportsmen Society, *4473*
Legislation aimed at protecting U.S. wetlands from drainage and development, *3148*
Lion listing under the U.S. Endangered Species Act, *2396*
National standards for air polluting motor vehicle emissions in the United States, *1406*
National Wild and Scenic River in New England, *3986*
Occupational safety coverage for U.S. workers, *3943*
Oil discovery of significance in the North Sea, *3661*
Oil spill of significance off Sweden, *3633*
Organization for the control of desert locusts, *3059*
Peregrine falcons listed under the U.S. Endangered Species Act, *2393*
Phosphate-free laundry detergent, *4405*
Pollution-related ban on commercial fishing in Lake St. Clair on the Canada-U.S. border, *4474*
Prairie dog listed under the U.S. Endangered Species Act, *2397*
Publication inspired by *The Whole Earth Catalog*, *3259*
Regular and systematic monitoring of air pollution levels in the United States, *1405*
Regulation of toy guns as a noise hazard in the United States, *3439*
Regulations by the U.S. Army Corps of Engineers requiring applicants for permits to fill wetlands to prove the work would be consistent with ecological considerations, *3149*
Reports of mercury contamination in tuna and swordfish, *4513*
Rhinoceros listings under the U.S. Endangered Species Act, *2398*
Single hailstone to measure 7.5 inches in diameter, *4219*
Snail listed under the U.S. Endangered Species Act, *2399*
State-level legislation in the United States mandating review of any development for impact on natural beauty, aesthetics, and historic sites, *3087*
Tortoise listed under the U.S. Endangered Species Act, *2400*
Turtles listed under the U.S. Endangered Species Act, *2401*
U.S. Environmental Protection Agency (EPA) role in wildlife conservation, *4779*
U.S. government listing of the nation's ten most polluted rivers, *4406*
U.S. National Sea Grant Program, *2289*
U.S. National Wilderness Area to enclose a bison sanctuary, *4650*
U.S. National Wilderness Areas in Alaska, *4651*
U.S. National Wildlife Refuge created specifically for watercress darters, *2334*
Water Bank Act in the United States, *3380*

1971

Ban on lead-based paint in the United States, *1358*
Banning of aircraft for hunting in the United States, *2959*
Catalogue of critical world wetlands, *3150*
Close-up evidence of volcanoes on Mars, *4383*
Conference on Third World environmental problems, *2291*
Doppler radar system to study storms, *1963*
Environmental treaty dealing with a particular ecosystem, *1744*
Example of a whole-systems approach to the environment, *3260*
Federal duck stamp in the United States to feature a cinnamon teal, *3208*
Federal management policies for wild horses in the United States, *4780*
Forestry volunteers for the U.S. Peace Corps, *2707*
Free investigations of indoor air pollution in U.S. workplaces, *1359*
Industrial pollution lawsuit in Japan that was successful, *2974*
International convention protecting wetlands, *3028*
Issue of the periodical *Acres U.S.A.*, *4042*
Large-scale epidemic of fungicide poisoning from treated grain seeds, *1292*
Law requiring the U.S. president to develop and conduct a comprehensive program of marine science activities, *3581*
Legislation in the United States limiting lead content in paint and providing funds for cleanup, *3944*
Local environmental political parties, *1105*
Mounted bird specimen to sell for more than £9,000 sterling, *2375*
Museum devoted to forestry education that started in a world's fair, *4943*
Museum established by a university to support its own programs, *4944*
Nobel Prize was awarded to ozone-layer researchers, *1972*
Nonreturnable bottle and can law enacted by a U.S. state, *4157*
Nuclear power-plant requirement in the United States to assess overall environmental impact, *3508*
Offshore oil produced in Norway, *3662*

Offshore oil-well fires of significance in the Gulf of Mexico, *3634*
Oil spill of significance off Japan, *3635*
Ozone-layer warning, *1973*
Presidential order prohibiting U.S. government assistance to concerns that violate air emissions standards, *1407*
Prohibition against nuclear weapons on the ocean floor, *3584*
Scale to measure tornado intensities, *4293*
Self-sustaining village experiment in Colombia, *4302*
Snowflakes reported to measure 8 by 12 inches, *4209*
State in the United States to pass a comprehensive noise-control law, *3440*
Tornadoes tracked by Doppler radar, *4294*
Tropical cyclone known to have lived for a month, *4262*
U.S. Justice Department lawsuit against U.S. Steel for failing to follow government recommendations to abate pollution from its Gary, IN, works, *4475*
United Nations agency to develop environmental education and research programs, *2282*
United Nations agreement on marine pollution from offshore industrial facilities, *3583*
United Nations compensation fund for marine oil pollution damage, *3586*
United Nations convention to define civil liability in nuclear accidents at sea, *3585*
Use of the term *green*, *1106*
Zoning law in the United States to encourage a mix of shops, offices, and residences, *3156*

1972

Automobile emissions requirements in Japan, *1578*
Automobile seat-belt buzzers, *1600*
Chemical and oil waste law enacted by the Danish government, *3945*
Climate hypothesis concerning animals and plants, *1828*
Comprehensive wildlife protection legislation in India, *4781*
Disabling smogs in Tokyo, *1351*
Disaster settlement to include provisions to address the long-term psychological needs of the victims, *2556*
Drought forcing the Soviet Union to buy U.S. grain, *2155*
Electric power in the United States from municipal garbage as a boiler fuel, *4130*
Extensive collection of quality data on fertility and family planning, *3882*
Federal duck stamp in the United States to be priced at five dollars, *3210*
Federal duck stamp in the United States to feature an empress goose, *3209*
Federal protection for bald eagles in the lower 48 states of the United States, *1694*
Geologist to go to the Moon, *2825*
Global conference on environmental issues, *1840*
Global environmental conference to spur activism, *1070*
Interagency government property transfers for wildlife conservation in the United States, *4769*
International compact on marine pollution from discharges of ships and aircraft, *4494*
International convention to preserve scenery, culture, and the environment, *3987*
International convention to specifically protect Antarctic seals, *3031*
International registry of data on chemicals in the environment, *2292*
Joint effort of major significance to solve pollution problems in the Great Lakes, *4476*
Landsat satellite, *2293*
Laws regulating space debris, *3032*
Legislation in the United States to allow noise pollution lawsuits, *3442*
Marine mammal conservation and management law in the United States, *4782*
Move to halt all fungicide uses of mercury in the United States, *1298*
National environmental political party, *1107*
National land-planning measure to receive U.S. congressional approval, *3088*
Noise control agency in the United States, *3429*
Noisc cmission standards for motor vehicles involved in U.S. interstate commerce, *3443*
Noise pollution legislation of significance in the United States, *3441*
Nontechnical text on population growth to be widely distributed, *3261*
Oil spill of significance in the Gulf of Oman, *3636*
Organization to regulate dumping of wastes in the high seas, *3061*
Park to commemorate the cattle industry of the American West, from its inception in the 1850s, *3807*
Political party to take an environmental issue to the polls, *1108*
Prohibition against nuclear weapons on the ocean floor, *3584*
Proposed concept of entitlement assistance for environmental conservation, *2294*
Public health hazard causing the U.S. government to buy a town, *3946*
Reintroduction of the term *acid rain*, *1469*
Satellite composite map of the U.S., *3381*
Substantial rainfall after more than 400 years at Calama, Chile, *2555*

1972—*continued*

Successful conservation organization to use paying volunteers to sponsor and assist scientists on research trips throughout the world, *1071*
Systematic dam inspection by the U.S. government, *2072*
Tax on bows and arrows in the United States, *2926*
Treaty to combine preservation of natural and cultural heritage sites, *3033*
U.S.–Canadian effort of significance to improve the water quality of the Great Lakes, *3029*
U.S. National Wildlife Refuge created specifically for Attwater prairie chickens, *2329*
U.S. National Wildlife Refuge created specifically for Columbian white-tailed deer, *2330*
United Nations agency for management of hazardous chemicals, *2975*
United Nations agreement to regulate dumping of waste into the oceans, *3587*
United Nations condemnation of French atmospheric nuclear weapons tests, *3509*
United Nations conference devoted to the global environment, *3030*
United Nations organization to establish its headquarters in an underdeveloped country, *1072*
Urban national park to exceed 75,000 acres, *3821*
Use of the phrase "think globally, act locally", *1073*
Water shortage in China causing the Yellow River to run dry, *2156*
Wild turkey restoration program in Georgia, *1671*
Zoning scheme to take time into account, *3157*

1973

Bird flight recorded at 37,000 feet, *1706*
Comprehensive reform of land-use planning laws in the United States, *3107*
Defense laboratory operated by the U.S. government to study wind power, *4853*
Ecotourism project started by the Indian government, *2241*
Embargo on oil supplies by the oil producers of the Middle East, *3669*
Expressway in France built with private funds, *1568*
Federal court ruling declaring clear-cutting in national forests a violation of U.S. law, *2639*
Federal duck stamp in the United States to feature a Steller's eider, *3211*
Genetic engineering, *2755*
Ice meteor to be subject to intensive study, *2296*
Law to protect future endangered fish and wildlife, *4727*
Manx shearwater nest discovered in the United States, *1705*
National Hunting and Fishing Day in the United States, *2927*
National political party in Europe committed primarily to environmental and ecological issues, *1109*
Nonviolent protest of significance by the modern Chipko movement in India, *1041*
Plant to use the pollution-to-homes process, *2295*
Pollution-related national health compensation law, *3947*
Sea turtle listing to include all species, *2402*
Tiger preserve in India, *4830*
Treaty protecting the polar bear and its natural habitat, *3034*
Volcano lava flow to be fought with sea water, *4384*
Wild horse adoptions in the United States, *4802*
Zoo with twilight conditions, *4945*

1974

Authentication of 148 tornadoes over a two-day period in the United States, *4295*
Clean drinking water standards in the United States that were legally binding, *4598*
Comprehensive U.S. legislation establishing minimum safety standards for chemical contaminants in water, *4478*
Condors reintroduced to Arizona, *2430*
Dam in India with a double curvature and parabolic arch, *2107*
Destruction of Michigan cattle, chickens, and eggs contaminated with the chemical compound PBB (polybrominated biphenyl), *2850*
Discovery of a marbled murrelet's nest, *1707*
Discovery room in a natural history museum in the United States, *4947*
Evidence of hazardous waste poisoning in the Michigan town of Hemlock, *2851*
Evidence of pollution damage to the earth's stratospheric ozone layer, *1470*
Federal legislation in the United States to tax hunting and fishing equipment in order to fund improvements in wildlife habitat, *4831*
Federal Sole Source Aquifer (SSA) Protection Program in the United States, *4599*
Flight of a solar-powered aircraft, *4113*
Hooded grebe to be discovered, *1672*
Indoor air quality conference in the United States, *1360*
International organization to manage breeding animals in captivity, *4946*
Law giving the U.S. Environmental Protection Agency authority to set water quality standards and monitor water supplies, *4477*
Laws stipulating the registration of objects sent into space, *3035*

Modern drinking water improvement legislation of significance in India, *4597*
Modern world conference on population, *3898*
National government publication emphasizing risk factors and lifestyle in public health, *3948*
National pollution control agency in India, *2976*
National trade organization for wind power in the United States, *4854*
National Wild and Scenic River in the southeastern United States, *3988*
Noise standards set by the United States Occupational Safety and Health Administration (OSHA), *3444*
Nuclear power organization of the U.S. government to emphasize public health, *3949*
Ocean Thermal Energy Conversion (OTEC) laboratory established by a U.S. state, *2297*
Ozone-layer data concerning damage by chlorofluorocarbons (CFCs), *1974*
Program for public acquisition of farmland development rights, *1313*
Recognition of "sick building syndrome", *1361*
Release of imported gypsy moth–killing nematodes in North America, *2607*
Research of significance pointing to a problem in the U.S. drinking water supply, *4514*
Satellite composite map of the U.S., *3381*
Significant forestry legislation in the United States in the second half of the 20th century, *2640*
Solar energy research funded by the U.S. Congress, *2298*
State in the United States to require nonsmoking sections in restaurants, *1362*
Study of a major U.S. water supply linking pollutants to high cancer rates, *4424*
Swampland designated as a U.S. National Wilderness Area, *4652*
U.S. law allowing consideration of recreation, fish, and wildlife enhancement as purposes of federal water development projects, *4561*
United Nations agreement on marine pollution from airborne sources, *3588*
United Nations agreement to regulate dumping of waste into the oceans, *3587*
Use of the term *ecofeminism*, *1042*
Wildlife preserve to protect golden lion tamarins, *4832*
Wolves reintroduced to Soviet Georgia, *2429*
Woman to direct an Antarctic research station, *3868*
World's fair with an environmental theme, *2310*

1975

Activist to popularize the term *speciesism*, *1019*
American Indian organization to collectively negotiate over control of natural resources in modern times, *3382*
Aquarium in North America to be accredited by the American Association of Zoological Parks and Aquariums, *4948*
Aquifer in the United States granted sole source protection from pollution, *4425*
Aquifer in the United States to be designated as a "sole source" under the Clean Water Act of 1974, *4600*
Australian Heritage Commission (AHC) Act, *4658*
California state agency to regulate wind power and other renewable energy sources, *4855*
Catalogue of critical world wetlands, *3150*
Cease-and-desist order to the U.S. Army to stop toxic contamination from the Rocky Mountain National Arsenal at Denver, CO, *2852*
Clean Indoor Air Act adopted by a U.S. state, *1408*
Coral reef off the coast of the United States protected by federal law, *3589*
Discovery of radioactive contamination in the town of Port Hope, Ontario, *2853*
Discovery of the monarch butterfly's winter home, *4729*
Endangered plants list issued by the United States, *2403*
EPA Journal, *1074*
Federal duck stamp in the United States to feature a waterfowl decoy, rather than a live bird, *3212*
Federal law to create wilderness areas in the eastern United States, *4653*
Flood limitation program of significance by the Canadian government, *2557*
Global agreement to protect plant and animal species from unregulated international trade, *3036*
Hazardous Materials Transportation Act in the United States, *2855*
Law controlling aircraft noise over the Grand Canyon, *3445*
Marine sanctuary in the United States, *3590*
Mediterranean fruit fly ("medfly") infestation in California, *1299*
National Wetlands Resource Center (NWRC) in the United States, *3383*
Natural gas well in Sudan, *2053*
Personal rapid transit (PRT) system in the United States, *1601*
Petroleum in Vietnam, *3623*
Plant in the United States to recover methane gas from garbage, *2054*
Ports in the United States to be monitored by U.S. Fish and Wildlife Service agents, *4730*
Report of the reappearance of the Campbell Island flightless brown teal, *1673*
Scale widely used to measure hurricanes, *4263*

1975—*continued*

State in the United States to ban the sale of aerosol products containing chlorofluorocarbons (CFCs), *1409*
U.S. government moves to limit permissible amounts of arsenic in drinking water, *2854*
U.S. National Wilderness Area in the Great Lakes region, *4654*
U.S. National Wilderness Area in the Mid-Atlantic states, *4655*
U.S. National Wilderness Area in the Ozark Plateau region, *4656*
U.S. National Wilderness Areas in the Great Plains, *4657*
U.S. National Wildlife Health Center, *4728*
U.S. National Wildlife Refuge created specifically for Mississippi sandhill cranes, *2331*
United Nations compensation policy on marine oil pollution, *3582*
United Nations convention to define civil liability in nuclear accidents at sea, *3585*
Valhalla Wilderness Society meeting, *1097*
Whooping crane artificial colony, *1726*
Whooping crane born in captivity, *4754*
World's fair with an ocean theme to be sanctioned by the Bureau of International Expositions (BIE), *2311*

1976

Agency in the United States to regulate solid waste, *4131*
Biosphere reserve in Asia, *1623*
Biosphere reserve in South America, *1624*
Biosphere reserves in Europe, *1625*
Biosphere reserves in the Middle East, *1626*
Biosphere reserves in the United States, *1627*
Biotechnology company in the United States, *2756*
Butterfly listing under the U.S. Endangered Species Act, *2407*
Coastal wetlands designated as a U.S. National Wilderness Area, *4659*
Comprehensive legislation in the United States banning PCBs and regulating their disposal, *2857*
Comprehensive legislation in the United States to regulate the generation, transport, and management of a wide range of hazardous wastes, *2858*
Comprehensive waste management legislation in the United States, *4132*
Earthquake in China with its epicenter directly under a city of 1 million people, *2231*
Ebola virus outbreak, *3950*
Elephant listed under the U.S. Endangered Species Act, *2406*
Ethanol promotion program in Brazil, *4328*
Federal "Adopt a Horse or Burro" program in the United States, *4803*
Federal guidelines for disposing of hazardous waste, *4143*
Federal law to restrict the dumping of hazardous waste in the United States, *2977*
Federal rules in the United States for closure of hazardous waste treatment, storage, and disposal facilities, *4144*
Fishing vessel buyback program in the United States, *2478*
Hawaiian monk seal listing under the U.S. Endangered Species Act, *2405*
Horse listed under the U.S. Endangered Species Act, *2404*
Land-use plan for a public desert area in the United States, *3089*
Law in the United States to specifically target toxic substances, *3951*
Legislation in the United States making animal fighting a federal offense, *1020*
Major release of the pesticide dioxin into the atmosphere, *1300*
Natural resource profits fund to benefit all Alaskans, *3384*
Nuclear energy research museum, *3510*
Offshore natural gas plant to separate liquids from the gas, *2055*
Offshore U.S. oil leases in the outer continental shelf of the Atlantic, *3663*
Oil spill of significance off Spain, *3637*
Ozone depletion warning about Freon, *1975*
PCB contamination to attract widespread national attention in the United States, *2856*
Pheasant restoration program in South Dakota, *1674*
Precipitation sampling network in Canada, *1471*
Productive mine above the Arctic Circle in Canada, *3335*
Radiometric survey in Antarctica, *3869*
Seismographs on Mars, *2230*
Steam well in a Hawaiian volcano, *4385*
Tax breaks for owners of historically significant commercial property in the United States, *3112*
Treaty protecting the polar bear and its natural habitat, *3034*
U.S. federal funding to preserve the Tule elk, *4833*
U.S. law allowing consideration of recreation, fish, and wildlife enhancement as purposes of federal water development projects, *4561*
U.S. National Wilderness Area in Hawaii, *4660*
World's fair with an ocean theme to be sanctioned by the Bureau of International Expositions (BIE), *2311*
Worldwide freshwater monitoring program, *4601*

1977

Activities of the Arctic Chemical Network, *1472*

Biosphere reserves in Africa, *1628*
Biosphere reserves in Australia, *1629*
Biosphere reserves in Mexico, *1630*
Canadian town to be partially demolished because of radioactive waste, *2859*
Chemical accident prevention law in France, *3953*
Commercial nuclear reactor that was gas-cooled, *3511*
Elephant and rhinoceros orphanage in Kenya, *4755*
Environmental group in Kenya to tie deforestation to poverty, *1043*
Federal guidelines governing research with genetically altered material in the United States, *1255*
Flow of oil through the Alaska Pipeline from Prudhoe Bay to Valdez on Prince William Sound, *3624*
Goose to lay an egg weighing 1.5 pounds, *1675*
Hailstorm blamed for the crash of a jetliner, *4220*
Indoor ventilation guidelines for commercial buildings, from the American Society of Heating, Refrigeration and Air-Conditioning Engineers, *1363*
Industrial-scale selective catalytic reduction system, *2299*
Intergovernmental conference on environmental education, *2300*
International meeting about desertfication, *4045*
Known trace of snow in the Miami, FL, area, *4210*
Large-scale reduction in levels of benzene permitted in U.S. factories, *3952*
Lava flow known to have attained a speed of 40 mph, *4386*
Learning lab in a national zoo, *4949*
Meteorology satellite launched by Europeans, *1964*
Nation to pass a constitutional amendment to protect its public health, forests and wildlife, *3386*
National amendment to restrict industrial polluters, *2978*
National Resources Inventory conducted by U.S. Natural Resources Conservation Service (NRCS), *3385*
Natural gas emergency legislation passed by the U.S. Congress, *2056*
Natural gas processing plant in the United Arab Emirates, *2057*
North American country to recognize acid rain as a serious environmental and transboundary problem, *1324*
Oil well blow-out in the North Sea, *3638*
Program in Colorado in which taxpayers could make a contribution to a wildlife program, *4731*
Significant federal legislation to ban the discharge of sewage sludge in U.S. waters, *4479*
Strip mining controls by the U.S. government, *3336*
Trans-Canada national foot trail, *4661*
Trees planted by the Green Belt Movement, *1098*
United Nations attempt to define civil liability for oil pollution resulting from seabed drilling, *3591*
United Nations conference on desertification, *2745*
United Nations Environment Programme Governing Council meeting on the ozone layer, *1976*

1978

Annual highway information for the U.S. government, *1569*
Appearance of the bayberry whitefly in California citrus groves, *1301*
Ban of chlorofluorocarbons (CFCs) by the U.S. government, *1978*
Ban on aerosol chlorofluorocarbons (CFCs), *1410*
Biomass feedstock research funded by the U.S. government, *4329*
Biosphere reserve in Canada, *1631*
Careful scientific monitoring of Lumsden Lake in Ontario, Canada, for acid rain, *1453*
Dangerous wastes law passed by the European Economic Community (EEC), *3954*
Elephant listed under the U.S. Endangered Species Act, *2406*
Evacuation of an entire chemically contaminated neighborhood in the United States, *4426*
Federal agency in the United States to monitor water use, *4602*
Federal ban of lead paint in the United States, *3955*
Federal law to allow designation of formerly developed lands as wilderness areas, *4662*
Federal legislation in the United States to grant federal funds for urban park rehabilitation and infrastructure maintenance, *3822*
Federal tax credits for wind power investments in the United States, *4856*
Flash flood national program in the United States, *2558*
Flood in the United States to cause $1 billion in damages, *2559*
Hailstorm known to cause a famine in Syria, *4221*
Hazardous waste treatment site to accept waste from 48 U.S. states, *2860*
Identification system for wild horses and burros by the U.S. Bureau of Land Management, *4804*

1978—*continued*

International holiday to promote solar energy, *4114*
Law in the United States deregulating the price of natural gas extracted from below 15,000 feet, *2058*
Law in the United States to require mine spoil waste land to be reclaimed, *3337*
Laws in Kenya against commercial trade in wildlife trophies or products, *2242*
Lawsuit on behalf of an endangered species to be decided by the U.S. Supreme Court, *2333*
Learning lab in the United States in a national zoo specifically related to birds, *4950*
Love Canal Homeowners Association meeting, *1099*
National network in the United States to analyze the effects of air pollution on precipitation chemistry, *1473*
National reserve in the United States, *3732*
National Scenic Trail in the Rocky Mountains, *3989*
National Wild and Scenic River in the Mid-Atlantic region, *3992*
Natural gas extraction in Ireland, *2059*
Natural gas price deregulation by the U.S. government, *2063*
Nuclear power plant in Taiwan, *3512*
Official "concentrated heavy snowstorm," defined as 20 inches or more within a 48-hour period, in Boston, MA, *4211*
Oil-tanker spill of significance off the coast of France, *3639*
Ozone-layer committee of the United Nations, *1977*
Ozone-layer measurements from space, *1979*
Quiet Communities Act in the United States, *3446*
Radioactive fallout of significance from a space program, *1326*
Recorded rainfall with a pH of 2, *1325*
Ruling by the U.S. Supreme Court that historic preservation was a form of community welfare, *3116*
Scientist to survive the eruption of an Antarctic volcano, *3870*
Species removed from the U.S. Endangered and Threatened Species List, *2408*
State nature preserve in Georgia, *4834*
Successful birth of an elephant in captivity in the Western Hemisphere, *4756*
Tax credit in the United States for ethanol, *4330*
Tornado in Great Britain to cause $1 million damage, *4296*
U.S. law exempting alternative energy producers from state and federal regulation, *4115*
U.S. National Wild and Scenic River in the Chihuahuan Desert, *3993*
U.S. National Wilderness Area in the Gulf of Mexico, *4663*
U.S. National Wildlife Refuge created specifically for wintering bald eagles, *2332*
U.S. Supreme Court ruling on state bans on imported waste, *4145*
United Nations agreement on marine pollution from airborne sources, *3588*
United Nations compensation fund for marine oil pollution damage, *3586*
Wilderness expeditions operator in Costa Rica, *2243*
World Heritage site established in the United States, *3808*
World Heritage site in Africa selected for environmental reasons, *3990*
World Heritage site in North America selected for environmental reasons, *3991*

1979

Alternative fuel measure adopted by the U.S. Postal Service, *4331*
Announcement that New York State would no longer accept out-of-state hazardous waste, *2861*
Aquarium with a salmon ladder, *4951*
Ban on chlorofluorocarbons (CFCs) on shipments between U.S. states, *1983*
Ban on the use of phosphate-containing laundry detergents in the United States, *4407*
Bilateral negotiations between the United States and Canada on acid rain, *3038*
Binding international treaty to address regional air pollution, *3039*
Catastrophic flood of the 20th century in Wales, *2560*
Dam safety guidelines issued by the U.S. government, *2073*
Ecological candidate elected to a national parliament, *1110*
Environmental organization to substantially practice "ecotage", *1075*
Federal government interagency study to evaluate methods by which states and localities in the United States could slow farmland losses, *3113*
Federal government move to relax water pollution controls for U.S. companies that discharged nontoxic substances, *4439*
Global warming report of significance made to the U.S. government, *1841*
Hurricane in the Atlantic with a masculine name, *4265*
Hurricane in the Gulf of Mexico with a masculine name, *4266*
Hurricanes to be given masculine names, *4264*
Intergovernmental organization devoted exclusively to protection of migratory species worldwide, *3062*

International agreement on cross-border air pollution, *1411*
International meeting of American Indians and environmentalists, *4408*
Introduction of the "bubble concept" in U.S. air pollution policy, *1412*
Jury award of damages in the U.S. to a victim of radiation contamination from a nuclear facility, *3513*
Known bird with a wingspan of 25 feet, *1676*
Law of significance to control acid rain in the United States, *1413*
Long-term study of the causes and effects of acid rain in the United States, *1474*
Major accident in a U.S. nuclear power plant, *3514*
Mandatory hunter education law in Idaho, *2960*
Modern study linking rain forest deforestation with accelerating species extinction due to habitat loss, *1632*
National surveys of wind as an energy source in the United States, *4857*
Oil platform blowout, *3640*
One-child-per-family policy in China, *3883*
Ozone depletion of major significance over Antarctica, *1980*
Ozone-layer vertical profile from space, *1981*
Patron saint of ecology, *1076*
Permanent Bureau of Land Management holding facility for wild horses and burros in the eastern United States, *4835*
Photographs of erupting volcanoes on the Jovian moon Io, *4387*
Photovoltaic (PV) village, *4116*
Politician from a Green party to be elected to a national parliament, *1111*
Successful offshore closed-cycle Ocean Thermal Energy Conversion (OTEC) operation, *4564*
Total ban of chlorofluorocarbons (CFCs) for aerosols by a country, *1982*
Treaty to insure humane slaughter of animals, *3037*
Underground animal rights organization to operate in the United States, *1021*
United Nations convention to protect the vicuña, *4777*
Use of solar energy at the White House, *4117*
Widely publicized scare in hazardous materials transport in Canada, *2862*
World Climate Conference, *1829*
World Heritage site in Asia, *3994*
World Heritage site in South America chosen for environmental reasons, *3995*
World Heritage sites in Europe chosen for environmental reasons, *3996*

1980s

Asian tiger mosquitoes in the United States, *1527*
Cool water or integrated gasification combined-cycle utility, *2301*
Emergence of wetlands science as a distinct discipline, *3151*
Environmentally conscious codes of ethics for whale-watching operators and participants in North America, *2244*
Hospitality ranch in Wyoming, *2245*
Introduction of early biologically based alternatives to chemical pesticides, *1302*
Recognition of radon as a dangerous indoor household pollutant in the United States, *1352*
Reports of possible extinction of the dusky seaside sparrow in Florida, *2377*
State in the United States to suspend vehicle registrations of litterers, *3276*
Use of the acronym SLAPP, *1077*
Warsaw Pact government to recognize that industrial wastes cause health problems, *2979*
Widespread damage to and deaths of trees in Germany from air pollution, *1475*
Widespread use of the acronym NIMBY, *1100*

1980

Aquaculture Development Act in the United States, *2497*
Biosphere reserve in Central America, *1633*
Biosphere reserve in Hawaii, *1634*
Colorado Wilderness Act, *4665*
Comprehensive hazardous waste cleanup program by the U.S. government, *3956*
Dam to reach 300 meters (984 feet) in height that is still in use, *2108*
Drought to cause damages of $1 billion or more in the United States, *2157*
Emperor penguins bred outside of Antarctica, *1678*
Eutrophication monitoring in the Baltic Sea, *4409*
Federal duck stamp in the United States to be priced at $7.50, *3213*
Federal government bailout of a major U.S. carmaker, *1548*
Federal trust fund for the cleanup of hazardous waste sites in the United States, *2863*
Geostationary Operational Environmental Satellite (GOES), *1965*
Giant panda born outside of China, *4757*
International organization to manage Antarctic wildlife, *4783*
Inventory and assessment of the world's forest resources, *2746*
Joint antipollution pact in the Mediterranean region, *3040*
Law in the United States to promote wind power as a renewable resource, *4858*
Legislation by the United States to give considerable authority for disposal of low-level radioactive waste to the states, *2865*

1980—*continued*

Long-term ecological research network in the United States, *2302*
Manned solar-powered aircraft flight, *4119*
National Scenic Trail in the United States devoted to glacial terrain, *3997*
National Scenic Trail to link the eastern and western United States, *3998*
Nuclear energy program to disassemble a damaged reactor, *3515*
Ohio State Scenic River, *3985*
Power plant using solar cells, *4120*
Public aquarium in the United States, *4895*
Refuge in the United States to protect endangered plants and insects, *4836*
Significant legislation in Massachusetts to address the shortage of waste treatment and storage capacity there, *2864*
Significant multinational agreement to curb pollution of the Mediterranean Sea, *4480*
Solar-powered aircraft that could carry a pilot, *4118*
"Superfund" environmental cleanup law in the United States, *2980*
Synthetic fuel plant in the United States built to operate on a commercial scale, *2060*
Tax exemptions in the United States for conservation easements, *4664*
Theory that an asteroid caused the extinction of dinosaurs, *2376*
Turtle excluder device, *2464*
U.S. legislation making states responsible for their nuclear waste, *3516*
U.S. legislation to promote Ocean Thermal Energy Conversion (OTEC) development for commercial use, *4565*
U.S. National Wild and Scenic Rivers in Alaska, *3999*
U.S. National Wildlife Refuge created specifically for watercress darters, *2334*
U.S. Supreme Court decision to rule that living things are patentable material, *2757*
Volcano eruption in the United States photographed by weather satellites, *4388*
Volcano known to claim human life in the contiguous 48 states, *4389*
Water Bank Act in the United States, *3380*
Wild horse range established in Colorado by the U.S. Bureau of Land Management, *4837*
Year Oklahoma required the purchase of a state duck stamp for hunting waterfowl, *1677*

1981

Bombing of a nuclear reactor, *3519*
Comprehensive legislation in New Jersey to deal with hazardous waste disposal, *2866*
Darwin Centenary Symposium on Earthworm Ecology, *4073*
Epidemiological evidence of the association between passive smoking and lung cancer, *1476*
Female golden-fronted bowerbird seen, *1708*
Firm evidence of the outbreak of AIDS in the United States, *3957*
Geostationary Operational Environmental Satellite (GOES), *1965*
Group of Ten meeting, *1078*
Law passed to protect black bass in the United States, *2458*
Massive spraying of the insecticide malathion to control the Mediterranean fruit fly in California, *1303*
National protest of significance sponsored by Earth First! *1079*
National zoo licensing legislation, *4952*
Natural gas field in Thailand, *2062*
"New Urbanism" community in the United States, *4303*
Nuclear power plant accident of significance in Japan, *3518*
Official confirmation of a powerful hydrogen bomb dropped accidentally from an aircraft, *3517*
Reproductive research foundation sponsored by a zoo, a university, and a wild animal park, *4953*
Snowfall in living memory in the Kalahari Desert in Namibia, *4212*
Species Survival Plan (SSP), *2335*
State-sponsored biotechnology initiative in the United States, *2758*
Study that defined the factors making old-growth Douglas fir forests unique, *2747*
Substantial cuts in the budget of the U.S. Environmental Protection Agency, *1080*
U.S. Supreme Court ruling on plastic milk containers, *4146*
Use of compressed natural gas to power a ship, *2061*
Volcano in Hawaii tapped to supply a geothermal power plant, *4390*
Wildlife biologists on the Northern Mariana Islands, *4732*
Wildlife Diversity Program in Oklahoma, *4784*
World Heritage Sites in Australia, *4000*
WorldWIDE Network Inc. meeting, *1044*
Year that North Sea oil production exceeded home demand in Great Britain, *3664*

1982

Automobile and truck ban in Greece to control emissions, *1579*
Bird known to have lived to a documented age of 80, *1679*
Comprehensive United Nations agreement on use of the oceans, *3592*
Elephant captive breeding program in Sri Lanka, *4758*

Federal law in the United States to withhold development funds for construction on coastal barrier islands, *3389*
Fish removed from the U.S. Endangered and Threatened Species List, *2378*
Flood barrier in Britain rotated from beneath the water, *2561*
Genetically engineered product to become commercially available, *2759*
Government program in the United States established to monitor waterborne toxic materials, *4603*
Hazardous substances protection law in Germany, *3958*
International body to govern use of the ocean, *3053*
Learning lab related to reptiles in a national zoo in the United States, *4954*
Limit on lead in gasoline initiated by the Italian government, *1580*
National Contingency Plan (NCP) and Hazard Ranking System (HRS) for hazardous wastes in the United States, *2981*
Noise control agency in the United States, *3429*
Nuclear-generated electricity exported from Canada, *3521*
Nuclear power plant shelled by environmentalists, *3520*
Offshore drilling-rig disaster near Canada, *3641*
Oryx reintroduction program, *2431*
Parliament to have a substantial Green bloc represented, *1112*
Pennsylvania law creating a scenic rivers system, *4001*
Permanent ban on oil and gas leasing on U.S. wilderness land, *4666*
Recovery plan approved by the U.S. Fish and Wildlife Service for a plant, *2336*
Report on the health of the oceans by the United Nations–sponsored Group of Experts on the Scientific Aspects of Marine Pollution (GESAMP), *4495*
Rewards offered by the U.S. Fish and Wildlife Service for information about wildlife crimes, *4733*
"Subsistence rights" legislation for Alaska, *3388*
Substantial cuts in the budget of the U.S. Environmental Protection Agency, *1080*
Survey of U.S. wetlands and wetlands losses that was comprehensive and statistically valid, *3387*
TBT (tributylin) restrictions, *4496*
Television network to broadcast weather programming exclusively, *1966*
Trilateral agreement to protect the Wadden Sea, *3010*
Unleaded gasoline campaign in Great Britain, *1549*
Urea-formaldehyde foam insulation ban in the United States, *1364*
Woman to graduate from the Pennsylvania School of Conservation, *2303*
World's fair in the American Southeast to be sanctioned by the Bureau of International Expositions (BIE), *2312*
World's fair to select energy as its theme, *2313*

1983

Amphibian removed from the U.S. Endangered and Threatened Species List, *2409*
Bilateral U.S.–Mexican agreement requiring American companies in Mexico to return hazardous waste to the United States, *2867*
Binding international treaty to address regional air pollution, *3039*
Bird removed from the U.S. Endangered and Threatened Species List due to extinction, *2379*
Chatham Islands black robin to continue breeding to the age of 13, *1680*
Costa Rican ecotourism organization whose purpose was to save an endangered part of the environment, *2246*
Dam constructed totally with roller-compacted concrete (RCC), *2109*
Dam safety organization among U.S. states, *2074*
Discovery of large numbers of dead and deformed birds in the Kesterson Refuge and other ponds containing irrigation runoff from farms in the Central Valley of California, *1221*
Federal government bailout of a major U.S. carmaker, *1548*
Federal law in the United States to withhold development funds for construction on coastal barrier islands, *3389*
Federal plan in the United States for developing biomass energy regionally, *4332*
Federal program to protect and restore an estuary in the United States, *3593*
Fossils of *Protoavis*, *1681*
Habitat Conservation Plan (HCP) in the United States, *2337*
Hurricane to cause damages of $1 billion or more in the United States, *4267*
Interagency Grizzly Bear Committee (IGBC), *4785*
Intergovernmental organization devoted exclusively to protection of migratory species worldwide, *3062*
National Priorities List (NPL) of hazardous waste sites in the United States, *2868*
National Scenic Trail in the eastern United States, *4004*
National Scenic Trail in the south central United States, *4002*
Rediscovery of MacGillivray's petrel, *1682*

1983—*continued*

Salmon caught by rod and line in the River Thames, *2509*
Spent nuclear fuel and high-level radioactive waste disposal law in the United States, *3522*
Subtropical National Scenic Trail in the United States, *4003*
Successful embryo transfer between females of endangered species, *4955*
Sustainable development organization for mountain regions, *4304*
Texas law to regulate game and fisheries statewide, *4786*
Toxic and hazardous wastes national policy of the Portuguese government, *3959*
U.S. Fish and Wildlife Service Environmental Contaminants Program, *4734*
U.S. National Wildlife Refuge created specifically for West Indian manatees, *2338*
United Nations organization concerned with global warming, *1842*
Use of the term *ecotourism*, *2247*

1984

African country to produce a National Conservation Strategy, *3390*
Aquaculture Development Act in the United States, *2497*
Biosphere reserve in Ecuador, *1635*
Butterfly listing under the U.S. Endangered Species Act, *2407*
Comprehensive legislation in the United States to regulate the generation, transport, and management of a wide range of hazardous wastes, *2858*
Drought to draw worldwide aid through television, *2158*
Effective international action to curb acid rain, *1414*
Establishment of what would become the U.S. Greens, *1113*
Extensive collection of quality data on fertility and family planning, *3882*
Government policy in the United States to deny aid to foreign family planning agencies that performed or promoted abortions as a method of population control, *3884*
Hazardous waste transportation law passed by the European Economic Community (EEC), *3960*
Insect removed from the U.S. Endangered and Threatened Species List, *2411*
Introduction of the concept of a "debt for nature" swap, *2719*
Jaguar preserve, *4838*
Joint document from major U.S. environmental groups to spell out a national environmental policy, *1081*
Large-scale boycott of a restaurant chain in an effort to halt imports of tropical beef, *2720*
Large settlement of Agent Orange lawsuits, *1304*
Law in the United States extending wilderness protection to land administered by the Bureau of Land Management (BLM), *4667*
Law in the United States to establish asbestos as a health hazard in schools, *2982*
Law in the United States to protect Atlantic striped bass, *2465*
Mollusk removed from the U.S. Endangered and Threatened Species List, *2380*
National Toxics Campaign in the United States, *1101*
National Wild and Scenic River in the Sonoran Desert, *4006*
Ohio highway beautification program to use wildflowers, *4005*
Pipeline to carry natural gas from the Soviet Union to a European country, *3708*
Radical "deep ecology" platform, *1082*
Recommendation from a California state waste management agency against siting hazardous waste incinerators in middle- to upper-income communities, *2869*
Reptile removed from the U.S. Endangered and Threatened Species List, *2410*
Successful effort to block a large hazardous waste disposal plant under the Massachusetts Hazardous Waste Facility Siting Act, *2870*
Successful embryo transfer between females of endangered species, *4955*
Tax on boats and motors in the United States to benefit wildlife habitats, *2510*
Tax on crossbow arrows in the United States, *2928*
Tidal power plant in North America, *4566*
Town plebiscite to approve a large hazardous waste treatment facility in Alberta, Canada, *2871*
U.S. government agency to adopt a formal environmental policy with American Indian tribes, *3391*
U.S. National Wildlife Refuge created specifically for endangered Nevada fish species, *2339*
World's fair to select rivers as its theme, *2314*

1985

Adopt-A-Highway program in the United States, *3277*
Agreement among nations to work together to create guidelines, policies, and procedures for protecting the ozone layer, *3041*
Call from a U.S. governors' group for remedial action on acid rain, *1327*
Carbon tetrachloride ban in the United States, *3961*
Chlorofluorocarbons (CFCs) limitation on an international level, *1985*

Discovery of large numbers of dead and deformed birds in the Kesterson Refuge and other ponds containing irrigation runoff from farms in the Central Valley of California, *1221*
Drought aid from a global rock music festival, *2159*
Federal legislation of significance to take erodable U.S. cropland out of production, *4057*
Global warming long-term policy of the government of the Soviet Union, *1843*
Government agency in the United States to regulate leaking underground storage tanks, *2872*
Handbook on ecological sabotage, *1083*
Hole in the earth's ozone layer, *1984*
International dam opposition group, *2075*
Law in New York City outlawing the sale of spray paint to minors, *3279*
Legislation by the United States to give considerable authority for disposal of low-level radioactive waste to the states, *2865*
Legislation in the United States to remove ecologically fragile farmland from production, *1229*
Mall to feature aquatic displays, *4956*
Mudslide in the 20th century to bury a city in Colombia, *2576*
National water use survey of the United States by county, *4604*
Natural gas price deregulation by the U.S. government, *2063*
Nuclear energy agreement between the U.S. and China, *3523*
Pére David's deer reintroduced to China, *2432*
Pipeline giving access to Canada's northern oil, *3625*
Radical environmental activist arrested and convicted for "monkeywrenching" in the United States, *1084*
Species removed from the U.S. Endangered and Threatened Species List due to recovery, *2412*
State in the United States to pass a beautification and antilitter act, *3278*
Time-constrained goals to be adopted by a group of nations to curb water pollution, *4410*
Volcano eruption causing massive deaths in Colombia, *4391*
World tropical timber marketing organization, *2687*
Zoo with a museum devoted to elephants, *4957*

1986

Accidental discharge of nuclear waste in British waters, *3524*
Aquatic theme park mixing education, research, and entertainment to be owned by a nonprofit organization, *4958*
Artificial flood of significance on the Colorado River in the western United States, *2562*
Asbestos legislation adopted by the U.S. government, *3962*
Asbestos mine rehabilitation in South Africa, *3963*
Ban on the use of lead pipes in public water systems in the United States, *4606*
Bayou in the U.S. National Wild and Scenic Rivers System, *4007*
Bilateral U.S.–Canadian agreement to waive "consent notification" for hazardous waste shipments from the United States into Canada, *2873*
Clean-coal joint program by the U.S. government and industry, *2064*
Compensation payments by the Australian government to Aborigines whose tribal lands were used for British nuclear weapons testing, *3525*
Environmental program to limit ecotourism damage in Nepal, *2248*
Federal duck stamp in the United States to feature a fulvous duck, *3214*
Federal policy in the United States on emergency removal of asbestos from schools, *4147*
Genetically altered plants, *2760*
Government report issued jointly by the United States and Canada to rate acid rain a serious environmental and transboundary problem for both countries, *1477*
Hazardous chemical public disclosure law in the United States, *3964*
Hazardous waste law enacted by the Spanish government, *3965*
Illinois State Scenic River, *4008*
Immediate ban on major uses of asbestos in the United States, *1328*
Known U.S. waste entrepreneurs to be convicted for fraudulent export of hazardous waste overseas, *2874*
Law in the United States mandating that all school systems check their buildings for asbestos-containing materials, *1365*
Law in the United States requiring environmental considerations to be assessed in the licensing of hydropower projects, *4567*
Law in the United States requiring public disclosure of pollutants through the Toxic Waste Inventory, *2984*
Law of importance in the United States requiring government and industry to inform the public about toxic chemicals, *1353*
Leaking Underground Storage Tank (LUST) trust fund in the United States, *2983*

1986—*continued*
Legislation to provide comprehensive planning for hazardous waste treatment in California, *2875*
Meteorological organization in Europe organized specifically to control weather satellites, *1967*
National memorial dedicated to farmers, *1135*
National Water Quality Laboratory in the United States, *4605*
National Wetlands Resource Center (NWRC) in the United States, *3383*
National Wilderness Inventory of Australia, *4668*
Nuclear agency established by the Chinese government for both civil and military activities, *3526*
Nuclear power plant disaster with worldwide fallout, *3528*
Person to be elected the head of a government after serving as a minister of the environment, *1114*
Reauthorization of the "Superfund" program, *2985*
Recognition by the U.S. government of the health hazards of passive smoking, *1366*
Recognition of radon as a dangerous indoor household pollutant in the United States, *1352*
Regional antipollution pact covering the South Pacific region, *4497*
Reported sighting of Jerdon's courser since 1900, *1683*
"Right to Know" law in the United States covering hazardous substances, *2986*
Town evacuated by the Environmental Protection Agency for dioxin levels 100 times emergency standards, *4427*
Toxic waste–carrying ship to draw international media attention, *2876*
Toxics Release Inventory (TRI) in the United States, *2987*
U.S.–Canadian bilateral report on acid rain, *3042*
United Nations educational center for African natural resource management, *3392*
Use of the term *biodiversity*, *1636*
Volcanic gas from a lake to cause massive deaths in Cameroon, *4392*
Wetlands Loan Act in the United States, *3373*
Wild horse training program in a U.S. prison, *4735*
Wind farm in Australia, *4859*
Workshop hosted by the European Community on the ozone layer, *1986*
Year that nuclear power plants numbered 100 in the U.S., *3527*

1987

Bali starling reintroductions, *2433*
Bird recorded at the North Pole, *1709*
Central American coral reef reserve, *3594*
Climate research agreement between the United States and Soviet Union, *1830*
Comprehensive hazardous waste treatment facility in the Canadian province of Alberta, *2877*
Comprehensive overview of the global environmental problem, *3262*
Congressional action on "nonpoint" source pollution in the U.S., *4481*
Coordinated assault on "nonpoint" sources of pollution in the United States, *4482*
Debt-for-nature swap, *3393*
Deliberate release of genetically altered organisms in the United States, *1256*
Extant species on the U.S. Endangered and Threatened Species List to become extinct, *2382*
Federal duck stamp in the United States to be priced at ten dollars, *3215*
Genetically altered bacterium tested outside of a laboratory in the environment, *2761*
Greenpeace campaign to stop exports of hazardous waste, *2878*
International treaty to protect the Earth's atmosphere, *3043*
Law in the United States banning the disposal of plastics at sea, *4148*
National Estuary Program (NEP) in the United States, *3595*
National Fishing Week in the United States, *2511*
Nuclear power plant in the United States to be totally decontaminated and decommissioned, *3529*
Nuclear waste site evaluation, *3530*
Ozone fund for developing nations, *1989*
Ozone pollutant agreement on an international level, *1987*
Proof that thunderstorms can push pollutants as high as the lower stratosphere, *1478*
Reported profit of Concord Resources, a U.S. company, for its hazardous waste disposal facility in Blainville, Quebec, *2879*
Reports of possible extinction of the dusky seaside sparrow in Florida, *2377*
Reptile removed from the U.S. Endangered and Threatened Species List due to recovery, *2413*
River protected by Iowa's Protected Water (PWA) program, *4669*
Scientific evidence that humans caused the Antarctic ozone hole, *1988*
Solar automobile race, *4121*
Species of bird to fall victim to space exploration, *2381*
Study of aircraft noise effecting U.S. national parks and forest wilderness, *3430*

Systematic legislation to clean up and restore Florida's polluted lakes, rivers, streams and bays, *4483*
Use of the term *sustainable development*, *4305*
Wolf to be reintroduced into the wild, *2434*
Workshop hosted by the European Community on the ozone layer, *1986*

1988

$100 million forest fire–fighting campaign in U.S. history, *2623*
African elephant protection law in the United States, *2340*
Agreement among nations to work together to create guidelines, policies, and procedures for protecting the ozone layer, *3041*
Beach closings in the United States due to medical wastes washing ashore, *4428*
Climate-change international panel, *1844*
Drought education and training center for developing nations, *2160*
Drought in the United States to strand boats on the Mississippi River, *2161*
Drought year producing worldwide environmental responses, *2162*
Earthquake in the Soviet Union allowed open media coverage, *2232*
Explosion of significance on an offshore oil rig, *3642*
Federal law in the United States to withhold development funds for construction on coastal barrier islands in the Great Lakes, *3394*
Federal laws in the United States to protect caves, *4670*
Federal legislation in the United States to protect caves located on federal lands, *3809*
Financial responsibility requirements for underground storage tanks (UST), *2988*
Formal promise from Brazil to take steps to protect its rain forests, *2721*
Genetically engineered mammal to be patented, *2762*
"Greenhouse effect" report to the U.S. Congress, *1845*
International convention attempting to regulate Antarctic mineral exploitation, *3044*
International core list of hazardous substances, *2880*
Joint biodiversity program between Nicaragua and Costa Rica, *1637*
Joint United Nations organization to assess scientific information on climate change, *1831*
Large-scale international private reforestation campaign aimed at individuals and communities, *2731*
Large-scale mercury pollution of streams in northern Brazil, *4440*
Law in the United States banning tin-based paint on boats, *2504*
Law in the United States requiring government agencies to purchase alternative-fueled vehicles, *4333*
Law in the United States to require the removal of lead from water coolers, *4607*
Medical waste disposal rules set by the U.S. Environmental Protection Agency (EPA), *4149*
Meeting of the Intergovernmental Program on Climate Change (IPCC), *1832*
Monsoon in Bangladesh to cause $1 billion in damage, *4189*
National Environmental Action Plan (NEAP) in Africa, *3395*
National population plan in Micronesia, *3885*
National Wetlands Policy Forum to call for a program of no net loss of wetlands, *4484*
Natural history museum in Egypt, *4959*
Negotiating session to prepare for a global convention on the control of transboundary movements of hazardous wastes, *2885*
Nuclear emergency plan by the Irish government, *3531*
Person to circumnavigate the North Pole on foot and solo, *3871*
Program of the National Park Service intended to help local groups undertake river and trail conservation projects, *3733*
Radon control legislation in the United States, *1367*
Reported disposal of hazardous waste by a "poison ship" from the United States, *2881*
Scientific evidence that humans caused the Antarctic ozone hole, *1988*
Significant expansion of federal presence in hazardous waste policy in Canada, *2882*
Student Environmental Action Coalition in the United States, *1085*
Swans sighted in the Antarctic, *1710*
U.S. legislation to regulate the disposal of hazardous medical materials, *2883*
U.S. National Fish and Wildlife Foundation, *4736*
U.S. National Wild and Scenic River managed in partnership with local governments, *4009*
Widely publicized epidemic of medical waste washing up on U.S. beaches, *2884*
Wildlife crime lab in the United States, *4787*
Woman to travel solo to either pole, *3872*
Year in which half of the agricultural counties in the United States were designated drought disaster areas, *2163*
Zebra mussels introduced in the United States, *1528*

1989

African-American organization to focus on pollution and hazardous waste in minority and poor communities, *2886*
Aquarium sleepovers, *4960*

1989—*continued*

Book on birds to sell for $3.96 million, *3263*
California condors to be bred in captivity, *1684*
Climate-impact forecasts by the U.S. government, *1833*
Conference held after the signing of the Protocol on Substances That Deplete the Ozone Layer (Montreal Protocol), *1990*
Dam safety organization in Canada, *2076*
Documented case of severe damage to a jetliner flying through a volcanic ash cloud, *4393*
Dolphin listing under the U.S. Endangered Species Act, *2414*
Earthquake to be broadcast on live television nationwide, *2233*
Evidence of volcanic activity on Triton, a moon of the planet Neptune, *4394*
Federal duck stamp in the United States to be priced at $12.50, *3216*
Federal duck stamp in the United States to feature a lesser scaup, *3217*
Global-change research coordination program established by the U.S. government, *1847*
Government commission to rent or purchase lands to be set aside for migratory birds in the United States, *1691*
Hybrid geopressure-geothermal power plant, *3396*
Illinois State Scenic River, *4008*
International accord requiring a reduction on nitrogen oxide levels, *1415*
International agreement on hazardous waste, *2887*
International climate conference held exclusively for developing countries, *1846*
International convention on transporting hazardous waste, *3045*
Introduction of the concept of "ecotheology", *1087*
Joint United Nations–World Council of Churches "Environmental Sabbath", *1086*
Mass burning of poached African ivory, *4788*
National Resources Defense Council study to show that the chemical Alar is a carcinogen, *1305*
Nuclear power plant to be closed by a popular state vote, *3532*
Ozone research coordination program of the European Community (EC), *1991*
Papal document to deal with ecology, *3264*
Parrot protection and welfare organization, *1717*
Plant removed from the U.S. Endangered and Threatened Species List, *2415*
Public aquarium in Japan, *4900*
Publication that outlined the goals of the "wise-use" movement, *1024*
Reformatted gasoline tested in the United States, *1479*
Reformulated gasoline sold in the United States, *3626*
Successful solid waste program in the Philippines, *4133*
Transatlantic flight powered by ethanol, *4334*
U.S. National Wildlife Refuge created specifically for Iowa pleistocene snails, *2341*
Wetlands protection and restoration law to provide funds for protection, *3090*
Woman to head a major U.S. conservation organization and receive a salary, *1039*
Year that tourism was Kenya's leading source of foreign currency, *2249*

1990s

Eco-industrial park in the United States, *4307*
Efforts to develop ecotourism in the Lake Kenyir region of Malaysia, *2250*
Evidence that the "ozone hole" had expanded outward to include New Zealand, Australia, Argentina, and Chile, *1480*
Introduction of a chemical "on-off switch" that would make seeds sterile in a generation, *1208*
State in the United States to suspend vehicle registrations of litterers, *3276*
Substantial growth of ecotourism in Hawaii, *2251*
Tourism restrictions on Ecuador's Galápagos Islands, *2252*
Virtual whale-watching tours, *2253*

1990

Amendments of the U.S. Clean Air act to regulate emissions of pollutants, including sulfur dioxide and nitrogen oxides, *1416*
Aquatic Nuisance Species Task Force, *1530*
Biosphere reserve in Madagascar, *1638*
Cleanup of an oil spill of significance in the United States with an oil-eating bacteria, *3644*
Comprehensive effort to introduce preventive pollution control in the United States, *1417*
Comprehensive upgrades of the 1970 and 1977 Clean Air Act and Amendments, *1418*
Discussion of the greenhouse effect and other environmental issues at a "post–cold war summit", *1848*
"Dolphin-safe" labels, *2479*
Earth Day to be celebrated around the world, *1089*
Eco-industrial park fueled by Ocean Thermal Energy Conversion (OTEC), *4306*
Environment minister in the ecologically critical country of Brazil, *2722*
Environmental activist to become a member of a women's hall of fame, *1045*
Environmental coalition of small island nations, *1090*
Extension of the offshore drilling ban, *3665*

Federal duck stamp in the United States to feature a black-bellied whistling duck, *3218*
Felony indictments under a U.S. environmental law for smuggling hazardous waste from California into Mexico, *2889*
Floods to cause damages of more than $1 billion in the United States, *2563*
Global-change research coordination program established by the U.S. government, *1847*
Government ban on the import of discarded batteries into Taiwan, *2888*
Human Genome Project, *2763*
Hybridization of the Norfolk Island boobook owl, *1685*
Intergovernmental agency devoted to marine and fisheries science of the Northern Pacific, *3596*
International compact barring the dumping of wastes into the oceans, *4485*
International horticultural show in Asia, *2315*
International journal devoted to protected areas issues, *3810*
"Killer bees" in the United States, *1529*
Law in the United States to establish federal and state standards for organic certification, *1264*
Legislation in the United States to mandate double hulls on oil tankers, *3627*
Mobilization and training of youth to participate in nonviolent environmental protest, *1088*
National database in the United States for pollution source reduction, *2989*
National organization formed to assist U.S. local, regional, and state grassroots environmental groups with their projects, *1102*
National population plan for Sudan, *3887*
National population program for Burundi, *3886*
National water use survey of U.S. aquifer systems, *4608*
New England Fishery Resources Restoration Act, *2466*
Northern spotted owl listing under the U.S. Endangered Species Act, *2416*
Nuclear power public safety organization of the South Korean government, *3966*
Office of Environmental Education in the United States, *2304*
Oil spill cleanup law in the United States, *3643*
Organization to study the North Pacific, *3063*
Organized resistance to environmental regulations, *1025*
Ozone-protection plan to phase out methyl chloroform, *1994*
Permits issued by the U.S. Fish and Wildlife Service to kill migratory birds at aquaculture sites, *2498*
State in the United States requiring retailers to obtain an annual litter-control permit, *3280*
State in which the U.S. Green political party gained ballot status, *1115*
U.S. federal source reduction policy, *2990*

1991

Agreement among the Big Three U.S. automakers to reduce the dumping of toxins into the Great Lakes, *4441*
Airline flight attendants' class action suit against tobacco companies, *1369*
All-natural pesticide made from the seeds of the neem tree, *1306*
Anadromous fish conservation law in the United States, *2463*
Antipollution device installed as a result of the 1977 law imposing special restrictions on pollution near national parks, *1419*
Arctic Environmental Protection Strategy (AEPS), *3873*
Automobile pollution device of significance in China, *1602*
"Bioprospecting" agreement, *1639*
Boycott of a major U.S. automaker in protest of the use of animals in crash tests, *1022*
Buses to use liquid natural gas (LNG) and diesel, *4335*
Canadian biodiversity government agency, *1640*
Cyclone to cause catastrophic loss of life in Bangladesh, *4268*
Ecological disaster caused by a deliberate oil spill, *3645*
Electric automobile battery research group in the United States, *1550*
Environmental accident insurance guarantee in India, *2991*
Federal duck stamp in the United States to be priced at $15, *3219*
Federal duck stamp in the United States to feature a king eider, *3220*
Federal standards in the United States for radon in drinking water, *4610*
Female oceanographer of distinction, *3597*
Flood to affect one-fifth of China's population, *2564*
Food irradiation plant in the United States, *1209*
Government report issued jointly by the United States and Canada to rate acid rain a serious environmental and transboundary problem for both countries, *1477*
Intentional oil-well fires of significance, *3646*
International conference on the connection between soil health and human health, *4027*
International convention attempting to regulate Antarctic mineral exploitation, *3044*
International organization to monitor the disappearance of amphibians, *4737*
Joint commitment to "zero discharge" of dangerous chemicals into Lake Superior by the United States and Canada, *4442*

1991—*continued*

Legislation in California to require fire-resistant roofing in that fire-prone state, *2624*
Litter and recycling strategy funded by fines and a statewide retail tax, *3281*
Massively lethal pesticide spill in the Sacramento River, *1307*
Minnesota wind power tax law, *4862*
Modern U.S. National Park Service (NPS) strategy for natural resource management, *3397*
National rules for reducing acid rain in the United States, *1421*
National Water Quality Assessment Program in the United States, *4609*
North American organization to monitor the disappearance of amphibians, *4738*
Offshore wind power plant, *4860*
Ozone measurements of chlorine monoxide from space, *1992*
Raven reintroduction in Italy, *2436*
Reintroductions of black-footed ferrets in the United States, *2435*
Report from the U.S. Environmental Protection Agency indicating a strong link between second-hand smoke and lung cancer, *1368*
Restoration to prairie of U.S. farmland previously reserved as a site for a nuclear power station, *3076*
Revision of the U.S. National Wetlands Policy to counter the "no net loss" theory, *4486*
Scenic Byways and All-American Roads program in the United States, *4010*
Specific treaty guidelines for Antarctic ecosystem protection, *3046*
State in the United States to set a time limit on accumulating litter near a public highway, *3282*
Sustainable development accord on the Alps, *4309*
Transportation act in the United States with significant mass transit funding, *1581*
U.S.–Canadian bilateral report on acid rain, *3042*
U.S. government program to reduce air pollution in the Grand Canyon, *1420*
U.S. program to use Christmas trees for wetlands restoration, *4308*
United Nations moratorium on drift net fishing, *2480*
Vacuum-powered prairie dog extraction device, *4805*
Wind farm in England, *4861*
Woman to win the U.S. federal duck stamp contest, *3221*
Women's international conference on the environment, *1046*

1992

Accord on transboundary effects of industrial accidents on the environment, *3047*
Aquarium devoted to freshwater life, *4961*
Arctic-subarctic biome surveyed by a U.S. National Park Service Inventory and Monitoring Program (I&M), *3398*
Asian long-horned beetle seen in North America, *2608*
Attack by loggers on a children's book, *2665*
Automobile emissions standards in Canada, *1582*
Automobile technology organization formed by large U.S. car companies, *1603*
Aye-aye born in captivity, *4759*
Ban on some U.S. uses of the EBDC fungicide class, *1308*
Biodiversity governmental agency in Mexico, *1641*
Biodome, *4964*
Biogeographic areas of the U.S. National Park Service Inventory and Monitoring Program (I&M), *3399*
Biotechnology museum, *4962*
Boycott of genetically engineered food products in the United States, *2764*
Canadian law to protect all endangered wildlife and flora, *2343*
Car fueled by methanol, *4336*
City in the United States to ban the sale of spray paint to reduce graffiti, *3284*
Claim by the U.S. House of Representatives that overestimates of reforestation efforts in the Pacific Northwest had led to dangerously high logging quotas, *2689*
Cooperative effort by 10 northeastern U.S. states to develop plans to reduce emissions of greenhouse gases, *1425*
Cooperative effort by northeastern states to curb nitrogen oxide emissions from coal-burning power plants, *1424*
Daily forecasts of ultraviolet radiation issued by the Canadian government, *1849*
Drought policy by Australia's national, state, and regional governments, *2164*
Eastern deciduous forest biomes surveyed by a U.S. National Park Service Inventory and Monitoring Program (I&M), *3400*
"Ecosystems management plan" for U.S. national forests, *2688*
Education center devoted exclusively to genetics and biotechnology, *2305*
Electric-car "solar carpark", *1604*
Embargo in the United States against tuna that was not "dolphin-safe", *2481*
Environmental health organization of European women, *3967*
European lynx reintroduced in Poland, *2437*
Evidence of the existence of the Vu Quang ox, *4739*
Farmland of significance in New York State to be permanently protected, *1136*

Federal duck stamp in the United States to feature a spectacled eider, *3222*
Federal income tax credit for wind energy use in the United States, *4863*
Federal law in the United States to reduce vessel sewage discharges, *3598*
Federal provision requiring companies that service air conditioning and refrigeration equipment to "capture and recycle" chlorofluorocarbons (CFCs), *1426*
Fertility control program for wild horses in the United States on public lands, *4806*
Field station for the U.S. National Wildlife Health Center, *4740*
General grants from the U.S. government to Indian tribes for solid waste disposal programs, *4151*
Global agreement on all aspects of biodiversity, *3048*
Government agency of India to promote wind power, *4864*
Guidelines issued by the U.S. Fish and Wildlife Service (FWS) on foregoing lawsuits in natural resource damage cases, *3403*
Guidelines issued in the United States by the Federal Trade Commission for environmental marketing claims, *1092*
Hurricane to cause more than $25 billion in damage, *4269*
Hybridization of a whooping crane and a sandhill crane, *1686*
Inter-American institute for research on global change, *3064*
International convention on biodiversity and sustainable development, *3049*
International convention on transporting hazardous waste, *3045*
Large-scale seizure and auction of cattle grazing on U.S. land without a permit, *1174*
Laundromat with an environmental theme, *4411*
Law in the United States to allow combining federal and private funds for wildlife program grants, *4840*
Law in the United States to require federal government agencies to obey environmental laws, *4134*
Law in the United States to specifically ban importation of endangered exotic birds, *2342*
Legally binding international document exclusively about global warming, *1850*
Low-emission automobiles, *1605*
Major apparel maker to make its entire line of clothing from organically grown cotton and nontoxic dyes, *1265*
Mandatory automobile-emissions tests in Canada, *1583*
Massive cleanup plan for the Florida Everglades, *4487*
Meeting of the contracting parties to the Basel Convention on the Control of Transboundary Movements of Hazardous Wastes and Their Disposal, *2891*
National Park Service unit including coral reef resources, *3799*
National park to focus on conservation history and the evolving nature of land use, *3812*
National policy on marketing genetically engineered food products in the United States, *2765*
National population guidelines for Bolivia, *3888*
National population policy for Indonesia, *3889*
National population policy for Jamaica, *3890*
National population policy for Niger, *3891*
Native American organization dedicated to preserving buffalo, *4839*
Natural gas fuel cell to produce electricity commercially, *2065*
New bird species established solely by DNA tests and photographs, *1687*
Nuclear waste site chosen by the U.S. Congress, *3533*
Oceanographic exhibit with more than 3,000 fish, *4963*
Officially sanctioned international horticultural fair in the United States, *2316*
Oil production in Papua New Guinea, *3709*
Organization to study the North Pacific, *3063*
Ozone fund for developing nations, *1989*
Ozone-protection plan to phase out methyl chloroform, *1994*
Ozone weekly summary by the Canadian government, *1993*
Pacific coast biome surveyed by a U.S. National Park Service Inventory and Monitoring Program (I&M), *3401*
Paleoclimatology research in the United States using a global network, *2015*
Park in the Western Hemisphere to commemorate metal mining, *3811*
Pencil made of recycled newspaper and cardboard fiber, *2716*
Pledge to halt all raw sewage flow from Victoria and Vancouver Island, British Columbia, into the Strait of Juan de Fuca, *4543*
Purchase of emissions allowances in the United States as permitted under the Clean Air Act Amendments, *1422*
Recommendations in California to shift responsibility, from businesses to drivers, for paying for Southern California air pollution abatement, *1427*
Rejections of state-mandated shoreland-zoning ordinances in coastal Maine, *3117*
Report that pollutants discharged into the Detroit River may have killed bald eagle hatchlings along Lake Erie, *4416*

1992—*continued*
Reported large-scale bird kills traced to the lawn care pesticide Diazinon, *1309*
Sockeye salmon listing under the U.S. Endangered Species Act, *2417*
State in the United States in which all electric utilities joined the U.S. Environmental Protection Agency's "Green Lights" energy efficiency program, *1423*
State in the United States to tax litter-producing industries, *3283*
State in which the U.S. Green political party gained ballot status, *1115*
Supercomputer used solely for environmental research, *2306*
Systematic research documenting the slow death of the sabal palm along Florida's Gulf of Mexico coastline, *2748*
Tax deduction in the United States for purchasers of Hybrid Electric Vehicles (HEVs), *4337*
Tax deductions in the United States for purchasers of Clean Fuel Vehicles (CFVs), *4338*
Tax on "gas-guzzling" automobiles by a U.S. state, *1584*
Trans-Canada multi-use trail, *4671*
Treaty setting the framework to address the production of greenhouse gases, *3050*
U.S. government research suggesting nitric acid (rather than sulfuric acid) is the chief cause of acid rain, *1481*
U.S. National Park Service (NPS) Inventory and Monitoring Program (I&M), *3402*
U.S. Supreme Court ruling on local bans on imported waste, *4150*
U.S. Supreme Court ruling striking down a federal requirement that states "take title" to all radioactive wastes after specified deadlines had passed, *2890*
United Nations sustainable development strategy, *4310*
Voluntary international principles to conserve the world's forests, *2715*
Widely acknowledged environmentalist to be chosen as a major-party U.S. vice presidential candidate, *1091*

1993

Administrator named to head the U.S. Forest Service who was not a career bureaucrat or a timber planter, *2708*
African-American president of the American Meteorological Society, *1834*
Automobile technology agreement between the U.S. government and industry, *1606*
Blizzard and winds to cause damages of $1 billion or more in the United States, *4213*
Blizzard to close all major U.S. airports on the East Coast, *4214*
Brownfields Economic Redevelopment Initiatives program in the United States, *2992*
Building industry consensus coalition for "green" construction in the United States, *4311*
Coniferous forest biome chosen for a U.S. National Park Service Inventory and Monitoring Program (I&M), *3404*
Death attributed to a "killer bee" attack in the United States, *3968*
Documentation of a *Mononychus* (one claw), a birdlike dinosaur, *1742*
Ecotourism research center in Australia, *2254*
Environmental health conference of U.S. physicians and environmentalists, *3969*
Environmental quality monitoring agency of the European Union (EU), *4312*
Executive Order requiring U.S. government agencies to use recycled paper, *4158*
Exposition with an environmental theme to be held in Korea, *2317*
Federal limits on shark fishing enacted by the United States, *2482*
Floods known to cover 10 U.S. Midwestern states, *2565*
Fly to be designated as an endangered species in the United States, *2344*
Forest Summit in the United States, *2690*
Global agreement on all aspects of biodiversity, *3048*
Institution to breed all 15 species of cranes, *1741*
International convention on biodiversity and sustainable development, *3049*
Law establishing the deepest cave in the United States as a protected area, *4672*
Law in Minnesota to treat as hazardous waste, *2892*
Law to discourage population growth in Iran, *3895*
Mexican spotted owl listing under the U.S. Endangered Species Act, *2418*
National population plan for Sierra Leone, *3894*
National population plan for Vietnam, *3892*
National population policy for Ethiopia, *3893*
Oil-tanker spill of major significance off Scotland, *3647*
Ozone-depletion Executive Order by a U.S. President, *1996*
Ozone-layer measurements from space, *1979*
Ozone-protecting household refrigerator, *1995*
Permanent United Nations sustainable development agency, *4313*
Permit system enacted in the United States for fishing vessels on the high seas, *2484*
Person in England to chain herself to a bulldozer to protest a highway, *1103*
State in the United States to impose a "hard to dispose" materials tax, *3285*

U.S. moratorium on building hazardous waste incinerators, *2893*
U.S. Natural Resource Inventory and Monitoring (I&M) mapmaking program, *3405*
U.S. Natural Resource Inventory and Monitoring (I&M) Program on water quality, *3406*
U.S. Navy report on the reintroduction of military dolphins to the wild, *4841*
U.S. President's Council on Sustainable Development, *4314*
Whooping crane artificial colony, *1726*
Woman to direct the U.S. Fish and Wildlife Service, *4741*
Woman to head the U.S. Environmental Protection Agency, *1116*

1994

Brownfields grants dispensed by the U.S. Environmental Protection Agency (EPA), *2993*
Cabinet-level environmental protection agency in Mexico, *2994*
Comprehensive database of natural resource information about U.S. national park lands, *3407*
Comprehensive inventory of vegetation in U.S. national parks, *3408*
Comprehensive United Nations agreement on use of the oceans, *3592*
Earthquake to occur directly under a large U.S. metropolitan area, *2234*
Ecosystem management plan upheld in a U.S. court, *3414*
Enforcement action by the U.S. Environmental Protection Agency (EPA) against a company for failure to comply with radionuclide emissions standards, *2997*
"Environmental justice" action by the U.S. government, *2995*
Exemption for U.S. military flyovers of National Wilderness deserts in California, *4673*
Feasibility studies of solar energy use in U.S. government buildings, *4122*
Feasibility studies of wind-generated energy use in U.S. government buildings, *4865*
Federal duck stamp in the United States to feature a red-breasted merganser, *3223*
Federal environmental justice grants in the United States, *3412*
Federal junior duck stamp in the U.S., *3224*
Genetic privacy legislation proposal in the United States, *2766*
Genetically altered food, *2767*
Government facility in the United States established to analyze chlorofluorocarbons (CFCs) in water and air, *4611*
Griffon vulture reintroduction in Italy, *2439*
Inter-American institute for research on global change, *3064*
International body to govern use of the ocean, *3053*
International body to recognize the obligations of countries to fight and prevent marine pollution, *3051*
International Fisheries Gene Bank (IFGB), *2499*
International organization to regulate use of the ocean floor, *3065*
Inventory of vegetation in U.S. national parks in Alaska, *3409*
Junior Duck Stamp program in the United States, *4789*
Lawsuit over industrial sewers initiated by the U.S. Resource Conservation and Recovery Act (RCRA), *2998*
Legal challenge to wolf reintroduction to Yellowstone National Park, *2440*
License plate in Massachusetts designed to benefit environmental education programs, *2307*
Mammal removed from the U.S. Endangered and Threatened Species List, *2419*
Mine gas extracted from coal fields in France, *2066*
National preserve in an American desert, *3813*
Natural resource damage settlement paid to New Jersey under the "Superfund" law, *2996*
North American conference on bird mortality and wind power, *4866*
Nuclear power in Great Britain generated from a pressurized water reactor, *3534*
Oil-pipeline spill of significance in Russia publicly reported, *3648*
Organization of African Unity (OAU) population agency, *3899*
Ostriches reintroduced to Arabia, *2438*
Penalties enacted by the United States for harassing hunters, *2961*
Prairies and grasslands biome surveyed by a U.S. National Park Service Inventory and Monitoring Program (I&M), *3410*
Rhinoceros and tiger protection law, *2346*
Rhinoceros preserve in Tanzania, *4760*
Sonic boom regulations for aircraft, deemed a noise pollutant and public health issue in the United States, *3447*
Standardized national vegetation classification system in the United States, *3413*
State in the United States to call for a moratorium on planned breeding of dogs and cats, *1023*
State in the United States to enact a "conspiracy to dump" law, *3286*
State in the United States to enact a "strict liability" antilitter law, *3287*
"Tornado chasers" systematically organized, *4297*
Trade sanctions imposed by the United States to protect endangered wildlife, *2345*
Treaty establishing an international 200-nautical-mile Exclusive Economic Zone (EEZ), *3052*

1994—*continued*
Treaty setting the framework to address the production of greenhouse gases, *3050*
U.S. Environmental Protection Agency (EPA) office for American Indian environmental affairs, *3411*
U.S. Fish and Wildlife Service (FWS) guidelines for distributing eagle parts, *4742*
U.S. Jobs Through Recycling grant to Arizona, *4159*
U.S. Jobs Through Recycling grant to California, *4160*
U.S. Jobs Through Recycling grant to Delaware, *4161*
U.S. Jobs Through Recycling grant to Maryland, *4162*
U.S. Jobs Through Recycling grant to Minnesota, *4163*
U.S. Jobs Through Recycling grant to Nebraska, *4164*
U.S. Jobs Through Recycling grant to New York, *4165*
U.S. Jobs Through Recycling grant to North Carolina, *4166*
U.S. Jobs Through Recycling grant to Ohio, *4167*
U.S. Jobs Through Recycling grant to Oregon, *4168*
U.S. Supreme Court decision that municipal incinerator ash be treated as hazardous waste, *2894*
United Nations conference to link the rights of women to population growth issues, *3900*
Uranium fuel bought from Russia by the United States, *3535*
Use of the Atlantic Coastal Fisheries Cooperative Management Act of 1993 against a state, *2467*
Vegetation inventory in the U.S. national parks system, *1797*
Vehicle recycling research center in the United States, *1607*
Wilderness protection act for California deserts, *4674*
World Health Organization (WHO) working group to address noise problems, *3431*
Zoo built for nocturnal viewing, *4965*

1995

Appearance of an oak tree blight previously unknown in the United States, *2609*
Asian elephant tracked with a satellite transmitter, *4744*
Drought research and development program nationally established in the United States, *2165*
Earthquake workshop canceled by an earthquake, *2235*
Elk reintroduced to Wisconsin, *2441*
Environmental lawsuit to challenge the United States under the North American Free Trade Agreement (NAFTA), *2348*
EPA Journal, *1074*
Female winner of the federal junior duck stamp contest in the United States, *3225*
Fishing vessel buyback program for New England, *2483*
Forest canopy crane used in North America, *2749*
Genetic recovery program for wildlife in the United States, *2347*
Genetically engineered corn, *2768*
Governmental tribunal in India for litigation relating to environmental accidents, *2999*
Green politician to join a national government as a minister, *1117*
High-resolution maps of the ocean floor, *3600*
Idaho Wildlife DNA Forensics Laboratory, *4743*
International airport to be re-developed as a community, *4315*
Land mammals removed from the U.S. Endangered and Threatened Species List, *2420*
Meteorological organization in Europe organized specifically to control weather satellites, *1967*
Natural gas production in Vietnam, *2067*
Natural Resources Inventory & Analysis Institute (NRIAI), *3415*
Nuclear power station to operate on a commercial basis, *3494*
Permit system enacted in the United States for fishing vessels on the high seas, *2484*
Proposed removal of bald eagles from the U.S. Threatened and Endangered Species List, *2425*
Remote-controlled submersible to reach the oceans' greatest depth, *3599*
Significant hog waste leakage into the New River of North Carolina, *4443*
Survey of U.S. biomass resources, *4339*
Sustainable development accord on the Alps, *4309*
Sustainable development governmental post in Canada, *4317*
Sustainable development handbook issued by the Canadian government, *4316*
Tests of leeches and mussels to monitor levels of freshwater contaminants, *4515*
U.S. Green Party presidential campaign, *1118*
U.S. Jobs Through Recycling grant to New Hampshire, *4169*
U.S. Jobs Through Recycling grant to Virginia, *4170*
U.S. President's Council on Sustainable Development, *4314*
"Universal Waste Rule" in the United States, *4135*

Wild dog breeding program in East Africa, *4761*

1996

All-American Road in the Appalachian Mountains, *4012*
All-American Road on the U.S. Pacific coast, *4013*
All-American Roads in the Rocky Mountains, *4014*
Atlantic–Gulf Coast biome surveyed by a U.S. National Park Service Inventory & Monitoring Program (I&M), *3416*
Automobile powered by fuel cells, *1608*
Biodiesel plant in Illinois, *4340*
Black-footed ferret reintroduction program to use acclimatization pens in the United States, *2443*
Canadian law to protect all endangered wildlife and flora, *2343*
Commercial fuel cell powered by landfill gas, *4341*
Dam safety code issued by the U.S. government, *2077*
Designation by the U.S. government of cigarettes and smokeless tobacco as nicotine delivery systems, *3970*
Electric car to be mass-produced using modern technology, *1609*
Environmental Quality Incentives Program (EQIP) in the United States, *3419*
Executive Order in the United States to protect the wildlife of the Midway Islands, *4843*
Federal Drinking Water State Revolving Fund (DWSRF) in the United States, *4614*
Federal duck stamp in the United States to feature a surf scoter, *3226*
Former hunting path designated as an All-American Road, *4015*
Genetic manipulation of hatchery fish stocks, *2500*
Greenhouse-effect warning about recycling paper, *1851*
Intergovernmental carnivore reintroduction strategy in Europe, *2442*
International Standards Organization (ISO) certification program for environmental management, *4318*
International treaty to specifically protect the sea turtles of the Western Hemisphere, *3054*
Landfill tax in England, *4153*
Law by the U.S. government to subject bottled water to the same standards as tap water, *4615*
Law in the United States to require phasing out of mercury in batteries, *4152*
Leaded gasoline total ban by the United States Congress, *1551*
Mammal to be cloned from an adult cell, *2769*
Nationwide survey of drinking water system infrastructures in the United States, *4612*
Noise control law in England with seizure provisions, *3448*
Nuclear byproduct technology capable of cleaning smokestack pollutants, *3000*
Nuclear power plant in Romania, *3536*
Oil-tanker spill of significance off Wales, *3649*
Permanent forum for coordinating sustainable development in the Arctic, *4319*
Scottish Environment Protection Agency (SEPA), *3001*
State Scenic Byway in Ohio, *4019*
Transfrontier protected area for marine turtles, *4842*
Tropical-subtropical biome surveyed by a U.S. National Park Service Inventory & Monitoring Program (I&M), *3417*
U.S. government declaration of cigarettes as a "drug delivery device", *1370*
U.S. National Scenic Byway in the Black Hills, *4011*
U.S. National Scenic Byway in the Great Lakes region, *4016*
U.S. National Scenic Byway on an American Indian reservation, *4017*
U.S. National Scenic Byway through Louisiana wetlands, *4018*
United Nations conference to address human habitation in a sustainable development context, *3901*
United Nations emergency response system for industrial pollution disasters, *3002*
United Nations training center for water supply issues, *4613*
Use of biodiesel fuel at a U.S. national political convention, *4342*
Wetlands Reserve Program (WRP) in the United States, *3420*
Wildlife Habitat Incentive Program (WHIP) in the United States, *4790*
World Wetlands Day, *3418*

1997

Agreement on International Humane Trapping Standards, *2929*
America Recycles Day, *4172*
American Heritage Rivers program, *3423*
Annual report on the environment and American foreign policy, *2308*
Automobile gasoline converted to hydrogen for future cars, *1613*
Ban on motorized watercraft on Wyoming's Snake River, *4675*
Battery recycling program in Canada, *4171*
Bioprospecting agreement in the United States, *2770*
Cars designed using computer-aided tools, *1610*
Clean Water Hardship Grant in the United States, *4617*

1997—*continued*
Congressional attempt to re-establish the Office of Noise Abatement and Control (ONAC), *3451*
Convention of people affected by the construction of dams, *2078*
Dam opposition day worldwide, *2081*
DNA test to detect whirling disease in trout and other salmonids, *2468*
Doctorate program in ecological economics in the United States, *2309*
Ecotourism appeal coordinated by governments, *2255*
Electric automobile to convert gas to hydrogen fuel, *1614*
Electric car from a major U.S. company, *1612*
Environmental justice complaint filed with the U.S. Environmental Protection Agency, *3003*
Environmental law challenged under the North American Free Trade Agreement (NAFTA), *3971*
Environmental lawsuit decided by the International Court of Justice, *3011*
Evidence that Martian volcanic rocks are in some respects similar to Earth's, *4395*
Executive Order issued to protect the Lake Tahoe region, *3421*
Fuel cell passenger car to use methane, *4343*
Geothermal building in Manhattan, *2826*
Global list of endangered plants, *2421*
Helicopter noise bill to reach a committee in the U.S. Congress, *3450*
Intentional poisoning of Lake Davis in California, *4807*
International treaty to ban the reproductive cloning of human beings, *3055*
International treaty to set binding limits on carbon dioxide emissions, *3056*
Jaguar Conservation Team in Arizona, *2349*
Leaf blower ban of significance in Los Angeles, CA, *3449*
Lemurs reintroduced to Madagascar, *2445*
Mexican wolf reintroduction, *2444*
Modern industrial pollution crime penalties in China, *3004*
Modern natural resource crime laws in China, *3422*
Modern solid waste disposal crime laws in China, *4154*
Modern wildlife resource crime laws in China, *4792*
National Environmental Action Plan (NEAP) in Africa, *3395*
Nuclear accident reported falsely in Japan, *3538*
Nuclear energy agreement between the U.S. and China, *3523*
Nuclear energy law allowing expropriation of power plants in Sweden, *3539*
Nuclear power plant in Turkey, *3544*
Phase of construction on the gigantic Three Gorges Dam on the Yangtze River in China, *2110*
Plant to produce methanol from coal, *4344*
Ranchers' organization to promote "predator-friendly" wool, *1175*
Record-high global temperatures in consecutive years, *1852*
Regulations enacted in the United States to limit swordfish harvesting according to size, *2485*
Self-sustaining greenhouse in a museum, *4966*
State commission in the United States funded by taxes on litter-producing corporations, *3288*
State in the United States to ban human cloning, *2771*
State in the United States to implement a multiple-species conservation plan statewide, *4791*
State in the United States to require utility companies to offer a minimum of electricity from renewable energy sources, *4320*
Test drive of the "veggie van," an experimental motor home powered by a diesel engine that ran on fuel made from vegetable oil, *1611*
U.S. government body devoted solely to mediating environmental disputes, *3066*
U.S. Sport Fishing and Boating Partnership Council, *2512*
U.S. state with a sustainable development agency, *4321*
United Nations convention on transboundary water supplies, *4616*
Year that nuclear power was the main source of British electricity, *3537*
Zoo in the United States to exhibit a pair of aye-ayes, *4967*

1998

All-American Road with glacial features, *4020*
All-American Road with volcanic features, *4021*
American Heritage Rivers program, *3423*
Australian indigenous species habitat listed as endangered, *2424*
Automobile emission cleaner than city air, *1615*
Biomass-to-ethanol conversion plant, *4346*
Comprehensive inventory of geologic resources in U.S. national parks, *3424*
Congressional attempt to re-establish the Office of Noise Abatement and Control (ONAC), *3451*
Dam cost analysis on a global scale, *2079*
Diamond mine in Canada, *3338*
Direct evidence that the West Antarctic Ice Sheet was once open sea, *3874*
Drought predictions for the 21st century by a U.S. government agency, *2166*
Elephant remote tracking program in Thailand, *4793*

Executive Order protecting coral reefs in the United States, *3601*
Genetically engineered papayas, *2772*
Human embryonic stem cells to be artificially cultivated, *2774*
International declaration on El Niño, *1968*
Logging company in California history to have its timber license suspended for environmental violations, *2666*
Mandatory international Atlantic bluefin tuna conservation plan, *2486*
Minnesota wind power tax law, *4862*
Modern environmental management policy in South Africa, *3425*
Music festival powered by biodiesel fuel, *4345*
National Green party to become part of a governing coalition in Europe, *1119*
National park to focus on conservation history and the evolving nature of land use, *3812*
National Threatened Species Day in Tasmania, *2350*
Nuclear waste lawsuit of significance against the U.S. government, *3540*
Record-high global temperatures in consecutive years, *1852*
Reintroduction of Aplomado falcons to Mexico, *2446*
Satellite station built to track "space litter", *3289*
Solar-powered vehicle to travel into deep space, *4123*
State in the United States to ban state funding for human cloning research, *2773*
State nature preserve in Massachusetts, *4844*
Steelhead trout listing under the U.S. Endangered Species Act, *2423*
Sturgeon listing by the Convention on International Trade in Endangered Species of Wild Fauna and Flora (CITES) to include all species, *2422*
Transportation act for the 21st century passed by the United States, *1585*
Trilateral biosphere reserve, *1642*
U.S. Government Paperwork Elimination Act, *4155*
U.S. National Scenic Byway below sea level, *4022*
U.S. National Scenic Byway in Alaska, *4023*
U.S. National Scenic Byway in an urban area, *4024*
U.S. National Scenic Byway in Grand Canyon National Park, *4025*
Wind farm in Africa, *4868*
Wind farm in Colorado, *4867*
Woman to reach the rank of major in the U.S. Park Police, *3814*

1999

Admission by the U.S. government that nuclear weapons production may have caused illnesses in thousands of workers, *3542*
American Indian tribal whale hunt in modern times during an International Whaling Commission (IWC) moratorium on whale hunting, *4628*
Appearance of the West Nile virus in the Western Hemisphere, *3972*
Ban of lead shot in Canada, *2962*
Bioprospecting clause in U.S. Fish and Wildlife Service Special Use specimen-collecting permits, *4676*
Brightfields initiative by the U.S. Department of Energy, *3005*
Building developments halted to protect an endangered species, *2351*
California waters in the state Heritage Trout Program (HTP), *2513*
Confirmation that peach oil kills fungi and other soil pests, *1310*
Cyber animal exhibit that allows people to view bears in their natural habitat, *4968*
Dam opposition day worldwide, *2081*
Deep-sea exploration of U.S. marine sanctuaries, *3602*
Deep underground nuclear depository, *3541*
Edict issued by the U.S. Environmental Protection Agency requiring the states to clean up air pollution in national parks, *1428*
Electric-vehicle project by the Canadian government and industry, *1616*
Evidence that a natural occurrence rather than toxins was causing deformities in frogs around the United States, *4417*
Government demolition of a U.S. dam without the owner's permission, *2080*
"Green" U.S. Post Office, *4322*
Haze regulations concerning U.S. wilderness areas and national parks, *4677*
International Adopt-a-Highway Day, *3290*
Large fish kill in Chesapeake Bay in the 1990s, *4498*
Light rail system in a national park in the United States, *3815*
Limits on the height of skyscrapers in New York City, *3158*
Lynx reintroduction in Colorado, *2447*
Major auction of confiscated wildlife products in the United States, *4794*
Major earthquake in the Izmit region of western Turkey, an area that holds nearly half the nation's population, *2236*
National assessment of biological resources in the United States, *3426*
Official designation of the "Mendocino Tree" as the world's tallest living thing, *2580*
Ozone depletion recorded over the Dead Sea, *1997*

1999—*continued*

Proposed removal of bald eagles from the U.S. Threatened and Endangered Species List, *2425*
Recorded appearance of a large haze of air pollution over the Indian Ocean, *1482*
Regional endangered-species plan to protect an aquatic ecosystem, *2352*
Scholarly article to identify a legal trend curbing the powers of environmental groups and private citizens to sue polluters in U.S. courts, *1093*
Substantial revisions in U.S. guidelines for approving new antibiotics for livestock and for monitoring the effects of existing drugs, *1176*
Town in the United States relocated due to airport noise, *3432*
Transfrontier wildlife park in Africa, *4845*
U.S. Interagency Council on Biobased Products and Bioenergy, *4347*
Use of an ancient weed-control technique by a modern American city, *1311*
Use of the term *genetic pollution, 1257*
Wind energy–generating station in southern Africa, *4870*
Wind-powered brewery in the United States, *4869*

2000

Comprehensive plan to redesign the water-flow system in and around the Everglades, *4678*
State in the United States to ban lead sinkers and jigs, *2514*

2001

Bill in California forcing developers to show that there would be an adequate supply of water for any proposed large subdivision, *3091*
Large-scale relocation of elephants, *4745*
Report on the impact of human predation upon coastal marine ecosystems dating back to prehistoric times, *3603*
Superpower to effectively withdraw from the Kyoto Protocol, *3012*
Systematic survey of the endangered Asiatic cheetah, *2353*

2002

Ban on fox hunting in Scotland, *2963*
Tax in Ireland on shoppers' plastic bags, *3291*
Working prototype of a commercially viable pebble-bed nuclear reactor, *3543*

2006

Nuclear power plant in Turkey, *3544*

2060

Year the normal ozone level will return to the stratosphere, *1998*

Index by Month and Day

The Index by Month and Day is a chronological listing of key information from the main text of the book, organized by month and day, starting with the earliest. Each index entry includes key information about the "first" and a 4-digit number in italics. That number directs you to the full entry in the main text, where entries are numbered in order, starting with 1001. To find the full entry, look in the main text for the entry tagged with that 4-digit number.

Note that some entries do not contain month or day information, so they are not included in this index. Entries that contain only month, but not day, information are listed first under each month.

For more information, see "How to Use This Book," on page ix.

January

1768 Botany professor at a college, *2257*

1821 Person to see the Antarctic territories of Peter I Island and Alexander I Island, *3831*

1840 Antarctic expedition funded by the U.S. Congress, *3835*

1888 Boone and Crockett Club meeting, *1030*

1919 Nature reserve in the Soviet Union, *3796*

1935 National organization dedicated specifically to preserving wilderness (as opposed to wildlife) in the United States, *1055*

1937 Sit-down strikes to cripple General Motors car production, *1544*

1946 Use of nuclear power to produce electricity in Great Britain, *3456*

1956 Nuclear power agency of the Japanese government, *3476*

1957 Colony of puna flamingos to be discovered, *1700*

1958 Nuclear accident of significance in the Soviet Union, *3482*

1959 Research park inside an "education triangle", *3726*

1963 Nuclear power barge, *3498*

1969 Environmental movement of significance in the Soviet Union, *1065*

Mudslides to cause property damage of over $135 million, *2575*

Regional wetlands development commission in New Jersey, *3377*

1970 Publication inspired by *The Whole Earth Catalog*, *3259*

1975 *EPA Journal*, *1074*

1978 Radioactive fallout of significance from a space program, *1326*

1986 U.S.–Canadian bilateral report on acid rain, *3042*

1994 Vehicle recycling research center in the United States, *1607*

1997 DNA test to detect whirling disease in trout and other salmonids, *2468*

Electric car from a major U.S. company, *1612*

January 1

1968 Law in California to limit motor-vehicle noise emissions on highways, *3435*

Law in California to limit motor-vehicle noise emissions through sales restrictions, *3436*

1972 Automobile seat-belt buzzers, *1600*

Extensive collection of quality data on fertility and family planning, *3882*

1975 California state agency to regulate wind power and other renewable energy sources, *4855*

1979 Total ban of chlorofluorocarbons (CFCs) for aerosols by a country, *1982*

1982 Hazardous substances protection law in Germany, *3958*

1985 Natural gas price deregulation by the U.S. government, *2063*

1991 Minnesota wind power tax law, *4862*

1993 State in the United States to impose a "hard to dispose" materials tax, *3285*

January 1—*continued*

1996 Leaded gasoline total ban by the United States Congress, *1551*
2000 State in the United States to ban lead sinkers and jigs, *2514*

January 2

1975 Discovery of the monarch butterfly's winter home, *4729*
1980 Water Bank Act in the United States, *3380*

January 3

1840 Deep-sea soundings, *3552*
1903 National park established to protect a cave, *3787*
1961 Deaths caused by an American nuclear reactor, *3495*
1975 Federal law to create wilderness areas in the eastern United States, *4653*
Hazardous Materials Transportation Act in the United States, *2855*
Law controlling aircraft noise over the Grand Canyon, *3445*
U.S. National Wilderness Area in the Great Lakes region, *4654*
U.S. National Wilderness Area in the Mid-Atlantic states, *4655*
U.S. National Wilderness Area in the Ozark Plateau region, *4656*
U.S. National Wilderness Areas in the Great Plains, *4657*
1977 Nation to pass a constitutional amendment to protect its public health, forests and wildlife, *3386*
National amendment to restrict industrial polluters, *2978*
1978 Tornado in Great Britain to cause $1 million damage, *4296*
1992 Sockeye salmon listing under the U.S. Endangered Species Act, *2417*

January 4

1935 National Park Service unit including coral reef resources, *3799*
1996 Greenhouse-effect warning about recycling paper, *1851*

January 5

1993 Oil-tanker spill of major significance off Scotland, *3647*

January 6

1806 European to lead an expedition into the area that later became Sequoia and Kings Canyon National Park, *3760*
1912 Cohesive theory of the concept of "continental drift", *2817*

January 7

1983 Spent nuclear fuel and high-level radioactive waste disposal law in the United States, *3522*

January 9

1908 National monument in the United States created from land donated by a private individual, *3747*
1959 Dam disaster of significance in Spain in the 20th century, *2117*
1984 Mollusk removed from the U.S. Endangered and Threatened Species List, *2380*

January 10

1800 Snowstorm known to have dropped 18 inches of snow on Savannah, GA, *4193*
1901 Oil well in the United States with a massive petroleum deposit, *3693*
1955 Nuclear energy nonmilitary joint projects of the U.S. government and industry, *3470*

January 11

1909 Canadian–U.S. International Boundary Treaties Act, *3366*
Canadian–U.S. treaty to regulate use of the Niagara River for water power, *4555*

January 12

1773 Natural history museum foundation in the United States, *4883*
1876 State forestry association, *2692*
1986 Reported sighting of Jerdon's courser since 1900, *1683*

January 15

1932 Officially reported measurable snowfall in downtown Los Angeles, *4204*
1955 House with solar heating and radiation cooling in the United States, *4109*
1982 Fish removed from the U.S. Endangered and Threatened Species List, *2378*

January 17

1773 Person to cross the Antarctic Circle, *3826*

1817 "Luminous" snowstorm in Vermont and Massachusetts, *4195*
1966 Oceanographic research vessel to recover a lost hydrogen bomb, *3580*
1994 Earthquake to occur directly under a large U.S. metropolitan area, *2234*
1995 Earthquake workshop canceled by an earthquake, *2235*

January 18

1921 Dam burst to cause a disaster by chemical contamination, *2114*
1933 National monument to commemorate the world's largest gypsum dune field, *3722*
1976 World's fair with an ocean theme to be sanctioned by the Bureau of International Expositions (BIE), *2311*

January 19

1977 Known trace of snow in the Miami, FL, area, *4210*

January 20

1943 Report of wild temperature variations on simultaneous readings from nearby stations in South Dakota, *1821*
1978 Official "concentrated heavy snowstorm," defined as 20 inches or more within a 48-hour period, in Boston, MA, *4211*
1999 Ozone depletion recorded over the Dead Sea, *1997*

January 21

1863 Heavy rainstorm and local flooding to cause the complete failure of an offensive operation in the American Civil War, *2529*
1880 Sewage-disposal system in the United States separate from the city water system, *4526*
1954 Nuclear-powered submarine, *3466*
1969 Nuclear power accident in Switzerland, *3507*
1981 Group of Ten meeting, *1078*
U.S. Supreme Court ruling on plastic milk containers, *4146*

January 22

1947 Winter in Great Britain in the 20th century during which snow fell somewhere in the country every day from January 22 to March 17, *4206*
1991 Environmental accident insurance guarantee in India, *2991*

January 23

1960 Manned descent to the greatest depth on earth, *3578*
1973 Volcano lava flow to be fought with sea water, *4384*
1991 Ecological disaster caused by a deliberate oil spill, *3645*

January 24

1556 List of earthquake deaths, *2182*
1895 Landing on Antarctica that initiated the "Heroic Age" of Antarctic exploration, *3839*
1935 Commercially marketed beer cans in the United States, *4127*

January 25

1945 Water supply in the United States to be fluoridated in order to reduce tooth decay, *4591*
1982 Unleaded gasoline campaign in Great Britain, *1549*
1999 Electric-vehicle project by the Canadian government and industry, *1616*

January 26

1531 Earthquake causing thousands of deaths in Portugal, *2181*

January 27

1967 Treaty to ban the introduction of nuclear weapons into orbit around the Earth, *3027*
1997 Mexican wolf reintroduction, *2444*

January 28

1951 Ice storm to cause $100 million in damages in the United States, *4272*
1969 Offshore oil spill to cause widespread damage, *3632*
1980 Ohio State Scenic River, *3985*

January 29

1999 Wind-powered brewery in the United States, *4869*

January 30

1997 State in the United States to implement a multiple-species conservation plan statewide, *4791*
1999 "Green" U.S. Post Office, *4322*

January 31

1982 Oryx reintroduction program, *2431*

February

1856 Underwater photograph, *3558*
1914 Underwater movie seen by the public, *3568*
1934 Commercial production of plants in water instead of soil, *1249*
1970 Conservation legal organization, *2290*
Oil discovery of significance in the North Sea, *3661*
1979 World Climate Conference, *1829*
1984 Pipeline to carry natural gas from the Soviet Union to a European country, *3708*
1985 Discovery of large numbers of dead and deformed birds in the Kesterson Refuge and other ponds containing irrigation runoff from farms in the Central Valley of California, *1221*
1988 Negotiating session to prepare for a global convention on the control of transboundary movements of hazardous wastes, *2885*
1989 International climate conference held exclusively for developing countries, *1846*
1991 Intentional oil-well fires of significance, *3646*
1997 Automobile gasoline converted to hydrogen for future cars, *1613*
Electric automobile to convert gas to hydrogen fuel, *1614*

February 1

1953 Flood leaving 1 million people homeless in the Netherlands, *2544*

February 2

1871 List of wetlands of international importance, *3016*
1902 Aerial photograph in Antarctica, *3845*
1937 Firearms tax in the United States to specifically benefit wildlife habi tats, *4820*
1952 Tropical storm to hit the United States in February, *4187*
1971 International convention protecting wetlands, *3028*
1996 World Wetlands Day, *3418*

February 3

1902 Polar explorer to ascend in a balloon in the Antarctic regions, *3846*
1931 Earthquake of significance in New Zealand, *2220*
1999 Lynx reintroduction in Colorado, *2447*

February 4

1797 Earthquake causing massive loss of life in Ecuador, *2196*
1985 Species removed from the U.S. Endangered and Threatened Species List due to recovery, *2412*
1990 Nuclear power public safety organization of the South Korean government, *3966*

February 5

1870 Government weather service in the United States, *1911*
1887 Snowfall recorded in excess of 3.5 inches in San Francisco, *4200*
1942 Hazardous substances transportation law in France, *3930*

February 7

1821 Explorer to land on Antarctica, *3832*

February 8

1889 Head of the Department of Agriculture to be a member of the President's Cabinet, *1129*
1965 List of wilderness areas in the U.S. National Wilderness Preservation System (NWPS), *4648*

February 9

1871 Conservation agency established by the U.S. government, *4763*
1909 Forestry school privately operated in the United States, *2681*

February 10

1972 Substantial rainfall after more than 400 years at Calama, Chile, *2555*

February 11

1808 Anthracite coal burned experimentally, *2036*
1971 Prohibition against nuclear weapons on the ocean floor, *3584*
1981 Nuclear power plant accident of significance in Japan, *3518*
1994 "Environmental justice" action by the U.S. government, *2995*
Federal environmental justice grants in the United States, *3412*

February 12

1820 Ship to bring home Antarctic sealskins, *3829*
1937 Description of hydroponic agriculture, *1247*

February 13

1960 Nuclear weapons test conducted by the French, *3492*
1992 Ban on some U.S. uses of the EBDC fungicide class, *1308*
2002 Ban on fox hunting in Scotland, *2963*

February 14

1872 Bird refuge established by a U.S. state, *1718*

February 15

1958 Blizzard to cause more than $500 million damage in the United States, *4208*
1972 United Nations agreement to regulate dumping of waste into the oceans, *3587*
1982 Offshore drilling-rig disaster near Canada, *3641*
1996 Oil-tanker spill of significance off Wales, *3649*

February 16

1938 Strong claim that industrial burning of fossil fuels increased carbon dioxide levels in the atmosphere, *1448*
1998 Dam cost analysis on a global scale, *2079*

February 17

1959 Satellite to transmit weather information to the earth, *1957*

February 18

1909 Intergovernmental conservation conference in North America, *3367*
Tri-national conference on conservation in North America, *1034*
1929 Government commission to rent or purchase lands to be set aside for migratory birds in the United States, *1691*

February 19

1884 Tornadoes in the United States reported to cause a great loss of life, *4283*
1912 U.S. Supreme Court decision regarding a sponge harvesting law, *2472*

February 20

1833 Man to reach 74 degrees 15 minutes south latitude, *3834*
1835 Earthquake to be described in detail by Charles Darwin, *2200*
1935 Woman to visit Antarctica, *3858*
1941 Proof that radar could be used to locate precipitation, *1944*

February 23

1883 National animal rights organization in the United States, *1010*
1993 Environmental health conference of U.S. physicians and environmentalists, *3969*

February 24

1898 Significant theory that Antarctica was a continent rather than a series of islands, *3841*
1978 Federal law to allow designation of formerly developed lands as wilderness areas, *4662*
1992 Automobile emissions standards in Canada, *1582*

February 26

1917 National park created after the establishment of the National Park Service, *3795*
1972 Disaster settlement to include provisions to address the long-term psychological needs of the victims, *2556*

February 28

1954 Phase-contrast cinemicrography film to be shown on television, *3266*

February 29

1748 So-called "Winter of the Deep Snow" in New England, *4192*
1984 Reptile removed from the U.S. Endangered and Threatened Species List, *2410*

March

1624 Cattle imported into a North American colony, *1156*
1788 Botanical society formed that is still in existence today, *1769*
1796 Coast survey book written and published in the United States, *3235*
1854 Parkland purchased by a city in the United States, *3817*
1944 Volcano eruption to end a World War II battle, *4379*
1951 World Meteorological Organization (WMO) Congress, *1822*

March—*continued*

1977 Commercial nuclear reactor that was gas-cooled, *3511*
1985 Chlorofluorocarbons (CFCs) limitation on an international level, *1985*
1989 Conference held after the signing of the Protocol on Substances That Deplete the Ozone Layer (Montreal Protocol), *1990*
1997 Convention of people affected by the construction of dams, *2078*
Dam opposition day worldwide, *2081*
Ecotourism appeal coordinated by governments, *2255*
2002 Tax in Ireland on shoppers' plastic bags, *3291*

March 1

1872 National park in the United States, *3772*
Superintendent of Yellowstone National Park, *3773*

March 2

1891 Gas conversion law by a state, *2047*
1931 Law in the United States to allow federal control of wildlife on both public and private lands, *4799*
1992 Report that pollutants discharged into the Detroit River may have killed bald eagle hatchlings along Lake Erie, *4416*

March 3

1849 Department of the Interior in the United States, *3358*
1867 Director of the U.S. Geological Survey (USGS), *2809*
1879 Agency created by U.S. government specifically for geological issues, *2811*
1885 U.S. government body to study the effects of insects on nature, *4687*
1891 Forest planted by the U.S. government, *2726*
1999 International Adopt-a-Highway Day, *3290*

March 4

1791 Drought recorded in Australia, *2133*
1881 Quarantine law for plants by a state, *1277*
1913 Bird hunting regulation enacted by the U.S. Congress, *2955*
Legislation to protect migratory game and insectivorous birds, *1689*
1962 Nuclear power plant in Antarctica, *3497*
1971 Mounted bird specimen to sell for more than £9,000 sterling, *2375*
1999 Wind energy–generating station in southern Africa, *4870*

March 5

1974 Discovery room in a natural history museum in the United States, *4947*

March 6

1875 State in the United States to restrict market hunting, *2951*
1948 International agency to address pollution of the oceans by ships, *3574*
1995 Sustainable development accord on the Alps, *4309*

March 7

1644 Whaling industry established by an American town, *4622*

March 8

1946 World Bank meeting, *4298*
1994 Feasibility studies of solar energy use in U.S. government buildings, *4122*
Feasibility studies of wind-generated energy use in U.S. government buildings, *4865*

March 9

1995 Land mammals removed from the U.S. Endangered and Threatened Species List, *2420*

March 10

1906 Coal mining accident in Europe to cause catastrophic deaths, *3322*
1982 Limit on lead in gasoline initiated by the Italian government, *1580*
1998 Elephant remote tracking program in Thailand, *4793*

March 11

1888 Blizzard to cause numerous deaths on the U.S. East Coast, *4201*
1932 Heath hen extinction report, *2370*
1967 Bat listing under the U.S. Endangered Species Act, *2386*
Bear listing under the U.S. Endangered Species Act, *2387*
Darter listed under the U.S. Endangered Species Act, *2388*
Deer listings under the U.S. Endangered Species Act, *2389*
Fox listing under the U.S. Endangered Species Act, *2390*

Parrot listed under the U.S. Endangered Species Act, *2391*
Snake listed under the U.S. Endangered Species Act, *2392*
1992 Cooperative effort by northeastern states to curb nitrogen oxide emissions from coal-burning power plants, *1424*
1997 Nuclear accident reported falsely in Japan, *3538*

March 12

1928 Dam of significance in the United States to collapse when completely filled, *2116*
1992 Organization to study the North Pacific, *3063*
1993 Blizzard and winds to cause damages of $1 billion or more in the United States, *4213*
Blizzard to close all major U.S. airports on the East Coast, *4214*
1994 Inter-American institute for research on global change, *3064*

March 13

1982 Woman to graduate from the Pennsylvania School of Conservation, *2303*
1992 Massive cleanup plan for the Florida Everglades, *4487*

March 14

1870 "Blizzard" designation for a snowstorm, *1912*
1903 Bird reservation established by the U.S. government, *1719*
1960 Sulfur mine offshore, *3331*
1999 Dam opposition day worldwide, *2081*
2001 Superpower to effectively withdraw from the Kyoto Protocol, *3012*

March 15

1887 Game warden in the United States to be paid a salary, *4688*
1958 Federal fish-agriculture rotation program in the United States, *2496*
1960 Undersea park established by the U.S. government, *3803*
1966 Oceanographic research vessel to recover a lost hydrogen bomb, *3580*
1992 Systematic research documenting the slow death of the sabal palm along Florida's Gulf of Mexico coastline, *2748*

March 16

1934 Annual license required of all U.S. waterfowl hunters 16 years of age or older, *2933*
Law in the United States to allow government fees to be levied to establish and maintain wildlife refuges, *4818*
Legislation that provided regular federal funding for waterfowl management in the United States, *1693*
1978 Oil-tanker spill of significance off the coast of France, *3639*
1992 Biodiversity governmental agency in Mexico, *1641*
1993 Mexican spotted owl listing under the U.S. Endangered Species Act, *2418*
1998 Automobile emission cleaner than city air, *1615*

March 17

1958 Satellite instruments powered by solar energy, *4112*
1992 Accord on transboundary effects of industrial accidents on the environment, *3047*

March 18

1543 Flood recorded by Europeans in North America, *2524*
1911 Hydropower facilities operated by the U.S. Bureau of Reclamation, *4556*
1925 Tornado in the United States with a death toll in excess of 650, *4285*
Tornado to last 3 hours in the United States, *4286*
1953 Accurate DNA structure description, *2753*
1967 Supertanker to founder and spill its cargo of oil, *3630*

March 19

1936 Prairie restoration in the U.S. National Parks System, *3749*

March 20

1970 Oil spill of significance off Sweden, *3633*

March 21

1969 International organization to regulate tuna fisheries, *2475*
1970 Earth Day to be celebrated nationwide in the United States, *1069*
1981 National protest of significance sponsored by Earth First! *1079*
1994 Treaty setting the framework to address the production of greenhouse gases, *3050*

March 22

1972 Move to halt all fungicide uses of mercury in the United States, *1298*

1985 Agreement among nations to work together to create guidelines, policies, and procedures for protecting the ozone layer, *3041*

1989 International convention on transporting hazardous waste, *3045*

March 23

1983 National Scenic Trail in the eastern United States, *4004*

March 24

1629 Hunting law enacted by an American colony, *2904*

1955 Seagoing oil drilling rig for drilling in water more than 100 feet deep, *3656*

1992 Mandatory automobile-emissions tests in Canada, *1583*

1995 Remote-controlled submersible to reach the oceans' greatest depth, *3599*

1997 Ban on motorized watercraft on Wyoming's Snake River, *4675*

March 25

1948 Tornado forecast for a specific U.S. location, *4288*

March 26

1980 Nuclear energy program to disassemble a damaged reactor, *3515*

1999 Deep underground nuclear depository, *3541*

March 27

1719 American colony to restrict oyster fishing, *2501*

1964 Use of elevation benchmarks to measure settling or uplifting of the earth, *2226*

1992 Guidelines issued by the U.S. Fish and Wildlife Service (FWS) on foregoing lawsuits in natural resource damage cases, *3403*

1996 Black-footed ferret reintroduction program to use acclimatization pens in the United States, *2443*

March 28

1865 Natural scenery protection state law connected with advertising, *3974*

1892 Eruption in modern times of the Mexican volcano El Chichon in the southeastern state of Chiapas, *4373*

1920 Killer tornado to cause $3 million in damages in Chicago, *4284*

1979 Major accident in a U.S. nuclear power plant, *3514*

March 29

1626 Forestry law enacted by a British colony, *2625*

1933 Employment program of significance for forestry and other natural resource projects, *2683*

March 30

1891 National forest in the United States, *3714*

National forest reserve in the United States, *2712*

1988 Woman to travel solo to either pole, *3872*

1992 Cooperative effort by 10 northeastern U.S. states to develop plans to reduce emissions of greenhouse gases, *1425*

1994 Natural resource damage settlement paid to New Jersey under the "Superfund" law, *2996*

U.S. Fish and Wildlife Service (FWS) guidelines for distributing eagle parts, *4742*

1999 Bioprospecting clause in U.S. Fish and Wildlife Service Special Use specimen-collecting permits, *4676*

April

1866 Society for animal welfare in the U.S., *1007*

1897 Ascent of Mount Elias, *3784*

1935 Use of the term *Dust Bowl* to describe the drought-devastated Southern Plains, *2150*

1971 Law requiring the U.S. president to develop and conduct a comprehensive program of marine science activities, *3581*

Offshore oil-well fires of significance in the Gulf of Mexico, *3634*

1972 Political party to take an environmental issue to the polls, *1108*

1974 Authentication of 148 tornadoes over a two-day period in the United States, *4295*

1978 Flood in the United States to cause $1 billion in damages, *2559*

1982 Nuclear-generated electricity exported from Canada, *3521*

1987 Use of the term *sustainable development*, *4305*

1988 Person to circumnavigate the North Pole on foot and solo, *3871*

Scientific evidence that humans caused the Antarctic ozone hole, *1988*
1990 International horticultural show in Asia, *2315*
1992 Aye-aye born in captivity, *4759*
Native American organization dedicated to preserving buffalo, *4839*
1993 Documentation of a *Mononychus* (one claw), a birdlike dinosaur, *1742*
Federal limits on shark fishing enacted by the United States, *2482*
National population policy for Ethiopia, *3893*
1997 Environmental law challenged under the North American Free Trade Agreement (NAFTA), *3971*
Jaguar Conservation Team in Arizona, *2349*
1998 Wind farm in Colorado, *4867*

April 1

1826 Internal combustion engine in the United States, *4323*
1897 Female pioneer in the field of mycology, the study of mushrooms, *1789*
1928 Zoo to use bar-less exhibits extensively, *4924*
1946 Tsunami of significance to hit a U.S. territory, *2543*
1960 All-weather satellite, *1960*
Satellite to provide cloud-cover photographs, *1825*
1973 Ecotourism project started by the Indian government, *2241*
Tiger preserve in India, *4830*
1992 Laundromat with an environmental theme, *4411*
1996 Scottish Environment Protection Agency (SEPA), *3001*
1998 Sturgeon listing by the Convention on International Trade in Endangered Species of Wild Fauna and Flora (CITES) to include all species, *2422*

April 2

1957 National park dedicated to the first transcontinental railroad, *3802*
1962 Establishment of a federal Bureau of Outdoor Recreation in the United States, *2686*
1973 Ice meteor to be subject to intensive study, *2296*
1993 National population plan for Sierra Leone, *3894*

April 3

1850 Meteorology society in Great Britain continuing through the 20th century, *1899*
1973 Zoo with twilight conditions, *4945*
1992 Officially sanctioned international horticultural fair in the United States, *2316*
1997 State in the United States to require utility companies to offer a minimum of electricity from renewable energy sources, *4320*

April 4

1947 Worldwide organization concerned with civil aviation safety, *3933*
1972 Electric power in the United States from municipal garbage as a boiler fuel, *4130*
1973 Nonviolent protest of significance by the modern Chipko movement in India, *1041*
1994 Trade sanctions imposed by the United States to protect endangered wildlife, *2345*
1996 Environmental Quality Incentives Program (EQIP) in the United States, *3419*
Wildlife Habitat Incentive Program (WHIP) in the United States, *4790*

April 6

1580 Death in England known to have been caused by an earthquake, *2183*
1909 Explorers credited with reaching the North Pole, *3849*

April 7

1948 Public health agency of the United Nations, *3934*
1966 Oceanographic research vessel to recover a lost hydrogen bomb, *3580*
1974 United Nations agreement to regulate dumping of waste into the oceans, *3587*
1976 Biotechnology company in the United States, *2756*
1999 Transfrontier wildlife park in Africa, *4845*

April 8

1997 Global list of endangered plants, *2421*

April 10

1872 Arbor Day celebration, *2724*

April 10—*continued*

Effective promoter of tree planting to prevent soil erosion on the U.S. Great Plains, *4049*

1963 Nuclear submarine from the U.S. Navy to be lost at sea, *3499*

1987 National Fishing Week in the United States, *2511*

April 11

1965 Set of tornadoes to cause $500 million in damages, *4291*

April 12

1887 Law in Wisconsin authorizing the pay of a game warden, *4689*

1992 Tax on "gas-guzzling" automobiles by a U.S. state, *1584*

April 14

1909 Oil discovery of significance in Persia, *3695*

1984 Large-scale boycott of a restaurant chain in an effort to halt imports of tropical beef, *2720*

April 15

1935 Arbitration ruling to assign national responsibility for air pollution damage, *1389*

1972 Joint effort of major significance to solve pollution problems in the Great Lakes, *4476*

U.S.–Canadian effort of significance to improve the water quality of the Great Lakes, *3029*

1979 Ban on chlorofluorocarbons (CFCs) on shipments between U.S. states, *1983*

April 16

1851 Storm that destroyed a lighthouse in the United States, *4181*

1996 Nuclear power plant in Romania, *3536*

April 17

1629 Commercial fishery in an American colony, *2470*

1861 Fire in a U.S. oil well, *3629*

1935 Drought rehabilitation program by the Canadian government, *2151*

1965 Flood crest on the Mississippi River at St. Paul, MN, to exceed the previous record by 4 feet, *2551*

April 18

1906 Earthquake in the United States believed to have surpassed 8.0 on the Richter scale, *2215*

1989 Ozone research coordination program of the European Community (EC), *1991*

1990 Permits issued by the U.S. Fish and Wildlife Service to kill migratory birds at aquaculture sites, *2498*

April 19

1920 International wildlife treaty upheld by the U.S. Supreme Court, *3018*

1999 California waters in the state Heritage Trout Program (HTP), *2513*

April 20

1973 National Hunting and Fishing Day in the United States, *2927*

1992 Pencil made of recycled newspaper and cardboard fiber, *2716*

1999 Limits on the height of skyscrapers in New York City, *3158*

April 21

1908 Explorers credited with reaching the North Pole, *3849*

1930 Radio broadcast from a zoo, *4928*

1993 Ozone-depletion Executive Order by a U.S. President, *1996*

April 22

1970 Earth Day to be celebrated nationwide in the United States, *1069*

1977 Oil well blow-out in the North Sea, *3638*

1990 Earth Day to be celebrated around the world, *1089*

1997 U.S. state with a sustainable development agency, *4321*

1999 Deep-sea exploration of U.S. marine sanctuaries, *3602*

Haze regulations concerning U.S. wilderness areas and national parks, *4677*

April 23

1969 Ohio State Scenic River, *3985*

April 24

1902 Volcano eruption recorded as a major disaster in the Western Hemisphere, *4374*

1987 Deliberate release of genetically altered organisms in the United States, *1256*

Genetically altered bacterium tested outside of a laboratory in the environment, *2761*

April 25

1896 U.S. government study into the effects of wildlife on natural resources, *3361*

1942 Coal mining accident in the Far East causing catastrophic deaths, *3328*

1954 Photovoltaic cell or solar battery to work effectively, *4108*

April 26

1986 Nuclear power plant disaster with worldwide fallout, *3528*

April 27

1935 U.S. Soil Conservation Service chief, *4056*

1987 Comprehensive overview of the global environmental problem, *3262*

1988 Woman to travel solo to either pole, *3872*

April 28

1976 Butterfly listing under the U.S. Endangered Species Act, *2407*

April 30

1864 Hunting license required by a state, *2913*

1921 Interstate planning and development agency in the United States, *3104*

1991 Cyclone to cause catastrophic loss of life in Bangladesh, *4268*

May

1934 Major windstorm of the Dust Bowl drought in the U.S. Midwest, *2149*

1960 Earthquake to register a magnitude of 9 or more on the Richter scale, *2225*

1962 Coal-mine fire to depopulate a U.S. city, *3332*

1974 World's fair with an environmental theme, *2310*

1982 World's fair in the American Southeast to be sanctioned by the Bureau of International Expositions (BIE), *2312*

World's fair to select energy as its theme, *2313*

1986 Workshop hosted by the European Community on the ozone layer, *1986*

1987 Bird recorded at the North Pole, *1709*

1990 Floods to cause damages of more than $1 billion in the United States, *2563*

1991 Flood to affect one-fifth of China's population, *2564*

1992 City in the United States to ban the sale of spray paint to reduce graffiti, *3284*

Daily forecasts of ultraviolet radiation issued by the Canadian government, *1849*

Low-emission automobiles, *1605*

1993 Ozone-layer measurements from space, *1979*

May 1

1854 Connecticut River flood crest at Hartford to reach 28 feet, 10.5 inches, *2528*

1900 Coal mining accident of significance in the United States, *3321*

1909 Hydroelectric power plant built by the U.S. government, *4554*

1977 United Nations attempt to define civil liability for oil pollution resulting from seabed drilling, *3591*

1998 Genetically engineered papayas, *2772*

May 3

1977 Goose to lay an egg weighing 1.5 pounds, *1675*

1978 International holiday to promote solar energy, *4114*

May 4

1493 East-west division of the world for exploration, *3008*

1822 Highway built in the United States entirely with federal funds, *1552*

1869 Oil drill offshore rig, *3650*

1875 Antivivisection legislation, *1008*

1992 U.S. government research suggesting nitric acid (rather than sulfuric acid) is the chief cause of acid rain, *1481*

1996 Canadian law to protect all endangered wildlife and flora, *2343*

May 5

1992 International convention on transporting hazardous waste, *3045*

May 6

1937 Brush charge of atmospheric electricity to destroy an airship, *4275*

May 6—*continued*

1978 United Nations agreement on marine pollution from airborne sources, *3588*

May 7

1840 Tornado in the United States to leave a city in ruins, *4281*

1857 Newspaper weather forecast in the U.S., *1904*

1982 Pennsylvania law creating a scenic rivers system, *4001*

May 8

1902 Volcano causing massive deaths in the West Indies, *4375*

1961 Seawater conversion plant in the United States on a practical scale, *4595*

1992 Pledge to halt all raw sewage flow from Victoria and Vancouver Island, British Columbia, into the Strait of Juan de Fuca, *4543*

May 9

1990 Felony indictments under a U.S. environmental law for smuggling hazardous waste from California into Mexico, *2889*

1992 Treaty setting the framework to address the production of greenhouse gases, *3050*

May 10

1752 Paper on the extinct volcanoes of Europe, *4363*

1933 Statewide construction and design standards for earthquakes in the United States, *2221*

1950 U.S. National Science Foundation (NSF), *2283*

1974 National Wild and Scenic River in the southeastern United States, *3988*

1979 Treaty to insure humane slaughter of animals, *3037*

May 11

1938 Federal law in the United States to develop salmon fisheries in the Columbia River Basin, *2495*

1989 Illinois State Scenic River, *4008*

1995 "Universal Waste Rule" in the United States, *4135*

May 12

1954 International convention to address oil pollution of the oceans, *3024*

1976 Oil spill of significance off Spain, *3637*

1978 Elephant listed under the U.S. Endangered Species Act, *2406*

1984 World's fair to select rivers as its theme, *2314*

May 13

1908 Governors' conference on U.S. conservation issues, *1033*

U.S. government conference devoted to conservation issues, *3365*

1930 Authenticated death by hail recorded by the U.S. Weather Bureau, *4216*

1992 Inter-American institute for research on global change, *3064*

1995 Environmental lawsuit to challenge the United States under the North American Free Trade Agreement (NAFTA), *2348*

1996 Law in the United States to require phasing out of mercury in batteries, *4152*

May 14

1966 International body to regulate conservation of Atlantic tunas, *3060*

International organization to regulate tuna fisheries, *2475*

1982 Genetically engineered product to become commercially available, *2759*

May 15

1885 Forest reserve set aside by a state, *2711*

Law establishing a protected wilderness in the Adirondack Mountains, *4631*

1928 Mississippi River flood control project of significance, *2537*

May 16

1873 Dam failure of major significance in the United States, *2111*

1874 American song written about a flood warning, *2530*

May 17

1913 Oil and gas conservation law by a state, *3618*

1995 Elk reintroduced to Wisconsin, *2441*

1999 American Indian tribal whale hunt in modern times during an International Whaling Commission (IWC) moratorium on whale hunting, *4628*

May 18

1937 U.S. legislation providing federal aid for up to 50 percent of the cost of reforestation of farmlands on the Great Plains, *2728*
1972 Prohibition against nuclear weapons on the ocean floor, *3584*
1980 Manned solar-powered aircraft flight, *4119*
Volcano eruption in the United States photographed by weather satellites, *4388*
Volcano known to claim human life in the contiguous 48 states, *4389*
1994 Genetically altered food, *2767*
1999 Regional endangered-species plan to protect an aquatic ecosystem, *2352*

May 19

1796 Protection of American Indian hunting grounds by the U.S. Congress, *2939*
1919 Conservation group to focus on the eastern United States in addition to western states, *1035*
1948 Interagency government property transfers for wildlife conservation in the United States, *4769*

May 20

526 Earthquake in which a massive loss of life was recorded, *2178*
1293 Earthquake in Japan causing massive deaths, *2180*
1844 Game protection society in the United States, *2967*
1919 Private organization to support national parks in the United States, *3718*
1926 Law passed to protect black bass in the United States, *2458*
1980 International organization to manage Antarctic wildlife, *4783*
1999 Use of the term *genetic pollution*, *1257*

May 21

1964 Nuclear-powered lighthouse, *3500*
1997 United Nations convention on transboundary water supplies, *4616*

May 22

1922 Volcano information network of significance set up internationally, *4378*
1963 General Fisheries Council for the Mediterranean, *2474*
1992 Natural gas fuel cell to produce electricity commercially, *2065*
1997 Clean Water Hardship Grant in the United States, *4617*

May 23

1908 Buffalo preserve in the United States, *4811*
Land bought by the U.S. government specifically for wildlife, *4812*
National wildlife refuge specially authorized by Congress, *4813*
1993 Law to discourage population growth in Iran, *3895*

May 24

1683 Public museum in Great Britain, *4881*
1911 Monument to commemorate the first major exploration of the American Southwest, *3748*

May 25

1900 Law in the United States to prohibit the transportation of wildlife across state lines if they were taken in violation of state law, *4691*
Wildlife management law in the United States, *4765*
1966 Proposed U.S. legislation regarding abandoned cars, *4142*

May 26

1908 Oil discovery in the Middle East, *3694*
Oil discovery of significance in Persia, *3695*
1976 Treaty protecting the polar bear and its natural habitat, *3034*
1994 Zoo built for nocturnal viewing, *4965*

May 27

1755 Water pumping plant in the United States, *4572*
1972 Chemical and oil waste law enacted by the Danish government, *3945*

May 28

1734 Fish protection law enacted by an American city, *2502*
1975 Whooping crane born in captivity, *4754*

May 30

1930 Legislation in Canada that preserved parks for future generations, *3735*

May 30—*continued*

1989 Dolphin listing under the U.S. Endangered Species Act, *2414*

May 31

1889 Dam disaster in the United States with a high death toll, *2113*
1949 Inter-American Tropical Tuna Commission, *2460*
1996 Transfrontier protected area for marine turtles, *4842*

June

1768 Geological survey of Russia, *2784*
1779 Oyster farming under state auspices, *2489*
1816 Summer blizzard recorded in the United States, *4194*
1860 Commercial oil refinery, *3687*
1899 Scientific expedition to what would later be designated the John Day Fossil Beds National Monument, *3744*
1943 Hydroelectric power plant in the United States to produce a million kilowatts, *4560*
1951 Plutonium-producing nuclear reactor in England, *3460*
1957 Classification of rivers, *4508*
1969 Aquatic environment built to revitalize urban waterfronts, *4942*
1971 Industrial pollution lawsuit in Japan that was successful, *2974*
1972 United Nations conference devoted to the global environment, *3030*
1974 Ozone-layer data concerning damage by chlorofluorocarbons (CFCs), *1974*
1977 Trees planted by the Green Belt Movement, *1098*
1979 Use of solar energy at the White House, *4117*
1980 Drought to cause damages of $1 billion or more in the United States, *2157*
1984 Successful embryo transfer between females of endangered species, *4955*
U.S. National Wildlife Refuge created specifically for endangered Nevada fish species, *2339*
1987 National Fishing Week in the United States, *2511*
1992 Car fueled by methanol, *4336*
Oil production in Papua New Guinea, *3709*
1993 Brownfields Economic Redevelopment Initiatives program in the United States, *2992*
Floods known to cover 10 U.S. Midwestern states, *2565*
Permanent United Nations sustainable development agency, *4313*
Person in England to chain herself to a bulldozer to protest a highway, *1103*
1994 Ostriches reintroduced to Arabia, *2438*
1995 Sustainable development handbook issued by the Canadian government, *4316*
1996 Commercial fuel cell powered by landfill gas, *4341*
United Nations emergency response system for industrial pollution disasters, *3002*
1998 All-American Road with glacial features, *4020*
All-American Road with volcanic features, *4021*
National park to focus on conservation history and the evolving nature of land use, *3812*
U.S. National Scenic Byway below sea level, *4022*
U.S. National Scenic Byway in Alaska, *4023*
U.S. National Scenic Byway in an urban area, *4024*
U.S. National Scenic Byway in Grand Canyon National Park, *4025*
2006 Nuclear power plant in Turkey, *3544*

June 1

1919 Forest fire air patrol, *2615*
1972 International convention to specifically protect Antarctic seals, *3031*
1978 U.S. Supreme Court ruling on state bans on imported waste, *4145*
1992 U.S. Supreme Court ruling on local bans on imported waste, *4150*

June 2

1889 Hydroelectric power plant in the United States transmitting alternating current over a long distance, *4552*
Irrigation advocacy group in the United States, *1031*
1964 Agreement governing treatment of Antarctic wildlife and plants, *3026*
1970 Crocodilians listed under the U.S. Endangered Species Act, *2394*
Gorilla listing under the U.S. Endangered Species Act, *2395*
Lion listing under the U.S. Endangered Species Act, *2396*
Prairie dog listed under the U.S. Endangered Species Act, *2397*

Rhinoceros listings under the U.S. Endangered Species Act, *2398*
Snail listed under the U.S. Endangered Species Act, *2399*
Tortoise listed under the U.S. Endangered Species Act, *2400*
Turtles listed under the U.S. Endangered Species Act, *2401*
1978 Hailstorm known to cause a famine in Syria, *4221*
1988 International convention attempting to regulate Antarctic mineral exploitation, *3044*

June 3

1886 U.S. government study into the effects of wildlife on natural resources, *3361*
1924 National wilderness area in the world, *3797*
1956 Waterfowl Depredations Prevention Act in the United States, *4800*
1979 Oil platform blowout, *3640*
1996 United Nations conference to address human habitation in a sustainable development context, *3901*
1998 State in the United States to ban state funding for human cloning research, *2773*

June 4

1825 Natural gas used to light a U.S. house, *2041*
1844 Report of extinction of the great auk, *2364*
1974 United Nations agreement on marine pollution from airborne sources, *3588*
1987 Reptile removed from the U.S. Endangered and Threatened Species List due to recovery, *2413*
1992 "Ecosystems management plan" for U.S. national forests, *2688*
Legally binding international document exclusively about global warming, *1850*
1997 Congressional attempt to re-establish the Office of Noise Abatement and Control (ONAC), *3451*
1999 Major auction of confiscated wildlife products in the United States, *4794*

June 5

1831 Complete state geological survey in the United States, *2802*
1992 Global agreement on all aspects of biodiversity, *3048*
International convention on biodiversity and sustainable development, *3049*

June 6

1932 Gasoline tax levied by the U.S. Congress, *1571*
1980 Theory that an asteroid caused the extinction of dinosaurs, *2376*

June 7

1494 Treaty between nations dividing the world for exploration, *3013*
1692 Archaeology site preserved by an earthquake, *2185*
1924 Fishing legally allowed on a U.S. government wildlife refuge, *2457*
Hunting legally allowed on a U.S. government wildlife refuge, *2919*
1980 Power plant using solar cells, *4120*

June 8

1783 Volcano known to have caused significant deaths by starvation, *4365*
1819 Campaign of significance against air pollution in London, *1319*
1937 International agreement on whaling regulations, *3020*
1981 Bombing of a nuclear reactor, *3519*
1990 Cleanup of an oil spill of significance in the United States with an oil-eating bacteria, *3644*

June 9

1953 Tornado in the eastern United States to claim 90 lives, *4290*
1998 Transportation act for the 21st century passed by the United States, *1585*

June 10

1895 Forest fire lookout tower, *2613*
1920 Hydroelectric power-plant licenses in the United States, *2095*
Law in the United States to license construction of commercial hydropower plants, *4557*

June 11

1998 Executive Order protecting coral reefs in the United States, *3601*

June 12

1951 Controls on the import and export of plants and produce, *3023*
1960 Law stating that U.S. forests would be exploited for a variety of recreational, commercial, and conservation purposes, *2685*

June 13

1968 Oil spill of significance off South Africa, *3631*

June 14

1854 Entomologist hired by the U.S. government, *1272*

1884 Burial of electric and other wires to be required by a state, *3921*

1903 Flash flood to cause $100 million in damages, *2534*

1976 Horse listed under the U.S. Endangered Species Act, *2404*

1991 Arctic Environmental Protection Strategy (AEPS), *3873*

1992 United Nations sustainable development strategy, *4310*

June 15

1971 Offshore oil produced in Norway, *3662*

1978 Lawsuit on behalf of an endangered species to be decided by the U.S. Supreme Court, *2333*

1992 Claim by the U.S. House of Representatives that overestimates of reforestation efforts in the Pacific Northwest had led to dangerously high logging quotas, *2689*

June 16

1947 Hurricane warning service to operate around the clock, *4257*

1970 Decision to build a highway into Brazil's Amazon wilderness, *2718*

1980 U.S. Supreme Court decision to rule that living things are patentable material, *2757*

1986 Aquatic theme park mixing education, research, and entertainment to be owned by a nonprofit organization, *4958*

1988 Law in the United States banning tin-based paint on boats, *2504*

1994 Mammal removed from the U.S. Endangered and Threatened Species List, *2419*

June 17

1859 Report of a Santa Ana wind to force temperatures to over 130 degrees, *4183*

1902 Law to provide for federal development of irrigated agriculture in the United States, *1220*

1960 Satellite to provide cloud-cover photographs, *1825*

1966 Law requiring the U.S. president to develop and conduct a comprehensive program of marine science activities, *3581*

1992 Tax on "gas-guzzling" automobiles by a U.S. state, *1584*

1995 Governmental tribunal in India for litigation relating to environmental accidents, *2999*

1998 Steelhead trout listing under the U.S. Endangered Species Act, *2423*

June 18

1969 Use of wrecked cars from a junkyard to make a barrier against a flood, *2554*

1989 Joint United Nations–World Council of Churches "Environmental Sabbath", *1086*

June 19

1884 Female awarded the Sage Fellowship in Entomology and Botany, *2266*

1932 Hailstorm of significance known to have hit the western Hunan province of China, *4217*

1965 State to own one million acres of game lands, *2923*

1969 Mass fish kill attributed to an insecticide spill, *1297*

1975 Australian Heritage Commission (AHC) Act, *4658*

United Nations compensation policy on marine oil pollution, *3582*

1986 Ban on the use of lead pipes in public water systems in the United States, *4606*

1992 Biodome, *4964*

June 20

1782 Bird officially designated as the national bird of the United States, *1688*

1930 Weather map to be telecast to a transatlantic steamer, *1937*

June 22

1936 Flood legislation by the U.S. government that called floods a national problem, *2539*

June 23

1938 Aquarium in the United States for large marine animals, *4934*

1961 International treaty regarding the use of Antarctica, *3025*

1979 Intergovernmental organization devoted exclusively to protection of migratory species worldwide, *3062*

1992 Farmland of significance in New York State to be permanently protected, *1136*

June 24

1976 Elephant listed under the U.S. Endangered Species Act, *2406*

June 25

1957 Hurricane reported to occur during early summer in the United States, *4261*

June 26

1797 Cast-iron plow, *1237*
1990 Northern spotted owl listing under the U.S. Endangered Species Act, *2416*
1992 Large-scale seizure and auction of cattle grazing on U.S. land without a permit, *1174*

June 27

1974 Noise standards set by the United States Occupational Safety and Health Administration (OSHA), *3444*
1977 Flow of oil through the Alaska Pipeline from Prudhoe Bay to Valdez on Prince William Sound, *3624*

June 28

1934 Law in the United States designed to stop overgrazing and its adverse effects on the soil, *4036*

June 29

1906 Cultural park set aside in the U.S. National Park System, *3788*
1925 Dam in the United States to fail during an earthquake, *2115*
1949 Television weather forecast in Great Britain, *1939*
1961 Satellite to use nuclear power for electricity, *3496*
1987 Systematic legislation to clean up and restore Florida's polluted lakes, rivers, streams and bays, *4483*
1993 U.S. President's Council on Sustainable Development, *4314*

June 30

1650 Hunting hounds in North America, *2907*
1864 State park in the United States, *3818*
1948 Water pollution law enacted by the U.S. Congress, *4458*
1949 Interagency government property transfers for wildlife conservation in the United States, *4769*
1954 Nuclear power station, *3467*
1984 Extensive collection of quality data on fertility and family planning, *3882*
1994 Vegetation inventory in the U.S. national parks system, *1797*

June 31

1940 Ross's goose breeding grounds, *1738*

July

1774 Geological survey of Russia, *2784*
1818 Geological periodical still published today, *2798*
1831 Successful harvesting machine, *1238*
1934 Federal duck stamp in the United States, *3165*
Federal duck stamp in the United States to feature the mallard, *3166*
1935 Scientist to use the term *ecosystem*, *1621*
1938 Federal duck stamp in the United States to feature the pintail, *3171*
1959 Artist to win the U.S. federal duck stamp contest three times, *3199*
1962 Smog chamber for air pollution research built by an industrial organization, *1462*
1967 Wandering albatross recorded ashore in the United States, *1703*
1972 Federal duck stamp in the United States to be priced at five dollars, *3210*
1975 Ports in the United States to be monitored by U.S. Fish and Wildlife Service agents, *4730*
1976 Major release of the pesticide dioxin into the atmosphere, *1300*
1979 Bilateral negotiations between the United States and Canada on acid rain, *3038*
1980 Federal duck stamp in the United States to be priced at $7.50, *3213*
1983 Dam constructed totally with roller-compacted concrete (RCC), *2109*
1984 Tax on boats and motors in the United States to benefit wildlife habitats, *2510*
1985 Law in New York City outlawing the sale of spray paint to minors, *3279*
1987 Federal duck stamp in the United States to be priced at ten dollars, *3215*
1989 Book on birds to sell for $3.96 million, *3263*
Federal duck stamp in the United States to be priced at $12.50, *3216*

July—*continued*

Federal duck stamp in the United States to feature a lesser scaup, *3217*

1990 Discussion of the greenhouse effect and other environmental issues at a "post–cold war summit", *1848*

1992 Government agency of India to promote wind power, *4864*

Reported large-scale bird kills traced to the lawn care pesticide Diazinon, *1309*

1994 Federal junior duck stamp in the U.S., *3224*

Griffon vulture reintroduction in Italy, *2439*

Organization of African Unity (OAU) population agency, *3899*

1995 High-resolution maps of the ocean floor, *3600*

July 1

1874 Zoo in the United States, *4896*

1908 National forest in the United States, *3714*

1916 Zoological Survey of India (ZSI), *4696*

1934 Federal law enforcement agency for wildlife in the United States, *4700*

1957 International Antarctic meteorological station, *3865*

1967 Satellite to transmit color photographs of the full Earth face, *2823*

1975 Endangered plants list issued by the United States, *2403*

Global agreement to protect plant and animal species from unregulated international trade, *3036*

1991 Litter and recycling strategy funded by fines and a statewide retail tax, *3281*

1992 Federal provision requiring companies that service air conditioning and refrigeration equipment to "capture and recycle" chlorofluorocarbons (CFCs), *1426*

1994 State in the United States to enact a "conspiracy to dump" law, *3286*

1997 Leaf blower ban of significance in Los Angeles, CA, *3449*

1999 Government demolition of a U.S. dam without the owner's permission, *2080*

July 2

1965 Meteorological satellite funded by the U.S. Weather Bureau, *1961*

1971 Nonreturnable bottle and can law enacted by a U.S. state, *4157*

1999 Proposed removal of bald eagles from the U.S. Threatened and Endangered Species List, *2425*

July 3

1918 Successful challenge to state-ownership doctrine, *1690*

1985 Global warming long-term policy of the government of the Soviet Union, *1843*

July 4

1893 Formal ascent of Devils Tower, *3743*

1997 Evidence that Martian volcanic rocks are in some respects similar to Earth's, *4395*

July 5

1994 Sonic boom regulations for aircraft, deemed a noise pollutant and public health issue in the United States, *3447*

1996 Mammal to be cloned from an adult cell, *2769*

July 6

1988 Explosion of significance on an offshore oil rig, *3642*

July 7

1932 School devoted exclusively to training wildlife conservation officers, *2278*

July 8

1869 State in the United States to protect an endangered plant species by law, *2320*

1916 National park east of the Mississippi River and the first located on an island, *3792*

1992 Recommendations in California to shift responsibility, from businesses to drivers, for paying for Southern California air pollution abatement, *1427*

July 9

1963 Bird known to fly at an altitude of 21,000 feet, *1701*

1965 U.S. law allowing consideration of recreation, fish, and wildlife enhancement as purposes of federal water development projects, *4561*

1979 Photographs of erupting volcanoes on the Jovian moon Io, *4387*

1992 Widely acknowledged environmentalist to be chosen as a major-party U.S. vice presidential candidate, *1091*

July 10

1930 U.S. federal law to protect a wilderness area, *4636*
1981 Massive spraying of the insecticide malathion to control the Mediterranean fruit fly in California, *1303*

July 11

1916 Highway program of significance in the United States granting federal funds for state roads, *1557*

July 12

1957 Nuclear power from a civilian reactor in the United States, *3477*
1970 National Wild and Scenic River in New England, *3986*

July 13

1985 Drought aid from a global rock music festival, *2159*

July 14

1922 Duck-billed platypus successfully transported outside Australia, *4699*
1991 Massively lethal pesticide spill in the Sacramento River, *1307*

July 15

1935 Sales of federal duck stamps to collectors in the United States, *3168*
1957 Oil discovery of major significance in Alaska, *3703*
1975 United Nations convention to define civil liability in nuclear accidents at sea, *3585*
1999 Admission by the U.S. government that nuclear weapons production may have caused illnesses in thousands of workers, *3542*

July 16

1945 Test of an atomic weapon, *3453*
1982 National Contingency Plan (NCP) and Hazard Ranking System (HRS) for hazardous wastes in the United States, *2981*

July 17

1943 Pilot to fly into a hurricane intentionally, *4252*
1955 Electric power from nuclear energy used to illuminate a town, *3471*

July 18

1955 Electric power generated by nuclear energy to be sold commercially, *3472*
1984 Tax on crossbow arrows in the United States, *2928*
1991 Federal standards in the United States for radon in drinking water, *4610*
1996 Noise control law in England with seizure provisions, *3448*

July 19

1784 Botanical expedition in the United States, *1768*
1948 U.S. preserve to feature indigenous wildlife in a natural setting, *4822*
1989 Mass burning of poached African ivory, *4788*
1998 Woman to reach the rank of major in the U.S. Park Police, *3814*

July 20

1975 World's fair with an ocean theme to be sanctioned by the Bureau of International Expositions (BIE), *2311*
1994 North American conference on bird mortality and wind power, *4866*

July 21

1959 Nuclear powered merchant ship, *3488*

July 22

1892 Oil tanker built for long-distance conveyance, *3613*
1968 International Council for the Exploration of the Sea (ICES), *3566*

July 23

1880 Hydroelectric power plant to furnish arc lighting service in the United States, *4548*
1938 Project after the 1938 appropriation by the U.S. Congress to assist state wildlife restoration projects, *4708*
1972 Satellite composite map of the U.S., *3381*

July 24

1934 Ptarmigan born and bred in captivity, *4749*

July 25

1916 Zoning ordinance in the U.S., *3153*

July 25—*continued*

1978 Species removed from the U.S. Endangered and Threatened Species List, *2408*
1997 Plant to produce methanol from coal, *4344*

July 26

1866 Irrigation law passed by U.S. Congress, *1216*
1997 Executive Order issued to protect the Lake Tahoe region, *3421*

July 27

1943 Hurricane report that was censored in the United States, *4253*
1963 Monument dedicated to a songbird, *3201*

July 28

1609 Hurricane that caused islands to be settled, *4229*
1976 Earthquake in China with its epicenter directly under a city of 1 million people, *2231*
1992 Guidelines issued in the United States by the Federal Trade Commission for environmental marketing claims, *1092*

July 29

1998 Reintroduction of Aplomado falcons to Mexico, *2446*

July 30

1879 American steam whaler, *4626*
1998 American Heritage Rivers program, *3423*

August

1588 Invasion of England to be halted by gale-force winds, *4176*
1927 International conference to address world population problems, *3896*
1937 Soil conservation district in the United States, *4040*
1944 Public service symbol created for the U.S. government, *3179*
1953 Bathyscaph, *3576*
1954 Time the Yangtze and Hwai rivers flooded to over 96 feet, *2545*
1955 Eight-day stretch of 100-plus degree heat in Los Angeles, CA, *1824*
1974 Significant forestry legislation in the United States in the second half of the 20th century, *2640*
1976 Offshore U.S. oil leases in the outer continental shelf of the Atlantic, *3663*
1977 Strip mining controls by the U.S. government, *3336*
1981 Natural gas field in Thailand, *2062*
1983 Successful embryo transfer between females of endangered species, *4955*
1988 Monsoon in Bangladesh to cause $1 billion in damage, *4189*
1992 Ozone-protection plan to phase out methyl chloroform, *1994*
1995 Fishing vessel buyback program for New England, *2483*

August 1

1916 U.S. national park in a volcanic range, *4377*
1943 Attack on the German-controlled oilfields at Ploesti, Romania, by the U.S. Air Force, *3667*
1977 Flow of oil through the Alaska Pipeline from Prudhoe Bay to Valdez on Prince William Sound, *3624*
1983 U.S. National Wildlife Refuge created specifically for West Indian manatees, *2338*

August 3

1968 Estuary protection funding in the United States, *4470*
1980 U.S. legislation to promote Ocean Thermal Energy Conversion (OTEC) development for commercial use, *4565*

August 4

1954 Law in the United States for watershed protection, *4593*

August 6

1945 Use of an atomic weapon in war, *3454*
1996 Federal Drinking Water State Revolving Fund (DWSRF) in the United States, *4614*
Law by the U.S. government to subject bottled water to the same standards as tap water, *4615*

August 7

1847 Plow for pulverizing the soil, *1241*
1974 Discovery of a marbled murrelet's nest, *1707*
1993 Exposition with an environmental theme to be held in Korea, *2317*

August 8

1955 United Nations meeting directly concerned with peaceful uses of nuclear energy, *3473*
1972 Systematic dam inspection by the U.S. government, *2072*
1993 Ecotourism research center in Australia, *2254*

August 9

1916 National park in the United States to contain an active volcano, *3793*
1944 Environmental public service symbol of the U.S. government, *2618*
1950 Tax on fishing equipment in the United States to specifically benefit wildlife habitats, *2507*
1965 Titan II silo explosion, *3502*
1995 Genetically engineered corn, *2768*

August 10

1984 Law in the United States extending wilderness protection to land administered by the Bureau of Land Management (BLM), *4667*

August 11

1965 Effort in the United States to address severe pollution in Lake Erie, *4539*
1979 One-child-per-family policy in China, *3883*
1980 Giant panda born outside of China, *4757*
1984 Law in the United States to establish asbestos as a health hazard in schools, *2982*

August 12

1883 Report of the extinction of the quagga, *2365*
1994 Oil-pipeline spill of significance in Russia publicly reported, *3648*
1995 Proposed removal of bald eagles from the U.S. Threatened and Endangered Species List, *2425*
1999 U.S. Interagency Council on Biobased Products and Bioenergy, *4347*

August 13

1970 Law in the United States allowing teenagers to work on federal conservation programs, *4778*

August 14

1937 National Trails system unit opened, *3723*
1946 Law to prevent damage to fish and wildlife, *4716*
1976 U.S. federal funding to preserve the Tule elk, *4833*

August 15

1635 Hurricane recorded in an American colony, *4231*
1775 Historic trail designated in the Western Region of the U.S. National Park Service, *3711*
1934 Bathysphere, *3572*

August 16

1916 Bird protection treaty, *3017*
International treaty to protect birds, *1743*
1969 United Nations convention to protect the vicuña, *4777*

August 17

1997 Bioprospecting agreement in the United States, *2770*
1999 Major earthquake in the Izmit region of western Turkey, an area that holds nearly half the nation's population, *2236*

August 18

1909 Reported period of more than 900 days without measurable precipitation in a southern California town, *2142*
1926 Weather map to be telecast from a land sending station to a land receiving station, *1934*
1983 Hurricane to cause damages of $1 billion or more in the United States, *4267*
1987 Study of aircraft noise effecting U.S. national parks and forest wilderness, *3430*

August 19

1992 National population policy for Niger, *3891*

August 20

1912 Quarantine law for plants enacted by the U.S. Congress, *1283*

August 22

1957 Potomac River pollution abatement conference, *4400*

August 23

1977 Trans-Canada national foot trail, *4661*

August 23—*continued*

1987 Scientific evidence that humans caused the Antarctic ozone hole, *1988*

August 24

79 Volcano eruption whose devastation was fully documented, *4356*

1804 Meteorological research done by balloon, *1881*

1964 River to be declared a national Scenic Riverway in the United States, *3805*

1966 Regulations on animal research in the United States, *2287*

1992 Hurricane to cause more than $25 billion in damage, *4269*

August 25

1916 Fisheries management in U.S. national parks, *2456*

Law in the United States requiring preservation of scenery within national parks, *3976*

National parks management agency in the United States, *3794*

1954 Hurricane of significance with a feminine name to hit the United States, *4260*

1979 Hurricane in the Atlantic with a masculine name, *4265*

August 26

1596 Europeans known to have wintered in the extreme north, *3824*

1721 Game law in Pennsylvania, *2908*

1883 Volcano eruption known to cause catastrophic tsunamis (giant sea waves), *4372*

1964 Uranium fuel ownership by private U.S. companies, *3501*

1996 Use of biodiesel fuel at a U.S. national political convention, *4342*

August 27

1667 Hurricane report in colonial Virginia, *4233*

1859 Oil well that was commercially productive, *3686*

1964 National Scenic Riverway law in the United States, *3978*

Rivers managed by the U.S. National Park Service, *4641*

August 28

1964 Satellite to provide high resolution night photographs of cloud-cover, *1827*

1984 National Wild and Scenic River in the Sonoran Desert, *4006*

1997 Zoo in the United States to exhibit a pair of aye-ayes, *4967*

August 29

1977 International meeting about desertfication, *4045*

1983 Texas law to regulate game and fisheries statewide, *4786*

August 30

1954 Nuclear technology access of significance by U.S. corporations, *3468*

August 31

1886 Severe earthquake to strike the U.S. east coast in modern times, *2206*

1955 Solar-powered car, *4110*

1984 Butterfly listing under the U.S. Endangered Species Act, *2407*

Insect removed from the U.S. Endangered and Threatened Species List, *2411*

September

1743 Hurricane known to be accurately tracked in colonial America, *4236*

1893 Water filtration system for bacterial purification of a U.S. city water supply, *4582*

1914 Carolina parakeet extinction report, *2367*

Extinction of a native American bird in modern times, *1656*

1922 Commercially marketed frozen fish, *2473*

1945 Hurricane entered intentionally with a large aircraft, *4255*

1958 Reproduction of U.S. federal duck stamps, *3196*

1961 Wildlife protection conference of modern African governments, *4773*

1965 International Conference on Polar Bears, *4775*

1971 Nuclear power-plant requirement in the United States to assess overall environmental impact, *3508*

1975 Aquarium in North America to be accredited by the American Association of Zoological Parks and Aquariums, *4948*

1987 Workshop hosted by the European Community on the ozone layer, *1986*

1988 Wildlife crime lab in the United States, *4787*

1989 Reformulated gasoline sold in the United States, *3626*

Successful solid waste program in the Philippines, *4133*
1991 "Bioprospecting" agreement, *1639*
Canadian biodiversity government agency, *1640*
Raven reintroduction in Italy, *2436*
1996 All-American Road in the Appalachian Mountains, *4012*
All-American Road on the U.S. Pacific coast, *4013*
All-American Roads in the Rocky Mountains, *4014*
Former hunting path designated as an All-American Road, *4015*
International Standards Organization (ISO) certification program for environmental management, *4318*
U.S. National Scenic Byway in the Great Lakes region, *4016*
U.S. National Scenic Byway on an American Indian reservation, *4017*
U.S. National Scenic Byway through Louisiana wetlands, *4018*
1997 Battery recycling program in Canada, *4171*
1998 State nature preserve in Massachusetts, *4844*

September 1

1869 Weather bulletin in the United States, *1910*
1914 Extinction caused by humans of an animal endemic to North America in modern times, *2368*
1916 International treaty to protect birds, *1743*
1972 Laws regulating space debris, *3032*
1981 Snowfall in living memory in the Kalahari Desert in Namibia, *4212*
1991 State in the United States to set a time limit on accumulating litter near a public highway, *3282*
1999 Ban of lead shot in Canada, *2962*

September 2

1775 Hurricane of significance recorded in North America, *4237*
1919 Hurricane to sink an ocean liner off the United States, *4249*
1937 Game preserve appropriation by Congress to assist state wildlife restoration projects, *4821*
Legislation for cooperative wildlife management, *4704*
1998 Congressional attempt to re-establish the Office of Noise Abatement and Control (ONAC), *3451*

September 3

1783 Treaty in which England recognized the fishing rights of the United States, *3014*
1899 Recorded incidence of an earthquake reversing the paths of two glaciers, *2209*
1930 Hurricane to cause massive deaths in the Dominican Republic, *4250*
1964 U.S. National Wilderness Area in the Superior Upland region, *4642*
U.S. National Wilderness Areas in the interior West, *4643*
U.S. National Wilderness Areas in the Pacific Northwest, *4644*
U.S. National Wilderness Areas in the Southeast, *4645*
U.S. National Wilderness Areas in the Southwest, *4646*
Wilderness Act in the United States, *4647*
1970 Single hailstone to measure 7.5 inches in diameter, *4219*

September 4

1618 Landslide known to kill more than 2,400 people, *2571*
1964 Orbiting geophysical observatory, *2822*

September 5

1925 Musk ox born in captivity, *4747*
1971 Tropical cyclone known to have lived for a month, *4262*
1994 United Nations conference to link the rights of women to population growth issues, *3900*

September 6

1869 Anthracite coal mine disaster of major significance in the United States, *2046*

September 7

1804 American explorer to describe prairie dogs, *4683*
Description of the "black-tailed deer" by William Clark, *4684*
1998 National Threatened Species Day in Tasmania, *2350*

September 8

1900 Hurricane causing massive loss of life in the United States, *4246*
1943 Indication of a new type of air pollution, *1449*

September 8—*continued*

Scientific paper demonstrating that hydrocarbons and nitric oxide react in sunlight to form ozone pollutants, *1454*

1980 Law in the United States to promote wind power as a renewable resource, *4858*

1983 National Priorities List (NPL) of hazardous waste sites in the United States, *2868*

September 9

1963 Giant panda born in captivity, *4752*

1972 Comprehensive wildlife protection legislation in India, *4781*

September 10

1875 National forestry association, *2691*

1899 Vertical surface displacement of more than 45 feet caused by an earthquake, *2210*

1993 Woman to direct the U.S. Fish and Wildlife Service, *4741*

September 11

1541 Mudslide catastrophe in Guatemala caused by a volcano, *2570*

1997 American Heritage Rivers program, *3423*

September 12

1979 Hurricane in the Gulf of Mexico with a masculine name, *4266*

September 13

1928 Catastrophic flood caused by a hurricane in the United States, *2538*

1994 Penalties enacted by the United States for harassing hunters, *2961*

September 14

1989 Plant removed from the U.S. Endangered and Threatened Species List, *2415*

September 15

1960 Wildlife conservation law for U.S. military property, *4772*

1968 United Nations convention on African natural resources, *3378*

1991 Ozone measurements of chlorine monoxide from space, *1992*

September 16

1804 Meteorological research done by balloon, *1881*

1971 United Nations agreement on marine pollution from offshore industrial facilities, *3583*

1980 Emperor penguins bred outside of Antarctica, *1678*

1987 International treaty to protect the Earth's atmosphere, *3043*

Ozone fund for developing nations, *1989*

Ozone pollutant agreement on an international level, *1987*

1990 National population plan for Sudan, *3887*

September 17

1999 National assessment of biological resources in the United States, *3426*

September 18

1906 Typhoon to cause massive loss of life in Hong Kong, *4247*

September 19

1898 Forestry school at a U.S. college, *2678*

1945 Kiwi bred in captivity, *1665*

1996 Permanent forum for coordinating sustainable development in the Arctic, *4319*

September 20

1947 U.S. Fish and Wildlife Service jurisdiction over sockeye salmon conservation, *2459*

September 21

1938 Flood in the United States to claim more than 600 lives, *2541*

1977 Chemical accident prevention law in France, *3953*

1994 World Health Organization (WHO) working group to address noise problems, *3431*

September 22

1908 Motorized transportation in a polar region, *3848*

1988 Agreement among nations to work together to create guidelines, policies, and procedures for protecting the ozone layer, *3041*

September 23

1885 Biology course offered at a U.S. college, *2267*

1998 Australian indigenous species habitat listed as endangered, *2424*

September 24

1906 Natural national monument in the United States, *3746*

September 25

1997 Environmental lawsuit decided by the International Court of Justice, *3011*

September 26

1542 European to set foot on what would later become the west coast of the United States., *3739*

1962 Flood of significance in Spain during the 20th century, *2548*

1972 Interagency government property transfers for wildlife conservation in the United States, *4769*

1980 Aquaculture Development Act in the United States, *2497*

1994 Enforcement action by the U.S. Environmental Protection Agency (EPA) against a company for failure to comply with radionuclide emissions standards, *2997*

1998 Music festival powered by biodiesel fuel, *4345*

September 27

1938 Vermont program authorized by the Federal Aid in Wildlife Restoration Act (Pittman-Robertson Act), *4709*

1997 State commission in the United States funded by taxes on litter-producing corporations, *3288*

September 28

1968 National Wilderness Area in New Jersey, *3730*

1976 Comprehensive legislation in the United States banning PCBs and regulating their disposal, *2857*

September 29

1993 Automobile technology agreement between the U.S. government and industry, *1606*

September 30

1882 Hydroelectric power plant in the United States to furnish incandescent lighting, *4550*

October

1780 Hurricane in the Atlantic causing massive deaths, *4238*

1938 Limited-access highway in the United States built like an interstate, *1563*

1947 Smog-related deaths recorded in the United States, *1341*

1950 Plutonium-producing nuclear reactor in England, *3460*

1966 Nuclear power commercial reactor in Japan, *3505*

1968 City in the U.S. to adopt a building code with internal noise suppression standards, *3437*

Report of a shorebird at latitude 71 degrees south, *1704*

1972 Urban national park to exceed 75,000 acres, *3821*

1976 Productive mine above the Arctic Circle in Canada, *3335*

1977 Intergovernmental conference on environmental education, *2300*

1978 Learning lab in the United States in a national zoo specifically related to birds, *4950*

Nuclear power plant in Taiwan, *3512*

1982 Learning lab related to reptiles in a national zoo in the United States, *4954*

1984 Drought to draw worldwide aid through television, *2158*

1990 "Killer bees" in the United States, *1529*

1992 Nuclear waste site chosen by the U.S. Congress, *3533*

Supercomputer used solely for environmental research, *2306*

1993 U.S. Navy report on the reintroduction of military dolphins to the wild, *4841*

1994 National preserve in an American desert, *3813*

1997 Nuclear power plant in Turkey, *3544*

October 1

1890 Weather service of the U.S. government not run by the military, *1917*

1913 Monument to a bird in the United States, *3163*

1940 Limited-access highway in the United States built like an interstate, *1563*

1957 Atomic energy organization of the United Nations for peaceful nuclear power, *3478*

1965 Automobile emissions bill approved by the U.S. Congress, *1574*

New York highway-noise emissions limits, *3434*

1974 Swampland designated as a U.S. National Wilderness Area, *4652*

October 1—*continued*

1983 Federal law in the United States to withhold development funds for construction on coastal barrier islands, *3389*

1991 Agreement among the Big Three U.S. automakers to reduce the dumping of toxins into the Great Lakes, *4441*

Boycott of a major U.S. automaker in protest of the use of animals in crash tests, *1022*

1996 Landfill tax in England, *4153*

State Scenic Byway in Ohio, *4019*

Wetlands Reserve Program (WRP) in the United States, *3420*

1997 Modern industrial pollution crime penalties in China, *3004*

Modern natural resource crime laws in China, *3422*

Modern solid waste disposal crime laws in China, *4154*

Modern wildlife resource crime laws in China, *4792*

October 2

1936 Fuel alcohol plant in the United States, *4327*

1968 Classification of rivers and federally-owned adjoining lands to determine their use by the public, *3731*

Waterways in the U.S. National Wild and Scenic Rivers System, *3983*

1969 National Wild and Scenic Rivers System in the United States, *3984*

1986 Federal policy in the United States on emergency removal of asbestos from schools, *4147*

1991 Joint commitment to "zero discharge" of dangerous chemicals into Lake Superior by the United States and Canada, *4442*

October 3

1841 "October Gale" to sink 40 Cape Cod fishing vessels off Nantucket Island, MA, *4241*

October 4

1961 Wetlands Loan Act in the United States, *3373*

1991 Specific treaty guidelines for Antarctic ecosystem protection, *3046*

1997 State in the United States to ban human cloning, *2771*

2001 Large-scale relocation of elephants, *4745*

October 5

1646 Wolf bounty enacted in what is now the United States, *4795*

1853 Female college professor with the same rights and privileges as her male colleagues, *2264*

1948 Version of the World Conservation Union (IUCN), *3058*

October 6

1947 Research of significance by the U.S. government on the peaceful uses of nuclear energy, *3457*

1992 Law in the United States to require federal government agencies to obey environmental laws, *4134*

1994 Junior Duck Stamp program in the United States, *4789*

October 7

1737 Typhoon to destroy 20,000 ships in India, *4235*

1905 Appearance of "The Great American Fraud," Samuel Hopkins Adams's article on patent medicines, *3250*

1988 African elephant protection law in the United States, *2340*

1994 Lawsuit over industrial sewers initiated by the U.S. Resource Conservation and Recovery Act (RCRA), *2998*

October 8

1871 Drought to create the conditions for a devastating fire in a major U.S. city, *2136*

Forest fire to cause a large number of deaths in the United States, *2612*

1957 Nuclear power-plant accident of significance, *3479*

1964 National recreation area designated in the United States, *3806*

1992 Embargo in the United States against tuna that was not "dolphin-safe", *2481*

October 9

1962 National fisheries center approved by Congress, *2462*

1963 Modern disaster blamed on faulty geologic analysis, *2118*

1969 United Nations convention on African natural resources, *3378*

2001 Bill in California forcing developers to show that there would be an adequate supply of water for any proposed large subdivision, *3091*

October 10

1950 Canadian–U.S. treaty to regulate use of the Niagara River for water power, *4555*
1995 Asian elephant tracked with a satellite transmitter, *4744*

October 12

1935 Commercial production of plants in water instead of soil, *1249*
1940 Treaty obliging the United States to maintain a list of endangered species, *3021*
1976 Federal law to restrict the dumping of hazardous waste in the United States, *2977*
1983 Bird removed from the U.S. Endangered and Threatened Species List due to extinction, *2379*
1988 Formal promise from Brazil to take steps to protect its rain forests, *2721*
Large-scale international private reforestation campaign aimed at individuals and communities, *2731*
1996 Dam safety code issued by the U.S. government, *2077*

October 13

1970 U.S. National Wildlife Refuge created specifically for watercress darters, *2334*

October 14

1907 Refuge established west of the Mississippi, *4810*
1941 Televised weather forecast in the United States, *1945*
1988 Law in the United States requiring government agencies to purchase alternative-fueled vehicles, *4333*
1998 Diamond mine in Canada, *3338*

October 15

1823 Meteorology society in Great Britain, *1886*
1862 Agriculture Bureau scientific publication, *3243*
1959 Nuclear power plant in the United States built without government funding, *3489*
1966 Law passed by U.S. Congress specifically aimed at preserving endangered birds, *1696*
Research grant program administered by the U.S. Department of the Interior, *2288*
1969 Canadian ban on hunting baby seals, *2958*
1978 Ban of chlorofluorocarbons (CFCs) by the U.S. government, *1978*
1997 Intentional poisoning of Lake Davis in California, *4807*

October 16

1971 United Nations agreement on marine pollution from offshore industrial facilities, *3583*
1978 United Nations compensation fund for marine oil pollution damage, *3586*
1986 Law in the United States requiring environmental considerations to be assessed in the licensing of hydropower projects, *4567*
1997 U.S. Sport Fishing and Boating Partnership Council, *2512*

October 17

1091 Written record of a devastating tornado in London, *4278*
1986 Law in the United States requiring public disclosure of pollutants through the Toxic Waste Inventory, *2984*
Reauthorization of the "Superfund" program, *2985*
"Right to Know" law in the United States covering hazardous substances, *2986*
Toxics Release Inventory (TRI) in the United States, *2987*
1989 Earthquake to be broadcast on live television nationwide, *2233*
1991 Anadromous fish conservation law in the United States, *2463*

October 18

1972 Noise pollution legislation of significance in the United States, *3441*
1982 Federal law in the United States to withhold development funds for construction on coastal barrier islands, *3389*

October 19

1934 National organization dedicated specifically to preserving wilderness (as opposed to wildlife) in the United States, *1055*
1941 Wind turbine, *4852*
1976 Coastal wetlands designated as a U.S. National Wilderness Area, *4659*

October 20

1818 Treaty to recognize the three-mile limit for fishing off the North American coast, *3015*
1916 International treaty to protect birds, *1743*
1969 Local private grassroots organization to successfully pressure governmental agencies to take steps to curb air pollution, *1404*
1976 U.S. National Wilderness Area in Hawaii, *4660*
1993 Executive Order requiring U.S. government agencies to use recycled paper, *4158*
1994 State in the United States to enact a "strict liability" antilitter law, *3287*
1998 Biomass-to-ethanol conversion plant, *4346*

October 21

1634 Storm on record to demolish an entire Dutch island, *4230*
1966 Major landslide of a slag heap in Wales, *2574*
1972 Marine mammal conservation and management law in the United States, *4782*
1976 Agency in the United States to regulate solid waste, *4131*
Comprehensive waste management legislation in the United States, *4132*
Federal rules in the United States for closure of hazardous waste treatment, storage, and disposal facilities, *4144*
1989 Transatlantic flight powered by ethanol, *4334*
1998 U.S. Government Paperwork Elimination Act, *4155*

October 22

1939 Snow cruiser for the Antarctic research, *3860*
1965 Federal law in the United States to restrict use of billboards in scenic areas, *3980*
1994 Rhinoceros and tiger protection law, *2346*

October 23

4004 BC Dating of the origin of Earth through biblical analysis, *2778*
1970 Federal aid to hunter education programs in the United States, *2925*
U.S. National Wilderness Area to enclose a bison sanctuary, *4650*
U.S. National Wilderness Areas in Alaska, *4651*
1992 Law in the United States to specifically ban importation of endangered exotic birds, *2342*

October 24

1978 Ozone-layer measurements from space, *1979*
1992 General grants from the U.S. government to Indian tribes for solid waste disposal programs, *4151*
Tax deduction in the United States for purchasers of Hybrid Electric Vehicles (HEVs), *4337*
Tax deductions in the United States for purchasers of Clean Fuel Vehicles (CFVs), *4338*
1998 Solar-powered vehicle to travel into deep space, *4123*

October 25

1972 Tax on bows and arrows in the United States, *2926*

October 26

1936 Dam to be constructed with a height greater than 152 meters (500 feet), *2099*
1974 Solar energy research funded by the U.S. Congress, *2298*
1988 Financial responsibility requirements for underground storage tanks (UST), *2988*
1992 National Park Service unit including coral reef resources, *3799*

October 27

1972 Legislation in the United States to allow noise pollution lawsuits, *3442*
Noise emission standards for motor vehicles involved in U.S. interstate commerce, *3443*
1992 Park in the Western Hemisphere to commemorate metal mining, *3811*

October 28

1891 Scientific formulation of the "fault theory" as we know it, *2208*
1982 Bird known to have lived to a documented age of 80, *1679*
1988 U.S. National Wild and Scenic River managed in partnership with local governments, *4009*

October 29

1947 Forest fire battled using artificial rain, *2620*

1959 Landslide in Mexico on record to kill more than 2,000 people, *2573*

October 30

1965 Anadromous fish conservation law in the United States, *2463*
1986 Bayou in the U.S. National Wild and Scenic Rivers System, *4007*
1991 National rules for reducing acid rain in the United States, *1421*

October 31

1969 Air Pollution Control Alert Warning System for New York City, *1350*
1984 Law in the United States to protect Atlantic striped bass, *2465*
1988 Law in the United States to require the removal of lead from water coolers, *4607*

November

1854 National storm warning service, *4182*
1892 Wood fossils found on the Antarctic, *3838*
1911 Japanese Antarctic expedition, *3851*
1927 Report of the extinction of the paradise bird of Australia, *2369*
1933 Dust storm known to carry dust from Montana all the way to the U.S. eastern seaboard, *2146*
1940 High altitude meteorological station in Antarctica, *3862*
1951 U.S. Soil Conservation Service chief, *4056*
1953 Oil refinery to achieve complete conversion of waste gasses into useful power with a carbon monoxide boiler, *3701*
1965 Federally sponsored interstate air pollution abatement conference, *1399*
1968 Publication that offered "access to tools" information for the energy-efficient and promoted ecologically aware lifestyles, *3258*
1972 Public health hazard causing the U.S. government to buy a town, *3946*
1974 Satellite composite map of the U.S., *3381*
1983 Amphibian removed from the U.S. Endangered and Threatened Species List, *2409*
1984 Jaguar preserve, *4838*
1988 Meeting of the Intergovernmental Program on Climate Change (IPCC), *1832*
1990 Environmental coalition of small island nations, *1090*
1991 Restoration to prairie of U.S. farmland previously reserved as a site for a nuclear power station, *3076*
1992 Ozone fund for developing nations, *1989*
1994 Mine gas extracted from coal fields in France, *2066*
Standardized national vegetation classification system in the United States, *3413*
1998 Logging company in California history to have its timber license suspended for environmental violations, *2666*

November 1

1755 Earthquake causing massive loss of life in Portugal, *2188*
Scientific investigation of an earthquake, *2189*
1936 Television weather forecast in Great Britain, *1939*
1959 British four-lane limited-access highway, *1567*
1983 Intergovernmental organization devoted exclusively to protection of migratory species worldwide, *3062*
1987 Solar automobile race, *4121*
1988 Medical waste disposal rules set by the U.S. Environmental Protection Agency (EPA), *4149*
1991 Airline flight attendants' class action suit against tobacco companies, *1369*
1994 Exemption for U.S. military flyovers of National Wilderness deserts in California, *4673*
Wilderness protection act for California deserts, *4674*
1996 Executive Order in the United States to protect the wildlife of the Midway Islands, *4843*

November 2

1966 Federal law in the United States to protect fur seals, *4776*

November 3

1995 Permit system enacted in the United States for fishing vessels on the high seas, *2484*
1998 Wind farm in Africa, *4868*

November 4

1974 Flight of a solar-powered aircraft, *4113*
1992 Federal law in the United States to reduce vessel sewage discharges, *3598*
Law in the United States to allow combining federal and private funds for wildlife program grants, *4840*

November 5

1889 Sewer district established by a city in the United States, *4527*
1913 Aqueduct to supply Los Angeles, CA, *4585*
1990 Comprehensive effort to introduce preventive pollution control in the United States, *1417*
National database in the United States for pollution source reduction, *2989*
U.S. federal source reduction policy, *2990*
1998 Human embryonic stem cells to be artificially cultivated, *2774*
Modern environmental management policy in South Africa, *3425*

November 6

1969 Regulation of toy guns as a noise hazard in the United States, *3439*

November 8

1870 U.S. Signal Corps storm warning, *4184*
1933 International treaty to propose African wildlife preserves, *3019*
1978 Quiet Communities Act in the United States, *3446*
1984 Aquaculture Development Act in the United States, *2497*
U.S. government agency to adopt a formal environmental policy with American Indian tribes, *3391*
1997 Helicopter noise bill to reach a committee in the U.S. Congress, *3450*

November 9

1921 Federal highway system in the United States, *1558*
1978 Federal tax credits for wind power investments in the United States, *4856*
U.S. law exempting alternative energy producers from state and federal regulation, *4115*
1998 International declaration on El Niño, *1968*

November 10

1948 International whaling commission, *3057*
1949 United States National Air Pollution Symposium, *1451*
1960 Nuclear power station to operate on a commercial basis, *3494*
1978 National Wild and Scenic River in the Mid-Atlantic region, *3992*
U.S. National Wild and Scenic River in the Chihuahuan Desert, *3993*
U.S. National Wilderness Area in the Gulf of Mexico, *4663*
1986 Wetlands Loan Act in the United States, *3373*
1997 Lemurs reintroduced to Madagascar, *2445*

November 11

1926 Highway-marking system in the United States, *1561*
1997 International treaty to ban the reproductive cloning of human beings, *3055*

November 12

1962 International conference on wetlands, *3374*

November 13

1870 Weather observations atop Mount Washington in New Hampshire, *1913*
1946 Artificial snowstorm, *1947*
1971 Banning of aircraft for hunting in the United States, *2959*
1979 Binding international treaty to address regional air pollution, *3039*
1985 Mudslide in the 20th century to bury a city in Colombia, *2576*
Volcano eruption causing massive deaths in Colombia, *4391*

November 14

1956 Interstate section in the United States, *1566*
1963 Volcanic island scientifically studied during its creation, *4381*

November 15

1896 Electricity generated by Niagara Falls waterpower, *4547*
1973 Treaty protecting the polar bear and its natural habitat, *3034*
1990 Comprehensive upgrades of the 1970 and 1977 Clean Air Act and Amendments, *1418*
1997 America Recycles Day, *4172*

November 16

1909 Environment and Natural Resources Division of the U.S. Department of Justice, *3368*
1945 United Nations agency to develop environmental education and research programs, *2282*

1972 Treaty to combine preservation of natural and cultural heritage sites, *3033*
1981 Law passed to protect black bass in the United States, *2458*
1990 New England Fishery Resources Restoration Act, *2466*
Office of Environmental Education in the United States, *2304*
1994 Comprehensive United Nations agreement on use of the oceans, *3592*
International body to govern use of the ocean, *3053*
International body to recognize the obligations of countries to fight and prevent marine pollution, *3051*
International organization to regulate use of the ocean floor, *3065*
Treaty establishing an international 200-nautical-mile Exclusive Economic Zone (EEZ), *3052*

November 17

1851 Postage stamps depicting the American bald eagle, *3161*

November 18

1755 Description of an earthquake or aftershock as a wave, *2190*
1960 Nuclear reactor to power space vehicles, *3493*

November 20

1811 Highway built in the United States entirely with federal funds, *1552*
1906 Earthquake organization formed because of a U.S. earthquake, *2216*
1995 U.S. President's Council on Sustainable Development, *4314*
1997 Regulations enacted in the United States to limit swordfish harvesting according to size, *2485*

November 22

1859 Synthetic fertilizer, *1127*

November 23

1972 International convention to preserve scenery, culture, and the environment, *3987*
1976 Hawaiian monk seal listing under the U.S. Endangered Species Act, *2405*

November 24

1993 Permit system enacted in the United States for fishing vessels on the high seas, *2484*
1998 Mandatory international Atlantic bluefin tuna conservation plan, *2486*

November 26

1921 Devastating ice storm in New England in the 20th century, *4270*
1947 Aqueduct to supply San Diego, CA, *4592*

November 27

1991 Scenic Byways and All-American Roads program in the United States, *4010*
Transportation act in the United States with significant mass transit funding, *1581*

November 28

1929 Airplane flight over the South Pole, *3856*
1988 Federal law in the United States to withhold development funds for construction on coastal barrier islands in the Great Lakes, *3394*

November 29

1929 Wildlife sanctuary in Tanganyika, *4817*
1969 United Nations compensation policy on marine oil pollution, *3582*
1973 Bird flight recorded at 37,000 feet, *1706*
1990 Aquatic Nuisance Species Task Force, *1530*

November 30

1971 Oil spill of significance off Japan, *3635*

December

1929 Description of hydroponic agriculture, *1247*
1936 Sit-down strikes to cripple General Motors car production, *1544*
1946 Hibernating bird discovered, *1740*
1952 Casualties of the great London killer fog of December 1952, *1343*
1957 Accident aboard the USS *Nautilus*, the world's first nuclear-powered submarine, *3480*
Nuclear accident of significance in the Soviet Union, *3482*
1966 Presentation of the essay "The Historical Roots of Our Ecological Crisis" by historian Lynn Townsend White Jr., *3257*

December—*continued*

1978 Scientist to survive the eruption of an Antarctic volcano, *3870*
1979 Catastrophic flood of the 20th century in Wales, *2560*
1980 U.S. legislation making states responsible for their nuclear waste, *3516*
1984 Government policy in the United States to deny aid to foreign family planning agencies that performed or promoted abortions as a method of population control, *3884*
1985 Zoo with a museum devoted to elephants, *4957*
1991 Wind farm in England, *4861*
1992 Fertility control program for wild horses in the United States on public lands, *4806*
Meeting of the contracting parties to the Basel Convention on the Control of Transboundary Movements of Hazardous Wastes and Their Disposal, *2891*
1996 Genetic manipulation of hatchery fish stocks, *2500*
1997 Nuclear energy law allowing expropriation of power plants in Sweden, *3539*

December 1

1892 Use of the term *oekology* (ecology) in the United States, *1052*
1913 Gas station in the United States, *3619*
1959 International treaty regarding the use of Antarctica, *3025*
1969 Aircraft noise certification standards in the U.S., *3438*
1978 Tax credit in the United States for ethanol, *4330*
1995 Meteorological organization in Europe organized specifically to control weather satellites, *1967*
1996 International treaty to specifically protect the sea turtles of the Western Hemisphere, *3054*

December 2

1946 International agency formed specifically to regulate whaling, *4627*
International whaling commission, *3057*
1957 Nuclear power plant exclusively used for peaceful purposes, *3481*
1970 U.S. Environmental Protection Agency (EPA) role in wildlife conservation, *4779*
1980 U.S. National Wild and Scenic Rivers in Alaska, *3999*
1982 "Subsistence rights" legislation for Alaska, *3388*
1993 Law establishing the deepest cave in the United States as a protected area, *4672*

December 3

1963 General Fisheries Council for the Mediterranean, *2474*
Organization for the control of desert locusts, *3059*

December 4

1952 Smog episode to cause widespread loss of life, *1344*
1987 Extant species on the U.S. Endangered and Threatened Species List to become extinct, *2382*
1996 Electric car to be mass-produced using modern technology, *1609*

December 5

1935 Commercial hydroponicum on a large scale, *1250*
1965 Recorded nuclear weapons accident involving a U.S. warship at sea, *3503*
1969 Law in the United States to protect other nations' wildlife, *4726*
1989 Introduction of the concept of "ecotheology", *1087*
Papal document to deal with ecology, *3264*
1994 Use of the Atlantic Coastal Fisheries Cooperative Management Act of 1993 against a state, *2467*

December 6

1907 Coal-mine explosion with massive loss of life in the United States, *3323*

December 7

1988 Earthquake in the Soviet Union allowed open media coverage, *2232*
1991 Sustainable development accord on the Alps, *4309*

December 8

1909 National U.S. bird banding association, *1716*
1916 International treaty to protect birds, *1743*
1953 Nuclear energy proposal at an international level, *3464*
1963 Jetliner to be destroyed in flight by lightning, *4277*

1987 Climate research agreement between the United States and Soviet Union, *1830*

December 9

1982 Trilateral agreement to protect the Wadden Sea, *3010*

December 10

1895 Meeting to create the New England Botanical Club, *1788*
1940 Treaty to establish national parks and preserves and to identify protected flora and fauna in the Western Hemisphere, *3022*
1982 Comprehensive United Nations agreement on use of the oceans, *3592*
International body to govern use of the ocean, *3053*
1997 International treaty to set binding limits on carbon dioxide emissions, *3056*

December 11

1919 Monument to an insect in the United States, *3164*
1972 Geologist to go to the Moon, *2825*
1980 "Superfund" environmental cleanup law in the United States, *2980*
1991 Food irradiation plant in the United States, *1209*
2000 Comprehensive plan to redesign the water-flow system in and around the Everglades, *4678*

December 12

1801 Automobile using steam power in England, *1587*
1959 U.S. submarine to carry nuclear-armed ballistic missiles, *3490*
1962 Recommendations for protecting landscapes issued by the United Nations, *3977*
1971 United Nations compensation fund for marine oil pollution damage, *3586*
1972 Organization to regulate dumping of wastes in the high seas, *3061*
1974 Condors reintroduced to Arizona, *2430*
1990 Intergovernmental agency devoted to marine and fisheries science of the Northern Pacific, *3596*
Organization to study the North Pacific, *3063*

December 13

1621 Furs exported from an American colony, *2902*
1916 Snow avalanche to kill Austrian and Italian troops during a war, *4202*
1950 Nuclear generator accident of significance, *3459*
1989 Wetlands protection and restoration law to provide funds for protection, *3090*

December 14

1911 Explorers to reach the South Pole, *3852*
1963 Televised dam collapse, *2119*
1998 Nuclear waste lawsuit of significance against the U.S. government, *3540*

December 15

1854 Street-cleaning machine of importance in the United States, *4175*
1971 Federal management policies for wild horses in the United States, *4780*
1983 Federal program to protect and restore an estuary in the United States, *3593*
1995 Sustainable development governmental post in Canada, *4317*
1997 Agreement on International Humane Trapping Standards, *2929*
1998 Drought predictions for the 21st century by a U.S. government agency, *2166*

December 16

1592 Great flood at Tombouctou on the River Niger in present-day Mali, *2525*
1811 Earthquake known to have changed the course of the Mississippi River, *2197*
1974 Clean drinking water standards in the United States that were legally binding, *4598*
Comprehensive U.S. legislation establishing minimum safety standards for chemical contaminants in water, *4478*
Federal Sole Source Aquifer (SSA) Protection Program in the United States, *4599*
1975 Aquifer in the United States granted sole source protection from pollution, *4425*
Aquifer in the United States to be designated as a "sole source" under the Clean Water Act of 1974, *4600*

December 17

1903 Monument erected to commemorate the first flight by an airplane, *3745*

December 17—*continued*

1921 Devastating storm of freezing rain to strike Illinois in the 20th century, *4271*

1949 Ban of aircraft access to Superior National Forest, *4639*

1963 Air pollution law of importance enacted by the U.S. Congress, *1398*

1971 United Nations convention to define civil liability in nuclear accidents at sea, *3585*

1980 Tax exemptions in the United States for conservation easements, *4664*

1992 Canadian law to protect all endangered wildlife and flora, *2343*

December 19

1970 Regulation of toy guns as a noise hazard in the United States, *3439*
Water Bank Act in the United States, *3380*

1972 Oil spill of significance in the Gulf of Oman, *3636*

1987 Law in the United States banning the disposal of plastics at sea, *4148*

December 20

1979 United Nations convention to protect the vicuña, *4777*

1991 United Nations moratorium on drift net fishing, *2480*

December 21

1872 Oceanographic expedition, *3560*

1994 Ecosystem management plan upheld in a U.S. court, *3414*
Legal challenge to wolf reintroduction to Yellowstone National Park, *2440*

December 22

1956 Gorilla born in captivity in the United States, *4751*

1980 Colorado Wilderness Act, *4665*

December 23

1985 Federal legislation of significance to take erodable U.S. cropland out of production, *4057*

December 24

1948 House completely heated by solar energy in the United States, *4107*

1970 Geothermal Steam Act in the United States, *3379*

December 25

1859 Rabbits in Australia, *1508*

December 27

1836 Disastrous snow avalanche in England, *4196*

December 28

1967 Evidence of prehistoric vertebrate life in Antarctica, *3866*
Fossil found in Antarctica, *3867*

1973 Law to protect future endangered fish and wildlife, *4727*
Sea turtle listing to include all species, *2402*

December 29

1993 Global agreement on all aspects of biodiversity, *3048*
International convention on biodiversity and sustainable development, *3049*

December 30

1969 Coal-mining law with penalty fines imposed by the U.S. government, *3334*

1982 Permanent ban on oil and gas leasing on U.S. wilderness land, *4666*

December 31

1970 National standards for air polluting motor vehicle emissions in the United States, *1406*

1971 Legislation in the United States limiting lead content in paint and providing funds for cleanup, *3944*

1982 Rewards offered by the U.S. Fish and Wildlife Service for information about wildlife crimes, *4733*

1985 Carbon tetrachloride ban in the United States, *3961*

1990 State in the United States requiring retailers to obtain an annual litter-control permit, *3280*

Personal Name Index

The Personal Name Index is a listing of personal names mentioned in the main text of the book, arranged alphabetically by last name. Each index entry includes key information about the "first" and a 4-digit number in italics. That number directs you to the full entry in the main text, where entries are numbered in order, starting with 1001. To find the full entry, look in the main text for the entry tagged with that 4-digit number.

For more information, see "How to Use This Book," on page ix.

A

Abbe, Cleveland: Weather bulletin in the United States, *1910*

Abbey, Edward: Environmental organization to substantially practice "ecotage", *1075*

Abbott, J. M.: Federal duck stamp in the United States to feature an American eider, *3194*

Abbott, William Hawkins: Commercial oil refinery, *3687*

Abert, James W.: Towhee named after an individual, *1731*

Adams, Julius: Dual sewer system in the United States for sewage and storm water, *4524*

Adams, Samuel Hopkins: Appearance of "The Great American Fraud," Samuel Hopkins Adams's article on patent medicines, *3250*

Adanson, Michel: Botanical classification based on the natural relationships between plants, *1767*

Aelianus, Claudius: Written description of fly fishing, *2448*

Agassiz, Alexander: U.S. oceanographic research ship, *3562*

Agassiz, Louis: Marine biology station in the United States, *3561*

Scientific evidence to support the theory of an ice age, *2010*

Statement of the theory of a world wide ice age, *2805*

Agricola, Georgius: Treatise on mining that was scholarly, detailed, and systematic, *3296*

Akeley, Carl Ethan: Cement gun for mounting wildlife specimens, *4693*

Wildlife dioramas exhibited in a museum that were accurate, *2281*

Wildlife films of mountain gorillas, *4698*

Albert I, Prince of Monaco: Oceanographic museum, *4916*

Wildlife films of mountain gorillas, *4698*

Aldrovani: Herbarium, *1751*

Alexander VI, Pope: East-west division of the world for exploration, *3008*

Treaty between nations dividing the world for exploration, *3013*

Allen, Arthur Augustus: Ptarmigan born and bred in captivity, *4749*

Allen, John: Agricultural book published in the United States, *3231*

Allen, Lewis Falley: Herd book for livestock compiled and published in the United States, *3240*

Alpini, Prospero: Western botanist to note the sexual identity of plants, *1752*

Alvarez, Luis: Theory that an asteroid caused the extinction of dinosaurs, *2376*

Alvarez, Walter: Theory that an asteroid caused the extinction of dinosaurs, *2376*

Amos: Biblical prophet to use a famous earthquake to date his prognostications, *2172*

Amundsen, Roald: Expedition to survive a winter in the Antarctic, *3840*

Explorers to reach the South Pole, *3852*

Landing on Antarctica that initiated the "Heroic Age" of Antarctic exploration, *3839*

Anderson, Neal: Federal duck stamp in the United States to feature a lesser scaup, *3217*

Federal duck stamp in the United States to feature a red-breasted merganser, *3223*

Anderson, W. W.: Description of the Virginia warbler, *1652*

Andronicus of Cyrrhus: Weather vane, *1854*

Annas, George J.: Genetic privacy legislation proposal in the United States, *2766*

Anza, Juan Bautista de: Historic trail designated in the Western Region of the U.S. National Park Service, *3711*

Appert, Nicolas: "Tin" can, *4124*

Appleton, Edward Victor: Ionosphere direct measurements, *1933*
Arber, Agnes Robertson: Female botanist to become a member of the Royal Society in England, *1796*
Aristotle: Attempt to explain the origin of thunder, *4273*
Botanical study of significance, *1746*
Mineralogy book, *2775*
Recorded scientific theories explaining earthquakes, *2174*
Written work of significance to use animals in captivity as subjects of a study, *4877*
Arkwright, Richard: Water frame, *4544*
Water-powered cotton mill in an American colony, *4545*
Arnold, Ron: Publication that outlined the goals of the "wise-use" movement, *1024*
Arrhenius, Svante: Mathematical equation to calculate atmospheric warming from carbon dioxide concentrations, *1438*
Warning about global warming, *1836*
Asaro, Frank: Theory that an asteroid caused the extinction of dinosaurs, *2376*
Ashe, W. W.: Recovery plan approved by the U.S. Fish and Wildlife Service for a plant, *2336*
Ashley, William H.: Non–Native Americans to enter what is now Dinosaur National Monument, *3740*
Ashmole, Elias: Public museum in Great Britain, *4881*
Atkins, Charles Grandison: Fish hatchery run by the U.S. government, *2492*
Atkinson, William: Research results in the United States on the use of sunlight through windows to generate heat, *4102*
Aubreville, André M. A.: Use of the term *desertification*, *4043*
Use of the term *desertification* to describe deforestation of tropical and subtropical Africa, *2743*
Audubon, John James: Bird banding in the United States, *1728*
Bird pictured in John James Audubon's *The Birds of America*, *3160*
Bird preservation organization in the United States, *1712*
Book on birds to sell for $3.96 million, *3263*
Lincoln's sparrow, *1654*
National Audubon Society meeting, *1715*
Rediscovery of MacGillivray's petrel, *1682*
Salaried employee at the Western Museum of Natural History, *4888*
Austin, Thomas: Rabbits in Australia, *1508*

B

Babbitt, Bruce: Regional endangered-species plan to protect an aquatic ecosystem, *2352*
Bache, Richard: Partridge propagation in the United States, *2930*
Backhouse, James: *Begonia sutherlandii* to flower in England, *1510*
Bagot, Charles: Copper mine in Australia, *3317*
Bailey, William J.: Solar water heater and hot water storage system to provide high-temperature water around the clock, *4101*
Baird, Spencer F.: Conservation agency established by the U.S. government, *4763*
Balboa, Vasco Núñez de: Natural resource exploited in Panama, *3346*
Balfour, Lady Evelyn: Comparative study of organic and conventional farming, *1260*
Organic farming association in the United Kingdom, *1262*
Woman to advocate organic gardening, *4065*
Balfour, Guy: Vacuum-powered prairie dog extraction device, *4805*
Banister, John: Botanical account, in detail, of the plants of North America, *1760*
Magnolias to arrive in England, *1491*
Banks, Joseph: Naturalist to describe the wildlife of Australia, *4682*
Barents, Willem: Europeans known to have wintered in the extreme north, *3824*
Bari, Judi: Mobilization and training of youth to participate in nonviolent environmental protest, *1088*
Barnett, M. A. F.: Ionosphere direct measurements, *1933*
Barnsdale, William: Commercial oil refinery, *3687*
Barrett, Peter J.: Evidence of prehistoric vertebrate life in Antarctica, *3866*
Fossil found in Antarctica, *3867*
Barton, Clara: Flood victims to receive aid from the hand of Clara Barton, *2531*
Barton, Otis: Bathysphere, *3572*
Bartram, John: Botanical garden in the United States, *1762*
Bartram, William: Breeding bird census, *1729*
Bascom, Florence: Female scientist employed by the U.S. Geological Survey (USGS), *2813*
Bath, Henry Frederick Thynne Bath, Marquess of: Safari park, *4940*
Bauer, Georg *See* Agricola, Georgius
Beattie, Mollie H.: Woman to direct the U.S. Fish and Wildlife Service, *4741*
Beatty, Martin: Oil well in the United States, *3674*
Beaufort, Francis: Scale used internationally for wind forces, *1882*

Becquerel, Antoine Henri: Woman to win a Nobel Prize, *2268*
Beebe, William: Bathysphere, *3572*
Beeton, Alfred M.: Scholarly article to show that Lake Erie was dying a slow death from pollution, *4510*
Behn, Mira: Nonviolent protest of significance by the modern Chipko movement in India, *1041*
Behn, Sarala: Nonviolent protest of significance by the modern Chipko movement in India, *1041*
Bellingshausen, Fabian Gottlieb von: Person to see the Antarctic territories of Peter I Island and Alexander I Island, *3831*
Sighting of Antarctica, *3828*
Bellmon, Henry: Water pollution national conference of significance sponsored by the U.S. Chamber of Commerce to address the issue of industrial waste, *4438*
Bennett, Hugh H.: U.S. Soil Conservation Service chief, *4056*
Benson, F. W.: Federal duck stamp in the United States to feature the canvasback, *3167*
Bentham, Jeremy: Published presentation of the philosophy that humans should not be cruel to animals because animals can feel pain, *1003*
Use of the phrase "The question is not, Can they reason?, not, Can they talk?, but, Can they suffer?", *1004*
Bentley, Wilson A.: Meteorologist to photograph ice crystals with a camera-equipped microscope, *4203*
Benz, Karl: Automobile that proved practical with an internal-combustion engine, *1589*
Bergh, Henry: Society for animal welfare in the U.S., *1007*
Bering, Vitus: Animal hunted to extinction in North America, *2358*
Spectacled cormorant description, *2357*
Berkeley, William: Wolf bounty enacted in what is now the United States, *4795*
Berners, Dame Juliana: Sport fishing guide written in English, *2505*
Berzilius, Jöns Jakob: Classification of minerals by their chemical composition, *2794*
Bien, Amos: Costa Rican ecotourism organization whose purpose was to save an endangered part of the environment, *2246*
Bierly, E. J.: Federal duck stamp in the United States to feature a merganser, *3193*
Federal duck stamp in the United States to feature a Ross's goose, *3207*
Bingham, Jesse: Recorded discovery of what would become Wind Cave National Park in South Dakota, *3779*
Biot, Jean-Baptiste: Meteorological research done by balloon, *1881*
Birch, Stephen: Oceanographic exhibit with more than 3,000 fish, *4963*
bird banding: Bird banding by a U.S. government agency, *1734*
Bishop, Bernice Pauahi: Natural history museum in the United States named for a descendant of royalty, *4904*
Bishop, Charles Reed: Natural history museum in the United States named for a descendant of royalty, *4904*
Bishop, R. E.: Federal duck stamp in the United States to feature the Canada goose, *3169*
Bjerknes, Jacob: Cyclone theory, *4248*
Bjerknes, Vilhelm: Cyclone theory, *4248*
Frontal air mass theory of forecasting, *1927*
Weather predictions using scientific methods, *1922*
Black, Frank Swett: Forestry school at a U.S. college, *2678*
Black, Richard: Permanent American base in Antarctica, *3861*
Blanford, Henry: El Niño known to have caused massive deaths, *1819*
Bliss, Raymond Whitcomb: House with solar heating and radiation cooling in the United States, *4109*
Bloodgood, Don E: Conference on industrial wastes, *2838*
Blue Devil (aye-aye): Aye-aye born in captivity, *4759*
Blytt, Axel: Use of pollen analysis and vegetation history to determine successive climate patterns, *2011*
Bohl, Walter E.: Federal duck stamp in the United States to feature the wood duck, *3177*
Bombardier, Joseph A.: Snowmobile, *4637*
Bonafede, Francesco: Professorship of botany in Europe, *2256*
Bond, Thomas: Hospital in the United States for the sick, injured, and insane, *3910*
Borcherdt, Victor: Zoo habitat constructed of simulated rock formations without bars, *4917*
Borchgrevnik, Carsten: Scientific record of an Antarctic winter, *3842*
Borlaug, Norman: High-yield crops were developed, *1132*
Variety of high-yield rice, *1133*
Boucher, Robert J.: Flight of a solar-powered aircraft, *4113*
Boulton, Matthew: Gas used to light a house, *2032*
Bowditch, E. B.: Oil well that was commercially productive, *3686*
Bowen, E. G.: Radar detection of air echoes from the lower atmosphere, *1938*
Bowers, Birdie: Emperor penguin egg collected for science, *1735*

Boyd, Louise Arner: Woman to fly over the North Pole, *3864*
Boyer, Herbert W.: Biotechnology company in the United States, *2756*
Genetic engineering, *2755*
Boyle, John: Solar power plant using a low-boiling-point liquid to power an engine, *4099*
Solar-powered energy plant capable of operating at night, *4100*
Braddock, Edward: Highway built in the United States entirely with federal funds, *1552*
Bradley, Guy: Wildlife warden killed in the line of duty in the United States, *4766*
Brand, Stewart: Publication that offered "access to tools" information for the energy-efficient and promoted ecologically aware lifestyles, *3258*
Brandes, Heinrich W.: Weather map, *1884*
Brandon, Ernest Alfred: Commercial hydroponicum on a large scale, *1250*
Brandsson, Jon: Report of extinction of the great auk, *2364*
Brandt, Johann Friedrich: Brandt's Cormorant description, *1730*
Bransfield, Edward: Sighting of Antarctica, *3828*
Brelaz, Daniel: Politician from a Green party to be elected to a national parliament, *1111*
Brewer, William: Scientific exploration of the area that later became Sequoia and Kings Canyon National, *3769*
Bridgeman, Thomas: Gardener's manual written and published in the United States, *3236*
Britton, Elizabeth Knight: Person to suggest starting the New York Botanical Garden, *1790*
Bromfield, Louis: Model conservationist farm in the United States, *1131*
Brooke, Robert: Hunting hounds in North America, *2907*
Brown, Arthur: Solar-heated school, *4106*
Brown, Janice: Manned solar-powered aircraft flight, *4119*
Browne, Charlie: Citation for heroism ever bestowed by the U.S. Department of the Interior, *3721*
Browner, Carol: Haze regulations concerning U.S. wilderness areas and national parks, *4677*
Woman to head the U.S. Environmental Protection Agency, *1116*
Bruce, Archibald: Geological publication in the United States, *2793*
Brueghel, Pieter: Severe winter said to have introduced a new artistic approach, *1800*
Brugger, Kenneth C.: Discovery of the monarch butterfly's winter home, *4729*
Brundtland, Gro Harlem: Comprehensive overview of the global environmental problem, *3262*
Person to be elected the head of a government after serving as a minister of the environment, *1114*
United Nations organization concerned with global warming, *1842*
Brush, Charles Francis: Windmill to generate electricity, *4850*
Buchan, Alexander: Map showing movement of cyclonic depression, *1908*
Buchanan, Patrick J.: Organized resistance to environmental regulations, *1025*
Bunsen, Robert Wilhelm: Natural gas invention for heating, *2045*
Burbank, Luther: Development of the Burbank or Idaho potato, *1207*
Burdin, Claude: Water-powered turbine, *4546*
Bush, George Herbert Walker: Law in the United States to specifically ban importation of endangered exotic birds, *2342*
Office of Environmental Education in the United States, *2304*
Proposal to dig a canal across the Florida peninsula, *3071*
Revision of the U.S. National Wetlands Policy to counter the "no net loss" theory, *4486*
U.S. government program to reduce air pollution in the Grand Canyon, *1420*
Bush, George W.: Government policy in the United States to deny aid to foreign family planning agencies that performed or promoted abortions as a method of population control, *3884*
Superpower to effectively withdraw from the Kyoto Protocol, *3012*
Byrd, Richard E.: Aerial mapping photographs over Antarctica, *3854*
Airplane flight over the South Pole, *3856*
Explosion seismology experiments in Antarctica, *3857*
Byrne, Ethel: Birth control clinic in the United States, *3879*

C

Cabeza de Vaca, Álvar Núñez *See* Núñez Cabeza de Vaca, Álvar
Cabrillo, Juan Rodriguez: European to set foot on what would later become the west coast of the United States., *3739*
Caesalpinus: Catalog of significance of plants in the ancient world, *1748*
Herbarium, *1751*

Plant taxonomist, *1753*
Caesar, Julius: Recorded drought that affected a Roman military campaign in Britain, *2122*
Callendar, George D.: Atmospheric carbon dioxide link to human activities, *1837*
Strong claim that industrial burning of fossil fuels increased carbon dioxide levels in the atmosphere, *1448*
Campanius, John: Weather records kept continuously in the United States, *1803*
Campbell, John Francis: Instrument to accurately measure the length of time the Earth's surface receives sunshine, *1901*
Campbell, Keith: Mammal to be cloned from an adult cell, *2769*
Candolle, Augustin Pyrame de: Use of the term *taxonomy* to describe the science of classification, *1774*
Cárdenas, García López de: European to see the Grand Canyon, *3753*
Carhart, Arthur H.: Full-time landscape architect hired by the U.S. Forest Service, *3717*
Landscape architect employed by the U.S. Forest Service (USFS), *4632*
Wilderness area protected by the U.S. Forest Service (USFS), *4633*
Carnot, Nicholas: Geothermal heat exchange theory, *2800*
Carson, Rachel: Commercial use of the pesticide DDT in the U.S., *1288*
Popular book to warn of pesticide threat, *3256*
Carter, Jimmy: Flood-plain protection advocated by a U.S. presidential executive order, *2553*
Long-term study of the causes and effects of acid rain in the United States, *1474*
National Hunting and Fishing Day in the United States, *2927*
Use of solar energy at the White House, *4117*
Cartier, Jacques: European to systematically record a North American winter, *1798*
Cass, Lewis: Pure copper deposits discovery in Michigan, *3309*
Catherine II, the Great, Empress of Russia: Geological survey of Russia, *2784*
Cato, Marcus Porcius, Censorius: Catalog of Roman plants, *1747*
Caton, John Dean: Game preserve in the United States, *2931*
Cavendish, Henry: Calculation of the Earth's mean density, *2791*
Caventou, Joseph Bienaimé: Use of *chlorophyll* as a term to describe the green substance that colors plants, *1775*
Ceausescu, Nicolae: Nuclear power plant in Romania, *3536*
Ceballos-Lascurain, Hector: Use of the term *ecotourism*, *2247*
Cervera, Fermin Z.: Species to be given a scientific name incorporating the first and last names of a naturalist, *1737*
Chadwick, Sir Edwin: Public health advocate of significance in Great Britain, *3915*
Public health legislation of significance, *3917*
Study in England of significance into the effects of industrialization on public health, *3916*
Chakrabarty, Ananda: U.S. Supreme Court decision to rule that living things are patentable material, *2757*
Chamberlin, Thomas Chrowder: Geologist to discovery that several ice ages had existed, *2012*
Champlain, Samuel de: Permanent French settlement in North America, *3127*
Chapin, Charles Willard: Animal disease of American origin, *3926*
Chapin, Daryl M.: Photovoltaic cell or solar battery to work effectively, *4108*
Chapin, James: Identification of the Congo peacock, *1697*
Chapman, Frank: National Christmas Bird Count in the United States, *1733*
Charles II, King of England: Major anti–air pollution tract, *1429*
Charles, Jules: Practical modern greenhouse, *4086*
Charlie (dog): Woman to travel solo to either pole, *3872*
Charney, Jule Gregory: Long-range numerical weather forecasts by computer, *1953*
Weather prediction system using modern numerical systems, *1949*
Charpentier, Johann von: Scientific evidence to support the theory of an ice age, *2010*
Chatterjee, Sankar: Fossils of *Protoavis*, *1681*
Cherry-Gerrard, Apsley: Emperor penguin egg collected for science, *1735*
Chesbrough, Ellis S.: Water tunnel to supply Chicago, *4579*
Chiang Kai-shek: Levees destroyed in China to halt a foreign invasion, *2569*
Chippenfield, Jimmy: Safari park, *4940*
Christiansen, Hans Christopher: Water pumping plant in the United States, *4572*
Chrysler, Walter: Car model with a high-compression six-cylinder engine, *1595*
Clark, (first name unknown): Men to winter in the Antarctic, *3830*
Clark, Roland: Federal duck stamp in the United States to feature the pintail, *3171*
Clark, William: American explorer to describe prairie dogs, *4683*
Description of the "black-tailed deer" by William Clark, *4684*

Clark, William:—*continued*
White Americans to systematically describe the weather of the western United States, *1812*
Claude, Georges: Ocean Thermal Energy Conversion (OTEC) power plant, *4559*
Clawson, Marion: Living historical farm in the United States, *1312*
Proposal for a national network of living historical farms in the United States, *1134*
Clinton, Bill: Admission by the U.S. government that nuclear weapons production may have caused illnesses in thousands of workers, *3542*
American Heritage Rivers program, *3423*
Comprehensive plan to redesign the waterflow system in and around the Everglades, *4678*
Embargo in the United States against tuna that was not "dolphin-safe", *2481*
"Environmental justice" action by the U.S. government, *2995*
Environmental Quality Incentives Program (EQIP) in the United States, *3419*
Executive Order in the United States to protect the wildlife of the Midway Islands, *4843*
Executive Order issued to protect the Lake Tahoe region, *3421*
Executive Order protecting coral reefs in the United States, *3601*
Executive Order requiring U.S. government agencies to use recycled paper, *4158*
Feasibility studies of solar energy use in U.S. government buildings, *4122*
Feasibility studies of wind-generated energy use in U.S. government buildings, *4865*
Federal environmental justice grants in the United States, *3412*
Forest Summit in the United States, *2690*
Government policy in the United States to deny aid to foreign family planning agencies that performed or promoted abortions as a method of population control, *3884*
Law by the U.S. government to subject bottled water to the same standards as tap water, *4615*
Law establishing the deepest cave in the United States as a protected area, *4672*
Law in the United States to require phasing out of mercury in batteries, *4152*
National preserve in an American desert, *3813*
Proposed removal of bald eagles from the U.S. Threatened and Endangered Species List, *2425*
Rhinoceros and tiger protection law, *2346*
Trade sanctions imposed by the United States to protect endangered wildlife, *2345*
U.S. Interagency Council on Biobased Products and Bioenergy, *4347*
U.S. President's Council on Sustainable Development, *4314*
Widely acknowledged environmentalist to be chosen as a major-party U.S. vice presidential candidate, *1091*
Wildlife Habitat Incentive Program (WHIP) in the United States, *4790*
Woman to direct the U.S. Fish and Wildlife Service, *4741*
Clusius, Carolus: Scientific botanical garden in Western Europe, *1754*
Tulips in Holland, *1755*
Clutius, Cornelis: Scientific botanical garden in Western Europe, *1754*
Cobb, William G.: Solar-powered car, *4110*
Cohen, Stanley N.: Genetic engineering, *2755*
Colbert, Charlie: Known U.S. waste entrepreneurs to be convicted for fraudulent export of hazardous waste overseas, *2874*
Colbert, Jack: Known U.S. waste entrepreneurs to be convicted for fraudulent export of hazardous waste overseas, *2874*
Colden, Jane: Female botanist of significance in the United States, *1761*
Coleman, John: Television network to broadcast weather programming exclusively, *1966*
Coleman, Norman Jay: Agricultural experiment stations established by the U.S. government in individual states, *1224*
Collinson, Peter: Hydrangea in England, *1494*
Collor de Mello, Fernando: Environment minister in the ecologically critical country of Brazil, *2722*
Colman, Norman Jay: Head of the Department of Agriculture to be a member of the President's Cabinet, *1129*
Colo (gorilla): Gorilla born in captivity in the United States, *4751*
Colter, John: Non–Native American to see the geysers of Yellowstone National Park, *3761*
Colton, Frank B.: Oral contraceptive, *3880*
Columbus, Christopher: Cattle in the Western Hemisphere, *1484*
Horses reintroduced in the Americas, *1485*
Hurricane reported in the Americas by a European explorer, *4225*
Hurricane to destroy a European colony in the Western Hemisphere, *4224*
Natural resource exploited in Cuba, *3345*
Pigs in the Western Hemisphere, *1486*
Sheep in the Western Hemisphere, *1487*
State nature preserve in Georgia, *4834*
Commoner, Barry: Example of a whole-systems approach to the environment, *3260*
Compton, Henry, Bishop of London: Magnolias to arrive in England, *1491*

Comstock, Henry: Comstock Lode exploration in Nevada, *3312*
Cook, Arthur M.: Federal duck stamp in the United States to feature an empress goose, *3209*
Cook, Frederick A.: Expedition to survive a winter in the Antarctic, *3840*
Explorers credited with reaching the North Pole, *3849*
Cook, James: Commercial sealing in the Antarctic region, *3827*
Naturalist to describe the wildlife of Australia, *4682*
Person to cross the Antarctic Circle, *3826*
Person to see the Antarctic territories of Peter I Island and Alexander I Island, *3831*
Cooper, Hugh L.: Major dam on the Mississippi River below St. Paul, MN, *2093*
Corbett, Jim: Tiger preserve in India, *4830*
Cordus, Valerius: Use of the word *pollen*, *1749*
Corky (cockatoo): Bird known to have lived to a documented age of 80, *1679*
Coronado, Francisco Vásquez de *See* Vázquez de Coronado, Francisco
Cortés, Hernán: Horses reintroduced to the American mainland, *1488*
Silkworms in the Western Hemisphere, *1490*
Costa, Jim: Bill in California forcing developers to show that there would be an adequate supply of water for any proposed large subdivision, *3091*
Cottrell, Frederick Gardner: Electrostatic device to remove particles from industrial smoke, *2970*
Cousteau, Jacques-Yves: Aqualung or diving tank, *3573*
Oceanographic museum, *4916*
Craighead, John: Classification of rivers, *4508*
Crary, Charlie: Person to enter into the cave at what would become Wind Cave National Park, *3778*
Crick, Francis: Accurate DNA structure description, *2753*
Crutzen, Paul: Nobel Prize was awarded to ozone-layer researchers, *1972*
Cubitt, William: Self-regulating windmill sails, *4848*
Cugnot, Nicholas Joseph: Automobile, *1586*
Curie, Marie: Woman to win a Nobel Prize, *2268*
Curie, Pierre: Woman to win a Nobel Prize, *2268*
Cushman, Robert: Furs exported from an American colony, *2902*
Custer, George Armstrong: Official exploration of the area that later became Mount Rushmore National Park, *3775*
Cuthbert, Saint: Duck sanctuary, *4808*
Cutler, Manasseh: Botanical expedition in the United States, *1768*
Cuvier, Georges: Natural catastrophe theory about animal extinctions, *2009*
Scientist to establish the fact of extinction, *2359*

D

da Vinci, Leonardo *See* Leonardo da Vinci
Dalhousie, Lord: Forest policy for British India, *2673*
Dali, Salvador: Flood of significance in Spain during the 20th century, *2548*
Dalton, John: Explanation of atmospheric condensation, *1879*
Daniell, John Frederic: Dew-point hygrometer, *1815*
Danvers, Henry: Botanic garden in England, *1756*
Darby, Abraham: Use of the process of smelting by coke, *2028*
D'Arcy, William Knox: Oil discovery of significance in Persia, *3695*
Darling, Jay Norwood ("Ding"): Federal duck stamp in the United States to feature the mallard, *3166*
Law in the United States to allow government fees to be levied to establish and maintain wildlife refuges, *4818*
Darlington, Harry: Permanent American base in Antarctica, *3861*
Darlington, Jennie: Permanent American base in Antarctica, *3861*
Women to spend the winter in the Antarctic, *3863*
D'Arsonval, Jacques: Theory of Ocean Thermal Energy Conversion (OTEC), *4549*
Darwin, Charles: Book on natural history written by Charles Darwin, *3238*
Darwin Centenary Symposium on Earthworm Ecology, *4073*
Earthquake to be described in detail by Charles Darwin, *2200*
First systematic weather forecasts, *1902*
Proposed theory that human population growth might create a global catastrophe, *3876*
Renowned scientist to ennoble the earthworm as one of the most important animals in the history of the world, *4070*
Theory that coral reefs are founded on sinking underwater land masses like subsiding volcanoes and continental shelves, *3553*
Use of the term *ecology*, *1049*
Weather maps in the modern version, *1907*
Zoo explicitly set up as a scientific institution, *4889*

Davey, John: Forestry school privately operated in the United States, *2681*
Davis, Gray: Bill in California forcing developers to show that there would be an adequate supply of water for any proposed large subdivision, *3091*
Davis, John: Explorer to land on Antarctica, *3832*
Davis, William: U.S.–Canadian bilateral report on acid rain, *3042*
Davy, Sir Humphry: Scientist to explain the reasons for the value of adding manure and ashes to the soil, *4068*
Dawn (crane): Whooping crane born in captivity, *4754*
de Doillier, Fadio: Fruit walls, *4087*
de Geer, Gerhard: Ice Age calculations done scientifically, *2004*
De la Beche, Sir Henry Thomas: British government inquiry into the practicability of laws to control smoke, *1377*
de Soto, Hernando: European mission of significance to the interior of North America, *3710*
Flood recorded by Europeans in North America, *2524*
Spanish ships built from pine trees felled along the southern reaches of the Mississippi, *2643*
d'Eaubonne, Françoise: Use of the term *ecofeminism*, *1042*
Delacour, Jean: Imperial pheasants collected from the wild, *1659*
Democritus: Recorded scientific theories explaining earthquakes, *2174*
Denny, O. N.: Successful introduction of the ring-neck pheasant in the United States, *1514*
Denys, Nicholas: Coal mined in Canada, *3299*
Des Voeux, H. A.: Use of the word *smog*, *1439*
Devall, Bill: Radical "deep ecology" platform, *1082*
Devi, Amrita: Ecological protests by women, *1040*
Mass death of "tree huggers", *1094*
Pacifist ecological protests in India, *2644*
Dick, John H.: Federal duck stamp in the United States to feature a harlequin duck, *3189*
Dickinson, Anna: Woman to climb Long's Peak in what is now Rocky Mountain National Park, *3774*
Dicoles: Geometric proof of the focal properties of parabolic and spherical mirrors, *4078*
Diesel, Rudolph: Diesel engine fuels, *4325*
Dittmar, William: Scientist to state the law of relative proportion in seawater, *3563*
Dohrn, Anton: Marine biology station of significance, *3559*
Dolly (sheep): Mammal to be cloned from an adult cell, *2769*
Dositheius: Western man to build a parabolic or "burning" mirror, *4077*
Douglas, Paul: Proposed U.S. legislation regarding abandoned cars, *4142*
Douglas-Hamilton, Iain: Significant modern studies of elephant behavior in the wild, *4724*
Douglass, Andrew Ellicott: Person to compare tree ring width to climate and droughts, *2143*
Use of dendroclimatology as a means of determining historic yearly rainfall and temperatures by tree rings, *2013*
Drake, E. R.: State forestry association, *2692*
Drake, Edwin Laurentine: Oil well that was commercially productive, *3686*
Dresser, Betsy: Reproductive research foundation sponsored by a zoo, a university, and a wild animal park, *4953*
Successful embryo transfer between females of endangered species, *4955*
Drowne, Shem: Weather vane maker in Colonial America, *1870*
Drummond-Hay, (first name unknown): Report of extinction of the great auk, *2364*
Du Pont, Pierre S.: Year in which General Motors vehicles outsold those of the Ford Motor Company, *1541*
Dubos, René: Use of the phrase "think globally, act locally", *1073*
Duckworth, Joseph P.: Pilot to fly into a hurricane intentionally, *4252*
Durand, Peter: "Tin" can, *4124*
Durant, William C.: Carmaking company to produce a range of models, *1537*
Year in which General Motors vehicles outsold those of the Ford Motor Company, *1541*
d'Urville, Dumont: Emperor penguin egg seen and collected by humans, *1649*
Duryea, Charles: Automobile race in the United States, *1532*
Mass-produced automobile in the United States, *1533*
Duryea, Frank: Automobile race in the United States, *1532*
Mass-produced automobile in the United States, *1533*
Dutrochet, Henri: Discovery and naming of the process of osmosis, *1777*
Dutton, Francis: Copper mine in Australia, *3317*
Dyer, John: Sierra Club chapter formed outside of California, *1057*
Dyer, Polly: Sierra Club chapter formed outside of California, *1057*

E

Earle, Sylvia A.: Female oceanographer of distinction, *3597*
Eastwood, B.: Cranberry treatise, *3241*
Eaton, Amos: Classification of the layers of the Earth's crust in North America, *2797*
Degree program in natural science in the United States, *2261*
Summer research term at a U.S. college, *2262*
Edison, Thomas: Electric generating station, *3359*
Edward I, King of England: Statute in England against air pollution, *1373*
Edward III, King of England: Female falconer of note, *2899*
Eights, James: American scientist to go to Antarctica, *3833*
Eisenhower, Dwight D.: Law stating that U.S. forests would be exploited for a variety of recreational, commercial, and conservation purposes, *2685*
Nuclear energy proposal at an international level, *3464*
Nuclear power plant exclusively used for peaceful purposes, *3481*
Undersea park established by the U.S. government, *3803*
Eliot, Jared: Distinctively American agricultural book, *3232*
Elizabeth I, Queen of England: Official "fish days" in England, *2449*
Ellis, William: Non-Hawaiian to observe Kilauea Volcano erupting in the area that later became Volcanoes National Park, *4367*
Elton, Charles S.: Introduction of the discipline of animal ecology, *1053*
Emerson, Ralph Waldo: Environmental movement of significance in the United States, *1047*
Eneas, Aubrey: Solar power company, *4098*
Erasistratus of Alexandra: Recorded evidence of the use of animals in research, *1002*
Ercker, Lazarus: Mining book detailing methods used by miners, *3297*
Ericsson, John: Parabolic trough solar reflector, *4095*
Solar-powered steam engine in the United States, *4094*
Espy, James Pollard: Meteorologist to claim that lighting large forest fires would produce widespread rain, *1889*
Meteorologist to discover the processes of cyclones and tornadoes, *4242*
Official meteorologist to work for the U.S. government, *1891*
Estes, Joel: Non–American Indian settler in what is now Rocky Mountain National Park, *3768*
Ethelbert II, King of Kent: English falconer of record, *2897*
Evelyn, John: Major anti–air pollution tract, *1429*
Everet, Mable: Death in England known to have been caused by an earthquake, *2183*
Ewing, Thomas: Cabinet-level agency concerned primarily with natural resource development and conservation in the United States, *3079*

F

Fabre, Jean Henri: Significant study of the behavior of insects, *4686*
Fabry, Charles: Discovery of the ozone layer, *1969*
Fahrenheit, Daniel Gabriel: Mercury thermometer, *1869*
Farlow, William G.: Meeting to create the New England Botanical Club, *1788*
Fawbush, Ernest J.: Tornado forecast for a specific U.S. location, *4288*
Fell, Jesse: Anthracite coal burned experimentally, *2036*
Ferdinand II, Grand Duke of Tuscany: Sealed thermometer, *1862*
Ferdinand V, King of Spain: East-west division of the world for exploration, *3008*
Treaty between nations dividing the world for exploration, *3013*
Fernow, Bernhard Eduard: Forestry school at a U.S. college, *2678*
Issue of *Forestry Quarterly*, *2736*
Ferrel, William: Tide-predicting machine, *3564*
Fisher, James L.: Federal duck stamp in the United States to feature a waterfowl decoy, rather than a live bird, *3212*
Fisher, John A.: Use of fuel oil to power warships, *3666*
Fitzroy, Robert: First systematic weather forecasts, *1902*
Use of the word *forecast* relating to meteorology, *1898*
Fjortoft, Ragner: Weather prediction system using modern numerical systems, *1949*
Flagler, Henry M.: Land boom of significance in Florida, *3141*
Flavell, George: Tourists to visit the Glen Canyon–Lake Powell area, *2238*
Fleay, David: Duck-billed platypus bred in captivity, *4750*
Fleming, Alexander: Use of penicillin in general clinical practice, *3932*

Ford, Gerald R.: Clean drinking water standards in the United States that were legally binding, *4598*
Solar energy research funded by the U.S. Congress, *2298*
Ford, Henry: Mass-produced car to run on ethanol, *4326*
U.S. manufacturer to make cars in great quantities to reduce costs, *1591*
Foreman, Dave: Handbook on ecological sabotage, *1083*
Radical environmental activist arrested and convicted for "monkeywrenching" in the United States, *1084*
Fortune, Robert: Forsythia in England, *1504*
Fourier, Jean Baptiste: Greenhouse effect description, *1835*
Fourier, Jean Baptiste Joseph: Meteorological theory saying human activities affect the climate, *1887*
Fourneyron, Benoît: Water-powered turbine, *4546*
The Fox (activist): Exploits of "The Fox," the anonymous Chicago-area environmentalist, *1058*
Foyn, Sven: Exploding grenade harpoon, *4625*
Francis of Assisi, Saint: Patron saint of ecology, *1076*
Francis I, Holy Roman Emperor: Surviving zoo, *4882*
Francis, Edward: Animal disease of American origin, *3926*
Franco, Raymond: Felony indictments under a U.S. environmental law for smuggling hazardous waste from California into Mexico, *2889*
François I, King of France: Trash collection tax in Paris, *4173*
Franklin, Benjamin: Charts of the Gulf Stream, *3548*
Edition of Benjamin Franklin's *Poor Richard's Almanac*, *1871*
Hurricane known to be accurately tracked in colonial America, *4236*
Lightning conductor in the United States, *4274*
Observer to propose a link between volcanic eruptions and the weather, *1809*
Partridge propagation in the United States, *2930*
Scientist to fly a kite in a thunderstorm in hopes of attracting a lightning bolt, *4180*
Street-sweeping service in the United States, *4174*
Treaty in which England recognized the fishing rights of the United States, *3014*
Franklin, Jerry: Study that defined the factors making old-growth Douglas fir forests unique, *2747*
Franklin, Rosalind: Accurate DNA structure description, *2753*
Frederick II, Holy Roman Emperor: Falconry book, *2898*
True book devoted to birds, *3227*
Fteley, Alphonse: Dam to have a maximum height of 295 feet, *2092*
Fujita, Tetsuya Theodore: Scale to measure tornado intensities, *4293*
Fuller, Calvin Souther: Photovoltaic cell or solar battery to work effectively, *4108*
Fuller, Fay: Woman to climb Mount Rainier, *3783*
Fuller, Kathryn: Woman to head a major U.S. conservation organization and receive a salary, *1039*
Fulton, Robert: Internal combustion engine in the United States, *4323*
Furlong, Lawrence: Coast survey book written and published in the United States, *3235*

G

Gagnan, Emil: Aqualung or diving tank, *3573*
Galilei, Galileo: Observer to realize that the atmosphere becomes colder at the upper levels nearer the heat-giving sun, *1861*
Open-air thermometer, *1860*
Gallatin, Albert: Treaty to recognize the three-mile limit for fishing off the North American coast, *3015*
Galton, Francis: Weather maps in the modern version, *1907*
Garrastazu, Emilio: Decision to build a highway into Brazil's Amazon wilderness, *2718*
Gay-Lussac, Joseph: Meteorological research done by balloon, *1881*
Geiger, Robert: Use of the term *Dust Bowl* to describe the drought-devastated Southern Plains, *2150*
Geiger, Rudolf: Microclimatology research, *1935*
Geldof, Bob: Drought aid from a global rock music festival, *2159*
George III, King of Great Britain: Treaty in which England recognized the fishing rights of the United States, *3014*
Gericke, William Frederick: Description of hydroponic agriculture, *1247*
Private hydroponic garden, *1248*
Use of the term *hydroponics*, *1251*
Gesner, Abraham: Natural history museum in Canada, *4891*
Patent for distilling kerosene from crude oil, *3679*
Gibbs, Lois: Love Canal Homeowners Association meeting, *1099*

Gibson, Janet: Central American coral reef reserve, *3594*
Giggenbach, Werner: Scientist to survive the eruption of an Antarctic volcano, *3870*
Gilbert, William: Terrestrial magnetism theory, *2777*
Glantz, Leonard H.: Genetic privacy legislation proposal in the United States, *2766*
Gleason, Kate: Tract housing development in the United States, *3074*
Glidden, Joseph Farwell: Barbed wire product on the market, *1163*
Glinka, K. D.: Establishment of the concept that soils are individual and natural and have their own form and structure, *4069*
Glover, Townend: Entomologist hired by the U.S. government, *1272*
Gmelin, Johann: Permafrost, *1806*
Goebel, Wilhelm: Federal duck stamp in the United States to feature a surf scoter, *3226*
Gooch, Sir William: Fox hunting pack in North America, *2910*
Goodhue, Lyle D.: Aerosol spray containers for insecticide that were portable, *1321*
Goodyear, Charles: Mass-produced condoms, *3878*
Gorbachev, Mikhail S.: Climate research agreement between the United States and Soviet Union, *1830*
Earthquake in the Soviet Union allowed open media coverage, *2232*
Gore, Albert, Jr.: Widely acknowledged environmentalist to be chosen as a major-party U.S. vice presidential candidate, *1091*
Gorman, Eville: Research showing acid-forming pollution could travel great distances, *1456*
Gossage, William: Effective treatment for stack emissions of hydrochloric acid in the alkali industry, *1434*
Gottlieb, Alan M.: Publication that outlined the goals of the "wise-use" movement, *1024*
Gottman, Jean: Formal identification of the line of continuous development along the U.S. eastern corridor from Washington to Boston, *3143*
Goulburn, Henry: Treaty to recognize the three-mile limit for fishing off the North American coast, *3015*
Grant, Ulysses S.: Conservation agency established by the U.S. government, *4763*
Government weather service in the United States, *1911*
Graunt, John: Systematic quantitative study of death in populations, *3908*
Use of London's weekly Bills of Mortality to establish a link between death rates and the burning of coal, *1331*
Graves, Collins: American song written about a flood warning, *2530*
Gray, Asa: Botany professorship in the United States, *2263*
"The Great Vance": Popular use of the word *zoo*, *4894*
Grey, Thomas: Death in England known to have been caused by an earthquake, *2183*
Griffiths, Martha Wright: Bill to require humane slaughter of animals in the United States, *1015*
Grinnell, George Bird: Bird preservation organization in the United States, *1712*
Boone and Crockett Club meeting, *1030*
Grinnell, Joseph: "Ecological niche" concept, *1620*
Gromme, Owen: Federal duck stamp in the United States to feature a shoveler, *3181*
Grotecloss, Elizabeth: Female awarded the Sage Fellowship in Entomology and Botany, *2266*
Grotius, Hugo: Freedom-of-the-seas doctrine, *3546*
Scholarly text on the international law of the sea, *3229*
Statement of the idea that nations should be held responsible for their actions just as individuals were, *3009*
Use of the term *mare liberum* or "freedom of the seas", *4488*
Grove, Sir William Robert: Fuel cell, *4324*
Guericke, Otto von: Weather forecast using a barometer, *1864*
Guettard, Jean Étienne: Paper on the extinct volcanoes of Europe, *4363*
Statement that rocks and minerals were not randomly scattered but deposited in bands, *2783*
Guminska, Maria: Warsaw Pact government to recognize that industrial wastes cause health problems, *2979*
Gutenberg, Beno: Scale to accurately measure earthquake strength, *2223*
Scientist to use records of seismic waves to demonstrate the existence of the earth's core and measure its distance from the surface, *2219*

H

Haagen-Smit, A. J.: Scientific paper demonstrating that hydrocarbons and nitric oxide react in sunlight to form ozone pollutants, *1454*
Haavisto, Pekka: Green politician to join a national government as a minister, *1117*

Haber, Fritz: Discovery of a laboratory process for extracting liquid ammonia from free nitrogen in the air, *4063*
Hadley, George: Trade-wind explanation, *1872*
Haeckel, Ernst: Use of the term *ecology*, *1049*
Hagen, Hermann August: Entomology professor at a U.S. college, *2265*
Hagenbeck, Carl: Zoo to display animals in a naturalistic setting, *4913*
Zookeeper to recognize the benefits of fresh air for animals in captivity, *4914*
Hall, James: Geological survey of Niagara Falls, *2807*
Hamilton, Alice: Female professor at Harvard, *2275*
Research conducted in the United States by a woman in the field of occupational health, *2270*
Statement of the theory that constant exposure over time to small doses of toxins causes poisoning, *2832*
Statewide survey of industrial poisons, *2272*
Textbook on industrial poisons in the United States, *2276*
Hammond, John: Parkland purchased by a city in the United States, *3817*
Hammond, Samuel: Significant statement about the value of wilderness that appeared in the United States, *3242*
Hanno: Description of chimpanzees, *4679*
Hansell, Martin: Cave discovered in American Forks Canyon, *3781*
Hanson, James: "Greenhouse effect" report to the U.S. Congress, *1845*
Harcourt, Mike: Pledge to halt all raw sewage flow from Victoria and Vancouver Island, British Columbia, into the Strait of Juan de Fuca, *4543*
Harrison, Benjamin: National forest reserve in the United States, *2712*
Harrison, John: Chronometer, *3545*
Hart, William: Natural gas well in the United States, *2040*
Harvey, Harold: Report on the effects of acid rain on fish populations, *1464*
Harvey, William Henry: Significant study of North American marine algae, *3555*
Hatfield, Charles M.: "Cloud compeller" to claim success in bringing rain to California with ill-smelling chemical cocktails set in pans atop high wooden towers, *2169*
Hatshepsut, Queen of Egypt: Acclimatization garden, *4875*
Importation of exotic animals for a royal collection, *4876*
Hautman, Jim: Federal duck stamp in the United States to feature a black-bellied whistling duck, *3218*
Federal duck stamp in the United States to feature a spectacled eider, *3222*
Havell, Robert, Jr.: Bird pictured in John James Audubon's *The Birds of America*, *3160*
Hay, M. W.: California quail in New Zealand, *1509*
Hayden, Ferdinand: Photographer to capture successfully the beauties of Yellowstone on film, *3265*
White men to see the great chasm at what later became Black Canyon of the Gunnison National Monument in Colorado, *3741*
Hayes, Rutherford B.: Agency created by U.S. government specifically for geological issues, *2811*
Public Land Review Commission in the United States, *3081*
Heaviside, Oliver: Prediction of the existence of the ionosphere, *1920*
Henderson, Hazel: Air pollution index broadcast on television, *1349*
Henry I, King of England: Royal menagerie known to exist in England, *4879*
Henry VIII, King of England: English law to support North American fisheries, *2469*
Henry, Joseph: Appeal for U.S. weather observers, *1894*
Use of the telegraph for reporting weather conditions, *1897*
Henson, Matthew: Explorers credited with reaching the North Pole, *3849*
Herbert, Lou: Treatise on mining that was scholarly, detailed, and systematic, *3296*
Hero of Alexandria: Person to demonstrate that air has weight, *1855*
Solar machine, *4081*
Herodotus: Description of an oil production industry, *3670*
Hess, Harry: Theory that seafloor spreading was caused by convective heat in the Earth's mantle, *2818*
Hess, Victor: Cosmic radiation discovery, *1925*
Hill, Alfred: Pronghorn antelope bred and reared in captivity, *4746*
Hill, David B.: Law establishing a protected wilderness in the Adirondack Mountains, *4631*
Hillman, John Wesley: European to "discover" the volcanic Crater Lake in Oregon, *4370*
Non–American Indian to see Crater Lake, *3767*
Hilton, William: Forest fire lookout tower, *2613*
Hines, Robert W.: Federal duck stamp in the United States to feature the redhead, *3182*
Hippocrates: Written description of the medical uses of plants, *1745*
Hirayama, T.: Epidemiological evidence of the association between passive smoking and lung cancer, *1476*

Hitchcock, Edward: Complete state geological survey in the United States, *2802*
Paper describing dinosaur footprints, *2803*
Hobart, Louis: Aquarium to be part of a natural history museum, *4919*
Hofmeister, Wilhelm: Demonstration of the alternation of generations, *1772*
Holmes, Arthur: Theory that seafloor spreading was caused by convective heat in the Earth's mantle, *2818*
Hooke, Robert: Cell description and use of the term *cell*, *2750*
Use of the term *cell* to describe microscopic units of plants, *1758*
Hooker, Joseph: Director of Kew Gardens, *1782*
Hooker, William: *Begonia sutherlandii* to flower in England, *1510*
Botanical chart to be published, *1781*
Director of Kew Gardens, *1782*
Hoover, Herbert Clark: Earthquake organization formed because of a U.S. earthquake, *2216*
National monument to commemorate the world's largest gypsum dune field, *3722*
Treatise on mining that was scholarly, detailed, and systematic, *3296*
Hoover, Lou Henry: Earthquake organization formed because of a U.S. earthquake, *2216*
Female college graduate in geology, *2814*
Hopkins, Cyril George: Soil survey of a state in the United States, *4071*
Test to identify critical elements missing in the soil, *4034*
Hough, Franklin Benjamin: Division of Forestry in the United States, *2674*
Forestry book written and published in the United States, *3247*
How, William: Flora of the British Isles, *1757*
Howard, Sir Albert: Instances of modern organic farming as a conscious rejection of traditional chemical methods, *1259*
Major opposition to chemical fertilizers, *4064*
Published warnings against using chemical fertilizers, *4066*
Woman to advocate organic gardening, *4065*
Howard, Ebenezer: Introduction of the "green belt" concept, *3099*
Use of the term *green belt*, *3100*
Howard, Luke: Cloud classification of scientific significance, *1880*
Howe, Nancy: Federal duck stamp in the United States to feature a king eider, *3220*
Woman to win the U.S. federal duck stamp contest, *3221*
Howells, Julius M.: Dam built using the hydraulic fill process, *2088*
Huang, Jie: Female winner of the federal junior duck stamp contest in the United States, *3225*
Hudson, Henry: Person to reach latitude 80 degrees north in the arctic region, *3825*
Humboldt, Alexander, Freiherr von: Map using isothermal lines to map temperatures, *1885*
Oilbird, *1647*
Humphrey, Hubert H.: Wilderness bill introduced in the U.S. Congress, *3736*
Hunt, Lynn B.: Federal duck stamp in the United States to feature the green-winged teal, *3172*
Hussey, Obed: Patent on a practical reaping machine, *1239*
Hutchison, Kay Bailey: Environmental lawsuit to challenge the United States under the North American Free Trade Agreement (NAFTA), *2348*
Hutton, James: Publication of the geological principle that the present is the key to the past, *2789*
Scientist to establish "uniformitarianism" as a principle of geology, *2790*
Hyat, John Wesley: Commercially available plastic, *4125*

I

Iacocca, Lee: Federal government bailout of a major U.S. carmaker, *1548*
Ibbetson, Agnes: Female botany experimentalist of significance in England, *1770*
Incas (parakeet): Carolina parakeet extinction report, *2367*
Isabella I, Queen of Spain: East-west division of the world for exploration, *3008*
Treaty between nations dividing the world for exploration, *3013*
Ishimori, Michiko: Minamata Disease or "cats dancing disease" cases, *2972*
Isleffson, Sigourer: Report of extinction of the great auk, *2364*
Iversen, J.: Detailed model for determining prevailing temperatures in prehistoric northern Europe, *2014*

J

Jackson, William Henry: Photographer to capture successfully the beauties of Yellowstone on film, *3265*
James I, King of England (James VI of Scotland): Proclamations against pollution caused by London starchmakers, *4445*
James, H. B.: Colony of puna flamingos to be discovered, *1700*

James, Tony: Instrument that could detect hazardous chemicals such as PCBs (polychlorinated biphenyls), *2843*
Jaques, Francis Lee: Federal duck stamp in the United States to feature the black duck, *3173*
Jay, John: Treaty in which England recognized the fishing rights of the United States, *3014*
Jefferson, Thomas: Acquisition of land of significance by the United States from a foreign government, *3135*
Simultaneous weather observations in the United States, *1875*
Soil conservation advocate of prominence in the United States, *4029*
Terraced or horizontal plowing in the United States, *4048*
Jellinek, Emil: Agreement to finance development of a new high-performance car on the condition that the model be named for the financier's daughter, *1535*
Jenner, Edward: Successful vaccination against smallpox, *3912*
Jensen, Soren: Research showing that PCBs (polychlorinated biphenyls) were present in human and wildlife tissues, *2847*
Jodhpur, Maharajah of: Pacifist ecological protests in India, *2644*
Johannsen, Wilhelm: Use of the word *gene* to describe units of heredity, *2752*
John II, King of Portugal: East-west division of the world for exploration, *3008*
Papal declaration regarding world exploration, *3007*
Treaty between nations dividing the world for exploration, *3013*
John Paul II, Pope: Introduction of the concept of "ecotheology", *1087*
Papal document to deal with ecology, *3264*
Patron saint of ecology, *1076*
Johnson, George Arthur: Drinking water supply in the United States to be chemically treated with chlorine compounds on a large scale, *4583*
Johnson, James B.: Fish hatchery to breed salmon in the United States, *2490*
Johnson, Lyndon B.: Air pollution law of importance enacted by the U.S. Congress, *1398*
Flood-danger evaluations required by a U.S. presidential executive order, *2552*
Public statement by a U.S. president that all of the country's major river systems were polluted, *4403*
Regulations on animal research in the United States, *2287*
U.S. National Sea Grant Program, *2289*
Waterways in the U.S. National Wild and Scenic Rivers System, *3983*
Wilderness bill introduced in the U.S. Congress, *3736*
Johnson, Martin: Wildlife films of the fauna of Borneo, *4697*
Johnson, Osa: Wildlife films of the fauna of Borneo, *4697*
Johnson, Samuel W.: Law in the United States requiring ingredient labels on fertilizers, *4062*
Johnston, Harold S.: Ozone-layer warning, *1973*
Johnston, Harry Hamilton: Okapi known to Western science, *4690*
Johnston, Velma: Federal law in the United States to ban the use of motorized vehicles in capturing wild horses, *4771*
Joliet, Louis: Coal discovery recorded in North America, *2026*
Jukes, Thomas: Use of antibiotics as an animal feed additive, *1172*
Julien, Denis: European to discover what is now Canyonlands National Park, *3764*
Junagarh, Nawab of: Law to protect the Asiatic lion, *4764*
Junge, Christian: Modern analysis of significance into the chemistry of rain, *1959*
Justinian I, Emperor of the East: Dam known to have a curved shape, *2084*
Recorded statement of the law of the sea, *3006*
"Sun rights" laws to guarantee buildings access to solar warmth and light, *4083*

K

Kahn, Adam: Botany professor at a college, *2257*
Kalimeris, Anthony: U.S. Supreme Court decision regarding a sponge harvesting law, *2472*
Kallet, Arthur: Major attack on the U.S. consumer products industry, *3251*
Kalmbach, Edwin Richard: Federal duck stamp in the United States to feature the ruddy duck, *3174*
Kamehameha I, King of the Hawaiian Islands: Natural history museum in the United States named for a descendant of royalty, *4904*
Kämpfer, Engelbert: Japanese flora description printed in English, *1763*
Karlsefni, Thorfinn: Commercial logger on the East Coast of North America, *2641*
Kaye, Michael: Wilderness expeditions operator in Costa Rica, *2243*
Keeley, Charles: Instrument to measure carbon dioxide in the atmosphere in parts per million, *1452*

Keeling, C. D.: Carbon dioxide monitoring station, *1458*
Keeling, David: Daily measurements of atmospheric carbon dioxide, *1839*
Keith, Sir William: Game law in Pennsylvania, *2908*
Kellogg, Albert: Systematic study of *Sequoia gigantea*, the California sequoia, *1784*
Kelly, Petra: Parliament to have a substantial Green bloc represented, *1112*
Kelvin, William Thomson, Baron: Geologist to discovery that several ice ages had existed, *2012*
Kemp, Clarence: Commercially marketed solar water heater, *4097*
Kennedy, John F.: Seawater conversion plant in the United States on a practical scale, *4595*
Kennelly, Arthur Edwin: Prediction of the existence of the ionosphere, *1920*
Kent, William: National monument in the United States created from land donated by a private individual, *3747*
Kepler, Johannes: Scientist to describe the characteristic six-sided symmetry of snow crystals, *4191*
Kessler, David A.: National policy on marketing genetically engineered food products in the United States, *2765*
Ketilsson, Ketil: Report of extinction of the great auk, *2364*
Kettering, Charles F.: Automatic starter for automobiles, *1592*
Khorana, Har Gobin: Artificial gene, *2754*
Kier, Samuel M.: Oil refinery in the world, *3681*
King, Clarence: Director of the U.S. Geological Survey (USGS), *2809*
Kirby, Mary: Complete flora of Leicestershire, England, *1783*
Kirkwood, James Pugh: Water filtration system for bacterial purification of a U.S. city water supply, *4582*
Klobassa, Mikolaj: Oil wells in Europe, *3680*
Knapp, J. D.: Federal duck stamp in the United States to feature the greater scaup, *3170*
Kölreuter, Josef Gottlieb: Artificial plant hybrid, *1765*
Scientific discovery of the role of bees in pollination, *1766*
Koto, Bunjiro: Scientific formulation of the "fault theory" as we know it, *2208*
Kouba, Leslie C.: Federal duck stamp in the United States to feature an old squaw duck, *3204*
Krapf, Johann Ludwig: Descriptions of Mount Kilimanjaro and Mount Kenya to reach Europe, *4630*
Kristensen, Leonard: Landing on Antarctica that initiated the "Heroic Age" of Antarctic exploration, *3839*
Kublai Khan, Mongol Emperor: Typhoon known to have prevented an invasion of Japan, *4222*
Kuehl, Sheila: Bill in California forcing developers to show that there would be an adequate supply of water for any proposed large subdivision, *3091*

L

La Cour, Poul: Aerodynamically efficient windmills to generate electricity, *4851*
Lady Jane (parakeet): Carolina parakeet extinction report, *2367*
Lafayette, Marie Joseph Paul Yves Roch Gilbert du Motier, Marquis de: Natural gas used to light a U.S. house, *2041*
Partridge propagation in the United States, *2930*
Landsberg, Helmut: Climatologist of repute, *1923*
Langford, Nathaniel Pitt: National park in the United States, *3772*
Superintendent of Yellowstone National Park, *3773*
Larsen, Carl: Whaling station in the Antarctic, *3847*
Wood fossils found on the Antarctic, *3838*
Latham, Increase: U.S. Signal Corps storm warning, *4184*
Laue, Max von: X-ray analysis of minerals, *2816*
Lawes, Sir John: Discovery that tricalcium phosphate could be converted into a soluble plant fertilizer by treatment with sulfuric acid, *4059*
Fertilizer factory, *4061*
Le Beau, Desirée: Female pioneer in the field of rubber recycling, *2284*
Le Blond de la Tour, Sieur: Levees along the Mississippi River, *2567*
Leakey, Richard: Mass burning of poached African ivory, *4788*
LeBlanc, Lee: Federal duck stamp in the United States to feature a Steller's eider, *3211*
L'Écluse, Charles de *See* Clusius, Carolus
Leder, Philip: Genetically engineered mammal to be patented, *2762*
Lederer, John: Written account of a journey through the area of the Blue Ridge Mountains that later became Shenandoah National Park, *3754*
Leffler, Ross: School devoted exclusively to training wildlife conservation officers, *2278*

Leister, Claude Willard: Radio broadcast from a zoo, *4928*
L'Enfant, Pierre: City planned and built as a seat of national government, *3095*
Lenin, Vladimir Ilich: Nature reserve in the Soviet Union, *3796*
Lenoir, Jean Joseph Étienne: Automobile with a practical internal-combustion engine, *1588*
Leonardo da Vinci: Person to imagine the possibilities of power-driven vehicles, *1531*
Statement of the theory of groundwater, *2776*
Leopold, Aldo: Federally designated wilderness area in the United States, *4634*
Game management chair at a U.S. college, *2932*
"Land Ethic" concept, *1054*
Textbook on wildlife management, *3252*
Levitt, William Jaird: Modern-day large-scale planned U.S. suburban development to use assembly-line methods of construction, *3075*
Lewis, Drew: U.S.–Canadian bilateral report on acid rain, *3042*
Lewis, John L.: Coal mines to be fully automated in the United States, *3329*
Lewis, Meriwether: Description of the "black-tailed deer" by William Clark, *4684*
White Americans to systematically describe the weather of the western United States, *1812*
Libby, Willard F.: Radiocarbon or C-14 dating, *2820*
Liebig, Justus von: Scientist to be dubbed the "father of chemical agriculture", *4060*
Likens, Gene: Reintroduction of the term *acid rain*, *1469*
Lincoln, Levi: Parkland purchased by a city in the United States, *3817*
Lindley, John: Professor of botany at the University of London, *1778*
Linnaeus, Carolus: Systematic use of binary nomenclature to describe plants, *1764*
Use of taxonomy to classify living organisms, *1617*
Linné, Carl von *See* Linnaeus, Carolus
Lisa, Manuel: Non–Native American to see the geysers of Yellowstone National Park, *3761*
Lister, Martin: Proposal to create a geological map, *2780*
Lodge, Oliver: Electrostatic precipitation application, *1446*
Logan, William Edmond: Geological Survey of Canada, *2806*
Geologist to classify rocks from Precambrian times, *2808*
Loncier, Adam: Record of using solar energy to make perfume, *4085*
Longman, James: Hotel in Mt. Rainier National Park, *3780*
Loomis, Elias: Scientist in the United States to conduct a detailed study of a single winter storm, *1888*
Lorentz, J. B.: Forestry school in France, *2733*
Louis XIV, King of France: Royal European menagerie with a research component, *4880*
Lovejoy, Thomas E. III: Introduction of the concept of a "debt for nature" swap, *2719*
Lovelock, James: Climate hypothesis concerning animals and plants, *1828*
Direct measurements of chlorofluorocarbon (CFC) levels at the earth's surface, *1466*
Expression of the concept that the earth is a living organism, *1050*
Measurements of chlorofluorocarbons (CFCs) in the atmosphere, *1971*
Lucas, Anthony F.: Oil well in the United States with a massive petroleum deposit, *3693*
Lukasiewicz, Ignacy: Oil refinery, *3685*
Oil wells in Europe, *3680*
Lutz, Frank Eugene: Outdoor museum with nature trails in the United States, *4920*
Lutzenburger, José: Environment minister in the ecologically critical country of Brazil, *2722*
Lyell, Charles: Geological survey of Niagara Falls, *2807*
Scientist to popularize uniformitarianism, *2801*
Lyon, Frank Farmington: Commercial hydroponicum on a large scale, *1250*

M

Maathai, Wangari: Environmental group in Kenya to tie deforestation to poverty, *1043*
Trees planted by the Green Belt Movement, *1098*
MacCready, Marshall: Manned solar-powered aircraft flight, *4119*
MacCready, Paul: Solar automobile race, *4121*
Solar-powered aircraft that could carry a pilot, *4118*
MacDonald, Gordon: Global warming report of significance made to the U.S. government, *1841*
MacGillivray, William: Rediscovery of MacGillivray's petrel, *1682*
Maclure, William: Geological map of the eastern United States, *2796*
Geological maps of the United States, *2792*
Geological organization in the United States, *2799*

Madison, James: Simultaneous weather observations in the United States, *1875*
Magnum, P. H.: Broadbase terrace for farming, *4031*
Magnus, Olaus: Person known to have studied snow crystals, *4190*
Mallet, Robert: Systematic catalog of earthquakes, *2201*
Theory stating that earthquakes come from the Earth's crust and not from outside forces working on the crust, *2202*
World seismic map of significance, *2203*
Malpighi, Marcello: Recognition of the correlation between cellular rings in trees and age, *1759*
Malthus, Thomas: Proposed theory that human population growth might create a global catastrophe, *3876*
Mapes, James Jay: Synthetic fertilizer, *1127*
Marcgrav, Georg: Reference work on Brazilian wildlife, *4680*
Reference work to identify Brazilian fish, *2450*
Maria Theresa, Empress of Austria: Surviving zoo, *4882*
Marie-Victorin, Brother: Person to lobby for a botanical garden in Montreal, *1794*
Markham, Clements R.: Cinchona trees in India, *1507*
Marquette, Jacques: Coal discovery recorded in North America, *2026*
Marsh, George Perkins: Comprehensive study of human impact on the environment, *3244*
Published presentation of a more holistic view of U.S. public land management, *3073*
Research showing that human activity could cause irretrievable damage to the earth, *1048*
Marshall, Robert: National organization dedicated specifically to preserving wilderness (as opposed to wildlife) in the United States, *1055*
Martha (pigeon): Extinction caused by humans of an animal endemic to North America in modern times, *2368*
Martin, Archer: Instrument that could detect hazardous chemicals such as PCBs (polychlorinated biphenyls), *2843*
Martin-Brown, Joan: *EPA Journal*, *1074*
Women's international conference on the environment, *1046*
WorldWIDE Network Inc. meeting, *1044*
Mason, Otis T.: Expression of the concept that the earth is a living organism, *1050*
Masters, Dick: Hospitality ranch in Wyoming, *2245*
Masters, Jean: Hospitality ranch in Wyoming, *2245*
Mather, Stephen T.: All-American Road with glacial features, *4020*
Director of the U.S. National Park Service, *3789*
Private organization to support national parks in the United States, *3718*
Matheson, Scott: Power plant using solar cells, *4120*
Matthews, Drummond: Plate tectonics theory verification, *2821*
Maury, Matthew Fontaine: Bathymetric map of the North Atlantic Ocean, *3556*
Charts of the Atlantic Ocean winds and currents, *3554*
Maps of the trade winds, *1895*
Oceanography textbook, *3557*
Text of modern oceanography detailing the trade winds and ocean currents, *3237*
Wind and current analysis of significance, *3551*
Mawson, Sir Douglas: Meteorite discovered in Antarctica, *3853*
Radio station in Antarctica, *3850*
Maxwell, George Hebard: Irrigation advocacy group in the United States, *1031*
Mazzeo, Peter: Recovery plan approved by the U.S. Fish and Wildlife Service for a plant, *2336*
McCall, Thomas: Comprehensive reform of land-use planning laws in the United States, *3107*
State governor in the United States to implement a doctrine of conservation in balance with economic concerns, *1104*
McCormick, Cyrus: Patent on a practical reaping machine, *1239*
Successful harvesting machine, *1238*
McCoy, George Walter: Animal disease of American origin, *3926*
McCrory, Colleen: Valhalla Wilderness Society meeting, *1097*
McCulloch, David: Use of elevation benchmarks to measure settling or uplifting of the earth, *2226*
McDonald, Marshall: Aquarium with an inland salt-water environment, *4907*
McDuffie, Bruce: Reports of mercury contamination in tuna and swordfish, *4513*
McGarrison, Sir Robert: Woman to advocate organic gardening, *4065*
McIntyre, Ray: Styrofoam, *4128*
McKee, Chris: Theory that an asteroid caused the extinction of dinosaurs, *2376*
McKinley, Ashley C.: Aerial mapping photographs over Antarctica, *3854*
McLean, George Payne: Bird hunting regulation enacted by the U.S. Congress, *2955*
McWhinnie, Mary Alice: Woman to direct an Antarctic research station, *3868*

Meigs, Montgomery Cunningham: Aqueduct to supply Washington, DC, *4577*
Mek Penawar (elephant): Asian elephant tracked with a satellite transmitter, *4744*
Mellon, Andrew: Oil well in the United States with a massive petroleum deposit, *3693*
Menzies, Archibald: Non-Hawaiian to climb Mauna Loa in the area that later became Volcanoes National Park, *3759*
Mercalli, Giuseppe: Scale widely used to measure earthquake intensity, *2212*
Mereruka: Illustrations of animals in a zoo-like setting, *4873*
Merkle, Gretchen W.: Woman to reach the rank of major in the U.S. Park Police, *3814*
Merle, Walter: Known journal about the weather, *1857*
Merriam, Clinton Hart: "Life zones" concept, *1619*
Merriam, John C.: Scientific expedition to what would later be designated the John Day Fossil Beds National Monument, *3744*
Merritt, Leonidas: Mining of significance in the iron-rich Mesabi Iron Range in northeast Minnesota, *3318*
Meserole, B. J.: Game protection society in the United States, *2967*
Michel, Helen: Theory that an asteroid caused the extinction of dinosaurs, *2376*
Michell, John: Calculation of the Earth's mean density, *2791*
Middleton, William: Confirmation that Dutch elm disease-carrying bark beetles deposit fungi spores on feeding wounds of trees, *2601*
Midgley, Thomas: Chemist to add tetraethyl lead to gasoline, *1339*
Chlorofluorocarbons (CFCs), *1970*
Miescher, Friedrich: Nucleic acids, *2751*
Mikkelson, Caroline: Woman to visit Antarctica, *3858*
Mikkelson, Klarius: Woman to visit Antarctica, *3858*
Miller, Robert C.: Tornado forecast for a specific U.S. location, *4288*
Mills, Hiram Francis: Water filtration system for bacterial purification of a U.S. city water supply, *4582*
Milne, John: International seismographic recording network, *2213*
Seismograph in its modern form, *2204*
Seismology association, *2205*
Mindell, Fania: Birth control clinic in the United States, *3879*
Ming Ming (panda): Giant panda born in captivity, *4752*
Mitchell, Chalmers: Modern animal park, *4929*
Mitchell, John: Description of two types of earthquake motion, *2191*
Theory on how to locate the epicenter of an earthquake, *2192*
Mohl, Hugo von: Use of the term *protoplasm* to describe the contents of plant cells, *1773*
Moi, Daniel Arap: Mass burning of poached African ivory, *4788*
Molina, Mario: Nobel Prize was awarded to ozone-layer researchers, *1972*
Ozone-layer data concerning damage by chlorofluorocarbons (CFCs), *1974*
Monroe, James: Highway built in the United States entirely with federal funds, *1552*
Montez, Ramón: Tourists to visit the Glen Canyon–Lake Powell area, *2238*
Moore, Burton E., Jr.: Federal duck stamp in the United States to feature a fulvous duck, *3214*
Moraga, Gabriel: European to lead an expedition into the area that later became Sequoia and Kings Canyon National Park, *3760*
Morey, Samuel: Internal combustion engine in the United States, *4323*
Morland, Sir Samuel: Angled barometer, *1865*
Morrell, Daniel: Warnings of structural problems at the South Fork Dam near Johnstown, PA, *2112*
Morris, E. A.: Federal duck stamp in the United States to feature a Pacific brant, *3200*
Morris, Henry M.: Widely publicized version of the Christian fundamentalist theory of flood geology, *2547*
Morton, Julius Sterling: Arbor Day celebration, *2724*
Effective promoter of tree planting to prevent soil erosion on the U.S. Great Plains, *4049*
Morton, Thomas George: National animal rights organization in the United States, *1010*
Mouchot, Augustin: Brandy distilled by the heat of the sun, *4091*
Combination of a glass heat trap and a burning mirror, *4092*
Solar motor, *4093*
Solar-powered steam engine in the United States, *4094*
Moynihan, Daniel Patrick: Law of significance to control acid rain in the United States, *1413*
Muir, John: Conservationist of note in the United States, *1026*
Dam project to provoke major opposition from environmentalists, *2090*
National environmental organization in the United States, *1051*

National monument in the United States created from land donated by a private individual, *3747*
Non–American Indian to discover Glacier Bay and the Muir Glacier, *3776*
Person to write a book promoting the idea of a U.S. national parks system, *3248*
Müller, Paul: DDT (Dichlorodiphenyl-Trichloroethane) synthesized for use as an insecticide, *1287*
Mulroney, Brian: U.S.–Canadian bilateral report on acid rain, *3042*
Murdock, William: Gas used to light a house, *2032*
Murray, Jack: Federal duck stamp in the United States to feature a snow goose, *3183*
Murray, John: Significant theory that Antarctica was a continent rather than a series of islands, *3841*
Must, Emma: Person in England to chain herself to a bulldozer to protest a highway, *1103*
Myers, Norman: Modern study linking rain forest deforestation with accelerating species extinction due to habitat loss, *1632*

N

Nader, Ralph: Automobile safety advocate of note in the United States, *1542*
Automobile safety legislation of significance passed by the U.S. Congress, *1575*
Automobile safety warning having national impact, *3940*
Compact car built by General Motors, *1547*
U.S. Green Party presidential campaign, *1118*
Naess, Arne: Radical "deep ecology" platform, *1082*
Nakaya, Ukichiro: Scientist to study snow with modern X-ray equipment, *4207*
Narváez, Pánfilo de: Known military expedition in North America sunk by a hurricane, *4226*
Naumann, Einar: Trophic-level system application to classify lakes, *4506*
Nearing, Helen: Noted "back to the land" conservationist in the United States, *1036*
Nearing, Scott: Noted "back to the land" conservationist in the United States, *1036*
Nelson, Gaylord: Reports of decreases in commercial fishermen's catches in Lake Superior, *4401*
Nelson, Murray: Sewer district established by a city in the United States, *4527*
Newbold, Charles: Cast-iron plow, *1237*
Newcomen, Thomas: Steam engine of high pressure used in a mine, *3300*
Newhouse, Sewell: Metal trap mass-produced in the U.S., *2912*
Nixon, Richard M.: Coal-mining law with penalty fines imposed by the U.S. government, *3334*
Presidential order prohibiting U.S. government assistance to concerns that violate air emissions standards, *1407*
U.S.–Canadian effort of significance to improve the water quality of the Great Lakes, *3029*
Nobile, Umberto: Explorers to reach the South Pole, *3852*
Norbeck, Peter: U.S. National Scenic Byway in the Black Hills, *4011*
Nordenskjöld, Nils Adolf Erik: Explorer to sail through the Northeast Passage from Norway to the Bering Sea, *3836*
Nordenskjöld, Nils Otto Gustaf: Long-distance sledge journey on the Antarctic continent, *3843*
Norton, Charles Eliot: National park in the United States made up entirely of lands donated by private citizens, *3791*
Núñez Cabeza de Vaca, Álvar: European discovery of Iguaçu Falls, *4629*
Nuttall, Thomas: Botanist to explore the American West, *1776*

O

O'Brien, Timothy: Systematic survey of the endangered Asiatic cheetah, *2353*
Ogle, Douglas: Recovery plan approved by the U.S. Fish and Wildlife Service for a plant, *2336*
O'Hair, Ralph: Pilot to fly into a hurricane intentionally, *4252*
Old Blue (robin): Chatham Islands black robin to continue breeding to the age of 13, *1680*
Olds, Ransom Eli: Automaker to use an assembly line, *1590*
Olmsted, Frederick Law: Botanical garden in a city opened to the public in the United States, *3819*
Plan of significance for a suburban community in the United States, *3097*
Orata, Sergius: Oyster farms, *2488*
O'Reilly, John Boyle: American song written about a flood warning, *2530*
Owen, David Dale: Geologist hired by the U.S. government, *2804*
Owen, Richard: Scientific description of a moa, *2362*
Use of the word *dinosaur*, *2363*

P

Pack, Christopher: Geological map, *2782*
Page, George: Plow for pulverizing the soil, *1241*
Pallas, Pierre Simon: Geological survey of Russia, *2784*
Palmer, Nathaniel: Sighting of Antarctica, *3828*
Palmer, W. C.: Index to scientifically measure droughts, *2154*
Park, John Finley: *Tornado* designation publicly used by the U.S. Weather Bureau, *4289*
Park, William H.: Municipal public health laboratory in the United States, *3922*
Parsons, Charles A.: Turbogenerator to produce hydroelectric power, *4551*
Passel, Charles: Research on the interaction of wind and cold on the human body, *3859*
Wind chill calculations, *1941*
Peabody, Amelia: House completely heated by solar energy in the United States, *4107*
Peale, Charles Willson: Animal exhibits in a U.S. museum to be in natural settings and groups, *4884*
Museum exhibits with labels and descriptions, *4885*
Natural history museum foundation in the United States, *4883*
Natural history museum in North America, *4886*
Pearson, Allan: Scale to measure tornado intensities, *4293*
Pearson, Gerald Leondus: Photovoltaic cell or solar battery to work effectively, *4108*
Peary, Matthew E.: Explorers credited with reaching the North Pole, *3849*
Pelletier, Pierre-Joseph: Use of *chlorophyll* as a term to describe the green substance that colors plants, *1775*
Penn, William: City plan in America, *3093*
Pennekamp, John: Underwater state park in the United States, *3728*
Pennell, Rebecca Mann: Female college professor with the same rights and privileges as her male colleagues, *2264*
Peter I, the Great, Emperor of Russia: Modern "artificial" capital built in Europe, *3094*
Peterson, Roger Tory: Field guide to birds, *3253*
Pfeiffer, Ehrenfried: Municipal composting plant in the United States, *4041*
Philippa of Hainault: Female falconer of note, *2899*
Phillip, Arthur: Drought recorded in Australia, *2133*
Phillips, Eleazar: Agricultural book published in the United States, *3231*
Phillips, James *See* The Fox (activist)
Picasso, Pablo: Flood of significance in Spain during the 20th century, *2548*
Piccard, Auguste: Bathyscaph, *3576*
Manned descent to the greatest depth on earth, *3578*
Piccard, Jacques: Bathyscaph, *3576*
Manned descent to the greatest depth on earth, *3578*
Piddington, H.: Appearance of the word *cyclone*, *4243*
Pignataro, Domenico: Earthquake intensity scale, *2195*
Pike, Nicholas: English sparrows imported into the United States, *1506*
Pike, Zebulon: Explorer to describe the High Plains of the western United States as the Great American Desert, *2135*
Pinchot, Gifford: U.S. government conference devoted to conservation issues, *3365*
Use of the expression *wise use* in discussing U.S. resource issues, *3364*
Pittman, Key: Game preserve appropriation by Congress to assist state wildlife restoration projects, *4821*
Legislation for cooperative wildlife management, *4704*
Playfair, Lyon, 1st Baron: British government inquiry into the practicability of laws to control smoke, *1377*
Pliny the Elder: Catalog of significance of plants in the ancient world, *1748*
Observer known to write about the toxicity of lead, *2827*
Oyster farms, *2488*
Recorded use of the word *asbestos*, *1317*
Pliny the Younger: Volcano eruption whose devastation was fully documented, *4356*
Podolinsky, Alex: Organic farming organization in Australia, *1263*
Polo, Marco: Recorded mention of oil in Persia, *3605*
Report to Europe of vast coal deposits in China, *2022*
Pombal, Sebastião José de Carvalho e Mello, Marquês de: Scientific investigation of an earthquake, *2189*
Porsche, Ferdinand: Subcompact car designed to be fuel-efficient, *1596*
Pott, Percival: Recognition of cancer as an occupational hazard, *2969*
Potter, Beatrix: Female pioneer in the field of mycology, the study of mushrooms, *1789*
Poulter, Thomas Charles: Explosion seismology experiments in Antarctica, *3857*
Snow cruiser for the Antarctic research, *3860*

Powell, John Wesley: Detailed geologic and topographic information about the area that later became Canyonlands National Park, *3770*
Director of the U.S. Geological Survey (USGS), *2809*
Geological observations of the Grand Canyon, *2810*
Geologist/explorer to state flatly that much of the western United States was too dry to farm, *2138*
Non–Native Americans to enter what is now Dinosaur National Monument, *3740*
Powers, William T.: Hydroelectric power plant to furnish arc lighting service in the United States, *4548*
Preuss, Roger E.: Federal duck stamp in the United States to be priced at two dollars, *3185*
Federal duck stamp in the United States to feature a goldeneye, *3186*
Pritchard, C. G.: Federal duck stamp in the United States to feature a hooded merganser, *3205*
Procopius: Dam known to have a curved shape, *2084*
Putnam, Palmer Cosslet: Wind turbine, *4852*
Pytheas: First recorded polar expedition, *3823*
Tin mines in ancient Britain, *3294*

Q

Quayle, Edward: Rainfall predictions in Australia using atmospheric pressure, *1924*
Quimper, Manuel: Discovery by Europeans of Mount Baker in Washington, *3757*

R

Raffles, Sir Stamford: Zoo explicitly set up as a scientific institution, *4889*
Ramazzini, Bernardino: Identification of some disabling industrial diseases, *1354*
Ramses III, Pharaoh of Egypt: Recorded use of pigeons as messengers, *1643*
Rasputin, Valentin: Environmental movement of significance in the Soviet Union, *1065*
Ray, John: Book devoted to birds and written in English, *3230*
Botanical account, in detail, of the plants of North America, *1760*
Rayleigh, John William Strutt, Baron: Scientist to describe the type of earthquake surface wave that acts like waves caused by a pebble tossed into a pond, *2211*
Raymond, Eleanor: House completely heated by solar energy in the United States, *4107*
Reagan, Ronald: Ban on the use of lead pipes in public water systems in the United States, *4606*
Climate research agreement between the United States and Soviet Union, *1830*
Congressional attempt to re-establish the Office of Noise Abatement and Control (ONAC), *3451*
Government policy in the United States to deny aid to foreign family planning agencies that performed or promoted abortions as a method of population control, *3884*
Government report issued jointly by the United States and Canada to rate acid rain a serious environmental and transboundary problem for both countries, *1477*
Group of Ten meeting, *1078*
Law in the United States banning tin-based paint on boats, *2504*
Medical waste disposal rules set by the U.S. Environmental Protection Agency (EPA), *4149*
National Fishing Week in the United States, *2511*
Noise control agency in the United States, *3429*
Nuclear waste site evaluation, *3530*
Permanent ban on oil and gas leasing on U.S. wilderness land, *4666*
"Right to Know" law in the United States covering hazardous substances, *2986*
Substantial cuts in the budget of the U.S. Environmental Protection Agency, *1080*
U.S.–Canadian bilateral report on acid rain, *3042*
Use of solar energy at the White House, *4117*
Rebmann, Johannes: Descriptions of Mount Kilimanjaro and Mount Kenya to reach Europe, *4630*
Redfield, William: Observer to describe tropical storms as large vortices of air rotating around an "eye" of low pressure, *4239*
Reece, Maynard: Artist to win the U.S. federal duck stamp contest three times, *3199*
Federal duck stamp in the United States to feature a bufflehead, *3184*
Federal duck stamp in the United States to feature a cinnamon teal, *3208*
Federal duck stamp in the United States to feature a gadwall, *3188*
Federal duck stamp in the United States to feature a white-winged scoter, *3206*
Federal duck stamp in the United States to feature an image other than waterfowl alone, *3197*

Reed, Walter: Evidence conclusively showing that yellow fever was transmitted by a variety of mosquito, *3923*
Reid, Harry Fielding: Seismologist to advance the elastic rebound theory to explain earthquakes in tectonic, *2218*
Reinier of Liège: Coal use documented in Europe, *2021*
Renaud, Claude B.: Tests of leeches and mussels to monitor levels of freshwater contaminants, *4515*
Ressegue, H.: Fur-bearing animals raised commercially, *1161*
Revelle, Roger: Daily measurements of atmospheric carbon dioxide, *1839*
Global warming report of significance made to the U.S. government, *1841*
Global warming report to gain international scientific attention, *1838*
Rey, Augustin: Modern experiments on the effects of sunlight on buildings, *4103*
Rhodes, James A.: "Save Lake Erie" petition, *4465*
Richards, Ellen Swallow: Board of health established in the United States by a state, *3919*
Richardson, Lewis Fry: Long-range weather forecasting attempt using equations, *1930*
Outline of the mathematical concepts that would lead to numerical modeling for accurate weather forecasting, *1931*
Richter, Charles F.: Scale to accurately measure earthquake strength, *2223*
Scientist to use records of seismic waves to demonstrate the existence of the earth's core and measure its distance from the surface, *2219*
Rifkin, Jeremy: Boycott of genetically engineered food products in the United States, *2764*
Riley, James Whitcomb: Ban on swimming in Indiana's Brandywine Creek because of bacterial contamination, *4471*
Ripley, Aiden Lassell: Federal duck stamp in the United States to feature the widgeon, *3176*
Ritter, William E.: U.S. institute for study and research on oceans, *2269*
Robertson, A. Willis: Game preserve appropriation by Congress to assist state wildlife restoration projects, *4821*
Legislation for cooperative wildlife management, *4704*
Robinson, Benjamin Lincoln: Meeting to create the New England Botanical Club, *1788*
Robinson, Frederick John: Treaty to recognize the three-mile limit for fishing off the North American coast, *3015*
Robinson, John: Cup anemometer, *1893*
Roche, Patricia A.: Genetic privacy legislation proposal in the United States, *2766*
Rockefeller, David: Farmland of significance in New York State to be permanently protected, *1136*
Rockefeller, John D., Jr.: Restoration of a colonial-era town to museum status, *3108*
Rockefeller, Margaret: Farmland of significance in New York State to be permanently protected, *1136*
Rockwell, Llewellyn H.: Organized resistance to environmental regulations, *1025*
Rodale, J. I.: Serial publication in the United States devoted to organic agriculture, *1261*
Rodale, Robert: International conference on the connection between soil health and human health, *4027*
Introduction of the concept of "regenerative farming", *1254*
Rogers, William: Formal ascent of Devils Tower, *3743*
Rolls, Charles S.: Long-lasting partnership to build luxury cars in Great Britain, *1536*
Ronne, Edith: Permanent American base in Antarctica, *3861*
Women to spend the winter in the Antarctic, *3863*
Ronne, Finn: Permanent American base in Antarctica, *3861*
Roosevelt, Franklin D.: Annual license required of all U.S. waterfowl hunters 16 years of age or older, *2933*
Federal agency charged with preventing soil erosion, *4055*
Firearms tax in the United States to specifically benefit wildlife habitats, *4820*
Legislation that provided regular federal funding for waterfowl management in the United States, *1693*
Proposal to dig a canal across the Florida peninsula, *3071*
Public health grants from the U.S. government to states, *3929*
Two states to adopt President Franklin Roosevelt's Standard State Soil Conservation District Laws, *4039*
Roosevelt, Theodore: Authorization of the U.S. Forest Service to make arrests for violations of its regulations, *2679*
Bird reservation established by the U.S. government, *1719*
Boone and Crockett Club meeting, *1030*
Failed U.S. Rocky Mountain mule deer management policy, *4797*
Governors' conference on U.S. conservation issues, *1033*
Hydropower facilities operated by the U.S. Bureau of Reclamation, *4556*
Intergovernmental conservation conference in North America, *3367*

Land bought by the U.S. government specifically for wildlife, *4812*
National Conservation Commission in the United States, *1032*
National forest in the United States, *3714*
National monument in the United States created from land donated by a private individual, *3747*
National park established to protect a cave, *3787*
National Wildlife Refuge in the United States, *4809*
Natural national monument in the United States, *3746*
Person to write a book promoting the idea of a U.S. national parks system, *3248*
Pronghorn antelope bred and reared in captivity, *4746*
Tri-national conference on conservation in North America, *1034*
U.S. government conference devoted to conservation issues, *3365*
U.S. National Wilderness Area to enclose a bison sanctuary, *4650*
Rosen, Walter: Use of the term *biodiversity*, *1636*
Ross, James Clark: Volcanoes discovered in Antarctica, *4369*
Ross, John: Evidence of life on the sea floor, *3550*
Rowland, F. Sherwood: Nobel Prize was awarded to ozone-layer researchers, *1972*
Rowland, Sherry: Ozone-layer data concerning damage by chlorofluorocarbons (CFCs), *1974*
Rowland, Thomas F.: Oil drill offshore rig, *3650*
Royce, Frederick H.: Long-lasting partnership to build luxury cars in Great Britain, *1536*
Ruckelshaus, William D.: Cabinet-level regulatory environmental agency in the United States, *1067*
U.S. government agency to adopt a formal environmental policy with American Indian tribes, *3391*
Ruffin, Edmund: Planter to recognize that some plants could put acid in the soil, *4030*
Rumboll, Mauricio: Hooded grebe to be discovered, *1672*
Rush, Richard: Treaty to recognize the three-mile limit for fishing off the North American coast, *3015*
Russell, F. A. R.: Air pollution book to dramatize London's exploding 19th-century air pollution problem, *3246*

S

Saffir, Herbert: Scale widely used to measure hurricanes, *4263*
Sage, Rufus: Account of the wonders of the Rocky Mountains to reach the East Coast of the United States, *3239*
Salk, Jonas: Mass polio vaccination campaign, *3937*
Salt, Henry: Use of the term *animal rights*, *1012*
Samuel, Marcus: Oil tanker built for long-distance conveyance, *3613*
Sandstrom, H. D.: Federal duck stamp in the United States to feature a ring-necked duck, *3191*
Sanger, Margaret: Birth control clinic in the United States, *3879*
International conference to address world population problems, *3896*
Sarney, José: Formal promise from Brazil to take steps to protect its rain forests, *2721*
Saussure, Horace Bénédict de: Experiments to measure the amount of heat trapped by glass, *4088*
Extensive geological study of the Alps, *2787*
Hygrometer utilizing human hair, *1877*
Solar collector, *4089*
Use of the term *geology*, *2785*
Savery, Thomas: Steam engine of high pressure used in a mine, *3300*
Suction pump to draw water from coal mine shafts, *2027*
Savoia-Aosta, Luigi Amedeo di: Ascent of Mount Elias, *3784*
Scalia, Antonin: Scholarly article to identify a legal trend curbing the powers of environmental groups and private citizens to sue polluters in U.S. courts, *1093*
Schaefer, Vincent J.: Artificial snowstorm, *1947*
Meteorologist to reproduce natural snow crystals in plastic, *4205*
Schaller, George B.: Significant studies of lions in the wild, *4725*
Systematic survey of the endangered Asiatic cheetah, *2353*
Schenck, Carl Alwin: "Master school" for forestry training in the United States, *2695*
Scherer, Reed: Direct evidence that the West Antarctic Ice Sheet was once open sea, *3874*
Schleinden, Matthias Jakob: Presentation of the cell theory, *1780*
Schlink, Frederick John: Major attack on the U.S. consumer products industry, *3251*
Schmitt, Harrison Hogan: Geologist to go to the Moon, *2825*

Schomburgk, Hans Hermann: Pygmy hippos known to Western science, *4695*
Schopf, Johann David: Systematic work on American geology, *2786*
Schroder, Gerhard: National Green party to become part of a governing coalition in Europe, *1119*
Schwann, Theodore: Presentation of the cell theory, *1780*
Schwartz, Marie Beatrice: Dutch elm disease description, *2600*
Schwender, Simon: Research showing that lichen are two organisms, *1785*
Scott, Robert Falcon: Aerial photograph in Antarctica, *3845*
Landing on Antarctica that initiated the "Heroic Age" of Antarctic exploration, *3839*
Polar explorer to ascend in a balloon in the Antarctic regions, *3846*
Scientific exploration of significance of Antarctica, *3844*
Scribner, J. W.: Appearance of the Scribner Log Rule for determining board-foot measurements of timber, *2652*
Seagers, C. B.: Federal duck stamp in the United State to feature a blue-winged teal, *3190*
Seal, Ulysses: International organization to manage breeding animals in captivity, *4946*
Selengut, Stanley: Resort powered by renewable energy, *4301*
Selikoff, Irving: Conclusive link between asbestos and lung cancer, *1348*
Semon, Waldo L.: Polyvinyl chloride (PVC), *2971*
Seneca, Lucius Annaeus, the Younger: Record of transparent glass windows in the West, *4082*
Shackleton, Ernest Henry: Aerial photograph in Antarctica, *3845*
Motorized transportation in a polar region, *3848*
Shauck, Maxwell: Transatlantic flight powered by ethanol, *4334*
Sheldrick, Daphne: Elephant and rhinoceros orphanage in Kenya, *4755*
Shipman, Herbert C.: Nene (Hawaiian goose) breeding program, *2426*
Shirase, Nobu: Japanese Antarctic expedition, *3851*
Shulgi, King of Ur: Large carnivores to be held in captivity, *4874*
Shuttleworth, John: Publication inspired by *The Whole Earth Catalog*, *3259*
Silkwood, Karen: Jury award of damages in the U.S. to a victim of radiation contamination from a nuclear facility, *3513*
Silliman, Benjamin: Geological organization in the United States, *2799*
Geological periodical still published today, *2798*
Silliman, Benjamin, Jr.: Published research suggesting a wide range of uses and products for distilled petroleum, *3682*
Simpson, James: Water filtration system, *4574*
Simpson, Robert: Scale widely used to measure hurricanes, *4263*
Sinclair, Mary: Environmental activist to become a member of a women's hall of fame, *1045*
Sinclair, Upton: Major exposé of unsanitary conditions in the U.S. meatpacking industry, *1168*
Serialization of Upton Sinclair's *The Jungle*, *3249*
Singer, Peter: Activist to popularize the term *speciesism*, *1019*
Siple, Paul: Research on the interaction of wind and cold on the human body, *3859*
Wind chill calculations, *1941*
Sixtus IV, Pope: Papal declaration regarding world exploration, *3007*
Slater, Samuel: Water-powered cotton mill in an American colony, *4545*
Sloan, Howard: Completely sun-oriented residential community in the United States, *4105*
Smith, J. D.: Arbor Day celebration, *2724*
Smith, Sir James Edward: Botanical society formed that is still in existence today, *1769*
Smith, Robert Angus: Reintroduction of the term *acid rain*, *1469*
Scientist to systematically study dispersal of industrial age air pollutants, *1436*
Treatise on the phenomenon of acid rain, *1437*
Use of the term *acid rain*, *1818*
Smith, William: Faunal succession theory, *2788*
Modern geological map, *2795*
Smith, William Alden: Game warden in the United States to be paid a salary, *4688*
Year in which salaried game wardens were appointed in the United States, *2914*
Smokey (bear): Environmental public service symbol of the U.S. government, *2618*
Snow, John: Researcher to prove the transmission of cholera via a contaminated water supply, *4420*
Somers, Sir George: Hurricane that caused islands to be settled, *4229*
Soyer, Alexis: Use of natural gas for cooking, *2043*
Spauling, Elizabeth: Women to pass along the Oregon Trail, *3765*
Speckle (goose): Goose to lay an egg weighing 1.5 pounds, *1675*

Spemann, Hans: Zoologist to win a Nobel Prize, *2280*
Spruce, Richard: Cinchona trees in India, *1507*
Staehle, Albert: Environmental public service symbol of the U.S. government, *2618*
Public service symbol created for the U.S. government, *3179*
Stalin, Joseph: Drought used for political purposes, *2145*
Widespread deaths of trees in Moscow attributed to air pollution, *1320*
Stauffer, Cheryl A.: Woman to graduate from the Pennsylvania School of Conservation, *2303*
Stearns, Stanley: Federal duck stamp in the United States to feature a blue goose, *3192*
Federal duck stamp in the United States to feature a Hawaiian nene goose, *3202*
Federal duck stamp in the United States to feature a whistling swan, *3203*
Steel, J. E.: Use of the term *bionics*, *2285*
Steenstrup, Johann Japetus: Description of the giant squid, *2451*
Steiner, Rudolf: Important scientific discussion of the negative impact of chemical fertilizers on food crops, *1258*
Municipal composting plant in the United States, *4041*
Organic farming organization in Australia, *1263*
Steinmetz, Charles P.: Warning of air pollution dangers from coal-fired electrical generation, *1337*
Steller, Georg Wilhelm: Animal hunted to extinction in North America, *2358*
Spectacled cormorant description, *2357*
Steno, Nicolaus: Formulation of the Principle of Original Horizontality, *2779*
Stratigraphy theory, *2781*
Stevens, Christine: Federal law in the United States closely regulating the use of animals in medical and commercial research, *1017*
Stevens, Hazard: Documented ascent of Mount Rainier, *3771*
Stewart, Timothy: Genetically engineered mammal to be patented, *2762*
Stiger, Albert: Attempts to cause rain with gun blasts into vapor-laden clouds, *2168*
Use of the Stiger gun as a defense against hail, *4215*
Stokes, Sir George: Instrument to accurately measure the length of time the Earth's surface receives sunshine, *1901*
Stopes, Marie Charlotte Carmichael: Woman on the science staff of Manchester University, *1791*
Strabo: Precise description of an earthquake, *2175*
Stuart, Robert: Recorded description of the area that would become the Fort Laramie National Historic Site at the mouth of the Laramie River, *3762*
Stuyvesant, Peter: Littering law in New York City, *3267*
Municipal garbage dumps in New York City, *4139*
Suess, Hans Eduard: Scientist to use the term *biosphere*, *1618*
Sullivan, Roy C.: Person known to survive being struck by lightning seven times, *4276*
Sullivan, William N.: Aerosol spray containers for insecticide that were portable, *1321*
Svobida, Lawrence: Edition of *Empire of Dust* by Lawrence Svobida, *3254*
Swallow, Ellen: Standards for water quality in the United States, *4421*
Use of the term *oekology* (ecology) in the United States, *1052*
Water purity tables, *4419*

T

Taft, William Howard: Automobile in the area of Mt. Rainier National Park, *3785*
Legislation to protect migratory game and insectivorous birds, *1689*
Tansley, Arthur George: Scientist to use the term *ecosystem*, *1621*
Society of professional ecologists, *2273*
Use of the term *ecosystem*, *1056*
Tarquinius Priscus, Lucius, King of Rome: Sewers of Rome, *4518*
Taylor, Ed: Law in the United States designed to stop overgrazing and its adverse effects on the soil, *4036*
Taylor, Michael Angelo: Campaign of significance against air pollution in London, *1319*
Taylor, Zachary: Cabinet-level agency concerned primarily with natural resource development and conservation in the United States, *3079*
Teisserenc de Bort, Léon: Stratosphere and troposphere discoveries, *1921*
Telkes, Maria: House completely heated by solar energy in the United States, *4107*
Tellier, Charles: Nonreflecting solar motor, *4096*
Thatcher, Margaret: Conference held after the signing of the Protocol on Substances That Deplete the Ozone Layer (Montreal Protocol), *1990*
Thayer, Helen: Person to circumnavigate the North Pole on foot and solo, *3871*
Woman to travel solo to either pole, *3872*

Theophrastus: Botanical study of significance, *1746*
Mineralogy book, *2775*
Thom, Earl C.: Index for temperature-humidity, *1955*
Thomas, Jack Ward: Administrator named to head the U.S. Forest Service who was not a career bureaucrat or a timber planter, *2708*
Thomas, Robert Bailey: Almanac with a continuous existence in the United States, *3234*
Thompson, David: Person to write about what later became Waterton-Glacier International Peace Park, *3756*
Thompson, Tommy G.: Car fueled by methanol, *4336*
Thompson, William: Underwater photograph, *3558*
Thoreau, Henry David: American naturalist to suggest that every town preserve a "primitive forest", *2710*
Environmental movement of significance in the United States, *1047*
Tiberius, Emperor of Rome: Greenhouses known in the West, *4079*
Timblon, Carlos: Ship to bring home Antarctic sealskins, *3829*
Todd, Charles: Drought patterns discovered in the Southern hemisphere, *2139*
Tohui (panda): Giant panda born outside of China, *4757*
Toit, Alexander du: Biological evidence that continents were once joined, *2819*
Tolmie, William: Non–American Indian to visit the area that later became Mount Rainier National Park, *3763*
Torres, David: Felony indictments under a U.S. environmental law for smuggling hazardous waste from California into Mexico, *2889*
Torricelli, Evangelista: Instrument to measure atmospheric pressure, *1863*
Towneley, Richard: Person to keep a continuous record of rainfall in England, *1867*
Rain gauge, *1866*
Tradescant, John: Public museum in Great Britain, *4881*
Treviranus, Gottfried Reinhold: Use of the term *biology*, *2259*
Trevithick, Richard: Automobile using steam power in England, *1587*
Trudeau, Pierre Elliott: U.S.–Canadian effort of significance to improve the water quality of the Great Lakes, *3029*
Truman, Harry S.: Ban of aircraft access to Superior National Forest, *4639*
U.S. Fish and Wildlife Service jurisdiction over sockeye salmon conservation, *2459*
U.S. National Science Foundation (NSF), *2283*
Truvelot, E. Leopold: Introduction of the gypsy moth in the United States, *1511*
Trzecieski, Titus: Oil wells in Europe, *3680*
Tulaczyk, Slawek: Direct evidence that the West Antarctic Ice Sheet was once open sea, *3874*
Turesson, Göte: Use of the term *genecology* to describe the study of genetic variations within individual plant species as they relate to the environment, *1793*
Turner, Frederick Jackson: Declaration that the American frontier had closed, *3140*
Turner, William: Book devoted entirely to birds, *3228*

U

Ui-te-Rangiora: Sighting of Antarctica, *3828*
Unwin, Raymond: Use of the term *green belt*, *3100*
Urquhart, Fred A.: Discovery of the monarch butterfly's winter home, *4729*
Ussher, James: Dating of the origin of Earth through biblical analysis, *2778*

V

Van Allen, James Alfred: Upper-atmosphere study with a rocket launched from a balloon, *1951*
van Straten, Florence W.: Cloud-seeding demonstration, *2170*
Van Trump, Philemon: Documented ascent of Mount Rainier, *3771*
Vancouver, George: Non-Hawaiian to climb Mauna Loa in the area that later became Volcanoes National Park, *3759*
Vanderbilt, George Washington: Forest management on a professional scale, *2675*
Vázquez de Coronado, Francisco: European to see the Grand Canyon, *3753*
Monument to commemorate the first major exploration of the American Southwest, *3748*
Vernadsky, Vladmir: Scientist to use the term *biosphere*, *1618*
Victoria, Queen of Great Britain: Meteorology society in Great Britain continuing through the 20th century, *1899*
Vidie, Lucien: Aneroid barometer, *1892*
Vignes, Jean Louis: Person to establish a vineyard in San Francisco, CA, *1500*
Vine, Frederick: Plate tectonics theory verification, *2821*

Vogt, William: Population control book published after World War II, *3255*
Vonnegut, Bernard: Artificial snowstorm, *1947*

W

Wakefield, Priscilla: Introduction to the botanical sciences written by a woman, *1771*
Walker, Hovenden: Storm to defeat a naval invasion of Canada, *4178*
Walsh, Don: Manned descent to the greatest depth on earth, *3578*
Walters, Charles: Issue of the periodical *Acres U.S.A.*, *4042*
Wankel, Felix: Automobile rotary engine, *1545*
Ward, Nathaniel B.: Terrarium, *1779*
Ward, Robert DeCourcy: Climatology professor at a U.S. college, *2271*
Waring, George Edwin: Rubbish-sorting facility in New York City, *4126*
Sewage-disposal system in the United States separate from the city water system, *4526*
Warren, Elizabeth: Botanical chart to be published, *1781*
Washington, George: Highway built in the United States entirely with federal funds, *1552*
Partridge propagation in the United States, *2930*
Washington, Warren M.: African-American organization to focus on pollution and hazardous waste in minority and poor communities, *2886*
African-American president of the American Meteorological Society, *1834*
Waterhouse, Benjamin: College instruction course in mineralogy, *2258*
Watson, James: Accurate DNA structure description, *2753*
Watson-Watt, Robert Alexander: Radar detection of air echoes from the lower atmosphere, *1938*
Watt, James: Coal-burning steam engine, *2030*
Gas used to light a house, *2032*
Patent on a device to enable a steam engine to consume its own smoke, *1433*
Watt, James G.: Permanent ban on oil and gas leasing on U.S. wilderness land, *4666*
Weber, Walter A.: Federal duck stamp in the United States to feature a trumpeter swan, *3187*
Federal duck stamp in the United States to feature the white-fronted goose, *3178*
Two-time winner of the U.S. federal duck stamp contest, *3180*
Weddell, James: Man to reach 74 degrees 15 minutes south latitude, *3834*
Weeks, John Wingate: Bird hunting regulation enacted by the U.S. Congress, *2955*
Wegener, Alfred: Biological evidence that continents were once joined, *2819*
Cohesive theory of the concept of "continental drift", *2817*
Theory of continental drift, *2217*
Wen-Wang, Ruler of Chou, in China: Chinese animal park, *4872*
Wetherill, Charles Mayer: Agriculture Bureau scientific publication, *3243*
Wetmore, Alexander: Bird banding by a U.S. government agency, *1734*
Whitcomb, John C., Jr.: Widely publicized version of the Christian fundamentalist theory of flood geology, *2547*
White, Caroline Earle: National animal rights organization in the United States, *1010*
White, Gilbert: Textbook on ecology, *3233*
White, Judith: Learning lab in a national zoo, *4949*
White, Lynn Townsend, Jr.: Presentation of the essay "The Historical Roots of Our Ecological Crisis" by historian Lynn Townsend White Jr., *3257*
Whitebread, Samuel Charles: Meteorology society in Great Britain continuing through the 20th century, *1899*
Whitman, Narcissa: Women to pass along the Oregon Trail, *3765*
Whitney, Eli: Widespread use of the cotton gin, *1236*
Whitney, Josiah Dwight: Scientific exploration of the area that later became Sequoia and Kings Canyon National, *3769*
Wilkes, Charles: Antarctic expedition funded by the U.S. Congress, *3835*
Wilkins, A. F.: Radar detection of air echoes from the lower atmosphere, *1938*
Wilkins, George Hubert: Flight across the Arctic Ocean, *3855*
William II, King of England: Royal menagerie known to exist in England, *4879*
Williams, Leonard: Woolly monkeys successfully bred in captivity, *4753*
Williamson, John E.: Underwater movie seen by the public, *3568*
Willkins, James Miller: Oil well drilled intentionally and successfully in North America, *3684*
Willsie, Henry E.: Solar power plant using a low-boiling-point liquid to power an engine, *4099*
Solar-powered energy plant capable of operating at night, *4100*
Willughby, Francis: Book devoted to birds and written in English, *3230*
Wilmut, Ian: Mammal to be cloned from an adult cell, *2769*

Wilson, Alexander (astronomer): Upper-air temperature measure of significance, *1874*
Wilson, Alexander (ornithologist): Breeding bird census, *1729*
Wilson, Des: Unleaded gasoline campaign in Great Britain, *1549*
Wilson, Edmund Beecher: Biology course offered at a U.S. college, *2267*
Wilson, Edward: Emperor penguin egg collected for science, *1735*
Wilson, Edward O.: Use of the term *biodiversity*, *1636*
Wilson, Woodrow: Bird protection treaty, *3017*
Highway program of significance in the United States granting federal funds for state roads, *1557*
International treaty to protect birds, *1743*
National park east of the Mississippi River and the first located on an island, *3792*
Successful challenge to state-ownership doctrine, *1690*
Windsor, Frank E.: Reservoir to cover four towns and affect seven others in the United States, *2101*
Winslow, Edward: Cattle imported into a North American colony, *1156*
Winthrop, John, IV: Description of an earthquake or aftershock as a wave, *2190*
Wolf, Hazel: International meeting of American Indians and environmentalists, *4408*
Wolke, Howie: Radical environmental activist arrested and convicted for "monkeywrenching" in the United States, *1084*
Woodwell, George: Global warming report of significance made to the U.S. government, *1841*
Wragge, Clement L.: Meteorologist to name hurricanes, *4245*
Wright, Orville: Monument erected to commemorate the first flight by an airplane, *3745*
Wright, Wilbur: Monument erected to commemorate the first flight by an airplane, *3745*

X

Xeng Li (panda): Giant panda born outside of China, *4757*

Y

Yard, Robert Sterling: Conservation group to focus on the eastern United States in addition to western states, *1035*
Private organization to support national parks in the United States, *3718*
Ying Ying (panda): Giant panda born outside of China, *4757*
Young, Brigham: Monument to a bird in the United States, *3163*
Young, James: Modern oil refining processes, *3678*
Young, Mahonri Mackintosh: Monument to a bird in the United States, *3163*
Yount, Harry: National park in the United States, *3772*
Ranger hired in Yellowstone National Park, *3777*

Z

Zannin, Grazia: Transatlantic flight powered by ethanol, *4334*
Zanthier, Hans Dietrich von: School of forest practice, *2732*
Zatz, Daniel: Cyber animal exhibit that allows people to view bears in their natural habitat, *4968*
Zedillo, Ernesto: Biodiversity governmental agency in Mexico, *1641*
Zhang Heng: Seismograph, *2177*

Geographical Index

The Geographical Index is a listing of key locations in the main text of the book, arranged alphabetically by nation, state or province, and city. Each index entry includes key information about the "first" and a 4-digit number in italics. That number directs you to the full entry in the main text, where entries are numbered in order, starting with 1001.

To find the full entry, look in the main text for the entry tagged with that 4-digit number.

Under each country's name in this index, entries that contain only the country's name are listed first. Following those are entries listed by state or province, and then by city, alphabetically.

Note that locations are generally identified by their modern names, i.e., Sri Lanka rather than Ceylon. However, in some cases, particularly those pertaining to the ancient world, entries are also listed under the name of the place at the time of the event—under Persia as well as Iran; under Carthage as well as Tunisia. Entries pertaining to ancient Rome are listed under Rome, Ancient as well as Italy/Rome, and there are also separate listings for the Byzantine Empire and the Holy Roman Empire, which have no precise modern equivalents.

Events that occured in the Soviet Union during the years when there was a Soviet Union will be listed under that name. Entries pertaining to England, Scotland, Wales, and Northern Ireland are to be found under the heading Great Britain.

In addition, because of the nature of the subjects covered here, there are numerous headings for geographical areas, such as Antarctica, Mesopotamia, and North America, as well as headings for Sea and Space.

For more information see "How to use this book," on page ix.

ADMIRALTY ISLANDS

Manus Island

Snail listed under the U.S. Endangered Species Act, *2399*

AFGHANISTAN

Organization for the control of desert locusts, *3059*

AFRICA

Biological evidence that continents were once joined, *2819*
Description of chimpanzees, *4679*
Drought aid from a global rock music festival, *2159*
Drought to draw worldwide aid through television, *2158*
Emergence of desertification as an international issue, *4044*
Gorilla listing under the U.S. Endangered Species Act, *2395*
International campaign to eradicate malaria, *3939*
International treaty to propose African wildlife preserves, *3019*
Introduction of manioc into Africa from Brazil, *1204*
Organization of African Unity (OAU) population agency, *3899*
Retreat of human settlement from the North African and Arabian deserts, *2008*
Rhinoceros listings under the U.S. Endangered Species Act, *2398*
United Nations convention on African natural resources, *3378*
United Nations educational center for African natural resource management, *3392*
Use of the term *desertification*, *4043*
Wildlife protection conference of modern African governments, *4773*

ALGERIA

Nuclear weapons test conducted by the French, *3492*

ANATOLIA

Flood known in Asia Minor, *2516*
Smelting of copper, *1329*

ANGOLA

Malange Province

National preserve created to protect the giant sable antelope, *4824*

ANTARCTICA

Aerial mapping photographs over Antarctica, *3854*
American scientist to go to Antarctica, *3833*
Antarctic expedition funded by the U.S. Congress, *3835*
Climate research agreement between the United States and Soviet Union, *1830*
Commercial whale-watching ventures, *2239*
Direct evidence that the West Antarctic Ice Sheet was once open sea, *3874*
Emperor penguin egg seen and collected by humans, *1649*
Evidence of prehistoric vertebrate life in Antarctica, *3866*
Expedition to survive a winter in the Antarctic, *3840*
Explorers to reach the South Pole, *3852*
Explosion seismology experiments in Antarctica, *3857*
Fossil found in Antarctica, *3867*
Hole in the earth's ozone layer, *1984*
International convention attempting to regulate Antarctic mineral exploitation, *3044*
International convention to specifically protect Antarctic seals, *3031*
International Polar Year, *3837*
International treaty regarding the use of Antarctica, *3025*
Long-distance sledge journey on the Antarctic continent, *3843*
Meteorite discovered in Antarctica, *3853*
Motorized transportation in a polar region, *3848*
Ozone depletion of major significance over Antarctica, *1980*
Ozone pollutant agreement on an international level, *1987*
Person to cross the Antarctic Circle, *3826*
Person to see the Antarctic territories of Peter I Island and Alexander I Island, *3831*
Polar explorer to ascend in a balloon in the Antarctic regions, *3846*
Radiometric survey in Antarctica, *3869*
Research on the interaction of wind and cold on the human body, *3859*
Scientific exploration of significance of Antarctica, *3844*
Scientific record of an Antarctic winter, *3842*
Scientist to survive the eruption of an Antarctic volcano, *3870*
Ship to bring home Antarctic sealskins, *3829*
Sighting of Antarctica, *3828*
Significant theory that Antarctica was a continent rather than a series of islands, *3841*
Specific treaty guidelines for Antarctic ecosystem protection, *3046*
Swans sighted in the Antarctic, *1710*
Tourists to visit the continent of Antarctica, *2240*
Volcanoes discovered in Antarctica, *4369*
Wind chill calculations, *1941*
Woman to direct an Antarctic research station, *3868*
Woman to visit Antarctica, *3858*
Year the normal ozone level will return to the stratosphere, *1998*

Alexander Island

Report of a shorebird at latitude 71 degrees south, *1704*

Antarctic Peninsula

High altitude meteorological station in Antarctica, *3862*

Bay of Whales

Airplane flight over the South Pole, *3856*
Japanese Antarctic expedition, *3851*

Cape Adare

Landing on Antarctica that initiated the "Heroic Age" of Antarctic exploration, *3839*

Cape Crozier

Emperor penguin egg collected for science, *1735*

Commonwealth Bay

Radio station in Antarctica, *3850*

Hughes Bay

Explorer to land on Antarctica, *3832*

Kainan Bay

International Antarctic meteorological station, *3865*

Kerguelen Islands

Rabbits in Antarctica, *1513*

King George's Island

Men to winter in the Antarctic, *3830*

McMurdo Sound

Aerial photograph in Antarctica, *3845*
Nuclear power plant in Antarctica, *3497*
Scientific evidence that humans caused the Antarctic ozone hole, *1988*

Seymour Island

Wood fossils found on the Antarctic, *3838*

South Georgia Island

Commercial sealing in the Antarctic region, *3827*
Magellan Goose on South Georgia Island, *1518*
Whaling station in the Antarctic, *3847*

Stonington Island

Permanent American base in Antarctica, *3861*
Women to spend the winter in the Antarctic, *3863*

Weddell Sea

Man to reach 74 degrees 15 minutes south latitude, *3834*

ARABIAN PENINSULA

Evidence that dromedary (one-humped) camels were domesticated in Arabia, *1146*

ARCTIC

Activities of the Arctic Chemical Network, *1472*
American steam whaler, *4626*
Arctic Environmental Protection Strategy (AEPS), *3873*
Bird recorded at the North Pole, *1709*
Climate research agreement between the United States and Soviet Union, *1830*
Europeans known to have wintered in the extreme north, *3824*
Explorers credited with reaching the North Pole, *3849*
First recorded polar expedition, *3823*
Flight across the Arctic Ocean, *3855*
International Polar Year, *3837*
Notice of the wintertime phenomenon known as "Arctic haze", *1468*
Oil-pipeline spill of significance in Russia publicly reported, *3648*
Permanent forum for coordinating sustainable development in the Arctic, *4319*
Person to circumnavigate the North Pole on foot and solo, *3871*
Reactor meltdown aboard a nuclear-powered ship, *3504*
Soviet law to ban polar bear hunting, *2956*
Tidal power plant north of the Arctic Circle, *4563*
Treaty protecting the polar bear and its natural habitat, *3034*
Weather satellite monitoring station north of the Arctic Circle, *1962*
Woman to fly over the North Pole, *3864*

Bering Sea

Animal hunted to extinction in North America, *2358*
Explorer to sail through the Northeast Passage from Norway to the Bering Sea, *3836*
Spectacled cormorant description, *2357*

ARGENTINA

European discovery of Iguaçu Falls, *4629*
Evidence that the "ozone hole" had expanded outward to include New Zealand, Australia, Argentina, and Chile, *1480*
Known bird with a wingspan of 25 feet, *1676*
Natural resource exploited in Argentina, *3342*
United Nations convention to protect the vicuña, *4777*

Buenos Aires

Ship to bring home Antarctic sealskins, *3829*

Patagonia

Laguna Las Escarchadas

Hooded grebe to be discovered, *1672*

ARMENIA

Earthquake in the Soviet Union allowed open media coverage, *2232*

ASIA

Climate change to lead to the peopling of a continent, *2001*
Extensive collection of quality data on fertility and family planning, *3882*
Humans to cross the Bering Land Bridge from Asia to North America, *3750*
International campaign to eradicate malaria, *3939*

ASIA—*continued*
Lion listing under the U.S. Endangered Species Act, *2396*

ASSYRIA

Evidence of humans using birds to hunt, *2896*

AUSTRALIA

Australian Heritage Commission (AHC) Act, *4658*
Australian indigenous species habitat listed as endangered, *2424*
Biological evidence that continents were once joined, *2819*
Biosphere reserves in Australia, *1629*
Cattle and sheep in Australia and New Zealand, *1492*
Compensation payments by the Australian government to Aborigines whose tribal lands were used for British nuclear weapons testing, *3525*
Drought patterns discovered in the Southern hemisphere, *2139*
Drought policy by Australia's national, state, and regional governments, *2164*
Drought recorded in Australia, *2133*
Duck-billed platypus bred in captivity, *4750*
Duck-billed platypus successfully transported outside Australia, *4699*
Evidence of long-term cooling and drying of Australia, *2005*
Evidence that the "ozone hole" had expanded outward to include New Zealand, Australia, Argentina, and Chile, *1480*
Global list of endangered plants, *2421*
International Adopt-a-Highway Day, *3290*
International organization to manage Antarctic wildlife, *4783*
Local environmental political parties, *1105*
Major destruction by starfish of coral in the Great Barrier Reef, *3579*
National Wilderness Inventory of Australia, *4668*
Naturalist to describe the wildlife of Australia, *4682*
Nuclear power organization of the Australian government, *3463*
Prickly pear in Australia, *1502*
Rainfall predictions in Australia using atmospheric pressure, *1924*
Symptoms of nutrient deficiencies in exotic pine plantations in Australia, *2740*
World Heritage Sites in Australia, *4000*

New South Wales

Newcastle

Coal discovery in Australia, *2031*
Coal exported from Australia, *2033*

Sydney Cove

Mining by European settlers in Australia, *3305*

Northern Territory

Darwin

Solar automobile race, *4121*

Queensland

Meteorologist to name hurricanes, *4245*
Report of the extinction of the paradise bird of Australia, *2369*

Brisbane

Attempts to cause rain with gun blasts into vapor-laden clouds, *2168*

Heron Island

Coral reef field research station in Australia, *3575*

Southport

Ecotourism research center in Australia, *2254*

South Australia

Adelaide

Lead mine in Australia, *3310*
Solar automobile race, *4121*
Use of compressed natural gas to power a ship, *2061*

Kangaroo Island

Emus on Kangaroo Island, *1521*

Kupunda

Copper mine in Australia, *3317*

Port Augusta

Ostrich in southeastern Australia, *1512*

Tasmania

National Threatened Species Day in Tasmania, *2350*

Hobart

Political party to take an environmental issue to the polls, *1108*

Macquarie Island

Radio station in Antarctica, *3850*

Victoria

Geelong

Rabbits in Australia, *1508*

Melbourne

International Antarctic meteorological station, *3865*

Powelltown

Organic farming organization in Australia, *1263*

Red Cliff

Ostrich in southeastern Australia, *1512*

Western Australia

Esperance

Wind farm in Australia, *4859*

AUSTRIA

Effective international action to curb acid rain, *1414*
Release of imported gypsy moth–killing nematodes in North America, *2607*
Scientist to use the term *biosphere*, *1618*

Salzburg

Sustainable development accord on the Alps, *4309*

Steinmark

Use of the Stiger gun as a defense against hail, *4215*

Vienna

Atomic energy organization of the United Nations for peaceful nuclear power, *3478*
Cosmic radiation discovery, *1925*
International forestry research agency, *2694*
International organization to promote the timber trade, *2702*
Meteorological organization on an international scale, *1914*
Surviving zoo, *4882*

AZERBAIJAN

Baku

Oil pipeline in Russia, *3614*
Oil well with a modern design, *3677*
Oil wells in Persia, *3672*
Recorded mention of oil in Persia, *3605*

BAHAMAS

Underwater movie seen by the public, *3568*

BANGLADESH

Cyclone to cause catastrophic loss of life in Bangladesh, *4268*
Monsoon in Bangladesh to cause $1 billion in damage, *4189*
Rhinoceros listings under the U.S. Endangered Species Act, *2398*

BARBADOS

Hurricane in the Atlantic causing massive deaths, *4238*

BELARUS

World Heritage sites in Europe chosen for environmental reasons, *3996*

BELGIUM

Flood leaving 1 million people homeless in the Netherlands, *2544*
International treaty to propose African wildlife preserves, *3019*
Major land reclamations in low-country Flanders and Zeeland, *3067*
Observation of the disease known as damping off, *2589*
Politician from a Green party to be elected to a national parliament, *1111*
Severe winter said to have introduced a new artistic approach, *1800*
Widespread cultivation of legumes that fixed nitrogen and thus improved soil fertility, *4058*

Brussels

Identification of the Congo peacock, *1697*

Liège

Coal use documented in Europe, *2021*

BELIZE

Central American coral reef reserve, *3594*
Volcano eruption mistaken for a war, *4368*

Cockscomb Basin

Jaguar preserve, *4838*

BERMUDA

Bathysphere, *3572*
Endangered species laws in the Western Hemisphere, *2319*
Hurricane that caused islands to be settled, *4229*
Introduction of the potato into North American colonies, *1205*

Castle Rocks

Report of the reappearance of the cahow, or Bermuda petrel, *1657*

BOLIVIA

United Nations convention to protect the vicuña, *4777*

La Paz

Debt-for-nature swap, *3393*
National population guidelines for Bolivia, *3888*

Laguna Colorado

Colony of puna flamingos to be discovered, *1700*

Potosi

Natural resource exploited in Bolivia, *3343*

BOTSWANA

Transfrontier wildlife park in Africa, *4845*

BRAZIL

Death attributed to a "killer bee" attack in the United States, *3968*
Environment minister in the ecologically critical country of Brazil, *2722*
Ethanol promotion program in Brazil, *4328*
European discovery of Iguaçu Falls, *4629*
Experimental rubber plantations in Malaya, *1180*
Formal promise from Brazil to take steps to protect its rain forests, *2721*
Introduction of manioc into Africa from Brazil, *1204*
"Killer bees" in the United States, *1529*
Large-scale mercury pollution of streams in northern Brazil, *4440*
Natural resource exploited in Brazil, *3344*
Wildlife preserve to protect golden lion tamarins, *4832*

Bahia

Mangos in the Western Hemisphere, *1493*

Brasilia

Announcement that Brazil would transfer its capital from Rio de Janeiro to a new city to be built in the undeveloped interior, *2717*

Rio de Janeiro

Announcement that Brazil would transfer its capital from Rio de Janeiro to a new city to be built in the undeveloped interior, *2717*
Coffee plantations in Brazil, *1178*

Amazon Basin

Decision to build a highway into Brazil's Amazon wilderness, *2718*
Program opening the Amazon region to economic development, *4299*
Woolly monkeys successfully bred in captivity, *4753*

BURMA

Widespread teak harvesting by British companies in Burma, *2584*

BURUNDI

National population program for Burundi, *3886*

BYZANTINE EMPIRE

Recorded statement of the law of the sea, *3006*
"Sun rights" laws to guarantee buildings access to solar warmth and light, *4083*

CAMAROON

Volcanic gas from a lake to cause massive deaths in Cameroon, *4392*

CANADA

Appearance of the alfalfa blotch leaf miner in the United States, *1194*
Arctic Environmental Protection Strategy (AEPS), *3873*
Automobile emissions standards in Canada, *1582*
Ban of lead shot in Canada, *2962*
Battery recycling program in Canada, *4171*
Bilateral negotiations between the United States and Canada on acid rain, *3038*
Bilateral U.S.–Canadian agreement to waive "consent notification" for hazardous waste shipments from the United States into Canada, *2873*
Bird protection treaty, *3017*
Canadian ban on hunting baby seals, *2958*

Canadian law to protect all endangered wildlife and flora, *2343*
Canadian–U.S. International Boundary Treaties Act, *3366*
Canadian–U.S. treaty to regulate use of the Niagara River for water power, *4555*
Commercial logger on the East Coast of North America, *2641*
Control method used on sea lampreys in the Great Lakes, *2461*
Daily forecasts of ultraviolet radiation issued by the Canadian government, *1849*
Dam safety organization in Canada, *2076*
Declaration of phosphates as a major source of water pollution issued jointly by two countries, *4404*
Drought rehabilitation program by the Canadian government, *2151*
Environmental law challenged under the North American Free Trade Agreement (NAFTA), *3971*
Environmentally conscious codes of ethics for whale-watching operators and participants in North America, *2244*
Evidence of life on the sea floor, *3550*
Flood limitation program of significance by the Canadian government, *2557*
Government report issued jointly by the United States and Canada to rate acid rain a serious environmental and transboundary problem for both countries, *1477*
Hurricane of significance recorded in North America, *4237*
Interagency Grizzly Bear Committee (IGBC), *4785*
Intergovernmental conservation conference in North America, *3367*
International Adopt-a-Highway Day, *3290*
International Joint Commission on Boundary Waters study of Lake Erie, Lake Ontario, and the upper St. Lawrence River, *4511*
International meeting of American Indians and environmentalists, *4408*
International treaty to protect birds, *1743*
Introduction of larger, meatier British cattle on U.S. ranches, *1164*
Introduction of the potato into North American colonies, *1205*
Joint commitment to "zero discharge" of dangerous chemicals into Lake Superior by the United States and Canada, *4442*
Joint effort of major significance to solve pollution problems in the Great Lakes, *4476*
Legislation in Canada that preserved parks for future generations, *3735*
Legislation in Canada to establish procedures for handling radioactive waste, *2840*
Mercury thermometer, *1869*
National government publication emphasizing risk factors and lifestyle in public health, *3948*
National professional organization for foresters in Canada, *2697*
National Scenic Trail in the Rocky Mountains, *3989*
North American country to recognize acid rain as a serious environmental and transboundary problem, *1324*
Organization to study the North Pacific, *3063*
Ozone weekly summary by the Canadian government, *1993*
Paralytic shellfish poisoning (PSP) cases, *4418*
Patent for distilling kerosene from crude oil, *3679*
Precipitation sampling network in Canada, *1471*
Radioactive fallout of significance from a space program, *1326*
Reports of sharp decreases in dissolved oxygen levels in Lake Erie, *4414*
Reports of the virtual disappearance of the mayfly in Lake Erie, *2374*
Reports of widespread fouling of Great Lakes beaches by ill-smelling seaweed, *4532*
Sea lamprey in Lake Erie, *4502*
Significant expansion of federal presence in hazardous waste policy in Canada, *2882*
Successful challenge to state-ownership doctrine, *1690*
Sustainable development governmental post in Canada, *4317*
Sustainable development handbook issued by the Canadian government, *4316*
TFM chemical releases into the lamprey-infested waters of Lake Superior, *1526*
Trans-Canada multi-use trail, *4671*
Tree species in North America to be widely harvested commercially, *2645*
Tri-national conference on conservation in North America, *1034*
U.S.–Canadian bilateral report on acid rain, *3042*
U.S.–Canadian effort of significance to improve the water quality of the Great Lakes, *3029*
U.S. Fish and Wildlife Service jurisdiction over sockeye salmon conservation, *2459*

Alberta

Drought rehabilitation program by the Canadian government, *2151*
Park to be designated an international peace park, *3798*

Banff

National park in Canada, *3782*
Natural history museum in western Canada, *4911*

CANADA—Alberta—*continued*

Edmonton

Mall to feature aquatic displays, *4956*

Lethbridge

Dam built by Canada's Prairie Farm Rehabilitation Administration, *2102*

Swan Hills

Comprehensive hazardous waste treatment facility in the Canadian province of Alberta, *2877*
Town plebiscite to approve a large hazardous waste treatment facility in Alberta, Canada, *2871*

British Columbia

Exploratory offshore oil wells in Canada, *3659*
Volcano eruption of significance in Canada, *4350*

Trail

Arbitration ruling to assign national responsibility for air pollution damage, *1389*

Valhalla Mountain Range

Valhalla Wilderness Society meeting, *1097*

Vancouver

Appearance of white pine blister rust on the Pacific coast of North America, *2597*
Aquarium in North America to be accredited by the American Association of Zoological Parks and Aquariums, *4948*
Aquarium sleepovers, *4960*
Attack by loggers on a children's book, *2665*
Environmental organization to pursue direct action as a primary strategy, *1063*
Mandatory automobile-emissions tests in Canada, *1583*

Victoria

International Fisheries Gene Bank (IFGB), *2499*
Pledge to halt all raw sewage flow from Victoria and Vancouver Island, British Columbia, into the Strait of Juan de Fuca, *4543*
Virtual whale-watching tours, *2253*

White Rock

Antilitter campaign in Canada, *3274*

Manitoba

Drought rehabilitation program by the Canadian government, *2151*
Drought to cause severe agricultural damage in Canada, *2134*

Churchill

Ptarmigan born and bred in captivity, *4749*

New Brunswick

Trans-Canada national foot trail, *4661*
Widespread use of DDT to spray for the tree-killing spruce budworm in Canada, *2605*

Fredericton

General hunting license in North America, *2954*

Point Lepreau

Nuclear-generated electricity exported from Canada, *3521*

St. Johns

Natural history museum in Canada, *4891*

Newfoundland

Commercial whaling in North America, *4621*
Ice Age calculations done scientifically, *2004*
Oil spill of significance recorded, *3628*
Transatlantic flight powered by ethanol, *4334*

Belle Isle

Storm to defeat a naval invasion of Canada, *4178*

Grand Falls

Exploratory offshore oil wells in Canada, *3659*

Labrador

English law to support North American fisheries, *2469*
Lincoln's sparrow, *1654*

St. John's

Offshore drilling-rig disaster near Canada, *3641*

Tors Cove

Exploratory offshore oil wells in Canada, *3659*

Northwest Territories

Ross's goose breeding grounds, *1738*
World Heritage site in North America selected for environmental reasons, *3991*

Baffin Island

Blue goose nest discovered, *1698*
Productive mine above the Arctic Circle in Canada, *3335*

Eureka

Joint arctic weather stations, *1948*

Norman Wells

Pipeline giving access to Canada's northern oil, *3625*

Yellowknife

Diamond mine in Canada, *3338*

Nova Scotia

Annapolis Royal

Tidal power plant in North America, *4566*

Louisbourg

Coal mine used on a commercial basis in Canada, *3302*

Port Morien

Coal mine used on a commercial basis in Canada, *3302*

Sydney Harbour

Coal mined in Canada, *3299*

Ontario

Canal connecting Lake Ontario with the other four of the Great Lakes, *3070*
Careful scientific monitoring of Lumsden Lake in Ontario, Canada, for acid rain, *1453*
Pollution-related ban on commercial fishing in Lake St. Clair on the Canada-U.S. border, *4474*
Report on the effects of acid rain on fish populations, *1464*
Report that pollutants discharged into the Detroit River may have killed bald eagle hatchlings along Lake Erie, *4416*
Sea lamprey positively identified in Lake Ontario, *1501*
Trans-Canada national foot trail, *4661*

Burlington

Worldwide freshwater monitoring program, *4601*

Hamilton

United Nations training center for water supply issues, *4613*

Niagara Escarpment

National Trail in Canada, *4649*

Ottawa

Intergovernmental agency devoted to marine and fisheries science of the Northern Pacific, *3596*

Port Hope

Canadian town to be partially demolished because of radioactive waste, *2859*
Discovery of radioactive contamination in the town of Port Hope, Ontario, *2853*

Sarnia

Oil well drilled intentionally and successfully in North America, *3684*

Toronto

Widely publicized scare in hazardous materials transport in Canada, *2862*

Prince Edward Island

Fur farm, *1011*

Quebec

Biosphere reserve in Canada, *1631*
Dam with a reservoir that could hold more than 100,000 cubic meters of water, *2106*
Large-scale widely reported outbreak of spruce budworm, *2599*
Snowmobile, *4637*
Tests of leeches and mussels to monitor levels of freshwater contaminants, *4515*
Timber trespass laws in French Canada, *2629*
Use of asbestos as an insulator on a large scale, *1333*

Blainville

Reported profit of Concord Resources, a U.S. company, for its hazardous waste disposal facility in Blainville, Quebec, *2879*

Chalk River

Nuclear generator accident of significance, *3459*

Hull

Canadian biodiversity government agency, *1640*

CANADA—Quebec—*continued*

Montreal

Biodome, *4964*
Electric-vehicle project by the Canadian government and industry, *1616*
Fox hunting club in Canada, *2965*
Geological Survey of Canada, *2806*
Geologist to classify rocks from Precambrian times, *2808*
Person to lobby for a botanical garden in Montreal, *1794*
Worldwide organization concerned with civil aviation safety, *3933*

Quebec

European to systematically record a North American winter, *1798*
Permanent French settlement in North America, *3127*

Saskatchewan

Drought rehabilitation program by the Canadian government, *2151*

Yukon Territory

Gold diggings discovery on Klondike Creek in the Yukon Territory of Canada, *3320*

CANARY ISLANDS

Introduction of bananas into the Americas, *1177*

CARIBBEAN REGION

European use of the term *hurricane*, *4223*

CARTHAGE

Description of chimpanzees, *4679*

CENTRAL AMERICA

Settled communities in Mesoamerica, *3123*

CHILE

Earthquake to register a magnitude of 9 or more on the Richter scale, *2225*
Evidence that the "ozone hole" had expanded outward to include New Zealand, Australia, Argentina, and Chile, *1480*
Natural resource exploited in Chile, *3351*
United Nations convention to protect the vicuña, *4777*
Volcano eruption of significance in Chile, *4351*

Calama

Substantial rainfall after more than 400 years at Calama, Chile, *2555*

Concepción

Earthquake causing thousands of deaths in Chile, *2193*
Earthquake to be described in detail by Charles Darwin, *2200*

Easter Island

Successful introduction of a tinamou, *1655*

Punta Arenas

Scientific evidence that humans caused the Antarctic ozone hole, *1988*

CHINA

Acute timber shortages in north China, *2582*
Appearance of wet-rice cultivation in paddy fields, *1198*
Chinese animal park, *4872*
Commercial use of coal, *2016*
Diversion of the Yellow River, *2566*
Domestication of millet, a type of cereal grass, *1196*
Domestication of the chicken, *1142*
Drought year producing worldwide environmental responses, *2162*
El Niño known to have caused massive deaths, *1819*
Farming communities in China, *3120*
Fish farms, *2487*
Flood death toll to exceed 1.5 million, *2532*
Floods known to have lasted 15 years in China, *2527*
Introduction of drought-resistant rice, *1122*
Known earthquake on the historical record, *2199*
Law in China to outlaw hunting giant panda, *2957*
Modern industrial pollution crime penalties in China, *3004*
Modern natural resource crime laws in China, *3422*
Modern solid waste disposal crime laws in China, *4154*
Modern wildlife resource crime laws in China, *4792*
Nuclear agency established by the Chinese government for both civil and military activities, *3526*
Nuclear energy agreement between the U.S. and China, *3523*

Oil wells, *3671*
One-child-per-family policy in China, *3883*
Ozone fund for developing nations, *1989*
People to keep regular records and collect data on earthquakes, *2171*
Phase of construction on the gigantic Three Gorges Dam on the Yangtze River in China, *2110*
Recorded flood on China's Yellow River, *2517*
Report to Europe of vast coal deposits in China, *2022*
Reversal of the post-glacial warming trend in China, *2002*
Seismograph, *2177*
Typhoon known to have prevented an invasion of Japan, *4222*
Use of Chrysanthemum powder as an insecticide, *1269*
Use of natural gas as a fuel, *2017*
Water shortage in China causing the Yellow River to run dry, *2156*
Written record of the use of a burning mirror in China, *4074*

Anhui

Flood to affect one-fifth of China's population, *2564*

Guangdong

Zhaoquin

Modern nature reserve in China, *4823*

Hebei

Beijing

Automobile pollution device of significance in China, *1602*
Giant panda born in captivity, *4752*
Pére David's deer reintroduced to China, *2432*

Tangshan

Earthquake in China with its epicenter directly under a city of 1 million people, *2231*

Henan

Kaifeng

Flood deaths of major significance deliberately caused by humans, *2526*
Levees destroyed in China to halt a foreign invasion, *2569*

Hunan

Hailstorm of significance known to have hit the western Hunan province of China, *4217*

Tungting Lake Region

Time the Yangtze and Hwai rivers flooded to over 96 feet, *2545*

Jiangsu

Flood to affect one-fifth of China's population, *2564*

Kwangtung

Hong Kong

Epidemiological evidence of the association between passive smoking and lung cancer, *1476*
Reservoir to supply Hong Kong, *4619*
Typhoon to cause massive loss of life in Hong Kong, *4247*
Worldwide organization to advocate family planning, *3897*

Manchuria

Coal mining accident in the Far East causing catastrophic deaths, *3328*
Drought known to have caused cannibalism in China, *2137*

Shaanxi

List of earthquake deaths, *2182*

Sichuan

Use of natural gas for illumination, *2019*

Zhejiang

Zhoushan

Forsythia in England, *1504*

COLOMBIA

Earthquake to register an 8.9 on the Richter scale, *2214*
Natural resources exploited in Colombia, *3350*

Armero

Mudslide in the 20th century to bury a city in Colombia, *2576*
Volcano eruption causing massive deaths in Colombia, *4391*

Gaviotas

Self-sustaining village experiment in Colombia, *4302*

COLOMBIA—*continued*

Nevada del Ruiz

Volcano eruption causing massive deaths in Colombia, *4391*

CONGO

Biosphere reserves in Africa, *1628*
Ebola virus outbreak, *3950*
Identification of the Congo peacock, *1697*
Lava flow known to have attained a speed of 40 mph, *4386*

Lake Kivu

Wildlife films of mountain gorillas, *4698*

COSTA RICA

"Bioprospecting" agreement, *1639*
Inter-American Tropical Tuna Commission, *2460*
Joint biodiversity program between Nicaragua and Costa Rica, *1637*
Natural resource exploited in Costa Rica, *3353*
Wilderness expeditions operator in Costa Rica, *2243*

Horquetas

Costa Rican ecotourism organization whose purpose was to save an endangered part of the environment, *2246*

CROATIA

World Heritage sites in Europe chosen for environmental reasons, *3996*

CUBA

Evidence conclusively showing that yellow fever was transmitted by a variety of mosquito, *3923*
Introduction of cattle and reintroduction of horses into the Americas, *1152*
Natural resource exploited in Cuba, *3345*
Species to be given a scientific name incorporating the first and last names of a naturalist, *1737*

Havana

Yellow fever epidemics in the New World, *3907*

Matanzas Bay

Ocean Thermal Energy Conversion (OTEC) power plant, *4559*

CYPRUS

Introduction of sugar cane from Mesopotamia into the Levant, Egypt, and Cyprus, *1200*

CZECH REPUBLIC

Prague

Mining book detailing methods used by miners, *3297*
Reported use of a refined oil for lighting, *3673*

CZECHOSLOVAKIA

Pipeline to carry natural gas from the Soviet Union to a European country, *3708*

DENMARK

Arctic Environmental Protection Strategy (AEPS), *3873*
Chemical and oil waste law enacted by the Danish government, *3945*
Description of the giant squid, *2451*
Detailed model for determining prevailing temperatures in prehistoric northern Europe, *2014*
International Conference on Polar Bears, *4775*
Offshore oil discovered off Denmark, *3660*
Trilateral agreement to protect the Wadden Sea, *3010*
United Nations agreement on marine pollution from offshore industrial facilities, *3583*

Askov

Aerodynamically efficient windmills to generate electricity, *4851*

Copenhagen

Environmental quality monitoring agency of the European Union (EU), *4312*
Intergovernmental agency devoted to marine and fisheries science, *3565*
Use of the word *gene* to describe units of heredity, *2752*

Kalundborg

Eco-industrial park, *4300*

Vindeby

Offshore wind power plant, *4860*

DOMINICAN REPUBLIC

Hurricane in the Atlantic with a masculine name, *4265*

Hurricane to destroy a European colony in the Western Hemisphere, *4224*
Introduction of cattle and reintroduction of horses into the Americas, *1152*

Santo Domingo

Hurricane to cause massive deaths in the Dominican Republic, *4250*

ECUADOR

Tourism restrictions on Ecuador's Galápagos Islands, *2252*
United Nations convention to protect the vicuña, *4777*

Ambato

Volcano eruption of significance recorded in today's Ecuador, *4362*

Quito

Earthquake causing massive loss of life in Ecuador, *2196*
Volcano eruption of significance recorded in today's Ecuador, *4362*

Galápagos Islands

Biosphere reserve in Ecuador, *1635*
Tortoise listed under the U.S. Endangered Species Act, *2400*
World Heritage site in South America chosen for environmental reasons, *3995*

EGYPT

Acclimatization garden, *4875*
Agricultural settlements in the fertile Nile Valley of Egypt, *3121*
Automatic water-driven irrigation wheel, *1213*
Drought recorded known to have caused a famine, *2120*
Human practice of sedentary agriculture, *1230*
Illustrations of animals in a zoo-like setting, *4873*
Importation of exotic animals for a royal collection, *4876*
International treaty to propose African wildlife preserves, *3019*
Introduction of sugar cane from Mesopotamia into the Levant, Egypt, and Cyprus, *1200*
Introduction of the animal-drawn water wheel in the Nile Valley of Egypt, *1232*
Known use of honey from honeybees as a food, *1144*
Recorded use of pigeons as messengers, *1643*
Surviving year-by-year record of flood levels in the Nile Valley in Egypt, *2007*
Use of artificial irrigation in the Nile Valley of Egypt, *1240*
Use of sulfur in an industrial process, *1314*
Use of the shaduf or bucket-and-pole system of watering fields in the Nile Valley of Egypt, *1231*
Western botanist to note the sexual identity of plants, *1752*

Alexandria

Person to demonstrate that air has weight, *1855*
Recorded evidence of the use of animals in research, *1002*
Solar machine, *4081*

Aswan

Dam built to totally control the annual flooding of Egypt's Nile River, *2105*
Treaty to combine preservation of natural and cultural heritage sites, *3033*

Cairo

Dam constructed in Ancient Egypt, *2083*
Natural history museum in Egypt, *4959*

Sinai Peninsula

Metal mines, *3292*

EL SALVADOR

Natural resource exploited in El Salvador, *3347*
Volcano eruption in Central America known to have moved a civilization, *4358*

ETHIOPIA

Drought to draw worldwide aid through television, *2158*
National population policy for Ethiopia, *3893*
Surviving year-by-year record of flood levels in the Nile Valley in Egypt, *2007*
World Heritage site in Africa selected for environmental reasons, *3990*

EUROPE

American vine aphids or *Phylloxera vastatrix* in Europe, *1271*
Appearance in Europe of American corn and potatoes, *1123*
Appearance of the Colorado potato beetle in Europe, *1519*
Appearance of the "Little Ice Age,", *1799*
Coal mining accident in Europe to cause catastrophic deaths, *3322*
Cultivation of vines, olives, and figs, *1197*

EUROPE—*continued*
Dangerous wastes law passed by the European Economic Community (EEC), *3954*
Development of the heavy ox-drawn plow, *1233*
Domestication of the common goose, *1145*
Evidence of the use of shoes for horses, *1149*
Extensive collection of quality data on fertility and family planning, *3882*
Hazardous waste transportation law passed by the European Economic Community (EEC), *3960*
Intergovernmental carnivore reintroduction strategy in Europe, *2442*
Meteorological organization in Europe organized specifically to control weather satellites, *1967*
Meteorology satellite launched by Europeans, *1964*
National Green party to become part of a governing coalition in Europe, *1119*
New technology to modify city design in late medieval Europe, *3092*
Nuclear power plant disaster with worldwide fallout, *3528*
Record of using solar energy to make perfume, *4085*
Treaty to insure humane slaughter of animals, *3037*
Use of fertilizers in forests, *2672*
Weathercock wind vanes, *1856*
Widespread use of canaries to detect deadly carbon monoxide gas in coal mines, *2034*
Widespread use of crop rotation to preserve soil health, *4028*
Widespread use of South American guano in European agriculture, *1128*
Widespread use of the horse as a plough animal in western Europe, *1151*
Windmills in Europe, *4847*

FIJI

Gua Island

Rediscovery of MacGillivray's petrel, *1682*

FINLAND

Arctic Environmental Protection Strategy (AEPS), *3873*
Green politician to join a national government as a minister, *1117*
United Nations agreement on marine pollution from offshore industrial facilities, *3583*

FLANDERS

Major land reclamations in low-country Flanders and Zeeland, *3067*

FRANCE

Adoption of the three-field rotation system, *1234*
Aneroid barometer, *1892*
Appearance of the American vine aphid in Europe, *1505*
Appearance of the Colorado potato beetle in Europe, *1519*
Aqualung or diving tank, *3573*
Automobile, *1586*
Automobile with a practical internal-combustion engine, *1588*
Botanical classification based on the natural relationships between plants, *1767*
Chemical accident prevention law in France, *3953*
Clearing of the natural forests of northern Europe, *2579*
Discovery and naming of the process of osmosis, *1777*
Effective international action to curb acid rain, *1414*
Fruit walls, *4087*
Fungicide considered effective, *1278*
General Fisheries Council for the Mediterranean, *2474*
Geothermal heat exchange theory, *2800*
Greenhouse effect description, *1835*
Hazardous substances transportation law in France, *3930*
Industrial distillation of oil from coal and shale, *3675*
International treaty to propose African wildlife preserves, *3019*
Mine gas extracted from coal fields in France, *2066*
Modern experiments on the effects of sunlight on buildings, *4103*
National storm warning service, *4182*
Nuclear weapons test conducted by the French, *3492*
Observation of the disease known as damping off, *2589*
Paper on the extinct volcanoes of Europe, *4363*
Pipeline to carry natural gas from the Soviet Union to a European country, *3708*
Scientist to establish the fact of extinction, *2359*
Significant legislation to manage French forests to assure a continuous yield of timber, *2626*
Statement that rocks and minerals were not randomly scattered but deposited in bands, *2783*
Supertanker to founder and spill its cargo of oil, *3630*
TBT (tributylin) restrictions, *4496*
Theory of Ocean Thermal Energy Conversion (OTEC), *4549*
United Nations condemnation of French atmospheric nuclear weapons tests, *3509*

Use of *chlorophyll* as a term to describe the green substance that colors plants, *1775*
Use of the term *desertification* to describe deforestation of tropical and subtropical Africa, *2743*
Use of the term *ecofeminism*, *1042*
Woman to win a Nobel Prize, *2268*

Arcachon

Notice of the effects of tributylin (TBT) on marine life, *4492*

Auteuil

Nonreflecting solar motor, *4096*

Chartres

Expressway in France built with private funds, *1568*

Courrieres

Coal mining accident in Europe to cause catastrophic deaths, *3322*

Creys-Malville

Nuclear power plant shelled by environmentalists, *3520*

Lascaux Cave

Artistic representation of a bird or bird parts, *3159*

Le Creuset

Water-powered turbine, *4546*

Marseilles

Discovery of the ozone layer, *1969*
First recorded polar expedition, *3823*

Montpellier

Use of the term *taxonomy* to describe the science of classification, *1774*

Nancy

Forestry school in France, *2733*

Paris

Dam organization of international significance, *2068*
Decision of the Paris authorities to ban the taking of water from the polluted Seine, *4522*
Expressway in France built with private funds, *1568*
Important scientific discussion of the negative impact of chemical fertilizers on food crops, *1258*
International convention to preserve scenery, culture, and the environment, *3987*
Meteorological research done by balloon, *1881*
Meteorological theory saying human activities affect the climate, *1887*
Meteorology satellite launched by Europeans, *1964*
Natural catastrophe theory about animal extinctions, *2009*
Observer to propose a link between volcanic eruptions and the weather, *1809*
Publication of modern weather maps, *1906*
Stratosphere and troposphere discoveries, *1921*
Transatlantic flight powered by ethanol, *4334*
Trash collection tax in Paris, *4173*
United Nations hydrological agency, *4596*

Portsall

Oil-tanker spill of significance off the coast of France, *3639*

Saint Malo

Dam to harness tidal power, *2104*
Electric power project to harness tidal motion, *4562*

Sérignan

Significant study of the behavior of insects, *4686*

Tours

Brandy distilled by the heat of the sun, *4091*
Combination of a glass heat trap and a burning mirror, *4092*
Solar motor, *4093*

Versailles

Rally to free animals in a menagerie in France, *4887*
Royal European menagerie with a research component, *4880*

GEORGIA

Wolves reintroduced to Soviet Georgia, *2429*

Batumi

Oil pipeline in Russia, *3614*

GERMANY

Agreement to finance development of a new high-performance car on the condition that the model be named for the financier's daughter, *1535*
Almanacs with yearlong weather forecasts, *1858*
Appearance of the American vine aphid in Europe, *1505*
Arachlor to be successfully synthesized, *2829*
Artificial plant hybrid, *1765*
Automobile powered by fuel cells, *1608*
Bird band recovery, *1727*
Book devoted entirely to birds, *3228*
Brandt's Cormorant description, *1730*
Brown trout in the United States, *2494*
Clearing of the natural forests of northern Europe, *2579*
Complete freeze of Lake Constance (the Bodensee) in West Germany in more than two centuries, *1826*
Controlled forest management, *2668*
Crude oil wells in Germany, *3683*
Demonstration of the alternation of generations, *1772*
Discovery of a laboratory process for extracting liquid ammonia from free nitrogen in the air, *4063*
Ecological candidate elected to a national parliament, *1110*
Epidemiological evidence of the association between passive smoking and lung cancer, *1476*
Eurasian tree sparrow in the United States, *1653*
Federal law in Germany to clean up the Ruhr river, *4453*
Forest experiment station to systematically study the effects of the removal of litter from the forest floor, *2735*
Forest fire ordinances, *2611*
Forestry experiment stations, *2734*
Formaldehyde synthesization, *1356*
Hazardous substances protection law in Germany, *3958*
Local environmental political parties, *1105*
Map using isothermal lines to map temperatures, *1885*
Microclimatology research, *1935*
National Green party to become part of a governing coalition in Europe, *1119*
Network of high-speed freeways in Germany, *1562*
Nucleic acids, *2751*
Observation of the disease known as damping off, *2589*
Oceanographic study of an entire ocean, *3569*
Ozone-protecting household refrigerator, *1995*
Parliament to have a substantial Green bloc represented, *1112*
Presentation of the cell theory, *1780*
Release of imported gypsy moth–killing nematodes in North America, *2607*
Scientific discovery of the role of bees in pollination, *1766*
Scientist to be dubbed the "father of chemical agriculture", *4060*
Scientist to describe the characteristic six-sided symmetry of snow crystals, *4191*
Scientist to use records of seismic waves to demonstrate the existence of the earth's core and measure its distance from the surface, *2219*
Subcompact car designed to be fuel-efficient, *1596*
Synthetic detergents for clothing, *4397*
Systematic work on American geology, *2786*
Theory of continental drift, *2217*
Trilateral agreement to protect the Wadden Sea, *3010*
Trophic-level system application to classify lakes, *4506*
United Nations attempt to define civil liability for oil pollution resulting from seabed drilling, *3591*
Use of sonar (sound navigation and ranging) to measure ocean depths, *3570*
Use of the term *biology*, *2259*
Use of the term *ecology*, *1049*
Use of the term *protoplasm* to describe the contents of plant cells, *1773*
Use of the word *pollen*, *1749*
Weather forecast using a barometer, *1864*
Weather map, *1884*
Widespread damage to and deaths of trees in Germany from air pollution, *1475*
Worldwide organization to advocate family planning, *3897*
Written reference to a mousetrap, *2901*

Berlin

Diesel engine fuels, *4325*
Ecotourism appeal coordinated by governments, *2255*
Nation to adopt building codes with noise standards, *3433*

Bingen

Mass fish kill attributed to an insecticide spill, *1297*

Chemnitz

Treatise on mining that was scholarly, detailed, and systematic, *3296*

Darmstadt

Meteorological organization in Europe organized specifically to control weather satellites, *1967*

Meteorology satellite launched by Europeans, *1964*

Frankfurt

Climatologist of repute, *1923*
Cohesive theory of the concept of "continental drift", *2817*
Fuel cell passenger car to use methane, *4343*

Freiburg

Zoologist to win a Nobel Prize, *2280*

Hamburg

Zoo to display animals in a naturalistic setting, *4913*

Heidelberg

Natural gas invention for heating, *2045*

Ilsenburg

School of forest practice, *2732*

Lindow

Automobile rotary engine, *1545*

Mannheim

Automobile that proved practical with an internal-combustion engine, *1589*
Meteorological society, *1876*

Munich

Earth Day to be celebrated around the world, *1089*
X-ray analysis of minerals, *2816*

Schleswig

Flood to cause massive deaths in what is now Germany, *2521*

Solenhofen

Discovery of *Archaeopteryx*, a fossil proto-bird, *1732*

Stallingen

Zookeeper to recognize the benefits of fresh air for animals in captivity, *4914*

GHANA

Introduction of cocoa into West Africa, *1179*
Natural resource exploited in Ghana, *3341*

Accra

United Nations educational center for African natural resource management, *3392*

GREAT BRITAIN

Alkali Act (1863) revisions of significance, *1387*
Appearance of the word *cyclone*, *4243*
Atmospheric carbon dioxide link to human activities, *1837*
Bird protection treaty, *3017*
British government attempt to restrict American colonial settlement to the area east of the Appalachian Mountains, *3131*
Climate hypothesis concerning animals and plants, *1828*
Coal-mine emergency legislation of significance passed by the British Parliament, *3326*
Comprehensive report to conclude that subtherapeutic use of antibiotics in animal feed posed a significant health risk to humans, *1173*
Criminal statutes against water pollution in Great Britain, *4447*
Darwin Centenary Symposium on Earthworm Ecology, *4073*
Direct measurements of chlorofluorocarbon (CFC) levels at the earth's surface, *1466*
Discovery that tricalcium phosphate could be converted into a soluble plant fertilizer by treatment with sulfuric acid, *4059*
Flood leaving 1 million people homeless in the Netherlands, *2544*
Flora of the British Isles, *1757*
General Fisheries Council for the Mediterranean, *2474*
Global list of endangered plants, *2421*
Instances of modern organic farming as a conscious rejection of traditional chemical methods, *1259*
Instrument to accurately measure the length of time the Earth's surface receives sunshine, *1901*
Intergovernmental conservation conference in North America, *3367*
International treaty to propose African wildlife preserves, *3019*
International treaty to protect birds, *1743*
International whaling commission, *3057*
Introduction of the discipline of animal ecology, *1053*
Law establishing comprehensive control of radioactive material by the British government, *3491*
Law in Great Britain in the modern era addressing hazardous waste, *2834*
Laws regulating space debris, *3032*

GREAT BRITAIN—*continued*
Local environmental political parties, *1105*
Long-lasting partnership to build luxury cars in Great Britain, *1536*
Long-standing survey of water quality conditions in the Western Approaches to the English Channel, *4399*
Measurements of chlorofluorocarbons (CFCs) in the atmosphere, *1971*
National act regulating painful experimentation on animals, *1009*
National antilitter organization in Great Britain, *3271*
National car industry to produce one million units a year, *1539*
National Health Service in Great Britain, *3935*
National zoo licensing legislation, *4952*
Oceanographic expedition, *3560*
Oil wells seized from foreign companies, *3699*
Outline of the mathematical concepts that would lead to numerical modeling for accurate weather forecasting, *1931*
Ozone-layer data concerning damage by chlorofluorocarbons (CFCs), *1974*
Parrot protection and welfare organization, *1717*
Patent on a device to enable a steam engine to consume its own smoke, *1433*
Plate tectonics theory verification, *2821*
Prohibition against nuclear weapons on the ocean floor, *3584*
Public health advocate of significance in Great Britain, *3915*
Public health legislation of significance, *3917*
Radar detection of air echoes from the lower atmosphere, *1938*
Radio gale warnings from Great Britain to ships in the eastern Atlantic, *4185*
Renowned scientist to ennoble the earthworm as one of the most important animals in the history of the world, *4070*
Scientist to describe the type of earthquake surface wave that acts like waves caused by a pebble tossed into a pond, *2211*
Standards for organic farming in Great Britain, *4067*
Storm tide warning service in Great Britain, *4188*
Successful challenge to state-ownership doctrine, *1690*
TBT (tributylin) restrictions, *4496*
Television weather forecast in Great Britain, *1939*
Treaty in which England recognized the fishing rights of the United States, *3014*
Treaty to ban the introduction of nuclear weapons into orbit around the Earth, *3027*
Treaty to recognize the three-mile limit for fishing off the North American coast, *3015*
United Nations attempt to define civil liability for oil pollution resulting from seabed drilling, *3591*
Use of carbolic acids to treat foul-smelling sewers, *4521*
Use of fuel oil to power warships, *3666*
Use of penicillin in general clinical practice, *3932*
Use of the term *green belt*, *3100*
Use of the word *dinosaur*, *2363*
Use of the word *forecast* relating to meteorology, *1898*
Weather forecasts available to the public in Great Britain, *1905*
Weather reports on radio in Britain, *1932*
Winter in Great Britain in the 20th century during which snow fell somewhere in the country every day from January 22 to March 17, *4206*
Year that nuclear power was the main source of British electricity, *3537*

England

Air pollution law in England granting British inspectors the right to enter factories during their investigations, *1382*
Air pollution law of significance in England aimed at actually ridding London of air pollution rather than making it easier to prosecute polluters, *1379*
Anglers' Co-operative Association request for an injunction against a water polluter under English Common Law, *4456*
Animal welfare organization started on a national level, *1006*
Antivivisection legislation, *1008*
Appearance of the "bosom friend," an article of warm clothing for women, *1811*
Authenticated use of coal by the Romans in England, *2018*
Avicultural organization founded, *1713*
Black Death epidemic in England, *3905*
Book on natural history written by Charles Darwin, *3238*
British government inquiry into the practicability of laws to control smoke, *1377*
Calculation of the Earth's mean density, *2791*
California quail in New Zealand, *1509*
Chronometer, *3545*
Cloud classification of scientific significance, *1880*
Coal mines in England to use steam engines, *3308*
Complete flora of Leicestershire, England, *1783*
Cup anemometer, *1893*
Development of the science of forestry, *2669*
Earthquake recorded in Great Britain, *2176*
Effective treatment for stack emissions of hydrochloric acid in the alkali industry, *1434*
English falconer of record, *2897*

English law forbidding night poaching, *2911*
English law to support North American fisheries, *2469*
Evidence of significant acid rain damage in the Great Britain, *1467*
Explanation of atmospheric condensation, *1879*
Exploitable petroleum "seepage" in England, *3676*
Exports of New England white pine to England, *2647*
Faunal succession theory, *2788*
Female falconer of note, *2899*
Fertilizer factory, *4061*
Forest replanting schemes promoted by the British Admiralty, *2723*
Formal government antipollution agency in England, *1381*
Fruit walls, *4087*
Game laws in England restricting hunting to the nobility, *2900*
Geological map, *2782*
Industrial distillation of oil from coal and shale, *3675*
Introduction of rabbits into England, *1150*
Introduction of the "green belt" concept, *3099*
Invasion of England to be halted by gale-force winds, *4176*
Landfill tax in England, *4153*
Law in England banning women from working underground, *2044*
Law to assert control of offshore waters, *3547*
Law to establish limits on emissions of hydrochloric acid, *1383*
Legislation explicitly barring industries in England from emitting "black smoke", *1384*
Map showing movement of cyclonic depression, *1908*
Meteorological data reported from airplanes in England, *1946*
Meteorological sensors on balloons in Great Britain, *1940*
Modern animal park, *4929*
Modern geological map, *2795*
Monastic and manorial records of the coal mining industry in England, *3295*
National political party in Europe committed primarily to environmental and ecological issues, *1109*
Noise control law in England with seizure provisions, *3448*
Official "fish days" in England, *2449*
Organized strike against a colliery in England, *2029*
Parliamentary committee to investigate air pollution in England, *1376*
Portland cement patent, *1332*
Proof that radar could be used to locate precipitation, *1944*
Published presentation of the philosophy that humans should not be cruel to animals because animals can feel pain, *1003*
Published warnings against using chemical fertilizers, *4066*
Rain gauge, *1866*
Recorded drought that affected a Roman military campaign in Britain, *2122*
Recorded imports of timber into England, *2642*
Regulations designed to curb the ill effects of coal smoke, *1372*
Research showing acid-forming pollution could travel great distances, *1456*
Scale used internationally for wind forces, *1882*
Scientist to explain the reasons for the value of adding manure and ashes to the soil, *4068*
Scientist to use the term *ecosystem*, *1621*
Self-regulating windmill sails, *4848*
Smoke density standard measurement in England, *1457*
Steam engine of high pressure used in a mine, *3300*
Strong claim that industrial burning of fossil fuels increased carbon dioxide levels in the atmosphere, *1448*
Study in England of significance into the effects of industrialization on public health, *3916*
Substantial waterborne export trade in British coal, *2023*
Successful application of electrostatic precipitation to scrub pollutants from cement kiln stacks in Britain, *1447*
Successful vaccination against smallpox, *3912*
Suction pump to draw water from coal mine shafts, *2027*
Systematic quantitative study of death in populations, *3908*
Terrarium, *1779*
Terrestrial magnetism theory, *2777*
Textbook on ecology, *3233*
"Tin" can, *4124*
Tin mines in ancient Britain, *3294*
Treatise on the phenomenon of acid rain, *1437*
Turbogenerator to produce hydroelectric power, *4551*
Use of the phrase "The question is not, Can they reason?, not, Can they talk?, but, Can they suffer?", *1004*
Use of the term *acid rain*, *1818*
Use of the term *animal rights*, *1012*
Use of the term *cell* to describe microscopic units of plants, *1758*
Use of the term *ecosystem*, *1056*
Waste disposal law in England, *4138*
Water frame, *4544*
Widely used chemical herbicide regarded as benign in the environment, *1291*
Widespread use of canaries to detect deadly carbon monoxide gas in coal mines, *2034*

GREAT BRITAIN—England—*continued*
Widespread use of oil cake as an animal feed in England, *1159*
Wind turbine ship to cross the Atlantic Ocean, *4849*

Acomb

Begonia sutherlandii to flower in England, *1510*

Aylesbury

Meteorology society in Great Britain continuing through the 20th century, *1899*

Bath

Safari park, *4940*

Birmingham

Gas used to light a house, *2032*

Burnley

Person to keep a continuous record of rainfall in England, *1867*

Camborne

Automobile using steam power in England, *1587*

Cambridge

Accurate DNA structure description, *2753*
Description of two types of earthquake motion, *2191*
Ozone research coordination program of the European Community (EC), *1991*
Theory on how to locate the epicenter of an earthquake, *2192*

Coalbrookdale

Use of the process of smelting by coke, *2028*

Delabole

Wind farm in England, *4861*

Dover

Cattle imported into a North American colony, *1156*

Durham

Drought in England known to have created slavery, *2124*

Exeter

Female botany experimentalist of significance in England, *1770*

Falmouth

Botanical chart to be published, *1781*

Farne Islands

Duck sanctuary, *4808*

Findon

Mines known in Great Britain, *3293*

Greenwich

Meteorological records officially kept in Great Britain, *1890*

Haughley

Comparative study of organic and conventional farming, *1260*
Organic farming association in the United Kingdom, *1262*
Woman to advocate organic gardening, *4065*

Isle of Wight

International seismographic recording network, *2213*

Leeds

British four-lane limited-access highway, *1567*
Research showing that the acidity of rain decreased the farther one traveled from a city center, *1441*

Letchworth

Planned green belt town, *3101*

Lewes

Disastrous snow avalanche in England, *4196*

Liverpool

Nuclear byproduct technology capable of cleaning smokestack pollutants, *3000*

London

Air pollution book to dramatize London's exploding 19th-century air pollution problem, *3246*
Air pollution measurements taken in London, *1443*

GEOGRAPHICAL INDEX

Air pollution prosecutions under the smoke clause of the City of London Sewers Bill, *1378*
Angled barometer, *1865*
Animal protection legislation, *1005*
Anti–air pollution activities by the English public interest group known as the Fog and Smoke Committee, *1430*
Bird known to have lived to a documented age of 80, *1679*
Bird pictured in John James Audubon's *The Birds of America*, *3160*
Book devoted to birds and written in English, *3230*
Botanical account, in detail, of the plants of North America, *1760*
Botanical society formed that is still in existence today, *1769*
British four-lane limited-access highway, *1567*
Campaign of significance against air pollution in London, *1319*
Cancellation of the British House of Commons sittings because of the stench of human waste in the River Thames, *4525*
Casualties of the great London killer fog of December 1952, *1343*
Commission of inquiry to investigate complaints about smoke levels in London, *1374*
Death in England known to have been caused by an earthquake, *2183*
Dew-point hygrometer, *1815*
Director of Kew Gardens, *1782*
Drought aid from a global rock music festival, *2159*
Drought to draw worldwide aid through television, *2158*
Electric generating station, *3359*
English police force given power to enforce smoke abatement laws, *1380*
Female botanist to become a member of the Royal Society in England, *1796*
Female pioneer in the field of mycology, the study of mushrooms, *1789*
First systematic weather forecasts, *1902*
Flood barrier in Britain rotated from beneath the water, *2561*
Forsythia in England, *1504*
"Frost Fair" on the frozen Thames river in London, *1802*
Fuel cell, *4324*
Gas lamps used to fully light a public thoroughfare, *2035*
Gas supplied by a private company, *2037*
Green belt legislation for London, *3083*
Greenhouse-effect warning about recycling paper, *1851*
Instrument that could detect hazardous chemicals such as PCBs (polychlorinated biphenyls), *2843*
International agency to address pollution of the oceans by ships, *3574*
Introduction to the botanical sciences written by a woman, *1771*
Ionosphere direct measurements, *1933*
Japanese flora description printed in English, *1763*
Large-scale air pollution problems in London, *1330*
Large-scale urban middle-class movement from townhouses to "suburbs of privilege", *3130*
Law making water closet connections to surface streams compulsory in London, *4446*
Live television reports of weather in England, *1952*
Long-range weather forecasting attempt using equations, *1930*
Magnolias to arrive in England, *1491*
Major anti–air pollution tract, *1429*
Meteorology society in Great Britain, *1886*
Modern public zoo, *4890*
Naming of the disease asbestosis, *1335*
Naturalist to describe the wildlife of Australia, *4682*
Ozone-protection plan to phase out methyl chloroform, *1994*
Popular use of the word *zoo*, *4894*
Proclamations against pollution caused by London starchmakers, *4445*
Professor of botany at the University of London, *1778*
Proposal to create a geological map, *2780*
Proposed theory that human population growth might create a global catastrophe, *3876*
Public aquarium for exhibition of fish, *4893*
Recognition of cancer as an occupational hazard, *2969*
Recognition of the correlation between cellular rings in trees and age, *1759*
Recorded evidence of serious pollution of the River Thames in London, *4520*
Researcher to prove the transmission of cholera via a contaminated water supply, *4420*
Royal proclamation in England to ban the burning of coal, *1375*
Salmon caught by rod and line in the River Thames, *2509*
Scientific description of a moa, *2362*
Smog episode to cause widespread loss of life, *1344*
Smog-related deaths reported in London, *1334*
Smoke reduction exhibition, *1431*
Smokeless coalite patent, *1440*
Society of professional ecologists, *2273*
Statute in England against air pollution, *1373*
Theory that coral reefs are founded on sinking underwater land masses like subsiding volcanoes and continental shelves, *3553*
Trade-wind explanation, *1872*

GREAT BRITAIN—England—London—*continued*

United Nations compensation fund for marine oil pollution damage, *3586*
Unleaded gasoline campaign in Great Britain, *1549*
Use of London's weekly Bills of Mortality to establish a link between death rates and the burning of coal, *1331*
Use of natural gas for cooking, *2043*
Use of the word *smog*, *1439*
Water filtration system, *4574*
Weather maps in the modern version, *1907*
Weather maps published for general sale, *1900*
Worldwide organization to advocate family planning, *3897*
Written record of a devastating tornado in London, *4278*
Zoo explicitly set up as a scientific institution, *4889*

Looe

Woolly monkeys successfully bred in captivity, *4753*

Manchester

Ice meteor to be subject to intensive study, *2296*
Municipal smokeless zones in England, *1392*
Scientist to systematically study dispersal of industrial age air pollutants, *1436*
Woman on the science staff of Manchester University, *1791*

Milton Keynes

International organization to monitor the disappearance of amphibians, *4737*

Newbury

International journal devoted to protected areas issues, *3810*

Newmarket

Tornado in Great Britain to cause $1 million damage, *4296*

Oxford

Botanic garden in England, *1756*
Cell description and use of the term *cell*, *2750*
Issue of *Forestry Abstracts*, *2741*
Known journal about the weather, *1857*
Public museum in Great Britain, *4881*
Use of nuclear power to produce electricity in Great Britain, *3456*
Weather observations in Britain that are still being taken, *1813*

Peckham

Hydrangea in England, *1494*

Reading

Meteorology society in Great Britain continuing through the 20th century, *1899*

Redruth

Gas used to light a house, *2032*

Saint Albans

Sport fishing guide written in English, *2505*

Scarborough

Recorded reference to a tornado as a "black horse", *4279*

Sellafield (Windscale)

Accidental discharge of nuclear waste in British waters, *3524*
Nuclear power-plant accident of significance, *3479*
Nuclear power plant in England, *3475*
Plutonium-producing nuclear reactor in England, *3460*

Seven Stones

Supertanker to founder and spill its cargo of oil, *3630*

Sizewell

Nuclear power in Great Britain generated from a pressurized water reactor, *3534*

Solway Moss

"Moss flood" of significance in England, *2572*

South Farnborough

Meteorological reports for British pilots, *1926*

Twyford Down

Person in England to chain herself to a bulldozer to protest a highway, *1103*

West Hartlepool

Oil tanker built for long-distance conveyance, *3613*

Weymouth

Underwater photograph, *3558*

Willesden

Electrostatic precipitation application, *1446*

Woodstock

Royal menagerie known to exist in England, *4879*

Northern Ireland

Famine as a result of potato blight, *1183*

Scotland

Ban on fox hunting in Scotland, *2963*
Coal-burning steam engine, *2030*
Evidence of significant acid rain damage in the Great Britain, *1467*
Law binding Scottish coal miners to their employers, *2024*
Long-term failure of the cod fishery in and around the Norwegian Sea, *1804*
Modern oil refining processes, *3678*
Natural gas piped ashore from the North Sea, *2052*
Oil discovery of significance in the North Sea, *3661*
Oil well blow-out in the North Sea, *3638*
Publication of the geological principle that the present is the key to the past, *2789*
Scientist to establish "uniformitarianism" as a principle of geology, *2790*
Scientist to popularize uniformitarianism, *2801*
Scottish Environment Protection Agency (SEPA), *3001*
Year that North Sea oil production exceeded home demand in Great Britain, *3664*

Aberdeen

Explosion of significance on an offshore oil rig, *3642*

Dounreay

Nuclear energy produced by a fast-breeder reactor in Great Britain, *3485*

Glasgow

Deaths caused by air pollution in Scotland, *1338*
Scientist to state the law of relative proportion in seawater, *3563*
Upper-air temperature measure of significance, *1874*

Roslin

Mammal to be cloned from an adult cell, *2769*

Saint Kilda

Report of the death of the last great auk in the British Isles, *2361*

Shetland Islands

Invasion of England to be halted by gale-force winds, *4176*
Oil-tanker spill of major significance off Scotland, *3647*

Stirling

Scottish Environment Protection Agency (SEPA), *3001*

Woodhead

British dam of significance built entirely of concrete, *2086*

Wales

Modern geological map, *2795*
Use of detergents and first-generation dispersants to clean up a beach after an oil spill, *4489*

Aberfan

Major landslide of a slag heap in Wales, *2574*

Glamorgan

Catastrophic flood of the 20th century in Wales, *2560*

Milford Haven

Oil-tanker spill of significance off Wales, *3649*

GREECE

Attempt to explain the origin of thunder, *4273*
Domestication of cattle, *1141*
Epidemiological evidence of the association between passive smoking and lung cancer, *1476*
General Fisheries Council for the Mediterranean, *2474*
Geometric proof of the focal properties of parabolic and spherical mirrors, *4078*
Mineralogy book, *2775*
Recorded scientific theories explaining earthquakes, *2174*
Signs of large-scale erosion because of deforestation in Greece, *4047*
Use of cattle for dairy products, *1147*
Use of solar architecture by the Greeks, *4075*
Western man to build a parabolic or "burning" mirror, *4077*

GREECE—*continued*

Written description of the medical uses of plants, *1745*
Written work of significance to use animals in captivity as subjects of a study, *4877*

Athens

Automobile and truck ban in Greece to control emissions, *1579*
Botanical study of significance, *1746*
Municipal dump in the Western world, *4136*
Weather vane, *1854*

Helice

Precise description of an earthquake, *2175*

Messene

Earthquake known to cause massive deaths throughout Greece, *2173*

Olynthus

Greek city to include solar designs in the original plans, *4076*

Sparta

Earthquake known to cause massive deaths throughout Greece, *2173*

Crete

Sanitary sewers, *4516*
Volcano eruption known on a Mediterranean island, *4353*

Knossos

Evidence of bullfighting as a sport, *1001*

Thíra (formerly Santorini)

Volcano eruption known on a Mediterranean island, *4353*

GUATEMALA

Volcano eruption recorded as a major disaster in the Western Hemisphere, *4374*

Antiqua

Earthquake to devastate Guatemala, *2194*

Ciudad Vieja

Mudslide catastrophe in Guatemala caused by a volcano, *2570*

Guatemala City

Earthquake to devastate Guatemala, *2194*

GUINEA-BISSAU

Kassa

Reported disposal of hazardous waste by a "poison ship" from the United States, *2881*

HISPANIOLA

Cases of smallpox to reach the Caribbean island of Hispaniola, *3906*
Cattle in the Western Hemisphere, *1484*
Horses reintroduced in the Americas, *1485*
Pigs in the Western Hemisphere, *1486*
Sheep in the Western Hemisphere, *1487*

Isabela

Hurricane to destroy a European colony in the Western Hemisphere, *4224*

HOLY ROMAN EMPIRE

Falconry book, *2898*
Flood to cause massive deaths in what is now Germany, *2521*
True book devoted to birds, *3227*

HONDURAS

Biosphere reserve in Central America, *1633*
Natural resource exploited in Honduras, *3349*

HUNGARY

Environmental lawsuit decided by the International Court of Justice, *3011*

Schemnitz

Mining school, *3303*

ICELAND

Colonization of Iceland from Norway, *3124*
Ice Age calculations done scientifically, *2004*
Long-term failure of the cod fishery in and around the Norwegian Sea, *1804*
Observer to propose a link between volcanic eruptions and the weather, *1809*
Recorded long-term blockade of Iceland by sea ice, *1805*
Volcano eruption of significance in Iceland, *4359*
Volcano known to have caused significant deaths by starvation, *4365*

Eldey Rock

Reykjavik

Surtsey

Vestmannaeyjar

INDIA

Calcutta

Assam

Digboi

Bengal

Calcutta

Gujarat

Gir forest

Surat

Hindustan

Kerala

Idikki

INDIA—*continued*

Rajasthan

Ecological protests by women, *1040*
Mass death of "tree huggers", *1094*
Pacifist ecological protests in India, *2644*

Jaipur

Oil well in India, *3689*

Uttar Pradesh

Fatepur Sikri

Abandonment of a city in India because of a failure of the water supply, *2127*

Gopeshwar

Nonviolent protest of significance by the modern Chipko movement in India, *1041*

Patlidun Valley

Tiger preserve in India, *4830*

INDONESIA

National population policy for Indonesia, *3889*
Natural resources exploited in Indonesia, *3339*
Rhinoceros listings under the U.S. Endangered Species Act, *2398*

Bali

Bali starling reintroductions, *2433*

Borneo

Commerce in cave swiftlet nests, *1645*

Java

Appearance of wet-rice cultivation in paddy fields, *1198*
National preserve created to protect the Javan rhinoceros, *4814*
Volcano eruption known to cause catastrophic tsunamis (giant sea waves), *4372*
Volcano eruption known to cause massive deaths in Indonesia, *4360*
Volcano eruption of significance in Java, *4364*

Jakarta

Offshore natural gas plant to separate liquids from the gas, *2055*

Krakatoa

Volcano eruption known to cause catastrophic tsunamis (giant sea waves), *4372*
Volcano eruption to be audible from 3,000 miles away, *4371*

Lesser Sundas

Sumbawa

Documented worldwide climate change from volcanic eruptions, *1318*
Volcano to cause deaths by famine in Indonesia, *4366*

Sulawesi (Celebes)

Zoo to exhibit rare black apes from the Indonesian island of Celebes (Sulawesi), *4899*

Sumatra

Petroleum discovery in Sumatra, *3691*
Volcano eruption known to cause catastrophic tsunamis (giant sea waves), *4372*

INTERNATIONAL

Accord on transboundary effects of industrial accidents on the environment, *3047*
Agreement among nations to work together to create guidelines, policies, and procedures for protecting the ozone layer, *3041*
Agreement governing treatment of Antarctic wildlife and plants, *3026*
Agreement on International Humane Trapping Standards, *2929*
Annual report on the environment and American foreign policy, *2308*
Arbitration ruling to assign national responsibility for air pollution damage, *1389*
Arctic Environmental Protection Strategy (AEPS), *3873*
Atomic energy organization of the United Nations for peaceful nuclear power, *3478*
Bilateral negotiations between the United States and Canada on acid rain, *3038*
Bilateral U.S.–Canadian agreement to waive "consent notification" for hazardous waste shipments from the United States into Canada, *2873*
Bilateral U.S.–Mexican agreement requiring American companies in Mexico to return hazardous waste to the United States, *2867*
Binding international treaty to address regional air pollution, *3039*
Canadian–U.S. International Boundary Treaties Act, *3366*
Canadian–U.S. treaty to regulate use of the Niagara River for water power, *4555*
Catalogue of critical world wetlands, *3150*
Chlorofluorocarbons (CFCs) limitation on an international level, *1985*

Climate-change international panel, *1844*
Climate research agreement between the United States and Soviet Union, *1830*
Comprehensive overview of the global environmental problem, *3262*
Comprehensive United Nations agreement on use of the oceans, *3592*
Conference held after the signing of the Protocol on Substances That Deplete the Ozone Layer (Montreal Protocol), *1990*
Conference on Third World environmental problems, *2291*
Controls on the import and export of plants and produce, *3023*
Convention of people affected by the construction of dams, *2078*
Dam opposition day worldwide, *2081*
Dam organization of international significance, *2068*
Dangerous wastes law passed by the European Economic Community (EEC), *3954*
Declaration of phosphates as a major source of water pollution issued jointly by two countries, *4404*
Drought education and training center for developing nations, *2160*
Drought to draw worldwide aid through television, *2158*
Drought year producing worldwide environmental responses, *2162*
Earth Day to be celebrated around the world, *1089*
Effective international action to curb acid rain, *1414*
Emergence of desertification as an international issue, *4044*
Environmental health organization of European women, *3967*
Environmental law challenged under the North American Free Trade Agreement (NAFTA), *3971*
Environmental lawsuit decided by the International Court of Justice, *3011*
Environmental quality monitoring agency of the European Union (EU), *4312*
Environmental treaty dealing with a particular ecosystem, *1744*
Extensive collection of quality data on fertility and family planning, *3882*
Felony indictments under a U.S. environmental law for smuggling hazardous waste from California into Mexico, *2889*
Freedom-of-the-seas doctrine, *3546*
General Fisheries Council for the Mediterranean, *2474*
Global agreement on all aspects of biodiversity, *3048*
Global agreement to protect plant and animal species from unregulated international trade, *3036*
Global conference on environmental issues, *1840*
Global environmental conference to spur activism, *1070*
Global list of endangered plants, *2421*
Global warming report to gain international scientific attention, *1838*
Government policy in the United States to deny aid to foreign family planning agencies that performed or promoted abortions as a method of population control, *3884*
Government report issued jointly by the United States and Canada to rate acid rain a serious environmental and transboundary problem for both countries, *1477*
Hazardous waste transportation law passed by the European Economic Community (EEC), *3960*
Inter-American institute for research on global change, *3064*
Inter-American Tropical Tuna Commission, *2460*
Interagency Grizzly Bear Committee (IGBC), *4785*
Intergovernmental agency devoted to marine and fisheries science, *3565*
Intergovernmental agency devoted to marine and fisheries science of the Northern Pacific, *3596*
Intergovernmental carnivore reintroduction strategy in Europe, *2442*
Intergovernmental conference on environmental education, *2300*
Intergovernmental conservation conference in North America, *3367*
Intergovernmental organization devoted exclusively to protection of migratory species worldwide, *3062*
International accord requiring a reduction on nitrogen oxide levels, *1415*
International Adopt-a-Highway Day, *3290*
International agency to address pollution of the oceans by ships, *3574*
International agreement on cross-border air pollution, *1411*
International agreement on hazardous waste, *2887*
International body to govern use of the ocean, *3053*
International body to recognize the obligations of countries to fight and prevent marine pollution, *3051*
International body to regulate conservation of Atlantic tunas, *3060*
International climate conference held exclusively for developing countries, *1846*
International compact barring the dumping of wastes into the oceans, *4485*
International compact on marine pollution from discharges of ships and aircraft, *4494*

INTERNATIONAL—*continued*
International conference on air pollution, *1461*
International Conference on Polar Bears, *4775*
International conference on the connection between soil health and human health, *4027*
International conference on wetlands, *3374*
International conference to address world population problems, *3896*
International conference to discuss global environmental issues, *1060*
International convention attempting to regulate Antarctic mineral exploitation, *3044*
International convention on biodiversity and sustainable development, *3049*
International convention on ocean pollution, *3577*
International convention on transporting hazardous waste, *3045*
International convention protecting wetlands, *3028*
International convention to address oil pollution of the oceans, *3024*
International convention to preserve scenery, culture, and the environment, *3987*
International core list of hazardous substances, *2880*
International Council for the Exploration of the Sea (ICES), *3566*
International declaration on El Niño, *1968*
International Fisheries Gene Bank (IFGB), *2499*
International forestry research agency, *2694*
International Joint Commission on Boundary Waters study of Lake Erie, Lake Ontario, and the upper St. Lawrence River, *4511*
International journal devoted to protected areas issues, *3810*
International meeting about desertfication, *4045*
International meeting of American Indians and environmentalists, *4408*
International monitoring system for air pollution, *1465*
International moratorium on above-ground nuclear weapons testing, *3484*
International organization of significance dedicated to protecting and conserving global resources, *1037*
International organization to manage Antarctic wildlife, *4783*
International organization to monitor the disappearance of amphibians, *4737*
International organization to promote the timber trade, *2702*
International organization to promote water quality monitoring, *4594*
International organization to regulate tuna fisheries, *2475*
International Ornithological Congress, *1711*
International professional forestry society, *2706*
International public health agency, *3925*
International registry of data on chemicals in the environment, *2292*
International Standards Organization (ISO) certification program for environmental management, *4318*
International treaty to ban the reproductive cloning of human beings, *3055*
International treaty to propose African wildlife preserves, *3019*
International treaty to protect birds, *1743*
International treaty to protect the Earth's atmosphere, *3043*
International wildlife treaty upheld by the U.S. Supreme Court, *3018*
Joint antipollution pact in the Mediterranean region, *3040*
Joint arctic weather stations, *1948*
Joint commitment to "zero discharge" of dangerous chemicals into Lake Superior by the United States and Canada, *4442*
Joint effort of major significance to solve pollution problems in the Great Lakes, *4476*
Joint United Nations organization to assess scientific information on climate change, *1831*
Joint United Nations–World Council of Churches "Environmental Sabbath", *1086*
Large-scale international private reforestation campaign aimed at individuals and communities, *2731*
Large-scale United Nations program to fund forestry projects, *2705*
Laws regulating space debris, *3032*
League of Nations conference of world timber experts, *2703*
Legally binding international document exclusively about global warming, *1850*
List of wetlands of international importance, *3016*
Mandatory international Atlantic bluefin tuna conservation plan, *2486*
Maps of the trade winds, *1895*
Meeting of the contracting parties to the Basel Convention on the Control of Transboundary Movements of Hazardous Wastes and Their Disposal, *2891*
Meeting of the Intergovernmental Program on Climate Change (IPCC), *1832*
Meteorological organization in Europe organized specifically to control weather satellites, *1967*
Meteorology satellite launched by Europeans, *1964*
Modern world conference on population, *3898*
Multilateral treaty to address an air pollution problem, *1397*
Multilingual forest terminology, *2742*
Negotiating session to prepare for a global convention on the control of transboundary movements of hazardous wastes, *2885*

North American organization to monitor the disappearance of amphibians, *4738*
Nuclear energy agreement between the U.S. and China, *3523*
Nuclear-generated electricity exported from Canada, *3521*
Organization for the control of desert locusts, *3059*
Organization of African Unity (OAU) population agency, *3899*
Organization to gather international forestry statistics, *2699*
Organization to regulate dumping of wastes in the high seas, *3061*
Organization to study the North Pacific, *3063*
Ozone fund for developing nations, *1989*
Ozone-layer committee of the United Nations, *1977*
Ozone pollutant agreement on an international level, *1987*
Ozone research coordination program of the European Community (EC), *1991*
Papal declaration regarding world exploration, *3007*
Park to be designated an international peace park, *3798*
Permanent forum for coordinating sustainable development in the Arctic, *4319*
Permanent United Nations sustainable development agency, *4313*
Prohibition against nuclear weapons on the ocean floor, *3584*
Proposed concept of entitlement assistance for environmental conservation, *2294*
Public health agency of the United Nations, *3934*
Recommendations for protecting landscapes issued by the United Nations, *3977*
Regional antipollution pact covering the South Pacific region, *4497*
Report on the health of the oceans by the United Nations–sponsored Group of Experts on the Scientific Aspects of Marine Pollution (GESAMP), *4495*
Scholarly text on the international law of the sea, *3229*
Significant multinational agreement to curb pollution of the Mediterranean Sea, *4480*
Specific treaty guidelines for Antarctic ecosystem protection, *3046*
Statement of the idea that nations should be held responsible for their actions just as individuals were, *3009*
Sturgeon listing by the Convention on International Trade in Endangered Species of Wild Fauna and Flora (CITES) to include all species, *2422*
Superpower to effectively withdraw from the Kyoto Protocol, *3012*
Time-constrained goals to be adopted by a group of nations to curb water pollution, *4410*
Transfrontier protected area for marine turtles, *4842*
Treaty between nations dividing the world for exploration, *3013*
Treaty in which England recognized the fishing rights of the United States, *3014*
Treaty obliging the United States to maintain a list of endangered species, *3021*
Treaty protecting the polar bear and its natural habitat, *3034*
Treaty setting the framework to address the production of greenhouse gases, *3050*
Treaty to ban the introduction of nuclear weapons into orbit around the Earth, *3027*
Treaty to combine preservation of natural and cultural heritage sites, *3033*
Treaty to establish national parks and preserves and to identify protected flora and fauna in the Western Hemisphere, *3022*
Treaty to insure humane slaughter of animals, *3037*
Treaty to recognize the three-mile limit for fishing off the North American coast, *3015*
Tri-national conference on conservation in North America, *1034*
Trilateral agreement to protect the Wadden Sea, *3010*
Trilateral biosphere reserve, *1642*
Truly international forestry literature, *2739*
U.S.–Canadian bilateral report on acid rain, *3042*
U.S.–Canadian effort of significance to improve the water quality of the Great Lakes, *3029*
United Nations agency for management of hazardous chemicals, *2975*
United Nations agency to develop environmental education and research programs, *2282*
United Nations agreement on marine pollution from airborne sources, *3588*
United Nations agreement on marine pollution from offshore industrial facilities, *3583*
United Nations agreement to regulate dumping of waste into the oceans, *3587*
United Nations attempt to define civil liability for oil pollution resulting from seabed drilling, *3591*
United Nations compensation fund for marine oil pollution damage, *3586*
United Nations compensation policy on marine oil pollution, *3582*
United Nations condemnation of French atmospheric nuclear weapons tests, *3509*
United Nations conference devoted to the global environment, *3030*
United Nations conference linking biodiversity and sustainable development, *1622*

INTERNATIONAL—*continued*
United Nations conference to address human habitation in a sustainable development context, *3901*
United Nations conference to link the rights of women to population growth issues, *3900*
United Nations convention on African natural resources, *3378*
United Nations convention on transboundary water supplies, *4616*
United Nations convention to define civil liability in nuclear accidents at sea, *3585*
United Nations convention to protect the vicuña, *4777*
United Nations emergency response system for industrial pollution disasters, *3002*
United Nations Environment Programme Governing Council meeting on the ozone layer, *1976*
United Nations hydrological agency, *4596*
United Nations meeting directly concerned with peaceful uses of nuclear energy, *3473*
United Nations moratorium on drift net fishing, *2480*
United Nations organization concerned with global warming, *1842*
United Nations organization to establish its headquarters in an underdeveloped country, *1072*
United Nations sustainable development strategy, *4310*
United Nations training center for water supply issues, *4613*
Uranium fuel bought from Russia by the United States, *3535*
Use of the term *mare liberum* or "freedom of the seas", *4488*
Version of the World Conservation Union (IUCN), *3058*
Volcano information network of significance set up internationally, *4378*
Voluntary international principles to conserve the world's forests, *2715*
Wildlife protection conference of modern African governments, *4773*
Women's international conference on the environment, *1046*
Workshop hosted by the European Community on the ozone layer, *1986*
World Charter for Nature, *1059*
World Climate Conference, *1829*
World Forestry Congress, *2701*
World Health Organization (WHO) working group to address noise problems, *3431*
World Meteorological Organization (WMO) Congress, *1822*
World Wetlands Day, *3418*
WorldWIDE Network Inc. meeting, *1044*
Worldwide organization concerned with civil aviation safety, *3933*
Worldwide organization to advocate family planning, *3897*

IRAN

Animal to be domesticated, *1138*
Biosphere reserves in the Middle East, *1626*
Dam known to have a curved shape, *2084*
Description of an oil production industry, *3670*
Domestication of the grain barley, *1120*
International oil cartel, *3622*
Law to discourage population growth in Iran, *3895*
Major oil producer to fully nationalize the oil industry, *3668*
Oil discovery in the Persian Gulf, *3654*
Organization for the control of desert locusts, *3059*
Systematic survey of the endangered Asiatic cheetah, *2353*
Windmills, *4846*

Mas jid-i-Suleiman

Oil discovery in the Middle East, *3694*
Oil discovery of significance in Persia, *3695*

IRAQ

Agricultural collapse caused by irrigation, *1212*
Domestication of pigs, *1140*
Domestication of sheep and goats, *1139*
Environmental damage of significance resulting from irrigation, *1211*
Human practice of sedentary agriculture, *1230*
Indications of private landowning, *3114*
Irrigation, as distinguished from dry farming, *1210*
Large carnivores to be held in captivity, *4874*
Large-scale epidemic of fungicide poisoning from treated grain seeds, *1292*
Oil pipeline of significance in Iraq, *3620*
Semi-permanent human settlements, *3118*
Surviving account of a great flood covering the Earth, *2518*

Baba Gurgur

Oil well that was productive in Iraq, *3698*

Baghdad

Bombing of a nuclear reactor, *3519*

IRELAND

Dating of the origin of Earth through biblical analysis, *2778*
Famine as a result of potato blight, *1183*
Instrument to accurately measure the length of time the Earth's surface receives sunshine, *1901*

Invasion of England to be halted by gale-force winds, *4176*
Nuclear emergency plan by the Irish government, *3531*
Tax in Ireland on shoppers' plastic bags, *3291*
United Nations attempt to define civil liability for oil pollution resulting from seabed drilling, *3591*

Dublin

Significant study of North American marine algae, *3555*
Systematic catalog of earthquakes, *2201*
Theory stating that earthquakes come from the Earth's crust and not from outside forces working on the crust, *2202*
World seismic map of significance, *2203*

Kinsale

Natural gas extraction in Ireland, *2059*

Waterford

Report of the death of the last great auk in the British Isles, *2361*

ISRAEL

Biblical prophet to use a famous earthquake to date his prognostications, *2172*
Bombing of a nuclear reactor, *3519*
Domesticated plants genuinely cultivated, *1195*
Human herding of semi-domesticated gazelles, *1137*
Ozone depletion recorded over the Dead Sea, *1997*

ITALY

Appearance of the American vine aphid in Europe, *1505*
Appearance of the Asiatic chestnut blight in Europe, *2602*
Earthquake intensity scale, *2195*
Formulation of the Principle of Original Horizontality, *2779*
General Fisheries Council for the Mediterranean, *2474*
Griffon vulture reintroduction in Italy, *2439*
Herbarium, *1751*
International treaty to propose African wildlife preserves, *3019*
Introduction of tobacco plants into Europe, *1124*
Introduction of tomatoes to Europe, *1201*
Limit on lead in gasoline initiated by the Italian government, *1580*
Modern disaster blamed on faulty geologic analysis, *2118*
Modern tollways, *1559*
Person to imagine the possibilities of power-driven vehicles, *1531*
Public health use of DDT, *3931*
Raven reintroduction in Italy, *2436*
Research showing that human activity could cause irretrievable damage to the earth, *1048*
Scale widely used to measure earthquake intensity, *2212*
Significant use of the natural heat of volcanic systems for electrical power in Italy, *4376*
Snow avalanche to kill Austrian and Italian troops during a war, *4202*
Statement of the theory of groundwater, *2776*
Stratigraphy theory, *2781*
Volcano eruption that was documented, *4354*
Written reference to a mousetrap, *2901*

Arcetri

Observer to realize that the atmosphere becomes colder at the upper levels nearer the heat-giving sun, *1861*

Assisi

Patron saint of ecology, *1076*

Bologna

Recognition of the correlation between cellular rings in trees and age, *1759*

Castellammare

Bathyscaph, *3576*

Chiavenna Valley

Landslide known to kill more than 2,400 people, *2571*

Florence

Instrument to measure atmospheric pressure, *1863*
Plant taxonomist, *1753*
Sealed thermometer, *1862*

Herculaneum

Volcano eruption whose devastation was fully documented, *4356*

Lardarello

Geothermal power plant, *2815*

Messina

Black Death (bubonic plague) outbreak in Europe, *3904*

ITALY—*continued*

Mirafiori

Mass-produced FIAT Topolino, *1543*

Modena

Identification of some disabling industrial diseases, *1354*

Naples

Marine biology station of significance, *3559*
Oyster farms, *2488*
Theory stating that earthquakes come from the Earth's crust and not from outside forces working on the crust, *2202*
Volcano eruption to end a World War II battle, *4379*
Volcano eruption whose devastation was fully documented, *4356*

Padua

Open-air thermometer, *1860*
Professorship of botany in Europe, *2256*

Paestum

Rapid deforestation of Italy, *2581*

Pisa

Botanical garden in Europe, *1750*

Pompeii

Volcano eruption whose devastation was fully documented, *4356*

Ravenna

Rapid deforestation of Italy, *2581*

Rome

Aedile office in Rome, *4519*
Aqueduct built to carry water into Rome, *4568*
Catalog of Roman plants, *1747*
Catalog of significance of plants in the ancient world, *1748*
Exotic animal spectacle in the Roman Empire, *4878*
Greenhouses known in the West, *4079*
Known exploitation of migratory swallows' homing instincts, *1644*
Municipal sanitation crews, *4137*
Observer known to write about the toxicity of lead, *2827*
Organization to gather international forestry statistics, *2699*
Public baths in the West to use solar energy for heat, *4080*
Published mentions of serious air pollution in classical Rome, *1316*
Record of transparent glass windows in the West, *4082*
Recorded flood in Europe, *2519*
Sewers of Rome, *4518*
Truly international forestry literature, *2739*

Seveso

Major release of the pesticide dioxin into the atmosphere, *1300*

Stabiae

Volcano eruption whose devastation was fully documented, *4356*

Venice

Signs of a timber shortage caused by the deforestation of western Europe, *2583*

Volterra

Known exploitation of migratory swallows' homing instincts, *1644*

IVORY COAST

Biosphere reserves in Africa, *1628*

Abidjan

Bird flight recorded at 37,000 feet, *1706*

JAMAICA

National population policy for Jamaica, *3890*

Kingston

Earthquake to cause massive loss of life in Jamaica, *2184*

Port Royal

Archaeology site preserved by an earthquake, *2185*
Earthquake to cause massive loss of life in Jamaica, *2184*

JAPAN

Appearance of wet-rice cultivation in paddy fields, *1198*
Automobile emissions requirements in Japan, *1578*

Coal mining accident in the Far East causing catastrophic deaths, *3328*
Discovery of "burning water" (petroleum) in Japan, *3604*
Earthquake to register a magnitude of 9 or more on the Richter scale, *2225*
Epidemiological evidence of the association between passive smoking and lung cancer, *1476*
Formal earthquake prediction research program in quake-prone Japan, *2227*
Japanese flora description printed in English, *1763*
Nuclear power agency of the Japanese government, *3476*
Nuclear power commercial reactor in Japan, *3505*
Observer to propose a link between volcanic eruptions and the weather, *1809*
Oil spill of significance off Japan, *3635*
Pollution-related national health compensation law, *3947*
Scientific formulation of the "fault theory" as we know it, *2208*
Scientist to study snow with modern X-ray equipment, *4207*
Seismology association, *2205*
Typhoon known to have prevented an invasion of Japan, *4222*

Chiba

Industrial-scale selective catalytic reduction system, *2299*

Chichi Shima

Report of the extinction of the Bonin Islands thrush, *2360*

Hiroshima

Use of an atomic weapon in war, *3454*

Honshu

Volcano eruption of significance in Japan, *4349*

Kamakura

Earthquake in Japan causing massive deaths, *2180*

Kii

Tsunami (giant wave) with great loss of life recorded in Japan, *2523*

Kobe

Earthquake workshop canceled by an earthquake, *2235*

Kyoto

International treaty to set binding limits on carbon dioxide emissions, *3056*

Minamata

Mass fatalities from eating poisoned fish and shellfish, *3936*
Minamata Disease or "cats dancing disease" cases, *2972*
Widespread incidence of poisoning on account of mercury pollution, *4434*

Mishima

Oil refinery construction halted by environmental protests in Japan, *3705*

Mutsu

Nuclear-powered ship launched by Japan, *3506*

Numazu

Oil refinery construction halted by environmental protests in Japan, *3705*

Osaka

International horticultural show in Asia, *2315*

Shimizu

Oil refinery construction halted by environmental protests in Japan, *3705*

Tochigi

Automobile emission cleaner than city air, *1615*

Tokai

Nuclear accident reported falsely in Japan, *3538*

Tokyo

Car produced by the firm that would become the Nissan Motor Company of Japan, *1538*
City to reach one million in population, *3877*
Disabling smogs in Tokyo, *1351*
Earthquake and tsunami (giant wave) to destroy Japan's capital, *2186*
Public aquarium in Japan, *4900*
Seismograph in its modern form, *2204*

Toyama

Industrial pollution lawsuit in Japan that was successful, *2974*

Tsuruga

Nuclear power plant accident of significance in Japan, *3518*

JAPAN—*continued*

Yokohama

World tropical timber marketing organization, *2687*

Izu Islands

Torishima

Short-tailed albatross protection law, *2383*

Ryukyu Islands

Recorded nuclear weapons accident involving a U.S. warship at sea, *3503*

Okinawa

Tsunami to cause more than 100,000 deaths in Japan, *4177*
World's fair with an ocean theme to be sanctioned by the Bureau of International Expositions (BIE), *2311*

JORDAN

Abandonment of villages because of deforestation and resulting soil erosion, *4046*
Ozone depletion recorded over the Dead Sea, *1997*

Jawa

Dams built whose remains have been discovered, *2082*

KENYA

Descriptions of Mount Kilimanjaro and Mount Kenya to reach Europe, *4630*
Earth Day to be celebrated around the world, *1089*
Environmental group in Kenya to tie deforestation to poverty, *1043*
Laws in Kenya against commercial trade in wildlife trophies or products, *2242*
Mass burning of poached African ivory, *4788*
Natural resource exploited in Kenya, *3363*
Nile perch in Lake Victoria, *1524*
Year that tourism was Kenya's leading source of foreign currency, *2249*

Kericho

Rocket trials for hailstone protection, *4218*

Nairobi

Elephant and rhinoceros orphanage in Kenya, *4755*
Modern study linking rain forest deforestation with accelerating species extinction due to habitat loss, *1632*
United Nations conference devoted to the global environment, *3030*
United Nations organization to establish its headquarters in an underdeveloped country, *1072*

Nyeri

Trees planted by the Green Belt Movement, *1098*

KUWAIT

Ecological disaster caused by a deliberate oil spill, *3645*
Intentional oil-well fires of significance, *3646*

LEBANON

General Fisheries Council for the Mediterranean, *2474*
Human herding of semi-domesticated gazelles, *1137*

LIBERIA

Pygmy hippos known to Western science, *4695*

MACEDONIA

World Heritage sites in Europe chosen for environmental reasons, *3996*
Written description of fly fishing, *2448*

MADAGASCAR

Biosphere reserve in Madagascar, *1638*
Lemurs reintroduced to Madagascar, *2445*
Natural resources exploited in Madagascar, *3362*

Masoala Peninsula

National Environmental Action Plan (NEAP) in Africa, *3395*

MADEIRA ISLANDS

Porto Santo

Rabbit infestation due to introduction, *1483*

MALAYSIA

Asian elephant tracked with a satellite transmitter, *4744*

Efforts to develop ecotourism in the Lake Kenyir region of Malaysia, *2250*
Experimental rubber plantations in Malaya, *1180*
Transfrontier protected area for marine turtles, *4842*
Zoo built for nocturnal viewing, *4965*

Sabah (North Borneo)

Commerce in cave swiftlet nests, *1645*
Wildlife films of the fauna of Borneo, *4697*

Sarawak

Miri

Oil well in Malaysia, *3696*

MALI

Tombouctou

Great flood at Tombouctou on the River Niger in present-day Mali, *2525*

MARSHALL ISLANDS

Atmospheric tests of a new, more powerful atomic weapon, the hydrogen or thermonuclear bomb, *1322*
Major public outcry on atmospheric nuclear-weapons testing, *1323*

MARTINIQUE

Hurricane in the Atlantic causing massive deaths, *4238*

St. Pierre

Volcano causing massive deaths in the West Indies, *4375*

MAURITIUS

Biosphere reserves in Africa, *1628*
Dodos encountered by Europeans, *2356*

MEDITERRANEAN REGION

Cultivation of vines, olives, and figs, *1197*
Earthquake to kill more than 1 million people, *2179*
Introduction of orange, lemon, and lime trees into Spain and the western Mediterranean, *1199*
Laws to locate sources of objectionable odors and smoke downwind of city walls, *1371*

MESOPOTAMIA

Agricultural collapse caused by irrigation, *1212*
Domestication of pigs, *1140*
Domestication of sheep and goats, *1139*
Human practice of sedentary agriculture, *1230*
Semi-permanent human settlements, *3118*

Babylonia

Surviving account of a great flood covering the Earth, *2518*

Khuzistan

Irrigation, as distinguished from dry farming, *1210*

Sumer

Environmental damage of significance resulting from irrigation, *1211*
Indications of private landowning, *3114*

Ur

Large carnivores to be held in captivity, *4874*

MEXICO

Appearance of modern sedentary ranching in the Western Hemisphere, *1157*
Appearance of the wooly whitefly in California, *1193*
Bilateral U.S.–Mexican agreement requiring American companies in Mexico to return hazardous waste to the United States, *2867*
Biodiversity governmental agency in Mexico, *1641*
Cabinet-level environmental protection agency in Mexico, *2994*
Dam with a height greater than 214 meters (800 feet), *2100*
Domesticated plants genuinely cultivated, *1195*
Felony indictments under a U.S. environmental law for smuggling hazardous waste from California into Mexico, *2889*
Horses reintroduced to the American mainland, *1488*
Intergovernmental conservation conference in North America, *3367*
Introduction of tomatoes to Europe, *1201*
Mexican spotted owl listing under the U.S. Endangered Species Act, *2418*
National Scenic Trail in the Rocky Mountains, *3989*
Oil wells seized from foreign companies, *3699*
Prairie dog listed under the U.S. Endangered Species Act, *2397*
Silkworms in the Western Hemisphere, *1490*
Tri-national conference on conservation in North America, *1034*
Use of the term *ecotourism*, *2247*

MEXICO—*continued*
Volcano eruption known to occur in Mexico, *4355*

Baja California Norte

Michoacán

Discovery of the monarch butterfly's winter home, *4729*

Chiapas

Eruption in modern times of the Mexican volcano El Chichon in the southeastern state of Chiapas, *4373*

Chihuahua

U.S. National Wild and Scenic River in the Chihuahuan Desert, *3993*

Mapimi

Biosphere reserves in Mexico, *1630*

Coahuila

Villa Acuña

Reservoir on the U.S.-Mexico border, *4620*

Colima

Isla San Benedicto

Species of bird to be eliminated by volcanic eruption, *2373*

Minatitlán

Landslide in Mexico on record to kill more than 2,000 people, *2573*

Federal District

Mexico City

Giant panda born outside of China, *4757*
High-yield crops were developed, *1132*

Hidalgo

Pachuca de Soto

Dam burst to cause a disaster by chemical contamination, *2114*

Michoacán

Michilia

Biosphere reserves in Mexico, *1630*

Tamaulipas

Reintroduction of Aplomado falcons to Mexico, *2446*

Ebano

Crude oil production in Mexico, *3615*

Veracruz

Veracruz

Hurricane to sink a fleet in the Gulf of Mexico, *4227*

Yucatan

Drought of major significance in Central America, *2123*
Oil platform blowout, *3640*
Yellow fever epidemics in the New World, *3907*

Zacatecas

Natural resource exploited in Mexico, *3352*

MICRONESIA

National population plan in Micronesia, *3885*

MIDDLE EAST

Embargo on oil supplies by the oil producers of the Middle East, *3669*
Oil discovery in the Middle East, *3694*

MONACO

Monte Carlo

Oceanographic museum, *4916*

MONGOLIA

Black Death (bubonic plague) outbreak, *3903*
Documentation of a *Mononychus* (one claw), a birdlike dinosaur, *1742*

MOROCCO

Tetuan

Wind farm in Africa, *4868*

MOZAMBIQUE

Large-scale relocation of elephants, *4745*

NAMIBIA

Lüderitz

Wind energy–generating station in southern Africa, *4870*

Windhoek

Snowfall in living memory in the Kalahari Desert in Namibia, *4212*

NEPAL

World Heritage site in Asia, *3994*

Annapurna

Environmental program to limit ecotourism damage in Nepal, *2248*

Kathmandu

Sustainable development organization for mountain regions, *4304*

NETHERLANDS

Area in Europe to become industrialized, *2968*
Clearing of the natural forests of northern Europe, *2579*
Development of the science of forestry, *2669*
Dutch elm disease description, *2600*
Flood leaving 1 million people homeless in the Netherlands, *2544*
Freedom-of-the-seas doctrine, *3546*
Hazardous waste exchange program in Europe, *2848*
Hunting privileges in a North American colony, *2906*
International core list of hazardous substances, *2880*
Local environmental political parties, *1105*
Major land reclamations in low-country Flanders and Zeeland, *3067*
Mercury thermometer, *1869*
Order to Dutch linen bleachers barring the dumping of wastes in canals, *4444*
Reference work to identify Brazilian fish, *2450*
Renovation of significance to the dike system in the Netherlands, *2546*
Scholarly text on the international law of the sea, *3229*
Statement of the idea that nations should be held responsible for their actions just as individuals were, *3009*
Stratosphere and troposphere discoveries, *1921*
Trilateral agreement to protect the Wadden Sea, *3010*
United Nations attempt to define civil liability for oil pollution resulting from seabed drilling, *3591*
Use of the term *mare liberum* or "freedom of the seas", *4488*
Widespread cultivation of buckwheat in northern Europe, *1202*
Worldwide organization to advocate family planning, *3897*

Amsterdam

Reference work on Brazilian wildlife, *4680*
Report of the extinction of the quagga, *2365*
Tide gauge, *1868*

Dordrecht

Flood to cause massive deaths in the Dordrecht region of Holland, *2522*

Leiden

Practical modern greenhouse, *4086*
Reference work on Brazilian wildlife, *4680*
Scientific botanical garden in Western Europe, *1754*
Tulips in Holland, *1755*

Nordstrand Island

Storm on record to demolish an entire Dutch island, *4230*

The Hague

Environmental lawsuit decided by the International Court of Justice, *3011*
Extensive collection of quality data on fertility and family planning, *3882*

Utrecht

Environmental health organization of European women, *3967*

Friesland

Flood known to cause massive loss of life in Europe, *2520*

NEW GUINEA

Evidence of widespread forest clearing by felling and the use of fire, *2667*

Foya Mountains

Female golden-fronted bowerbird seen, *1708*

NEW ZEALAND

Cattle and sheep in Australia and New Zealand, *1492*

NEW ZEALAND—*continued*

Evidence that the "ozone hole" had expanded outward to include New Zealand, Australia, Argentina, and Chile, *1480*
Ice Age calculations done scientifically, *2004*
Local environmental political parties, *1105*
National environmental political party, *1107*
Scientific description of a moa, *2362*
Volcano eruption in New Zealand, *4357*

Campbell Island

Report of the reappearance of the Campbell Island flightless brown teal, *1673*

Chatham Islands

Reported new sighting of the taiko, *1670*

Dent Island

Report of the reappearance of the Campbell Island flightless brown teal, *1673*

North Island

Hastings

Earthquake of significance in New Zealand, *2220*

Napier

Earthquake of significance in New Zealand, *2220*
Kiwi bred in captivity, *1665*

Papakura

California quail in New Zealand, *1509*

Wairoa

Earthquake of significance in New Zealand, *2220*

South Island

Fiordland

Rediscovery of the takahe, *1667*

Kaikoura Mountains

Discovery of the Hutton's shearwater breeding grounds, *1702*

Taiaroa Head

Albatross colony located on a mainland, *1663*

Stephen Island

Species of bird to be eliminated by a single cat, *2366*

NICARAGUA

Banana plantations in Central America started by a U.S. company, *1181*
Joint biodiversity program between Nicaragua and Costa Rica, *1637*
Volcano eruption mistaken for a war, *4368*

NIGER

National population policy for Niger, *3891*

NIGERIA

Biosphere reserves in Africa, *1628*
Natural resource exploited in Nigeria, *3356*

Oloibiri

Oil discovered in Nigeria, *3702*

NORTH AMERICA

Appearance of the "Little Ice Age,", *1799*
Bird species native to North America to be extinguished, *1650*
Climate change to lead to the peopling of a continent, *2001*
Extinctions of North American animals caused by humans, *2355*
Flood recorded by Europeans in North America, *2524*
Humans to cross the Bering Land Bridge from Asia to North America, *3750*
Hurricane reported in the Americas by a European explorer, *4225*
Introduction of bananas into the Americas, *1177*
Natural gas sighting by Europeans in North America, *2025*
Natural resource exploited in North America, *3354*
Naturalization of the honeybee in North America, *1158*
North American conference on bird mortality and wind power, *4866*
North American organization to monitor the disappearance of amphibians, *4738*
Purple loosestrife in North America, *1498*
Significant study of North American marine algae, *3555*
Tree species in North America to be widely harvested commercially, *2645*
Tri-national conference on conservation in North America, *1034*
Wildlife conference in North America, *4701*

NORTH KOREA

Appearance of wet-rice cultivation in paddy fields, *1198*

NORTHERN HEMISPHERE

NORWAY

Bergen

Vardö

Svalbard

Spitsbergen

OMAN

PAKISTAN

PALESTINE

PANAMA

Canal Zone

PAPUA NEW GUINEA

PERSIA

Mas jid-i-Suleiman

PERU

Callao

PERU—*continued*

Lima

Earthquake to cause massive loss of life in Peru, *2187*
Natural resource exploited in Peru, *3348*

PHILIPPINES

Widespread commercial logging in the Philippines, *2586*

Los Banos

Variety of high-yield rice, *1133*

Luzon Island

Typhoon to sink U.S. warships, *4254*

Olangapo

Successful solid waste program in the Philippines, *4133*

Turtle Islands

Transfrontier protected area for marine turtles, *4842*

POLAND

Biosphere reserves in Europe, *1625*
Elk reintroduced in Europe, *2428*
European bison reintroduction, *2427*
European lynx reintroduced in Poland, *2437*
Trilateral biosphere reserve, *1642*
Warsaw Pact government to recognize that industrial wastes cause health problems, *2979*
World Heritage sites in Europe chosen for environmental reasons, *3996*

Bobraka

Oil wells in Europe, *3680*

Jaslo

Oil refinery, *3685*

Krakow

Warsaw Pact government to recognize that industrial wastes cause health problems, *2979*

Krosno

Street lamps lighted by using oil, *3606*

Skawina

Warsaw Pact government to recognize that industrial wastes cause health problems, *2979*

PORTUGAL

East-west division of the world for exploration, *3008*
International treaty to propose African wildlife preserves, *3019*
Papal declaration regarding world exploration, *3007*
Toxic and hazardous wastes national policy of the Portuguese government, *3959*
Treaty between nations dividing the world for exploration, *3013*

Lisbon

Description of two types of earthquake motion, *2191*
Earthquake causing massive loss of life in Portugal, *2188*
Earthquake causing thousands of deaths in Portugal, *2181*
Scientific investigation of an earthquake, *2189*
Theory on how to locate the epicenter of an earthquake, *2192*
Transatlantic flight powered by ethanol, *4334*

Azores Islands

Transatlantic flight powered by ethanol, *4334*

Madeiras Islands

Sugar cane cultivation in the Madeiras Islands of the Atlantic, *1235*

ROMANIA

Cernavoda

Nuclear power plant in Romania, *3536*

Ploesti

Attack on the German-controlled oilfields at Ploesti, Romania, by the U.S. Air Force, *3667*

ROME, ANCIENT

Aedile office in Rome, *4519*
Aqueduct built to carry water into Rome, *4568*
Catalog of Roman plants, *1747*
Catalog of significance of plants in the ancient world, *1748*
Exotic animal spectacle in the Roman Empire, *4878*
Greenhouses known in the West, *4079*
Known exploitation of migratory swallows' homing instincts, *1644*
Known use of the substance later called asbestos, *1315*

Municipal sanitation crews, *4137*
Observer known to write about the toxicity of lead, *2827*
Oyster farms, *2488*
Public baths in the West to use solar energy for heat, *4080*
Published mentions of serious air pollution in classical Rome, *1316*
Record of transparent glass windows in the West, *4082*
Recorded flood in Europe, *2519*
Recorded use of the word *asbestos*, *1317*
Sewers of Rome, *4518*
Smallpox epidemic of significance recorded, *3902*
Volcano eruption whose devastation was fully documented, *4356*
Written description of fly fishing, *2448*

RUSSIA

Arctic Environmental Protection Strategy (AEPS), *3873*
Establishment of the concept that soils are individual and natural and have their own form and structure, *4069*
Geological survey of Russia, *2784*
Hydroelectric dams on Russia's Volga river, *2069*
Nuclear accident of significance in the Soviet Union, *3482*
Oil pipeline of significance in Russia, *3616*
Recorded imports of timber into England, *2642*
Uranium fuel bought from Russia by the United States, *3535*

Baku

Oil pipeline in Russia, *3614*
Oil well with a modern design, *3677*

Batumi

Oil pipeline in Russia, *3614*

Moscow

Sewer in Moscow, *4529*

Saint Petersburg

Modern "artificial" capital built in Europe, *3094*

Ursinsk

Oil-pipeline spill of significance in Russia publicly reported, *3648*

Komandorskiye Islands

Discovery of the spectacled cormorant, *1646*

Novaya Zemlya

Europeans known to have wintered in the extreme north, *3824*

Siberia

Permafrost, *1806*

SAINT EUSTATIUS

Hurricane in the Atlantic causing massive deaths, *4238*

SAINT HELENA

Introduction of goats on St. Helena Island in the South Atlantic, *1499*

SAINT KITTS AND NEVIS

Basseterre

Hurricane to kill thousands in the Lesser Antilles, *4232*

SAUDI ARABIA

Mahazat as-Sayd Protected Area

Ostriches reintroduced to Arabia, *2438*

SEA

Animal hunted to extinction in North America, *2358*
Bathymetric map of the North Atlantic Ocean, *3556*
Bathyscaph, *3576*
Bathysphere, *3572*
Charts of the Atlantic Ocean winds and currents, *3554*
Charts of the Gulf Stream, *3548*
Comprehensive United Nations agreement on use of the oceans, *3592*
Coral reef field research station in Australia, *3575*
Deep-sea exploration of U.S. marine sanctuaries, *3602*
Deep-sea soundings, *3552*
Earthquake to register an 8.9 on the Richter scale, *2214*
Eutrophication monitoring in the Baltic Sea, *4409*
Evidence of life on the sea floor, *3550*
Executive Order protecting coral reefs in the United States, *3601*
Explorer to sail through the Northeast Passage from Norway to the Bering Sea, *3836*
Explosion of significance on an offshore oil rig, *3642*

SEA—*continued*
Extension of the offshore drilling ban, *3665*
Federal law in the United States to reduce vessel sewage discharges, *3598*
Federal program to protect and restore an estuary in the United States, *3593*
Female oceanographer of distinction, *3597*
Flight across the Arctic Ocean, *3855*
Flood leaving 1 million people homeless in the Netherlands, *2544*
Freedom-of-the-seas doctrine, *3546*
General Fisheries Council for the Mediterranean, *2474*
High-resolution maps of the ocean floor, *3600*
Incineration of liquid organohalogen compounds at sea, *4490*
Intergovernmental agency devoted to marine and fisheries science, *3565*
Intergovernmental agency devoted to marine and fisheries science of the Northern Pacific, *3596*
International agency formed specifically to regulate whaling, *4627*
International agreement on whaling regulations, *3020*
International body to govern use of the ocean, *3053*
International body to recognize the obligations of countries to fight and prevent marine pollution, *3051*
International body to regulate conservation of Atlantic tunas, *3060*
International compact barring the dumping of wastes into the oceans, *4485*
International compact on marine pollution from discharges of ships and aircraft, *4494*
International convention on ocean pollution, *3577*
International convention to address oil pollution of the oceans, *3024*
International Council for the Exploration of the Sea (ICES), *3566*
International whaling commission, *3057*
Joint antipollution pact in the Mediterranean region, *3040*
Law in the United States banning the disposal of plastics at sea, *4148*
Law requiring the U.S. president to develop and conduct a comprehensive program of marine science activities, *3581*
Law to assert control of offshore waters, *3547*
Legislation in the United States to mandate double hulls on oil tankers, *3627*
Long-standing survey of water quality conditions in the Western Approaches to the English Channel, *4399*
Long-term failure of the cod fishery in and around the Norwegian Sea, *1804*
Major destruction by starfish of coral in the Great Barrier Reef, *3579*
Manned descent to the greatest depth on earth, *3578*
Maps of the trade winds, *1895*
Marine biology station in the United States, *3561*
Marine biology station of significance, *3559*
Multi-purpose oceanographic research vessel in the United States, *3571*
Multilateral treaty to address an air pollution problem, *1397*
National Estuary Program (NEP) in the United States, *3595*
Notice of the effects of tributylin (TBT) on marine life, *4492*
Nuclear-powered submarine, *3466*
Nuclear submarine from the U.S. Navy to be lost at sea, *3499*
Ocean Thermal Energy Conversion (OTEC) laboratory established by a U.S. state, *2297*
Oceanographic expedition, *3560*
Oceanographic research vessel to recover a lost hydrogen bomb, *3580*
Oceanographic study of an entire ocean, *3569*
Oceanography institution, *3567*
Oil discovery of significance in the North Sea, *3661*
Oil spill cleanup law in the United States, *3643*
Oil spill of significance off Spain, *3637*
Oil-tanker spill of major significance off Scotland, *3647*
Oil-tanker spill of significance off Wales, *3649*
Oil well blow-out in the North Sea, *3638*
Organization to regulate dumping of wastes in the high seas, *3061*
Organization to study the North Pacific, *3063*
Permit system enacted in the United States for fishing vessels on the high seas, *2484*
Prohibition against nuclear weapons on the ocean floor, *3584*
Recorded appearance of a large haze of air pollution over the Indian Ocean, *1482*
Regional antipollution pact covering the South Pacific region, *4497*
Remote-controlled submersible to reach the oceans' greatest depth, *3599*
Report on the health of the oceans by the United Nations–sponsored Group of Experts on the Scientific Aspects of Marine Pollution (GESAMP), *4495*
Report on the impact of human predation upon coastal marine ecosystems dating back to prehistoric times, *3603*
Scientist to state the law of relative proportion in seawater, *3563*
Significant federal legislation to ban the discharge of sewage sludge in U.S. waters, *4479*
Significant multinational agreement to curb pollution of the Mediterranean Sea, *4480*

Significant study of North American marine algae, 3555
Spectacled cormorant description, 2357
Storm tide warning service in Great Britain, 4188
Supertanker to founder and spill its cargo of oil, 3630
Survey of the coast of the United States, 3549
Theory that coral reefs are founded on sinking underwater land masses like subsiding volcanoes and continental shelves, 3553
Theory that seafloor spreading was caused by convective heat in the Earth's mantle, 2818
Tide-predicting machine, 3564
Time-constrained goals to be adopted by a group of nations to curb water pollution, 4410
Trilateral agreement to protect the Wadden Sea, 3010
U.S. institute for study and research on oceans, 2269
U.S. institute for study and research on oceans, 2269
U.S. legislation to promote Ocean Thermal Energy Conversion (OTEC) development for commercial use, 4565
U.S. National Sea Grant Program, 2289
U.S. oceanographic research ship, 3562
U.S. submarine to carry nuclear-armed ballistic missiles, 3490
Underwater movie seen by the public, 3568
Underwater photograph, 3558
United Nations agreement on marine pollution from airborne sources, 3588
United Nations agreement on marine pollution from offshore industrial facilities, 3583
United Nations agreement to regulate dumping of waste into the oceans, 3587
United Nations attempt to define civil liability for oil pollution resulting from seabed drilling, 3591
United Nations compensation fund for marine oil pollution damage, 3586
United Nations compensation policy on marine oil pollution, 3582
United Nations convention to define civil liability in nuclear accidents at sea, 3585
United Nations moratorium on drift net fishing, 2480
Use of sonar (sound navigation and ranging) to measure ocean depths, 3570
Wind and current analysis of significance, 3551
World's fair with an ocean theme to be sanctioned by the Bureau of International Expositions (BIE), 2311

SIERRA LEONE

National population plan for Sierra Leone, 3894

SINGAPORE

Oil refinery in Singapore, 3704
Toxic waste–carrying ship to draw international media attention, 2876
Worldwide organization to advocate family planning, 3897

SLOVAKIA

Environmental lawsuit decided by the International Court of Justice, 3011
Trilateral biosphere reserve, 1642

SOMALIA

New bird species established solely by DNA tests and photographs, 1687

SOUTH AFRICA

Biological evidence that continents were once joined, 2819
Commercial ostrich farm, 1648
Diamond in South Africa, 3313
Global list of endangered plants, 2421
International treaty to propose African wildlife preserves, 3019
Large-scale relocation of elephants, 4745
Modern environmental management policy in South Africa, 3425
Oil spill of significance off South Africa, 3631
Transfrontier wildlife park in Africa, 4845
Working prototype of a commercially viable pebble-bed nuclear reactor, 3543

Cape Town

Asbestos mine rehabilitation in South Africa, 3963
Botanical garden dedicated to the native flora of a single country, 1792
Dam cost analysis on a global scale, 2079
Natural resource exploited in South Africa, 3355

SOUTH AMERICA

Biological evidence that continents were once joined, 2819
Domestication of the root vegetable potato, 1121
Extensive collection of quality data on fertility and family planning, 3882
Introduction of bananas into the Americas, 1177
Wheat plants grown in the Americas, 1489
Widespread use of South American guano in European agriculture, 1128

SOUTH KOREA

Appearance of wet-rice cultivation in paddy fields, *1198*
Nuclear power agency of the South Korean government, *3486*
Nuclear power public safety organization of the South Korean government, *3966*

Chungnam

Oil refinery in South Korea built with private funds, *3706*

Taejon

Exposition with an environmental theme to be held in Korea, *2317*

SOVIET UNION

Appearance of the Colorado potato beetle in Europe, *1519*
Climate research agreement between the United States and Soviet Union, *1830*
Drought forcing the Soviet Union to buy U.S. grain, *2155*
Drought used for political purposes, *2145*
Drought year producing worldwide environmental responses, *2162*
Earthquake in the Soviet Union allowed open media coverage, *2232*
Environmental movement of significance in the Soviet Union, *1065*
Global warming long-term policy of the government of the Soviet Union, *1843*
International Conference on Polar Bears, *4775*
International moratorium on above-ground nuclear weapons testing, *3484*
Laws regulating space debris, *3032*
Military uses of marine mammals, *4723*
Nature reserve in the Soviet Union, *3796*
Nuclear accident of significance in the Soviet Union, *3482*
Pipeline to carry natural gas from the Soviet Union to a European country, *3708*
Prohibition against nuclear weapons on the ocean floor, *3584*
Radioactive fallout of significance from a space program, *1326*
Reactor meltdown aboard a nuclear-powered ship, *3504*
Soviet law to ban polar bear hunting, *2956*
Treaty to ban the introduction of nuclear weapons into orbit around the Earth, *3027*
Wolves reintroduced to Soviet Georgia, *2429*

Chernobyl

Nuclear power plant disaster with worldwide fallout, *3528*

Moscow

Widespread deaths of trees in Moscow attributed to air pollution, *1320*

Murmansk

Tidal power plant north of the Arctic Circle, *4563*

Obninsk

Nuclear power station, *3467*

Sverdlovsk

Generally known instance of a river catching fire because of pollution, *4437*

Siberia

Bratsk

Snowflakes reported to measure 8 by 12 inches, *4209*

Irkutsk

Environmental movement of significance in the Soviet Union, *1065*

SPACE

All-weather satellite, *1960*
Close-up evidence of volcanoes on Mars, *4383*
Evidence of volcanic activity on Triton, a moon of the planet Neptune, *4394*
Evidence that Martian volcanic rocks are in some respects similar to Earth's, *4395*
Geologist to go to the Moon, *2825*
Landsat satellite, *2293*
Laws regulating space debris, *3032*
Meteorological satellite funded by the U.S. Weather Bureau, *1961*
Meteorology satellite launched by Europeans, *1964*
Multilateral treaty to address an air pollution problem, *1397*
Orbiting geophysical observatory, *2822*
Ozone-layer vertical profile from space, *1981*
Ozone measurements of chlorine monoxide from space, *1992*
Photographs of erupting volcanoes on the Jovian moon Io, *4387*
Recorded evidence of quakes on the Moon, *2229*
Satellite instruments powered by solar energy, *4112*
Satellite station built to track "space litter", *3289*
Satellite to offer useful weather data, *1956*

Satellite to provide cloud-cover photographs, *1825*
Satellite to transmit color photographs of the full Earth face, *2823*
Satellite to transmit weather information to the earth, *1957*
Satellite to use nuclear power for electricity, *3496*
Seismographs on Mars, *2230*
Solar-powered vehicle to travel into deep space, *4123*
Treaty to ban the introduction of nuclear weapons into orbit around the Earth, *3027*

SPAIN

East-west division of the world for exploration, *3008*
Evidence of bullfighting as a sport, *1001*
Hazardous waste law enacted by the Spanish government, *3965*
International treaty to propose African wildlife preserves, *3019*
Introduction of orange, lemon, and lime trees into Spain and the western Mediterranean, *1199*
Introduction of the potato into Europe, *1203*
Introduction of tobacco plants into Europe, *1124*
Introduction of tomatoes to Europe, *1201*
Silkworms in the Western Hemisphere, *1490*

Almanza

Masonry dam built that is still in use, *2085*

Barcelona

Flood of significance in Spain during the 20th century, *2548*

Cadiz

Earthquake causing massive loss of life in Portugal, *2188*

Corunna

Oil spill of significance off Spain, *3637*

Madrid

International organization to regulate tuna fisheries, *2475*

Palomares

Oceanographic research vessel to recover a lost hydrogen bomb, *3580*

Rivaldelago

Dam disaster of significance in Spain in the 20th century, *2117*

Tordesillas

Treaty between nations dividing the world for exploration, *3013*

SRI LANKA

Use of biological control against an unwanted plant, *1273*
Widespread damage from coffee leaf disease in Ceylon (Sri Lanka), *1187*

Pinnawela

Elephant captive breeding program in Sri Lanka, *4758*

SUDAN

International treaty to propose African wildlife preserves, *3019*
National population plan for Sudan, *3887*
Natural gas well in Sudan, *2053*

SWEDEN

Acid rain measurement monitoring systems, *1450*
Arctic Environmental Protection Strategy (AEPS), *3873*
Balloon logging experiments, *2663*
Ban on mercury fungicides in agriculture, *1296*
Ban on phenylmercury as a slimicide in the pulp and paper industry in Sweden, *4468*
Carefully documented reports of a link between acid rain and damage to fish populations, *1460*
Local environmental political parties, *1105*
Mathematical equation to calculate atmospheric warming from carbon dioxide concentrations, *1438*
Nuclear energy law allowing expropriation of power plants in Sweden, *3539*
Oil spill of significance off Sweden, *3633*
PCB (polychlorinated biphenyl) testing to show that the chemical was highly toxic and dangerous, *2846*
Reference to the problem of acid rain, *1435*
Total ban of chlorofluorocarbons (CFCs) for aerosols by a country, *1982*
United Nations agreement on marine pollution from offshore industrial facilities, *3583*
United Nations attempt to define civil liability for oil pollution resulting from seabed drilling, *3591*
Use of the term *genecology* to describe the study of genetic variations within individual plant species as they relate to the environment, *1793*
Worldwide organization to advocate family planning, *3897*

SWEDEN—*continued*

Kiruna

Weather satellite monitoring station north of the Arctic Circle, *1962*

Stockholm

Classification of minerals by their chemical composition, *2794*
Ice Age calculations done scientifically, *2004*
Nobel Prize was awarded to ozone-layer researchers, *1972*
Research showing that PCBs (polychlorinated biphenyls) were present in human and wildlife tissues, *2847*
Use of the phrase "think globally, act locally", *1073*
Warning about global warming, *1836*
Weather predictions using scientific methods, *1922*

Uppsala

Person known to have studied snow crystals, *4190*
Systematic use of binary nomenclature to describe plants, *1764*
Use of taxonomy to classify living organisms, *1617*

SWITZERLAND

Appearance of the Asiatic chestnut blight in Europe, *2602*
Dam with a height greater than 273 meters (900 feet), *2103*
DDT (Dichlorodiphenyl-Trichloroethane) synthesized for use as an insecticide, *1287*
Ecological candidate elected to a national parliament, *1110*
Effective international action to curb acid rain, *1414*
Extensive geological study of the Alps, *2787*
Local environmental political parties, *1105*
Politician from a Green party to be elected to a national parliament, *1111*
Research showing that lichen are two organisms, *1785*

Geneva

Climate-change international panel, *1844*
Comprehensive overview of the global environmental problem, *3262*
Experiments to measure the amount of heat trapped by glass, *4088*
Global agreement to protect plant and animal species from unregulated international trade, *3036*
Hygrometer utilizing human hair, *1877*
International monitoring system for air pollution, *1465*
International Standards Organization (ISO) certification program for environmental management, *4318*
Public health agency of the United Nations, *3934*
Solar collector, *4089*
United Nations agency for management of hazardous chemicals, *2975*
United Nations emergency response system for industrial pollution disasters, *3002*
Use of the term *geology*, *2785*
World Health Organization (WHO) working group to address noise problems, *3431*

Gland

International convention protecting wetlands, *3028*

Lucens Vad

Nuclear power accident in Switzerland, *3507*

Neubuhl

Modern planned solar community in Switzerland, *4104*

Neuchâtel

Scientific evidence to support the theory of an ice age, *2010*
Statement of the theory of a world wide ice age, *2805*

SYRIA

Human herding of semi-domesticated gazelles, *1137*
Large-scale human settlement, *3119*

Antioch

Earthquake in which a massive loss of life was recorded, *2178*

Jabalah

Hailstorm known to cause a famine in Syria, *4221*

TAIWAN

Government ban on the import of discarded batteries into Taiwan, *2888*
Trade sanctions imposed by the United States to protect endangered wildlife, *2345*

Taipei County

Nuclear power plant in Taiwan, *3512*

TAJIKISTAN

Vakhsh

Dam to reach 300 meters (984 feet) in height that is still in use, *2108*

TANZANIA

Natural resource exploited in Tanzania, *3340*
Nile perch in Lake Victoria, *1524*
Significant modern studies of elephant behavior in the wild, *4724*
Wildlife sanctuary in Tanganyika, *4817*

Mkomazi

Rhinoceros preserve in Tanzania, *4760*
Wild dog breeding program in East Africa, *4761*

Mweka

Wildlife resource management college in Africa, *4774*

Seronera

Significant studies of lions in the wild, *4725*

THAILAND

Biosphere reserve in Asia, *1623*
Elephant remote tracking program in Thailand, *4793*
National park in Thailand, *3727*

Gulf of Thailand

Natural gas field in Thailand, *2062*

TRINIDAD AND TOBAGO

Offshore oil wells in Trinidad, *3655*
Oil deposits in Trinidad, *3688*

TUNISIA

Biosphere reserves in Africa, *1628*

Carthage (now Tunis)

Description of chimpanzees, *4679*

TURKEY

Accidental distribution in Turkey of feed grain treated with toxic hexacholorobenzene, *2841*
Bird band recovery, *1727*
Earthquake in the Soviet Union allowed open media coverage, *2232*
Flood known in Asia Minor, *2516*
General Fisheries Council for the Mediterranean, *2474*
Human herding of semi-domesticated gazelles, *1137*
Nuclear power agency established by the Turkish government, *3474*
Nuclear power plant in Turkey, *3544*
Smelting of copper, *1329*

Izmit

Major earthquake in the Izmit region of western Turkey, an area that holds nearly half the nation's population, *2236*

UGANDA

Nile perch in Lake Victoria, *1524*
Okapi known to Western science, *4690*

UKRAINE

Domestication of the horse, *1143*
Drought used for political purposes, *2145*
Trilateral biosphere reserve, *1642*

Askania-Nova

Wildlife preserve in Ukraine, *4815*

Chernobyl

Nuclear power plant disaster with worldwide fallout, *3528*

UNITED ARAB EMIRATES

Offshore oil drilling of significance in the Trucial States (now United Arab Emirates), *3657*

Abu Dhabi

Das Island

Natural gas processing plant in the United Arab Emirates, *2057*

UNITED STATES

Acquisition of land of significance by the United States from a foreign government, *3135*
Activist to popularize the term *speciesism*, *1019*
Administrator named to head the U.S. Forest Service who was not a career bureaucrat or a timber planter, *2708*
Admission by the U.S. government that nuclear weapons production may have caused illnesses in thousands of workers, *3542*
Aerial surveys of large parts of the United States to support soil conservation programs, *4072*

UNITED STATES—*continued*
Aerosol spray containers for insecticide that were portable, *1321*
African-American organization to focus on pollution and hazardous waste in minority and poor communities, *2886*
African elephant protection law in the United States, *2340*
Agency created by U.S. government specifically for geological issues, *2811*
Agency in the United States created to deal exclusively with the problem of water pollution, *4460*
Agency in the United States to regulate solid waste, *4131*
Agency to administer national water pollution legislation in the United States, *4535*
Agricultural bureau established by the U.S. government, *1222*
Agricultural experiment stations established by the U.S. government in individual states, *1224*
Agriculture Bureau scientific publication, *3243*
Air pollution law of importance enacted by the U.S. Congress, *1398*
Air pollution study sponsored by the U.S. government, *1455*
Aircraft noise certification standards in the U.S., *3438*
Airline flight attendants' class action suit against tobacco companies, *1369*
All-natural pesticide made from the seeds of the neem tree, *1306*
Alternative fuel measure adopted by the U.S. Postal Service, *4331*
Amendments of the U.S. Clean Air act to regulate emissions of pollutants, including sulfur dioxide and nitrogen oxides, *1416*
America Recycles Day, *4172*
American Forest Congress, *2693*
American Heritage Rivers program, *3423*
American Indian organization to collectively negotiate over control of natural resources in modern times, *3382*
American naturalist to suggest that every town preserve a "primitive forest", *2710*
Amphibian removed from the U.S. Endangered and Threatened Species List, *2409*
Anadromous fish conservation law in the United States, *2463*
Animal to be domesticated, *1138*
Announcement that the U.S. Army Corps of Engineers would require mandatory disclosure of industrial effluent, *4472*
Annual highway information for the U.S. government, *1569*
Annual license required of all U.S. waterfowl hunters 16 years of age or older, *2933*
Annual report on the environment and American foreign policy, *2308*
Appeal for U.S. weather observers, *1894*
Appearance of modern sedentary ranching in the Western Hemisphere, *1157*
Appearance of the alfalfa blotch leaf miner in the United States, *1194*
Appearance of the alfalfa weevil in the United States, *1188*
Appearance of "The Great American Fraud," Samuel Hopkins Adams's article on patent medicines, *3250*
Appearance of the potato-destroying Colorado beetle on the east coast of the U.S., *1185*
Appearance of the Scribner Log Rule for determining board-foot measurements of timber, *2652*
Appearance of the two-man crosscut saw for felling timber, *2656*
Appearance of the two-wheeled sulky plow, *1242*
Aquaculture Development Act in the United States, *2497*
Aquatic Nuisance Species Task Force, *1530*
Arbitration ruling to assign national responsibility for air pollution damage, *1389*
Arctic Environmental Protection Strategy (AEPS), *3873*
Artist to win the U.S. federal duck stamp contest three times, *3199*
Asbestos legislation adopted by the U.S. government, *3962*
Authorization of the U.S. Forest Service to make arrests for violations of its regulations, *2679*
Automobile emissions bill approved by the U.S. Congress, *1574*
Automobile safety advocate of note in the United States, *1542*
Automobile safety legislation of significance passed by the U.S. Congress, *1575*
Automobile seat-belt buzzers, *1600*
Automobile technology agreement between the U.S. government and industry, *1606*
Ban of chlorofluorocarbons (CFCs) by the U.S. government, *1978*
Ban on aerosol chlorofluorocarbons (CFCs), *1410*
Ban on chlorofluorocarbons (CFCs) on shipments between U.S. states, *1983*
Ban on fencing the public domain in the United States, *1166*
Ban on lead-based paint in the United States, *1358*
Ban on some U.S. uses of the EBDC fungicide class, *1308*
Ban on the use of lead pipes in public water systems in the United States, *4606*
Banning of aircraft for hunting in the United States, *2959*
Barbed wire product on the market, *1163*

Bat listing under the U.S. Endangered Species Act, *2386*
Bathymetric map of the North Atlantic Ocean, *3556*
Beach closings in the United States due to medical wastes washing ashore, *4428*
Bear listing under the U.S. Endangered Species Act, *2387*
Bilateral negotiations between the United States and Canada on acid rain, *3038*
Bilateral U.S.–Canadian agreement to waive "consent notification" for hazardous waste shipments from the United States into Canada, *2873*
Bilateral U.S.–Mexican agreement requiring American companies in Mexico to return hazardous waste to the United States, *2867*
Bill to require humane slaughter of animals in the United States, *1015*
Biogeographic areas of the U.S. National Park Service Inventory and Monitoring Program (I&M), *3399*
"Bioprospecting" agreement, *1639*
Bioprospecting clause in U.S. Fish and Wildlife Service Special Use specimen-collecting permits, *4676*
Biosphere reserves in the United States, *1627*
Bird hunting regulation enacted by the U.S. Congress, *2955*
Bird officially designated as the national bird of the United States, *1688*
Bird pictured in John James Audubon's *The Birds of America*, *3160*
Bird protection treaty, *3017*
Bird removed from the U.S. Endangered and Threatened Species List due to extinction, *2379*
Bird reservation established by the U.S. government, *1719*
Blizzard and winds to cause damages of $1 billion or more in the United States, *4213*
Botanist to explore the American West, *1776*
Boycott of a major U.S. automaker in protest of the use of animals in crash tests, *1022*
Boycott of genetically engineered food products in the United States, *2764*
British government attempt to restrict American colonial settlement to the area east of the Appalachian Mountains, *3131*
Broad-based animal welfare organization to oppose the slaughter of fur seals, *1016*
Brownfields Economic Redevelopment Initiatives program in the United States, *2992*
Butterfly listing under the U.S. Endangered Species Act, *2407*
Cabinet-level agency concerned primarily with natural resource development and conservation in the United States, *3079*
Cabinet-level regulatory environmental agency in the United States, *1067*
Call from a U.S. governors' group for remedial action on acid rain, *1327*
Camels imported into the United States, *1160*
Canadian–U.S. International Boundary Treaties Act, *3366*
Canadian–U.S. treaty to regulate use of the Niagara River for water power, *4555*
Car model with a high-compression six-cylinder engine, *1595*
Carbon tetrachloride ban in the United States, *3961*
Cement gun for mounting wildlife specimens, *4693*
Census of fish kills by pollution in the United States, *4415*
Census of public roads in the United States, *1555*
Charter for the National Trust for Historic Preservation in the United States, *3109*
Charts of the Atlantic Ocean winds and currents, *3554*
Charts of the Gulf Stream, *3548*
Chemist to add tetraethyl lead to gasoline, *1339*
Chestnut blight in the United States, *2592*
Citizen lawsuit provisions allowing people to file legal actions in U.S. courts to protect the environment, *1096*
Civil works project undertaken by the U.S. Army Corps of Engineers, *3069*
Claim by the Centers for Disease Control that malaria had been eradicated in the United States, *3941*
Claim by the U.S. House of Representatives that overestimates of reforestation efforts in the Pacific Northwest had led to dangerously high logging quotas, *2689*
Classification of rivers and federally-owned adjoining lands to determine their use by the public, *3731*
Clean-coal joint program by the U.S. government and industry, *2064*
Clean drinking water standards in the United States that were legally binding, *4598*
Climate-impact forecasts by the U.S. government, *1833*
Climate research agreement between the United States and Soviet Union, *1830*
Climatologist of repute, *1923*
Cloud-seeding demonstration, *2170*
Coal discovery recorded in North America, *2026*
Coal-mine inspection law by the U.S. government, *3327*
Coal mines to be fully automated in the United States, *3329*
Coal-mining agency in the U.S. government, *3325*
Coal-mining law with penalty fines imposed by the U.S. government, *3334*

UNITED STATES—*continued*
Coal-mining safety law in the United States, *3319*
Collection of weather adages gathered under the aegis of the U.S. Army Signal Corps, *1916*
Colorado potato beetles, *1182*
Colorado Wilderness Act, *4665*
Commercial land disposal of low level nuclear waste in the United States, *2845*
Commercial use of the ion exchange method for treating industrial wastewater in the United States, *4505*
Commercial use of the pesticide DDT in the U.S., *1288*
Company forests in the United States, *2653*
Company to voluntarily limit production of PCBs (polychlorinated biphenyls), *2849*
Comprehensive effort to introduce preventive pollution control in the United States, *1417*
Comprehensive hazardous waste cleanup program by the U.S. government, *3956*
Comprehensive inventory of geologic resources in U.S. national parks, *3424*
Comprehensive inventory of vegetation in U.S. national parks, *3408*
Comprehensive legislation in the United States banning PCBs and regulating their disposal, *2857*
Comprehensive legislation in the United States to regulate the generation, transport, and management of a wide range of hazardous wastes, *2858*
Comprehensive national policies for protecting the environment in the United States, *1062*
Comprehensive scientific effort to conserve soil and water resources on a national scale, *4035*
Comprehensive study of human impact on the environment, *3244*
Comprehensive U.S. legislation establishing minimum safety standards for chemical contaminants in water, *4478*
Comprehensive U.S. national fish and wildlife policy, *4770*
Comprehensive upgrades of the 1970 and 1977 Clean Air Act and Amendments, *1418*
Comprehensive waste management legislation in the United States, *4132*
Conclusive link between asbestos and lung cancer, *1348*
Confirmation that Dutch elm disease-carrying bark beetles deposit fungi spores on feeding wounds of trees, *2601*
Confirmation that peach oil kills fungi and other soil pests, *1310*
Congressional action on "nonpoint" source pollution in the U.S., *4481*
Congressional attempt to re-establish the Office of Noise Abatement and Control (ONAC), *3451*
Conservation agency established by the U.S. government, *4763*
Conservation group to focus on the eastern United States in addition to western states, *1035*
Consolidation of various types of lands into a single U.S. National Wildlife Refuge System, *4828*
Control method used on sea lampreys in the Great Lakes, *2461*
Cooperative effort by 10 northeastern U.S. states to develop plans to reduce emissions of greenhouse gases, *1425*
Cooperative effort by northeastern states to curb nitrogen oxide emissions from coal-burning power plants, *1424*
Coordinated assault on "nonpoint" sources of pollution in the United States, *4482*
Coordinated weather observations in the United States, *1896*
Corporate effort at pollution control at Du Pont de Nemours chemical plants in the United States, *4429*
Crocodilians listed under the U.S. Endangered Species Act, *2394*
Crop dusting by an airplane, *1245*
Dam safety code issued by the U.S. government, *2077*
Dam safety guidelines issued by the U.S. government, *2073*
Darter listed under the U.S. Endangered Species Act, *2388*
Death attributed to a "killer bee" attack in the United States, *3968*
Declaration of phosphates as a major source of water pollution issued jointly by two countries, *4404*
Declaration that the American frontier had closed, *3140*
Deer listings under the U.S. Endangered Species Act, *2389*
Department of the Interior in the United States, *3358*
DES hormone ban by the U.S. government, *4422*
Description of hydroponic agriculture, *1247*
Description of the "black-tailed deer" by William Clark, *4684*
Designation by the U.S. government of cigarettes and smokeless tobacco as nicotine delivery systems, *3970*
Detailed five-day U.S. Weather Bureau forecasts, *1942*
Detergent industry research for an answer to the problem of foam in tap water, *4507*
Development by the United States of small nuclear reactors for use in space, *3469*
Diebacks of significance to western white pine forests in the United States, *2588*

Director of the U.S. Geological Survey (USGS), *2809*
Director of the U.S. National Park Service, *3789*
Disposable diapers mass-marketed in the United States, *4537*
Division of Forestry in the United States, *2674*
Dolphin listing under the U.S. Endangered Species Act, *2414*
"Dolphin-safe" labels, *2479*
Doppler radar system to study storms, *1963*
Drought forcing the Soviet Union to buy U.S. grain, *2155*
Drought in modern United States to affect 28 percent of the population, *2153*
Drought of significance recorded in the American colonies, *2130*
Drought predictions for the 21st century by a U.S. government agency, *2166*
Drought to cause damages of $1 billion or more in the United States, *2157*
Drought year producing worldwide environmental responses, *2162*
Duck stamps for waterfowl hunters in the United States, *1692*
Dumping of highly toxic PCBs (polychlorinated biphenyls) in waterways, *4430*
Earth Day to be celebrated nationwide in the United States, *1069*
Earthquake data center established by the U.S. government, *2228*
"Ecological niche" concept, *1620*
"Ecosystems management plan" for U.S. national forests, *2688*
Edict issued by the U.S. Environmental Protection Agency requiring the states to clean up air pollution in national parks, *1428*
Effort of significance by the United States government to develop soil conservation and rehabilitation programs for the southern Great Plains during the Dust Bowl, *4037*
Electric automobile battery research group in the United States, *1550*
Elephant listed under the U.S. Endangered Species Act, *2406*
Embargo in the United States against tuna that was not "dolphin-safe", *2481*
Emergence of wetlands science as a distinct discipline, *3151*
Employment program of significance for forestry and other natural resource projects, *2683*
Endangered plants list issued by the United States, *2403*
Endangered species list by the U.S. government, *2385*
Endangered species list to include birds, *1669*
Entomologist hired by the U.S. government, *1272*
Environment and Natural Resources Division of the U.S. Department of Justice, *3368*
Environmental health organization of significance started by the U.S. government, *3942*
"Environmental justice" action by the U.S. government, *2995*
Environmental law challenged under the North American Free Trade Agreement (NAFTA), *3971*
Environmental law in the American colonies, *2627*
Environmental lawsuit to challenge the United States under the North American Free Trade Agreement (NAFTA), *2348*
Environmental public service symbol of the U.S. government, *2618*
Environmental Quality Incentives Program (EQIP) in the United States, *3419*
Environmentally conscious codes of ethics for whale-watching operators and participants in North America, *2244*
Epidemiological evidence of the association between passive smoking and lung cancer, *1476*
Establishment of a federal Bureau of Outdoor Recreation in the United States, *2686*
Estuary protection funding in the United States, *4470*
European to set foot on what would later become the west coast of the United States., *3739*
Evidence of solar architecture, *4084*
Example of a whole-systems approach to the environment, *3260*
Executive Order issued to protect the Lake Tahoe region, *3421*
Executive Order protecting coral reefs in the United States, *3601*
Executive Order requiring U.S. government agencies to use recycled paper, *4158*
Exemption for U.S. military flyovers of National Wilderness deserts in California, *4673*
Explorer to describe the High Plains of the western United States as the Great American Desert, *2135*
Exports of Datsun cars from Japan to the U.S., *1546*
Express nationwide highway system in the United States, *1565*
Expression of the concept that the earth is a living organism, *1050*
Extant species on the U.S. Endangered and Threatened Species List to become extinct, *2382*
Extensive collection of quality data on fertility and family planning, *3882*
Extensive peacetime drought relief project in the United States, *2144*
Extensive U.S. government standards for quality and labeling of foods, *1227*
Feasibility studies of solar energy use in U.S. government buildings, *4122*

UNITED STATES—*continued*
Feasibility studies of wind-generated energy use in U.S. government buildings, *4865*
Federal "Adopt a Horse or Burro" program in the United States, *4803*
Federal agency charged with comprehensively monitoring and protecting all wildlife in the United States, *4768*
Federal agency charged with preventing soil erosion, *4055*
Federal agency in the United States to monitor water use, *4602*
Federal aid to hunter education programs in the United States, *2925*
Federal ban of lead paint in the United States, *3955*
Federal chemical standards for drinking water in the United States, *4588*
Federal drinking water safety regulations in the United States, *4584*
Federal drinking water standards in the United States, *4586*
Federal Drinking Water State Revolving Fund (DWSRF) in the United States, *4614*
Federal duck stamp in the United State to feature a blue-winged teal, *3190*
Federal duck stamp in the United States, *3165*
Federal duck stamp in the United States to be priced at $12.50, *3216*
Federal duck stamp in the United States to be priced at $15, *3219*
Federal duck stamp in the United States to be priced at $7.50, *3213*
Federal duck stamp in the United States to be priced at five dollars, *3210*
Federal duck stamp in the United States to be priced at ten dollars, *3215*
Federal duck stamp in the United States to be priced at two dollars, *3185*
Federal duck stamp in the United States to feature a black-bellied whistling duck, *3218*
Federal duck stamp in the United States to feature a blue goose, *3192*
Federal duck stamp in the United States to feature a bufflehead, *3184*
Federal duck stamp in the United States to feature a cinnamon teal, *3208*
Federal duck stamp in the United States to feature a fulvous duck, *3214*
Federal duck stamp in the United States to feature a gadwall, *3188*
Federal duck stamp in the United States to feature a goldeneye, *3186*
Federal duck stamp in the United States to feature a harlequin duck, *3189*
Federal duck stamp in the United States to feature a Hawaiian nene goose, *3202*
Federal duck stamp in the United States to feature a hooded merganser, *3205*
Federal duck stamp in the United States to feature a king eider, *3220*
Federal duck stamp in the United States to feature a lesser scaup, *3217*
Federal duck stamp in the United States to feature a merganser, *3193*
Federal duck stamp in the United States to feature a Pacific brant, *3200*
Federal duck stamp in the United States to feature a red-breasted merganser, *3223*
Federal duck stamp in the United States to feature a ring-necked duck, *3191*
Federal duck stamp in the United States to feature a Ross's goose, *3207*
Federal duck stamp in the United States to feature a shoveler, *3181*
Federal duck stamp in the United States to feature a snow goose, *3183*
Federal duck stamp in the United States to feature a spectacled eider, *3222*
Federal duck stamp in the United States to feature a Steller's eider, *3211*
Federal duck stamp in the United States to feature a surf scoter, *3226*
Federal duck stamp in the United States to feature a trumpeter swan, *3187*
Federal duck stamp in the United States to feature a waterfowl decoy, rather than a live bird, *3212*
Federal duck stamp in the United States to feature a whistling swan, *3203*
Federal duck stamp in the United States to feature a white-winged scoter, *3206*
Federal duck stamp in the United States to feature an American eider, *3194*
Federal duck stamp in the United States to feature an empress goose, *3209*
Federal duck stamp in the United States to feature an image other than waterfowl alone, *3197*
Federal duck stamp in the United States to feature an old squaw duck, *3204*
Federal duck stamp in the United States to feature the black duck, *3173*
Federal duck stamp in the United States to feature the Canada goose, *3169*
Federal duck stamp in the United States to feature the canvasback, *3167*
Federal duck stamp in the United States to feature the greater scaup, *3170*
Federal duck stamp in the United States to feature the green-winged teal, *3172*
Federal duck stamp in the United States to feature the mallard, *3166*
Federal duck stamp in the United States to feature the pintail, *3171*
Federal duck stamp in the United States to feature the redhead, *3182*
Federal duck stamp in the United States to feature the ruddy duck, *3174*

Federal duck stamp in the United States to feature the white-fronted goose, *3178*
Federal duck stamp in the United States to feature the widgeon, *3176*
Federal duck stamp in the United States to feature the wood duck, *3177*
Federal duck stamp in the United States to sell for $3.00, *3198*
Federal effort to regulate hazardous materials on railroads in the United States, *2830*
Federal environmental justice grants in the United States, *3412*
Federal fish-agriculture rotation program in the United States, *2496*
Federal government action granting the U.S. states sovereignty over swamplands, *3144*
Federal government bailout of a major U.S. carmaker, *1548*
Federal government interagency study to evaluate methods by which states and localities in the United States could slow farmland losses, *3113*
Federal government move to relax water pollution controls for U.S. companies that discharged nontoxic substances, *4439*
Federal government program to finance construction of flood control levees in the United States, *2568*
Federal grants-in-aid to U.S. municipalities for pollution cleanup and control, *4461*
Federal guidelines for disposing of hazardous waste, *4143*
Federal guidelines governing research with genetically altered material in the United States, *1255*
Federal health program in the United States, *3913*
Federal highway system in the United States, *1558*
Federal income tax credit for wind energy use in the United States, *4863*
Federal junior duck stamp in the U.S., *3224*
Federal land-grant program to promote irrigation in western U.S. states, *1219*
Federal land grants for colleges that would further the "agricultural and mechanical arts", *1223*
Federal law enforcement agency for wildlife in the United States, *4700*
Federal law in the United States closely regulating the use of animals in medical and commercial research, *1017*
Federal law in the United States protecting and preserving rivers, *4469*
Federal law in the United States to ban the use of motorized vehicles in capturing wild horses, *4771*
Federal law in the United States to develop salmon fisheries in the Columbia River Basin, *2495*
Federal law in the United States to protect fur seals, *4776*
Federal law in the United States to reduce vessel sewage discharges, *3598*
Federal law in the United States to restrict use of billboards in scenic areas, *3980*
Federal law in the United States to withhold development funds for construction on coastal barrier islands, *3389*
Federal law in the United States to withhold development funds for construction on coastal barrier islands in the Great Lakes, *3394*
Federal law to allow designation of formerly developed lands as wilderness areas, *4662*
Federal law to create wilderness areas in the eastern United States, *4653*
Federal law to protect the bald eagle, *2384*
Federal law to restrict the dumping of hazardous waste in the United States, *2977*
Federal laws in the United States to protect caves, *4670*
Federal legislation in the United States authorizing recreational uses of wildlife conservation areas managed by the Department of the Interior, *4825*
Federal legislation in the United States to grant federal funds for urban park rehabilitation and infrastructure maintenance, *3822*
Federal legislation in the United States to protect caves located on federal lands, *3809*
Federal legislation in the United States to tax hunting and fishing equipment in order to fund improvements in wildlife habitat, *4831*
Federal legislation of significance to take erodable U.S. cropland out of production, *4057*
Federal legislation to establish the national parks system, *3790*
Federal limits on shark fishing enacted by the United States, *2482*
Federal Lumber Code Authority in the United States, *2635*
Federal management policies for wild horses in the United States, *4780*
Federal management policy for public grazing lands in the United States, *1169*
Federal plan in the United States for developing biomass energy regionally, *4332*
Federal policy in the United States on emergency removal of asbestos from schools, *4147*
Federal program of significance to purchase forestland for watershed protection in the United States, *2713*
Federal program of significance to seal up abandoned mines in the United States to prevent water entry and polluting drainage, *4431*
Federal program to protect and restore an estuary in the United States, *3593*
Federal protection for bald eagles in the lower 48 states of the United States, *1694*

UNITED STATES—*continued*

Federal provision requiring companies that service air conditioning and refrigeration equipment to "capture and recycle" chlorofluorocarbons (CFCs), *1426*
Federal regulation in the United States for designation and management of wild areas, *4635*
Federal requirement in the United States for bacteriological analysis of drinking water systems, *4590*
Federal research on ways to limit the release of smoke into the atmosphere in the United States, *1442*
Federal rules in the United States for closure of hazardous waste treatment, storage, and disposal facilities, *4144*
Federal Sole Source Aquifer (SSA) Protection Program in the United States, *4599*
Federal standards in the United States for radon in drinking water, *4610*
Federal studies of the effects on the atmosphere of carbon monoxide from automobile exhaust in the United States, *1445*
Federal tax credits for wind power investments in the United States, *4856*
Federal trust fund for the cleanup of hazardous waste sites in the United States, *2863*
Federally sponsored interstate air pollution abatement conference, *1399*
Federally sponsored water resource research in the United States, *4402*
Federation of Sewage Works Association meetings, *4534*
Female scientist employed by the U.S. Geological Survey (USGS), *2813*
Female winner of the federal junior duck stamp contest in the United States, *3225*
Field guide to birds, *3253*
Financial responsibility requirements for underground storage tanks (UST), *2988*
Firearms tax in the United States to specifically benefit wildlife habitats, *4820*
Firm evidence of the outbreak of AIDS in the United States, *3957*
Fish removed from the U.S. Endangered and Threatened Species List, *2378*
Fisheries management in U.S. national parks, *2456*
Fishing legally allowed on a U.S. government wildlife refuge, *2457*
Fishing vessel buyback program for New England, *2483*
Flash flood national program in the United States, *2558*
Floating hatchery run by the U.S. government in ocean waters, *2493*
Flood control legislation of significance in the United States, *2542*
Flood-danger evaluations required by a U.S. presidential executive order, *2552*
Flood legislation by the U.S. government that called floods a national problem, *2539*
Flood legislation by the U.S. government to emphasize the environment, *2550*
Flood-plain protection advocated by a U.S. presidential executive order, *2553*
Flood victims to receive aid from the hand of Clara Barton, *2531*
Food additive regulation passed by the U.S. Congress, *3938*
Forest conservation legislation in the United States, *2631*
"Forest preserves" legislation in the United States, *3734*
Forestry volunteers for the U.S. Peace Corps, *2707*
Formal identification of the line of continuous development along the U.S. eastern corridor from Washington to Boston, *3143*
Fox listing under the U.S. Endangered Species Act, *2390*
Free investigations of indoor air pollution in U.S. workplaces, *1359*
Full-time landscape architect hired by the U.S. Forest Service, *3717*
Funding for multisector wildlife research, *2279*
Game animals on U.S. postage stamps in 1956, *2922*
Game preserve appropriation by Congress to assist state wildlife restoration projects, *4821*
Gasoline tax levied by the U.S. Congress, *1571*
General grants from the U.S. government to Indian tribes for solid waste disposal programs, *4151*
Genetic engineering, *2755*
Genetically altered plants, *2760*
Geological maps of the United States, *2792*
Geological periodical still published today, *2798*
Geological publication in the United States, *2793*
Geologist/explorer to state flatly that much of the western United States was too dry to farm, *2138*
Geologist hired by the U.S. government, *2804*
Geostationary Operational Environmental Satellite (GOES), *1965*
Geothermal Steam Act in the United States, *3379*
Global-change research coordination program established by the U.S. government, *1847*
Global list of endangered plants, *2421*
Global warming report of significance made to the U.S. government, *1841*
Gorilla listing under the U.S. Endangered Species Act, *2395*
Government agency in the United States to keep track of the sale of federal land, *3137*

Government agency in the United States to regulate leaking underground storage tanks, *2872*
Government-backed crop insurance in the United States, *1228*
Government-built saline water conversion demonstration plants in the United States, *4462*
Government commission to rent or purchase lands to be set aside for migratory birds in the United States, *1691*
Government demolition of a U.S. dam without the owner's permission, *2080*
Government land grant to a railroad by the United States, *3139*
Government policy in the United States to deny aid to foreign family planning agencies that performed or promoted abortions as a method of population control, *3884*
Government program in the United States established to monitor waterborne toxic materials, *4603*
Government program in the United States to screen plant and animal species for natural products with anticancer properties, *2744*
Government programs in the United States to assist migrant workers and their families, *1252*
Government publication of importance issued by the United States telling farmers how to protect their soil, *4032*
Government report issued jointly by the United States and Canada to rate acid rain a serious environmental and transboundary problem for both countries, *1477*
Government Soil Survey in the United States, *4033*
Government study on U.S. coal industry conditions, *2049*
Government survey of wildlife-related recreation activities by the U.S., *4719*
Government weather service in the United States, *1911*
Governors' conference on U.S. conservation issues, *1033*
Grazing fees assessment for use of U.S. federal lands, *1167*
"Greenhouse effect" report to the U.S. Congress, *1845*
Greenpeace campaign to stop exports of hazardous waste, *2878*
Groundwater contamination inventory in the United States, *2844*
Group of Ten meeting, *1078*
Guidelines issued by the U.S. Fish and Wildlife Service (FWS) on foregoing lawsuits in natural resource damage cases, *3403*
Guidelines issued in the United States by the Federal Trade Commission for environmental marketing claims, *1092*
Habitat Conservation Plan (HCP) in the United States, *2337*
Handbook on ecological sabotage, *1083*
Hawaiian monk seal listing under the U.S. Endangered Species Act, *2405*
Hazardous chemical public disclosure law in the United States, *3964*
Hazardous Materials Transportation Act in the United States, *2855*
Haze regulations concerning U.S. wilderness areas and national parks, *4677*
Head of the Department of Agriculture to be a member of the President's Cabinet, *1129*
Heat wave in the United States known to have lasted for two months, *1823*
Helicopter logging experiments in the United States, *2664*
Helicopter noise bill to reach a committee in the U.S. Congress, *3450*
High-resolution maps of the ocean floor, *3600*
Highway law in the United States for federal-state funding, *1570*
Highway-marking system in the United States, *1561*
Highway program of significance in the United States granting federal funds for state roads, *1557*
Horse listed under the U.S. Endangered Species Act, *2404*
Housing program for U.S. military veterans subsidized by the federal government, *3084*
Hunting legally allowed on a U.S. government wildlife refuge, *2919*
Hurricane of significance recorded in North America, *4237*
Hurricane warning, *4244*
Hurricanes named by the U.S. Weather Bureau, *4258*
Hurricanes to be given feminine names, *4259*
Hurricanes to be given masculine names, *4264*
Hydroelectric power-plant licenses in the United States, *2095*
Hydropower agency of the U.S. government, *4553*
Identification system for wild horses and burros by the U.S. Bureau of Land Management, *4804*
Immediate ban on major uses of asbestos in the United States, *1328*
Important federal legislation to provide technical forestry services to private nonindustrial landowners, *2637*
Index for temperature-humidity, *1955*
Index to scientifically measure droughts, *2154*
Indoor ventilation guidelines for commercial buildings, from the American Society of Heating, Refrigeration and Air-Conditioning Engineers, *1363*
Insect removed from the U.S. Endangered and Threatened Species List, *2411*

UNITED STATES—*continued*
Inter-American Tropical Tuna Commission, *2460*
Interagency government property transfers for wildlife conservation in the United States, *4769*
Interagency Grizzly Bear Committee (IGBC), *4785*
Intergovernmental conservation conference in North America, *3367*
International Adopt-a-Highway Day, *3290*
International Conference on Polar Bears, *4775*
International conference on the connection between soil health and human health, *4027*
International convention on ocean pollution, *3577*
International Joint Commission on Boundary Waters study of Lake Erie, Lake Ontario, and the upper St. Lawrence River, *4511*
International Log Rule for measuring timber yield, *2658*
International meeting of American Indians and environmentalists, *4408*
International professional forestry society, *2706*
International treaty to protect birds, *1743*
International wildlife treaty upheld by the U.S. Supreme Court, *3018*
Interstate gas pipelines in the United States, *2050*
Introduction into the United States of the method of timber cruising called "variable plot cruising", *2662*
Introduction of a chemical "on-off switch" that would make seeds sterile in a generation, *1208*
Introduction of blended tobacco, *1130*
Introduction of early biologically based alternatives to chemical pesticides, *1302*
Introduction of kudzu into the United States for erosion control, *4050*
Introduction of the "bubble concept" in U.S. air pollution policy, *1412*
Introduction of the concept of a "debt for nature" swap, *2719*
Introduction of the pH scale to measure acidity in rain, *1444*
Introduction of the rectangular system of land surveys in the United States, *3132*
Involvement of the U.S. Department of Commerce with commercial fisheries, *2477*
Iron mines in the American colonies, *3298*
Irrigation advocacy group in the United States, *1031*
Irrigation ditches in what is now the United States, *4569*
Irrigation law passed by U.S. Congress, *1216*
Irrigation project of significance in what is now the United States, *1214*
Issue of *Forestry Quarterly*, *2736*
Joint commitment to "zero discharge" of dangerous chemicals into Lake Superior by the United States and Canada, *4442*
Joint document from major U.S. environmental groups to spell out a national environmental policy, *1081*
Joint effort of major significance to solve pollution problems in the Great Lakes, *4476*
Junior Duck Stamp program in the United States, *4789*
Land bought by the U.S. government specifically for wildlife, *4812*
"Land Ethic" concept, *1054*
Land mammals removed from the U.S. Endangered and Threatened Species List, *2420*
Land-use plan for a public desert area in the United States, *3089*
Large-scale federal legislation aimed at purifying salty waters in the arid western United States, *4459*
Large-scale peripheral suburban growth in the United States, *3142*
Large-scale reduction in levels of benzene permitted in U.S. factories, *3952*
Large settlement of Agent Orange lawsuits, *1304*
Law by the U.S. government to subject bottled water to the same standards as tap water, *4615*
Law giving the U.S. Environmental Protection Agency authority to set water quality standards and monitor water supplies, *4477*
Law in the United States allowing teenagers to work on federal conservation programs, *4778*
Law in the United States banning the disposal of plastics at sea, *4148*
Law in the United States banning tin-based paint on boats, *2504*
Law in the United States calling for ways of controlling jet aircraft emissions, *1401*
Law in the United States deregulating the price of natural gas extracted from below 15,000 feet, *2058*
Law in the United States designed to stop overgrazing and its adverse effects on the soil, *4036*
Law in the United States extending wilderness protection to land administered by the Bureau of Land Management (BLM), *4667*
Law in the United States for watershed protection, *4593*
Law in the United States mandating that all school systems check their buildings for asbestos-containing materials, *1365*
Law in the United States regulating explosives and flammable materials, *2828*
Law in the United States requiring environmental considerations to be assessed in the licensing of hydropower projects, *4567*

Law in the United States requiring government agencies to purchase alternative-fueled vehicles, *4333*
Law in the United States requiring preservation of scenery within national parks, *3976*
Law in the United States requiring public disclosure of pollutants through the Toxic Waste Inventory, *2984*
Law in the United States that protected the public from hazardous substances, *1281*
Law in the United States to allow combining federal and private funds for wildlife program grants, *4840*
Law in the United States to allow federal control of wildlife on both public and private lands, *4799*
Law in the United States to allow government fees to be levied to establish and maintain wildlife refuges, *4818*
Law in the United States to establish asbestos as a health hazard in schools, *2982*
Law in the United States to establish federal and state standards for organic certification, *1264*
Law in the United States to license construction of commercial hydropower plants, *4557*
Law in the United States to make shipment of wild animals a federal offense if the animals were taken in violation of state laws, *1013*
Law in the United States to prohibit the transportation of wildlife across state lines if they were taken in violation of state law, *4691*
Law in the United States to promote wind power as a renewable resource, *4858*
Law in the United States to protect Atlantic striped bass, *2465*
Law in the United States to protect other nations' wildlife, *4726*
Law in the United States to require an environmental impact statement for all major federal projects that could significantly affect the environment, *3086*
Law in the United States to require federal government agencies to obey environmental laws, *4134*
Law in the United States to require laboratories to furnish adequate food and shelter for test animals, *1018*
Law in the United States to require mine spoil waste land to be reclaimed, *3337*
Law in the United States to require phasing out of mercury in batteries, *4152*
Law in the United States to require the removal of lead from water coolers, *4607*
Law in the United States to require warning labels on hazardous substances, *1285*
Law in the United States to specifically ban importation of endangered exotic birds, *2342*
Law in the United States to specifically target toxic substances, *3951*
Law making it illegal to dump waste in U.S. navigable waters except by special permission, *4449*
Law of importance in the United States regulating pesticide use, *1290*
Law of importance in the United States requiring government and industry to inform the public about toxic chemicals, *1353*
Law of significance to control acid rain in the United States, *1413*
Law of significance to permit timber cutting in U.S. forest preserves, *2676*
Law passed by U.S. Congress specifically aimed at preserving endangered birds, *1696*
Law passed to protect black bass in the United States, *2458*
Law requiring the U.S. president to develop and conduct a comprehensive program of marine science activities, *3581*
Law setting up the Cooperative Extension Service in the United States, *1266*
Law stating that U.S. forests would be exploited for a variety of recreational, commercial, and conservation purposes, *2685*
Law to open vast areas of the Great Plains in the United States to systematic settlement, *3080*
Law to prevent damage to fish and wildlife, *4716*
Law to prohibit the dumping of solid waste in U.S. waters, *4141*
Law to protect future endangered fish and wildlife, *4727*
Law to provide for federal development of irrigated agriculture in the United States, *1220*
Law to regulate the discharge of petroleum products into coastal waters, *4455*
Laws in the United States to include a cost-benefit analysis of industrial wastes, *2836*
Laws regulating space debris, *3032*
Laws requiring catalytic converters for automobiles in the United States, *1576*
Leaded gasoline total ban by the United States Congress, *1551*
Leafy spurge in the United States, *1497*
Leaking Underground Storage Tank (LUST) trust fund in the United States, *2983*
Legal basis for U.S. federal drinking water standards, *4581*
Legislation aimed at protecting U.S. wetlands from drainage and development, *3148*
Legislation allowing exchanges of U.S. public domain forest land for other lands of equal value, *2633*
Legislation allowing settlers and miners to cut timber on U.S. public lands for their own use, *2630*
Legislation authorizing federal-state cooperation in U.S. farm forestry, *2636*

UNITED STATES—*continued*
Legislation by the United States barring private acquisition of American Indian lands, *3078*
Legislation by the United States to give considerable authority for disposal of low-level radioactive waste to the states, *2865*
Legislation for cooperative wildlife management, *4704*
Legislation in the United States granting a homesteader land at bargain rates in return for a promise to irrigate, *1217*
Legislation in the United States limiting lead content in paint and providing funds for cleanup, *3944*
Legislation in the United States making animal fighting a federal offense, *1020*
Legislation in the United States to allow noise pollution lawsuits, *3442*
Legislation in the United States to mandate double hulls on oil tankers, *3627*
Legislation in the United States to provide federal funds to states for cooperation in forest fire control, *2614*
Legislation in the United States to provide for use of fees from offshore oil drilling for land purchases and historic preservation, *3658*
Legislation in the United States to remove ecologically fragile farmland from production, *1229*
Legislation in the United States to specifically control the disposal of nuclear waste, *3465*
Legislation in the United States to withdraw substantial amounts of land from cultivation, *1225*
Legislation linking soil conservation and commodity policy in the United States, *4038*
Legislation of significance by the U.S. government promoting tree farming, *2725*
Legislation of significance in the United States promoting and funding mass public housing, *3085*
Legislation of significance in the United States to promote reforestation and soil conservation as flood control measures, *2536*
Legislation on health issues involving water pollution passed by the U.S. government, *4452*
Legislation permitting settlers to live on U.S. public land until they could pay for it or until someone else bought it, *3136*
Legislation that provided regular federal funding for waterfowl management in the United States, *1693*
Legislation to protect migratory game and insectivorous birds, *1689*
Legislation to regulate Indian grazing ranges in the United States, *1226*
"Life zones" concept, *1619*
Lion listing under the U.S. Endangered Species Act, *2396*
List of protected rivers in the United States, *4463*
List of wilderness areas in the U.S. National Wilderness Preservation System (NWPS), *4648*
Local land-sales offices in the United States, *3134*
Long-term study of the causes and effects of acid rain in the United States, *1474*
Major apparel maker to make its entire line of clothing from organically grown cotton and nontoxic dyes, *1265*
Major attack on the U.S. consumer products industry, *3251*
Major private organization to campaign against air pollution in the United States, *1336*
Major public outcry on atmospheric nuclear-weapons testing, *1323*
Major windstorm of the Dust Bowl drought in the U.S. Midwest, *2149*
Mammal removed from the U.S. Endangered and Threatened Species List, *2419*
Marine geology vessel, *2824*
Marine mammal conservation and management law in the United States, *4782*
Mass-membership animal rights organization in the United States, *1014*
Mass polio vaccination campaign, *3937*
Mass-produced automobile in the United States, *1533*
Massively destructive eastward migration of the Rocky Mountain grasshopper, *1186*
Maximum Allowable Concentrations (MACs) for hazardous substances in the United States, *2837*
Medical waste disposal rules set by the U.S. Environmental Protection Agency (EPA), *4149*
Mercury thermometer, *1869*
Meteorologist to claim that lighting large forest fires would produce widespread rain, *1889*
Meteorologist to discover the processes of cyclones and tornadoes, *4242*
Mexican spotted owl listing under the U.S. Endangered Species Act, *2418*
Military uses of marine mammals, *4723*
Mine safety law in the United States applying to non-coal mines, *3333*
Miners' health protection agency established by the U.S. government, *3324*
Mining law that freely released federal land for public exploitation, *3316*
Mississippi River flood control project of significance, *2537*
Modern herbicide in the United States, *1289*
Modern U.S. National Park Service (NPS) strategy for natural resource management, *3397*
Mollusk removed from the U.S. Endangered and Threatened Species List, *2380*

Move to halt all fungicide uses of mercury in the United States, *1298*
Multistate clean streams campaign for the Ohio Valley, *4457*
National Air Pollution Control Act in the United States, *1395*
National and state efforts to eradicate the foliage-destroying gypsy moth in the United States, *2591*
National assessment of biological resources in the United States, *3426*
National association for city planning in the United States, *3103*
National Audubon Society meeting, *1715*
National car industry to produce one million units a year, *1539*
National Conservation Commission in the United States, *1032*
National Contingency Plan (NCP) and Hazard Ranking System (HRS) for hazardous wastes in the United States, *2981*
National database in the United States for pollution source reduction, *2989*
National Disaster Relief Act in the United States, *2198*
National Estuary Program (NEP) in the United States, *3595*
National fisheries center approved by Congress, *2462*
National Fishing Week in the United States, *2511*
National forecasting map in the United States, *1903*
National guidelines for local zoning ordinances in the United States, *3154*
National Historic Preservation Act, *3111*
National Hunting and Fishing Day in the United States, *2927*
National insect quarantine program in the United States, *1282*
National land-planning measure to receive U.S. congressional approval, *3088*
National law regulating the surveying and sale of U.S. public lands, *3077*
National legislation mandating protection of timber and water resources in U.S. national forests, *2632*
National natural resource inventory document in the United States, *3371*
National network in the United States to analyze the effects of air pollution on precipitation chemistry, *1473*
National organization formed to assist U.S. local, regional, and state grassroots environmental groups with their projects, *1102*
National parks management agency in the United States, *3794*
National policy on marketing genetically engineered food products in the United States, *2765*
National Priorities List (NPL) of hazardous waste sites in the United States, *2868*
National Resources Defense Council study to show that the chemical Alar is a carcinogen, *1305*
National Resources Inventory conducted by U.S. Natural Resources Conservation Service (NRCS), *3385*
National rules for reducing acid rain in the United States, *1421*
National Scenic Riverway law in the United States, *3978*
National Scenic Trail in the Rocky Mountains, *3989*
National Scenic Trails in the United States, *3981*
National soil erosion survey in the United States, *4054*
National standards for air polluting motor vehicle emissions in the United States, *1406*
National survey of wetlands in the United States, *3147*
National timber industry trade group in the United States, *2704*
National trade group to promote the U.S. timber industry's positions on environmental issues, *2700*
National trade organization for wind power in the United States, *4854*
National Trails System in the United States, *3982*
National Water Quality Assessment Program in the United States, *4609*
National water use survey of the United States by county, *4604*
National water use survey of U.S. aquifer systems, *4608*
National Wetlands Policy Forum to call for a program of no net loss of wetlands, *4484*
National Wild and Scenic Rivers System in the United States, *3984*
National wildlife refuge specially authorized by Congress, *4813*
Nationwide survey of drinking water system infrastructures in the United States, *4612*
Natural energy directly regulated by the U.S. government, *2051*
Natural gas emergency legislation passed by the U.S. Congress, *2056*
Natural gas price deregulation by the U.S. government, *2063*
Natural Resources Conservation Service of the U.S. Department of Agriculture, *3370*
Natural Resources Inventory & Analysis Institute (NRIAI), *3415*
Natural resources inventory to employ statistical sampling in the United States, *3372*
New England Fishery Resources Restoration Act, *2466*
Noise control agency in the United States, *3429*

UNITED STATES—*continued*
Noise emission standards for motor vehicles involved in U.S. interstate commerce, *3443*
Noise pollution legislation of significance in the United States, *3441*
Noise standards set by the United States Occupational Safety and Health Administration (OSHA), *3444*
Nontechnical text on population growth to be widely distributed, *3261*
Northern spotted owl listing under the U.S. Endangered Species Act, *2416*
Nuclear energy agreement between the U.S. and China, *3523*
Nuclear energy development agency of the U.S. government, *3455*
Nuclear energy nonmilitary joint projects of the U.S. government and industry, *3470*
Nuclear energy program to disassemble a damaged reactor, *3515*
Nuclear power organization of the U.S. government to emphasize public health, *3949*
Nuclear reactor to power space vehicles, *3493*
Nuclear submarine from the U.S. Navy to be lost at sea, *3499*
Nuclear technology access of significance by U.S. corporations, *3468*
Nuclear waste lawsuit of significance against the U.S. government, *3540*
Nuclear waste site chosen by the U.S. Congress, *3533*
Nuclear waste site evaluation, *3530*
Nuclear-weapons tests in the atmosphere in the Pacific, *1342*
Observer to describe tropical storms as large vortices of air rotating around an "eye" of low pressure, *4239*
Occupational safety coverage for U.S. workers, *3943*
Oceanography textbook, *3557*
Office of chief forester in the United States, *2696*
Office of Environmental Education in the United States, *2304*
Official meteorologist to work for the U.S. government, *1891*
Official use of the term *wilderness* to classify U.S. federal lands, *4638*
Offshore oil to be accessed from a modern platform out of sight of land, *3653*
Offshore U.S. oil leases in the outer continental shelf of the Atlantic, *3663*
Oil spill cleanup law in the United States, *3643*
Oil wells seized from foreign companies, *3699*
Organized resistance to environmental regulations, *1025*
Ozone-depletion Executive Order by a U.S. President, *1996*
Ozone-layer measurements from space, *1979*
Paleoclimatology research in the United States using a global network, *2015*
Parabolic trough solar reflector, *4095*
Parrot listed under the U.S. Endangered Species Act, *2391*
Penalties enacted by the United States for harassing hunters, *2961*
Pencil made of recycled newspaper and cardboard fiber, *2716*
Peregrine falcons listed under the U.S. Endangered Species Act, *2393*
Permanent ban on oil and gas leasing on U.S. wilderness land, *4666*
Permit system enacted in the United States for fishing vessels on the high seas, *2484*
Permits issued by the U.S. Fish and Wildlife Service to kill migratory birds at aquaculture sites, *2498*
Person known to survive being struck by lightning seven times, *4276*
Phosphate-free laundry detergent, *4405*
Plant removed from the U.S. Endangered and Threatened Species List, *2415*
Popular book to warn of pesticide threat, *3256*
Population control book published after World War II, *3255*
Postage stamps depicting the American bald eagle, *3161*
Postcard weather-forecasting service in the United States, *1918*
Prairie dog listed under the U.S. Endangered Species Act, *2397*
Pre-emption act in the United States involving lands settled before the government had purchased and surveyed them, *3138*
Presentation of the essay "The Historical Roots of Our Ecological Crisis" by historian Lynn Townsend White Jr., *3257*
Presidential order prohibiting U.S. government assistance to concerns that violate air emissions standards, *1407*
Private and nonprofit land conservation organization in the United States, *1029*
Private environmental organization of significance in the United States to purchase and preserve substantial natural habitats, *1038*
Private organization to support national parks in the United States, *3718*
Program of the National Park Service intended to help local groups undertake river and trail conservation projects, *3733*
Prohibition against nuclear weapons on the ocean floor, *3584*
Project after the 1938 appropriation by the U.S. Congress to assist state wildlife restoration projects, *4708*
Proof that radar could be used to locate precipitation, *1944*
Proof that thunderstorms can push pollutants as high as the lower stratosphere, *1478*

Proposal for a national network of living historical farms in the United States, *1134*
Proposals for a federal tax on water polluters in the United States, *4464*
Proposed removal of bald eagles from the U.S. Threatened and Endangered Species List, *2425*
Proposed U.S. legislation regarding abandoned cars, *4142*
Protection of American Indian hunting grounds by the U.S. Congress, *2939*
Provision for federal cooperation in fire control on private timber or forest-producing lands, *2616*
Public aquarium in the United States, *4895*
Public health department created by the U.S. government, *3920*
Public health grants from the U.S. government to states, *3929*
Public Land Review Commission in the United States, *3081*
Public outcry of significance in the United States about cancer-causing chemicals in food, *1294*
Public service symbol created for the U.S. government, *3179*
Public statement by a U.S. president that all of the country's major river systems were polluted, *4403*
Publication inspired by *The Whole Earth Catalog*, *3259*
Publication that offered "access to tools" information for the energy-efficient and promoted ecologically aware lifestyles, *3258*
Published presentation of a more holistic view of U.S. public land management, *3073*
Quarantine law for plants enacted by the U.S. Congress, *1283*
Quiet Communities Act in the United States, *3446*
Radical "deep ecology" platform, *1082*
Radical environmental activist arrested and convicted for "monkeywrenching" in the United States, *1084*
Radon control legislation in the United States, *1367*
Rankings by the U.S. government of the nation's most polluted cities, *1432*
Reauthorization of the "Superfund" program, *2985*
Recognition by the U.S. government of the health hazards of passive smoking, *1366*
Recognition of radon as a dangerous indoor household pollutant in the United States, *1352*
Recognition of "sick building syndrome", *1361*
Recommendation to sell U.S. public lands in large parcels for grazing, *1165*
Recorded "year without summer" in New England, *1814*
Refuge established west of the Mississippi, *4810*
Regular and systematic monitoring of air pollution levels in the United States, *1405*
Regulation of toy guns as a noise hazard in the United States, *3439*
Regulations by the U.S. Army Corps of Engineers requiring applicants for permits to fill wetlands to prove the work would be consistent with ecological considerations, *3149*
Regulations enacted in the United States to limit swordfish harvesting according to size, *2485*
Regulations on animal research in the United States, *2287*
Report from the U.S. Environmental Protection Agency indicating a strong link between second-hand smoke and lung cancer, *1368*
Reports of sharp decreases in dissolved oxygen levels in Lake Erie, *4414*
Reports of the virtual disappearance of the mayfly in Lake Erie, *2374*
Reports of widespread fouling of Great Lakes beaches by ill-smelling seaweed, *4532*
Reproduction of U.S. federal duck stamps, *3196*
Reptile removed from the U.S. Endangered and Threatened Species List, *2410*
Reptile removed from the U.S. Endangered and Threatened Species List due to recovery, *2413*
Requirement for an environmental impact statement for development projects in the United States, *1064*
Research and development program for water pollution control sponsored by the U.S. government, *4509*
Research grant program administered by the U.S. Department of the Interior, *2288*
Research of significance by the U.S. government on the peaceful uses of nuclear energy, *3457*
Research showing that human activity could cause irretrievable damage to the earth, *1048*
Return of normal rainfall to the area of the central United States hit by the Dust Bowl, *2152*
Revision of the U.S. National Wetlands Policy to counter the "no net loss" theory, *4486*
Rewards offered by the U.S. Fish and Wildlife Service for information about wildlife crimes, *4733*
Rhinoceros and tiger protection law, *2346*
Rhinoceros listings under the U.S. Endangered Species Act, *2398*
"Right to Know" law in the United States covering hazardous substances, *2986*
Rivers and harbors antipollution law in the United States, *4450*
Rivers managed by the U.S. National Park Service, *4641*

UNITED STATES—*continued*
Road inventory in the United States, *1556*
Ruling by the U.S. Supreme Court that historic preservation was a form of community welfare, *3116*
Saltcedar or tamarisk in the United States, *1503*
Scenic Byways and All-American Roads program in the United States, *4010*
Scholarly article to identify a legal trend curbing the powers of environmental groups and private citizens to sue polluters in U.S. courts, *1093*
Scholarly article to show that Lake Erie was dying a slow death from pollution, *4510*
Scientist in the United States to conduct a detailed study of a single winter storm, *1888*
Sea lamprey in Lake Erie, *4502*
Sea turtle listing to include all species, *2402*
Serialization of Upton Sinclair's *The Jungle*, *3249*
Serious warnings about a timber shortage in the United States, *2585*
Set of tornadoes to cause $500 million in damages, *4291*
Severe drought year affecting nearly one-half of the United States, *2148*
"Shellfish clause" in the U.S. Clean Water Amendments, *4467*
Significant conference on the risks of chemical contaminants to future generations to be sponsored by the U.S. government, *4423*
Significant federal legislation attacking forest pests in the United States, *2604*
Significant federal legislation to ban the discharge of sewage sludge in U.S. waters, *4479*
Significant forestry legislation in the United States in the second half of the 20th century, *2640*
Small all-purpose tractor in the United States, *1246*
Snail listed under the U.S. Endangered Species Act, *2399*
Snake listed under the U.S. Endangered Species Act, *2392*
Society emphasizing conservation in the United States, *1028*
Society for animal welfare in the U.S., *1007*
Society of professional ecologists in the United States, *2274*
Sockeye salmon listing under the U.S. Endangered Species Act, *2417*
Solar energy research funded by the U.S. Congress, *2298*
Solar-powered aircraft that could carry a pilot, *4118*
Solar-powered steam engine in the United States, *4094*
Solar water heaters on record, *4090*
Sonic boom regulations for aircraft, deemed a noise pollutant and public health issue in the United States, *3447*
Spanish ships built from pine trees felled along the southern reaches of the Mississippi, *2643*
Species removed from the U.S. Endangered and Threatened Species List, *2408*
Species removed from the U.S. Endangered and Threatened Species List due to recovery, *2412*
Spent nuclear fuel and high-level radioactive waste disposal law in the United States, *3522*
Stabilization of climates of the present-day United States to their current levels, *2003*
Standardized national vegetation classification system in the United States, *3413*
State park in the United States, *3818*
Statement of the theory that constant exposure over time to small doses of toxins causes poisoning, *2832*
Steelhead trout listing under the U.S. Endangered Species Act, *2423*
Strip mining controls by the U.S. government, *3336*
Study of aircraft noise effecting U.S. national parks and forest wilderness, *3430*
Study that defined the factors making old-growth Douglas fir forests unique, *2747*
"Subsistence rights" legislation for Alaska, *3388*
Substantial cuts in the budget of the U.S. Environmental Protection Agency, *1080*
Substantial federal effort to promote industrial forestry on private land in the United States, *2677*
Substantial federal legislation to provide funds for research into reforestation in the United States, *2634*
Substantial revisions in U.S. guidelines for approving new antibiotics for livestock and for monitoring the effects of existing drugs, *1176*
Substantial U.S. government funding for a study of causes of soil erosion and methods of erosion control, *4051*
Successful challenge to state-ownership doctrine, *1690*
Successful control of the cotton-destroying boll weevil, *1284*
"Superfund" environmental cleanup law in the United States, *2980*
Superpower to effectively withdraw from the Kyoto Protocol, *3012*
Survey of recreational fishing on a nationwide basis in the United States, *2508*
Survey of recreational hunting on a nationwide basis in the United States, *2921*
Survey of the coast of the United States, *3549*
Survey of U.S. biomass resources, *4339*

Survey of U.S. wetlands and wetlands losses that was comprehensive and statistically valid, *3387*
Survey of wildlife-related recreation on a nationwide basis in the United States, *4720*
Systematic dam inspection by the U.S. government, *2072*
Systematic work on American geology, *2786*
Tax breaks for owners of historically significant commercial property in the United States, *3112*
Tax credit in the United States for ethanol, *4330*
Tax deduction in the United States for purchasers of Hybrid Electric Vehicles (HEVs), *4337*
Tax deductions in the United States for purchasers of Clean Fuel Vehicles (CFVs), *4338*
Tax exemptions in the United States for conservation easements, *4664*
Tax on boats and motors in the United States to benefit wildlife habitats, *2510*
Tax on bows and arrows in the United States, *2926*
Tax on crossbow arrows in the United States, *2928*
Tax on fishing equipment in the United States to specifically benefit wildlife habitats, *2507*
Teletype communications of regularity between U.S. weather stations, *1936*
Television network to broadcast weather programming exclusively, *1966*
Test drive of the "veggie van," an experimental motor home powered by a diesel engine that ran on fuel made from vegetable oil, *1611*
Text of modern oceanography detailing the trade winds and ocean currents, *3237*
Textbook on wildlife management, *3252*
TFM chemical releases into the lamprey-infested waters of Lake Superior, *1526*
Tide-predicting machine, *3564*
Time that natural gas was overtaken by electricity as the primary power source for lighting in the United States, *2048*
Tornado designation publicly used by the U.S. Weather Bureau, *4289*
Tornado observation instructions in the United States, *4282*
Tornadoes in the United States reported to cause a great loss of life, *4283*
Tortoise listed under the U.S. Endangered Species Act, *2400*
Toxics Release Inventory (TRI) in the United States, *2987*
Trade sanctions imposed by the United States to protect endangered wildlife, *2345*
Transportation act for the 21st century passed by the United States, *1585*
Transportation act in the United States with significant mass transit funding, *1581*
Treaty in which England recognized the fishing rights of the United States, *3014*
Treaty obliging the United States to maintain a list of endangered species, *3021*
Treaty to ban the introduction of nuclear weapons into orbit around the Earth, *3027*
Treaty to recognize the three-mile limit for fishing off the North American coast, *3015*
Tree-killing white pine blister rust in the United States, *2596*
Tree species in North America to be widely harvested commercially, *2645*
Tri-national conference on conservation in North America, *1034*
Turtles listed under the U.S. Endangered Species Act, *2401*
Two-day U.S. Weather Bureau forecasts, *1919*
Two-time winner of the U.S. federal duck stamp contest, *3180*
U.S.–Canadian bilateral report on acid rain, *3042*
U.S.–Canadian effort of significance to improve the water quality of the Great Lakes, *3029*
U.S. census to collect information on a natural resource, *3357*
U.S. congressional "treaty" acquiring American Indian lands for the government, *3133*
U.S. Environmental Protection Agency (EPA) office for American Indian environmental affairs, *3411*
U.S. Environmental Protection Agency (EPA) role in wildlife conservation, *4779*
U.S. federal funding to preserve the Tule elk, *4833*
U.S. federal law to protect a wilderness area, *4636*
U.S. federal source reduction policy, *2990*
U.S. Fish and Wildlife Service Environmental Contaminants Program, *4734*
U.S. Fish and Wildlife Service (FWS) guidelines for distributing eagle parts, *4742*
U.S. Fish and Wildlife Service jurisdiction over sockeye salmon conservation, *2459*
U.S. Forest Service study of recreational opportunities and values in the national forests, *2682*
U.S. government agency to adopt a formal environmental policy with American Indian tribes, *3391*
U.S. government body devoted solely to mediating environmental disputes, *3066*
U.S. government body to study the effects of insects on nature, *4687*
U.S. government conference devoted to conservation issues, *3365*
U.S. government declaration of cigarettes as a "drug delivery device", *1370*
U.S. government listing of the nation's ten most polluted rivers, *4406*

UNITED STATES—*continued*
U.S. government manual with guidelines for placement of public wells and instructions about possible contamination from industrial sites, *2839*
U.S. government moves to limit permissible amounts of arsenic in drinking water, *2854*
U.S. Government Paperwork Elimination Act, *4155*
U.S. government program to reduce air pollution in the Grand Canyon, *1420*
U.S. government research suggesting nitric acid (rather than sulfuric acid) is the chief cause of acid rain, *1481*
U.S. government study into the effects of wildlife on natural resources, *3361*
U.S. Green Party presidential campaign, *1118*
U.S. institute for study and research on oceans, *2269*
U.S. Interagency Council on Biobased Products and Bioenergy, *4347*
U.S. Jobs Through Recycling grant to Arizona, *4159*
U.S. Jobs Through Recycling grant to California, *4160*
U.S. Jobs Through Recycling grant to Delaware, *4161*
U.S. Jobs Through Recycling grant to Maryland, *4162*
U.S. Jobs Through Recycling grant to Minnesota, *4163*
U.S. Jobs Through Recycling grant to Nebraska, *4164*
U.S. Jobs Through Recycling grant to New Hampshire, *4169*
U.S. Jobs Through Recycling grant to New York, *4165*
U.S. Jobs Through Recycling grant to North Carolina, *4166*
U.S. Jobs Through Recycling grant to Ohio, *4167*
U.S. Jobs Through Recycling grant to Oregon, *4168*
U.S. Jobs Through Recycling grant to Virginia, *4170*
U.S. law allowing consideration of recreation, fish, and wildlife enhancement as purposes of federal water development projects, *4561*
U.S. law exempting alternative energy producers from state and federal regulation, *4115*
U.S. legislation making states responsible for their nuclear waste, *3516*
U.S. legislation providing federal aid for up to 50 percent of the cost of reforestation of farmlands on the Great Plains, *2728*
U.S. legislation to promote Ocean Thermal Energy Conversion (OTEC) development for commercial use, *4565*
U.S. legislation to regulate the disposal of hazardous medical materials, *2883*
U.S. moratorium on building hazardous waste incinerators, *2893*
U.S. national census, *3875*
U.S. National Fish and Wildlife Foundation, *4736*
U.S. National Park Service (NPS) Inventory and Monitoring Program (I&M), *3402*
U.S. National Science Foundation (NSF), *2283*
U.S. National Sea Grant Program, *2289*
U.S. National Wild and Scenic Rivers in Alaska, *3999*
U.S. Natural Resource Inventory and Monitoring (I&M) mapmaking program, *3405*
U.S. Natural Resource Inventory and Monitoring (I&M) Program on water quality, *3406*
U.S. Navy report on the reintroduction of military dolphins to the wild, *4841*
U.S. nuclear-powered surface warship, *3487*
U.S. oceanographic research ship, *3562*
U.S. President's Council on Sustainable Development, *4314*
U.S. Soil Conservation Service chief, *4056*
U.S. Sport Fishing and Boating Partnership Council, *2512*
U.S. Supreme Court decision regarding a sponge harvesting law, *2472*
U.S. Supreme Court decision that municipal incinerator ash be treated as hazardous waste, *2894*
U.S. Supreme Court decision to rule that living things are patentable material, *2757*
U.S. Supreme Court ruling affirming the legality of zoning, *3155*
U.S. Supreme Court ruling on local bans on imported waste, *4150*
U.S. Supreme Court ruling on plastic milk containers, *4146*
U.S. Supreme Court ruling on state bans on imported waste, *4145*
U.S. Supreme Court ruling striking down a federal requirement that states "take title" to all radioactive wastes after specified deadlines had passed, *2890*
Underground animal rights organization to operate in the United States, *1021*
"Universal Waste Rule" in the United States, *4135*
Uranium fuel bought from Russia by the United States, *3535*
Uranium fuel ownership by private U.S. companies, *3501*
Urea-formaldehyde foam insulation ban in the United States, *1364*
Use by Smokey Bear of the fire prevention slogan "Remember—only you can prevent forest fires", *2619*
Use of a locomotive in logging operations, *2654*
Use of antibiotics as an animal feed additive, *1172*

Use of asphalt for street paving in the United States, *1553*
Use of copper sulphate to destroy the charlock weed, *1280*
Use of high-speed electronic computers for U.S. weather forecasts, *1943*
Use of organic chemicals for weed control, *1286*
Use of rapid sand filtration for treating industrial wastewater in the United States, *4504*
Use of the acronym SLAPP, *1077*
Use of the Atlantic Coastal Fisheries Cooperative Management Act of 1993 against a state, *2467*
Use of the expression *wise use* in discussing U.S. resource issues, *3364*
Use of the phrase *tornado watch* by the U.S. Weather Bureau, *4292*
Use of the telegraph for reporting weather conditions, *1897*
Use of the term *bionics*, *2285*
Use of the term *Dust Bowl* to describe drought-seared and wind-eroded U.S. southern Great Plains, *4052*
Use of the term *Dust Bowl* to describe the drought-devastated Southern Plains, *2150*
Use of the term *genetic pollution*, *1257*
Use of the term *green*, *1106*
Use of the term *primitive area* by the U.S. Forest Service to describe wilderness, *2714*
Use of tractors in timber harvesting, *2659*
Vegetation inventory in the U.S. national parks system, *1797*
Volunteer tornado spotters network established by the U.S. government, *4287*
Warnings against tapping groundwater near industrial developments, *2835*
Water Bank Act in the United States, *3380*
Water pollution law enacted by the U.S. Congress, *4458*
Water pollution legislation passed by Congress, *4448*
Water pollution national conference of significance sponsored by the U.S. Chamber of Commerce to address the issue of industrial waste, *4438*
Water quality standards established at the U.S. federal level, *4466*
Water resources inventory of the arid western U.S., *1218*
Waterfowl Depredations Prevention Act in the United States, *4800*
Waterfowl species to appear twice on U.S. federal duck stamps, *3195*
Waterways in the U.S. National Wild and Scenic Rivers System, *3983*
Weather bulletins for U.S. farmers, *1915*
Weather observations by the U.S. military, *1883*
Weather service of the U.S. government not run by the military, *1917*
Wetlands conservation book published by the U.S. government for a general audience, *3375*
Wetlands Loan Act in the United States, *3373*
Wetlands protection and restoration law to provide funds for protection, *3090*
Wetlands Reserve Program (WRP) in the United States, *3420*
White Americans to systematically describe the weather of the western United States, *1812*
Widely acknowledged environmentalist to be chosen as a major-party U.S. vice presidential candidate, *1091*
Widely publicized epidemic of medical waste washing up on U.S. beaches, *2884*
Widely publicized version of the Christian fundamentalist theory of flood geology, *2547*
Widespread and catastrophic drought to afflict farmers in the U.S. Great Plains, *2141*
Widespread attempts to grow yeast on pulp wastes from paper manufacturing in the United States, *4432*
Widespread replacement of railroad logging by trucks in the United States, *2660*
Widespread spraying of the pesticide DDT to eradicate gyspy moths in the United States, *2603*
Widespread use of the acronym NIMBY, *1100*
Wild horses appeared in America, *1153*
Wilderness Act in the United States, *4647*
Wilderness bill introduced in the U.S. Congress, *3736*
Wilderness protection act for California deserts, *4674*
Wilderness refuge system in the United States, *4827*
Wildlife conference in North America, *4701*
Wildlife conservation law for U.S. military property, *4772*
Wildlife Habitat Incentive Program (WHIP) in the United States, *4790*
Wildlife management law in the United States, *4765*
Wind and current analysis of significance, *3551*
Wind turbine ship to cross the Atlantic Ocean, *4849*
Woman to direct the U.S. Fish and Wildlife Service, *4741*
Woman to head a major U.S. conservation organization and receive a salary, *1039*
Woman to head the U.S. Environmental Protection Agency, *1116*
Woman to win the U.S. federal duck stamp contest, *3221*
Workshops on hunter safety in the United States, *2924*
World Heritage site established in the United States, *3808*

UNITED STATES—*continued*
Worldwide organization to advocate family planning, *3897*
Year all the states in the United States had at least one law regulating hazardous wastes, *2831*
Year in which General Motors vehicles outsold those of the Ford Motor Company, *1541*
Year in which half of the agricultural counties in the United States were designated drought disaster areas, *2163*
Year oil became the major energy source in the United States, *3621*
Yellow fever vaccine, *3928*

Alabama

Appearance of alligatorweed in the United States, *4500*
Catfish farms in the United States, *2476*
Effort for organized wildlife protection in Alabama, *4692*
Former hunting path designated as an All-American Road, *4015*
Human resident in what is now Russell Cave National Monument in Alabama, *3738*
Hurricane reported to occur during early summer in the United States, *4261*
Legal proceedings against Alabama polluters by the Bass Anglers Sportsmen Society, *4473*
State in the United States to mandate citizen participation in refuse collection systems, *3275*
Tornadoes in the United States reported to cause a great loss of life, *4283*

Bessemer

U.S. National Wildlife Refuge created specifically for watercress darters, *2334*

Emelle

Hazardous waste treatment site to accept waste from 48 U.S. states, *2860*

Enterprise

Monument to an insect in the United States, *3164*

Mobile

Hurricane in the Gulf of Mexico with a masculine name, *4266*

Muscle Shoals

Dams on the Tennessee River near Muscle Shoals, AL, *2094*

Alaska

Arctic-subarctic biome surveyed by a U.S. National Park Service Inventory and Monitoring Program (I&M), *3398*
Ascent of Mount Elias, *3784*
Conservationist of note in the United States, *1026*
Discovery of a bristle-thighed curlew's nest, *1699*
Extension of the offshore drilling ban, *3665*
Humans to cross the Bering Land Bridge from Asia to North America, *3750*
Legislation in the United States to mandate double hulls on oil tankers, *3627*
National park created after the establishment of the National Park Service, *3795*
Natural resource profits fund to benefit all Alaskans, *3384*
Non–American Indian to discover Glacier Bay and the Muir Glacier, *3776*
Oil discovery of major significance in Alaska, *3703*
Recorded wildcat oil drilling team in Alaska, *3692*
Road link between Alaska and the lower 48 states, *1564*
State in which the U.S. Green political party gained ballot status, *1115*
"Subsistence rights" legislation for Alaska, *3388*
Tsunami to devastate the U.S. West Coast, *2549*
U.S. National Scenic Byway in Alaska, *4023*
U.S. National Wild and Scenic Rivers in Alaska, *3999*
U.S. National Wilderness Areas in Alaska, *4651*

Anchorage

Documented case of severe damage to a jetliner flying through a volcanic ash cloud, *4393*
Inventory of vegetation in U.S. national parks in Alaska, *3409*
U.S. National Scenic Byway in Alaska, *4023*

Bering Strait

Sea otters encountered by Europeans, *4681*

Homer

Cyber animal exhibit that allows people to view bears in their natural habitat, *4968*

Klawock

Cannery in Alaska, *2471*

Oil Bay

Recorded oil seepages in Alaska, *3612*

Point Barrow

Flight across the Arctic Ocean, *3855*

Point Hope

Harbor created by an atomic bomb, *3483*

Prudhoe Bay

Flow of oil through the Alaska Pipeline from Prudhoe Bay to Valdez on Prince William Sound, *3624*
Oil discovery at Prudhoe Bay on Alaska's North Slope, *3700*

Seward

U.S. National Scenic Byway in Alaska, *4023*

Valdez

Flow of oil through the Alaska Pipeline from Prudhoe Bay to Valdez on Prince William Sound, *3624*
Use of elevation benchmarks to measure settling or uplifting of the earth, *2226*

Yakuta Bay

Recorded incidence of an earthquake reversing the paths of two glaciers, *2209*
Vertical surface displacement of more than 45 feet caused by an earthquake, *2210*

Arizona

Artificial flood of significance on the Colorado River in the western United States, *2562*
Coal use known in North America, *2020*
Condors reintroduced to Arizona, *2430*
Dam to be constructed with a height greater than 152 meters (500 feet), *2099*
Documented journey through the Glen Canyon–Lake Powell area, *3755*
Electric car from a major U.S. company, *1612*
European to see the Grand Canyon, *3753*
Evidence of water reservoir-induced earthquakes, *2222*
Failed U.S. Rocky Mountain mule deer management policy, *4797*
Forest experiment station in the United States, *2737*
Geological observations of the Grand Canyon, *2810*
Irrigation ditches in what is now the United States, *4569*
Irrigation project of significance in what is now the United States, *1214*
Jaguar Conservation Team in Arizona, *2349*
Large-scale seizure and auction of cattle grazing on U.S. land without a permit, *1174*
Law controlling aircraft noise over the Grand Canyon, *3445*
Law in the United States extending wilderness protection to land administered by the Bureau of Land Management (BLM), *4667*
National protest of significance sponsored by Earth First! *1079*
National recreation area designated in the United States, *3806*
National Wild and Scenic River in the Sonoran Desert, *4006*
Plant to use the pollution-to-homes process, *2295*
U.S. government program to reduce air pollution in the Grand Canyon, *1420*
U.S. Jobs Through Recycling grant to Arizona, *4159*
U.S. National Wilderness Areas in the Southwest, *4646*
Written record of the discovery of Glen Canyon, *3245*

Apache National Forest

Mexican wolf reintroduction, *2444*

Ashfork

Steel dam in the United States, *2089*

Flagstaff

Person to compare tree ring width to climate and droughts, *2143*

Hereford

Monument to commemorate the first major exploration of the American Southwest, *3748*

Jacob Lake

U.S. National Scenic Byway in Grand Canyon National Park, *4025*

Nogales

Historic trail designated in the Western Region of the U.S. National Park Service, *3711*

Page

Antipollution device installed as a result of the 1977 law imposing special restrictions on pollution near national parks, *1419*

Phoenix

Hydropower facilities operated by the U.S. Bureau of Reclamation, *4556*

Schuchuli

Photovoltaic (PV) village, *4116*

UNITED STATES—Arizona—*continued*

Seligman

Black-footed ferret reintroduction program to use acclimatization pens in the United States, *2443*

Tucson

House with solar heating and radiation cooling in the United States, *4109*
Solar-heated school, *4106*

Tusayan

Light rail system in a national park in the United States, *3815*

Arkansas

Catfish farms in the United States, *2476*
Floods to cause damages of more than $1 billion in the United States, *2563*
Interstate forest-fire protection agreement in the southern United States, *2622*
National park in the United States, *3772*
State in the United States to restrict market hunting, *2951*
Two states to adopt President Franklin Roosevelt's Standard State Soil Conservation District Laws, *4039*
U.S. National Wilderness Area in the Ozark Plateau region, *4656*

Searcy

Titan II silo explosion, *3502*

California

All-American Road on the U.S. Pacific coast, *4013*
American migrants to be forced into cannibalism after being trapped in the High Sierra by heavy snow, *4197*
Appearance of an oak tree blight previously unknown in the United States, *2609*
Appearance of the bayberry whitefly in California citrus groves, *1301*
Appearance of the environmental movement known as bioregionalism, *1066*
Automobile crankcase emissions law in the United States, *1573*
Automobile emissions control law of significance in the United States, *1577*
Automobile emissions standards set by a U.S. state, *1572*
Bill in California forcing developers to show that there would be an adequate supply of water for any proposed large subdivision, *3091*
California condor sanctuary, *1723*
California state agency to regulate wind power and other renewable energy sources, *4855*
California waters in the state Heritage Trout Program (HTP), *2513*
Closed season on antelope and elk hunting in the United States, *2944*
Commercial whale-watching ventures, *2239*
Conservationist of note in the United States, *1026*
Deep-sea exploration of U.S. marine sanctuaries, *3602*
Discovery of a marbled murrelet's nest, *1707*
Discovery of large numbers of dead and deformed birds in the Kesterson Refuge and other ponds containing irrigation runoff from farms in the Central Valley of California, *1221*
Ecosystem management plan upheld in a U.S. court, *3414*
Electric car from a major U.S. company, *1612*
European to lead an expedition into the area that later became Sequoia and Kings Canyon National Park, *3760*
Executive Order issued to protect the Lake Tahoe region, *3421*
Exemption for U.S. military flyovers of National Wilderness deserts in California, *4673*
Extension of the offshore drilling ban, *3665*
Federal recommendation that areas of the San Joaquin Valley in California be taken out of agricultural use because of naturally high selenium concentrations, *4026*
Felony indictments under a U.S. environmental law for smuggling hazardous waste from California into Mexico, *2889*
Firm evidence dating a major U.S. forest fire, *2610*
Fly to be designated as an endangered species in the United States, *2344*
Fox listing under the U.S. Endangered Species Act, *2390*
Game commission in the United States, *2953*
Genetically altered bacterium tested outside of a laboratory in the environment, *2761*
Gold deposit discovery of significance in California, *3311*
Hibernating bird discovered, *1740*
Hunting law in California, *2945*
Introduction of larger, meatier British cattle on U.S. ranches, *1164*
Introduction of the citrus-damaging cottony-cushion scale insect in California, *1184*
Introduction of the scale insect–destroying Vedalia beetle in California, *1279*
Known successful introduction of a fish species within the United States, *2491*
Land-use plan for a public desert area in the United States, *3089*

Law in California to limit motor-vehicle noise emissions on highways, *3435*
Law in California to limit motor-vehicle noise emissions through sales restrictions, *3436*
Law in the United States to establish federal and state standards for organic certification, *1264*
Legislation in California to require fire-resistant roofing in that fire-prone state, *2624*
Legislation preserving a significant extent of coastal redwood forest in the United States, *2638*
Legislation to provide comprehensive planning for hazardous waste treatment in California, *2875*
Logging company in California history to have its timber license suspended for environmental violations, *2666*
Low-emission automobiles, *1605*
Massive spraying of the insecticide malathion to control the Mediterranean fruit fly in California, *1303*
Mediterranean fruit fly infestation in North America, *1522*
Mediterranean fruit fly ("medfly") infestation in California, *1299*
Mobilization and training of youth to participate in nonviolent environmental protest, *1088*
National environmental organization in the United States, *1051*
National insect quarantine program in the United States, *1282*
National monument in the United States created from land donated by a private individual, *3747*
National park in the United States to contain an active volcano, *3793*
National preserve in an American desert, *3813*
Pacific coast biome surveyed by a U.S. National Park Service Inventory and Monitoring Program (I&M), *3401*
Person to write a book promoting the idea of a U.S. national parks system, *3248*
Private organization for protecting original redwood forests, *2698*
Public preserve in the United States, *3713*
Quarantine law for plants by a state, *1277*
Recognition of the citrus-attacking California red scale insect as a serious pest in the United States, *1275*
Recommendation from a California state waste management agency against siting hazardous waste incinerators in middle- to upper-income communities, *2869*
Refuge in the United States to protect endangered plants and insects, *4836*
Scientific exploration of the area that later became Sequoia and Kings Canyon National, *3769*
Sierra Club chapter formed outside of California, *1057*
Sierra Club outing to a national park, *3786*
Signs of damage to trees far from automobile pollution sources, *2606*
Snake listed under the U.S. Endangered Species Act, *2392*
State fishing commission in the United States, *2503*
State government concerns about public landfills, *2833*
State in the United States to ban human cloning, *2771*
State in the United States to enact legislation to curtail aircraft emissions, *1403*
State in the United States to permanently employ wildlife pathologists, *4713*
State in which the U.S. Green political party gained ballot status, *1115*
State wildlife conservation agency in the United States, *4762*
Statewide construction and design standards for earthquakes in the United States, *2221*
Steelhead trout listing under the U.S. Endangered Species Act, *2423*
Systematic study of *Sequoia gigantea*, the California sequoia, *1784*
Tourists in Yosemite National Park, *2237*
Tsunami to devastate the U.S. West Coast, *2549*
U.S. federal funding to preserve the Tule elk, *4833*
U.S. Jobs Through Recycling grant to California, *4160*
Use of the term *hydroponics*, *1251*
Waterways in the U.S. National Wild and Scenic Rivers System, *3983*
Wilderness protection act for California deserts, *4674*

Arcata

Artificial wetlands treatment of wastewater in the United States, *4540*

Bagdad

Reported period of more than 900 days without measurable precipitation in a southern California town, *2142*

Berkeley

Electrostatic device to remove particles from industrial smoke, *2970*
International dam opposition group, *2075*
Ozone-layer warning, *1973*
Private hydroponic garden, *1248*
Theory that an asteroid caused the extinction of dinosaurs, *2376*

UNITED STATES—California—*continued*

Capitola

Commercial production of plants in water instead of soil, *1249*

Carmel

All-American Road on the U.S. Pacific coast, *4013*

Colton

Building developments halted to protect an endangered species, *2351*

Dagget

Cool water or integrated gasification combined-cycle utility, *2301*

Davis

DNA test to detect whirling disease in trout and other salmonids, *2468*
Genetically altered food, *2767*

Death Valley

U.S. National Scenic Byway below sea level, *4022*

Diamond Bar

Electric-car "solar carpark", *1604*
Natural gas fuel cell to produce electricity commercially, *2065*

Dunsmuir

Massively lethal pesticide spill in the Sacramento River, *1307*

Edwards

Flight of a solar-powered aircraft, *4113*
Manned solar-powered aircraft flight, *4119*

Fontana

Building developments halted to protect an endangered species, *2351*

Hemet

Building developments halted to protect an endangered species, *2351*

Irvine

Ozone-layer data concerning damage by chlorofluorocarbons (CFCs), *1974*

La Jolla

Daily measurements of atmospheric carbon dioxide, *1839*
Global warming report to gain international scientific attention, *1838*
Oceanographic exhibit with more than 3,000 fish, *4963*
Oceanography institution, *3567*
U.S. institute for study and research on oceans, *2269*

Lassen Peak

Volcano eruption in North America for which a date can be estimated, *4361*

Long Beach

Statewide construction and design standards for earthquakes in the United States, *2221*

Los Angeles

Air pollution control director position for a U.S. city, *1391*
Air pollution control districts in a U.S. state, *1393*
Aqueduct to supply Los Angeles, CA, *4585*
"Cloud compeller" to claim success in bringing rain to California with ill-smelling chemical cocktails set in pans atop high wooden towers, *2169*
Dam of significance in the United States to collapse when completely filled, *2116*
District zoning in the United States, *3152*
Earthquake to occur directly under a large U.S. metropolitan area, *2234*
Eight-day stretch of 100-plus degree heat in Los Angeles, CA, *1824*
Indication of a new type of air pollution, *1449*
Instrument to measure carbon dioxide in the atmosphere in parts per million, *1452*
Leaf blower ban of significance in Los Angeles, CA, *3449*
Mudslides to cause property damage of over $135 million, *2575*
Officially reported measurable snowfall in downtown Los Angeles, *4204*
Ports in the United States to be monitored by U.S. Fish and Wildlife Service agents, *4730*
Recommendations in California to shift responsibility, from businesses to drivers, for paying for Southern California air pollution abatement, *1427*
Reformatted gasoline tested in the United States, *1479*
Reformulated gasoline sold in the United States, *3626*
Science and technology museum that began as a state agricultural exhibition hall, *4937*

Smog emergency alert plan in Los Angeles County, *1347*
Successful embryo transfer between females of endangered species, *4955*
Televised dam collapse, *2119*

Monrovia

Solar water heater and hot water storage system to provide high-temperature water around the clock, *4101*

Montebello

Commercial hydroponicum on a large scale, *1250*

Mount Hamilton

Seismograph in the United States, *2207*

Needles

Solar-powered energy plant capable of operating at night, *4100*

Oakland

Bird refuge established by a U.S. state, *1718*
Deliberate release of genetically altered organisms in the United States, *1256*
Municipal composting plant in the United States, *4041*

Ontario

Building developments halted to protect an endangered species, *2351*

Palo Alto

Female college graduate in geology, *2814*
Research park in the United States, *3724*

Palos Verdes

Marine mammals studied by the U.S. military, *4722*
Plant in the United States to recover methane gas from garbage, *2054*

Pasadena

Scientific paper demonstrating that hydrocarbons and nitric oxide react in sunlight to form ozone pollutants, *1454*
Solar power company, *4098*
United States National Air Pollution Symposium, *1451*

Petaluma

Zoning scheme to take time into account, *3157*

Point Arguello

Satellite to provide high resolution night photographs of cloud-cover, *1827*

Portola

Intentional poisoning of Lake Davis in California, *4807*

Rancho Cucamonga

Building developments halted to protect an endangered species, *2351*

Rialto

Building developments halted to protect an endangered species, *2351*

Riverside

Forest fire air patrol, *2615*

Sacramento

Nuclear power plant to be closed by a popular state vote, *3532*

San Bruno Mountain

Habitat Conservation Plan (HCP) in the United States, *2337*

San Diego

Appearance of the wooly whitefly in California, *1193*
Aqueduct to supply San Diego, CA, *4592*
California condors to be bred in captivity, *1684*
Emperor penguins bred outside of Antarctica, *1678*
European to set foot on what would later become the west coast of the United States., *3739*
Major pollution damage to West Coast oyster beds, *4396*

San Francisco

Aquarium to be part of a natural history museum, *4919*
Biotechnology company in the United States, *2756*
Building industry consensus coalition for "green" construction in the United States, *4311*
Dam project to provoke major opposition from environmentalists, *2090*
Earth Day to be celebrated nationwide in the United States, *1069*

UNITED STATES—California—San Francisco—*continued*
Earthquake in the United States believed to have surpassed 8.0 on the Richter scale, *2215*
Earthquake organization formed because of a U.S. earthquake, *2216*
Earthquake to be broadcast on live television nationwide, *2233*
Establishment of U.S. Forest Service Field Districts, *2680*
Historic trail designated in the Western Region of the U.S. National Park Service, *3711*
Major pollution damage to West Coast oyster beds, *4396*
Organizational expression of the bioregional movement, *1068*
Person to establish a vineyard in San Francisco, CA, *1500*
Phase-contrast cinemicrography film to be shown on television, *3266*
Ports in the United States to be monitored by U.S. Fish and Wildlife Service agents, *4730*
Snowfall recorded in excess of 3.5 inches in San Francisco, *4200*
Urban national park to exceed 75,000 acres, *3821*
Wandering albatross recorded ashore in the United States, *1703*
Woman to reach the rank of major in the U.S. Park Police, *3814*
Zoo in the United States to exhibit a pair of aye-ayes, *4967*

San Luis Obispo

All-American Road on the U.S. Pacific coast, *4013*

Santa Ana

Use of wrecked cars from a junkyard to make a barrier against a flood, *2554*

Santa Barbara

Dam in the United States to fail during an earthquake, *2115*
Offshore oil spill to cause widespread damage, *3632*
Report of a Santa Ana wind to force temperatures to over 130 degrees, *4183*

Santa Rosa

Development of the Burbank or Idaho potato, *1207*

Santa Susana

Nuclear power from a civilian reactor in the United States, *3477*

Summerland

Oil wells in the ocean floor to be successfully drilled, *3651*

Tulane County

Animal disease of American origin, *3926*

Ukiah

Official designation of the "Mendocino Tree" as the world's tallest living thing, *2580*

Vallejo

Aquatic theme park mixing education, research, and entertainment to be owned by a nonprofit organization, *4958*

Yosemite Valley

State park in the United States, *3818*

Colorado

Account of the wonders of the Rocky Mountains to reach the East Coast of the United States, *3239*
All-American Roads in the Rocky Mountains, *4014*
Application of the wilderness concept to U.S. Forest Service areas, *3719*
Coffinite mineral, *3330*
Colorado Wilderness Act, *4665*
Cultural park set aside in the U.S. National Park System, *3788*
Evidence of toxic chemicals in groundwater near the U.S. government's Rocky Mountain arsenal in Colorado, *2842*
Insecticide that was effective in the United States, *1274*
Landscape architect employed by the U.S. Forest Service (USFS), *4632*
Lynx reintroduction in Colorado, *2447*
Non–American Indian settler in what is now Rocky Mountain National Park, *3768*
Non–Native Americans to enter what is now Dinosaur National Monument, *3740*
Program in Colorado in which taxpayers could make a contribution to a wildlife program, *4731*
U.S. National Wilderness Areas in the interior West, *4643*
White men to see the great chasm at what later became Black Canyon of the Gunnison National Monument in Colorado, *3741*
Wilderness area protected by the U.S. Forest Service (USFS), *4633*
Wind farm in Colorado, *4867*
Woman to climb Long's Peak in what is now Rocky Mountain National Park, *3774*

World Heritage site in the United States selected for environmental reasons, *3973*

Boulder

African-American president of the American Meteorological Society, *1834*
Earthquake data center established by the U.S. government, *2228*

Canon City

Wild horse training program in a U.S. prison, *4735*

Cortez

Vacuum-powered prairie dog extraction device, *4805*

Denver

Cease-and-desist order to the U.S. Army to stop toxic contamination from the Rocky Mountain National Arsenal at Denver, CO, *2852*
Commercial nuclear reactor that was gas-cooled, *3511*
Establishment of U.S. Forest Service Field Districts, *2680*
International airport to be re-developed as a community, *4315*
Major auction of confiscated wildlife products in the United States, *4794*
National Water Quality Laboratory in the United States, *4605*
Pronghorn antelope bred and reared in captivity, *4746*
Snow goose born and bred in captivity, *4748*
Use of an ancient weed-control technique by a modern American city, *1311*
Zoo habitat constructed of simulated rock formations without bars, *4917*

Fort Collins

Wind-powered brewery in the United States, *4869*

Golden

Earthquake data center established by the U.S. government, *2228*

Grand Junction

Wild horse range established in Colorado by the U.S. Bureau of Land Management, *4837*

Mesa Verde

North American civilization known to end due to a drought, *2125*

Connecticut

Account of the wonders of the Rocky Mountains to reach the East Coast of the United States, *3239*
Automobile safety warning having national impact, *3940*
Devastating ice storm in New England in the 20th century, *4270*
Flood in the United States to claim more than 600 lives, *2541*
Introduction of gypsy moth-attacking parasites into the United States from Europe and Asia, *2594*
Law in the United States requiring ingredient labels on fertilizers, *4062*
National Trails system unit opened, *3723*
State in the United States to forbid the posting of advertising flyers on telephone poles, *3269*
State in the United States to protect an endangered plant species by law, *2320*
State in the United States to require nonsmoking sections in restaurants, *1362*
Summer blizzard recorded in the United States, *4194*

Bridgeport

Brownfields grants dispensed by the U.S. Environmental Protection Agency (EPA), *2993*

Groton

Accident aboard the USS *Nautilus*, the world's first nuclear-powered submarine, *3480*
Commercial fuel cell powered by landfill gas, *4341*
Nuclear-powered submarine, *3466*
U.S. submarine to carry nuclear-armed ballistic missiles, *3490*

Hartford

City planning commission in the United States, *3102*
Connecticut River flood crest at Hartford to reach 28 feet, 10.5 inches, *2528*

Middlefield

Clean Water Hardship Grant in the United States, *4617*

Middletown

"October Gale" to sink 40 Cape Cod fishing vessels off Nantucket Island, MA, *4241*

New Haven

Geological organization in the United States, *2799*

UNITED STATES—Connecticut—New Haven—*continued*
Published research suggesting a wide range of uses and products for distilled petroleum, *3682*

New Milford

Pumped storage hydroelectric plant of significance in the United States, *4558*

Stamford

National antilitter organization in the United States, *3270*

Delaware

Colony in America to require the killing of wolves, *4796*
Reports of high concentrations of toxic metals in bivalves from the U.S. Mussel Watch Program, *4493*
U.S. Jobs Through Recycling grant to Delaware, *4161*

Selbyville

Federally sponsored interstate air pollution abatement conference, *1399*

Wilmington

Effective refrigerant for mass consumer use, *1340*
Weather records kept continuously in the United States, *1803*

District of Columbia

Federal program to protect and restore an estuary in the United States, *3593*

Washington

Annual report on the environment and American foreign policy, *2308*
Aqueduct to supply Washington, DC, *4577*
Biosphere reserves in the United States, *1627*
City planned and built as a seat of national government, *3095*
Coordinated weather observations in the United States, *1896*
Discovery room in a natural history museum in the United States, *4947*
Environmental health conference of U.S. physicians and environmentalists, *3969*
Environmental public service symbol of the U.S. government, *2618*
EPA Journal, *1074*
Governors' conference on U.S. conservation issues, *1033*
Group of Ten meeting, *1078*
Heat wave in the United States known to have lasted for two months, *1823*
International public health agency, *3925*
Lawsuit on behalf of an endangered species to be decided by the U.S. Supreme Court, *2333*
Learning lab in a national zoo, *4949*
Learning lab in the United States in a national zoo specifically related to birds, *4950*
Learning lab related to reptiles in a national zoo in the United States, *4954*
Maps of the trade winds, *1895*
National fisheries center approved by Congress, *2462*
National forecasting map in the United States, *1903*
National Herbarium in the United States, *1787*
National organization formed to assist U.S. local, regional, and state grassroots environmental groups with their projects, *1102*
National trade organization for wind power in the United States, *4854*
Newspaper weather forecast in the U.S., *1904*
Ozone depletion warning about Freon, *1975*
Plantings of ornamental Japanese cherry trees along the Potomac River in Washington, DC, *2727*
Plow for pulverizing the soil, *1241*
Potomac River pollution abatement conference, *4400*
Private organization to support national parks in the United States, *3718*
Public aquarium in the United States, *4895*
Railroad traffic in the United States, *3766*
Scale to accurately measure earthquake strength, *2223*
Secondary sewage treatment program in the District of Columbia, *4536*
Severe storm forecasting service of significance in the United States, *4186*
Society of professional ecologists in the United States, *2274*
Storm sewer building project of significance in the District of Columbia, *4533*
Surviving zoo established by the U.S. Congress, *4905*
Use of solar energy at the White House, *4117*
Use of the term *biodiversity*, *1636*
Weather map to be telecast from a land sending station to a land receiving station, *1934*
Weather service of the U.S. government not run by the military, *1917*

Florida

Acquisition of fragile wetlands on a large scale by the United States, *3146*
Appearance of alligatorweed in the United States, *4500*
Australian melaleuca in the United States, *1517*
Australian pine in the United States, *1495*

Brazilian pepper in the United States, *1496*
Comprehensive plan to redesign the water-flow system in and around the Everglades, *4678*
Eradication of the flesh-eating screwworm in the southeastern U.S., *1293*
Food irradiation plant in the United States, *1209*
Genetic recovery program for wildlife in the United States, *2347*
Hurricane experiment to reduce cloud intensity, *4256*
Hurricane in the Atlantic with a masculine name, *4265*
Hurricane in the Gulf of Mexico with a masculine name, *4266*
Hurricane in the United States rated Category 5 (most intense), *4251*
Hurricane to cause more than $25 billion in damage, *4269*
Hurricane to sink a fleet off North America, *4234*
Hurricane warning service to operate around the clock, *4257*
Infestation in the United States of the Mediterranean fruit fly, *1191*
Introduction of alligatorweed-eating adult flea beetles in Florida, *4503*
Land purchased for what would become Everglades National Park in Florida, *1095*
Large-scale dredging effort to reclaim land in Florida's Everglades for farming, *1244*
Massive cleanup plan for the Florida Everglades, *4487*
Mediterranean fruit fly infestation in North America, *1522*
National Hurricane Center in the United States, *1954*
Proposal to dig a canal across the Florida peninsula, *3071*
Regional endangered-species plan to protect an aquatic ecosystem, *2352*
Reports of high concentrations of toxic metals in bivalves from the U.S. Mussel Watch Program, *4493*
Reports of possible extinction of the dusky seaside sparrow in Florida, *2377*
Scarlet ibis introduced into the United States, *1668*
Screwworm infestations in Florida and Georgia, *1171*
Sri Lankan hydrilla in the United States, *1525*
Subtropical National Scenic Trail in the United States, *4003*
Systematic legislation to clean up and restore Florida's polluted lakes, rivers, streams and bays, *4483*
Systematic research documenting the slow death of the sabal palm along Florida's Gulf of Mexico coastline, *2748*
Tropical storm to hit the United States in February, *4187*
Undersea park established by the U.S. government, *3803*
Whooping crane artificial colony, *1726*

Big Pine Key

U.S. National Wildlife Reserve created specifically for Key deer, *2327*

Bradenton

European mission of significance to the interior of North America, *3710*

Cape Canaveral

All-weather satellite, *1960*
Landsat satellite, *2293*
Meteorological satellite funded by the U.S. Weather Bureau, *1961*
Orbiting geophysical observatory, *2822*
Ozone-layer vertical profile from space, *1981*
Ozone measurements of chlorine monoxide from space, *1992*
Satellite instruments powered by solar energy, *4112*
Satellite to offer useful weather data, *1956*
Satellite to provide cloud-cover photographs, *1825*
Satellite to transmit color photographs of the full Earth face, *2823*
Satellite to transmit weather information to the earth, *1957*
Satellite to use nuclear power for electricity, *3496*
Solar-powered vehicle to travel into deep space, *4123*
Species of bird to fall victim to space exploration, *2381*

Cape Sable

Bird of a new type discovered in the 20th century in the United States, *1658*
Wildlife warden killed in the line of duty in the United States, *4766*

Crystal River

U.S. National Wildlife Refuge created specifically for West Indian manatees, *2338*

Florida Straits

Hurricane to sink an ocean liner off the United States, *4249*

Homestead

Known trace of snow in the Miami, FL, area, *4210*

UNITED STATES—Florida—*continued*

Key Largo

Coral reef off the coast of the United States protected by federal law, *3589*
Underwater state park in the United States, *3728*

Key West

National Park Service unit including coral reef resources, *3799*

Lake Buena Vista

Man-made salt-water environment to contain over 5.5 million gallons of water, *4871*

Lake Okeechobee

Catastrophic flood caused by a hurricane in the United States, *2538*

Marineland

Underwater action picture studio and marine attraction, *4932*

Merritt Island

U.S. National Wildlife Refuge created specifically for the dusky seaside sparrow, *2328*

Miami

Known trace of snow in the Miami, FL, area, *4210*
Land boom of significance in Florida, *3141*
Ports in the United States to be monitored by U.S. Fish and Wildlife Service agents, *4730*
Women's international conference on the environment, *1046*

Pelican Island

Bird reservation established by the U.S. government, *1719*
National Wildlife Refuge in the United States, *4809*

Pensacola

Hurricane to wreck a colonists' ship in North America, *4228*

Saint Augustine

Aquarium in the United States for large marine animals, *4934*
Permanent Spanish settlement in what is now the United States, *3125*

Seaside

"New Urbanism" community in the United States, *4303*

Tallahassee

Known military expedition in North America sunk by a hurricane, *4226*

Georgia

Detailed life-history investigation of a major wildlife species, *1736*
Eradication of the flesh-eating screwworm in the southeastern U.S., *1293*
Federally funded wildlife management area in Georgia, *4715*
Gold discovery in Georgia, *3307*
National Trails system unit opened, *3723*
National Wild and Scenic River in the southeastern United States, *3988*
Programs in Georgia authorized by the Federal Aid in Wildlife Restoration Act (Pittman-Robertson Act), *1695*
Screwworm infestations in Florida and Georgia, *1171*
State nature preserve in Georgia, *4834*
State to require the registration of foresters, *2684*
Summer blizzard recorded in the United States, *4194*
Swampland designated as a U.S. National Wilderness Area, *4652*
Wild turkey restoration program in Georgia, *1671*

New Hope

Hailstorm blamed for the crash of a jetliner, *4220*

Saint Simon's Island

Hurricane to kill many slaves in the United States, *4240*

Savannah

Agricultural experimental farm in an American colony, *1126*
Snowstorm known to have dropped 18 inches of snow on Savannah, GA, *4193*
Widespread use of the cotton gin, *1236*
World Bank meeting, *4298*

Guam

Program in Guam authorized by the Federal Aid in Wildlife Restoration Act (Pittman-Robertson Act), *4721*

Hawaii

Biosphere reserve in Hawaii, *1634*
Chestnut-winged chachalaca introduced in the United States, *1660*
Earthquake to register a magnitude of 9 or more on the Richter scale, *2225*
Genetically engineered papayas, *2772*
Great currasows introduced into the United States, *1662*
Hawaiian monk seal listing under the U.S. Endangered Species Act, *2405*
Large-scale astronomical observatory on a volcano, *4382*
Measurement of carbon dioxide levels in the atmosphere, *1459*
Mediterranean fruit fly infestation in North America, *1522*
Non-Hawaiian to climb Mauna Loa in the area that later became Volcanoes National Park, *3759*
Non-Hawaiian to observe Kilauea Volcano erupting in the area that later became Volcanoes National Park, *4367*
Steam well in a Hawaiian volcano, *4385*
Substantial growth of ecotourism in Hawaii, *2251*
U.S. national park in a volcanic range, *4377*
U.S. National Wilderness Area in Hawaii, *4660*
Utilization of the Federal Aid in Wildlife Restoration Act (Pittman-Robertson Act) in Hawaii, *1664*

Hilo

Nene (Hawaiian goose) breeding program, *2426*
Seismic Tidal Wave Warning System, *2224*
Tsunami of significance to hit a U.S. territory, *2543*

Honolulu

Field station for the U.S. National Wildlife Health Center, *4740*
Natural history museum in the United States named for a descendant of royalty, *4904*
Ports in the United States to be monitored by U.S. Fish and Wildlife Service agents, *4730*
Seismic Tidal Wave Warning System, *2224*

Keahole Point

Eco-industrial park fueled by Ocean Thermal Energy Conversion (OTEC), *4306*
Ocean Thermal Energy Conversion (OTEC) laboratory established by a U.S. state, *2297*
Successful offshore closed-cycle Ocean Thermal Energy Conversion (OTEC) operation, *4564*

Laysan Island

Species of bird to be resuscitated from a single live specimen, *2323*

Maui

Common quail introduced to the Hawaiian Islands, *1520*

Mauna Loa

Carbon dioxide monitoring station, *1458*

Mauna Loa Observatory

Daily measurements of atmospheric carbon dioxide, *1839*

Oahu

Female oceanographer of distinction, *3597*

Puna

Volcano in Hawaii tapped to supply a geothermal power plant, *4390*

Waikiki

Aquarium in Hawaii, *4912*

Idaho

Dam safety guidelines issued by the U.S. government, *2073*
Hydroelectric power plant built by the U.S. government, *4554*
Law in the United States extending wilderness protection to land administered by the Bureau of Land Management (BLM), *4667*
Mandatory hunter education law in Idaho, *2960*
National park in the United States, *3772*
Nuclear energy program to disassemble a damaged reactor, *3515*
Sockeye salmon listing under the U.S. Endangered Species Act, *2417*
State in the United States to protect bison, *2947*
U.S. National Wilderness Areas in the interior West, *4643*
Waterways in the U.S. National Wild and Scenic Rivers System, *3983*
World Heritage site in the United States selected for environmental reasons, *3973*

Arco

Electric power from nuclear energy used to illuminate a town, *3471*

Boise

Geothermal district heating system, *2812*

UNITED STATES—Idaho—*continued*

Caldwell

Idaho Wildlife DNA Forensics Laboratory, *4743*

Grays Lake

Whooping crane artificial colony, *1726*

Idaho Falls

Deaths caused by an American nuclear reactor, *3495*
Electric power from nuclear energy, *3461*

Illinois

Conference on industrial wastes, *2838*
Devastating storm of freezing rain to strike Illinois in the 20th century, *4271*
Eurasian tree sparrow in the United States, *1653*
Female pioneer in the field of rubber recycling, *2284*
Floods known to cover 10 U.S. Midwestern states, *2565*
Hunting legally allowed on a U.S. government wildlife refuge, *2919*
Illinois State Scenic River, *4008*
Large U.S. states to file civil suits charging airlines with violating air pollution control, *1402*
Nuclear power plant in the United States built without government funding, *3489*
Set of tornadoes to cause $500 million in damages, *4291*
Soil survey of a state in the United States, *4071*
State government concerns about public landfills, *2833*
Statewide survey of industrial poisons, *2272*
Test to identify critical elements missing in the soil, *4034*
Tornado in the United States with a death toll in excess of 650, *4285*

Argonne

Electric power from nuclear energy, *3461*

Aurora

Exploits of "The Fox," the anonymous Chicago-area environmentalist, *1058*

Chicago

Air pollution ordinance enacted in Chicago, *1396*
Aquarium with an inland salt-water environment, *4907*
Asian long-horned beetle seen in North America, *2608*
Automobile race in the United States, *1532*
Biodiesel plant in Illinois, *4340*
Brightfields initiative by the U.S. Department of Energy, *3005*
Building code in the United States to require new factories to reduce smoke production, *1388*
City in the United States to ban the sale of spray paint to reduce graffiti, *3284*
Completely sun-oriented residential community in the United States, *4105*
Drought to create the conditions for a devastating fire in a major U.S. city, *2136*
Dust storm known to drop massive quantities of Great Plains topsoil on Chicago, *2147*
Exploits of "The Fox," the anonymous Chicago-area environmentalist, *1058*
Geologist to discovery that several ice ages had existed, *2012*
Killer tornado to cause $3 million in damages in Chicago, *4284*
Lawsuit against the state of Illinois and the city of Chicago claiming the diversion of Lake Michigan water to flush sewage down the Chicago Sanitary and Ship Canal was lowering water levels in the Great Lakes system, *4531*
Local ordinances curbing air pollution in the United States, *1386*
Major exposé of unsanitary conditions in the U.S. meatpacking industry, *1168*
National forestry association, *2691*
National urban beautification movement in the United States, *3098*
Ports in the United States to be monitored by U.S. Fish and Wildlife Service agents, *4730*
Radiocarbon or C-14 dating, *2820*
Research conducted in the United States by a woman in the field of occupational health, *2270*
Reversal of course of the Chicago River to keep raw human waste from washing up on beaches in front of Chicago Gold Coast apartment buildings, *4530*
Sewage-disposal system in the United States separate from the city water system, *4526*
Sewer district established by a city in the United States, *4527*
Snow cruiser for the Antarctic research, *3860*
Solar-powered car, *4110*
U.S. Signal Corps storm warning, *4184*
Underground city sewer system in the United States, *4523*
Use of biodiesel fuel at a U.S. national political convention, *4342*
Water tunnel to supply Chicago, *4579*

Hamilton

Major dam on the Mississippi River below St. Paul, MN, *2093*

Lockport

Jointure of the Gulf of Saint Lawrence and the Gulf of Mexico, *3712*

Ottawa

Game preserve in the United States, *2931*

Riverside

Plan of significance for a suburban community in the United States, *3097*

Rock Island

Fishing legally allowed on a U.S. government wildlife refuge, *2457*

West Frankfort

Tornado to last 3 hours in the United States, *4286*

Indiana

Gas conversion law by a state, *2047*
Geologist hired by the U.S. government, *2804*
Ice jam to close the Ohio River for nearly two months, *1820*
Indiana program authorized by the Federal Aid in Wildlife Restoration Act (Pittman-Robertson Act), *4710*
Tornado in the United States with a death toll in excess of 650, *4285*
Tornado to last 3 hours in the United States, *4286*
Tornadoes in the United States reported to cause a great loss of life, *4283*

Gary

U.S. Justice Department lawsuit against U.S. Steel for failing to follow government recommendations to abate pollution from its Gary, IN, works, *4475*

Greenfield

Ban on swimming in Indiana's Brandywine Creek because of bacterial contamination, *4471*

Indianapolis

Genetically engineered product to become commercially available, *2759*

Iowa

Bag limit on game bird hunting in the United States, *2952*
"Blizzard" designation for a snowstorm, *1912*
Floods known to cover 10 U.S. Midwestern states, *2565*
Hunting legally allowed on a U.S. government wildlife refuge, *2919*
Iowa State Preserves System, *3376*
Prairies and grasslands biome surveyed by a U.S. National Park Service Inventory and Monitoring Program (I&M), *3410*
Restoration to prairie of U.S. farmland previously reserved as a site for a nuclear power station, *3076*
River protected by Iowa's Protected Water (PWA) program, *4669*
Set of tornadoes to cause $500 million in damages, *4291*
U.S. National Wildlife Refuge created specifically for Iowa pleistocene snails, *2341*

Ames

Natural Resources Inventory & Analysis Institute (NRIAI), *3415*
Position of director of game research, *2920*

Des Moines

Living historical farm in the United States, *1312*

Iowa City

Upper-atmosphere study with a rocket launched from a balloon, *1951*

Keokuk

Major dam on the Mississippi River below St. Paul, MN, *2093*

Sioux City

Public hearing on meatpackers' discharge of offal into the Missouri River sponsored by the U.S. Public Health Service, *4435*

Kansas

Drought to depopulate the Great Plains state of Nebraska by half, *2140*
Edition of *Empire of Dust* by Lawrence Svobida, *3254*
Floods known to cover 10 U.S. Midwestern states, *2565*
Issue of the periodical *Acres U.S.A.*, *4042*
Use of the term *Dust Bowl* to describe drought-seared and wind-eroded U.S. southern Great Plains, *4052*

UNITED STATES—Kansas—*continued*

Atchinson

Fuel alcohol plant in the United States, *4327*

Bonner Springs

National memorial dedicated to farmers, *1135*

Coffeyville

Single hailstone to measure 7.5 inches in diameter, *4219*

Topeka

Interstate section in the United States, *1566*

Wichita

Irrigation advocacy group in the United States, *1031*

Kentucky

Flash flood national program in the United States, *2558*
Floods known to cover 10 U.S. Midwestern states, *2565*
Ice jam to close the Ohio River for nearly two months, *1820*
Tornadoes in the United States reported to cause a great loss of life, *4283*

Heritage Creek

Town in the United States relocated due to airport noise, *3432*

Lexington

Dam safety organization among U.S. states, *2074*

Louisville

Great flood to maroon the entire city of Louisville, KY, *2540*

Minor Lane Heights

Town in the United States relocated due to airport noise, *3432*

Monticello

Oil well in the United States, *3674*

Louisiana

Bayou in the U.S. National Wild and Scenic Rivers System, *4007*
Catfish farms in the United States, *2476*
Coastal wetlands designated as a U.S. National Wilderness Area, *4659*
Floods to cause damages of more than $1 billion in the United States, *2563*
Hurricane to cause more than $25 billion in damage, *4269*
Interstate forest-fire protection agreement in the southern United States, *2622*
Offshore oil well, *3652*
Offshore oil-well fires of significance in the Gulf of Mexico, *3634*
U.S. National Scenic Byway through Louisiana wetlands, *4018*

Cameron

Hurricane reported to occur during early summer in the United States, *4261*

Convent

Environmental justice complaint filed with the U.S. Environmental Protection Agency, *3003*

Grand Isle

Sulfur mine offshore, *3331*

Harahan

U.S. program to use Christmas trees for wetlands restoration, *4308*

Jennings

Biomass-to-ethanol conversion plant, *4346*

Lafayette

National Wetlands Resource Center (NWRC) in the United States, *3383*

New Orleans

Introduction of the water hyacinth in the United States, *4501*
Levees along the Mississippi River, *2567*
Natural resource exploited in Honduras, *3349*
Nuclear power-plant requirement in the United States to assess overall environmental impact, *3508*
Ports in the United States to be monitored by U.S. Fish and Wildlife Service agents, *4730*
Recorded instance of ice floes blocking the Mississippi River at New Orleans, LA, *1810*
Research of significance pointing to a problem in the U.S. drinking water supply, *4514*
Steam-powered sawmill, *2649*
Study of a major U.S. water supply linking pollutants to high cancer rates, *4424*

World's fair to select rivers as its theme, *2314*
Zoo that started as a sugar plantation, *4901*

Maine

Case of winter cold causing an English colony in New England to fail, *1801*
Closed season on caribou hunting in the United States, *2948*
Closed season on moose hunting in the United States, *2941*
Devastating ice storm in New England in the 20th century, *4270*
Flood in the United States to claim more than 600 lives, *2541*
Indoor air quality conference in the United States, *1360*
Interstate forest-fire protection agreement in the United States, *2621*
National park in the United States made up entirely of lands donated by private citizens, *3791*
National Trails system unit opened, *3723*
National Wild and Scenic River in New England, *3986*
Nuclear-generated electricity exported from Canada, *3521*
Rejections of state-mandated shoreland-zoning ordinances in coastal Maine, *3117*

Augusta

Fur trading post in the American colonies, *2903*
Government demolition of a U.S. dam without the owner's permission, *2080*

Bangor

Great "lumber town" in the United States, *2650*

Bucksport

Fish hatchery run by the U.S. government, *2492*

Greenville

Forest fire lookout tower, *2613*

Mount Desert Island

National park east of the Mississippi River and the first located on an island, *3792*

Maryland

Federal program to protect and restore an estuary in the United States, *3593*
Hunting hounds in North America, *2907*
Hunting law establishing "rest days" for waterfowl in the United States, *2949*
Instigation in the United States of the use of easements to preserve parklands from urban expansion, *3820*
Large fish kill in Chesapeake Bay in the 1990s, *4498*
National Trails system unit opened, *3723*
Reports of high concentrations of toxic metals in bivalves from the U.S. Mussel Watch Program, *4493*
Statewide statutes authorizing preferential tax assessment of farmland in the United States, *3110*
Tax on "gas-guzzling" automobiles by a U.S. state, *1584*
U.S. Jobs Through Recycling grant to Maryland, *4162*

Baltimore

Boards of health in American cities, *3909*
Commercially marketed solar water heater, *4097*
Heat wave in the United States known to have lasted for two months, *1823*
Human embryonic stem cells to be artificially cultivated, *2774*
Seismologist to advance the elastic rebound theory to explain earthquakes in tectonic, *2218*
Use of natural gas for street lamps in a U.S. city, *2039*

Beltsville

Plant Materials Center (PMC) in the United States, *1795*

Bethesda

Human Genome Project, *2763*
Professional association of fisheries scientists in the United States, *2452*

Bishop

Federally sponsored interstate air pollution abatement conference, *1399*

Chesapeake Bay

Nuclear-powered lighthouse, *3500*

Cumberland

Railroad traffic in the United States, *3766*

Elkton

Jetliner to be destroyed in flight by lightning, *4277*

Greenbelt

Ozone measurements of chlorine monoxide from space, *1992*

UNITED STATES—Maryland—*continued*

Hyattsville

Satellite composite map of the U.S., *3381*

Laurel

North American organization to monitor the disappearance of amphibians, *4738*
Whooping crane born in captivity, *4754*

Rockville

Earthquake data center established by the U.S. government, *2228*

Massachusetts

Atlantic–Gulf Coast biome surveyed by a U.S. National Park Service Inventory & Monitoring Program (I&M), *3416*
Board of health established in the United States by a state, *3919*
Closed hunting season in North America, *2934*
Closed hunting term in North America, *2937*
Colonial American game warden system, *2909*
Complete state geological survey in the United States, *2802*
Court ruling in colonial New England barring settlers from direct purchase of American Indian lands, *3129*
Devastating ice storm in New England in the 20th century, *4270*
Edict setting aside for Royal Navy use all Massachusetts trees more than 24 inches in diameter, *2670*
Flood in the United States to claim more than 600 lives, *2541*
Hunting law to protect non-game birds in the United States, *2940*
Introduction of gypsy moth-attacking parasites into the United States from Europe and Asia, *2594*
Law in Massachusetts to protect an endangered plant species, *2322*
License plate in Massachusetts designed to benefit environmental education programs, *2307*
"Luminous" snowstorm in Vermont and Massachusetts, *4195*
National parkland acquired with congressionally authorized funds, *3804*
National Trails system unit opened, *3723*
Nuclear-generated electricity exported from Canada, *3521*
Post–American Revolution "broad arrow" policy marking trees for naval use, *2671*
Prohibition against using camouflaged hunting boats in North America, *2936*
Significant legislation in Massachusetts to address the shortage of waste treatment and storage capacity there, *2864*
Standards for water quality in the United States, *4421*
Tunnel in the Massachusetts water system, *4589*

Amherst

Paper describing dinosaur footprints, *2803*

Boston

Agricultural book published in the United States, *3231*
Almanac with a continuous existence in the United States, *3234*
American song written about a flood warning, *2530*
Aquatic environment built to revitalize urban waterfronts, *4942*
Bark beetle in the United States, *2593*
Boards of health in American cities, *3909*
Botanical garden in a city opened to the public in the United States, *3819*
Court ruling upholding Boston's right to regulate the boundary line for wharves, *3115*
Description of an earthquake or aftershock as a wave, *2190*
Distinctively American agricultural book, *3232*
Draining and filling of the marshy area of the Charles River near Boston, MA, *3145*
Drinking water conduit in the United States to be built underwater, *4576*
Genetic privacy legislation proposal in the United States, *2766*
Modern aquarium, *4941*
National Toxics Campaign in the United States, *1101*
Official "concentrated heavy snowstorm," defined as 20 inches or more within a 48-hour period, in Boston, MA, *4211*
Public arboretum affiliated with a university, *1786*
Research results in the United States on the use of sunlight through windows to generate heat, *4102*
Solar power company, *4098*
Storm that destroyed a lighthouse in the United States, *4181*
Street railway in Boston, *3072*
Urban park in the United States, *3816*
Use of the term *oekology* (ecology) in the United States, *1052*
Water supply system built for a U.S. city, *4570*
Weather vane maker in Colonial America, *1870*

Cambridge

Climatology professor at a U.S. college, *2271*
Continuous daily weather observations in the United States, *1873*
Entomology professor at a U.S. college, *2265*

Female professor at Harvard, *2275*
Genetically engineered mammal to be patented, *2762*
Meeting to create the New England Botanical Club, *1788*
Modern analysis of significance into the chemistry of rain, *1959*
Museum established by a university to support its own programs, *4944*
Prediction of the existence of the ionosphere, *1920*
Textbook on industrial poisons in the United States, *2276*
Use of dendroclimatology as a means of determining historic yearly rainfall and temperatures by tree rings, *2013*
Water purity tables, *4419*
Weather reports kept continuously by a U.S. college, *1807*

Concord

Environmental movement of significance in the United States, *1047*

Dana

Reservoir to cover four towns and affect seven others in the United States, *2101*

Dover

House completely heated by solar energy in the United States, *4107*

Enfield

Reservoir to cover four towns and affect seven others in the United States, *2101*

Greenwich

Reservoir to cover four towns and affect seven others in the United States, *2101*

Holden

State nature preserve in Massachusetts, *4844*

Ipswich

Botanical expedition in the United States, *1768*

Lawrence

Water filtration system for bacterial purification of a U.S. city water supply, *4582*

Martha's Vineyard

Heath hen extinction report, *2370*

Medford

Commercial fishery in an American colony, *2470*
Gypsy moth infestation in a U.S town, *2590*
Introduction of the gypsy moth in the United States, *1511*

Nantucket

"October Gale" to sink 40 Cape Cod fishing vessels off Nantucket Island, MA, *4241*
Recorded whaling expedition from an American colony, *4624*
Sperm whale captured at sea by an American ship, *4623*

New Bedford

American steam whaler, *4626*

Newburyport

Coast survey book written and published in the United States, *3235*

Penikese Island

Marine biology station in the United States, *3561*

Plymouth

Cattle imported into a North American colony, *1156*
Christian rainmaking ceremony in New England, *2167*
Drought affecting English colonists in Massachusetts, *2128*
Forestry law enacted by a British colony, *2625*
Furs exported from an American colony, *2902*
Hurricane recorded in an American colony, *4231*

Prescott

Reservoir to cover four towns and affect seven others in the United States, *2101*

Rowe

Nuclear power station to operate on a commercial basis, *3494*

Salem

National historic site in the United States National Park System, *3801*
So-called "Winter of the Deep Snow" in New England, *4192*

South Dartmouth

Self-sustaining greenhouse in a museum, *4966*

UNITED STATES—Massachusetts—*continued*

Sterling

Almanac with a continuous existence in the United States, *3234*

Truro

Legislation in an American colony regulating timber cutting to protect the soil, *2628*

Ware

Reservoir to cover four towns and affect seven others in the United States, *2101*

Warren

Successful effort to block a large hazardous waste disposal plant under the Massachusetts Hazardous Waste Facility Siting Act, *2870*

Watertown

Successful conservation organization to use paying volunteers to sponsor and assist scientists on research trips throughout the world, *1071*

Williamsburg

American song written about a flood warning, *2530*
Dam failure of major significance in the United States, *2111*

Williamstown

Classification of the layers of the Earth's crust in North America, *2797*

Woburn

Enforcement action by the U.S. Environmental Protection Agency (EPA) against a company for failure to comply with radionuclide emissions standards, *2997*
Mass-produced condoms, *3878*

Woods Hole

Fish research laboratory operated by the U.S. government on a permanent basis, *2454*
Marine aquarium built primarily for education and research, *4902*
Multi-purpose oceanographic research vessel in the United States, *3571*
Oceanographic research vessel to recover a lost hydrogen bomb, *3580*
Vessel constructed solely for fisheries research, *2453*

Worcester

Parkland purchased by a city in the United States, *3817*
Sewage disposal system in the United States by chemical precipitation, *4528*
Tornado in the eastern United States to claim 90 lives, *4290*

Michigan

Cars designed using computer-aided tools, *1610*
Destruction of Michigan cattle, chickens, and eggs contaminated with the chemical compound PBB (polybrominated biphenyl), *2850*
Documentation of industrial pollutants in a lake on the Lake Superior wilderness preserve Isle Royale, *4512*
Evidence that human pollution had reduced fish spawns in the Saginaw River and Saginaw Bay of Lake Huron, *4412*
Federal law in the United States to withhold development funds for construction on coastal barrier islands in the Great Lakes, *3394*
Logging crews to form the advance guard of what would become known as the Big Cut in the U.S. Great Lakes region, *2651*
National lakeshore in the U.S., *3729*
Probable appearance of the cereal leaf beetle in the United States, *1192*
Pure copper deposits discovery in Michigan, *3309*
Report that pollutants discharged into the Detroit River may have killed bald eagle hatchlings along Lake Erie, *4416*
Set of tornadoes to cause $500 million in damages, *4291*
Skyline lead for logging, *2657*
Soo Lock, to carry shipping over the rapids at Sault Ste. Marie from Lake Huron to Lake Superior, *3068*
State in the United States to ban state funding for human cloning research, *2773*
U.S. manufacturer to make cars in great quantities to reduce costs, *1591*
U.S. Supreme Court ruling on local bans on imported waste, *4150*
Use of the "Big Wheel" in logging, *2655*

Ann Arbor

Botany professorship in the United States, *2263*
School of conservation, *2277*

Auburn Hills

Automobile gasoline converted to hydrogen for future cars, *1613*

Detroit

Advanced pollution-control system introduced by a U.S. car manufacturer, *1599*

Agreement among the Big Three U.S. automakers to reduce the dumping of toxins into the Great Lakes, *4441*
Automaker to use an assembly line, *1590*
Automobile technology organization formed by large U.S. car companies, *1603*
Car fueled by methanol, *4336*
Chlorofluorocarbons (CFCs), *1970*
Compact car built by General Motors, *1547*
Electric automobile to convert gas to hydrogen fuel, *1614*
Electric car to be mass-produced using modern technology, *1609*
Mass-produced car to run on ethanol, *4326*
Mass-produced car with a fiberglass body, *1597*
Pollution-control data swapped by U.S. car companies, *1598*
U.S. city to decide to widen roads rather than build a subway system, *3106*
Use of tetraethyl lead for automobiles, *1594*
Zebra mussels introduced in the United States, *1528*
Zoo to use bar-less exhibits extensively, *4924*

Flint

Carmaking company to produce a range of models, *1537*
Sit-down strikes to cripple General Motors car production, *1544*

Grand Rapids

Game warden in the United States to be paid a salary, *4688*
Hydroelectric power plant to furnish arc lighting service in the United States, *4548*
Water supply in the United States to be fluoridated in order to reduce tooth decay, *4591*
Year in which salaried game wardens were appointed in the United States, *2914*

Hemlock

Evidence of hazardous waste poisoning in the Michigan town of Hemlock, *2851*

Highland Park

Vehicle recycling research center in the United States, *1607*

Keweenaw

Park in the Western Hemisphere to commemorate metal mining, *3811*

Lansing

Successful U.S. passenger car manufacturer, *1534*

Midland

Environmental activist to become a member of a women's hall of fame, *1045*
Styrofoam, *4128*

Mio

Monument dedicated to a songbird, *3201*

Muskegon

Large-scale natural waste disposal system in the United States, *4541*

Seney

U.S. National Wildlife Reserve created specifically for Kirtland's warbler, *2325*

Warren

Smog chamber for air pollution research built by an industrial organization, *1462*

Midway Islands

Executive Order in the United States to protect the wildlife of the Midway Islands, *4843*

Minnesota

Ban of aircraft access to Superior National Forest, *4639*
"Blizzard" designation for a snowstorm, *1912*
Clean Indoor Air Act adopted by a U.S. state, *1408*
Floods known to cover 10 U.S. Midwestern states, *2565*
Hunting legally allowed on a U.S. government wildlife refuge, *2919*
Landscape architect employed by the U.S. Forest Service (USFS), *4632*
Large-scale widely reported attack of the larch sawfly, *2598*
Law in Minnesota to treat as hazardous waste, *2892*
Law in the United States to establish federal and state standards for organic certification, *1264*
Mining of significance in the iron-rich Mesabi Iron Range in northeast Minnesota, *3318*
Minnesota wind power tax law, *4862*
National forest established east of the Mississippi, *3716*
Prairies and grasslands biome surveyed by a U.S. National Park Service Inventory and Monitoring Program (I&M), *3410*
U.S. federal law to protect a wilderness area, *4636*
U.S. Jobs Through Recycling grant to Minnesota, *4163*

UNITED STATES—Minnesota—*continued*

U.S. National Scenic Byway in an urban area, *4024*
U.S. National Wilderness Area in the Superior Upland region, *4642*
U.S. Supreme Court ruling on plastic milk containers, *4146*
Waterways in the U.S. National Wild and Scenic Rivers System, *3983*

Apple Valley

International organization to manage breeding animals in captivity, *4946*

Minneapolis

Establishment of what would become the U.S. Greens, *1113*

Saint Paul

Flood crest on the Mississippi River at St. Paul, MN, to exceed the previous record by 4 feet, *2551*
State forestry association, *2692*

Mississippi

Catfish farms in the United States, *2476*
Flood in the United States to cause $1 billion in damages, *2559*
Hurricane entered intentionally with a large aircraft, *4255*
Hurricane in the Gulf of Mexico with a masculine name, *4266*
Hurricane reported to occur during early summer in the United States, *4261*
Interstate forest-fire protection agreement in the southern United States, *2622*
State in the United States to control littering on levees, *3268*
State in the United States to enact a "conspiracy to dump" law, *3286*
Tornadoes in the United States reported to cause a great loss of life, *4283*
U.S. National Wilderness Area in the Gulf of Mexico, *4663*

Gauthier

U.S. National Wildlife Refuge created specifically for Mississippi sandhill cranes, *2331*

Natchez

Former hunting path designated as an All-American Road, *4015*
National Scenic Trail in the south central United States, *4002*
Tornado in the United States to leave a city in ruins, *4281*

Pascagoula

Turtle excluder device, *2464*

Stennis

National Wetlands Resource Center (NWRC) in the United States, *3383*

Missouri

Floods known to cover 10 U.S. Midwestern states, *2565*
National Scenic Riverway law in the United States, *3978*
Prairies and grasslands biome surveyed by a U.S. National Park Service Inventory and Monitoring Program (I&M), *3410*
River to be declared a national Scenic Riverway in the United States, *3805*
Rivers managed by the U.S. National Park Service, *4641*
Tornado in the United States with a death toll in excess of 650, *4285*
Tornado to last 3 hours in the United States, *4286*
Waterfowl management area developed by the Missouri Conservation Commission, *1725*
Waterways in the U.S. National Wild and Scenic Rivers System, *3983*

Kansas City

Large shopping center built on the outskirts of an urban area specifically to serve people with cars, *3105*
Scale to measure tornado intensities, *4293*
Severe storm forecasting service of significance in the United States, *4186*

New Madrid

Earthquake known to have changed the course of the Mississippi River, *2197*
National Disaster Relief Act in the United States, *2198*

Saint Joseph

Formal complaint about Missouri River pollution from stockyards and municipal sewers in St. Joseph, MO, *4451*

Saint Louis

Effective control measures for solid-fuel stationary sources of air pollution in the United States, *1390*
Electric power in the United States from municipal garbage as a boiler fuel, *4130*
Eurasian tree sparrow in the United States, *1653*

Garbage incinerator, *4140*
Local ordinance in the United States to regulate chimney height as a means of curbing air pollution, *1385*
National meteorological organization in the United States, *1928*
Solar power plant using a low-boiling-point liquid to power an engine, *4099*

Times Beach

Public health hazard causing the U.S. government to buy a town, *3946*
Town evacuated by the Environmental Protection Agency for dioxin levels 100 times emergency standards, *4427*

Montana

Brucellosis in the Yellowstone National Park buffalo herd, *4798*
Classification of rivers, *4508*
Dust storm known to carry dust from Montana all the way to the U.S. eastern seaboard, *2146*
Introduction of larger, meatier British cattle on U.S. ranches, *1164*
Large-scale practice of dry farming methods in the United States, *1243*
Montana program authorized by the Federal Aid in Wildlife Restoration Act (Pittman-Robertson Act), *4714*
National park in the United States, *3772*
Park to be designated an international peace park, *3798*
Person to write about what later became Waterton-Glacier International Peace Park, *3756*
Ranchers' organization to promote "predator-friendly" wool, *1175*
State law to revoke camping permits for littering convictions, *3272*
U.S. National Wilderness Areas in the interior West, *4643*
Wild horse adoptions in the United States, *4802*
Wild horse range established by the U.S. Bureau of Land Management, *4829*
World Heritage site in the United States selected for environmental reasons, *3973*

Deer Lodge

Park to commemorate the cattle industry of the American West, from its inception in the 1850s, *3807*

Flathead Valley

Land bought by the U.S. government specifically for wildlife, *4812*

Missoula

Environmental organization to substantially practice "ecotage", *1075*
Establishment of U.S. Forest Service Field Districts, *2680*
Glacial floods of major significance known in North America, *2515*

Moiese

Buffalo preserve in the United States, *4811*

Nebraska

Arbor Day celebration, *2724*
Drought to depopulate the Great Plains state of Nebraska by half, *2140*
Effective promoter of tree planting to prevent soil erosion on the U.S. Great Plains, *4049*
Floods known to cover 10 U.S. Midwestern states, *2565*
Litter and recycling strategy funded by fines and a statewide retail tax, *3281*
Nebraska program authorized by the Federal Aid in Wildlife Restoration Act (Pittman-Robertson Act), *4711*
Prairies and grasslands biome surveyed by a U.S. National Park Service Inventory and Monitoring Program (I&M), *3410*
U.S. Jobs Through Recycling grant to Nebraska, *4164*
Use of the term *Dust Bowl* to describe drought-seared and wind-eroded U.S. southern Great Plains, *4052*

Beatrice

Prairie restoration in the U.S. National Parks System, *3749*

Lincoln

Drought education and training center for developing nations, *2160*
Drought research and development program nationally established in the United States, *2165*

Omaha

Zoo in Nebraska, *4908*

Swan

Forest planted by the U.S. government, *2726*

Nevada

Closed season on hunting mountain goats and mountain sheep in the United States, *2946*

UNITED STATES—Nevada—*continued*
Comstock Lode exploration in Nevada, *3312*
Dam to be constructed with a height greater than 152 meters (500 feet), *2099*
Evidence of water reservoir-induced earthquakes, *2222*
Executive Order issued to protect the Lake Tahoe region, *3421*
Fertility control program for wild horses in the United States on public lands, *4806*
Nevada program authorized by the Federal Aid in Wildlife Restoration Act (Pittman-Robertson Act), *4717*
Open hunting season on the chukar in the United States, *1666*
Program to stop extinction of trumpeter swans, *2324*
U.S. National Scenic Byway on an American Indian reservation, *4017*

Amargosa Valley

U.S. National Wildlife Refuge created specifically for endangered Nevada fish species, *2339*

Boulder City

Hydroelectric power plant in the United States to produce a million kilowatts, *4560*

Clark County

Amphibian to become extinct in U.S. history, *2371*

Elko

Bird known to fly at an altitude of 21,000 feet, *1701*

Las Vegas

Underground nuclear weapons test explosion at the Nevada Test Site (NTS) northwest of Las Vegas, *3462*

Nellis Air Force Base

Wild horse range overseen by the U.S. government, *4801*

Reno

Publication that outlined the goals of the "wise-use" movement, *1024*

New Hampshire

Devastating ice storm in New England in the 20th century, *4270*
Exports of New England white pine to England, *2647*
Flood in the United States to claim more than 600 lives, *2541*
Game commission in the United States, *2953*
National Trails system unit opened, *3723*
New England pines for Royal Navy masts, *2648*
Serious evidence of acid rain in the United States, *1463*
State in the United States to ban lead sinkers and jigs, *2514*
State law to revoke fishing, hunting, and trapping licenses for littering convictions, *3273*
U.S. Jobs Through Recycling grant to New Hampshire, *4169*
U.S. National Wilderness Area in New England, *4640*
Weather observations atop Mount Washington in New Hampshire, *1913*

Concord

Forest fire battled using artificial rain, *2620*

Dublin

Farmer's Almanac, *1878*

Jackson

U.S. National Wild and Scenic River managed in partnership with local governments, *4009*

Newington

Community forest, *2709*

Ordford

Internal combustion engine in the United States, *4323*

New Jersey

American colony to restrict oyster fishing, *2501*
Amphibian removed from the U.S. Endangered and Threatened Species List, *2409*
Comprehensive legislation in New Jersey to deal with hazardous waste disposal, *2866*
Interstate planning and development agency in the United States, *3104*
Large U.S. states to file civil suits charging airlines with violating air pollution control, *1402*
National reserve in the United States, *3732*
National Trails system unit opened, *3723*
National Wild and Scenic River in the Mid-Atlantic region, *3992*
National Wilderness Area in New Jersey, *3730*
New Jersey program authorized by the Federal Aid in Wildlife Restoration Act (Pittman-Robertson Act), *4702*

Nonresident hunting license in the United States, *2950*
Public disclosure of a U.S. Army proposal to dump nerve gas and live explosives 200 miles off the New Jersey coast, *4491*
"Refreshment taverns" in the middle of the frozen Hudson River, *1816*
Release of imported gypsy moth–killing nematodes in North America, *2607*
Reports of high concentrations of toxic metals in bivalves from the U.S. Mussel Watch Program, *4493*
State highway department in the United States, *1554*
State in the United States in which all electric utilities joined the U.S. Environmental Protection Agency's "Green Lights" energy efficiency program, *1423*
State in the United States to pass a comprehensive noise-control law, *3440*
U.S. National Wilderness Area in the Mid-Atlantic states, *4655*
U.S. state with a sustainable development agency, *4321*
U.S. Supreme Court ruling on state bans on imported waste, *4145*
Use of the Atlantic Coastal Fisheries Cooperative Management Act of 1993 against a state, *2467*

Beverly

Partridge propagation in the United States, *2930*

Burlington County

Cast-iron plow, *1237*

Camden

Nuclear-powered merchant ship, *3488*

Fairfield

Natural resource damage settlement paid to New Jersey under the "Superfund" law, *2996*

Hackensack

Regional wetlands development commission in New Jersey, *3377*

Jersey City

Drinking water supply in the United States to be chemically treated with chlorine compounds on a large scale, *4583*

Lakehurst

Brush charge of atmospheric electricity to destroy an airship, *4275*

Murray Hill

Photovoltaic cell or solar battery to work effectively, *4108*

Newark

Ports in the United States to be monitored by U.S. Fish and Wildlife Service agents, *4730*
Synthetic fertilizer, *1127*

Princeton

Attempts at climate modeling, *1958*
Long-range numerical weather forecasts by computer, *1953*
Numerical weather forecast on an electronic computer, *1950*
Weather prediction system using modern numerical systems, *1949*

Riverton

Japanese beetles in the United States, *1189*

New Mexico

Description of the Virginia warbler, *1652*
Environmental public service symbol of the U.S. government, *2618*
European to see the Grand Canyon, *3753*
Federally designated wilderness area in the United States, *4634*
National wilderness area in the world, *3797*
State in the United States to pass a beautification and antilitter act, *3278*
Test of an atomic weapon, *3453*
Towhee named after an individual, *1731*
U.S. Forest Service study of recreational opportunities and values in the national forests, *2682*
U.S. National Wilderness Areas in the Southwest, *4646*
Waterways in the U.S. National Wild and Scenic Rivers System, *3983*
Whooping crane artificial colony, *1726*
Wilderness area in New Mexico, *3720*

Alamagordo

National monument to commemorate the world's largest gypsum dune field, *3722*

Albuquerque

Defense laboratory operated by the U.S. government to study wind power, *4853*
Establishment of U.S. Forest Service Field Districts, *2680*
Long-term ecological research network in the United States, *2302*

UNITED STATES—New Mexico—Albuquerque—*continued*

Native American organization dedicated to preserving buffalo, *4839*
Nuclear energy research museum, *3510*
Official confirmation of a powerful hydrogen bomb dropped accidentally from an aircraft, *3517*
Solar-heated office building, *4111*

Bosque del Apache Refuge

Hybridization of a whooping crane and a sandhill crane, *1686*

Carlsbad

Deep underground nuclear depository, *3541*
Nuclear waste site chosen by the U.S. Congress, *3533*

Carlsbad Caverns

Law establishing the deepest cave in the United States as a protected area, *4672*

New York

Announcement that New York State would no longer accept out-of-state hazardous waste, *2861*
Ban on the use of phosphate-containing laundry detergents in the United States, *4407*
Bounty for tree-damaging porcupines, *2595*
Burial of electric and other wires to be required by a state, *3921*
Closed season on bird hunting in North America, *2935*
Colonial American game warden system, *2909*
Dam to have a maximum height of 295 feet, *2092*
Dust storm known to carry dust from Montana all the way to the U.S. eastern seaboard, *2146*
Environmental law case in which citizens were granted standing for a noneconomic interest, *3737*
Evidence that a natural occurrence rather than toxins was causing deformities in frogs around the United States, *4417*
Farmland of significance in New York State to be permanently protected, *1136*
Female botanist of significance in the United States, *1761*
Fish hatchery to breed salmon in the United States, *2490*
Fish-stocking program failures in acidic lakes in upstate New York, *4413*
Flash flood national program in the United States, *2558*
Forest reserve set aside by a state, *2711*
Fruit spraying in the United States, *1276*
Fur-bearing animals raised commercially, *1161*
Genetically engineered papayas, *2772*
Geological survey of Niagara Falls, *2807*
Government declaration of large stretches of the Hudson River as dangerous for recreation, *4398*
Hunting license required by a state, *2913*
Interstate forest-fire protection agreement in the United States, *2621*
Interstate planning and development agency in the United States, *3104*
Known successful introduction of a fish species within the United States, *2491*
Lands in the United States to have constitutional protection, *3715*
Large U.S. industrial state to ban the burning of high-sulfur fuels, *1400*
Law establishing a protected wilderness in the Adirondack Mountains, *4631*
Lawsuit against the state of Illinois and the city of Chicago claiming the diversion of Lake Michigan water to flush sewage down the Chicago Sanitary and Ship Canal was lowering water levels in the Great Lakes system, *4531*
National Scenic Trail to link the eastern and western United States, *3998*
National Trails system unit opened, *3723*
National Wild and Scenic River in the Mid-Atlantic region, *3992*
Natural scenery protection state law connected with advertising, *3974*
New York highway-noise emissions limits, *3434*
PCB contamination to attract widespread national attention in the United States, *2856*
Program for public acquisition of farmland development rights, *1313*
Prohibition against the use of gun batteries in hunting in the United States, *2942*
"Refreshment taverns" in the middle of the frozen Hudson River, *1816*
Reports of mercury contamination in tuna and swordfish, *4513*
Sea lamprey positively identified in Lake Ontario, *1501*
Tests by the military of the defoliant Agent Orange, *1295*
U.S. Jobs Through Recycling grant to New York, *4165*
U.S. National Scenic Byway in the Great Lakes region, *4016*
Wildlife conservation group in the U.S., *1027*
Written mention of petroleum in what would become the United States, *3607*

Albany

Commercially available plastic, *4125*

Significant statement about the value of wilderness that appeared in the United States, *3242*

Buffalo

Electricity generated by Niagara Falls waterpower, *4547*
Herd book for livestock compiled and published in the United States, *3240*

Cold Spring

Brown trout in the United States, *2494*

Cold Spring Harbor

Biotechnology museum, *4962*
Education center devoted exclusively to genetics and biotechnology, *2305*

Cornwall-on-Hudson

Lawsuit by a conservation group to overrule a decision by a U.S. federal regulatory agency, *3979*

Cuba

Map showing a "fountain of bitumen" (petroleum) in what would become the U.S., *3608*

Fredonia

Natural gas used to light a U.S. house, *2041*
Natural gas well in the United States, *2040*

Greenpoint

Oil drill offshore rig, *3650*

Ithaca

Female awarded the Sage Fellowship in Entomology and Botany, *2266*
Forestry school at a U.S. college, *2678*
Ptarmigan born and bred in captivity, *4749*
Reintroduction of the term *acid rain*, *1469*

Levittown

Modern-day large-scale planned U.S. suburban development to use assembly-line methods of construction, *3075*

Long Island

Pekin duck brought to the United States, *1162*

Mount Vernon

Known U.S. waste entrepreneurs to be convicted for fraudulent export of hazardous waste overseas, *2874*

New York

Air Pollution Control Alert Warning System for New York City, *1350*
Air pollution index broadcast on television, *1349*
Appearance of the West Nile virus in the Western Hemisphere, *3972*
Aqueduct to supply New York City, *4575*
Asian long-horned beetle seen in North America, *2608*
Bird preservation organization in the United States, *1712*
Birth control clinic in the United States, *3879*
Blizzard to cause numerous deaths on the U.S. East Coast, *4201*
Blizzard to close all major U.S. airports on the East Coast, *4214*
Boards of health in American cities, *3909*
Book on birds to sell for $3.96 million, *3263*
Boone and Crockett Club meeting, *1030*
Building code to be adopted in the United States, *1355*
City in the U.S. to adopt a building code with internal noise suppression standards, *3437*
Commercially marketed frozen fish, *2473*
Comprehensive public health housing law in New York City, *3924*
Conservation legal organization, *2290*
Contemporary American science and technology museum, *4927*
Cranberry treatise, *3241*
District zoning in the United States, *3152*
Dual sewer system in the United States for sewage and storm water, *4524*
Duck-billed platypus successfully transported outside Australia, *4699*
Electric generating plant in the United States, *3360*
English sparrows imported into the United States, *1506*
European starlings introduced to the United States, *1516*
Fish protection law enacted by an American city, *2502*
Game protection society in the United States, *2967*
Gardener's manual written and published in the United States, *3236*
Geothermal building in Manhattan, *2826*
Heat wave in the United States known to have lasted for two months, *1823*
House (English) sparrows to thrive in the United States, *1651*
Hunting privileges in a North American colony, *2906*
Hurricane of significance with a feminine name to hit the United States, *4260*
International organization to regulate use of the ocean floor, *3065*

UNITED STATES—New York—New York—*continued*

Introduction of the famous "grid pattern" of New York City, *3096*
Laundromat with an environmental theme, *4411*
Law in New York City outlawing the sale of spray paint to minors, *3279*
Laws stipulating the registration of objects sent into space, *3035*
Limits on the height of skyscrapers in New York City, *3158*
Littering law in New York City, *3267*
Municipal garbage dumps in New York City, *4139*
Municipal public health laboratory in the United States, *3922*
Musk ox born in captivity, *4747*
National Christmas Bird Count in the United States, *1733*
National guidelines for local zoning ordinances in the United States, *3154*
National U.S. bird banding association, *1716*
Noise control commission in the United States, *3427*
Nuclear energy proposal at an international level, *3464*
Occasion of ice closing the Erie Canal for an entire month, *1817*
Parkway in the United States, *1560*
Person to suggest starting the New York Botanical Garden, *1790*
Ports in the United States to be monitored by U.S. Fish and Wildlife Service agents, *4730*
Public aquarium in the United States funded as a municipal facility, *4909*
Public health housing law in a large U.S. city, *3918*
Public photography at the Bronx Zoo, *4935*
Public water supply in New York City, *4571*
Radio broadcast from a zoo, *4928*
Recorded blanket of killer smog to cover New York City, *1345*
Reservoir to supply New York City, *4618*
Rubbish-sorting facility in New York City, *4126*
Ruling by the U.S. Supreme Court that historic preservation was a form of community welfare, *3116*
Society for animal welfare in the U.S., *1007*
Televised weather forecast in the United States, *1945*
U.S. Supreme Court ruling affirming the legality of zoning, *3155*
United Nations Fund for Population Activities (UNFPA), *3881*
Use of the term *sustainable development*, 4305
Warden hired by a bird conservation society in the United States, *1714*
Warning of air pollution dangers from coal-fired electrical generation, *1337*
Water pollution legislation passed by Congress, *4448*
Water tunnel to supply New York City, *4587*
Weather map to be telecast to a transatlantic steamer, *1937*
Wildlife dioramas exhibited in a museum that were accurate, *2281*
Winter in which all the waters surrounding New York City froze solid, *1808*
Zoning law in the United States to encourage a mix of shops, offices, and residences, *3156*
Zoning ordinance in the U.S., *3153*
Zoo with a veterinary hospital, *4918*

Niagara Falls

Electricity generated by Niagara Falls waterpower, *4547*
Evacuation of an entire chemically contaminated neighborhood in the United States, *4426*
Love Canal Homeowners Association meeting, *1099*

Oneida

Metal trap mass-produced in the U.S., *2912*

Poughkeepsie

Water filtration system for bacterial purification of a U.S. city water supply, *4582*

Rochester

Lawsuit over industrial sewers initiated by the U.S. Resource Conservation and Recovery Act (RCRA), *2998*
Tract housing development in the United States, *3074*

Saratoga Springs

Commercially bottled water in the United States, *4580*

Schenectady

Artificial snowstorm, *1947*
Meteorologist to reproduce natural snow crystals in plastic, *4205*

Southampton

Whaling industry established by an American town, *4622*

Theresa

Reinforced concrete buttress dam, *2091*

Troy

Degree program in natural science in the United States, *2261*

Doctorate program in ecological economics in the United States, *2309*
Substantial radiation "hot spot" to be observed far from the site of an atmospheric nuclear test, *1346*
Summer research term at a U.S. college, *2262*

Tuxedo Park

Outdoor museum with nature trails in the United States, *4920*

West Milton

Electric power generated by nuclear energy to be sold commercially, *3472*

West Point

Reports that latrines at the U.S. Military Academy at West Point were seriously polluting the Hudson River, *4538*

North Carolina

All-American Road in the Appalachian Mountains, *4012*
Appearance of alligatorweed in the United States, *4500*
Blizzard to close all major U.S. airports on the East Coast, *4214*
Broadbase terrace for farming, *4031*
Corporation to join the newly created Research Triangle Institute in North Carolina, *3725*
Eastern deciduous forest biomes surveyed by a U.S. National Park Service Inventory and Monitoring Program (I&M), *3400*
Flash flood national program in the United States, *2558*
Gold discovery of significance in the United States, *3306*
"Master school" for forestry training in the United States, *2695*
National seashore in the United States, *3800*
National Trails system unit opened, *3723*
National Wild and Scenic River in the southeastern United States, *3988*
Significant hog waste leakage into the New River of North Carolina, *4443*
Soil conservation district in the United States, *4040*
State nature preserve in North Carolina, *4826*
Tornadoes in the United States reported to cause a great loss of life, *4283*
U.S. Jobs Through Recycling grant to North Carolina, *4166*

Asheville

Forest management on a professional scale, *2675*

Cape Hatteras

Marine sanctuary in the United States, *3590*

Chapel Hill

Student Environmental Action Coalition in the United States, *1085*

Durham

Aye-aye born in captivity, *4759*
Mass production of cigarettes, *1357*

Greensboro

Genetically engineered corn, *2768*

Kitty Hawk

Monument erected to commemorate the first flight by an airplane, *3745*

Pamlico Sound

Wolf to be reintroduced into the wild, *2434*

Research Triangle Park

Research park inside an "education triangle", *3726*
State-sponsored biotechnology initiative in the United States, *2758*
Supercomputer used solely for environmental research, *2306*

Roanoke Island

Drought of significance in colonial America, *2126*

North Dakota

Blizzards to completely halt railroad traffic in the Great Plains for many weeks, *4199*
Floods known to cover 10 U.S. Midwestern states, *2565*
Large-scale practice of dry farming methods in the United States, *1243*
Ledger-art drawings of the Plains Indians, *3162*
National Scenic Trail to link the eastern and western United States, *3998*
U.S. National Wilderness Areas in the Great Plains, *4657*
Wildlife research center in the United States, *4819*

Beulah

Synthetic fuel plant in the United States built to operate on a commercial scale, *2060*

UNITED STATES—*continued*

Northern Mariana Islands

Wildlife biologists on the Northern Mariana Islands, *4732*

Ohio

Asian long-horned beetle seen in North America, *2608*
Ice jam to close the Ohio River for nearly two months, *1820*
Large-scale flood control project by a single U.S. state, *2535*
Ohio highway beautification program to use wildflowers, *4005*
Ohio State Scenic River, *3985*
"Save Lake Erie" petition, *4465*
Set of tornadoes to cause $500 million in damages, *4291*
State commission in the United States funded by taxes on litter-producing corporations, *3288*
State in the United States to enact a "strict liability" antilitter law, *3287*
State Scenic Byway in Ohio, *4019*
U.S. Jobs Through Recycling grant to Ohio, *4167*
U.S. National Wilderness Area in the Great Lakes region, *4654*

Akron

Polyvinyl chloride (PVC), *2971*

Cincinnati

Carolina parakeet extinction report, *2367*
Disposable diapers, *4129*
Dutch elm disease in the United States, *1523*
Extinction caused by humans of an animal endemic to North America in modern times, *2368*
Extinction of a native American bird in modern times, *1656*
Forestry book written and published in the United States, *3247*
Great flood to maroon the entire city of Louisville, KY, *2540*
Local ordinances curbing air pollution in the United States, *1386*
Patent on a practical reaping machine, *1239*
Reproductive research foundation sponsored by a zoo, a university, and a wild animal park, *4953*
Salaried employee at the Western Museum of Natural History, *4888*
Surviving zoo in Ohio, *4897*
Weather bulletin in the United States, *1910*
Weather charts in the United States, *1909*

Cleveland

Brownfields grants dispensed by the U.S. Environmental Protection Agency (EPA), *2993*
Dutch elm disease in the United States, *1523*
Effort in the United States to address severe pollution in Lake Erie, *4539*
State Scenic Byway in Ohio, *4019*
Windmill to generate electricity, *4850*

Columbus

Gorilla born in captivity in the United States, *4751*
Officially sanctioned international horticultural fair in the United States, *2316*

Dayton

Automatic starter for automobiles, *1592*
Ethyl gasoline with tetraethyl lead, *1540*
Gasoline additive tried, *1593*

Dover

State Scenic Byway in Ohio, *4019*

Goshem

Goose to lay an egg weighing 1.5 pounds, *1675*

Kent

Forestry school privately operated in the United States, *2681*

Mansfield

Model conservationist farm in the United States, *1131*

Perry

Year that nuclear power plants numbered 100 in the U.S., *3527*

Rocky River

Municipality to adapt the Federal Water Quality Administration's tertiary treatment system to raw sewage rather than effluent that had undergone secondary treatment, *4542*

Xenia

Authentication of 148 tornadoes over a two-day period in the United States, *4295*

Yellow Springs

Female college professor with the same rights and privileges as her male colleagues, *2264*

Oklahoma

Floods to cause damages of more than $1 billion in the United States, *2563*
Game laws in Oklahoma, *2915*
Interstate forest-fire protection agreement in the southern United States, *2622*
Jury award of damages in the U.S. to a victim of radiation contamination from a nuclear facility, *3513*
Oil and gas conservation law by a state, *3618*
"Tornado chasers" systematically organized, *4297*
Two states to adopt President Franklin Roosevelt's Standard State Soil Conservation District Laws, *4039*
U.S. National Wilderness Area to enclose a bison sanctuary, *4650*
Use of the term *Dust Bowl* to describe drought-seared and wind-eroded U.S. southern Great Plains, *4052*
Wildlife Diversity Program in Oklahoma, *4784*
Year Oklahoma required the purchase of a state duck stamp for hunting waterfowl, *1677*

Norman

Tornadoes tracked by Doppler radar, *4294*

Oklahoma City

Tornado forecast for a specific U.S. location, *4288*

Oregon

All-American Road with volcanic features, *4021*
Comprehensive reform of land-use planning laws in the United States, *3107*
Ecosystem management plan upheld in a U.S. court, *3414*
Emergence of Oregon as the leading U.S. timber-producing state, *2661*
European to "discover" the volcanic Crater Lake in Oregon, *4370*
Evidence that a natural occurrence rather than toxins was causing deformities in frogs around the United States, *4417*
Federal law in the United States to develop salmon fisheries in the Columbia River Basin, *2495*
Human beings to witness volcanic eruptions of Mount Mazama in what is now Crater Lake National Park, *3751*
Hydroelectric dams on the Columbia River in the U.S. Pacific Northwest, *2097*
Non–American Indian to see Crater Lake, *3767*
Nonreturnable bottle and can law enacted by a U.S. state, *4157*
Program to stop extinction of trumpeter swans, *2324*
Scenic highway in the United States, *3975*
Scientific expedition to what would later be designated the John Day Fossil Beds National Monument, *3744*
Significant air pollution control program to be adopted in the United States, *1394*
State governor in the United States to implement a doctrine of conservation in balance with economic concerns, *1104*
State in the United States to ban the sale of aerosol products containing chlorofluorocarbons (CFCs), *1409*
Tsunami to devastate the U.S. West Coast, *2549*
U.S. Jobs Through Recycling grant to Oregon, *4168*
U.S. National Wilderness Areas in the Pacific Northwest, *4644*
U.S. National Wildlife Refuge created specifically for Columbian white-tailed deer, *2330*
Volcano eruption to create a lake in North America, *4352*

Ashland

Wildlife crime lab in the United States, *4787*

Heppner

Dam constructed totally with roller-compacted concrete (RCC), *2109*
Flash flood to cause $100 million in damages, *2534*

Oregon City

Dam built in the United States to run a hydroelectric plant, *2087*
Hydroelectric power plant in the United States transmitting alternating current over a long distance, *4552*

Portland

Establishment of U.S. Forest Service Field Districts, *2680*
Forest Summit in the United States, *2690*
Hydroelectric power plant in the United States transmitting alternating current over a long distance, *4552*
Museum devoted to forestry education that started in a world's fair, *4943*
Zoo with a museum devoted to elephants, *4957*

Willamette Valley

Successful introduction of the ring-neck pheasant in the United States, *1514*

Worden

U.S. National Wildlife Refuge created specifically for wintering bald eagles, *2332*

UNITED STATES—*continued*

Pennsylvania

Bird banding in the United States, *1728*
Coal-mine safety laws in a U.S. state, *3314*
Federal program to protect and restore an estuary in the United States, *3593*
Flash flood national program in the United States, *2558*
Game law in Pennsylvania, *2908*
Hydroelectric dam of significance in the United States financed by private capital, *2096*
Introduction of the concept of "regenerative farming", *1254*
Introduction of the potato into North American colonies, *1205*
Mining law in Pennsylvania, *3315*
National Scenic Trail in the eastern United States, *4004*
National Trails system unit opened, *3723*
National Wild and Scenic River in the Mid-Atlantic region, *3992*
Nuclear energy program to disassemble a damaged reactor, *3515*
Oil company in the United States, *3609*
Oil-tank railroad car purposely built in the United States, *3611*
Pennsylvania law creating a scenic rivers system, *4001*
Pennsylvania state law prohibiting aliens from owning firearms, *2918*
State in the United States to ban use of automatic weapons in hunting, *2917*
State to own one million acres of game lands, *2923*
Strict standards for maintaining stream temperatures in a U.S. state, *4436*
Woman to graduate from the Pennsylvania School of Conservation, *2303*

Allentown

Blizzard to cause more than $500 million damage in the United States, *4208*
Hot-blast iron furnace fueled by anthracite coal in the United States, *2042*

Avondale

Anthracite coal mine disaster of major significance in the United States, *2046*

Bethlehem

Seagoing oil drilling rig for drilling in water more than 100 feet deep, *3656*
Water pumping plant in the United States, *4572*

Brockway

School devoted exclusively to training wildlife conservation officers, *2278*

Bryn Mawr

Biology course offered at a U.S. college, *2267*
Female scientist employed by the U.S. Geological Survey (USGS), *2813*

Centralia

Coal-mine fire to depopulate a U.S. city, *3332*

Donora

Smog-related deaths recorded in the United States, *1341*

East Liberty

Gas station in the United States, *3619*

Emmaus

Serial publication in the United States devoted to organic agriculture, *1261*

Germantown

Recycled paper in the United States, *4156*

Hanover

Entomologist in the United States, *4685*

Harrisburg

Bird sanctuary to protect migrating hawks and eagles, *1720*
Major accident in a U.S. nuclear power plant, *3514*
Raptor sanctuary, *1721*
School devoted exclusively to training wildlife conservation officers, *2278*

Johnstown

Dam disaster in the United States with a high death toll, *2113*
Warnings of structural problems at the South Fork Dam near Johnstown, PA, *2112*

Middlesex

Limited-access highway in the United States built like an interstate, *1563*

Oil City

Flood covered with burning oil in the United States, *2533*

Oil Creek Valley

Commercial oil refinery, *3687*

Philadelphia

Animal exhibits in a U.S. museum to be in natural settings and groups, *4884*
Archery club in the United States, *2966*
Boards of health in American cities, *3909*
Botanical garden in the United States, *1762*
Botany professor at a college, *2257*
Breeding bird census, *1729*
City plan in America, *3093*
Continuously operating natural sciences institution in North America, *2260*
Drought aid from a global rock music festival, *2159*
Edition of Benjamin Franklin's *Poor Richard's Almanac*, *1871*
Federal game law in the United States, *2938*
Fishing club in the United States of any duration, *2506*
Fox hunting club in the United States, *2964*
Geological map of the eastern United States, *2796*
Geological maps of the United States, *2792*
Heat wave in the United States known to have lasted for two months, *1823*
Hospital in the United States for the sick, injured, and insane, *3910*
Hurricane known to be accurately tracked in colonial America, *4236*
Lightning conductor in the United States, *4274*
Museum exhibits with labels and descriptions, *4885*
Music festival powered by biodiesel fuel, *4345*
National animal rights organization in the United States, *1010*
National public health agency in the United States, *3914*
Natural history museum foundation in the United States, *4883*
Natural history museum in North America, *4886*
Scientist to fly a kite in a thunderstorm in hopes of attracting a lightning bolt, *4180*
Street-cleaning machine of importance in the United States, *4175*
Street-sweeping service in the United States, *4174*
Toxic waste–carrying ship to draw international media attention, *2876*
Use of natural gas for commercial lighting in the United States, *2038*
Waterworks system in the United States to use cast-iron pipes, *4573*
Yellow fever outbreak to force as much as half the population of Philadelphia, PA, to flee to the countryside, *3911*
Zoo in the United States, *4896*
Zoo to breed a Bornean orangutan, *4922*
Zoo to breed a captive chimpanzee, *4923*
Zoo to breed a cheetah, *4938*
Zoo to breed a Chilean flamingo, *4939*
Zoo to exhibit a great Indian rhinoceros, *4898*
Zoo to exhibit a pair of African buffalo, *4925*
Zoo to exhibit adult bull elephants, *4903*
Zoo to exhibit rare black apes from the Indonesian island of Celebes (Sulawesi), *4899*
Zoo to have a "baby pet zoo", *4933*
Zoo to have a model dairy barn, *4930*
Zoo to have a regular radio program, *4931*
Zoo to have a research institution, *4910*
Zoo to have an African forest elephant, *4921*
Zoo to have an eagle aviary, *4892*
Zoo to have an outdoor gibbon cage, *4926*

Pittsburgh

Local private grassroots organization to successfully pressure governmental agencies to take steps to curb air pollution, *1404*
Oil refinery in the world, *3681*
Zoo with twilight conditions, *4945*

Rouseville

Fire in a U.S. oil well, *3629*

Shippingport

Nuclear power plant exclusively used for peaceful purposes, *3481*
Nuclear power plant in the United States to be totally decontaminated and decommissioned, *3529*

Titusville

Oil production on a large scale in the United States, *3610*
Oil well that was commercially productive, *3686*

Wilkes Barre

Anthracite coal burned experimentally, *2036*

Puerto Rico

Tests by the military of the defoliant Agent Orange, *1295*

Rhode Island

Flood in the United States to claim more than 600 lives, *2541*
Law in Rhode Island to protect endangered plants, *2326*
Law in the United States to outlaw spring shooting, *2943*
Oyster farming under state auspices, *2489*
Rhode Island project initiated by the Federal Aid in Wildlife Restoration Act (Pittman-Robertson Act), *4712*

UNITED STATES—Rhode Island—*continued*
State in the United States requiring retailers to obtain an annual litter-control permit, *3280*
State in the United States to impose a "hard to dispose" materials tax, *3285*
State in the United States to suspend vehicle registrations of litterers, *3276*

Newport

Closed season on deer hunting in the United States, *2905*

Pawtucket

Water-powered cotton mill in an American colony, *4545*

Providence

College instruction course in mineralogy, *2258*
Zoo in Rhode Island to continuously operate, *4906*

South Carolina

National Wild and Scenic River in the southeastern United States, *3988*
Tornadoes in the United States reported to cause a great loss of life, *4283*

Charleston

Continuous daily weather observations in the United States, *1873*
Natural history museum foundation in the United States, *4883*
Protected historic district in the United States, *3082*
Severe earthquake to strike the U.S. east coast in modern times, *2206*
Snowstorm known to have dropped 18 inches of snow on Savannah, GA, *4193*
Tornado to cause a major death toll in a U.S. city, *4280*

South Dakota

American explorer to describe prairie dogs, *4683*
Blizzards to completely halt railroad traffic in the Great Plains for many weeks, *4199*
Floods known to cover 10 U.S. Midwestern states, *2565*
Large-scale practice of dry farming methods in the United States, *1243*
Ledger-art drawings of the Plains Indians, *3162*
Mountain range to become a work of art, *3175*
National park established to protect a cave, *3787*
Official exploration of the area that later became Mount Rushmore National Park, *3775*
Person to enter into the cave at what would become Wind Cave National Park, *3778*
Pheasant restoration program in South Dakota, *1674*
Program to stop extinction of trumpeter swans, *2324*
Recorded discovery of what would become Wind Cave National Park in South Dakota, *3779*
South Dakota program authorized by the Federal Aid in Wildlife Restoration Act (Pittman-Robertson Act), *4703*

Black Hills

U.S. National Scenic Byway in the Black Hills, *4011*

Deadwood

Report of wild temperature variations on simultaneous readings from nearby stations in South Dakota, *1821*

Lead

Report of wild temperature variations on simultaneous readings from nearby stations in South Dakota, *1821*

Rapid City

Native American organization dedicated to preserving buffalo, *4839*

Tennessee

Eastern deciduous forest biomes surveyed by a U.S. National Park Service Inventory and Monitoring Program (I&M), *3400*
Flash flood national program in the United States, *2558*
Lawsuit on behalf of an endangered species to be decided by the U.S. Supreme Court, *2333*
Multi-purpose dam development of significance in the United States, *2070*
National Trails system unit opened, *3723*
Purchase of emissions allowances in the United States as permitted under the Clean Air Act Amendments, *1422*
Tennessee program authorized by the Federal Aid in Wildlife Restoration Act (Pittman-Robertson Act), *4705*
Tornadoes in the United States reported to cause a great loss of life, *4283*

Boston

Former hunting path designated as an All-American Road, *4015*

Chattanooga

Aquarium devoted to freshwater life, *4961*

Cross Plains

Permanent Bureau of Land Management holding facility for wild horses and burros in the eastern United States, *4835*

Elgin

Dam built as part of the Tennessee Valley Authority (TVA), *2098*

Kingsport

Plant to produce methanol from coal, *4344*

Knoxville

National organization dedicated specifically to preserving wilderness (as opposed to wildlife) in the United States, *1055*
Successful birth of an elephant in captivity in the Western Hemisphere, *4756*
World's fair in the American Southeast to be sanctioned by the Bureau of International Expositions (BIE), *2312*
World's fair to select energy as its theme, *2313*

Memphis

Drought in the United States to strand boats on the Mississippi River, *2161*
Sewage-disposal system in the United States separate from the city water system, *4526*

Nashville

National Scenic Trail in the south central United States, *4002*

Oak Ridge

"Atomic city" in the United States, *3452*
Biomass feedstock research funded by the U.S. government, *4329*
Energy museum that began in a cafeteria, *4936*
Museum dedicated to explaining the peaceful uses of atomic energy to the public, *3458*

Texas

Appearance of alligatorweed in the United States, *4500*
Appearance of the cotton-destroying boll weevil in Virginia, *1190*
Asian tiger mosquitoes in the United States, *1527*
Boll weevils in the United States, *1515*
Floods to cause damages of more than $1 billion in the United States, *2563*
Fossils of *Protoavis*, *1681*
Hurricane reported to occur during early summer in the United States, *4261*
Hurricane to cause damages of $1 billion or more in the United States, *4267*
Ice storm to cause $100 million in damages in the United States, *4272*
Interstate forest-fire protection agreement in the southern United States, *2622*
Introduction of larger, meatier British cattle on U.S. ranches, *1164*
"Killer bees" in the United States, *1529*
Offshore oil well, *3652*
Refuge for whooping cranes, *1724*
State agency in Texas to regulate game, *2916*
State in the United States to set a time limit on accumulating litter near a public highway, *3282*
Tests by the military of the defoliant Agent Orange, *1295*
Texas law to regulate game and fisheries statewide, *4786*
Texas program authorized by the Federal Aid in Wildlife Restoration Act (Pittman-Robertson Act), *4706*
Texas state agency to regulate fishing, *2455*
U.S. National Wild and Scenic River in the Chihuahuan Desert, *3993*
Use of the term *Dust Bowl* to describe drought-seared and wind-eroded U.S. southern Great Plains, *4052*
Use of the term *wildcatter* in oil exploration, *3697*

Beaumont

Oil well in the United States with a massive petroleum deposit, *3693*

Bryant

Pilot to fly into a hurricane intentionally, *4252*

Corpus Christi

Hurricane to sink an ocean liner off the United States, *4249*

Del Rio

Reservoir on the U.S.-Mexico border, *4620*

Eagle Lake

U.S. National Wildlife Refuge created specifically for Attwater prairie chickens, *2329*

Fort Worth

Continuously operated zoo in Texas, *4915*
"Green" U.S. Post Office, *4322*

Freeport

Seawater conversion plant in the United States on a practical scale, *4595*

UNITED STATES—Texas—*continued*

Galveston

Cleanup of an oil spill of significance in the United States with an oil-eating bacteria, *3644*
Hurricane causing massive loss of life in the United States, *4246*
Hurricane report that was censored in the United States, *4253*

Houston

Buses to use liquid natural gas (LNG) and diesel, *4335*
Discussion of the greenhouse effect and other environmental issues at a "post–cold war summit", *1848*
Hurricane report that was censored in the United States, *4253*
Oil-drilling equipment to penetrate hard rock, *3617*
Oil refinery to achieve complete conversion of waste gasses into useful power with a carbon monoxide boiler, *3701*

Lubbock

Authenticated death by hail recorded by the U.S. Weather Bureau, *4216*

Pleasant Bayou

Hybrid geopressure-geothermal power plant, *3396*

San Antonio

Aquifer in the United States granted sole source protection from pollution, *4425*
Aquifer in the United States to be designated as a "sole source" under the Clean Water Act of 1974, *4600*

Tyler

Adopt-A-Highway program in the United States, *3277*
Dam built using the hydraulic fill process, *2088*

Waco

Transatlantic flight powered by ethanol, *4334*

Utah

Bird banding by a U.S. government agency, *1734*
Cave discovered in American Forks Canyon, *3781*
Detailed geologic and topographic information about the area that later became Canyonlands National Park, *3770*
Discovery by non–American Indians of the Horsecollar Ruins in what would become Natural Bridges National Monument in Utah, *3742*
Documented journey through the Glen Canyon–Lake Powell area, *3755*
European to discover what is now Canyonlands National Park, *3764*
Human occupation of the area that later became Zion National Park, *3752*
National protest of significance sponsored by Earth First! *1079*
Non–Native Americans to enter what is now Dinosaur National Monument, *3740*
Oil exploration and production in a U.S. National Forest, *3707*
Power plant using solar cells, *4120*
Project after the 1938 appropriation by the U.S. Congress to assist state wildlife restoration projects, *4708*
Project in Utah approved under the Federal Aid in Wildlife Restoration Act (Pittman-Robertson Act), *1722*

Alta

Snow avalanche to destroy a mining camp near Alta, UT, *4198*

Ogden

Establishment of U.S. Forest Service Field Districts, *2680*

Promontory

National park dedicated to the first transcontinental railroad, *3802*

Salt Lake City

Monument to a bird in the United States, *3163*

Scofield

Coal mining accident of significance in the United States, *3321*

Vermont

Dust storm known to carry dust from Montana all the way to the U.S. eastern seaboard, *2146*
Law in Vermont to protect endangered plants, *2321*
"Luminous" snowstorm in Vermont and Massachusetts, *4195*
Meteorologist to photograph ice crystals with a camera-equipped microscope, *4203*

National Trails system unit opened, *3723*
Noted "back to the land" conservationist in the United States, *1036*
State in the United States to require utility companies to offer a minimum of electricity from renewable energy sources, *4320*
State-level legislation in the United States mandating review of any development for impact on natural beauty, aesthetics, and historic sites, *3087*
Summer blizzard recorded in the United States, *4194*
Vermont program authorized by the Federal Aid in Wildlife Restoration Act (Pittman-Robertson Act), *4709*

Grandpa's Knob

Wind turbine, *4852*

Woodstock

National park to focus on conservation history and the evolving nature of land use, *3812*

Virgin Islands

Saint John

Resort powered by renewable energy, *4301*

Virginia

Appearance of the cotton-destroying boll weevil in Virginia, *1190*
Assembly of significance of environmental lawyers, law school faculty and environmental leaders in the United States, *1061*
Botanical account, in detail, of the plants of North America, *1760*
Comprehensive study on wild turkeys, *1739*
Eastern deciduous forest biomes surveyed by a U.S. National Park Service Inventory and Monitoring Program (I&M), *3400*
Federal program to protect and restore an estuary in the United States, *3593*
Flash flood national program in the United States, *2558*
Fox hunting pack in North America, *2910*
Honey bees carried into the Western Hemisphere, *1155*
Hunting law enacted by an American colony, *2904*
Land grants of 100 acres to each Virginia colony settler, *3128*
National Scenic Trail in the eastern United States, *4004*
National Trails system unit opened, *3723*
Recovery plan approved by the U.S. Fish and Wildlife Service for a plant, *2336*
Reported large-scale bird kills traced to the lawn care pesticide Diazinon, *1309*
Soil conservation advocate of prominence in the United States, *4029*
U.S. Jobs Through Recycling grant to Virginia, *4170*
Wolf bounty enacted in what is now the United States, *4795*
Written account of a journey through the area of the Blue Ridge Mountains that later became Shenandoah National Park, *3754*

Alexandria

Boone and Crockett Club meeting, *1030*

Arlington

Standardized national vegetation classification system in the United States, *3413*
U.S. National Science Foundation (NSF), *2283*
Weather map to be telecast from a land sending station to a land receiving station, *1934*

Cape Charles

Eco-industrial park in the United States, *4307*

Chesapeake

Planetarium constructed by a school system in reponse to a space launch, *4969*

Fredericksburg

Heavy rainstorm and local flooding to cause the complete failure of an offensive operation in the American Civil War, *2529*

Hot Springs

Conference to discuss the scope of the newly established international Food and Agriculture Organization, *1267*

Isle of Wight

Planter to recognize that some plants could put acid in the soil, *4030*

Jamestown

Commercial cultivation of tobacco, *1125*
Hurricane report in colonial Virginia, *4233*
Permanent English settlement in America, *3126*
Sawmill established in America, *2646*
Sheep imported into an American colony, *1154*

Manakin

Coal deposits recorded in the American colonies, *3301*

Monticello

Simultaneous weather observations in the United States, *1875*

UNITED STATES—Virginia—Monticello—*continued*
Terraced or horizontal plowing in the United States, *4048*

Reston

Government facility in the United States established to analyze chlorofluorocarbons (CFCs) in water and air, *4611*

Richmond

Bituminous coal commercially mined in the American colonies, *3304*
Brownfields grants dispensed by the U.S. Environmental Protection Agency (EPA), *2993*
Coal deposits recorded in the American colonies, *3301*
Commercially marketed beer cans in the United States, *4127*

Walnut Grove

Successful harvesting machine, *1238*

Williamsburg

Restoration of a colonial-era town to museum status, *3108*
Simultaneous weather observations in the United States, *1875*

Washington

All-American Road with glacial features, *4020*
Automobile in the area of Mt. Rainier National Park, *3785*
Citation for heroism ever bestowed by the U.S. Department of the Interior, *3721*
Coniferous forest biome chosen for a U.S. National Park Service Inventory and Monitoring Program (I&M), *3404*
Discovery by Europeans of Mount Baker in Washington, *3757*
Documented ascent of Mount Rainier, *3771*
Emergence of Oregon as the leading U.S. timber-producing state, *2661*
European settlement in the area that later became Olympic National Park, *3758*
Federal law in the United States to develop salmon fisheries in the Columbia River Basin, *2495*
Fishing vessel buyback program in the United States, *2478*
Forest canopy crane used in North America, *2749*
Glacial floods of major significance known in North America, *2515*
Hotel in Mt. Rainier National Park, *3780*
Hydroelectric dams on the Columbia River in the U.S. Pacific Northwest, *2097*
Large-scale fire prevention publicity campaign in the United States, *2617*
Non–American Indian to visit the area that later became Mount Rainier National Park, *3763*
Sierra Club chapter formed outside of California, *1057*
State in the United States to call for a moratorium on planned breeding of dogs and cats, *1023*
State in the United States to tax litter-producing industries, *3283*
Tree farm in the United States, *2730*
Tree farm movement in the United States, *2729*
Tsunami to devastate the U.S. West Coast, *2549*
U.S. National Wilderness Areas in the Pacific Northwest, *4644*
U.S. National Wildlife Refuge created specifically for Columbian white-tailed deer, *2330*
Volcano eruption in the United States photographed by weather satellites, *4388*
Widespread, widely reported blowdown in the United States, *2587*

Marysville

Genetic manipulation of hatchery fish stocks, *2500*

Mount Saint Helens

Volcano known to claim human life in the contiguous 48 states, *4389*

Neah Bay

American Indian tribal whale hunt in modern times during an International Whaling Commission (IWC) moratorium on whale hunting, *4628*

Olympia

Woman to climb Mount Rainier, *3783*

Richland

National surveys of wind as an energy source in the United States, *4857*
Nuclear power plant in the United States to be totally decontaminated and decommissioned, *3529*

Seattle

Aquarium with a salmon ladder, *4951*
Comprehensive database of natural resource information about U.S. national park lands, *3407*
Ecosystem management plan upheld in a U.S. court, *3414*

Industrial waste treatment program at an airport, *2973*
Ports in the United States to be monitored by U.S. Fish and Wildlife Service agents, *4730*
Science center originally designed for a world's fair in the United States, *2286*
State in the United States to implement a multiple-species conservation plan statewide, *4791*

Spokane

World's fair with an environmental theme, *2310*

West Virginia

Federal court ruling declaring clear-cutting in national forests a violation of U.S. law, *2639*
Flash flood national program in the United States, *2558*
Ice storm to cause $100 million in damages in the United States, *4272*
National Trails system unit opened, *3723*
Wildlife management project in West Virginia, *4707*

Belle

Significant pollution abatement at the giant Du Pont de Nemours chemical plant on the Kanawha River in Belle, WV, *4433*

Buffalo Creek

Disaster settlement to include provisions to address the long-term psychological needs of the victims, *2556*

Logan City

Disaster settlement to include provisions to address the long-term psychological needs of the victims, *2556*

Man

Disaster settlement to include provisions to address the long-term psychological needs of the victims, *2556*

Monongah

Coal-mine explosion with massive loss of life in the United States, *3323*

Morgantown

Personal rapid transit (PRT) system in the United States, *1601*

Wheeling

Recorded rainfall with a pH of 2, *1325*

Species Survival Plan (SSP), *2335*

Wisconsin

Elk reintroduced to Wisconsin, *2441*
Floods known to cover 10 U.S. Midwestern states, *2565*
Hunting legally allowed on a U.S. government wildlife refuge, *2919*
Law in Wisconsin authorizing the pay of a game warden, *4689*
National Scenic Trail in the United States devoted to glacial terrain, *3997*
Purchase of emissions allowances in the United States as permitted under the Clean Air Act Amendments, *1422*
Refuge acquired by the U.S. National Wildlife Refuge System specifically for the management of waterfowl, *4816*
Reports of decreases in commercial fishermen's catches in Lake Superior, *4401*
Set of tornadoes to cause $500 million in damages, *4291*
State law in the United States to protect wild and scenic rivers from development, *4454*
Use of the name Green Bay for an arm of Lake Michigan along the Wisconsin shore, *4499*
Waterways in the U.S. National Wild and Scenic Rivers System, *3983*
Year in which salaried game wardens were appointed in the United States, *2914*

Appleton

Hydroelectric power plant in the United States to furnish incandescent lighting, *4550*

Baraboo

Institution to breed all 15 species of cranes, *1741*

Coon Creek Valley

Watershed erosion control demonstration project, *4053*

Green Bay

Forest fire to cause a large number of deaths in the United States, *2612*

Madison

Artificial gene, *2754*
Forest products laboratory of the U.S. Forest Service, *2738*
Game management chair at a U.S. college, *2932*
Human embryonic stem cells to be artificially cultivated, *2774*

UNITED STATES—Wisconsin—Madison—*continued*
U.S. National Wildlife Health Center, *4728*

Peshtigo

Forest fire to cause a large number of deaths in the United States, *2612*

Wyoming

$100 million forest fire–fighting campaign in U.S. history, *2623*
Bioprospecting agreement in the United States, *2770*
Federal protection for bald eagles in the lower 48 states of the United States, *1694*
Formal ascent of Devils Tower, *3743*
Grazing district on public lands in the United States, *1170*
Introduction of larger, meatier British cattle on U.S. ranches, *1164*
Large-scale practice of dry farming methods in the United States, *1243*
Legal challenge to wolf reintroduction to Yellowstone National Park, *2440*
National forest in the United States, *3714*
National forest reserve in the United States, *2712*
National park in the United States, *3772*
Natural national monument in the United States, *3746*
Non–Native American to see the geysers of Yellowstone National Park, *3761*
Photographer to capture successfully the beauties of Yellowstone on film, *3265*
Ranger hired in Yellowstone National Park, *3777*
Recorded description of the area that would become the Fort Laramie National Historic Site at the mouth of the Laramie River, *3762*
Superintendent of Yellowstone National Park, *3773*
U.S. National Wilderness Areas in the interior West, *4643*
Volcano eruption known in North America, *4348*
Women to pass along the Oregon Trail, *3765*
World Heritage site established in the United States, *3808*
World Heritage site in the United States selected for environmental reasons, *3973*

Jackson

Ban on motorized watercraft on Wyoming's Snake River, *4675*
Elk refuge in the United States, *4694*

Medicine Bow

Reintroductions of black-footed ferrets in the United States, *2435*

Moran

U.S. preserve to feature indigenous wildlife in a natural setting, *4822*

Ranchester

Hospitality ranch in Wyoming, *2245*

Wheatland

Wildlife research unit in the United States, *4718*

URUGUAY

Biosphere reserve in South America, *1624*

VATICAN

East-west division of the world for exploration, *3008*
Introduction of the concept of "ecotheology", *1087*
Papal declaration regarding world exploration, *3007*
Papal document to deal with ecology, *3264*
Patron saint of ecology, *1076*

VENEZUELA

Oilbird, *1647*

Caracas

International treaty to specifically protect the sea turtles of the Western Hemisphere, *3054*

VIETNAM

Appearance of wet-rice cultivation in paddy fields, *1198*
Evidence of the existence of the Vu Quang ox, *4739*
Imperial pheasants collected from the wild, *1659*
Introduction of drought-resistant rice, *1122*
National population plan for Vietnam, *3892*
Petroleum in Vietnam, *3623*
Tests by the military of the defoliant Agent Orange, *1295*

Vung Tau

Natural gas production in Vietnam, *2067*

VIRGIN ISLANDS

Tropical-subtropical biome surveyed by a U.S. National Park Service Inventory & Monitoring Program (I&M), *3417*

WESTERN HEMISPHERE

Large animals domesticated in the Western Hemisphere, *1148*
Treaty obliging the United States to maintain a list of endangered species, *3021*
Treaty to establish national parks and preserves and to identify protected flora and fauna in the Western Hemisphere, *3022*

YUGOSLAVIA (FORMER)

Biosphere reserves in Europe, *1625*
General Fisheries Council for the Mediterranean, *2474*

ZAMBIA

African country to produce a National Conservation Strategy, *3390*
Major hydroelectric power-producing dam on the Zambesi River in Africa, *2071*

Lusaka

United Nations educational center for African natural resource management, *3392*

ZIMBABWE

Large-scale relocation of elephants, *4745*
Major hydroelectric power-producing dam on the Zambesi River in Africa, *2071*